BROKEN SYMMETRY

SELECTED PAPERS OF Y. NAMBU

World Scientific Series in 20th Century Physics

Published

Vol. 20 The Origin of the Third Family
edited by O. Barnabei, L. Maiani, R. A. Ricci and F. R. Monaco

Vol. 21 Spectroscopy with Coherent Radiation — Selected Papers of Norman F. Ramsey (with Commentary)
edited by N. F. Ramsey

Vol. 22 A Quest for Symmetry — Selected Works of Bunji Sakita
edited by K. Kikkawa, M. Virasoro and S. R. Wadia

Vol. 23 Selected Papers of Kun Huang (with Commentary)
edited by B.-F. Zhu

Vol. 24 Subnuclear Physics — The First 50 Years: Highlights from Erice to ELN by A. Zichichi
edited by O. Barnabei, P. Pupillo and F. Roversi Monaco

Vol. 25 The Creation of Quantum Chromodynamics and the Effective Energy by V. N. Gribov, G. 't Hooft, G. Veneziano and V. F. Weisskopf
edited by L. N. Lipatov

Vol. 26 A Quantum Legacy — Seminal Papers of Julian Schwinger
edited by K. A. Milton

Vol. 27 Selected Papers of Richard Feynman (with Commentary)
edited by L. M. Brown

Vol. 28 The Legacy of Léon Van Hove
edited by A. Giovannini

Vol. 29 Selected Works of Emil Wolf (with Commentary)
edited by E. Wolf

Vol. 30 Selected Papers of J. Robert Schrieffer — In Celebration of His 70th Birthday
edited by N. E. Bonesteel and L. P. Gor'kov

Vol. 31 From the Preshower to the New Technologies for Supercolliders — In Honour of Antonino Zichichi
edited by B. H. Wiik, A. Wagner and H. Wenninger

Vol. 32 In Conclusion — A Collection of Summary Talks in High Energy Physics
edited by J. D. Bjorken

Vol. 33 Formation and Evolution of Black Holes in the Galaxy — Selected Papers with Commentary
edited by H. A. Bethe, G. E. Brown and C.-H. Lee

Vol. 35 A Career in Theoretical Physics, 2nd Edition
by P. W. Anderson

Vol. 36 Selected Papers (1945–1980) with Commentary
by Chen Ning Yang

Vol. 37 Adventures in Theoretical Physics — Selected Papers with Commentaries
by Stephen L. Adler

Vol. 38 Matter Particled — Patterns, Structure and Dynamics — Selected Research Papers of Yuval Ne'eman
edited by R. Ruffini and Y. Verbin

Vol. 39 Searching for the Superworld — A Volume in Honour of Antonino Zichichi on the Occasion of the Sixth Centenary Celebrations of the University of Turin, Italy
edited by S. Ferrara and R. M. Mössbauer

For information on Vols. 1–19, please visit http://www.worldscibooks.com/series/wsscp_series.shtml

World Scientific Series in 20th Century Physics – Vol. 13

BROKEN SYMMETRY

SELECTED PAPERS OF Y. NAMBU

Editors

T. Eguchi
University of Tokyo, Japan

K. Nishijima
Chuo University, Japan

World Scientific
Singapore • New Jersey • London • Hong Kong

Published by

World Scientific Publishing Co. Pte. Ltd.
5 Toh Tuck Link, Singapore 596224
USA office: 27 Warren Street, Suite 401-402, Hackensack, NJ 07601
UK office: 57 Shelton Street, Covent Garden, London WC2H 9HE

Library of Congress Cataloging-in-Publication Data
Nambu, Y., 1921–
Broken symmetry : selected papers of Y. Nambu / edited by T. Eguchi, K. Nishijima.
p. cm. -- (World Scientific series in 20th century physics ; vol. 13)
ISBN-13 978-981-02-2356-4 -- ISBN-10 981-02-2356-0
ISBN-13 978-981-02-2420-2 (pbk) -- ISBN-10 981-02-2420-6 (pbk)
1. Broken symmetry (Physics) 2. Particles (Nuclear physics)
3. Mathematical physics. I. Eguchi, T. (Toru), 1948– .
II. Nishijima, K. (Kazuhiko), 1926– . III. Title. IV. Series.
QC793.3.S9N34 1995
539.7'2--dc20

95-21985
CIP

British Library Cataloguing-in-Publication Data
A catalogue record for this book is available from the British Library.

The editors and publisher would like to thank the following publishers of the various journals and books for their assistance and permissions to include the selected reprints found in this volume:

Academic Press (*Ann. Phys.*);
The American Physical Society (*Phys. Rev., Phys. Rev. Lett.*);
Elsevier Science Publishers B.V. (*Nucl. Phys., Phys. Lett., Phys. Rep.*);
Gordon and Breach Publishing;
Institute of Physics Publishing;
John Wiley & Sons;
Kyoto University (*Prog. Theor. Phys.*);
Purdue Research Foundation;
Società Italiana di Fisica (*Il Nuovo Cimento*);
Springer-Verlag.

FOREWORD

It is our great pleasure to publish a Collection of Papers of Professor Yoichiro Nambu who has been one of the most renowned masterminds among our contemporaries. We have chosen the title "Broken Symmetry" for this volume since we think it represents the best-known contributions of his repertoire.

Many of us are clients of his papers and subscribe to his messages which convey his foresights always leading to a new frontier of physics. It is not necessarily easy to follow his publications, however, since some of his ideas are beyond immediate comprehension. Indeed, on the occasion of his 65th birthday, one of the leading theorists made the following remark: "He is always 10 years ahead of us, so I tried to understand his works in order to contribute to a new area which will flourish 10 years later. Contrary to my expectation, however, it took me 10 years to understand them."

For those who try to grasp the entire flow of his ideas it is gratifying and helpful to have an access to a collection of 40 papers selected by Professor Nambu himself. Furthermore, it should be emphasized that some of his unpublished papers and lecture notes, unavailable otherwise, have been reproduced and included in this volume. They are the famous unpublished lecture notes on string theory and the first paper on spontaneous symmetry breaking read at the Midwest Conference on Theoretical Physics held in 1960.

In addition to his scientific publications, a memoir on his research career (*Research in Elementary Particle Theory*) prepared by himself and published first in "Butsuri", the Bulletin of the Physical Society of Japan, in 1977 (in Japanese) is also translated in English and included in this volume. This self-portrait of his offers a detailed description of how and why he was motivated to develop and pursue many of his pioneering works.

The chronology of this list of selected papers would eventually enable us to trace back some of his ideas to his earlier publications with the help of his memoir.

We hope that the readers would treasure this volume for years to come as a constant source of inspiration.

Editors, May 1995

RESEARCH IN ELEMENTARY PARTICLE THEORY*

1. It has been more than thirty years since I entered my career as a physicist immediately after the end of the War. As I now look back and try to comment on my work done in all these years, I cannot help an uneasy feeling because I have never undertaken this before.[1] Admitting, however, that my career has been a little unusual, I think it may be worthwhile to contribute an article to this special issue even at the risk of inaccuracies. I will therefore pick and discuss below some representative periods and physics problems from my life.

I graduated from the University of Tokyo during the War under the old system (more or less at the MS level, cut short from three to two and a half years), and entered military life. After the War I was fortunate to be able to go back to the University of Tokyo as a kind of research associate, a position which had been secured for me at graduation time. This was probably around the beginning of 1946. It is difficult now to realistically describe the miserable environment we lived in after the War, but it is possible that such an environment was in fact beneficial to me in laying the foundations of my career. This period probably corresponded to the present PhD level, but for professors and students alike the overwhelming concern was merely to survive. One was content with satisfying our basic needs, and what extra energy one had was spent in digesting important physical concepts and trying to develope one's own ideas. It was ideal in this respect that I literally lived in Rm 305 of the Physics Department building for three years. My colleagues in the same situation had occupied several large offices, and we used to visit each other at night, freed from the daytime crowding, and discuss physics for long hours. (The life at the University of Chicago is not very different because professors and students all live within walking distance from the campus.)

Among those people around me, I was particularly influenced by my roommates Giichi Iwata and Ziro Koba.[2] From Iwata I learned some mathematics as well as a taste for French literature and Latin. From Koba I absorbed the Tomonaga theory. Ryogo Kubo was at that time a young leader of the solid state theory group which was intensely studying the Onsager solution of the Ising model. This problem also cast a spell on me, and after a while I found a simple algebraic derivation of the solution. This was my first success, which happened in the spring of 1947. But I did not regard it worth publishing. Moreover, it

*Translated from the Bulletin of the Physical Society of Japan, **32** (1977), No. 10, pp. 773–778.

was immediately followed by the experiment of Lamb and Retherford, and I found myself fully involved in the Lamb shift problem. Later I managed to write my work on the Ising model after receiving encouragement from Prof. K. Husimi of Osaka University who also developed a similar line of work with H. Syozi about the same time. As an added benefit of this, I started correspondence with Bruria Kaufman. She was at that time an assistant of Einstein, and arranged for me a meeting with Einstein when I came to Princeton in 1952. I owe Husimi not only the publication of the work on the Ising model, but also my 1973 paper on Generalized Hamiltonian Dynamics. The idea actually goes back to my Osaka City University days, when Husimi took quite an interest in this.

In the office Koba and I were sitting face to face. A very serious man, he showed up earlier than anybody else, and I was often embarrassed when I was still sleeping using both of our desks. He was closely collaborating with Tomonaga at Bunrika University nearby. At first I did not know what the Tomonaga theory was about. But as I started to commute regularly to Tomonaga's seminars, I was gradually initiated into it, and also made acquaintance with other people in the Tomonaga group. There was an episode involving Koba, that he shaved his head after discovering a serious error in his calculations. I was a witness to this story which was probably too well known to be repeated here in detail. Among other people at the University of Tokyo, I also became close to Seitaro Nakamura who was a resident across the hall. His did unforgettable services to all of us by finding moonlighting jobs for us, initiating a movement to improve our status and salary, and starting the semi-official journal Soryushiron Kenkyu.[3] Occasionally some well known figures came to see Nakamura and Koba. Once Mitsuo Taketani and Satosi Watanabe showed up together, and I watched with fascination the lively sparrings that went on between them.

I am not qualified to discuss the development of the Tomonaga theory, but I can relate some happenings around me. At that time current issues of foreign journals like Physical Review were not generally available; one had to go to the library of the Allied Occupation Forces or to special individuals like Tomonaga and Taketani. I think I first learned of the Lamb–Retherford experiment, Bethe's work and the Schwinger theory from the Time magazine. (The Lamb–Retherford experiment was announced in June 1947.) After seeing Bethe's paper I tried to find an intuitive interpretation of radiative corrections. Together with Ken-ichi Ono, a friend and fellow resident, we were able to report the result at the fall meeting (November 1947) of the Physical Society of Japan. At the same session Tomonaga presented his ideas of how to handle the radiative corrections with his renormalization theory. I did not understand his talk well, except that I remember the concept of normal ordering was introduced there. Anyway we were overwhelmed by his talk, and ended up not publishing our own work. But in fact ours was very similar to that of T. A. Welton which was published in Physical Review later. As for the anomalous magnetic moment of the electron, I had seen the phenomenological theory of G. Breit. But one day I realized that it might come out of radiative corrections, when I saw a letter in the Physical Review by Schwinger to my great chagrin, as I remember. (Schwinger's paper apppeared in February 1949.) We were thus always one step behind the Americans, which was regrettable but to be expected. In his calculations Koba had introduced a diagrammatic method, albeit primitive compared to Feynman's; Tomonaga had originated the normal product expansion of the S-matrix which Koba and I tried to develop, but the essential results were preempted by F. Dyson and G. C. Wick. Sometimes, however, we were able to catch up with the Americans and almost pass them. As an example I recall the work of H. Fukuda and Y. Miyamoto (1949) on the gamma decay of the π^0, which was simultaneous with that of J. Steinberger.

I moved to the new Osaka City University, probably in the fall of 1949. Prof. Yuzuru Watase, Chairman of the Physics Department, was instrumental in bringing this about; from Tokyo came, besides me, Satio Hayakawa, Yoshio Yamaguchi, and Kazuhiko Nishijima, and from Osaka came Tadao Nakano. It was a thrilling event for all of us who, until then mere assistants at best, suddenly occupied the whole ranks of an elementary particle theory group, from assistant to full professor according to our age. (Around the same time Koba became an associate professor at Osaka University.) The science and engineering division of the university took up a temporary residence in a war-damaged building that had previously been Oogimachi Elementary School, and our group was housed in a wooden barrack built on the schoolyard. It was also around this time that the practice of so-called *mushashugyo* (honing of the samurai's skills by training pilgrimage) got started. Thus we were visited by young and fresh graduates like Hironari Miyazawa and Masatoshi Koshiba from Tokyo, who would sleep at night in a sofa in our office.

During the three years spent in Osaka, one of the major happenings was the problem of the V particles (strange particles). We were in fact ahead of Americans in theoretical work, due to our City University group and other people like S. Oneda, and were able to keep pace until it culminated in the Nakano–Nishijima–Gell-Mann law. It was easy for us to plunge into such new phenomena, perhaps because we had been exposed to comsic ray physics through contact with the Nishina–Tomonaga group, and were used to order-of-magnitude thinking. I also believe that Fermi's theory of multiple production and his lectures on nuclear physics, both of which came out at that time, gave me considerable impact. We were also helped by the fact that Hayakawa, who had gone for postdoctoral study in America in the meantime, corresponded regularly with us, and kept us informed of developments there. As for the Nakano–Nishijima–Gell-Mann law, I was already in Princeton by then. I learned it from Murray Gell-Mann first, then heard of the independent work of Nakano and Nishijima from Hayakawa who was back in Osaka.

Perhaps a word is in order about the so-called Bethe–Salpeter equation. It is true that I was the first to publish it, but it was written down without derivation or application, at the end of a paper which discussed the derivation of two-particle potentials from field theory. (Some more details were given in Soryushiron Kenkyu. People in Kyoto, Hideji Kita, Yasuo Munakata, and Chushiro Hayashi, also took up the topic.) The reason why I did not follow it up is partly because of my illness, but more importantly because I did not find it very useful. In fact, the ladder approximation, although fully relativistic, does not even give the correct Breit formula for electron-electron interaction. I was not satisfied with mere mathematical elegance.

Another topic comes to mind that belongs to this period: the WKB approximation to the S-matrix. It was motivated by the "(functional) diffusion equation" of Tomonaga and N. Fukuda (Soryushiron Kenkyu **3** (1951), No. 2). I wrote an article in Soryushiron Kenkyu, but did not remember the details. Later I published a paper in English (1968) when the concept turned out to be useful in chiral dynamics. (Today the method goes by the name of loop expansion.)

2. In 1952 I came to the Institute for Advanced Study for a two-year stay, together with Toichiro Kinoshita, at the recommendation of Tomonaga to Oppenheimer. Thereafter I moved to Chicago, where I remain to this day. The Princeton days were my second trial period, as I came abroad for the first time and started to compete among the first-rate people of the world. I would say that these were the Golden Days of the Institute, and I got to know W. Pauli, A. Pais, F. Dyson, C. N. Yang, T. D. Lee, L. Van Hove, G. Kallen, W. Thirring, etc. Quantum electrodynamics and V particles were the biggest topics of the day, but somehow I had been drawn into many-body problems. I was preoccupied with the Bohm–Pines theory of plasmas, the origins of nuclear saturation and spin-orbit coupling, the equation of state for nuclear matter, and the like. In collaboration with Kinoshita, I managed to write a paper on generalized Hartree–Fock method, but I could not convince myself that I could solve real nuclear physics problems. Failing other ideas, too, I was driven to desperation. It was M. L. (Murph) Goldberger who saved me out of this personal crisis by bringing me to Chicago. There, still on leave from Osaka, I joined the ranks of research associates with Reinhard Oehme and Hironari Miyazawa. The theory of dispersion relations, originated by Goldberger, and the analytic properties of the S-matrix were the vogue of the late 1950s. I too was completely taken in by these topics and wrote papers myself as well as in collaboration, but I must say they were a mixed bag of gems and pebbles. Some of them give me a chill even now, for they were of such a quality as to fit the famous remark of Pauli's, "It is not even wrong." The point is that I had made wild assumptions about the analyticity of Green's functions with many variables. When I subsequently discovered the error myself, I was so embarrassed and disappointed that I threw away a gem in the paper as well. In fact Kurt Symanzik, who was visiting our group for a while, showed the viability of a special case of my formulas after proper amendment, and was eager to write a paper with me, but I was in no mood to agree. That formula is known as the Mandelstam representation. (This is only meant as a belated word of appreciation and apology to Symanzik.)

It was not all blunders, after all. There were meaningful contributions like the discovery of the anomalous threshold and the parametric representation of the Green's functions. But I suppose I came to be well recognized by the physics community as a result of the "CGLN" (Chew, Goldberger, Low and Nambu) papers on dispersion theory of π-N scattering and γ-π processes. To tell the truth, however, these papers were written by the first three authors in Princeton. If I did contribute anything, it is perhaps that I had told Goldberger some of my ideas beforehand.

There were of course other more fundamental issues like parity nonconservation and the symmetries of strong interactions that were developing at that time, but my hands were full. An event happened, in the meantime, for which my long-standing interest in condensed matter dating back to my Tokyo days proved useful. It is the famous BCS (Bardeen, Cooper and Schrieffer) theory of superconductivity. The University of Illinois, with a strong tradition in condensed matter physics, is located not far from Chicago, and I had known people like J. Bardeen and D. Pines there. One day, before publication of the BCS paper, Bob Schrieffer, still a student, came to Chicago to give a seminar on the BCS theory in progress. Gregor Wentzel, our senior professor, looked rather skeptical. I was also very much disturbed by the fact that their wave function did not conserve electron number. It did not make sense, I thought, to discuss electromagnetic properties of superconductors with such an approximation. At the same time I was impressed by their boldness, and tried to understand the problem. I ended up becoming captive to the BCS theory. In the meantime Schrieffer took an assistant professor job in Chicago, so we were in close touch. But my major concern was in the purely theoretical issues like gauge invariance, and it took me two years before I was able to convince myself of its resolution and write a paper. While in the meantime experts like N. N. Bogoliubov and P. W. Anderson were refining the BCS theory, I deliberately tried to keep independence.

The name "spontaneous breakdown of symmetry" is due to M. Baker and S. Glashow. This idea germinated in me during the above course of investigation. The concept actually consists of three elements: 1) degeneracy of the vacuum (ground state) 2) the associated massless mode (Goldstone or zero mode) 3) subsuming of the zero mode and the Coulomb field (long range field) into the plasmon mode. It was my contention that the BCS theory contained all these three elements. I also thought of applying the idea to other phenomena like crystal lattice and ferromagnetism but, not being an expert in condensed matter physics, hesitated to announce a general principle of spontaneous symmetry breaking, of which they were examples. The fact that I devoted my efforts to the analysis of gauge invariance of the Meissner effect, however, subsequently turned out useful in understanding the Goldberger–Treiman relation in $\pi \rightarrow \mu - \nu$ decay, for I realized that it was the same relation that showed gauge invariance of the Meissner effect. The next step was an easy one, to reach the analogy between the energy gap in superconductors and the mass in the Dirac equation. (I first made a comment to this effect at the 1959 Kiev High Energy Physics Conference.) It was before the quark model, so the main question was how to choose a model in such a way as to show that symmetry breaking can actually happen. One possibility would be to take an orthodox field theory like quantum electrodynamics and solve the Dyson–Schwinger-BCS equation. It seemed clear that one could get a solution if one ignored the photon renormalization effects (i.e. the running of the coupling constant), but I was not sure I could handle the problem honestly. (Later the group of K. Johnson tackled it.) So I chose a second model, namely a nonlinear spinor field theory which was a straightforward relativistic version of the BCS theory. To tell the truth, I did not like the fact that it belonged to the type of Heisenberg's nonlinear theory, for I had never believed that a theory like that, with no elegance at all, should be the ultimate law of physics. Nevertheless it compelled me to study Heisenberg's papers seriously. I was duly impressed when I found the concept of spontaneous breakdown invoked there. (I have to say, however, that his way of doing looks too formal.) I adopted the nonlinear model because it was mathematically simple and clean provided that one did not resort to gimmicks like indefinite metric but instead simply introduced a cutoff. Thus I started to proceed carefully with the help of Gianni Jona-Lasinio who had just arrived from Rome as a research associate. Our preliminary report was given at the annual Midwest Theoretical Physics Conference which took place at Purdue University. Because of an illness in my family, it was presented by Jona-Lasinio. Soon after the preprint version of our talk had gone out, I received a preprint from J. Goldstone. This is the paper that made him famous. He started by quoting me, then discussed a simple example in field theory, and finally conjectured the generality of the existence of zero modes. As for the last point, I had been debating with myself how to write a paper addressing it as a general phenomenon, so I felt as if a prize catch had been stolen from under my nose. Eventually I decided to concentrate on the details of the relationship between gamma-5 invariance and massless pion.

Such dynamical models, however, were far removed from the prevalent paradigm of the S-matrix popularized by the Chew–Low school. There was also the formal approach of Gell-Mann and Levy on the Goldberger-Treiman relation. So for the next few years I tried to find expermimental confirmation of my ideas with the help of a research associate David Lurie and my student Ely Shrauner. This was the so-called soft-pion theorem, and it also contained an element of the current algebra which Gell-Mann would formulate later. I also tried to apply it to K-meson decays and the like, but lacking the quark model still, there was too much arbitrariness for making useful statements. After the quark model and current algebra had become established, S. Adler and others further developed the theorem, and ushered in the days of chiral dynamics. It was in this period that the work of H. Sugawara and M. Suzuki, fresh PhD's at Tokyo, appeared. In our group, too, Yasuo Hara, Joseph Schechter, and my students Jeremiah Cronin and Louis Clavelli were active.

Apart from the above line of work, there is another one having to do with a three-color generalization of the quark model (integral charge color triplet, 1965). This and the hadronic string model, begun around 1969, are still alive today, but I will skip them because of space, and devote the rest of the article to general observations from the perspective of the history of Japanese physics.

3. Japan has a unique tradition in elementary physics represented by such names as H. Nagaoka, H. Yukawa, S. Tomonaga, and S. Sakata. But I have not seen an account of how this has come about, and I wish those interested in the history of physics will take this up someday. It seems that the mode of thinking typified by Yukawa and Sakata looks rather alien when seen from the outside, as I often hear comments to that effect.[4] But I have to say that their influence on me has been overwhelming even though I was not trained in their school. I am sure everyone of the immediate postwar generation will agree with me on this. As I have stated already, I got familiar with the work of the Tomonaga group through my friends and through attending the seminars conducted by Nishina and Tomonaga at Riken (Institute of Physico-Chemical Research) and Bunrika University. (I attended them occasionally as a student even before the War.) It is remarkable that Tomonaga's theory group and Nishina's experimental group collaborated very closely with each other. Unfortunately this tradition stopped after the war. It must be due to the disruption of experimental physics caused by the postwar economic condition, and they are still struggling to recover from it. Especially these days the scale of experiment has grown so large that a need for interational cooperation has become greater. Japan is geograpically handicapped in this respect, and its sense of international cooperation is very weak. I would say this applies to theory as well. To us the postwar generation the influence of the methodology articulated by leaders like S. Sakata and M. Taketani was like sunlight; it was strong even if we were not conscious of it. In fact it was too strong, I would say. Methodology is not physics itself. The failure of particle physics in Japan to keep up its momentum after the initial successes is, in my opinion, because there were no competing strong groups. The Yukawa school of particle physics is noteworthy in that it adopted the guiding principle: To each field of force, a particle. It opened, on the one hand, a path leading to such ideas as vector dominance (J. J. Sakurai), bootstrap (G. Chew) and the Regge poles. On the other hand, it established the powerful method of "inventing" new particles on demand.[5] It is true that this mode of thinking is found already in Pauli's neutrino hypothesis, but the fact is that after Yukawa it has become a routine to theorists. It may be said, for example, Z. Maki used it, consciously or subconsciously, in his prediction of charmed particles, as did Glashow, Iliopoulos, and Maiani. Furthermore, there is a conservative side to western science in that it tries to stick to existing framework as long as possible, and this tends to make the Yukawa-type approach difficult to accept. An example is the tortuous way the cosmic-ray meson problem was resolved by the two-meson theory of Sakata and Inouye and of Marshak and Bethe. To quote my own experiences, my neutral vector meson hypothesis (1957) met with rather cool reception, and my 3-color quark model was considered irrelevant till the 1970s.

The path from the Sakata model to the quark model is not a simple one, historically as well as methodologically, for a new layer of matter is introduced in the latter. Nevertheless, it should not be surprising that the concept of SU(3) was first taken up by Ikeda, Ogawa and Ohnuki of the Sakata group (as well as by Yamaguchi). Sakata's influence is also detected in the SU(6) theory of B. Sakita.

At present the main line of particle physics is based on quarks and leptons as its fundamental material particles and gauge theory as its dynamical principle. Since the gauge concept encompasses gravity as well, we have come one step closer to the dream of a true unification of forces. In my opinion this was made possible because of the analogy between particle and condensed matter physics, i.e. between the vacuum of the world and ordinary media. In fact this analogy has advanced to a degree not contemplated when I first started the superconductivity model. It is remarkable that one sees this analogy to superconductivity and superfluidity in phenomena ranging from the W and Z particles of the Salam–Weinberg theory, hadronic strings to monopoles.

It seems fair to say, however, that we have not completely understood particle physics yet even at the present level. For there still exist two major elements of arbitrariness. One is that one does not know the number of families of quarks and leptons. The other is that one does not know the fundamental mechanism for giving mass to quarks, leptons, W and Z particles, etc. The present theory is only phenomenology in these respects. Of course there is no reason to insist that the quarks and leptons are the ultimate units. If more particles should be found experimentally, it might become necessary to go one more layer down. At the same time, if the contention that a quark can never exist in isolation can be firmly established, it may also necessitate a revision of the concept of elementarity in a different sense.[6]

One day which was more than ten years ago, E. C. G. Sudarshan and I were in a car together near Chicago. In our conversation he started asking why I had not worked on gravity. "Don't you think", he confronted me in the next breath, "that the goal of physics is to find general principles?" Thereupon I answered, "Right now I am more interested in substance than principle." For example, the levels of the hydrogen atom might be understood in terms of the SO(4, 2) group and an infinite component wave equation, but it would not be the ultimate way; one has to pursue the entities behind it, namely electron and proton. By this I also meant to reveal my true intention when I had taken up the infinite component equations some time earlier. Suddenly I recalled Heisenberg's remark somewhere to the effect that symmetry principle is the essence of physics, and I also realized that probably Sudarshan was under the influence of Indian philosophy whereas I was under the influence of the philosophy of the Yukawa–Sakata shool.

The fact is, though, that both elements, substance and principle, are necessary for particle physics. It is remarkable that great progress has been made on both counts within recent years. These two get merged into one in the approaches of Heisenberg and bootstrap, but the present mainstream view is still based on dualism. Monistic synthesis may belong, to borrow a term from theorists, in the realm of asymptopia.

References

[1] There is a kind of record, however: Y. Nambu, Symmetry breakdown and small mass bosons, *Fields and Quanta* **1** (1970) 33; *The Past Decade in Particle Theory,* ed. E. C. G. Sudarshan and Y. Ne'eman (Gordon and Breach, 1973), p. 33.

[2] Reminiscences about the late Ziro Koba exist in the literature: M. Nogami *et al.*, Soryushiron Kenkyu **48** (1973) No. 3; S. Tomonaga, Kagaku **44** (1974) No. 6; S. Umehara, Igaku Geijutsu (1975), Nos. 5 and 6, but I was not able to consult them all.

[3] "Soryushiron Kenkyu" is one of my favorite journals. Old fragments of it at the University of Chicago Library have been useful in writing this article.

[4] For an example of western critique of the Sakata-Taketani school, see Y. Ne'eman, "Concrete versus abstract theoretical models" (submitted to the Jerusalem Conference on the Historical Interaction between Science and Philosophy, 1971).

[5] I owe much of the following observations to conversations I have had with Laurie Brown, for which I would like to thank him.

[6] Even as I wrote this article, I heard reports about the discovery of a new heavy lepton and of a fractionally charged particle (quark), but these results do not yet seem to be of high reliability.

CONTENTS

*Numbers in brackets refer to the list of publications on pp. 455–467.

Progress of Theoretical Physics Vol. V, No. 1, Jan.~Feb., 1950.

A Note on the Eigenvalue Problem in Crystal Statistics.

Yôichirô NAMBU

Department of Physics, Tokyo University.

(Received September 1, 1949)

§ 1. Introduction.

In recent years there have been some remarkable developments in the mathematical theory of order-disorder transition. The problem of determining the partition function for a lattice with simple interaction model was found, by several authors,[1] to be reduced to an eigenvalue problem which is easier to handle with, and which, with suitable approximations, would reveal more refined details of the phase transition than the former theories. This method of attack culminated in the work of Onsager,[2] who succeeded in giving the rigorous solution of the above mentioned eigenvalue problem for two-dimensional simple square Ising lattice. The mathematical tools involved, however, are so hopelessly complicated that one would quite simply lose sight in the jungle of hypercomplex numbers. A considerable improvement on this mathematical point has been attained by Husimi and Syôzi[3] who used it to solve the eigenvalue problem for the honeycomb type lattice. The present author, independently of them, has reached more or less similar ideas and considerations which it is the purpose of the present paper to expose in brief detail. Though as yet no substantial applications has been attempted, nor anything physically new has been derived, it may be hoped that it will do some profit for those who are interested in such problems.

§ 2. Preliminary Considerations.

Let us take N identical particles 1, 2,$\cdots N$, arranged in order, and consider the interchange $P_{n,n+1}$ of two adjacent particles n and $n+1$ ($n=1, 2,\cdots N$). Such operators are met with in the theory of ferromagnetism in which the eigenvalues of the operator (for one-dimensional case)

$$P=\sum_n P_{n,n+1}=\sum \frac{1+\sigma_n\cdot\sigma_{n+1}}{2} \tag{1}$$

are asked. Here σ_n means the spin matrix $(\sigma_x, \sigma_y, \sigma_z)$ at the nth site. As P commutes with $\sum_n \sigma_{nz}=2S-N$, where S is the total number of spins oriented to $+z$ direction, we can work within a subspace in which S is kept fixed. Then

P is equivalent to an operator affecting the arrangement of S particles over N sites. Each site is either occupied or not occupied by one of these particles. If we introduce such operators, that change the unoccupied nth site into occupied, and *vice versa*, by

$$a_n^+ = \begin{pmatrix} 0 & 0 \\ 1 & 0 \end{pmatrix}, \quad a_n = \begin{pmatrix} 0 & 1 \\ 0 & 0 \end{pmatrix}, \tag{2}$$

then P can be written in terms of them as follows:

$$\begin{aligned} P &= \sum (a_n^+ a_{n+1} + a_{n+1}^+ a_n + a_n^+ a_n a_{n+1}^+ a_{n+1} + a_n a_n^+ a_{n+1} a_{n+1}^+) \\ &= \sum (a_n^+ a_{n+1} + a_{n+1}^+ a_n - 2a_n^+ a_n + 2a_n^+ a_n a_{n+1}^+ a_{n+1}) + N. \end{aligned} \tag{3}$$

The operators (2) remind one of the operators familiar in the second quantization, the only difference being the lack of anticommutativity between different sites. But we can easily see that the introduction of the sign functions necessary to make the a_n's and a_n^+'s anticommute does not impair the relation (3), so that we are permitted to put

$$[a_n, a_m^+]_+ = \delta_{nm}, \quad [a_n, a_m]_+ = [a_n^+, a_m^+]_+ = 0. \tag{4}$$

Now we make use of the Fourier transformation to rewrite (3) as

$$\begin{aligned} P &= \sum_{n,m} (a_n^+ a_m \delta(n-m+1) - a_n^+ a_m \delta(n-m) + a_n^+ a_n a_m^+ a_m \delta(n-m+1) + \text{conj.} \\ &= \frac{1}{N} \sum_{n,m} \sum_k \left(a_n^+ e^{-\frac{2\pi i}{N} nk} a_m e^{+\frac{2\pi i}{N} mk} e^{-\frac{2\pi i}{N} k} - a_n^+ e^{-\frac{2\pi i}{N} nk} a_m e^{+\frac{2\pi i}{N} mk} + \text{conj.} \right) \\ &\quad + \sum_{n\,m} a_n^+ a_n a_m^+ a_m \delta(n-m\pm 1) \\ &= \sum_k \left(a_k^+ a_k e^{-\frac{2\pi i}{N} k} - a_k^+ a_k + \text{conj.} \right) + \sum a_n^+ a_n a_m^+ a_m \delta(n-m\pm 1), \end{aligned} \tag{5}$$

where

$$a_k = \frac{1}{\sqrt{N}} \sum_n a_n e^{\frac{2\pi i}{N} nk}, \quad a_k^+ = \frac{1}{\sqrt{N}} \sum_n a_n e^{-\frac{2\pi i}{N} nk}. \tag{5'}$$

There is a striking analogy to the ordinary quantum mechanics. P is just the Hamiltonian of interacting identical systems for which the "kinetic energy" and the "interaction potential" are given respectively by

$$2\left(\cos \frac{2\pi}{N} k - 1\right) \quad \text{and} \quad \delta(x_i - x_j \pm 1), \tag{6}$$

where k and x take discrete values. The kinetic terms alone lead to the well known Bloch spin waves whereas the interaction terms cause scattering of them. The latter can be neglected when the number of particles are small because the interaction force is of short range, which corresponds to the fact that the notion of spin wave is a good approximation when the magnetization is nearly complete (low temperature).

As the above example shows, the introduction of a set of anticommuting operators is particularly convenient because of its orthogonal property[4]. Instead of the creation and annihilation operators a_n^+ and a_n, we can also use the quantities $a_n^+ + a_n$ and $i(a_n^+ - a_n)$, or in other words, a set of anticommuting quantities with the commutation relations

$$[x_r,\ x_s]_+ = 2\delta_{rs}, \quad r,\ s=1,\cdots 2N. \tag{7}$$

According to the theory of spinors, such quantities can be considered as making up an orthogonal basis for a $2N$-dimentional vectors space[5]. A product of two vectors $x_r x_s$ commutes with x_t if $t \neq r,\ s$. Thus we have a means for expressing commutable quantities in terms of anticommutable quantities.

As a next example we shall take the eigenvalue problem for the operator

$$P' = \sum_{n=1}^{N} (\sigma_{nx}\sigma_{n+1,x} + \sigma_{ny}\sigma_{n+1,y}) \equiv \sum (A_n + B_n), \tag{8}$$

which is a part of the operator P defined in (1). The commutation relations for A and B are

$$\begin{aligned} &[A_n,\ A_m]_- = [B_n,\ B_m]_- = 0, \\ &[A_n,\ B_m]_- = 0 \quad \text{if} \quad n=m \quad \text{or} \quad |\,n-m\,| \neq 1, \\ &[A_n,\ B_{n\pm 1}]_+ = 0, \quad A_n^2 = B_n^2 = 1. \end{aligned} \tag{9}$$

That is, A_n and B_n anticommute when they are adjacent, otherwise simply commuting. Such quantities can equally be composed of totally anticommuting operators. Indeed, let us introduce $2N$ basic vectors $x_1,\ x_2, \cdots x_{2N}$ with the commutation relations (7), and define (Fig. 1), for N=even,

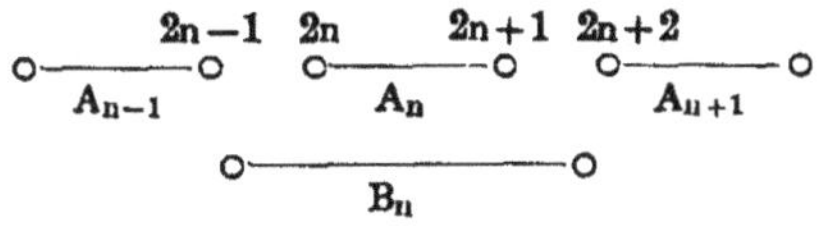

Fig. 1.

$$\begin{aligned} &A_n = i x_{2n} x_{2n+1}, \\ &\qquad n=1,\ 2, \cdots N;\ 2N+1 \equiv 1. \\ &B_n = i x_{2n-1} x_{2n+2}. \end{aligned} \tag{10}$$

Obviously these quantities satisfy all the relations (8). The only difference is as follows. From (8) we have

$$\prod_{n=1}^{N} A_n = \prod_{n=1}^{N} B_n = 1, \tag{11}$$

while from (10), on the other hand,

$$\begin{aligned} \Pi A_n &= i^N x_2 x_3 \cdots x_{2N} x_1 = -i^N x_1 x_2 \cdots x_{2N} \equiv x = \pm 1, \\ \Pi B_n &= i^N x_1 x_4 \cdot x_3 x_6 \cdot x_5 x_8 \cdots x_{2N-1} x_2 \\ &= i^N (-1)^{N-1} x_1 x_2 \cdots x_{2N} = (-1)^N x = x, \end{aligned} \tag{11'}$$

so that the operators in (8) cover only a subspace of the operator domains for (10). We have deliberately to select those solutions which lie in the subspace $x=1$.

Now (8) can be written as

$$P'=i\sum_{n=1}^{N}(x_{2n}x_{2n+1}+x_{2n-1}x_{2n+2})=i\sum_{n=1}^{N}(x_{2n}x_{2m+1}+x_{2n-1}x_{2m+2})\delta_{nm}$$

$$=\frac{i}{N}\sum_{k=1}^{N}\sum_{n,m}(x_{2n}x_{2m+1}+x_{2n-1}x_{2m+2})e^{\frac{2\pi k}{N}i(n-m)}=2\sum_{k=1}^{N}(x_k y_{-k}+y_k x_{-k}e^{\frac{2\pi k}{N}2i})$$

$$=2i\sum_{k=1}^{N/2}\{x_k y_{-k}(1-e^{-\frac{2\pi k}{N}2i})+x_{-k}y_k(1-e^{\frac{2\pi k}{N}2i})\}$$

$$=-4\sum_{k=1}^{N/2}(x_k y_{-k}e^{-\frac{2\pi k}{N}i}-x_{-k}y_k e^{\frac{2\pi k}{N}i})\sin\frac{2\pi k}{N}\equiv-4\sum_{k=1}^{N/2}z_k\sin\frac{2\pi k}{N}, \tag{12}$$

where

$$x_k\equiv\frac{1}{\sqrt{2N}}\sum_{n=1}^{N}x_{2n}e^{\frac{2\pi k}{N}ni},\quad y_k\equiv\frac{1}{\sqrt{2N}}\sum_{n=1}^{N}x_{2n+1}e^{\frac{2\pi k}{N}ni},$$

$$[x_k,\ x_{-l}]_+=[y_k,\ y_{-l}]_+=\delta_{kl},\quad k,\ l=-N,\ -(N-1),\cdots+N, \tag{12'}$$

$$[z_k,\ z_l]_-=0,\quad [x_k,\ y_{-l}]_+=0.$$

Thus we see that the eigenvalue of P' is a sum of the contributions from individual terms with z_k, which have four eigenvalues 0, 0, ±1 as shown by matrix representation:

$$x_k=\begin{pmatrix} & 1\\ 0 & \end{pmatrix}\times\begin{pmatrix}1 & \\ & -1\end{pmatrix},\quad x_{-k}=\begin{pmatrix} & 0\\ 1 & \end{pmatrix}\times\begin{pmatrix}1 & \\ & -1\end{pmatrix},$$

$$y_k=\begin{pmatrix}1 & \\ & 1\end{pmatrix}\times\begin{pmatrix} & 1\\ 0 & \end{pmatrix},\quad y_{-k}=\begin{pmatrix}1 & \\ & 1\end{pmatrix}\times\begin{pmatrix} & 0\\ 1 & \end{pmatrix}, \tag{13}$$

$$\therefore\quad z_k=x_k y_{-k}e^{-\frac{2\pi k}{N}i}-x_{-k}y_k e^{\frac{2\pi k}{N}i}=-\begin{pmatrix}0 & & e^{\frac{2\pi k}{N}i}\\ & & \\ e^{-\frac{2\pi k}{N}i} & & 0\end{pmatrix}.$$

Then

$$P'=4\sum_{k=1}^{N/2}\epsilon_k\sin\frac{2\pi k}{N},\quad \epsilon_k=0,\ 0,\ 1\ \text{ or }\ -1. \tag{14}$$

The restriction $x=1$ becomes as follows. We write

$$\prod_{n=1}^{N}(\cos\theta+\sin\theta x_{2n}x_{2n+1})=\prod\exp(\theta x_{2n}x_{2n+1})=\exp[\theta\sum x_{2n}x_{2n+1}] \tag{15}$$

$$=\exp[2\theta\sum_{k=1}^{N/2}(x_k y_{-k}+x_{-k}y_k)]\equiv\prod_{k=1}^{N/2}[2\theta iR_k],$$

$$=\Pi\{1+R_k^2(\cos 2\theta-1)+i\sin 2\theta R_k\},\tag{15'}$$

since
$$R_k^3=R_k.\tag{16}$$

Putting $\theta=\pi/2$, we have

$$(-1)^{\frac{N}{2}}x=\prod_{k=1}^{N/2}(1-2R_k^2).\tag{17}$$

R_k turns out to commute with z_k and has the eigenvalues ε_k^2. Hence we get the condition

$$(-1)^{N/2}x=\Pi(1-2\varepsilon_k^2)=(-1)^{N/2}.\tag{18}$$

which determines the number of non-zero ε_k.

Thus we see that the use of anticommuting quantities is successful for the operator (8). The problem of diagonalizing the operator has been reduced to diagonalizing a quadratic form. However, this ceases to be the case when we take instead of (8) the operator P defined in (1), which calls for a quartic form rather than a quadratic.

§ 3. Onsager's Problem.

Now let us take up the case of Onsager[2], that is, a plane square lattice with interaction according to Ising model:

$$E=J\sum s_i s_i'+J'\sum s_i s_{i+1},\tag{19}$$

where s_i represents the spin $(=\pm 1)$ of the ith site on a certain row and s_i' that of the neighboring one on the next row. The summation extends over all the sites in the lattice. According to Onsager, the problem of obtaining the partition function $F=Tr\exp(-E/kT)$ is reduced to asking for the largest eigenvalue of the operator

$$H=\exp[\beta\sum_{n=1}^{N}s_n s_{n+1}]\exp[\alpha\sum_{n=1}^{N}c_n],\quad N+1\equiv 1,\tag{20}$$

where
$$s_n^2=c_n^2=1,$$
$$[s_n,c_n]_+=0,\tag{20'}$$
$$[s_n,s_m]_-=[s_n,c_m]_-=[c_n,c_m]_-=0,\quad n\neq m.$$

We shall not occupy ourselves with the deduction of the formula (20), but suppose it as given, and try to solve the eigenvalue problem for H. Writing (20) simply as

$$H=\exp[\beta\sum S_n]\exp[\alpha\sum C_n],$$
$$S_n\equiv s_n s_{n+1},\quad C_n\equiv c_n,\tag{21}$$

we see that S_n and C_n have the properties

$$[S_n, C_{n\pm1}]_+=0, \quad \text{otherwise all commuting, and}$$
$$S\equiv S_1 S_2 \cdots S_N=1, \quad C\equiv C_1 C_2 \cdots C_N=\pm 1. \tag{21'}$$

In view of the relations (20) and (21′), a natural idea which suggests itself is to express H in terms of the operators of the second quantization as was illustrated in the last section. Indeed, we can put

$$s_n s_{n+1}=i(a_n^+-a_n)(a_{n+1}^++a_{n+1}), \quad c_n=2a_n^+a_n-1,$$
$$a_n^++a_n=c_1c_2\cdots c_{n-1}s_n, \quad i(a_n^+-a_n)=c_1c_2\cdots c_n s_n, \tag{22}$$

with $$[a_n, a_m^+]_+=\delta_{nm}.$$

Here c_n plays just the rôle of the sign function. Such a procedure has been independently devised by Husimi. Instead of (22) we may directly start from the anticommuting orthogonal basis $\{x_n\}$. Thus we replace S_n and C_n by

$$S_n=ix_{2n}x_{2n+1}, \quad C_n=ix_{1n-1}x_{2n},$$
$$[x_r, x_s]_+=2\delta_{rs}, \quad r, s=1, 2, \cdots 2N,$$
$$S=\Pi S_n=i^N x_2 x_3 \cdots x_{2N} x_1=-X, \tag{23}$$
$$C=\Pi C_n=i^n x_1 x_2 \cdots x_{2N}\equiv X.$$

This substitution corresponds to the case $C=-1$ if we confine ourselves to the subspace $X=1$. The case $C=1$ obtains when one of the defining equations, say for C_1, is substituted by

$$C_1=-ix_1x_2. \tag{23'}$$

However, such minor modifications do not affect the final result for large N except some delicate problems like the boundary tension, so that we shall look apart from the eigenvalues of S and C in order to avoid unnecessary complexity. With the choice (20), the operator H can now be written as

$$H=\exp[i\beta\textstyle\sum x_{2n}x_{2n+1}]\exp[iu\textstyle\sum x_{2n-1}x_{2n}]\equiv H_2H_1. \tag{24}$$

As mentioned before, the quantities $x_1, x_2, \cdots x_{2N}$ may be regarded as orthogonal basic vectors in a $2N$-dimensional space. In this space, a rotation of angle θ in the plane spanned by x_r and x_s is expressed by the operator

$$x_n \to U^{-1}x_nU,$$
$$U=\exp[(\theta/2)x_rx_s]=\cos(\theta/2)+\sin(\theta/2)x_rx_s. \tag{25}$$

In fact it is easily verified that

$$U^{-1}x_rU=\cos\theta x_r+\sin\theta x_s,$$
$$U^{-1}x_sU=-\sin\theta x_r+\cos\theta x \;, \tag{26}$$
$$U^{-1}x_nU=x_n, \quad n\neq r, s.$$

Then (24) means nothing but a product of rotations, first by angle $2\alpha i$, in the planes $x_{2n-1}x_{2n}$, and then by angle $2\beta i$, in the planes $x_{2n}x_{2n+1}$. Let us suppose that in a suitable coordinate system $\{x_n'\}$ the rotation H is brought to Jordan's standard form, so that

$$H=\exp\left[i\sum x_{2n}'x_{2n+1}'\gamma_n\right]. \tag{27}$$

Since $|ix_n'x_{n+1}'|=1$, the eigenvalue of the operator H that operates on a eigenvector Ψ is found in the expression

$$H\Psi=\exp\left[\sum\varepsilon_n\gamma_n\right]\Psi,\quad \varepsilon_n=\pm 1. \tag{28}$$

We have only to solve the problem in the $2N$-dimensional subspace $\{x_n\}$ of basic vectors, instead of the whole space of Ψ with 2^N dimensions. By the first rotation H_1, a vector (x_n) is transformed into another vector (y_n), which, by the second rotation H_2, again changes into a third (z_n). Thus

$$\begin{aligned} y_{2n-1}&=\cos 2\alpha' x_{2n-1}+\sin 2\beta' x_{2n}, \\ y_{2n}&=-\sin 2\alpha' x_{2n-1}+\cos 2\beta' x_{2n}, \end{aligned} \tag{28}$$

$$\begin{aligned} z_{2n}&=\cos 2\beta' y_{2n}+\sin 2\beta' y_{2n+1}, \\ z_{2n+1}&=-\sin 2\beta' y_{2n}+\cos 2\beta' y_{2n+1}, \end{aligned} \tag{28'}$$

or

$$\begin{aligned} z_{2n}&=-bx_{2n-1}+dx_{2n}+ax_{2n+1}+cx_{2n+2}; \\ z&=-ax_{2n}+dx_{2n+1}+bx_{2n+2}+cx_{2n-1}. \end{aligned} \tag{28''}$$

Here $\alpha'=\alpha i$, $\beta'=\beta i$,

$$\begin{aligned} a&=\cos 2\alpha' \sin 2\beta', \quad b=\sin 2\alpha' \cos 2\beta', \\ c&=\sin 2\alpha' \sin 2\beta', \quad d=\cos 2\alpha' \cos 2\beta'. \end{aligned} \tag{28'''}$$

In matrix form,

$$H=\begin{pmatrix} \ddots & & & & & & \\ -b & d & a & c & & & \\ & & \ddots & & & & \\ c & -a & d & b & & & \\ & & & \ddots & & & \\ & & -b & d & a & c & \\ & & & & \ddots & & \\ & & c & -a & d & b & \\ & & & & & \ddots & \end{pmatrix} \begin{matrix} \\ \\ \\ \cdots 2n \\ \\ \cdots 2n+1 \\ \\ \\ \\ \end{matrix} \tag{29}$$

The eigenvalue equation for H becomes

$$\begin{aligned} -bx_{2n-1}+(d-\lambda)x_{2n}+ax_{2n+1}+cx_{2n+2}&=0, \\ cx_{2n-1}-ax_{2n}+(d-\lambda)x_{2n+1}+bx_{2n+2}&=0, \end{aligned} \tag{30}$$

or

$$\begin{aligned} ax_{2n+1}+cx_{2n+2}&=bx_{2n-1}+(\lambda-d)x_{2n}, \\ (\lambda-d)x_{2n+1}-bx_{2n+2}&=cx_{2n-1}-ax_{2n}. \end{aligned} \tag{30'}$$

Let us write

$$(x_{2n+1},\ x_{2n+2})=\psi_{n+1}\,,\quad (x_{2n-1},\ x_{2n})=\psi_n\,, \tag{31}$$

and

$$A=\begin{pmatrix} a, & c \\ \lambda-d, & -b \end{pmatrix},\quad B=\begin{pmatrix} b, & \lambda-d \\ c, & -a \end{pmatrix}. \tag{31'}$$

Then (30′) means

$$A\psi_{n+1}=B\psi_n\,,\quad \text{or}\quad \psi_{n+1}=A^{-1}B\psi_n\equiv C\psi_n\,, \tag{32}$$

hence $$\psi_n=C^{n-1}\psi_1\,. \tag{32'}$$

The periodicity condition requires that $\psi_{N+1}=\psi_1$, or

$$(E-C^N)\psi_1=0,\quad \det|E-C^N|=0. \tag{33}$$

If we denote the Nth roots of unity by $\xi_k=\exp(2\pi ki/N)=\exp(\varphi_k i)$, $k=1, 2,\cdots N$, then

$$\det|\xi_k E-C|=0,\quad \text{or}\quad |\xi_k A-B|=0, \tag{33'}$$

that is,

$$\begin{vmatrix} \xi_k a-b, & \xi_k c-(\lambda-d) \\ \xi_k(\lambda-d)-c, & -\xi_k b+a \end{vmatrix}=0. \tag{33''}$$

This gives a quadratic equation for λ:

$$\xi\lambda^2-\lambda[2\xi d+\xi^2 c+c]+\xi(a^2+b^2+c^2+d^2)-(1+\xi^2)(ab-cd)=0. \tag{34}$$

Substituting (28‴),

$$a^2+b^2+c^2+d^2=1,\quad ab-cd=0,$$

$$\lambda^2-\lambda[2d+(\xi+\xi^{-1})c]+1=0,$$

or $$\lambda^2-2\lambda(\cos 2\alpha'\cos 2\beta'+\sin 2\alpha'\sin 2\beta'\cos\varphi_k)+1=0. \tag{34'}$$

Putting the two roots as $\exp(\pm 2\gamma_k' i)$,

$$(1/2)(e^{2\gamma' i}+e^{-2\gamma' i})=\cos 2\gamma_k'=\cos 2\alpha'\cos 2\beta'+\sin 2\alpha'\sin 2\beta'\cos\varphi_k. \tag{35}$$

Returning to the original constants α and β, and writing $\gamma=\gamma' i$, we get just the relation given by Onsager:

$$\cosh 2\gamma_k=\cosh 2\alpha\cosh 2\beta-\sinh 2\alpha\sinh 2\beta\cos\varphi_k,\quad \varphi_k=2\pi k/N. \tag{35'}$$

The largest eigenvalue of H becomes then, according to (28),

$$H_{max}=\exp[\Sigma|\gamma_k|]. \tag{36}$$

Such a procedure also applies, *mutatis mutandis*, to certain variants of Onsager model, e.g. the honeycomb lattice treated by Husimi and Syôzi. It fails, however, in case of those lattices with too many nearest neighbours, in particular the three-dimensional lattices. For we cannot express a quantity which anticommutes with more than two neighbors by a product of two totally anticommuting vectors. In fact the three-dimensional model necessitates us to resort to products of four x's as illustrated below:

$$H=\exp[\beta\sum(s_{nm}s_{n,m+1}+s_{nm}s_{n+1,m})\exp[a\sum c_{nm}]$$
$$=\exp[i\beta\sum h_r]\exp[a\sum k_s], \tag{37}$$
$$h=x_1x_3',\ x_4x_2'',\quad k=x_1x_2x_3x_4.\quad \text{(See Fig. 2.)}$$

Supplementary conditions are also to be considered. The occurence of quartic forms may look reasonable when we note the so-called dual transformation. Such transformations do really exist even in the three-dimensional case. In fact let us take a cubic lattice with atoms located at the middle point of each edge of the cell, and let each set of the four atoms lying on a side plane determine a four-body interaction energy $\mu s_1s_2s_3s_4$, $s_i=\pm1$. Then it is easy to show that this lattice is dual to a simple cubic lattice made up of the body center sites of the former with ordinary two-body interactions. Perhaps the four-body force may have some bearing on the quantity k_s in (37).

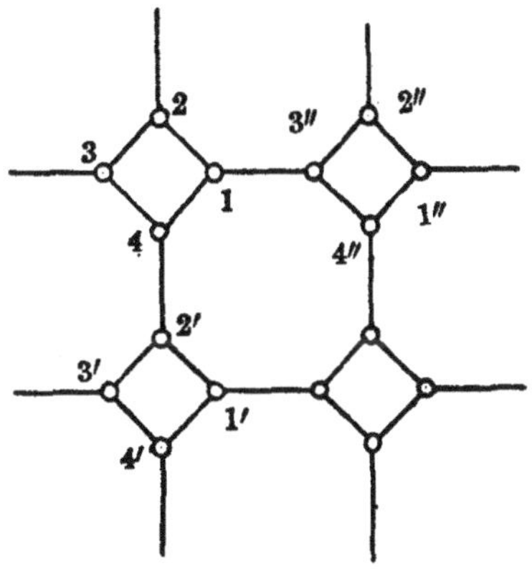

Fig. 2.

§ 4. Kramers-Wannier's method of approach.

Our next investigations concern the screw lattice model adopted by Kramers and Wannier for solving the two-dimensional case. This model seems more convenient for general purposes than that used by Onsager. Now let us try to transform the Onsager model into the K-W model as follows. Rewrite (20) (after reversing the order of H_1 and H_2) as

$$H_1H_2=\exp[iax_1x_2]\exp[iax_3x_4]\cdots\exp[i\beta x_2x_3]\exp[i\beta x_4x_5]\cdots$$
$$=\exp[iax_1x_2]\exp[iax_3x_4]\exp[i\beta x_2x_3]\exp[iax_5x_6]\exp[i\beta x_4x_5]\cdots$$
$$=\exp[iax_1x_2]\prod_{n=1}^{N-1}(\exp[iax_{2n+1}x_{2n+2}]\exp[i\beta x_{2n}x_{2n+1}]\exp[i\beta x_{2N}x_1], \tag{38}$$

which can be modified, without grave consequences, to

$$H=\prod_{n=1}^{N}\exp[iax_{2n+1}x_{2n+2}]\exp[i\beta x_{2n}x_{2n+1}]\equiv\Pi H_n,$$
$$H_n\equiv\exp[iax_{2n+1}x_{2n+2}]\exp[i\beta x_{2n}x_{2n+1}]. \tag{38'}$$

It turns out more convenient to consider an operator which has its component factors H_n in the reverse order to (51) (which is also equivalent to a renumbering of the order). Then define a displacement operator P, which changes x_n into x_{n+1}, and put

$$\Psi_n=H_nH_{n-1}\cdots H_1\Psi_0 \tag{39}$$

with obvious relations

$$H_n=P^nH_0P^{-n},\quad H_0=\exp\left[i\beta x_{2N}x_1\right]\exp\left[i\alpha x_1x_2\right]. \tag{39'}$$

The relation (a Schrödinger equation for discrete time variable !)

$$\Psi_{n+1}=H_n\Psi_n \tag{40}$$

is turned into

$$\Psi_{n+1}'=P^{-1}H_0\Psi_n'\equiv A\Psi_n' \tag{40'}$$

by putting

$$P^{-n}\Psi_n=\Psi_n' \quad \text{or} \quad \Psi_n=P^n\Psi_n'. \tag{40''}$$

But $P^{-1}H_0$ is now independent of n, hence the eigenvalue problem for Ψ_n is equivalent to

$$\lambda\Psi'=A\Psi'=P^{-1}H_0\Psi',\quad \lambda^N=E,\quad (H\Psi_0=\Psi_N=E\Psi_0). \tag{41}$$

P can be expressed in terms of the basic vectors as

$$\begin{aligned}P=&\exp\left[\frac{\pi}{4}x_1x_{2N-1}\right]\exp\left[-\frac{\pi}{4}x_1x_{2N-3}\right]\cdots\exp\left[\pm\frac{\pi}{4}x_1x_3\right]\\ &\times\exp\left[\frac{\pi}{4}x_2x_{2N}\right]\exp\left[-\frac{\pi}{4}x_2x_{2N-2}\right]\cdots\exp\left[\pm\frac{\pi}{4}x_2x_4\right]\end{aligned} \tag{42}$$

because of the relations

$$\begin{aligned}&\exp\left[\frac{\pi}{4}x_1x_2\right]x_1\exp\left[-\frac{\pi}{4}x_1x_2\right]\\ &=\frac{1}{2}[1-x_1x_2]x_1[1+x_1x_2]=\frac{1}{2}(x_1-x_1+2x_2)=x_2,\\ &\exp\left[\frac{\pi}{4}x_1x_2\right]x_2\exp\left[-\frac{\pi}{4}x_1x_2\right]=\frac{1}{2}(x_2-x_2-2x_1)=-x_1.\end{aligned} \tag{42'}$$

Thus the operator A can again be regarded as a product of rotation operators. Let us look for an eigenvector of the form

$$x=\sum_{s=1}^{N-1}\lambda^{n-1}(ax_{2n-1}+bx_{2n})+cx_{2N}+\lambda^{N-1}ax_{2N-1}. \tag{43}$$

Substituting in the equation $AxA^{-1}=\lambda x$, we get the following result:

$$\left.\begin{aligned}&\cos 2a'\cos 2\beta' x-\sin 2a'\cos 2\beta' y+\sin 2\beta' z=\lambda^N x\,,\\&\sin 2a' x+\cos 2a' y=\lambda z\,,\\&-\sin 2\beta'\cos 2a' x+\sin 2a'\sin 2\beta' y+\cos 2\beta' z=\lambda^{N-1}y\,,\end{aligned}\right\}\qquad(44)$$

or

$$\left.\begin{aligned}&\cosh 2a\cosh 2\beta x+i\sinh 2a\cosh 2\beta y-i\sinh 2\beta z=\lambda^N x\,,\\&-i\sinh 2a x+\cosh 2a y=\lambda z\,,\\&i\sinh 2\beta\cosh 2a x-\sinh 2a\sinh 2\beta y+\cosh 2\beta z=\lambda^{N-1}y\,.\end{aligned}\right\}\qquad(44')$$

Eliminating z by virture of the second equation,

$$\left.\begin{aligned}&(\cosh 2a\cosh 2\beta-\sinh 2a\sinh 2\beta\lambda^{-1})x+i(\sinh 2a\cosh 2\beta\\&\qquad-\sinh 2\beta\cosh 2a\lambda^{-1}y=\lambda^N x\,,\\&i(\sinh 2\beta\cosh 2a-\sinh 2a\cos 2\beta\lambda^{-1})x-(\sinh 2a\sinh 2\beta\\&\qquad-\cosh 2\beta\cosh 2a\lambda^{-1}y=\lambda^{N-1}y\,,\end{aligned}\right.\qquad(44'')$$

which determine λ according to

$$\begin{aligned}&\lambda^{2N-1}+\sinh 2a\sinh 2\beta\lambda^N+(-\cosh 2\beta\cosh 2a-\cosh 2a\cosh 2\beta)\lambda^{N-1}\\&\quad+\sinh 2a\sinh 2\beta\lambda^{N-2}+(\sinh^2 2a\sinh^2 2\beta+\cosh^2 2\beta\cosh^2 2a\\&\quad-\sinh^2 2a\cosh^2 2\beta-\sinh^2 2\beta\cosh^2 2a\lambda^{-1}\\&\quad+(-\sinh 2a\cosh 2a\sinh 2\beta\cosh 2\beta+\sinh 2a\cosh 2a\sinh 2\beta\cosh 2\beta)\lambda^{-2}\\&\quad-\cosh 2a\sinh 2a\cosh 2\beta\cosh 2\beta+\sinh 2a\cosh 2a\sinh 2\beta\cosh 2\beta=0.\end{aligned}$$

This in turn reduces to

$$\lambda^{2N}+\sinh 2a\sinh 2\beta(\lambda+\lambda^{-1})-2\cosh 2a\cosh 2\beta+\lambda^{-N}=0.\qquad(45')$$

Putting $\lambda=\exp 2\gamma$,

$$\cosh 2N\gamma=\cosh 2a\cosh 2\beta-\sinh 2a\sinh 2\beta\cosh\gamma.\qquad(46)$$

This is the equation for γ. To solve this, we assume that $2\gamma=2\gamma_0+\omega i$, and substitute in (46), thereby making use of the relation

$$\cosh(2\gamma_0+\omega i)=\cosh 2\gamma_0\cos\omega+i\sinh 2\gamma_0\sin\omega\,.\qquad(47)$$

Comparison of real and imaginary parts yields the relations

$$\begin{aligned}&\cosh 2N\gamma_0\cos N\omega=\cosh 2a\cosh 2\beta-\sinh 2a\sinh 2\beta\cosh\gamma_0\cos\omega\,,\\&\sinh 2N\gamma_0\sin N\omega=\sinh 2a\sinh 2\beta\sinh\gamma_0\sin\omega\,.\end{aligned}\qquad(48)$$

Choose $\omega=k\pi/N$, $k=1,2,\cdots 2N$, so that $\sin N\omega=0$, $\cos N\omega=(-1)^k$. In the limit $N\to\infty$ and $N\gamma_0\to\Gamma$=finite, we have $\gamma_0=0$, and both equations become satisfied by this choice of ω. Thus

$$\pm\cosh 2\Gamma=\cosh 2a\cosh 2\beta-\sinh 2a\sinh 2\beta\cos\omega\,,\qquad(49)$$

with $\omega=k\pi/N$. (49′)

But the minus sign is impossible, hence the allowed values for ω are

$$\omega=2k\pi/N, \quad k=1, 2, \cdots N. \tag{49''}$$

This completely agrees with the previous results. The case of the three-dimensional lattice can be treated analogously, but we shall not give here the details.

§ 5. Additional Remarks.

The mathematical tricks employed in the preceding sections seem to allow a rather natural explanation from a somewhat more general point of view. The essential point is that we consider, instead of an ordinary eigenvalue problem, a different one of the form

$$[H, X]_-=\lambda X \tag{50}$$

for the "eigenoperator" X[6]. Important consequences drawn from this equation are: (a) λ is the difference of two eigenvalues of H: $\lambda=E_n-E_m$; (b) X transforms certain eigenvector Ψ_m of H with E_m into another one Ψ_n with eigenvalue $E_n=E_m+\lambda$; and (c) Product of two eigenoperators X_2X_1 is again an eigenoperator, with eigenvalue $\lambda=\lambda_1+\lambda_2$, transforming Ψ_m into a third eigenvector Ψ_l. If we integrate the relation (50) according to

$$\frac{d}{ds}X=[H, X], \tag{51}$$

we arrive at the equation

$$e^{sH}Xe^{-sH}=e^{\lambda}X. \tag{51}$$

The equation (51) just corresponds to the problem of obtaining the eigenoperator $\{x_n'\}$ for the rotation H encountered in Section 3 and 4. When the operator H has a simple structure, a general eigenoperator X will be factorized into a product of prime eigenoperators:

$$X=X_1X_2\cdots X_k,$$

with $$e^{\lambda}=e^{\lambda_1+\lambda_2+\cdots\lambda_k}. \tag{52}$$

The largest λ is the difference of the largest and smallest eigenvalues of sH in (52). If the eigenvalues E_i are symmetrically distributed around zero, E_{max} will itself be given by $E_{max}=\lambda_{max}/2$, a fact which casts a sidelight on the procedure used in the previous sections.

The main part of the present work had been completed nearly two years ago. It is through the kindness of Professor Husimi and Mr. Syôzi of Osaka University that the author enjoys the opportunity of publishing this note. He wishes to express on this occasion his cordial thanks to them and also to Professor Kubo of Tokyo University for helpful discussions.

References.

(1) H. A. Kramers and G. H. Wannier, Phys. Rev. **60** (1941), 252, 263.
E. N. Lassetre and J. P. Howe, J. Chem. Phys. **9** (1941), 747 and 801.
E. W. Montroll and J. E. Mayer, J. Chem. Phys. **9** (1941), 626.

(2) L. Onsager, Phys. Rev. **65** (1944), 117.

(3) Husimi and Syôzi, Lectures at the Annual Meeting of the Phys. Soc., 1948 and 1949; Husimi and Syozi, Prog. Theor. Phys. in press.

(4) An alternative way for solving (3) is to use commutative, rather than anticommutative, operators which corresponds to Bose statistics, together with some supplementary conditions. This leads to Bethe's rigorous solution for one-dimensional model (ZS. f. Phys. **71** (1931), 205; *Handbuch d. Phys.* XXIV/2, 604). We shall not here enter into details about it.

(5) Cf. Cartan: *Leçons sur la théorie des spineurs*, 1938.

(6) A deeper research on this subject will be found in: Y. Nambu, *On the method of the third quantization*, Prog. Theor. Phys. **4** (1949), 331 and 399.

Progress of Theoretical Physics, Vol. V, No. 1, Jan.～Feb., 1950.

The Use of the Proper Time in Quantum Electrodynamics I.

Yôichirô NAMBU

*Department of Physics, University of City Osaka**

(Received November 8, 1949)

I. Introduction

The space-time approach to quantum electrodynamics, as has been developed by Feynman,[1] seems to offer a very attractive and useful idea to this domain of physics. His ingenious method is indeed attractive, not only because of its intuitive procedure which enables' one to picture to oneself the complicated interactions of elementary particles, its ease and relativistic correctness with which one can calculate the necessary matrix elements or transition probabilities, but also because of its way of thinking which seems somewhat strange at first look and resists our minds that are accustomed to causal laws. According to the new standpoint, one looks upon the world in its four-dimensional entirety. A phenomenon that will come into play in this theatre is now laid out beforehand in full detail from immemorial past to ultimate future and one investigates the whole of it at glance. The time itself loses sense as the indicator of the development of phenomena; there are particles which flow down as well as up the stream of time; the eventual creation and annihilation of pairs that may occur now and then, is no creation nor annihilation, but only a change of directions of moving particles, from past to future, or from future to past; a virtual pair, which, according to the ordinary view, is foredoomed to exist only for a limited interval of time, may also be regarded as a single particle that is circulating round a closed orbit in the four-dimensional theatre; a real particle is then a particle whose orbit is not closed but reaches to infinity. . .

In such a view, a state with prescribed number of particles including real as well as virtual does not exactly correspond to a four-dimensional state in the ordinary sense, that is, a state represented by the wave function satisfying the time dependent Schroedinger equation. But the former is rather a part of the latter in which any number of virtual particles may be allowed to occur. To obtain an idea of the actual state we shall have to sum over all possibilities as to the number of virtual particles.

The interpretation of the four-dimensional state in the present sense becomes also somewhat different from the conventional one as giving the transition pro-

*) Now staying temporarily at Tokyo University.

bability or amplitude from a given state A to a final state F in the three-dimensional space. We can rather ask for the relative probability that a four-dimensional state A-I-F with prescribed real as well as virtual (intermediate) particles be realized in nature. This will be zero unless the arbitrary chosen A-I-F is not such as is an actually possible transition under the Schroedinger equation.

The above-mentioned view of the entire space-time behavior of nature ***sub specie aeternitatis***, however, might not appeal to a reason which is liable to think in the language of differential equations and pursue the development of things along a certain parameter. In fact we find it hard to regard the world line of a particle as a mere status of that particle, but are unconciously following the motion of an imaginary mass point along the world line. Thus, in Feynman's theory where the ordinary time loses its rôle as the indicator of the development of the world, it would still be convenient to introduce some parameter with which the four-dimensional world is going to shape itself. How this is possible to a certain extent we shall see in what follows.

2. Formal introduction of the proper time

Let us consider a wave function obeying the ordinary Schroedinger equation

$$i\frac{\partial}{\partial t}\psi(t,x)=H(x)\psi(t,x). \tag{1}$$

The scalar product (ψ,φ) of two wave functions ψ and φ at a given time carries the meaning of the probability amplitude for finding a state characterized by ψ when we know that the system is in the state φ. This interpretation is based upon the mathematical fact that the length of the wave function vector is constant in time according to (1). If we go over to the four-dimensional standpoint and regard the behavior of ψ in both t and x for a finite interval of time as characterizing the state $\Psi(t,x)$, we shall naturally have to define the norm of a wave function by

$$(\Psi,\Psi)=\int_{t_0}^{t_0+T}\int_{-\infty}^{\infty}(\psi(tx),\psi(tx))dx\,dt=T\int(\psi(x),\psi(x))\,dx, \tag{2}$$

which is a multiple of the ordinary norm of the state ψ. Thus we have only to alter the normalization of the wave function for the transition from ψ to Ψ, though this means an infinite factor when the time interval is extended indefinitely. The probability amplitude (ψ,φ) mentioned above is then in the new standpoint merely a probability amplitude density for a cross section of time from which the full amplitude is derived by integration with respect to t:

$$(\Psi,\Phi)=\int(\psi,\phi)\,dt. \tag{3}$$

Eq. (3) allows the following interpretation. Suppose we take for Ψ and Φ two stationary states for a system, then (3) will be zero unless Ψ and Φ represents the same state. But if some perturbation is introduced in the system, the system will no longer remain in the original (three-dimensional) state φ, but the space-time behavior of Φ will be expressed by decomposing it with respect to unperturbed eigenfunctions,

$$\Phi=\sum_k a_k(t)\,\Phi_k(t), \qquad a_k(t_0)=\delta_{k0}. \tag{4}$$

The probability that we find the system after an infinite lapse of time in the state k is given by

$$\lim_{t\to\infty}|a_k(t)|^2=\lim_{t\to\infty}|(\psi_k,\Phi)|^2. \tag{5}$$

If we displace the time scale and shift the initial point to $-\infty$, we obtain the probability for finding a state Ψ_k or ψ_k (referred to unperturbed coordinates) irrespective of time when we know the system was in the state φ_0 and the perturbation has been, say adiabatically, switched on, by

$$|a_k(\infty)|^2=\left|\int_{-\infty}^{\infty}(\Psi_k(t),\Phi(t))\,dt\right|^2=|(\Psi_k,\Phi)|^2, \tag{6}$$

for we may suppose that $a_k(t)$ has reached its stationary value for any finite t. In this way the four-dimensional scalar product acquires a physical meaning.

Now let us investigate the problem from a different point of view. The Schroedinger equation (1) implies, when regarded four-dimensionally, a sort of supplementary condition imposed on Ψ:

$$(i\partial/\partial t-H)\Psi=0 \tag{7}$$

since we no longer look upon t as a parameter along which Ψ develops itself. Consequently a Ψ which corresponds to the real world must be of the form

$$\Psi_{\text{real}}=\delta\left(i\frac{\partial}{\partial t}-H\right)\Psi_0. \tag{8}$$

If we introduce here a redundant variable τ and assume an equation of the type

$$i\frac{\partial}{\partial\tau}\Psi=\left(i\frac{\partial}{\partial t}-H\right), \tag{9}$$

and an accompanying eigenvalue problem

$$\lambda\Psi=\left(i\frac{\partial}{\partial t}-H\right)\Psi, \tag{9'}$$

(8) becomes

$$\Psi_{\text{real}}=\delta(\lambda)\Psi_0=\frac{1}{2\pi}\int_{-\infty}^{\infty}\Psi(\tau)\,d\tau. \tag{10}$$

The four-dimensional view has a static character in that a state is defined once for all t and after that a condition is invoked in order that it correspond to any reality. But now that τ is introduced and (9) has regained the aspect of an ordinary Schroedinger equation, the new variable must play a similar rôle as that the ordinary time played in (1). As τ goes on, a wave packet localized over certain four-dimensional volume will move to and fro and change its shape gradually. Then we may think of τ as something like the proper time of the particle represented by the wave packet. If we directed a camera on to that particle with the shutter open for an infinitely long time, we should obtain a vague strip of world line as the locus of the particle, which would correspond to the real wave function Ψ_{real}. Eq. (10) tells us just this situation in the mathematical language.

Next we shall consider the transition probability. If a perturbation term is inserted, (9) becomes

$$i\frac{\partial}{\partial t}\Psi=\left(i\frac{\partial}{\partial t}-H_0-H_1\right)\Psi, \tag{11}$$

while the free particle wave functions to which we refer the initial and final state satisfy the equation

$$\left(i\frac{\partial}{\partial t}-H_0\right)\Psi_A=\left(i\frac{\partial}{\partial t}-H_0\right)\Psi_F=0. \tag{12}$$

Let us write $i\partial/\partial t-H_0\equiv L$, and perform the transformation

$$\Psi=\exp[-iL\tau]\Psi_1. \quad H_1'=\exp[iL\tau]H_1\exp[-iL\tau], \tag{13}$$

(11) then goes over into

$$i\frac{\partial}{\partial t}\Psi_1=-H_1'\Psi_1. \tag{14}$$

Starting at $\tau=\tau_0$ from a free state Ψ_A, the transition amplitude at $\tau=\tau$ to a free state Ψ_F will be expressed by

$$P(AF)=\delta(L)(F|U(\tau\tau_0)|A), \quad \Psi(\tau)=U(\tau\tau_0)\Psi(\tau_0), \tag{15}$$

since H_1' would in general bring a real particle into unreal one for which $L\Psi\neq 0$. Eq. (15) admits a twofold interpretation: a) formally, $P(AF)$ is given from $(\Psi_F, \Psi(\tau))$ by taking the zero-frequency component of $\Psi(\tau)$:

$$P(AF)=\frac{1}{2\pi}\int_{-\infty}^{\infty}(\Psi_F, \Psi(\tau))\,d\tau\,; \tag{16}$$

b) physically, it is the accumulated amplitude for the transition starting from the state A and arriving at F after an infinite lapse of time:

$$P(AF)=\int_{-\infty}^{\infty}(\Psi_F, \Psi(\tau))\,d\tau=\int_{-\infty}^{\infty}(F|U(\tau,-\infty)|A)\,d\tau. \tag{17}$$

(16) and (17) differ only by a normalization factor.

Thus we see that the formal introduction of a redundant parameter and its identification with the proper time carries a bit of mathematical convenience as well as physical plausibility. In order to convince ourselves further on this point, we shall next recall Fock's theory and go ahead on his line.

3. Theory of Fock and its extension

Fock[2] once introduced the concept of the proper time in the Dirac electron in parallel with the classical theory and proved its correspondence to the ordinary proper time. It will be briefly recapitulated below. The classical Lagrangian for an electron interacting with the electromagnetic field is given by

$$L_0 = -mc\dot{x}_\mu^2 - mc^2/2 - (e/c)(\dot{x}_\mu A_\mu), \quad \dot{x}_\mu = dx_\mu/d\tau, \tag{18}$$

and the equation of motion follows from the variational principle

$$\delta S = 0, \quad S = \int^\tau L_0 d\tau, \tag{19}$$

together with the supplementary condition

$$\dot{x}_\mu^2 = \text{const.} = -c^2. \tag{20}$$

Eliminating τ from (19) and (20) we get the ordinary action function $S(x_\mu)$ for the system. The Hamilton-Jacobi partial differential equation which follows from above is

$$\frac{\partial S}{\partial \tau} + \frac{1}{2m}\left[\left(\text{grad } S + \frac{e}{c}A\right)^2 - \frac{1}{c}\left(\frac{\partial S}{\partial t} - e\Phi\right)^2\right] + m^2c^2 = 0, \tag{21}$$

with the condition

$$\partial S/\partial \tau = 0. \tag{21'}$$

Now turning to quantum theory, the Dirac electron obeys the wave equation

$$(\gamma_\mu D_\mu + \varkappa)\psi = 0, \quad D_\mu = \partial/\partial x_\mu - e/\hbar c \cdot A_\mu. \tag{22}$$

Such a ψ may be expressed as

$$\psi = (\gamma_\mu D_\mu - \varkappa)\Psi, \tag{23}$$

where Ψ in turn shall obey the second order differential equation

$$(\gamma D + \varkappa)(\gamma D - \varkappa)\Psi = \left[\left(\Box - \frac{ei}{\hbar c}A\right)^2 - \varkappa^2 + \frac{e}{\hbar c}\sigma_{\mu\nu}F_{\mu\nu}\right]\Psi \equiv \Lambda\Psi = 0. \tag{24}$$

Fock proposed to introduce a function F which satisfies instead of (24) an equation closely analogous to (21):

$$i\hbar\frac{\partial}{\partial \tau}F = \frac{\hbar^2}{2mc}\Lambda F, \tag{25}$$

and express Ψ as a suitable integral of F over τ:

$$\Psi=\int_C F d\tau, \tag{26}$$

the contour being taken in such a way that

$$\frac{\partial}{\partial\tau}\Psi=\int_C\frac{\partial F}{\partial\tau}d\tau=F|_C=0, \tag{27}$$

which corresponds to the condition (21′).

When we neglect temporarily the electromagnetic interaction, the electron becomes free and (25) is solved by

$$F(\tau)=\exp\left(-\frac{i\hbar\tau}{2mc}\Lambda\right)F(0). \tag{28}$$

Given an initial distribution of the wave packet over space-time, it will change with τ by the relation (28). But only those stationary states with zero eigenvalue can ever correspond to a real world:

$$\Lambda F=\lambda F=0. \tag{29}$$

In classical language, a preassigned pattern of streamlines representing the motion of an aggregate of electrons will change according to the dynamical law (21). Only the stationary stream is the true state of the world where individual electrons follow invariant paths.

Now let us determine the behavior of the wave packet which starts from a delta function $F(0)=\delta(x)$ at $\tau=0$. The answer is easily given by the Fourier representation

$$\begin{aligned}F(\tau,x)&=\left(\frac{1}{2\pi}\right)^4\int e^{i\lambda\tau}e^{ik_\mu x_\mu}\,\delta(\lambda+(k^2+\varkappa^2)/2\varkappa)\,dk_\mu d\lambda\\&=\frac{\varkappa}{\pi}\int\Delta^{(1)}(x^2,\varkappa^2+2\varkappa\lambda)e^{i\lambda\tau}d\tau\equiv\varkappa\Delta_\varkappa(\tau,x^2)\\&=-\frac{i}{4\pi^2}\frac{\varkappa^2}{\tau^2}\exp\left(\frac{\varkappa}{2\tau}x^2i+\frac{\varkappa}{2}\tau i\right),\qquad(\tau>0).\end{aligned}\tag{30}$$

$F(\tau,x)$ is the probability amplitude for finding the particle at x when we know that it was at the origin a time τ ago. According to the previous argument the real observed amplitude will be obtained when we sum $F(\tau,x)$ for all $\tau>0$, which results in

$$\begin{aligned}\int_0^\infty F(\tau,x)\,d\tau&=\frac{2\varkappa}{(2\pi)^3}\int_0^\infty\cdot\delta_+(k^2+\varkappa^2)\,e^{ik_\mu x_\mu}\,dk_\mu\\&=\frac{2\varkappa}{i}\left(\bar\Delta(x)+\frac{i}{2}\Delta^{(1)}(x)\right)=\varkappa\Delta_F.\end{aligned}\tag{31}$$

Here $\bar{\Delta}$, $\Delta^{(1)}$, and Δ_F bear the meaning as defined in Schwinger,[3] Feynman,[1] and Dyson[4]:

$$\bar{\Delta}(x)=\frac{1}{(2\pi)^4}\int\frac{1}{k^2+\kappa^2}\,e^{ik_\mu x_\mu}\,dk_\mu,$$

$$\Delta^{(1)}(x)=\frac{1}{(2\pi)^3}\int\delta(k^2+\kappa^2)\,e^{ik_\mu x_\mu}\,dk_\mu,$$

$$\Delta_F(x)=-2i\bar{\Delta}(x)+\Delta^{(1)}(x)=\frac{2}{(2\pi)^3}\int\delta_+(k^2+\kappa^2)\,e^{ik_\mu x_\mu}\,dk_\mu. \tag{32}$$

Thus we see that we shall be able to arrive at Δ_F, the fundamental quantity in positron theory, if we take $F(\tau,x)$ for the (five-dimensional) commutation relation between quantized wave functions satisfying (25). The proper time τ is nothing but what has been a mere parameter in the integral representation of the Δ-functions. The integration with respect to τ followed above, not from $-\infty$ to $+\infty$ but only for positive τ, is a departure from the standpoint expounded before, and corresponds to the fact that four-dimensional outgoing waves divergent from a source into past and future are exclusively considered in Feynman's theory.

Now the starting equation (25) seems somewhat artificial and impairing the simplicity of the original Dirac equation. In fact one would be tempted to introduce the proper time as a fifth coordinate in a linearized form. Thus one may put for instance

$$\gamma_5\frac{\partial}{\partial\tau}\psi=\left(\gamma_\mu\frac{\partial}{\partial x_\mu}+\kappa\right)\psi,\quad \gamma_5\frac{\partial}{\partial\tau}\psi^\dagger=\left(\gamma_\mu\frac{\partial}{\partial x_\mu}-\kappa\right)\psi^\dagger, \tag{33}$$

which, by iteration, yields

$$\left(\frac{\partial^2}{\partial\tau^2}+\Box^2-\kappa^2\right)\psi=\left(\frac{\partial^2}{\partial\tau^2}+\Box^2-\kappa^2\right)\psi^\dagger=0. \tag{34}$$

A solution of (24) can be written as

$$\psi=\left(\gamma_5\frac{\partial}{\partial\tau}+L^\dagger\right)\Phi,\quad (\Box_\sigma^2-\kappa^2)\,\Phi=0,$$

$$L=\gamma_\mu\frac{\partial}{\partial x_\mu}+\kappa,\quad L^\dagger=\gamma_\mu\frac{\partial}{\partial x_\mu}-\kappa,\quad \sigma=1,\ldots\ldots5,\ x_5\equiv\tau. \tag{35}$$

For the commutation relation we adopt the expression

$$\{\psi,\psi^\dagger\}=\left(\gamma_5\frac{\partial}{\partial\tau}+L^\dagger\right)f(\tau,x),$$

$$f(\tau,x)=-\frac{2i}{(2\pi)^4}\int_c e^{ik_\sigma x_\sigma}\,\delta(k_\sigma^2+\kappa^2)\,dk_\sigma, \tag{36}$$

where C means the path of integration for k_5: $+\infty i\to 0\to +\infty$. First, when integrated over $\tau=x_5$ from 0 to ∞, (36) gives

$$\int_0^\infty f(\tau, x)d\tau = -\frac{2i}{(2\pi)^4}\int_c e^{ik_\mu x_\mu}\,\delta(k_0^2+\varkappa^2)\Big[\pi\delta(k_5)-\frac{1}{ik_5}\Big]dk_a$$

$$= -\frac{2i}{(2\pi)^4}\int e^{ik_\mu x_\mu}\Big[\frac{\pi}{2}\,\delta(k_\mu^2+\varkappa^2)+\frac{1}{2i(k_\mu^2+\varkappa^2)}\Big]dk_\mu = -\frac{i}{2}\,\Delta_F(x), \quad (37)$$

and (36) reduces to

$$\int_0^\infty \{\psi(\tau),\ \psi^{\dagger\prime}(0)\}\,d\tau = -\frac{i}{2}L^\dagger\Delta_F(x), \quad (38)$$

the term with $\partial/\partial\tau$ being put to zero on integration. The initial value of f for $\tau=0$ is calculated in a similar way, and yields the result

$$\Big(\frac{\partial}{\partial\tau}\gamma_5 - L^\dagger\Big)f\Big|_{\tau=0} = \frac{-i}{(2\pi)^4}\int\Big(i\gamma_5-\frac{L^\dagger}{(-k_\mu^2-\varkappa^2)^{1/2}}\Big)e^{ik_\mu x_\mu}\,dk_\mu. \quad (39)$$

But the factor in the bracket becomes

$$i\gamma_5-\frac{L^\dagger}{(-k_\mu^2-\varkappa^2)^{1/2}} = i\gamma_5(1+\varepsilon),\quad \varepsilon=\gamma_5 L^\dagger/(-k_\mu^2-\varkappa^2)^{1/2}, \quad (40)$$

$$\varepsilon^2=1,\quad \varepsilon=\pm 1.$$

Thus, (39) is not of the form const. $\delta(x)$, as it should be. This is because we are selecting only those waves which are outgoing from the source point. Neglecting ε, or taking its average, we get the reasonable result

$$<\{\psi(x),\ \gamma_5\psi^\dagger(x')\}>_{\tau=\tau'} \sim \delta(x). \quad (41)$$

In this point the linearized form does not prove to be much preferable to the original Fock equation. Though it may be convenient when we try to make use of the interaction representation, we shall follow hereafter Fock's procedure which seems most natural after all.

4. Problem of vacuum polarization

Now we shall turn to the interpretation of the various terms of the S matrix. The starting point is that the probability amplitude for an electron going from x to x' in a time τ is assumed to be given by

$$(\Psi^*,\ \psi(x,\tau)) = \Big(\gamma_\mu\frac{\partial}{\partial x_\mu}-\varkappa\Big)\Delta(\tau, x-x') \equiv S(\tau, x-x'). \quad (42)$$

If, for example, we consider the self-energy of the electron, we have to do with the process: an electron and a photon start simultaneously at x and afterwards meet again at x', thereby giving rise to the matrix element

$$-\frac{e^2}{\hbar c}\gamma_\mu S(\tau, x-q')\gamma_\mu D_r(x-x'). \tag{43}$$

Integrating over τ we get the usual self-energy element. We could also introduce another proper time for the photon at least formally, modifying $D(x-x')$ to $D(\tau, x-x')$. But then the radiation field no more remains real.

The problem of the vacuum polarization is a more interesting subject. Here we shall confine ourselves to the external polarization only. An electron, starting from x, suffers a scattering by the field at x' (time τ), then comes back to the origin (time $\tau+\tau'$) and there produces a polarization current δj_μ. This will be given by

$$\delta j_\mu = i\frac{e}{\hbar c}\, Tr[\gamma_\mu S(\tau', x-x')\, A_\nu \gamma_\nu S(\tau, x'-x)]. \tag{44}$$

Integration with respect to τ and τ' leads to the usual. exppession. If we consider the possibility that an electron can be scattered by the field over and over again before return, then we are dealing with an electron moving under continuous influence of the field. Such an electron will be described by the wave function satisfying

$$\lambda\Psi_\lambda = \Lambda\Psi_\lambda. \tag{45}$$

Its contribution to the induced current is

$$i\int_0^\infty (\Psi^*(x)\, e^{-i\lambda\tau}\gamma_\mu\Psi_\lambda(x))\, d\tau = 2\pi i\delta_+(\lambda)\,\Psi^*_\lambda(x)\,\gamma_\mu(\gamma_\nu D_\nu - \varkappa)\,\Psi_\lambda(x). \tag{46}$$

Summing over all stationary states (and adjusting the normalization factor), we get the induced current expression

$$\delta j_\mu = i\sum\nolimits_\lambda \delta_f(\lambda)\,\Psi^*_\lambda\gamma_\mu(\gamma_\nu D_\nu - \varkappa)\,\Psi_\lambda. \tag{47}$$

Letting the external field vanish, (47) reduces simply to

$$\delta j^0_\mu = i\,\mathrm{Tr}\left[\gamma_\mu\left(\gamma_\lambda\frac{\partial}{\partial x_\lambda} - \varkappa\right)\frac{1}{(2\pi)^4}\int \delta_+(k^2+\varkappa^2)\, e^{ik_\mu x_\mu}\, dk_\mu\right]_{x=0} = \frac{i}{2}\,\mathrm{Tr}\,[\gamma_\mu S_r(0)], \tag{48}$$

which may be regarded as zero by virtue of symmetry. If, however, (47) is expanded in powers of the coupling constant, an expression of the form

$$\delta j_\mu = c_1 A_\mu + c_2\Box^2 A_\mu + \cdots\cdots \tag{49}$$

will be obtained. The first and second term are divergent, corresponding to the self-energy of photon and the renormalization of the external charge respectively. We shall show, however, that there is a different method of approach to determine the form of δj_μ.

Fock resorted to a kind of the W.K.B. method to solve the proper time wave equation (25), setting

$$F = e^{\frac{i}{\hbar}S} f, \tag{50}$$

where S is the classical action function satisfying (21). This method is applied to obtain the Riemann function R for the Dirac electron, which is expressed as

$$R=\oint F d\tau, \tag{51}$$

the contour encircling the origin $\tau=0$ in the complex domain. F can be rigorously solved when the electron is free or at least the external field is constant. In the former case, in particular,

$$S=+\frac{m}{2\tau}x^2-\frac{1}{2}mc^2\tau, \qquad f=-\frac{m}{8\pi^2\hbar c}\frac{1}{\tau^2},$$

$$R=-\frac{m}{8\pi^2\hbar c}\oint e^{\frac{im}{2\hbar\tau}x^2-\frac{imc^2}{2\hbar}\tau}\frac{d\tau}{\tau^2}=-\frac{m}{4\pi\hbar\sqrt{-x^2}}J_1\left(\frac{mc}{\hbar}\sqrt{-x^2}\right). \tag{52}$$

On the other hand, Schwinger's $\bar{\Delta}$-function has the integral representation

$$\bar{\Delta}(x)=\frac{1}{8\pi^2}\int_{-\infty}^{\infty}e^{-idx^2+i\frac{x^2}{4a}}du=\frac{\kappa}{16\pi^2}\int_{-\infty}^{\infty}e^{-\frac{ix^2}{2\tau}\kappa+\frac{i\kappa\tau}{2}}\frac{d\tau}{\tau^2}. \tag{53}$$

When $-x^2$ is positive, we may take a contour integral going round the upper half plane at infinity and deviating slightly below the real axis near the origin; when $-x^2$ is negative, we go round the lower half plane at infinity. Then the former contour shrinks to a circle around zero, giving just (43), and the latter integral vanishes. Thus

$$\bar{\Delta}(x)=-\frac{1}{2}R(x)\gamma(-x^2), \tag{54}$$

where

$$\gamma(\lambda)=\begin{cases}1, & \lambda>0\\ 0, & \lambda<0\end{cases}$$

From this result we see that the function $\Delta_F^{(e)}$ in the presence of the external field will be obtained if we choose in (42) a straight integration path from 0 to ∞ instead of the circle around zero. When there is a constant magnetic and electric field H and E parallel to the z axis, $F=\exp(iS/\hbar)f$ is given by

$$S=S_0-\frac{1}{2}mc^2\tau+\frac{eE}{4c}[(z-z')^2-c^2(t-t')^2]\operatorname{cth}\frac{eE\tau}{2mc}$$
$$+\frac{eH}{4c}[(x-x')^2+(y-y')^2]\operatorname{ctg}\frac{eH\tau}{2mc},$$

$$S_0=\frac{1}{2}eE(z+z')(t-t')+\frac{eH}{2c}(x'y-y'x), \tag{55}$$

$$f=\frac{m}{8\pi^2\hbar c}\frac{eH}{2mc}\frac{eE}{2mc}f_0\Big/\sin\frac{eH\tau}{2mc}\operatorname{sh}\frac{eE\tau}{2mc},$$

$$f_0=\exp\left[\frac{ie}{2mc}\sigma_z H\tau+\frac{e}{2mc}\alpha_z E\tau\right],$$

and

$$-\frac{1}{2}\Delta_F^{(e)}(x,x')=\int_0^\infty F(x,x')\,d\tau. \tag{56}$$

The polarization current is then

$$\delta j_\mu=i\,Tr\left[\gamma_\mu\left(\gamma_\lambda\frac{\partial}{\partial x_\lambda}-\varkappa-\frac{ei}{\hbar c}\gamma_\lambda{}'A_\lambda\right)\Delta_F^{(e)}(x,x')\right]_{x=x'}, \tag{57}$$

where

$$A_x=-\frac{1}{2}Hy,\quad A_y=\frac{1}{2}Hx,\quad A_z=0,\quad A_4=Ezi. \tag{58}$$

Let us consider the problem of the gauge invariance. When A_μ is replaced by $A_\mu+\partial\chi/\partial x_\mu$, S acquires an additional term $(e/c)(\chi-\chi')$ so that $\Delta_F^{(e)}$ is multiplied by a factor $\exp[i(\chi(x)-\chi(x')/\hbar]$. But the current (48) turns out to suffer no modification:

$$\delta j_\mu{}'=e^{i(\chi-\chi')/\hbar}\,\delta j_\mu|_{x=x'}=\delta j_\mu. \tag{59}$$

The gauge invariance is thus guaranteed. The charge conservation law also holds since

$$\left(\gamma\Box+\varkappa-\frac{ei}{\hbar c}A\gamma\right)S_F^{(e)}=S_F^{(e)}\left(-\gamma\Box'+\varkappa-\frac{ie}{\hbar c}A'\gamma\right)=-\delta(x-x'). \tag{60}$$

These relations are a consequence of the equation (25) for F and the initial condition: $F(x)=\delta(x)$ at $\tau=0$.

On the other hand the usual perturbation formula (40) gives in general non-gauge invariant, divergent results when evaluated numerically. In our expression (56) for the particular constant field, however, δj_μ turns out to vanish if we regard expressions of the type $(x_\mu-x_\mu')\Delta_F^{(e)}(x,x')|_{x=x'}$ to be zero. This may be approved because here appears only one singular function and not a product of two or more, so that it is a consistent condition compatible with other requirements. It is also shown in the appendix that no essential difficulties arise in case of an arbitrary constant field. The self-energy of the photon is then zero at least in the order e^2, for the first term in the expansion (40) vanishes according to the above result.

Appendix. Solution for an arbitrary constant external field.[5]

We take the constant external field $F_{\lambda\mu}=-F_{\mu\lambda}$, and the vector potential for it:

$$A_\mu=\frac{1}{2}(x_\lambda-x_{0\lambda})F_{\lambda\mu}, \tag{A1}$$

where $x_{0\lambda}$ are arbitrary constants corresponding to an initial condition in the subsequent calculation. The classical equation of motion is then

$$\frac{d}{d\tau}\dot{x}_\mu=-\frac{e}{mc}F_{\mu\lambda}\dot{x}_\lambda$$

or

$$\frac{d}{d\tau}\dot{x}=-\frac{e}{mc}\boldsymbol{F}\dot{x},\qquad \boldsymbol{F}=(F_{\mu\lambda}). \tag{A2}$$

The Hamilton-Jacobi equation is, on the other hand, given by

$$\frac{\partial S}{\partial\tau}+\frac{1}{2m}\left[\left(\frac{\partial S}{\partial x_\mu}+\frac{e}{2c}(x_\lambda-x_{0\lambda})F_{\lambda\mu}\right)^2+m^2c^2\right]=0. \tag{A3}$$

Instead of solving (A3) directly, we shall first handle the equation of motion (A2) which can easily be integrated because $\boldsymbol{F}$ is a constant matrix. Thus,

$$\dot{\boldsymbol{x}}=\exp\left[-\frac{e}{mc}\boldsymbol{F}\tau\right]\dot{\boldsymbol{x}}_0\equiv\boldsymbol{S}\dot{\boldsymbol{x}}_0, \tag{A4}$$

and further

$$(\boldsymbol{x}-\boldsymbol{x}_0)=\left[(1-\boldsymbol{S}(\tau))\Big/\frac{e}{mc}\boldsymbol{F}\right]\dot{\boldsymbol{x}}_0,$$

$$\therefore\quad \dot{\boldsymbol{x}}_0=\left[\frac{e\boldsymbol{F}}{mc}\Big/(1-\boldsymbol{S}(\tau))\right](\boldsymbol{x}-\boldsymbol{x}_0), \tag{A5}$$

$$\dot{\boldsymbol{x}}=\boldsymbol{S}\dot{\boldsymbol{x}}_0=\left[\frac{e\boldsymbol{F}}{mc}\boldsymbol{S}\Big/(1-\boldsymbol{S})\right](\boldsymbol{x}-\boldsymbol{x}_0).$$

The conjugate momenta p_μ and the Hamiltonian $H=(p\dot{x})-L$ become

$$\begin{aligned}\boldsymbol{p}&=m\dot{\boldsymbol{x}}-(e/c)\boldsymbol{A}=m\dot{\boldsymbol{x}}+(e/2c)\boldsymbol{F}(\boldsymbol{x}-\boldsymbol{x}_0)\\&=\left[\frac{e\boldsymbol{F}}{c}\boldsymbol{S}\Big/(1-\boldsymbol{S})+\frac{e\boldsymbol{F}}{2c}\right](\boldsymbol{x}-\boldsymbol{x}_0)=\frac{e\boldsymbol{F}}{2c}(1+\boldsymbol{S})/(1-\boldsymbol{S})\cdot(\boldsymbol{x}-\boldsymbol{x}_0)\\&=\frac{e}{2c}\boldsymbol{F}\,\mathrm{cth}\left(\frac{e}{2mc}\boldsymbol{F}\tau\right)\cdot(\boldsymbol{x}-\boldsymbol{x}_0),\end{aligned} \tag{A6}$$

$$\begin{aligned}H&=\frac{1}{2}mc^2+\frac{1}{2m}\left(\boldsymbol{p}+\frac{e}{c}\boldsymbol{A}\right)^2=\frac{1}{2}mc^2+\frac{1}{2}m\dot{\boldsymbol{x}}^2\\&=\frac{1}{2}mc^2+\frac{1}{2}mc^2\cdot\frac{e^2}{m^2c^2}\left(\frac{\boldsymbol{F}\boldsymbol{S}}{1-\boldsymbol{S}}\boldsymbol{x}-\boldsymbol{x}_0,\ \frac{\boldsymbol{F}\boldsymbol{S}}{1-\boldsymbol{S}}\boldsymbol{x}-\boldsymbol{x}_0\right)\\&=\frac{1}{2}mc^2+\frac{e^2}{2m}\left(\boldsymbol{x}-\boldsymbol{x}_0,\ \frac{-\boldsymbol{F}^2\boldsymbol{S}}{(1-\boldsymbol{S})^2}\boldsymbol{x}-\boldsymbol{x}_0\right),\end{aligned} \tag{A7}$$

using the relation for the transposed:

$$\tilde{\boldsymbol{S}}=\boldsymbol{S}^{-1},\qquad \tilde{\boldsymbol{F}}=-\boldsymbol{F}. \tag{A8}$$

The action function S, which satisfies $\delta S = p\delta x - H\delta\tau$, is now obtained by putting

$$S = -\frac{1}{2} mc^2\tau + \frac{e}{4c}(x - x_0, \ G(x - x_0)), \tag{A9}$$

$$G = F \operatorname{cth}\left(\frac{e}{2mc} F\tau\right).$$

The remaining function f is determined by the equation

$$2m\frac{df}{d\tau} + \left(\Box^2 S + \frac{ei}{2c}\sigma_{\mu\nu}F_{\mu\nu}\right)f = 0 \tag{A10}$$

From (A9),

$$\Box^2 S = \frac{e}{2c} Tr\, G,$$

so that

$$f = \exp\left[-\frac{1}{2} Tr\left\{\ln \operatorname{sh}\left(\frac{e}{2mc} F\tau\right) + \frac{ei}{2mc}(\sigma F)\tau\right\}\right] f_0. \tag{A11}$$

In these expressions, functions of the matrix F bear symbolical meaning. When it happens that det $|F| = 0$ and consequently no inverse can be defined directly, we should go back to the beginning for the correct interpretation.

Now that the functions $F = \exp(iS/\hbar)f$ and A_μ are both even in the coordinates $x - x'$, we see that the argument proposed in Section 4 for the discussion of the induced current holds also in this general case.

References

1) R. P. Feynman, a manuscript kindly sent to Prof. Tomonaga, which appeared later in Phys. Rev. **76** (1949), 749 and 769.
2) V. Fock, Phys. Zeit. Sow. Un. **12** (1937), 404.
3) J. Schwinger, Phys. Rev. **75** (1949), 651.
4) F. J. Dyson, Phys. Rev. **75** (1949), 484.
5) A detailed analysis on this subject was made by A. H. Taub, Phys. Rev. **73** (1948), 786.

Progress of Theoretical Physics, Vol. 5, No. 4, July~August, 1950

Force Potentials in Quantum Field Theory*

Yoichiro NAMBU
Osaka City University

(Received July 31, 1950)

1. Introduction and Summary

As early as 1935 Yukawa conjectured that the nuclear force, which ties together the component nucleons into a solid nucleus, could be attributed to an intermediate field with an intrinsic mass corresponding to the range of the nuclear force. The general success of the meson theory that followed the discovery of such particles in cosmic rays, has been so great that one cannot now discuss the nuclear and cosmic phenomena without the help of Yukawa's idea on the nature of mesons. One must admit, however, that the meson theory in the present stage is far from satisfactory almost in every detail regarding its quantitative predictions on various phenomena. This may be due to our insufficient knowledge about the nature of the real mesons as well as to a more profound crisis of the present quantum field theory which precludes us from drawing reasonable conclusions out of physical assumptions. Both difficulties are intrinsically connected with each other and their complete solution does not seem to be an easy one.

Here we will pick up one characteristic feature of the Yukawa theory, i.e. the theory of nuclear forces. Up to now, however, there is a wide gap between the nuclear potentials predicted on the basis of various meson theories and the one that is obtained empirically. No simple assumptions have ever succeeded in explaining the quantitative behavior of the internucleonic interactions. The theoretical side of the difficulty seems to consist of two facts. One is the high singularity of the nuclear potential which is more or less of a common origin in the present quantum theory, while the other is that we have no established formalism of the relativistic many-body problem, a peculiar situation encountered in the theory of the deuteron. At least formally, the S-matrix theory of Heisenberg affords a means to deal equally with scattering processes and bound systems. But the latter case is more helpless since the higher order terms in the coupling constant play an essential role in the determination of energy levels. In the non-relativistic limit, such higher order effects were properly represented in the interaction potential appearing in the Schrödinger eigenvalue

* Preliminary report: Prog. Theor. Phys. 5 (1950), 321. Some errors are corrected in the text.

problem, a typical case being that of the hydrogen atom. At the genesis of the meson theory, Yukawa borrowed this successful idea from the electromagnetic theory. There are, however, some remarkable distinctions between the electromagnetic and meson field. The latter has mass and charge, together with strong coupling constants. The Coulomb potential, being not quantized, was an exact theoretical consequence, while there is no such classical potential in the mesonic case. These situations exhibit themselves in the relativistic effect, or the retardation on one hand, and in the quantum effect, or the recoil and higher order forces on the other. Before deciding between the existing " korrespondenzmässigen " theories and any more revolutionary ones, we probably have to re-examine the old concept of potential and improve it so as to conform more closely to the rigorous field theoretical view.

In the following sections we apply the recent method of quantum electrodynamics to the analysis of these problems. At the present stage, however, we have to abandon the complete relativistic covariance, which is one of the most beautiful achievements of the new theory, but rather confine ourselves to more crude and approximate considerations. In Section 2 we begin with the derivation of the second order potential with the aid of the covariant field theory, and point out the ambiguities that naturally arise. The latter leads to the consideration of the fourth order potential in the next section. An extension of these methods to still higher order terms seems almost impracticable, but very simplified arguments are presented in the last section. The examination, however, of the existing individual meson models in the light of the obtained results will be carried out only on a later occasion.

2. The Second Order Potential

The derivation of the second order interaction potential between two Fermi particles has been carried out in various ways. Originally the Coulomb potential for charged particles was directly imported from the classical theory. Its relativistic correction was first derived by Breit[1] also on the basis of the correspondence principle. He afterwards developed his principle of the approximate relativistic covariance[2] to obtain some general expressions for electromagnetic as well as nuclear potentials with relativistic corrections. On the other hand, perturbation calculations based on the quantized field theory were carried out by many authors[3], with the results that were essentially the same as that expected from the correspondence arguments. In the meson theory an attempt was also made[4] to use canonical transformations instead of the ordinary perturbation method. Very recently this was replaced by the more elegant and perfect theory of Tomonaga and Schwinger[5], from which standpoint some papers have already been published on the meson potentials[6]. All these results seem to be in general, but not complete, agreement. The correct interpretation of them will only be given by a careful inspection and reflection on the meaning of the potential.

Here we start from the Tomonaga-Schwinger equation (in natural units):

$$i\frac{\delta}{\delta\sigma}\Psi[\sigma]=H(x)\Psi[\sigma], \tag{1}$$

$$H(x)=-\varepsilon j_\lambda \varphi_\lambda \tag{2}$$

for the interaction of a fermion field ψ with a boson field φ. ε is the coupling constant, and j_λ a quantity bilinear in $\bar{\psi}$ and ψ. λ represents the concurrent tensor and isotopic spin indices. By the routine transformation[11]

$$\Psi(\sigma)=\exp\left[-\frac{1}{2}\int_{-\infty}^{\infty}H(x')\varepsilon(\sigma\sigma')(dx')\right]\Psi_1[\sigma], \tag{3}$$

we obtain

$$i\frac{\delta}{\delta\sigma}\Psi_1[\sigma]=-\frac{1}{4}i[H(x),\int_{-\infty}^{\infty}H(x')\varepsilon(\sigma\sigma')(dx')]\Psi_1[\sigma] \tag{4}$$

to the second order in ε. For a system in which real bosons are absent, this gives

$$i\frac{\delta}{\delta\sigma}\Psi_1[\sigma]=\left[-\frac{1}{2}\varepsilon^2 j_\lambda(x)\int_{-\infty}^{\infty}j_\mu(x')\bar{\Delta}_{\lambda\mu}(x-x')(dx')\right]\Psi_1[\sigma]\equiv H^{(2)}(x)\Psi_1[\sigma]$$

using the commutation relation

$$[\varphi_\lambda(x),\varphi_\mu(x')]=i\Delta_{\lambda\mu}(x-x')=iO_{\lambda\mu}\Delta(x-x'),$$
$$\bar{\Delta}_{\lambda\mu}(x)=-\frac{1}{2}\varepsilon(x)\Delta_{\lambda\mu}(x)\ \text{ or }\ -\frac{1}{2}O_{\lambda\mu}(\varepsilon(x)\Delta(x)), \tag{6}$$

where $\Delta_{\lambda\mu}(x)$ contains the essential factor $\Delta(x)$ as defined by Schwinger. In order to get the usual potential form we make use of the Fourier decomposition of $\bar{\Delta}$, or more conveniently the operational relation

$$\int_{-\infty}^{\infty}\bar{\Delta}(x)f(x)dx_0=\frac{1}{4\pi}\frac{1}{r}\exp\left(-r\sqrt{\mu^2+(d/dx_0)^2}\right)f(x)|_{x_0=0},$$
$$r^2=\sum_{i=1}^{3}x_i^2,\qquad x_0=t, \tag{7}$$

where μ characterizes the mass of the field. In case $\mu=0$,

$$\int\bar{D}(x)f(x)dx_0=\frac{1}{4\pi}\frac{1}{2r}\left(\exp\left(-r\frac{d}{dx_0}\right)+\exp\left(r\frac{d}{dx_0}\right)\right)f(x)|_{x_0=0}$$
$$=\frac{1}{4\pi}\frac{1}{2r}(f(\boldsymbol{r},t-r)+f(\boldsymbol{r},t+r)) \tag{8}$$
$$=\frac{1}{4\pi}\frac{1}{2r}(f(\boldsymbol{r})_{\text{ret}}+f(\boldsymbol{r})_{\text{adv}}).$$

We should apply these formulas after making the space-like surface σ flat. Thus in Eq. (5) the interaction becomes

$$H^{(2)}(x) = -\frac{1}{2}\frac{\varepsilon^2}{4\pi} j_\lambda(\boldsymbol{r}) \int O_{\lambda\mu} V(\boldsymbol{r}-\boldsymbol{r}') j_\mu(\boldsymbol{r}') d\boldsymbol{r}', \tag{9}$$

$$V(\boldsymbol{r}-\boldsymbol{r}') = \begin{cases} \dfrac{1}{|\boldsymbol{r}-\boldsymbol{r}'|} \exp(-|\boldsymbol{r}-\boldsymbol{r}'|\sqrt{\mu^2+(d/dt)^2}), \quad \text{or} \\ \dfrac{1}{|\boldsymbol{r}-\boldsymbol{r}'|}\dfrac{1}{2}\left\{\exp\left(-|\boldsymbol{r}-\boldsymbol{r}'|\dfrac{d}{dt}\right) + \exp\left(|\boldsymbol{r}-\boldsymbol{r}'|\dfrac{d}{dt}\right)\right\}. \end{cases} \tag{10}$$

$H^{(2)}(x)$ means the potential energy density for the particle at x due to other particles. In the two-particle system in which we are interested, the total potential energy will be written down in configuration space as

$$H^{(2)}(12) = -\frac{1}{2}\frac{\varepsilon^2}{4\pi}\{j_\lambda(1)\, O_{\lambda\mu} V(12) j_\mu(2) + j_\mu(2)\, O_{\mu\lambda} V(21) j_\lambda(1)\}, \tag{11}$$

where 1, 2 refer to the numbering of the particles, and j_λ means the quantity introduced before, but represented in the configuration space. V is an operator operating on the quantity that comes after it. We may regard (11) as the potential energy to be inserted in the ordinary Schrödinger equation, in which j_λ is a constant operator and d/dt is replaced by

$$\frac{d}{dt_i} = i\,[H_i^{(0)},\ ], \qquad i=1,2, \tag{12}$$

$H^{(0)}$ being the free state Hamiltonian for each particle. We note here that the r-dependent potential itself should always be thrown under the operation (12). Eq. (10) tells us that V corresponds to the attached field introduced by Dirac and others[8]. Its appearance is a consequence of the Schwinger transformation (3). When the real boson field is absent (no incoming and outgoing waves), the attached field is equivalent to the retarded field which follows from the transformation

$$\Psi[\sigma] = \exp\left[-i\int_{-\infty}^{\sigma} H(x')(dx')\right] \Psi_1[\sigma]. \tag{13}$$

Though they should be equivalent under the above condition, the form of the potential derived from these transformations is apparently not equivalent. Since such a condition is eliminated in the Schrödinger equation, we here meet with an ambiguity. On physical grounds, however, the symmetrical form may be preferred to the retarded form which is not invariant under the interchange of past and future. It is to be noted that the ambiguity does not arise in the non-relativistic limit for which the familiar potential $1/r$ or $\exp(-\mu r)/r$ results.

Instead of using the rather unesthetical one-time theory to derive the potential, we may also invoke the formula[11]

$$\psi(x)=\int_\sigma S(x-x')\,\gamma_\mu\,\psi(x')\,d\sigma'_\mu,$$

$$\bar\psi(x)=\int_\sigma d\sigma_\mu'\,\bar\psi(x')\,\gamma_\mu\,S(x'-x), \tag{14}$$

with
$$\frac{1}{i}S(x-x')=\{\psi(x),\,\bar\psi(x')\},$$

and express the retarded and advanced quantities $j_\lambda(x)$ in terms of ψ and $\bar\psi$ on the surface σ. The resulting interaction is non-local, or in other words, velocity dependent. The Schrödinger equation with such an interaction is then an integro-differential equation, and its solution will never be an easy task. The non-local effect, however, will set in only at very close distances for which large momentum transfers between the particles become important. Thus if we will take account of the retardation in an approximate manner, we may usually be allowed to expand (10) in powers of d/dt and retain only the first few terms. But if j_λ contains a Zitterbewegung (or non-static) part, we cannot do this, for it means the transition from positive to negative state or vice versa, and $d/dt\sim\pm 2mi$, where m is the mass of the particle. If $2m$ is much larger than μ, (10) gives a rapidly oscillating function of the form $\cos(2mr)/r$ which had rather better be substituted by a delta function:

$$\frac{1}{4\pi}\frac{1}{r}\cos 2mr\sim-\frac{1}{4m^2}\left(1-\frac{1}{4m^2}\Delta+\cdots\right)\delta(\boldsymbol{r}). \tag{15}$$

The problem of ambiguity mentioned above, however, is not yet settled at all. We show this with the example of the Breit interaction. In this case,

$$H(x)=-\varepsilon j_\lambda A_\lambda,\qquad j_\lambda=i\bar\psi\gamma_\lambda\psi, \tag{16}$$

and

$$H^{(2)}(x)=-\frac{1}{2}\frac{\varepsilon^2}{4\pi}\int j_\lambda(x)\,\frac{\cos\left(\frac{d}{dt'}\right)}{r}\,j_\lambda(x')\,d\boldsymbol{r}'. \tag{17}$$

This is the Møller formula.[3] The fourth component of the product $j_\lambda j_\lambda'$ corresponds to the Coulomb force with small retardation effect. An expansion in d/dt, which is allowed here, yields the potential in configuration space

$$H^{(2)}(12)=\frac{\varepsilon^2}{4\pi}\left[(\gamma_\lambda)_1\frac{1}{r}(\gamma_\lambda)_2+\frac{1}{4}\left(\frac{d^2}{dt_1^2}+\frac{d^2}{dt_2^2}\right)(\gamma_4)_1 r(\gamma_4)_2\right]+\cdots \tag{18}$$

When this is evaluated, we do not obtain the Breit formula, contrary to our expectation! However, (18) may also be written as

$$H^{(2)}(12)=\frac{\varepsilon^2}{4\pi}\left[(\gamma_\lambda)_1\frac{1}{r}(\gamma_\lambda)_2-\frac{1}{2}\frac{d}{dt_1}\frac{d}{dt_2}(\gamma_4)_1 r(\gamma_4)_2\right], \tag{19}$$

for the energy conservation means

$$\frac{d}{dt_1} = -\frac{d}{dt_2} \tag{20}$$

neglecting the interaction term which amounts to a higher order correction. From (19) now follows the desired formula:*

$$H^{(2)}(12) = \left[(\gamma_4)_1 (\gamma_4)_2 - \frac{1}{2} \left\{ (\gamma_i)_1 (\gamma_i)_2 + \frac{((\gamma_i)_1, \boldsymbol{r})((\gamma_i)_2, \boldsymbol{r})}{r^2} \right\} \right] \frac{1}{r}. \tag{21}$$

The assumption (20) was also made by Bethe and Fermi[8], and more explicitly by Toyoda[9] in meson theory. It is true that an ambiguity does not influence the second order S-matrix corresponding to the elastic scattering of two particles, since in this case (20) holds exactly. But a potential in its proper sense should contain those matrix elements in which energy is not conserved and which brings the system into a virtual state. From the above result we see that the potential is indeterminate to within an arbitrary matrix whose diagonal elements with respect to energy are zero. Let us try to find out how this problem can be settled.

We make here a requirement that the two-body problem shall be rendered to a relativistic form as far as possible. Thus we assume a pair of the Dirac many-time equations for the wave function $\psi(x_1, x_2)$ describing a two-particle system under action at a distance:

$$\begin{aligned} \left\{ i\frac{\partial}{\partial t_1} - H_1 - V(12) \right\} \psi &= 0, \\ \left\{ i\frac{\partial}{\partial t_2} - H_2 - V(21) \right\} \psi &= 0. \end{aligned} \tag{22}$$

H_1 and H_2 being the Hamiltonians for free state. In general the potentials $V(12)$ and $V(21)$ may be different. By a standard transformation (22) can be brought to the interaction representation

$$\begin{aligned} \left\{ i\frac{\partial}{\partial n_1} - V(12) \right\} \psi &= 0, \\ \left\{ i\frac{\partial}{\partial n_2} - V(21) \right\} \psi &= 0. \end{aligned} \tag{23}$$

where the time-like parameters n may point to different directions, but for the time being assumed to be parallel. Now an integrability condition should be enounced:

$$i\left(\frac{\partial V(21)}{\partial n_1} - \frac{\partial V(12)}{\partial n_2} \right) = [V(12), V(21)]. \tag{24}$$

---

* An alternative method to derive (21) is to separate the pure Coulomb interaction beforehand and treat only the transversal part in the above way.

Suppose the potentials V to be defined in terms of the variables of the two particles which stand in a definite relation to each other with respect to their space-time position, thus taking account of the retardation effect. Then, since the relative position of the particles should remain invariant in V and n_1 is parallel to n_2, we have

$$\frac{\partial V}{\partial n_1}=\frac{\partial V}{\partial n_2}, \tag{25}$$

so that

$$i\frac{\partial}{\partial n}(V(21)-V(12))=[V(12),\ V(21)], \tag{26}$$

which in turn requires that

$$V(12)=V(12), \tag{27}$$

as will be seen by developing V in a power series in the coupling constant. (27) may be regarded as expressing Newton's third law of motion. Let us test this condition for the above mentioned example. The Møller interaction gives, in view of (17) and (18),

$$\begin{aligned} V(12)\,|_{t_1=t_2}&=\frac{\varepsilon^2}{4\pi}\cdot\frac{1}{2}\left\{(\gamma_\lambda)_1\frac{1}{r}(\gamma_\lambda)_2+\frac{1}{2}(\gamma_\lambda)_1\frac{d^2}{dt_2^2}r(\gamma_4)_2\right\},\\ V(21)\,|_{t_1=t_2}&=\frac{\varepsilon^2}{4\pi}\cdot\frac{1}{2}\left\{(\gamma_\lambda)_1\frac{1}{r}(\gamma_\lambda)_2+\frac{1}{2}(\gamma_4)_2\frac{d^2}{dt_1^2}r(\gamma_4)_1\right\}, \end{aligned} \tag{28}$$

hence clearly $V(12)\neq V(21)$! To make the equation of motion (23) integrable, we must therefore supplement the potentials by some additional terms. This can be performed by a transformation independent of past history:

$$\psi=\exp\left[\frac{i}{4}\frac{\varepsilon^2}{4\pi}\left(\frac{d}{dt_1}+\frac{d}{dt_2}\right)(\gamma_4)_1 r(\gamma_4)_2\right]\psi', \tag{29}$$

with the result

$$V'(12)=V'(21)=\frac{\varepsilon^2}{4\pi}\frac{1}{2}\left[(\gamma_\lambda)_1\frac{1}{r}(\gamma_\lambda)_2-\frac{1}{2}\frac{d}{dt_1}\frac{d}{dt_2}(\gamma_4)_1 r(\gamma_4)_2\right]+O(\varepsilon^4), \tag{30}$$

which just leads to (21).

The meaning of the above condition may be illustrated as follows. Let us describe the retarded interactions V by the Feynman diagram. So long as these interactions occur successively with sufficiently large intervals of time, there is nothing difficult. But when they come nearer and cross each other, there arises a difference in the order of emission and absorption of quanta represented by the process $V(21)\,V(12)$ or $V(12)\,V(21)$ and the one corresponding to the actual process, which always occurs in the chronological order. Such a situation is characteristic of the relativistic interaction, for the range of the time interval in which the crossing of the lines prevails will tend to zero as $c\to\infty$. In the expression of the S-matrix, however, these discrepancies are properly compensated, giving the correct result in the fourth order term. Thus we may suppose that in amending the second order potential so as to satisfy the condition (27) we are automatically taking account, at least partially, of the higher order interactions. The ambiguities in the second order potential will then be reflected in the higher order ones in such a way that the overall effect turns out identical. In other words, they will correspond simply to our freedom of canonical transformation of the whole representation.

Anticipating the justification of the above considerations in later sections, we shall next derive an alternative form of the potential which seems more convenient for our purpose. From the point of view of the S-matrix, the transformation (3) is a rather clumsy procedure. We had better write down the S-matrix after Dyson[10] as

$$S=1+\int H(x)\,dx+\frac{1}{2!}\iint P(H(x_1),\,H(x_2))\,(dx_1)\,(dx_2)+\cdots, \tag{31}$$

and identify its sub-matrix corresponding to the two-particle system with

$$1+\int V(x_1,x_2)\,(dX)+\frac{1}{2!}\iint P(V(x_1,x_2),\,V(x_1',x_2'))\,(dX)\,(dX)'+\cdots, \tag{32}$$

where X means a set of coordinates describing the motion of the system as a whole, and the symbol P is defined with respect to these coordinates. Indeed, if a relativistic two-body problem could actually be formulated, we could consider a generalization of the center of gravity to which belongs the constants of motion, or the momentum-energy vector. In case the two particles are identical, it is natural to adopt for X

$$X_\mu=\frac{1}{2}\,((x_\mu)_1+(x_\mu)_2), \tag{33}$$

the remaining variables being the relative coordinates

$$x_\mu=(x_\mu)_1-(x_\mu)_2, \tag{34}$$

so that

$$x_1=X+\frac{x}{2},\qquad x_2=X-\frac{x}{2}. \tag{35}$$

In terms of these variables V becomes, to the second order,

$$H^{(2)}\equiv V(X)=\frac{1}{2}\int P\Big(H\Big(X+\frac{x}{2}\Big),H\Big(X-\frac{x}{2}\Big)\Big)(dx). \tag{36}$$

With the expression (2), it is

$$H^{(2)}=-\frac{i}{4}\varepsilon^2\int j_\lambda\left(X+\frac{x}{2}\right)\Delta_{F\lambda\mu}(x)j_\mu\left(X-\frac{x}{2}\right)(dx),$$
$$\Delta_F(x)=\Delta^{(1)}(x)-2i\bar{\Delta}(x). \tag{37}$$

The appearance of the function $\Delta_F(x)$ as the vacuum expectation value is a common feature of the Feynman-Dyson theory.[5][10] This makes, at first glance, the potential function non-Hermitic, but actually the term with $\Delta^{(1)}(x)$ does not contribute to the process and may be omitted. In case the retardation expansion is allowed, (37) gives

$$\begin{aligned}H^{(2)}=&-\frac{\varepsilon^2}{4\pi}\Big[j_\lambda(\boldsymbol{r}_1)\,O_{\lambda\mu}\frac{e^{-\mu r}}{r}j_\mu(\boldsymbol{r}_2)\\&-\frac{1}{2\mu}\left(\frac{d}{dx_0}\right)^2\left\{j_\lambda\left(\boldsymbol{r}_1,X_0+\frac{x_0}{2}\right)O_{\lambda\mu}e^{-\mu r}j_\mu\left(\boldsymbol{r}_2,X_0-\frac{x_0}{2}\right)\right\}_{x_0=0}\Big]\\=&-\frac{\varepsilon^2}{4\pi}\Big[j_\lambda(1)\,O_{\lambda\mu}\frac{e^{-\mu r}}{r}j_\mu(2)\\&-\frac{1}{2\mu}\left\{\frac{1}{4}\left(\frac{d}{dt_1}\right)^2-\frac{1}{2}\frac{d}{dt_1}\frac{d}{dt_2}+\frac{1}{4}\left(\frac{d}{dt_2}\right)^2\right\}j_\lambda(1)e^{-\mu r}j_\mu(2)\,O_{\lambda\mu}\Big].\end{aligned}\tag{38}$$

For $\mu=0$, replace $-\exp(-\mu r)/\mu$ by r. This formula applies, being an expansion in $d/dt_1-d/dt_2=i[H_1-H_2,\quad]$, to those transitions in which the two particles possess the same energy sign both in the initial and final state, a wider condition than the previous one. It is to be noted, however, that in this way the separate derivation of $V(12)$ and $V(21)$ in (23) is impossible, and the physical insight into the condition on V is accordingly obscured.

3. The Fourth Order Potential

As was shown in the preceding analysis, we cannot give in general an unambiguous second order interaction unless the fourth order one is specified. They are related to each other in such a way that the second and fourth order (and possibly higher order) terms in the S-matrix come out correctly as the combined effect of these potentials. In other words, the fourth order terms of the S-matrix are the sum of the first order effect of the fourth order potential and the second order effect of the second order potential. In the S-matrix theory a similar circumstance arises when S is expressed as[11]

$$S=\frac{1-iK}{1+iK}. \tag{39}$$

Expanding S and K:

$$S=1+S_2+S_4+\cdots,\qquad K=K_2+K_4+\cdots, \tag{40}$$

we see that

$$S_2^+=-S_2,\quad K_2=\frac{i}{2}S_2,\quad K_4=\frac{i}{2}S_4+i(K_2)^2=\frac{i}{2}S_4-i\left(\frac{S_2}{2}\right)^2. \tag{41}$$

Thus $2K_4$ is the difference of S_4 and a two-fold iteration of S_2, and is the real (Hermitic) part of iS_4. In this case, however, the term $(S_2)^2$ corresponds to the process in which the particles are scattered twice but the energy is conserved in the intermediate state (the damping part). In our case, on the other hand, a second order potential is meant to include the reactive effect as well and consequently the fourth order potential is of a nature different from K.

Now from the standpoint of the Feynman theory, the fourth order processes consist of two main graphs (a) and (b). Other graphs contain self-energy effects, and we shall neglect them for the present purpose, for it only gives a

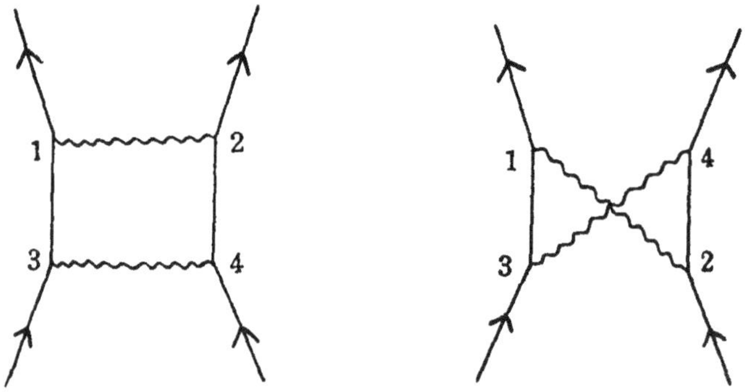

(a) (b)

small smearing out effect to the source if divergences can be avoided. On iterating the second order graph, we get a graph corresponding to (a) if the two boson lines do not cross, and a graph similar to (b) if they do. But the latter is not exactly (b), for the order of occurrence of the events at 1, 2, 3 and 4 is not chronological, and it does not give the correct matrix element corresponding to (b). Hence the difference between the true fourth order term and the one arizing from the iteration of the second order term can be written as

$$-\frac{1}{8}(-i\varepsilon^2)^2\int j_\mu(1)j_\nu(3)[j_\lambda(4),j_\rho(2)]\,\Delta_{F\mu\rho}(12)\Delta_{F\nu\lambda}(34)$$
$$\times\frac{1+\varepsilon(13)}{2}\,\frac{1+\varepsilon(42)}{2}(dx_1)(dx_2)(dx_3)(dx_4), \tag{42}$$

where 1, 3 and 4, 2 are understood to refer to different particles. A more symmetrical form is provided by

$$\frac{1}{32}\varepsilon^4\int\Big\{\frac{1+\varepsilon(13)\varepsilon(42)}{2}[j_\mu(1)\cdot j_\nu(3)][j_\lambda(4),j_\rho(2)]$$
$$+\frac{\varepsilon(13)+\varepsilon(42)}{4}(\{j_\mu(1),j_\nu(3)\}[j_\lambda(4),j_\rho(2)]+[j_\mu(1),j_\nu(3)]\{j_\lambda(4),j_\rho(2)\})\Big\}$$
$$\times\Delta_{F\mu\rho}(12)\,\Delta_{F\nu\lambda}(34)\,(dx_1)(dx_2)(dx_3)(dx_4). \tag{43}$$

(41) and (42) contain the sign function $\varepsilon(x)$ which depends on the surface parametrization. In fact, on writing down the matrix element explicitly, the second term of (43) is seen to contain a factor $\varepsilon(x)\,S^{(1)}(x)$. This shows that the potential energy and accompanying eigenvalue problem do not have in general the ordinary covariant character. As it will not be of much importance to obtain the rigorous form of the potential, we shall be content with rough estimations. For this purpose we specify the time axis and proceed with v/c expansions. Then two cases discriminate themselves. The first is the one in which the quantities j are static so that in an expression like $j_\mu(1)\,j_\nu(3)$ we can take account of the difference of t_1 and t_3 by a simple Taylor expansion in t_1-t_3. If, on the other hand, the j are non-static ($\sim v/c$), then such an expansion fails because of the large Zitterbewegung frequency, and we consequently face the second case. The cross term, in which one j is static and the other non-static, will be considered afterwards.

a) The static case. We use the expression (43), not explicitly writing down the S-functions for the particles, but going with the j_λ themselves. We make beforehand a remark that the first term in the bracket in (43) gives a zeroth order ($\sim(v/c)^0$) potential while the second is responsible for the retardation correction to it. Thus in the first term we put $j_\lambda(t)\sim 1$ with respect to the time dependence. The Δ-function is, in Fourier representation,

$$\Delta_F(x)=\frac{-2i}{(2\pi)^4}\int e^{i(kx)}\,\frac{(dk)}{k^2+\mu^2}, \tag{44}$$

with the understanding that μ^2 is furnished with a small negative imaginary part. The first term of Eq. (43), which we call (A), becomes then

$$\begin{aligned}&\frac{1}{32}\,\varepsilon^4\,\frac{(-2i)^2}{(2\pi)^8}\int\cdots\int\frac{1+\varepsilon(13)\,\varepsilon(42)}{2}\,[j_\mu(\boldsymbol{r}_1),j_\nu(\boldsymbol{r}_3)]\,[j_\lambda(\boldsymbol{r}_4),j_\rho(\boldsymbol{r}_2)]\,O_{\mu\rho}O_{\lambda\nu}\\ &\quad\times\frac{1}{k^2+\mu^2}\cdot\frac{1}{k'^2+\mu^2}\cdot e^{-ik_0(t_1-t_2)}\,e^{-ik_0'(t_3-t_4)}\,e^{i(\boldsymbol{k},\boldsymbol{r}_1-\boldsymbol{r}_2)}\,e^{i(\boldsymbol{k}',\boldsymbol{r}_3-\boldsymbol{r}_4)}\\ &\quad\times(d\boldsymbol{k})\,(d\boldsymbol{k}')\,dt_1\,dt_2\,dt_3\,dt_4\cdot d\boldsymbol{r}_1\,d\boldsymbol{r}_2\,d\boldsymbol{r}_3\,d\boldsymbol{r}_4.\end{aligned} \tag{45}$$

Making use of the fact that $j_\mu(r)$ and $j_\nu(r')$ commute (for equal t) if $r\neq r'$, we can identify r_1 and r_2 with r_3 and r_4 respectively, and put $r_1-r_2=r_3-r_4=r$, thereby reducing the quadruple integral $\int\Pi dr$ to a double one. As regards the integration in t, we invoke the centers of time

$$T=\frac{t_1+t_2+t_4+t_3}{4}=\frac{T'+T''}{2},\quad T'=\frac{t_1+t_2}{2},\quad T''=\frac{t_3+t_4}{2},$$

together with

$$t=T'-T'',\quad t'=t_1-t_2,\quad t''=t_3-t_4, \tag{46}$$

and leave out the integration in T to obtain the potential energy. Further putting

$$t'-t''=u, \qquad t=\frac{1}{2}uv, \tag{47}$$

we find

$$\int\cdots\int\frac{1+\varepsilon(13)\varepsilon(42)}{2}e^{-ik_0(t_1-t_2)-ik_0'(t_3-t_4)}dt_1\cdots dt_4$$

$$=\int dT\int_{-\infty}^{\infty}e^{-i(k_0-k_0')u}\frac{1}{2}|u|du\int_{-1}^{1}dv\cdot 2\pi\delta(k_0+k'_0)$$

$$=-\frac{8}{(k_0-k_0')^2}2\pi\delta(k_0+k_0')\int dT. \tag{48}$$

The potential (A) is given by

$$H_A^{(4)}=\left(\frac{\varepsilon^2}{4\pi}\right)^2\frac{1}{2}\int[j_\mu(\boldsymbol{r}_1),j_\nu(\boldsymbol{r}_1)]\,[j_\lambda(\boldsymbol{r}_2),j_\rho(\boldsymbol{r}_2)]\,O_{\mu\rho}\,O_{\lambda\nu}\,V(\boldsymbol{r}_1-\boldsymbol{r}_2)\,d\boldsymbol{r}_1\,d\boldsymbol{r}_2,$$

$$V(\boldsymbol{r})=\frac{i}{16\pi^5}\int\frac{1}{k_0^2}\cdot\frac{e^{-i\boldsymbol{k}\boldsymbol{r}}}{k^2+\mu^2}\cdot\frac{e^{i\boldsymbol{k}'\boldsymbol{r}}}{k'^2+\mu^2}\delta(k_0+k'_0)\,(dk)\,(dk'). \tag{49}$$

V can be evaluated, with the above mentioned definition at the poles, as

$$V(\boldsymbol{r})=\frac{i}{16\pi^5}\iint\left[-\pi i\left\{\frac{1}{(k^2+\mu^2)^{1/2}}\frac{1}{k'^2-k^2}+\frac{1}{(k'^2+\mu^2)^{1/2}}\frac{1}{k^2-k'^2}\right\}e^{i\boldsymbol{k}\boldsymbol{r}}\,e^{i\boldsymbol{k}'\boldsymbol{r}}\right]d\boldsymbol{k}\,d\boldsymbol{k}'$$

$$=\frac{i}{16\pi^5}\int(-\pi i)\frac{e^{i\boldsymbol{k}\boldsymbol{r}}}{(k^2+\mu^2)^{1/2}}(-2\pi^2)\frac{\cos kr}{r}\times 2\,d\boldsymbol{k}$$

$$=\frac{1}{2\pi r^2}\int\frac{\sin 2kr}{(k^2+\mu^2)^{1/2}}k\,dk=\frac{1}{\pi r}K_0(2\mu r) \tag{50}$$

where K_ν is the modified Bessel function of order ν:

$$K_\nu(x)=\frac{1}{2}\pi i e^{-\frac{1}{2}\pi i\nu}H_{-\nu}^{(1)}(ix). \tag{51}$$

Asymptotically, V behaves as

$$V(r)\sim-(1/\pi r)\ln(\gamma\mu r), \quad \gamma=1.781 \text{ for } 2\mu r\ll 1$$

$$\sim(1/2r)(1/\pi\mu r)^{1/2}e^{-2\mu r} \quad \text{for } 2\mu r\gg 1. \tag{52}$$

The range of the force is thus half of that in the second order potential, corresponding to the fact that two mesons are interchanged between the particles.

The retardation correction has been neglected in the above calculation, but it can be shown that the $(v/c)^1$-effect is zero. In view of the approximate nature of the calcultion, we shall not investigate the $(v/c)^2$ -correction to the fourth order potential which is already small compared to the second order one.

On the other hand, the second term of Eq. (43), called (B), gives a $(v/c)^1$-effect. We accordingly put

$$j(1)\,j(3) \sim e^{i\Delta p_0(1)t_1} \cdot e^{i\Delta p_0(3)t_3}. \tag{53}$$

An integral of the form

$$\int \frac{\varepsilon(13)+\varepsilon(42)}{4}\, e^{-i(k_0-k'_0)\frac{t_1-t_3-t_2+t_4}{2}}\, e^{ip(t_1-t_3)}\, e^{iq(t_4-t_2)}\, dt_1 \ldots dt_4 \tag{54}$$

is, by a similar procedure, found to be

$$= -\frac{\pi i}{2}\,\frac{1}{p-q}\,\{\delta(k_0-k_0'-2p)-\delta(k_0-k_0'-2q)\}. \tag{55}$$

Using this, the part (B) of the potential can be calculated, giving

$$H_B^{(4)} = \frac{i}{16}\left(\frac{\varepsilon^2}{4\pi}\right)^2 \int\!\!\int \Big\{ \Big[\frac{d}{dt}\,(j_\mu(\boldsymbol{r}_1)\,j_\lambda(\boldsymbol{r}_2)),\, j_\nu(\boldsymbol{r}_1)\,j_\rho(\boldsymbol{r}_2) \Big] - \Big[j_\mu(\boldsymbol{r}_1) j_\lambda(\boldsymbol{r}_2),$$

$$\frac{d}{dt}\,(j_\nu(\boldsymbol{r}_1) j_\rho(\boldsymbol{r}_2)) \Big] \Big\}\, O_{\mu\rho}\, O_{\nu\lambda}\, \frac{e^{-\mu r}}{r}\, \frac{e^{-\mu r}}{2\mu}\, d\boldsymbol{r}_1\, d\boldsymbol{r}_2$$

$$= \frac{i}{8}\left(\frac{\varepsilon^2}{4\pi}\right)^2 \int\!\!\int \Big[\frac{d}{dt}(j_\mu(\boldsymbol{r}_1) j_\lambda(\boldsymbol{r}_2))\, V_2(r),\, j_\nu(\boldsymbol{r}_1) j_\rho(\boldsymbol{r}_2)\, V_0(r) \Big]\, O_{\mu\rho}\, O_{\lambda\rho}\, d\boldsymbol{r}_1,\, d\boldsymbol{r}_2, \tag{56}$$

$$V_0 = \frac{e^{-\mu r}}{r}, \qquad V_2 = \frac{e^{-\mu r}}{2\mu} = -\frac{1}{2\mu}\,\frac{\partial}{\partial\mu}\, V_0. \tag{57}$$

(b) The non-static case. We make the approximation

$$j(1)\,j(3) \sim e^{2mi(t_1-t_3)} \tag{58}$$

since the main contribution comes from processes in which the initial and final state is positive, while the intermediate is negative, in energy. Here we must be careful about the sign of m in the exponential. It is determined by the order of the operators $j(1)$ and $j(3)$, but not by the order of the times t_1 and t_3. The integral corresponding to (48) is then

$$-\left\{ \frac{1}{\left(2m-\frac{1}{2}(k_0-k_0')\right)^2} + \frac{1}{\left(2m+\frac{1}{2}(k_0-k_0')\right)^2} \right\} 2\pi\delta(k_0+k_0') \int dT$$

for the order 13·42,

and

$$\left\{ -\frac{2\pi i}{m}\,\delta(k_0-k_0') + \frac{2}{(2m)^2-\left(\frac{k_0-k_0'}{2}\right)^2} \right\} 2\pi\delta(k_0+k_0') \int dT \tag{59}$$

for the order 13·24.

We neglect k_0 and k_0' compared to m, since the important contribution comes from values of k_0 of the order μ, which is $\ll m$ for actual cases. The interaction (A) is thus given by

$$H_A^{(4)}=\left(\frac{\varepsilon^2}{4\pi}\right)^2\frac{1}{2}\int\{j_\mu(1),O_{\mu\rho}j_\rho(2)\}^2\cdot\frac{1}{4m}\,V_0(r)^2\,d\boldsymbol{r}_1\,d\boldsymbol{r}_2$$

$$+\left(\frac{\varepsilon^2}{4\pi}\right)^2\frac{1}{2}\int\{j_\mu(1),\,j_\nu(1)\}\{j_\lambda(2),\,j_\rho(2)\}O_{\mu\rho}\,O_{\nu\lambda}\,\frac{1}{8\pi}\,\frac{\mu}{m^2r^2}\,K_1(2\mu r)\,d\boldsymbol{r}_1\,d\boldsymbol{r}_2. \tag{60}$$

The interaction (B), on the other hand, becomes zero in the present approximation. We shall not, however, enter into more detailed treatments.

c) Cross terms. When some j are static and others non-static, there occur in the expression (43) cross terms of these quantities. They may be calculated in a similar manner. But the resulting fourth order potentials will be non-static with radial dependence essentially the same as in the case b). Consequently their order of magnitude will be smaller than the preceding ones. We shall not investigate these terms here.

In the above results, a finite field mass μ has been assumed. If $\mu=0$, they cannot be applied at once since some of the expressions diverge. In this case we can show that

in (57) $\qquad V_2\rightarrow\frac{r}{2},$

and in (60) $\qquad \frac{1}{8\pi}\,\frac{\mu}{m^2r^2}\,K_1(2\mu r)\rightarrow\frac{1}{16\pi}\,\frac{1}{m^2r^3}. \qquad (61)$

A reasonable evaluation of (50) for $\mu=0$ is not known, but it may be left out of consideration since fortunately such a case does not occur actually.

4. The Relation between the Second and Fourth Order Potential

Now we have at hand the second and fourth order potential $H^{(2)}$ and $H^{(4)}$. Their sum should be taken as the potential energy for the system which is rigorous up to the fourth order in the coupling contant. $H^{(2)}$ is a familiar interaction which can be derived by various methods, except that the retardation effect should be taken account of in a manner specified above. In case j is static, a retardation expansion is allowed. But for general j the situation is different. In fact, for transitions in which either of the particles changes its energy sign but not both, (odd transition), the interaction is virtually of a delta-function type which will not fit in the category of the ordinary potential. In other words, we had better decompose every quantity j into a static part $\bar{j}$ and a non-static (Zitterbewegung) part* $\tilde{j}$, and apply the retarded potential (38) only for the

* The terms static and non-static are somewhat deviating from ordinary usage. They refer to the change of the energy sign.

even part $\bar{j}_\lambda \bar{j}_\mu + \tilde{j}_\lambda \tilde{j}_\mu$. In this sense the Pauli approximation plays a major role in the two-body problem.

The fourth order potential, on the other hand, exhibits several interesting features. It takes on different forms according to the v/c-dependence of j as above. Moreover, it has an additional dependence on the spin variables through commutator expressions. Suppose that all the j's commute with each other, then the term (49) or the first term of (43) vanishes, leaving only the retardation correction (56) in the static case. Thus there remain no fouıth order forces except for a small retardation correction.

Let us see how these fourth order characteristics show up in the electromagnetic interaction of two electrons. Here we have one static and three non-static interactions, namely

$$j_4(\boldsymbol{r}_1)\,j_4(\boldsymbol{r}_2) = -(\bar{\psi}\gamma_4\psi)_1\,(\bar{\psi}\gamma_4\psi)_2 \tag{62a}$$

and

$$j_i(\boldsymbol{r}_1)\,j_i(\boldsymbol{r}_2) = -(\bar{\psi}\gamma_i\psi)_1\,(\bar{\psi}\gamma_i\psi)_2, \quad i=1,2,3. \tag{62b}$$

We apply the formula (49) and (50) for (62a) and (60) for (62b). The former, however, vanishes, showing that the Coulomb potential is exact even to the fourth order effect. The latter yields a relatively large spin-spin and a smaller spin-independent interaction of the electrons, but the spin-spin term is just cancelled by a second order perturbation of the second order potential with $j_i \check{j}_i$ *via* negative state. The retardation correction (56) can be applied only to (62a). We can, however, again eliminate it by a canonical transformation (which is equivalent to the previously considered one (29)):

$$\Psi = \exp\left[\frac{i}{8}\frac{d}{dt}\int j_4(\boldsymbol{r}_1)\;V_2 j_4(\boldsymbol{r}_2)\;d\boldsymbol{r}_1\,d\boldsymbol{r}_2\right]\Psi'. \tag{63}$$

In fact the transformed interaction becomes

$$\begin{aligned} &-\frac{1}{2}\int j_\lambda(1)\,j_\lambda(2)\;V - \frac{i}{8}\iint\left[\frac{d}{dt}\,(j_4(1)\,j_4(2)),\,j_4(1)\,j_4(2)\right]V_0\,V_2 \\ &+\frac{i}{8}\iint\left[\frac{d}{dt}\,(j_4(1)\,j_4(2)),\,j_4(1)\,j_4(2)\right]V_0 V_2 + \frac{1}{8}\int\frac{d^2}{dt^2}\,(j_4(1)\,j_4(2))\,V_2 \\ &+\text{higher order terms.} \end{aligned} \tag{64}$$

The last term contributes

$$\frac{1}{8}\int\left\{\left(\frac{d}{dt_1}\right)^2 j_4(1)\,j_4(2) + 2\,\frac{d}{dt_1}j_4(1)\frac{d}{dt_2}j_4(2) + j_4(1)\left(\frac{d}{dt_2}\right)^2 j_4(2)\right\}V_2, \tag{65}$$

which combines with the first teım with the expansion (38). Thus we get the corrected Breit interaction, with $j_\lambda j_\lambda$ replaced by $\bar{j}_\lambda \bar{j}_\lambda + \tilde{j}_\lambda \tilde{j}_\lambda \equiv \overline{j_\lambda j_\lambda}$,

$$H^{(2)\prime}=\frac{\varepsilon^2}{4\pi}\,\frac{1}{2}\int\left\{\overline{j_\lambda(1)\,j_\lambda(2)}\,V_0-\frac{1}{2}\,\overline{\frac{dj_4(1)}{dt}}\;\overline{\frac{dj_4(2)}{dt}}\,V_2\right\}d\boldsymbol{r}_1\,d\boldsymbol{r}_2$$

$$+\left(\frac{\varepsilon^2}{4\pi}\right)^2\frac{1}{2}\int\{\tilde{j}_i(1)\,\tilde{j}_i(2)\}^2\cdot\frac{1}{4m}\,V_0^2\,d\boldsymbol{r}_1\,d\boldsymbol{r}_2+\text{smaller terms.} \qquad (66)$$

There remains no net fourth order effect of detectable magnitude. On the other hand, if we try to perform a second order perturbation using the original Breit formula, a result is obtained which contradicts the experimental facts about the hyperfine structure of helium[12]. This is due to the lack of the fourth order corrections which cancel a part of the iteration of the second order potential, leaving no spin-dependent fourth order effects. The same circumstance is observed in the (symmetrical or charged) pseudoscalar meson theory with pseudoscalar coupling investigated by Watson and Lepore[18], in which the isotopic spin variables replace the ordinary ones.

5. Further Discussions

In the preceding analysis we have given a general form of the potential which is correct up to the fourth order in the coupling constant and takes an approximate account of the retardation effect. The result shows that the corrections, which should be made to the potential directly derivable from non-relativistic classical theory, is essentially dependent on the nature of the interaction. Those interactions which contain only classical variables give in the quantum theory also potentials not much different from classical ones. On the other hand, if the interaction contains quantum variables, such as the Dirac spin and isotopic spin matrices, the higher order corrections in general affect the classical potential. Although these corrections have been considered only in the fourth approximation, we can in principle obtain the higher order ones step by step, in such a manner that the resulting potential yield exactly the S-matrix derived from the field theory. The origin of these correction lies, as was seen above, in the fact that the order of the emission and absorption of quanta exchanged between the particles affect the matrix elements. As an illustration, let us consider a charged meson theory, which contains the non-commutative isotopic spin variables τ_λ. The physical meaning of the non-commutativity is that the charge must be conserved at any instant, so that a nucleon cannot emit or absorb successively more than one positive or negative mesons. In other words, there is a strong correlation between the partaking quanta, which produces deviations in the higher order S-matrix terms from those obtained by a mere iteration of the second order process. For scattering problems these higher order corrections may be neglected. But for the bound system they would give in general a large influence on the eigenvalues and eigenfunctions. It is very difficult, however, to give an explicit expression for these corrections. We give below only an idea about these circumstances on a simple model, which may rather be too rough an approximation.

Take a symmetrical meson theory with static interaction. The main nth order processes consist of those graphs in which n meson lines connect the two nucleon lines, the order of the joints on the particle lines being different for each graph. Every graph gives rise to a matrix element of the form

$$\sum_{\tau}(\tau_1\,\tau_2\ldots\tau_n)_1\,(\tau_{i1}\,\tau_{i2}\ldots\tau_{in})_2\,f_n(p_1\,q_1\,p_2\,q_2,\,(i_1\ldots i_n)), \tag{67}$$

where p_1, q_1, p_2, q_2 are the initial and final momenta of the particles. Now we *assume* that the function f virtually does not depend on the permutation symbol $(i_1\ldots i_n)$ for the *irreducible* graphs, in which every meson line crosses some other meson lines.

For these graphs we put

$$f_n=\langle f_n\rangle_{Av}\sim-\frac{1}{C_n}\,v_0^{n-1}v=-kv_0^n, \tag{68}$$

$$v=ce^{-\mu r}/r,\qquad v_0\sim ce^{-\mu r}/\sqrt{\mu r},\quad k\sim\sqrt{\mu/r},$$

where C_n is the number of irreducible graphs. This form may be suggested by observing that

$$(-i)^2\,C_mf_m\cdot C_{n-m}f_{n-m}\cdot T\sim C_nf_n \tag{69}$$

and

$$\int\frac{e^{ikr}}{k^2+\mu^2}\,dk=4\pi K_0(\mu r)\sim\frac{1}{\sqrt{\mu r}}e^{-\mu r}, \tag{70}$$

where $T\sim 1/k$ is the time interval in which a mth and an $(n-m)$th order graph overlap, giving an nth order irreducible graph. Then we extend for simplicity the average over the irreducible graphs to all graphs, thereby using the formula

$$\frac{1}{n\,!}\sum_{(i_1\ldots i_n)}(\tau_i\ldots\tau_n)_1\,(\tau_{i_1}\ldots\tau_{i_n})_2=\begin{cases}n+1 & n=\text{even},\\ \dfrac{n+2}{3}\,\tau_1\cdot\tau_2 & n=\text{odd}.\end{cases} \tag{71}$$

Next, starting with an effective potential

$$\{(\tau_1\cdot\tau_2)+a\}V=-k\sum s_nv_0^n, \tag{72}$$

we get similar irreducible terms of S-matrix:

$$\{(\tau_1\cdot\tau_2)+a\}^nF_n,\quad F_n\sim-VV_0^{n-1}=-kV_0^n. \tag{73}$$

Equating both S-matrices, we obtain an equation determining the effective potential, which is solved as

$$1+\frac{1}{k}V=\frac{(A-2B)-B(\tau_1\cdot\tau_2)}{(A+B)(A-3B)},$$

$$A=\frac{1+v_0^2}{(1-v_0^2)^2},\quad B=\frac{1-v_0^2/3}{(1-v_0^2)^2}v_0. \tag{74}$$

This is a rather strange result. For large r and small coupling constant, V is nearly equal to $-v(\tau_1\cdot\tau_2)$. But for the contrary case v_0 will be large, and the isotopic dependence is of a linear combination of $\tau_1\cdot\tau_2$ and 1. Of course the above form of the functional dependence is not a reliable one, because of the nature of the present investigation. But it may be sufficient to make one modest in asserting the validity of lower order approximations.

Next we make a remark on the relation between the results obtained here and those which follows from the ordinary perturbation calculation. Writing the momentum-energy of the particle and emitted quantum as (p, p_0) and (k, k_0) respectively, the second order matrix element is

$$\left(\frac{1}{2\pi}\right)^3\left\{\frac{1}{-(\Delta p_0)_1-k_0}+\frac{1}{-(\Delta p_0)_2-k_0}\right\}O_1O_2\frac{1}{2k_0}. \tag{75}$$

The energy denominators can be rewritten as

$$\frac{1}{-(\Delta p_0)_1-k_0}+\frac{1}{-(\Delta p_0)_2-k_0}=\frac{1}{-(\Delta p_0)_1-k_0}+\frac{1}{(\Delta p_0)_1-k_0}$$
$$+\frac{1}{-(\Delta p_0)_2-k_0}-\frac{1}{(\Delta p_0)_1-k_0}$$
$$=\frac{-2k_0}{k_0^2-(\Delta p_0)_1^2}+\frac{(\Delta p_0)_1+(\Delta p_0)_2}{\{k_0-(\Delta p)_1\}\{k_0+(\Delta p_0)_2\}}. \tag{76}$$

The first term corresponds to the Eqs. (9) and (10), while the second is, being of the form $i(d/dt)V$ (non-Hermitic), contributes only to a higher order effect. The symmetrical form (37), however, does not follow from the perturbation theory. The fourth order interaction is therefore also different. In fact the perturbation theory will not give a fourth order potential as simple as the present one.

Thus far we have been confining ourselves to essentially non-relativistic treatments. Those regions where the distance of the particles is small and relativistic effects play a predominant role are out of consideration in the present investigation. Moreover, the difficulty of high singularities of the potential which haunts these regions is of a more profound origin and will only be solved with the general difficulties of the present quantum theory. A complete relativistic eigenvalue problem itself, if it could be formulated, would not give much help to these fundamental difficulties.

As for the relativistic formulation of the two-body problem, our results seem to present a slight progress over the older ones in that the many-time theory has been invoked as a guiding principle which requires a certain integrability condition for the interaction term. In view of the preceding analysis, we easily find that a pair of equations of the form (in the interaction representation)

$$\left\{i\frac{\partial}{\partial t_2}-V\left(\frac{t_1+t_2}{2}\right)\right\}\psi=0, \qquad \left\{i\frac{\partial}{\partial t_2}-V\left(\frac{t_1+t_2}{2}\right)\right\}\psi=0 \tag{77}$$

just meet our purpose. Transforming back to the Schrödinger representation, we have

$$\left\{i\frac{\partial}{\partial t_1}-H_1-V(t_1-t_2)\right\}\psi=0, \quad \left\{i\frac{\partial}{\partial t_2}-H_2-V(t_1-t_2)\right\}\psi=0\,; \tag{78}$$

or

$$\left\{i\frac{\partial}{\partial T}-(H_1+H_2)-2V(t)\right\}\psi=0, \qquad \left\{i\frac{\partial}{\partial t}-(H_1-H_2)\right\}\psi=0, \tag{79}$$

with

$$\begin{aligned} &i\frac{\partial V}{\partial t}-[H_1-H_2,\ V]=0, \qquad T=\frac{t_1+t_2}{2}, \qquad t=t_1-t_2. \\ &V=e^{-i(H_1-H_2)t}\,V(0)\,e^{i(H_1-H_2)t}, \end{aligned} \tag{80}$$

The space-time behavior of the wave function ψ is such that it is essentially determined at $t_1=t_2$, and extended to $t_1 \neq t_2$ simply as free of interaction. The latter procedure means nothing but a canonical transformation of the whole representation by a function $\exp[i(H_1-H_2)t]$. Thus the above many-time equations do not really present a purely relativistic (four-dimensional) formulation. They are essentially the same as the ordinary one-time equation. However, we notice here a fact which seems to have been overlooked. That is the freedom of canonical transformation just mentioned. Indeed any equation of the form

$$\left(i\frac{\partial}{\partial t}-H-V\right)\psi=0 \tag{81}$$

allows a transformation $\exp[if(H)]\psi$ with the result that only the interaction potential suffers a modification. In this sense the potential is not uniquely determined though the ensuing S-matrix is unique[14].

Finally a remark may be added concerning the Eqs. (31) and (32). The S-matrix (31) is unitary, but its sub-matrix under consideration is not. For if two particles collide, the Bremsstrahlung will occur, which diminishes the norm of the final elastic state. This means that V is in general complex. In fact an atom in an excited state will "decay" by emitting photons, and this should be properly represented by the complex potential. We may, however, regard V approximately Hermitic, at least at low energies. The preceding results give it some justification.

In conclusion, we may say that the concept of potential energy cannot enjoy a wide and practical extension much beyond the classical and non-relativistic form. Consequently a more ambitious attempt at a relativistic eigenvalue problem for closed systems will have to deviate considerably from the present one. An example will be furnished by the equation

$$\left(\gamma_\lambda \frac{\partial}{\partial x_\lambda} - \varkappa\right)_1 \left(\gamma_\mu \frac{\partial}{\partial x_\mu} - \varkappa\right)_2 \psi = -\frac{\varepsilon^2}{2} (\gamma_\nu)_1 (\gamma_\nu)_2 D_F(12) \psi. \tag{82}$$

This equation can be deduced, as a reasonable one, from the many-time equation with intermediate field at the cost of some assumption and alteration on the meaning of state vector. We might expect that it determines an eigen-momentum-energy vector of the system from some boundary conditions at (space-time) infinity. Eq. (81) only takes account of the second order effect. Inclusion of higher order effects is not possible in the above form and can only be achieved by an integral equation. If higher order effects play an important part on the interaction, we shall therefore have to resort to still different methods, such as the *S*-matrix theory, the strong coupling theory, direct tackling with integral equations,[15] and so on.

References

1) G. Breit, Phys. Rev. **34** (1929), 553.
2) G. Breit, Phys. Rev. **51** (1937), 248, 778; **53** (1938), 153.
3) G. Breit, Phys. Rev. **39** (1932), 616. C. Møller, ZS. f. Phys. **78** (1931), 786.
H. Bethe and E. Fermi, Zeit. Phys. **77** (1932), 296.
4) C. Møller and L. Rosenfeld, Det Kgl. Danske Wid. Selsk. Mat. Fys. Med. XVII, No. 11 (1940).
5) S. Tomonaga, Prog. Theor. Phys. **1** (1946), 27.
J. Schwinger, Phys. Rev. **74** (1948), 1439; **75** (1949), 651; **76** (1949), 790.
R. P. Feynman, Phys. Rev. **76** (1949), 749, 769.
6) L. Van Hove, Phys. Rev. **75** (1949), 1519.
Y. Nambu, Prog. Theor. Phys. **3** (1948), 444.
7) Y. Nambu, Ref. 6)
8) P. A. M. Dirac, Proc. Roy. Soc. **136** (1932), 453.
9) T. Toyoda, Phys. Rev. **77** (1950), 353; Soryûsiron-Kenkyû **1** (1949), No. 3, 129 (in Japanese).
10) F. J. Dyson, Phys. Rev. **75** (1949), 486.
11) J. Schwinger, Ref. 5).
12) G. Breit, Phys. Rev. **36** (1930), 303; **39** (1932), 616.
13) K. M. Watson and J. V. Lepore, Phys. Rev. **76** (1949), 1157.
14) C. Møller, Det Kgl. Danske Vid. Selsk. Mat. Fys. Med. XXIII, No. 1 (1945).
15) S. M. Dancoff, Phys. Rev. **78** (1950), 382.

On the Nature of V-Particles, I

Y. Nambu, K. Nishijima and
Y. Yamaguchi

*Department of Physics,
Osaka City University*

June 19, 1951

Recently Butler et al.[1] presented further evidences concerning the nature of the so-called V-particles which had been observed by Rochester and Butler[2] and by Anderson and collaborators[3]. According to these authors: i) The V-particles are found among penetrating showers with a rate of the order of 1%; ii) Their decay life is estimated to be about 10^{-10} sec; iii) There are two kinds of them, charged and neutral, the latter being about 5 to 10 times more abundant than the former, presumably due to their difference in lives; iv) According to Butler et al., moreover, they can be

Reprinted from Prog. Theor. Phys. 6 (1951) pp. 615–622.

classified in two groups of different masses. The heavier particles, with mass about 2200 to 2300 m_e, decay into a nucleon and a (π) meson, hence fermions. The lighter ones, with mass around 1000 m_e, decay into two (π) mesons, and hence bosons; v) One case is reported which shows a successive decay of a charged V through neutral V into three charged particles (three mesons or two mesons plus one proton), though no suggestion is given as to whether they belong to the heavier or the lighter group.

Then the arguments go on as follows. First of all, there is a possibility that the decay products of V and τ might include μ-mesons as well. This, however, seems rather improbable in view of the fairly large nuclear interaction of the decay mesons reported by Anderson et al.[3], and the relatively long life of the parents in contrast to their sizable production rate. A three particle decay, such as $V \rightarrow P+\mu+\nu$, is all the more unlikely according to Anderson et al.'s arguments about the coplanar character of the parent and daughter particles. It is also to be noted that soft showers are not associated with the V or τ events, which shows that decays giving rise to photon or electron must be rare compared to the main processes. The above mentioned contradiction between production and decay may be lifted if we postulate, as was just the case in the π-μ decay, that the observed V (though not necessarily including τ) are decay products of some unknown particles with sufficiently short life and strong nuclear interaction. But at the present stage we will try to solve this problem using only the observed particles, assuming the decay products to be π's and nucleons (of course excluding the case V)). This can be done as follows:

Now we face two alternative interpretations of the point v) above. One is to assume that the unknown particles involved in this reaction are V's, decaying according to the scheme:

$$V_{\pm} \rightarrow V_0+\pi_{\pm}, \quad V_0 \rightarrow N+\pi \quad (N\text{—nucleon}); \qquad (1)$$

while the other is to assume them to be τ's:

$$\tau_{+} \rightarrow \tau_0+\pi_{\pm}, \quad \tau_0 \rightarrow 2\pi \qquad (2)$$

According to (1), the V's may be regarded as excited states of nucleons which make transitions to the lower states by emitting mesonic radiation. (τ-emission can be forbidden energetically.) Such a situation has long since been anticipated in the strong coupling theory of nucleon-meson interaction. This theory however, does not just seem to be very useful for our analysis. For it is too crude and incomplete to be relied upon quantitatively, especially in view of the extremely small width of the actual levels ($\sim 10^{-5}$ev). In the present stage we had rather better treat them as different elementary particles, obeying Fermi statistics and having half-odd spins, possibly higher than 1/2, and introduce formal interactions which cause the observed transitions. Thus we assume the following scheme

$$\begin{array}{lcc} \text{Interaction} & \underbrace{V_{\pm}V_0\pi,\ V_0N\pi} & \underbrace{V_{\pm}N\tau,\ V_0N\tau} \\ \text{Coupling const.} & G_1 & G_2 \end{array}$$

$$\begin{array}{cccc} VV\pi & NN\pi & VV\tau & NN\tau \\ G_3 & g_3(\text{known}) & G_4 & g_4 \end{array} \qquad (3)$$

The conditions to be considered are: a) decay life roughly all of order 10^{-10} sec.; b) competition among various possible decay modes (especially for the τ decay process); and c) production mechanism, to give a yield of $\sim 10^{-2}$ times that of ordinary mesons.

For the calculation of $\tau \rightarrow \pi+x (x=\pi,\gamma,\cdots)$ processes, the results of covariant calculation by many authors[6] can directly be applied with only a few alterations. Unfortunately, however, most of the calculation involved are not free from the diverging ambiguities which have to be disposed of with the aid of the Pauli regulator. Accordingly, in the present order of magnitude consideration, we check them with more rough and intuitive estimation which only takes account of the coupling constants and the volume of phase space —in a manner more or less similar to Fermi's.[7)] On the other hand, the various selection rules, such as described by Fukuda et al.,[6e)] are more reliable and can be used to narrow down the possibilities.

In this way we get the following results:

$$G_1^2 \sim 10^{-11}-10^{-13},\ G_2^2 \sim 10^{-2}-10^{-3},$$
$$g_4^2 \sim 10^{-7}-10^{-9}$$
$$(G_3^2 \gtrsim g_3^2,\ G_4^2 \lesssim g_4^2,\ \text{not necessary}). \quad (4)$$

The transformation property of τ(or at least τ_0) must be either scalar or vector, since it is very likely that the π mesons are pseudoscalar.[8)] The range of values in (4) correspond to different assumptions as to Fermi's reaction volume (or alternatively the general trend of the covariant calculation), as well as the assignment of spin values for the V's*. Thus discrepancies with experiment of the order, say, 100, should be tolerated.

Next let us examine the assumption (2). In this case, $\tau_\pm$ must be pseudoscalar, while τ_0 scalar or vector.** This choice, which may also be adopted in the first model (1), excludes the process $\tau_\pm \to 2\pi$ in favor of $\tau_\pm \to \tau_0+\pi$ and $\tau_\pm \to 3\pi$ (Powell's case).[5e)] There arise the following nine couplings:

$$\begin{array}{lll} VN\pi, & VN\tau_\pm, & VN\tau_0, \\ VV\pi, & VV\tau_\pm, & VV\tau_0, \\ NN\pi, & NN\tau_\pm, & NN\tau_0. \end{array} \quad (5)$$

Of these, $NN\pi$ is known, and $NN\pi_\pm$ and $NN\tau_0$ and $VN\tau_0$ turn out either unnecessary or harmful. Among many possible combinations of the remaining couplings few yield consistent results. Thus we take as a relatively reasonable choice (see Appendix):

$$\begin{array}{lll} VN\tau_0 & VV\tau_\pm & VV\tau_0 \\ G_1^2 \sim 10^{-2}, & G_2^2 \sim 10^{-1.5}, & G_3^2 \sim 10^{-6}, \end{array}$$
$$\begin{array}{ll} VV\pi & VN\pi \\ G_4^2 \sim 10^{-1.5}, & G_5^2 \sim 10^{-12}, \end{array} \quad \text{others}=0. \quad (6)$$

These estimations are of course susceptible to fluctuations, by as large a factor as $\sim 10^2$, depending on different assumptions on the calculational procedure.

At present we cannot tell with confidence which of the above two alternatives (1) and (2) is the more preferable. The former is relatively free from ambiguities and conflictions in determining the coupling constants. The idea of the existence of a series of excited levels of nucleons also attracts us. On the other hand, however, the latter cannot be excluded and even seems natural in view of the other evidences concerning τ-mesons, e.g. Powell's $\tau_\pm \to 3\pi$ decay, which can well compete with $\tau_\pm \to \tau_0+\pi$ on this model. It is hoped that the forthcoming paper of Butler et al. will settle this question.

In conclusion, we should like to call attention to some effects which could be related to the V and τ mesons. First, the nuclear force would be modified. But the range being at most only $\sim 1/m_\tau \sim 2/m_\pi$, the effect would be too small to account for the existing N-N scattering data. Second, the conventional strong coupling theory, which allows the nucleon isobars of both charges to occur not as anti-particles, could be tested by the experiments, in which the stability of nuclei should also be taken account of. (According to our models, the anti-particles can only appear as pairs.)*** Third, the new particles would play some part in the anomalous magnetic moment of nucleons, though we do not know how and to what extent. Fourth, the production of these particles, though depending on the model, would occur mainly a V-τ pairs and V-V pairs, and to a lesser degree as single (or multiple) τ's. The corresponding threshold energy would be about 1.1 Bev ($\gamma+N \to V+\tau$) and 0.8 Bev ($N+N \to V+V$) respectively in the laboratory system. So far these predictions are not definitely at variance with experiments.[9)]

A more detailed account of the present analysis will be given later.

Appendix

On the second model (Eq. (2)), the possible competing processes which need consideration are as follows**

$$\tau_0 \to 2\pi, \quad \tau_0 \to \pi_0+\gamma, \quad \tau_0 \to 2\gamma;$$

$$\tau_\pm \to \tau_0+\pi_\pm \quad \tau_\pm \to 3\pi, \quad \tau_\pm \to \pi_\pm+2\gamma;$$

$$V \to N+\pi.$$

From these processes we can first determine the relative magnitude of the coupling constants so as to fit the experimental facts, and then normalize them by some process such as the production rate. A crude life-time formula is, for example, furnished by

$$1/t \sim (G_3^2)(G_4^2)^2 I_2/m_\pi \sim (G_3^2)(G_4^2)^2 \cdot 10^{23}\text{sec}^{-1} \quad \text{for} \quad \tau_0 \to 2\pi,$$

where I_2 means the availabe volume per unit energy of momentum space. An additional factor $\sim 10^{-3}(\sim(m_\pi/m_V)^3)$ may be introduced if we take account of the coordinate space volume corresponding to the third order process in which virtual V-pairs are created (see Figure). This estimation also agrees roughly with the results of covariant calculation using regulators. The numerical values given in (6) were obtained by the latter refined method assuming scalar and pseudoscalar coupling for the scalar τ_0 and pseudoscalar $\tau_\pm$ respectively.

The first model (1) may be treated analogously.

*) Spin 3/2 for V would make the decay $V \to N+\pi$ forbidden of first order if it were allowed for spin 1/2.

**) Actually pseudovector $\tau_\pm$ is to be discarded since they favor $\tau_\pm \to \pi_\pm+\gamma$ (assuming π to be pseudoscalar) For τ_0, either pure neutral scalar or symmetrical neutral vector must be taken in order to allow the process $\tau_\pm \to \tau_0+\pi_\pm$

***) For example, the process: anti-$V_0' \to P_-+\pi_+$ would be much rarer than $V_0 \to P_-+\pi_+$ in our case. This is consistent with experiments.

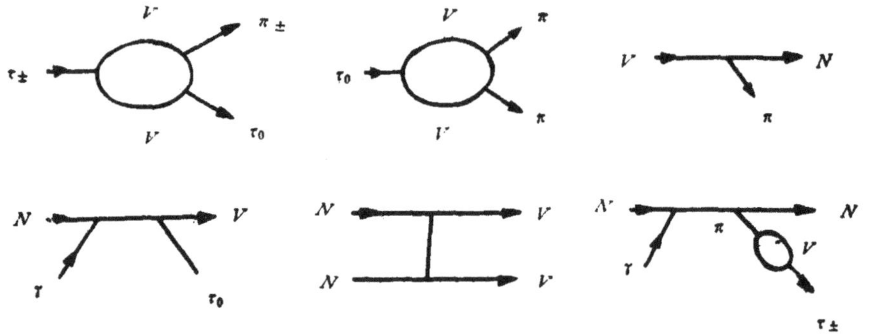

Feynman diagrams for decay and production on the second model (2). Those on the first model are analogous.

1) R. Armenteros, K. H. Barker, C. C. Butler, A. Cachon and A. H. Chapman, Nature **167** (1951), 501.

2) G. D. Rochester and C. C. Butler, Nature **160** (1947), 885.

3) A. J. Seriff, B. B. Leighton, C. Hsiao, E. W. Cowan and C. D. Anderson, Phys. Rev. **78** (1950), 290.
Also see: C. C. Butler, W. G. V. Rosser and K. H. Barker, Proc. Phys. Soc. A **63** (1950), 145; K. H. Barker and C. C. Butler, Proc. Phys. Soc. A **64** (1951), 4.

4) V. D. Hopper and S. Biswas, Phys. Rev. **80** (1950), 1099.
5) Cloud chamber data:
a) L. Leprince-Ringuet and M. L'heritier, C. R. **219** (1944), 616.
b) R. Brode, Rev. Mod. Phys. **21** (1949), 37.
c) H. S. Bridge and M. Annis, Phys. Rev. **81** (1951), 445.
Also see refs. 1), 2), 3) and 8).
Photographic emulsion data:
d) R. Brown, U. Camerini, P. H. Fowler, H. Muirhead, C. F. Powell and D. N. Ritson, Nature **163** (1949), 48 and 82.
e) A. I. Alikhanian, E. M. Samoilovich, I. I. Grevich and Kh. P. Babayan, J. Exp. Theor. Phys. **19** (1949), 667.
f) L. Leprince-Ringuet, Rev. Mod. Phys. **21** (1949), 42.
g) H. H. Forster, Phys. Rev. **77** (1950), 733.
h) N. Wagner and D. Cooper, Phys. Rev. **76** (1949), 449.
Also ref. 4).
6) Neutral mesons:
a) J. R. Finkelstein, Phys. Rev. **72** (1947), 414.
b) H. Fukuda and Y. Miyamoto, Prog. Theor. Phys. **4** (1949), 347, 394.
c) J. Steinberger, Phys. Rev. **76** (1949), 1180.
τ-mesons:
d) H. Fukuda, S. Hayakawa and Y. Miyamoto, Prog. Theor. Phys. **5** (1950), 283 and 352.
e) S. Ôneda, S. Sasaki and S. Ozaki, Prog. Theor. Phys. **4** (1949), 524; ibid. 5 (1950), 25 and 165.
f) H. Fukuda and S. Hayakawa, Prog. Theor. Phys. **5** (1950), 993.
7) E. Fermi, Prog. Theor. Phys. **5** (1950), 570.
8) E. g., see K. Aizu et al., Prog. Theor. Phys. **5** (1950), 931.
9) C. B. A. McCusker and D. D. Millar, Nuovo Cim. **8** (1951), 289.

On the Nature of V-Particles, II

Y. Nambu, K. Nishijima
and Y. Yamaguchi

Osaka City University

July 31, 1951

In a preceding letter[1] we have proposed two possible interpretations of the phenomena associated with V-particles, mainly on the basis of Butler et al.'s recent observations.[2] Of course these interpretations have not been claimed to be the only and definite, but merely possible and more or less natural ones. Also the whole argumentation itself was not complete and exhaustive enough to get full insight into the nature of the V-particles. In view of these points, we will here supplement the previous letter with some more general considerations, while adding to our knowledge the experimental results of American groups[3] in so far as they are accessible to us.

Of the various properties of the V-particles which are known to us, the most remarkable seems to be their large yield and long life, two apparently contradicting properties on the basis of simple detailed balance consideration. They suggest that production and decay are not inverse processes and/or some kind of selection rules (in a very general sense) are at work in the decay reaction. On the other hand, among what are not yet clarified experimentally, there are such important things as the identification of the decay products of the V-particles and their decay modes. Thus at least some of the decay mesons could be μ-mesons (e.g. $\tau \rightarrow \pi + \mu$), and some of the decays could or should involve three or more product particles[3d] (e.g. $\tau + \pi_0, \rightarrow 2\pi_\pm$ $V \rightarrow N + \mu + \nu$). Since no other crucial evidences are known (such as the mode of V production), we are led to a wide variety of possible interpretations if we will take all these points into account, and they can be recommended or disfavored only after a closer examination.

Following the above considerations, let us

first summarize as follows the various conceivable assumptions which shall serve to find out and characterize systematically the possible individual models: 1) Assume the V-particles as elementary entities obeying some basic field equations; or $\bar{1}$) Regard them as composite substance with substructure. 2) Regard V and τ as separate and independent events; or $\bar{2}$) Regard them as inherently related. 3) Assume the product mesons to be π's only; or $\bar{3}$) They may involve μ's as well. 4) Assume only two-particle decays; or $\bar{4}$) Allow of three or more particle decays. 5) In order to account for the large difference in production and decay rate: a) Use the relation in mass values (e.g. for production, $\tau+N\to V+\tau$, but for decay the inverse is forbidden energetically). b) Assume some unknown parents which, directly produced in nuclear events, decay instantly into the observed V's (analogy of π and μ mesons). c) Assume an interaction mechanism such that V's are generated in a nucleon-nucleon impact, but are hard to decay singly (e.g. $N+N\to V+V$). d) Take advantage of large internal angular momenta for V, which make the decay highly forbidden. e) Take advantage of parity and other selection rules related to transformation properties (including Furry's theorem). f) Other special devices. These various assumptions are not mutually exclusive, but may be appropriately combined. 6) Use the conventional perturbation theory (weak coupling); or $\bar{6}$) Exploit the strong interaction (like Fermi[4]) or do not depend on the details of coupling at all (e.g. selection rules which can be enunciated by observing only the initial and final states).

By suitable combinations of these characteristic assumptions, we shall be able to arrive at a large number of models (or interpretations) which can explain more or less consistently the essential features of V events. We give below some of the representative models that follow in this way from our present general consideration, and make comments on their merits and defects.

1) All observed processes are assumed to be direct ones. For example, we introduce the following couplings:

a) $NNNV\tau$, $VN\pi$, $\tau\pi\pi$ ($1\bar{2}345$c6)*.
b) $NV'\tau$, $V'V\pi$, $VN\pi$, $\tau\pi\pi$ ($m_{V'}>m_V+m_\pi$) ($1\bar{2}345$b6)[5],
c) $NNVV$, $NN\tau\tau$, $VN\pi$, $\tau\pi\pi$ or $\tau\pi\mu$ (12(3)45c6).

This standpoint is equivalent to determining all the relevant terms of the S-matrix independently in so far as they do not lead to inconsistencies. Though formally possible, it is not a very attractive procedure since we little understand the nature of the events by such a highly phenomenological approach. The introduction of a short-lived parent V's as in the above second model, seems to be an unnecessary complication unless some definite evidence on the existence of such a particle is presented.

2) Our two models proposed in I ($1\bar{2}3$(4)5ace6).[6] They were specially designed to account for the successive decays of V or τ. But the pairwise production which they predict does not seem to be favored by experiments, if not yet rejected.

3) A three particle decay is assumed for V: $V\to N+\mu+\nu$, while the production occurs, for example, as $V+\pi$ pairs ($12\bar{3}\bar{4}5$a6). Since some evidences show that the decay does not necessarily follow a two-body scheme, such a possibility will not be excluded as responsible for at least part of the decays. (But a universal Fermi-type coupling ($g\sim10^{-49}$) would lead to a rather long life ($>10^{-6}$ sec).

4) Gamow-type model. Turning to the structural theories, the most naive one may be to regard V as a bound $N+\pi$ system ($\bar{1}2345$cd$\bar{6}$). This, however, would require an unusually high potential barrier (or the order of several Bev) to assure the long life, and if this barrier were supplied by the centrifugal force, the angular momentum would have to be$\sim$8, which seems too high to be easily attained by nucleon-nucleon collisions

$(N+N\to V+V)$.

5) We may modify the above model so that V is a bound $N+\tau$ system. The assumed interaction is, for example,

$\tau\tau NN$ (for τ and V production), $\tau\pi\pi$ (for τ and V decay) ($\bar{1}\bar{2}345ac6$),
or τNN, (for τ and V production), $\tau\pi\pi$ (for τ and V decay) („),

The decay of both V and τ are controlled by the same coupling $\tau\pi\pi$. The first model will predict the pairwise production, whereas on the second (τ must be a boson), the process $V\to N+\pi$ is threatened by $V\to N+\gamma$.

6) Isomeric transition. Alternatively we amend the defects of model 4 by assuming that in the decay of V the π meson is radiated from a small volume of the dimensions of nucleon Compton wave length ($\bar{1}2345cdf\bar{6}$). Using the well-known multiple radiation formula, the angular momentum L can be lowered to ~ 4, which is not unreasonable to be realized by nuclear collisions. But in order to suppress the γ-emission, the bound meson field which is responsible for the excitation energy would have to be extremely rigid. If we calculate ad hoc the moment of inertia of the bound meson field around a nucleon and quantize this rigid body we obtain the observed excitation of ~ 200 Mev for $L=4\frac{1}{2}$ with the coupling constant $g^2\sim 10$, and only the transitions to $L=1/2$ or $3/2$ are energetically allowed for π emission. Although this model is interesting, it is hard to be founded upon orthodox field theory. The conventional strong coupling theory, on the other hand, which is similar to above in physical ideas, does not seem to give the required long life of the excited states.

Some remarks may be added on the spin (or transformation property) of V and τ particles. We did not take it into account in the above classification since it did not seem essential in the characterization of the models. In general both fermion and boson property may be allowed for these particles. But in any way the stability of protons and neutrons should be guaranteed. Thus, for instance, a coupling like $VN\mu$ (V=boson elementary particle), leading to $V\to\mu+N$, cannot be admitted.

In this way we have seen that at the present stage several different model can assert themselves just about as good. But, in general, we may say that the elementary particle theories (Alternative 1) are liable to predict pairwise production which remains to be confirmed by experiment, while the structural theories (Alternative $\bar{1}$) tend to suffer from high probabilities of radiative decay. It will be premature and useless to demand anything more definite from what we know at present about the V particles.

In concluding, we express deep gratitude to Prof. S. Hayakawa for informing us of the recent activities of American groups as his own opinions. The models mentioned here include those which have been proposed in America, but no published papers being available, we have refrained from explicit citation.[7] Thanks are also due to Messrs. G. Takeda, S. Takagi, H. Fukuda, K. Aizu, T. Kinoshita, H. Miyazawa and S. Ôneda for valuable discussions.

References

1) Y. Nambu, K. Nishijima and Y. Yamaguchi, Prog. Theor. Phys. **6** (1951), this number. Cited in the text as I. The notations used here follow those of I.

2) C. C. Butler et al., Nature **167** (1951), 501. See I.

3)
a) H. S. Bridge and M. Annis, Phys. Rev. **82** (1951), 445.
b) R. W. Thompson, H. O. Cohn and R. S. Flum, Bull. Am. Phys. Soc. **26** (1951), No. 3, p. 6.
c) W. B. Fretter, ibid. No. 5, p. 16.
d) R. B. Leighton, A. J. Seriff and C. D. Anderson, ibid.
d) R. B. Leighton, A. J. Seriff and C. D. Anderson, ibid.
e) E. J. Althaus, ibid.

4) E. Fermi, Prog. Theor. Phys. **5** (1950), 570.
5) This type of model was suggested by S. Takagi.
6) The first of these was also proposed by S. Ôneda.
7) According to Prof. Hayakawa, theories have been given by R. P. Feynman, E. Fermi, R. E. Marshak, R. F. Christy and others.

*) This notation signifies the nature of the model according to the above mentioned criteria.

POSSIBLE EXISTENCE OF A HEAVY NEUTRAL MESON*

YOICHIRO NAMBU
The Enrico Fermi Institute for Nuclear Studies, The University of Chicago, Chicago, Illinois

(Received April 25, 1957)

In an attempt to account for the charge distributions of the proton and the neutron as indicated by the electron scattering experiments,[1] we would like to consider the possibility that there may be a heavy neutral meson which can contribute to the form factor of the nucleon. We assume that this meson, ρ^0, is a vector field with isotopic spin zero and a mass two to three times that of the ordinary pion, coupled strongly to the nucleon field. An isolated ρ^0 would decay through virtual nucleon pair formation according to the following schemes:

$$\begin{aligned}&\text{(a)}\quad \rho^0 \to \pi^0 + \gamma, \qquad 2\pi^0 + \gamma, \qquad \pi^+ + \pi^- + \gamma;\\ &\text{(b)}\quad \rho^0 \to e^+ + e^-, \qquad \mu^+ + \mu^-;\\ &\text{(c)}\quad \rho^0 \to \pi^+ + \pi^-.\end{aligned}$$

The process (a) would have a decay probability roughly of the order of $P_a \sim (\mu c^2/\hbar)(G^2/\hbar c)(e^2/\hbar c)$ $\times(\mu/M)^2$, where G is the nuclear coupling constant, μ and M the ρ^0 and the nucleon masses, respectively. For the process (b), the probability would be $P_b \sim (\mu c^2/\hbar)(G^2/\hbar c)(e^2\hbar c)^2(\mu/M)^2$. The process (c) is a forbidden transition, so that it can take place only in violation of the isotopic spin conservation, with a decay rate comparable to that for (b).

Now the process (b) gives rise to a short-range interaction between a nucleon and an electron (or a muon) by exchange of a ρ^0. This will contribute a form factor $F'(k^2) \sim Gg/(\mu^2+k^2)$ to the electron-nucleon scattering, where g is the effective ρ^0-electron coupling, $g^2/\hbar c \sim (G^2/\hbar c)(e^2/\hbar c)^2(\mu/M)^2$. Since ρ^0 is an isotopic scalar, F' has the same sign for both proton and neutron, whereas the corresponding form factor F due to the pion cloud should change sign. Relativistic field theory shows on general grounds that $F(k^2)$ has the form

$$F(k^2) = \int_{2m_\pi}^{\infty} \frac{\rho(m)}{m^2+k^2} dm,$$

where the lower limit of integration corresponds to the threshold for pion pair creation by an external electromagnetic field. With our assumptions about ρ^0, it is thus possible that the two form factors F and F' cancel approximately for the neutron but reinforce for the proton, in agreement with observation. If we equate tentatively the mean square radius of the proton with the one due to ρ^0:

$$Gg/\mu^2 \sim e^2\langle a^2\rangle/b,$$

we get

$$(G^2/\hbar c)(g^2/\hbar c) \sim [(e^2/\hbar c)/b]^2 \sim 10^{-6},$$

which checks with the previous estimate since $G^2/\hbar c$ would be of the order one. The decay lives become, very approximately,

$$\tau_a \sim 10^{-19}\text{–}10^{-20} \text{ sec},$$
$$\tau_b \sim \tau_c \sim 10^{-17}\text{–}10^{-18} \text{ sec}.$$

Published in *Phys. Rev.* **106**, 1366, ©1957.

We can pursue further consequences of our assumption.

(1) ρ^0 could be produced by any strong nuclear reactions, but it would instantly decay mostly into a high-energy $\gamma(\gtrsim 140$ MeV) and a ρ^0. The ratio of charged to neutral components in high-energy reactions should accordingly be influenced.

(2) The second maximum of the pion-nucleon scattering around 1 BeV[2] could be attributed to the reaction

$$\pi^- + p \rightarrow n + \rho^0,$$

if a resonance should occur for such a system.

(3) ρ^0 would contribute a repulsive nuclear force of Wigner type and short range ($\lesssim 0.7 \times 10^{-13}$ cm), more or less similar to the phenomenological hard core.

(4) The anomalous moment of the nucleon[3] should be affected by ρ^0. The main effect seems to be that ρ^0 and the usual pion give opposite contributions to the isotopic scalar part of the core moment, thus tending to bring better agreement between theory and experiment.

(5) If it is energetically possible, we ought to expect that K mesons and hyperons would sometimes decay by emitting a ρ^0.

It should perhaps be added that the neutral meson considered here is similar in nature to the one introduced by Teller for quite different purposes.[4]

* This work was supported by the U. S. Atomic Energy Commission.

[1] R. Hofstadter, Revs. Modern Phys. **28**, 214 (1956). Other references are given in this paper. Also E. E. Chambers and R. Hofstadter, Phys. Rev. **103**, 1456 (1956); J. A. McIntyre, Phys. Rev. **103**, 1464 (1956); R. W. McAllister, Phys. Rev. **104**, 1494 (1956); Hughes, Harvey, Goldberg, and Stafne, Phys. Rev. **90**, 497 (1953); Melkonian, Rustad, and Havens, Bull. Am. Phys. Soc. Ser. II, **1**, 62 (1956). For theoretical interpretations, see Yennie, Lévy, and Ravenhall, Revs. Modern Phys. **29**, 144 (1957).

[2] Shapiro, Leavitt, and Chen, Phys. Rev. **92**, 1073 (1953); Cool, Madansky, and Piccioni, Phys. Rev. **93**, 637 (1954); O. Piccioni and other authors, *Proceedings of the Sixth Annual Rochester Conference on High-Energy Physics 1956* (Interscience Publishers, Inc., New York, 1956). See also F. Dyson, Phys. Rev. **99**, 1037 (1955); G. Takeda, Phys. Rev. **100**, 440 (1955).

[3] H. Miyazawa, Phys. Rev. **101**, 1564 (1956); see also G. Sandri, Phys. Rev. **101**, 1616 (1956).

[4] M. H. Johnson and E. Teller, Phys. Rev. **98**, 783 (1955); H. Duerr and E. Teller, Phys. Rev. **101**, 494 (1956); E. Teller, *Proceedings of the Sixth Annual Rochester Conference on High-Energy Physics 1956* (Interscience Publishers, Inc., New York, 1956); H. Duerr, Phys. Rev. **103**, 469 (1956).

Y. NAMBU
1957, Novembre
Il Nuovo Cimento
Serie X, Vol. 6, pag. 1064-1083

Parametric Representations of General Green's Functions (*).

Y. NAMBU

The Enrico Fermi Institute for Nuclear Studies
The University of Chicago - Chicago, Ill.

(ricevuto il 23 Giugno 1957)

Summary. — Parametric representations for the Green's function of field theory are derived in perturbation theory. These representations are valid for each term in the perturbation series that corresponds to a Feynman diagram, and reflect its analytic property and threshold characteristics. As an example, the three-body (vertex) function is shown to satisfy a dispersion relation when two of the three momenta are fixed, with the correct location of the singularities expected from the thresholds of the competing real processes.

1. – Introduction.

Attempts to investigate the structure of the present field theory have been pursued by various authors making use of only such fundamental properties as causality and unitarity. The so-called Green's functions are of particular interest in this respect as well as for the fact that their asymptotic part gives the S-matrix elements. In previous papers ([1]) we proposed specific representations of the Green's functions which seemed to follow from simple physical considerations and indicate their analytic property. It has been found, however, that these formulae are not necessarily satisfied by the actual Green's

(*) This work was supported by the United States Atomic Energy Commission.
([1]) Y. NAMBU: *Phys. Rev.*, **100**, 394 (1955); **101**, 459 (1956).

functions calculated by perturbation theory ([2]); it seems that although the representations have the correct properties we would expect from physical considerations, they are too restricted to correspond to reality.

In this paper a different, more general kind of representation is proposed and proved in perturbation theory. The main difference is that in the new formula there is only a single denominator determining the dependence of the function on all the momenta, the kind of denominator which one would obtain in actual calculations by combining various denominators according to Feynman's method. In this respect it resembles another formula proposed before for scattering matrices ([3]). As in the case of scattering matrices, the dispersion relations for general Green's functions are expected to reflect two properties, namely that the functions have certain analyticity in appropriate variables, and that the so-called absorptive part has a spectrum determined by the various reaction thresholds, where branch points or poles are located. The existence of branch points at threshold of the S matrix were shown in general by EDEN ([4]), and the general dispersion relations for causal amplitudes were recently discussed by POLKINGHORNE ([5]). The present formulas have properties which seem to be closely related to these points. But an explicit demonstration of the conventional dispersion relations has been possible so far only for the vertex (three-field) function, which will be illustrated in Sect. **3**. The relation of the present representation to the previous more restricted one will be discussed.

2. – Derivation of the representation.

In this section we will derive a representation that applies to every Feynman diagram in the perturbation series of a Green's function or related functions. For definiteness we work with quantum electrodynamics. But we ignore here the problem of ultraviolet as well as infrared divergences, and assume all the integrals are convergent. Renormalization is not an important point in our consideration.

In terms of the interaction representation, the Green's functions, as de-

---

([2]) A simplest counter-example wuold be the first radiative correction to the three-body Green's function (vertex function). The main point has to do with the range of the integration parameters which, according to the results of Ref. ([1]), carried a meaning of the square of the mass of intermediate states, and thus were supposed to take positive values only. The present paper leads only to a much weaker statement.

([3]) Y. NAMBU: *Phys. Rev.*, **98**, 803 (1955).

([4]) R. J. EDEN: *Proc. Roy. Soc.*, A **210**, 388 (1951).

([5]) J. C. POLKINGHORNE: *Nuovo Cimento*, **4**, 216 (1956); *Proc. Camb. Phil. Soc.*, **53** Part I, 260 (1957).

fined in I, can be expressed as

$$(1)\quad \begin{cases} \varrho_{2.0}(x|y|) &= \langle T(\psi(x),\,\overline{\psi}(y),\,S)\rangle_0 \\ \varrho_{0.2}(||zz') &= \langle T(A(z),\,A(z'),\,S)\rangle_0\,, \\ \varrho_{2.1}(x|y|z) &= \langle T(\psi(x),\,\overline{\psi}(y),\,A(z),\,S)\rangle_0\,, \quad \text{etc.} \\ S = T\exp\left[+ie\int i\overline{\psi}\gamma_\mu\psi A_\mu \mathrm{d}^4x\right]. \end{cases}$$

Expanding S and rearranging the T-product into a normal product, one gets a series of terms which can be constructed according to the rules of Dyson and corresponds to various Feynman diagrams. The contribution of a Feynman diagram to $\varrho_{2r,m}$ is of the form

$$(2)\quad F(x|y|z) = (-e)^R(-1)^l\int\!\!\dots\!\!\int \prod_{i,r} S_F(x_i-u_r)\prod_{r,s}\gamma_r S_F(u_k-u_s)\gamma_s\,\cdot$$
$$\cdot\prod_{s,j} S_F(u_s-y_j)\prod_{k,t} D_F(z_k-u_t)\prod_{r,s} D_F(u_r-u_s)\cdot \mathrm{d}^4u_1\dots\mathrm{d}\,u_R\,,$$

where R is the number of internal points (vertices), and l the number of closed electron loops. The rules as to the multiplication of γ-matrices are not explicitly indicated. Hereafter the letter x will be used to denote all the external points x, y and z indiscriminately. Our theorems in this section will hold for any quantity if it has the structure shown in Eq. (2), namely if it consists of Δ_{F-} functions and their derivatives, being integrated over internal points.

We begin by observing that S_F and D_F have the representation

$$(3)\quad \begin{cases} D_F(x) = \dfrac{-i}{8\pi^2}\displaystyle\int_0^\infty \exp[iax^2/2]\,\mathrm{d}a\,, \\ S_F(x) = (-\gamma\cdot\partial/\partial x+m)\dfrac{-i}{8\pi^2}\displaystyle\int_0^\infty \exp[iax^2/2-im^2/2a]\,\mathrm{d}a\,, \end{cases}$$

while the corresponding Fourier coefficients are

$$(4)\quad \begin{cases} \displaystyle\int D_F(x)\exp[-ik\cdot x]\,\mathrm{d}^4x = \frac{1}{2}\int_0^\infty \exp[-ibk^2/2]\,\mathrm{d}b = -i/(k^2-i\varepsilon)\,, \\ \displaystyle\int S_F(x)\exp[-ik\cdot x]\,\mathrm{d}^4x = \frac{1}{2}\int_0^\infty(-i\gamma\cdot k+m)\exp[-ib(k^2+m^2)/2]\,\mathrm{d}b = \\ \qquad = -i(-i\gamma\cdot k+m)/(k^2+m^2-i\varepsilon)\,. \end{cases}$$

We will substitute the above representations into Eq. (2), and carry out the integrations one by one. The first integrations to be done will be of the type

$$\int S_F(x_1 - u)\gamma_\mu S_F(u - x_2) D_F(u - x_3)\, d^4u \,. \tag{5}$$

We shall make use of the following lemma.

Lemma 1. Let I be a quadratic form in the N vectors x_i:

$$I = \tfrac{1}{2}\sum_{i>j} a_{ij}(x_i - x_j)^2 = \tfrac{1}{2}\sum_{j} A_{\,j} x_i \cdot x_j \qquad a_{ij} \geqslant 0, \qquad \sum_{i,j} a_{ij} > 0 \tag{6}$$

a and A are related by

$$A_{ji} = A_{ij} = - a_{ij} \leqslant 0\,, \qquad A_{ij} = \sum_{j=1}^{N} a_{ij} > 0\,, \qquad \sum_{j=1}^{N} A_{ij} = 0\,. \qquad i \neq j \tag{7}$$

Then

$$\exp[iI]\, d^4x_1 = i(4\pi^2/A_{11}^2) \exp[iI']\,, \tag{8}$$

with

$$\left\{ \begin{array}{ll} I' = \tfrac{1}{2}\sum\limits_{i>j>1} a'_{ij}(x_i - x_j)^2 = \sum\limits_{i,j\neq 1}{}' {}_{ij}A\, x_i \cdot x_j\,, & \\ a'_{ij} = a_{ij} + a_{i1}a_j \Big/ \sum\limits_{R} a_{k1} \geqslant a_{ij}, & i, j > 1 \\ A'_{ij} = A_{ij} - A_{ii}A_{ji}/A_{11}\,, & \sum\limits_{j=2}^{N} A'_{ij} = 0\,. \end{array} \right. \tag{9}$$

Eq. (8) can be derived by observing

$$\sum_{i=2}^{N} a_{i1}(x_i - x_1)^2 = \Big(\sum_i a_{i1}\Big)\Big(x_1 - \sum_i a_{i1}x_i \Big/ \sum_i a_{i1}\Big)^2 + \tag{10}$$

$$+ \Big(1 \Big/ \sum_i a_{i1}\Big) \sum_{i,j\neq 1} a_{i1}a_{j1}(x_i - x_j)^2$$

and

$$\int \exp[iAx^2/2]\, d^4x = 4\pi^2 i/A^2\,, \qquad (A > 0).$$

The above lemma implies that the basic structure of the exponential $\exp[iI]$, as given by Eqs. (6) and (7), is preserved after integration over some of its co-ordinates. Eq. (5) becomes, for example,

$$(11)\quad \begin{cases} (\gamma\cdot\partial/\partial x_1 - m)\gamma_\mu(\gamma\cdot\partial/\partial x_2 - m)[i/(8\pi^2)^3]\int_0^\infty\int_0^\infty\int_0^\infty\int \exp[(i/2)\{a_1(x_1-u)^2 + \\ + a_2(x_2-u)^2 + a_3(x_3-u)^2\}]\cdot\exp[(-im^2/2)(1/a_1+1/a_2]\, da_1\, da_2\, da_3\, d^4u = \\ \qquad = -(\gamma\cdot\partial/\partial x_1 - m)\gamma_\mu(\gamma\cdot\partial/\partial x_2 - m)(1/128\pi^4)\cdot \\ \cdot\int_0^\infty\int_0^\infty\int_0^\infty \exp[(i/2)\{a_{12}(x_1-x_2)^2 + a_{23}(x_2-x_3)^2 + a_{31}(x_3-x_1)^2\}]f(a_{ij})\, da_{12}\, da_{23}\, da_{31}, \\ f(a_{ij}) = \int\int\int \exp[(-im^2/2)(1/a_1+1/a_2)]\cdot \\ \qquad \cdot[\prod_{i>j}\delta\{a_{ij} - a_ia_j/(a_1+a_2+a_3)\}](a_1+a_2+a_3)^{-2}\, da_1\, da_2\, da_3\,. \end{cases}$$

One can then proceed to the next integration by using Lemma 1 again. It should be noted that, if a differential operator arising from an S_F-function acts on an internal point over which to integrate, it can always be replaced by a linear combination of differential operators with respect to the remaining points. This will be shown in the appendix. One can thus exhaust all the integrations and arrive at the theorem:

Theorem 1. The function $F_{2n,m}$ of Eq. (2), which corresponds to a certain Feynman diagram, has a parametric representation

$$(12)\quad \begin{cases} F_{2n,m} = (-i)^{(N+R)/2}\sum_\alpha\int\ldots\int O^{(\alpha)}(\gamma,(1/m)\partial/\partial x)\exp[iI]f^{(\alpha)}(a_{ij})\prod da_{ij}\,, \\ I = \frac{1}{2}\sum_{i>j}a_{ij}(x_i-x_j)^2 = \frac{1}{2}\sum_{i,j}A_{ij}x_ix_j\,, \qquad i,j = 1,\ldots 2n+m = N \\ a_{ij} = -A_{ij} = -A_{ji} \geqslant 0\,, \qquad \sum_{i=1}^N A_{ij} = 0\,, \end{cases}$$

$O^{(\alpha)}$ here designates a product of Dirac matrices, photon polarization vectors, and a number $n_\alpha \leqslant \text{Min}\,(n+R,\, n(N-1)+m)$ of differential operators with respect to the external points x (see Appendix); $f^{(\alpha)}$ is a complex function of the positive parameters a_{ij}. The power of i in front of the integral is the combined result of all the i's coming from D_F, S_F and integrations over the internal points.

As regards the function $f^{(\alpha)}$ one can further make the following statement.

Theorem 2. Introduce a common scaling factor t with the dimensions of m^2 for all the variables such that

$$(13)\quad a_{ij} = t\alpha_{ij}\,, \qquad t > 0.$$

Then $f_{(\alpha)}(a_{ij})$ of Eq. (12) has the form

$$(14)\qquad f^{(\alpha)}(t\alpha_{ij})t^{N(N-1)/2} = m^{3n+m}\sum_{\nu=0}^{(N+R-N\alpha)/2}\int_0^\infty \exp\left[-iM^2/2t\right]i_\nu\tau_\nu^{\alpha}(\alpha_{ij},\, M^2/m^2)\cdot \cdot (m^2/t)^{((R-N)/2)-\nu}\,\mathrm{d}M^2/m^2\,,$$

where $\tau_\nu^{(\alpha)}(\alpha_{ij},\, M^2/m^2)$ is real. In other words, $f^{(\alpha)}$ as a function of t has a continuation which is analytic for $\operatorname{Im} t > 0$.

To prove the theorem, first observe that both sides of the transformation Eq. (9) are homogeneous functions of order one with respect to the variables a'_{ij} or a_{ij} respectively, so that the relation does not depend on t; it simply tells how the ratios $a_{12} : a_{13} : \ldots$ are transformed. Now each S_F in Eq. (1), according to Eq. (4), contributes a factor of the form

$$\exp\left[-im^2/2a_l\right] = \exp\left[-im^2/2t\alpha_l\right]\,, \qquad l = 1, \ldots, (n+R)/2,$$

and after integration over the internal points α_j will eventually be expressed as functions of α_{ij}. Further, one may write

$$(14')\quad \exp\left[(-im^2/2t)\sum 1/\alpha_l\right] = \int_0^\infty \exp\left]-iM^2/2t\right]\delta\left[M^2/m^2 - \sum 1/\alpha_l\right]\mathrm{d}M^2/m^2\,.$$

The factor m^{3n+m} in front of Eq. (14) is put simply for dimensional reasons. The factor $(m^2/t)^{((R-N)/2)-\nu}$ comes from the change of variables and the fact that each integration over a vertex yields a factor of the form $1/A_{ii}^2$ according to Eq. (8), which altogether gives $t^{-2R+3R/2+N/2} = t^{-R/2+N/2}$; in addition, as is shown in the appendix, the differential operators in S_F can bring about extra t's up to $t^{-(n+R-n\alpha)/2}$ when $F_{2n,m}$ is brought into the form Eq. (12).

Let us next introduce the Fourier transform of the function according to

$$(15)\qquad \delta(\sum_i k_i)\, G(k_i) = \frac{1}{(2\pi)^4}\int\!\ldots\!\int \exp\left[-i\sum k_i\cdot x_i\right] F(x_i)\prod_i \mathrm{d}^4x_i\,.$$

The conservation law $\sum_i k_i = 0$ is implied in Eq. (12) by the relation $\sum_i A_{ij} = 0$, which also means that the rank of matrix A is $N-1$ ([6]), and $\det A = 0$. This brings about a little inconvenience in carrying out the integration in

([6]) The rank is even smaller if the Feynman diagram contains disconnected graphs, which is not the case here by virtue of the definition of the Green's functions.

Eq. (15). If $\det A \neq 0$, one could use the formula

$$\text{(16)} \qquad \int \cdots \int \exp\left[-i \sum k_i \cdot x_i\right] \exp\left[(i/2) \sum A_{ij} x_i \cdot x_j'\right] \prod d^4 x_i = \frac{(4\pi^2 i)^N}{[\det A]^2} \exp\left[(-i/2) \sum B_{ij} k_i \cdot k_j\right],$$

where B_{ij} is the inverse matrix of A. When $\det A = 0$, there is no unique way of defining an $N \times N$ matrix B. Two matrices B_{ij} and $B'_{ij} = B_{ij} + X_i + X_j$ are equivalent for $\sum k_i = 0$ since

$$\sum B'_{ij} k_i \cdot k_j = \sum B_{ij} k_i \cdot k_j + 2 \left(\sum X_i k_i\right) \cdot \left(\sum k_j\right).$$

We will adopt here a more or less arbitrary prescription. We modify A into $A' = A + \lambda E$, so that $\det A' \neq 0$, and take an appropriate limit $\lambda \to 0$ according to

$$\text{(17)} \qquad \begin{cases} \lim\limits_{\lambda \to 0} \int \cdots \int \exp\left[-i \sum k_i \cdot x_i\right] \exp\left[(i/2) \sum A'_{ij} x_i \cdot x_j\right] \prod d^4 x_i = \\ = \lim\limits_{\lambda \to 0} \left[(4\pi i)^N / \lambda^2 D^2\right] \exp\left[-i \left(\sum k_i\right)^2 / 2N\lambda\right] \exp\left[(-i/2) \sum B_{ij} k_i \cdot k_j\right] = \\ \qquad = (2\pi)^4 \delta\left(\sum k_i\right) \left[(4\pi^2 i)^{N-1} / D^2\right] \exp\left[(-i/2) \sum B_{ij} k_i \cdot k_j\right], \\ D \equiv \left[d(\det A')/d\lambda\right]_{\lambda=0}, \\ B_{ij} \equiv \lim\limits_{\lambda \to 0} \dfrac{\det A_{(ij)}(\lambda) - \det A_{(ij)}(0)}{\det A(\lambda)} = \dfrac{\sum_k \det A_{(ik,jk)}}{D} \equiv \dfrac{b_{ij}}{D}. \end{cases}$$

Here $A_{(ij)}$ and $A_{(ik,jk)}$ mean submatrices of A where the designated rows and columns are omitted. This particular choice of B_{ij} has an interesting feature which will be stated in the following theorem.

Theorem 3. The Fourier transform $G(k)$ of $F(x)$, as defined in Eq. (15), has a representation

$$\text{(18)} \qquad \begin{cases} G_{2n,m}(k_i) = \sum\limits_{\alpha} \int \cdots \int O^{(\alpha)}_{2n,m}(\gamma, ik/m) \exp[-iJ] g^{\alpha}(a_{ij}) \prod da_{ij}, \\ g^{(\alpha)}(a_{ij}) = (-i)^{(N+R)/2} \left[(4\pi^2 i)^{N-1} / D^2(a_{ij})\right] f^{(\alpha)}(a_{ij}), \\ J = \frac{1}{2} \sum\limits_{i,j=1}^{N} B_{ij} k_i \cdot k_j, \\ \quad = \frac{1}{2} \sum\limits_{\mu=1}^{N(N-1)/2} b_\mu l_\mu^2, \qquad b_\mu \geqslant 0. \end{cases}$$

l_μ is a sum of up to $[N/2]$ vectors

$$k_i\,,\quad k_i+k_j\,,\quad k_i+k_j+k_l\,,\quad \text{etc.}$$

taken from $k_1, \dots, k_N$. The summation need be extended only up to $[N/2]$ momenta since any l_μ may be replaced by $-l'_\mu$ where l'_μ is the sum of the complementary momenta.

The last expression for J can be derived by noting that $b_{ij} = \sum_k A_{(ik,jk)}$ is a positive, multi-linear function of the coefficients a_{lm}:

$$b_{ij} = \sum a_{lm} a_{np} \dots \quad \text{and} \quad b_{ii} \geqslant b_{ij} \geqslant 0\,. \tag{19}$$

Each term in the summation is a product of $N-2$ different a_{lm}'s taken appropriately out of $N(N-1)/2$ parameters $a_{12}, \dots, a_{N,N-1}$. By reason of symmetry, the sum must run over the indices in such a way that it is symmetric in i and j, and that if a particular combination of the a_{lm}'s occurs, it also includes all the terms obtained from it by permutation of the indices except i and j. It is then possible to break up the sum for b_{ij}, $i \neq j$, into various groups:

$$b_{ij} = b_{(ij)} + \sum_l b_{(ijl)} + \sum_{l,m} b_{(ijlm)} + \dots\,, \tag{20}$$

where $b_{(ij)}$, $b_{(ijl)}$, $b_{(ijlm)}$, etc., are partial sums which are symmetric with respect to the indices ij, ijl, $ijlm$, etc. This means that $b_{(ijl)}$, for example, occurs equally in b_{ij}, b_{jl} and b_{li}. Since furthermore any term in b_{ij}, $i \neq j$, is contained also in b_{ii} and b_{jj}, these partial sums $b_{(ij)}$, $b_{(ijl)}$, $b_{(ijlm)}$, etc., are coefficients of $(k_i+k_j)^2$, $(k_i+k_j+k_l)^2$, $(k_i+k_j+k_l+k_m)^2$, etc., respectively. The remaining part b_i of b_{ii}, which does not belong to any of these terms, will then give the coefficient of k_i^2. In other words

$$J = \frac{1}{2D}\left\{\sum_i b_i k_i^2 + \sum_{(ij)} b_{(ij)}(k_i+k_j)^2 + \sum_{(ijl)} b_{(ijl)}(k_i+k_j+k_l)^2 + \dots\right\}. \tag{21}$$

Since B is essentially the inverse of the matrix A, it depends on the scaling factor t of Theorem 2 as

$$B_{ij} = \frac{1}{t}\beta_{ij}\,, \qquad J = \frac{1}{2t}\sum \beta_{ij} k_i \cdot k_j = \frac{1}{2t}\sum \beta_\mu l_\mu^2\,,$$

where β_{ij} or β_μ is a function of the variables α_{ij}. Let us now choose t as one of the integration parameters by normalizing α_{ij} or β_{ij} in a suitable way. For example, adopting the convention

$$\sum_{i>j} \alpha_{ij} = 1 \qquad \text{or} \qquad t = \sum_{i>j} a_{ij}\,,$$

one may write for any $\mathcal{F}(a_{ij})$

$$\int_0^\infty \!\!\dots\!\! \int \mathcal{F}(a_{ij}) \prod \mathrm{d}a_{ij} = \int_0^\infty \!\!\dots\!\! \int \mathcal{F}(t\alpha_{ij})\, \delta(\textstyle\sum\alpha_{ij}) \prod \mathrm{d}\alpha \; t^{(N(N-1)/2)-1}\mathrm{d}t\,. \tag{22}$$

t will then run from 0 to ∞. Substituting this expression into Eq. (18), the t integration may be carried out first. Noting that

$$\int_0^\infty \exp\,[-ib/t]\,\mathrm{d}t/t^{n+2} = (i\partial/\partial b)^n \frac{-i}{b-i\varepsilon}\,,$$

one gets the following result.

Theorem 4. G has the representation

$$(23)\quad \begin{cases} G(k) = -im^{3n+m} \sum\limits_{\alpha,\nu} \int\!\dots\!\int O^{(\alpha)}(\gamma, ik/m)\sigma_\nu^{(\alpha)}(\alpha_{ij}, M^2/m^2)\cdot \\ \qquad \cdot m^{R-N-2\nu}(\partial/\partial M^2)^{((3N+R)/2)-\nu-3} \dfrac{1}{\frac{1}{2}\sum\beta_\mu l_\mu^2 + M^2 - i\varepsilon}\,\mathrm{d}\left(\dfrac{M^2}{m^2}\right)\prod \mathrm{d}\alpha_{ij}\,, \\ \sigma_\nu^{(\alpha)} = [(4\pi^2)^{N-1}/D^2(\alpha_{ij})]\,\tau_\nu^{(\alpha)}(\alpha_{ij}, M^2/m^2)\delta(\sum\alpha_{ij}-1)\,, \end{cases}$$

where all the integration variables run only over positive values:

$$\alpha_{ij} \geqslant 0\,, \qquad M^2 \geqslant 0\,.$$

This means that G, regarded as a function of $N(N-1)/2$ *independent* variables l_μ^2, is analytic if all the variables lie in the upper half of the complex plane. If the differential operator in the integrand ($(3N+R)/2-\nu-3 \geqslant 0$ always) is incorporated into the weight function, Eq. (23) may also be written

$$(23')\quad \begin{cases} G(k) = -im^{\varepsilon-3n-3m} \sum\limits_{\alpha} \int\!\dots\!\int O^{(\alpha)}(\gamma, ik/m)\sigma^{(\alpha)}(\alpha_{ij}, M^2/m^2)\cdot \\ \qquad\qquad \cdot \dfrac{1}{\frac{1}{2}\sum\limits_\mu \beta_\mu l_\mu^2 + M^2 - i\varepsilon}\,\mathrm{d}\left(\dfrac{M^2}{m^2}\right)\prod \mathrm{d}\alpha_{ij}\,, \\ \sigma^{(\alpha)} = \sum\limits_\nu \sigma_\nu^{(\alpha)}(\alpha_{ij}, M^2/m^2)m^{R-N-2\nu}(\partial/\partial M^2)^{(3N+R)/2-\nu-3} \\ \qquad = \sum\limits_\nu (-m^2\,\partial/\partial M^2)^{(3N+R)/2-\nu-3}\sigma_\nu^{(\alpha)}(\alpha_{ij}, M^2/m^2)\,. \end{cases}$$

The last expression obtains if integration by parts is admitted. The fact that the weight functions $\sigma_\nu^{(\alpha)}$ or $\sigma^{(\alpha)}$ are real (with the proper definition of $O^{(\alpha)}$) is essentially a consequence of the time revesal invariance of our underlying field theory.

Our next task is to look closely into the range of the variables α_{ij} or β_μ, which is necessary to exhibit the dispersion relations of the desired nature for G. For this purpose some more lemmas have to be introduced

Lemma 2. The contents of lemma 1 can be restated as follows: If one replaces all the vectors x_i by real numbers z_i (or vectors in an Euclidean space of any dimensions), so that $\sum_i z_i^2 > 0$, then I' in Eq. (9) is numerically equal to the minimum of the positive definite form I with respect to the variation of z_1.

This is obvious in view of Eq. (10). But it has also an interesting implication that the factor $\exp[iI]$, which governs the analytic behavior of the Green's functions, is determined by a classical action principle. One may regard $1/a_{ij}$ as a parameter similar to the proper time (or its square) required by a particle to travel between x_i and x_j. One may also adopt the following mechanical model. Take N points $\mathbf{z}_1, \ldots, \mathbf{z}_N$ (vectors now) in space, and between each pair of points $\mathbf{z}_i$ and $\mathbf{z}_j$ hook a spring with a force constant a_{ij}. The potential energy of such a system is then equal to I, and as one sets the point $\mathbf{z}_1$ free, it will settle down to an equilibrium position $\mathbf{z}_1 = \sum a_{i1} z_i / \sum a_{i1}$ with energy I'.

Lemma 3. Two quadratic forms

$$I = \tfrac{1}{2}a(z_1 - z_2)^2 , \qquad I' = \tfrac{1}{2}a'(z_1 - z)^2 + \tfrac{1}{2}a''(z - z_2)^2 , \tag{24}$$

with

$$1/a = 1/a' + 1/a'' ,$$

satisfy the inequality

$$I' \geqslant I .$$

This can be establihsed by finding the minimum of I' as a function of z which turns out to be

$$\text{Min } I' = (z_1 - z_2)^2 / (1/a' + 1/a'') .$$

In terms of the above mechanical model, it means that the energy of the system always increases by picking any point P on a spring stretched between z_1 and z_2, and moving it to any new fixed position z by force. The spring constants of the resulting two sub-springs $z_1 z$ and $z z_2$ are given by the relation

$$\frac{1}{a'} = \frac{|z_1 - P|}{|z_1 - z_2|}\frac{1}{a} , \quad \frac{1}{a''} = \frac{|P - z_2|}{|z_1 - z_2|}\frac{1}{a} , \quad \frac{1}{a'} + \frac{1}{a''} = \frac{1}{a} . \tag{25}$$

This observation immediately leads to the following conclusion.

Lemma 4. Let $I(z_i; a_l)$ and $M^2(a_l)$ correspond to a Feynman diagram $\mathcal{G}$, where all the parameters a_l carried by the lines are fixed. If $\mathcal{G}$ is changed to $\mathcal{G}'$ by inserting an internal photon line (radiative correction), then for any $I'(z_i, a_l')$ and $M'^2(a_l')$ for $\mathcal{G}'$ there exist such $I(z_i; a_l)$ and $M^2(a_l)$ for $\mathcal{G}$ that

$$(26)\qquad \begin{cases} I'(z_i; a_l') \geqslant I(z_i; a_l) & \text{for all } z_i, \\ M'^2(a_l') = M^2(a_l), & \end{cases}$$

and vice versa. The corresponding quantities $J'(q; a_l')$ and $J(q; a_l)$ in the momentum space $(z_i \to q_i)$ satisfy the relation

$$(26')\qquad J'(q_i; a_l') \leqslant J(q_i; a_l) \qquad \text{for all } q_i, \ \sum q_i = 0,$$

since they are inverse quadratic forms of I' and I. If we do not insist on fixing M^2 and M'^2, Eqs. (26) and (27) may alternatively be stated respectively as

$$(27)\qquad I'M'^2 \geqslant IM^2 \quad \text{and} \quad J'/M'^2 \leqslant J/M^2,$$

since these quantities are scaling-invariant. One can always change the scale of a_l or a_l' according to Theorem 2 to make $M^2 = M'^2$.

In general, radiative corrections to a Feynman diagram are built up by 1) attaching photon lines to electron lines, and 2) inserting electron closed loops in (single or groups of) photon lines. The effect of the first process is covered by the above lemma. The effect of the second process may be understood again in terms of the mechanical model. Take first a diagram with an electron loop correction. We keep temporarily all the vertices on the loop fixed in space, and replace the spring representing the electron loop by a set of photon springs which connect the vertices in pairs, and which have the same potential energy as the former. Then the vertices are set free, and the whole system is allowed to settle down to an equilibrium position that corresponds to a diagram without the radiative correction. It is clear that by this operation both the potential energy I and the mass M^2 decrease. Thus:

Lemma 5. The relations of Lemma 4 hold true also for insertion of an electron loop.

Combining these results, we are led to the important theorem.

Theorem 4. The weight functions $\sigma_\nu^{(\alpha)}(\alpha_{ij}, M^2/m^2)$ in Eq. (23) have a property such that J/M^2 is not larger than a suitable J_0/M_0^2 for a lower order diagram from which the diagram under consideration can be constructed by insertion of electron and photon lines:

$$0 < J(q;\beta)/M^2 \leqslant J_0(q;\beta^{(0)})/M_0^2 \,. \tag{28}$$

for all q_i with $\sum_i q_i = 0$, $\sum_i q_i^2 > 0$.

The lemmas and theorems in this section have been derived under the assumption that the Bose field is massless. In case a meson field is considered, most of the results remain essentially unaltered, except that Theorem 4 and the preceding lemmas need a careful examination. Thus if a meson line is inserted, both I and M^2 will increase. In the case of inserting a closed loop in meson lines, it is necessary for the theorem to hold that M^2 does not decrease by the operation. This sets an upper limit $2m/n$ to the meson mass that depends on the number n of the meson lines connected to the loop. For example, if a loop is inserted in a single meson line (meson self-energy), the meson and nucleon masses have to satisfy the relation $\mu \leqslant 2m$. Physically it means that the meson must be energetically stable against decay into a pair. The proof will be given in the appendix.

3. - Application to the three-field Green's function.

We will here apply the results of the preceding section to exhibit the structure of the three-field Green's function or the vertex function.

The three-field Green's function $\varrho_{2,1}(p|p'|k)$ may be written in terms of eight scalar functions as

$$\varrho_{2,1}(p|p'|k) = i\sum_{\lambda,\lambda'=0}^{1}(ip\cdot\gamma)^{\lambda}[H_{\lambda\lambda'}(pp'k)\gamma_\mu + K_{\lambda\lambda'}(pp'k)\sigma_{\mu\nu}k_\nu](ip'\cdot\gamma)^{\lambda'}\,, \tag{29}$$

$$\sigma_{\mu\nu} = (\gamma_\mu\gamma_\nu - \gamma_\nu\gamma_\mu)/2i\,.$$

In the (lowest) order e,

$$\left\{\begin{aligned} H^{(0)}_{\lambda\lambda'} &= eC_{\lambda\lambda'}\frac{1}{p^2+m^2-i\varepsilon}\,\frac{1}{p'^2+m^2-i\varepsilon}\,\frac{1}{k^2-i\varepsilon} = \\ &= 2eC_{\lambda\lambda'}\iiint_0^1\frac{\delta(1-\alpha_0-\beta_0-\gamma_0)\,\mathrm{d}\alpha_0\,\mathrm{d}\beta_0\,\mathrm{d}\gamma_0/(1-\gamma_0)^3}{[(\alpha_0/1-\gamma_0)p^2+(\beta_0/(1-\gamma_0))p'^2+(\gamma_0/(1-\gamma_0))k^2+m^2-i\varepsilon]^3}\,, \\ K^{(0)}_{\lambda\lambda'} &= 0\,, \\ C_{\lambda\lambda'} &= \begin{pmatrix} m^2 & -m \\ -m & 1\end{pmatrix}. \end{aligned}\right. \tag{30}$$

Accordingly to Theorem 4, $H_{\lambda\lambda'}$ in general should have a form

$$H_{\lambda\lambda'} = \int\!\cdots\!\int \frac{\sigma_{\lambda\lambda'}(\alpha\beta\gamma)\,d\alpha\,d\beta\,d\gamma}{[\alpha p^2 + \beta p'^2 + \gamma k^2 + m^2 - i\varepsilon]^n}\,, \qquad (n \geqslant 3), \tag{31}$$

where we have normalized M^2 to m^2 instead of normalizing $\alpha+\beta+\gamma$. Theorem 5 then says that there exists such J_0 that

$$J_0 = (\alpha_0/(1-\gamma_0))q_1^2 + (\beta_0/(1-\gamma_0))q_2^2 + (\gamma_0/(1-\gamma_0))q_3^2 \geqslant J = \alpha q_1^2 + \beta q_2^2 + \gamma q_3^2 \geqslant 0\,, \tag{32}$$

if three real numbers q_1, q_2 and q_3 satisfy $q_1+q_2+q_3=0$. Writing also

$$\begin{cases} \alpha \equiv (\alpha_0/(1-\gamma_0)) - \xi \geqslant 0\,, \\ \beta \equiv (\beta_0/(1-\gamma_0)) - \eta \geqslant 0\,, \\ \gamma \equiv (\gamma_0/(1-\gamma_0)) - \zeta \geqslant 0\,, \end{cases} \tag{33}$$

Eq. (32) becomes

$$\xi q_1^2 + \eta q_2^2 + \zeta(q_1+q_2)^2 \geqslant 0\,, \tag{34}$$

which entails the inequalities

$$\xi+\eta \geqslant 0\,, \quad \eta+\zeta \geqslant 0\,, \quad \zeta+\xi \geqslant 0\,, \quad \xi\eta+\eta\zeta+\zeta\xi \geqslant 0\,. \tag{35}$$

Eqs. (33) and (35) are the conditions that essentially determine the spectrum of the absorptive part of H. For example, if one makes the electron legs free: $p^2 = p'^2 = -m^2$, then

$$\alpha p^2 + \beta p'^2 + \gamma k^2 = (\xi+\eta-1)m^2 + \gamma k^2\,. \tag{36}$$

As a function of $-k^2$, the zero of the denominator $J+m^2$ in Eq. (31) is given by

$$-k^2 = (\xi+\eta)m^2/\gamma \geqslant 0\,. \tag{37}$$

Thus in this case the singularities of $H(-k^2)$ lie on the positive real axis, and their smallest value is indeed zero, which comes from $H^{(0)}$. In the same way, if $p'^2 = -m^2$, $k^2 = 0$, one can show that the singularities (absorptive part) of $H(-p^2)$ are restricted to $-p^2 \geqslant 0$.

It is possible to recognize the thresholds for various real processes by further classification of the Feynman diagrams. In the case of $p^2 = p'^2 = -m^2$, the lowest threshold, apart from the incoming single photon state, is at $k^2=0$, corresponding to the creation of three photons. Now if one starts from the lowest order vertex (Eq. (30)) minus the external photon line (a V-shaped graph with electron lines only), and inserts all radiative corrections to it, the above kind of many-photon processes do not occur in such diagrams, but the first reaction that can take place is creation of a pair: $-k \to p+p'$; p_0, $p_0' \geqslant m$, which is possible if $-k^2 \geqslant 4m^2$. That this is indeed so can be seen as follows: The lowest order H is now

$$(38)\qquad \frac{1}{p^2+m^2-i\varepsilon}\,\frac{1}{p'^2+m^2-i\varepsilon} = \iint_0^1 \frac{\delta(1-\alpha_0-\beta_0)\,d\alpha_0\,d\beta_0}{[\alpha_0 p^2+\beta_0 p'^2+m^2-i\varepsilon]^2},$$

so that γ_0 of Eq. (33) is zero, and $-\zeta \geqslant 0$. Putting $p^2=p'^2=-m^2$,

$$(39)\qquad J+m^2=(\xi+\eta)m^2-\zeta k^2 .$$

Its zero corresponds to

$$(40)\qquad -k^2=m^2(\xi+\eta)/(-\zeta) \geqslant m^2(\xi+\eta)(1/\xi+1/\eta) \geqslant 4m^2 .$$

where use was made of the relation

$$(41)\qquad -1/\zeta \geqslant 1/\xi+1/\eta \geqslant 0 ,$$

a consequence of Eq. (35).

As for the anomalous moment part K of $\varrho_{2,1}$, the lowest order is e^3, which will not be discussed here. But it is obvious that for any diagram all the functions have, in general, the same thresholds unless they vanish accidentally or by selection rules.

It is difficult to generalize these considerations to Green's functions of higher order. In particular, the dispersion relations of the familiar variety for the scattering matrix do not seem to follow in a simple way from our formulas. The main reason is as follows. The lowest order Compton scattering, for example, consists of the familiar uncrossed and crossed diagrams. As a function of $-P\cdot K$, where $P(K)$ is the sum of the initial and final electron (photon) momenta, they have a singularity at $-P\cdot K=-\Delta^2$ and $-P\cdot K=+\Delta^2$ respectively, Δ being the momentum transfer. In higher orders, it is expected that one class of matrix elements has singularities for $-P\cdot K \geqslant -\Delta^2$ and the other $-P\cdot K \leqslant \Delta^2$ so that each can be treated separately. But these classes cannot in general be related to the crossed and uncrossed diagrams. In fact some higher order diagrams contain both branches of the singularities simultaneously, which are difficult to unscramble.

4. – Remarks.

The derivation of the representations in Sect. 2 was made on the assumption that all the parametric integrations converge and the order of integration can be freely interchanged. This is not actually true because of the various divergences. These divergences are hidden in the weight functions $\sigma_\nu^{(\alpha)}(\alpha_{ji}, M^2/m^2)$ of Eq. (23), the explicit evaluation of which was not attempted. To remove the ultraviolet divergences, renormalization has to be carried out by introducing into the Hamiltonian appropriate counter terms. This will not affect the results of Sect. 2 since the counter terms do not destroy the nature of the representation, but the weight functions will be modified in such a way that they themselves are finite and give convergent integrals (7), apart from the infrared divergences characteristic of the radiation field. The presence of the infrared divergences in the Green's functions is a reflection of the fact that these functions do not necessarily correspond to physical quantities. Formally the difficulty is removed by giving a finite mass to the photon, and our theorems will still be valid.

The results of Sect. 2 exhibit the analytic property and characteristic thresholds of each function corresponding to a Feynman diagram; but obviously they cannot be carried over to a complete Green's function which is an infinite sum of such functions. In particular, our approach will certainly fail when bound states of particles exist since they affect the thresholds. It would be interesting if our results, after suitable modifications, could be derived without the use of perturbation. The variational properties of the functions observed in Sect. 2 may offer a clue to this problem.

Finally we will discuss the relation of the present representation to the previously proposed one of more restricted type. The latter had essentially the form

$$G(k_\mu^2) = \int \dots \int_0^\infty \prod_{\mu=1}^{N(N-1)/2} \frac{1}{l_\mu^2 + s_\mu - i\varepsilon} \sigma(s_\mu) \pi \, \mathrm{d}s_\mu \,, \tag{54}$$

which should be compared with the new form

$$G'(k) = \int \dots \int \frac{1}{[\sum_\mu \beta_\mu l_\mu^2 + M^2 - i\varepsilon]^n} \sigma'(M^2, \beta_\mu) \pi \, \mathrm{d}\beta_\mu \mathrm{d}M^2 \,, \tag{55}$$

(7) The problem of renormalization was discussed in detail by Bogoljubow and Shirkow using the same kind of representation as in the present work. Bogoljubow and Shirkow: *Fortschritte d. Phys.*, **4**, 438 (1956).

where the l_μ's are defined in Theorem 3. As was pointed out in Sect. 2, Eq. (55) shows that G' is the boundary value of a function analytic for $\mathrm{Im}\,(-l_\mu^2) > 0$. Thus it can be expressed as

$$G'(l_\mu^2) = \frac{1}{2\pi i}\int_{-\infty}^{\infty} \frac{G'(-s_1, l_2^2, \dots)}{l_1^2 + s_1 - i\varepsilon}\, \mathrm{d}s_1 \,. \tag{56}$$

The same procedure can then be applied to $G'(-s_1, l_2^2, \dots)$ in the integrand with respect to the variable l_2^2, and so on.

One gets finally

$$\left\{\begin{aligned} G'(l_\mu^2) &= \frac{1}{(2\pi i)^{N(N-1)/2}}\int_{-\infty}^{\infty}\!\!\dots\!\int \prod \frac{1}{l_\mu^2 + s_\mu - i\varepsilon}\, G'(-s_\mu)\pi\, \mathrm{d}_\mu s \,, \\ &= \frac{2i}{(2\pi i)^{N(N-1)/2}}\int_{-\infty}^{\infty}\!\!\dots\!\int \prod \frac{1}{l_\mu^2 + s_\mu - i\varepsilon}\, \mathrm{Im}\, G'(-s_\mu)\pi\, \mathrm{d}s_\mu \,, \end{aligned}\right. \tag{57}$$

where in the last step only the imaginary part was used in the integrand to express $G'(-S_1, \dots, -S_{N(N-1)/2-1}, l^2_{N(N-1)/2})$. Eq. (57) is similar to Eq. (54), but the parameters S_μ here run all the way from $-\infty$ to $+\infty$. Thus we see that $G(k_i)$ must have an additional special character if it is at all representable in the restricted form Eq. (54).

* * *

The author expresses his gratitude to Prof. M. GELL-MANN for his discussions when the former was staying at the California Institute of Technology in the summer of 1956.

APPENDIX I.

Take a diagram with $2n$ external electron lines, m photon lines, and R internal vertices. The total number of electron lines is $n+R$, and each electron line connecting two points x_i and x_k contributes a

$$\begin{aligned} S_F(x_i - x_k) &= (-\gamma\cdot\partial/\partial x + m)\Delta_F(x) = \\ &= -\frac{i}{8\pi^2}\int_0^\infty \{-ia\gamma\cdot(x_i - x_k) + m\} \exp\left[ia(x_i - x_k)^2/2 - im^2/2a\right] \mathrm{d}a \,. \end{aligned} \tag{A1}$$

When one is supposed to integrate over x_i or x_k, the differentiation may be regarded as $-\partial/\partial x_k$ or $+\partial/\partial x_i$ respectively, and taken out of the integral sign Suppose in general one has a form

(A2) $$(x_1-x_2)_\mu \exp[iI], \qquad I=\tfrac{1}{4}\sum_{i,k=1}^{N} a_{ik}(x_i-x_k)^2,$$

which has to be integrated over x_1. Our problem is to express $(x_1-x_2)_\mu$ in terms of derivatives with respect to the other $N-1$ co-ordinates, $x_2, \ldots, x_N$. Each $-i\partial/\partial x_i$ on $\exp[iI]$ gives $r_i \equiv \sum_k a_{ik}(x_i-x_k)$, so that one requires

(A3) $$\begin{cases} x_1-x_2=\sum_i c_i r_i = \sum_{i,k} x_i(\delta_{ik}\sum_j a_{ij}-a_{ik})c_k \\ \qquad =\sum_{i,k} A_{ik}x_i c_k, \end{cases}$$

where A_{ik} is the matrix introduced in Eq. (7). Since this equation must hold for all x_i, one gets

(A4) $$\begin{cases} \sum_k A_{ik}c_k = C_i, \\ C_i=\begin{cases} 1 & i=1 \\ -1 & i=2 \\ 0 & i>2 \end{cases} \end{cases}$$

Both sides become zero on summing over i, so that one equation and one c_i is redundant. Thus there exists a solution with $c_1=0$, which enables x_1-x_2 to be expressed in terms of the derivatives $\partial/\partial x_2, \ldots \partial/\partial x_N$.

Now each derivative in $S_F(x_i-x_k)$ is accompanied by a γ-matrix so that in our diagram there will be up to $(n+R)$ γ-matrices coupled with derivatives. On the other hand, our Green's function contains $2n$ spinor indices which can span a direct product of n γ-spaces. In each space there are $N-1$ independent scalars $\gamma\cdot r_i$ $(\sum_{i=1}^{N} r_i=0)$. Thus the number of γ-matrices coupled with derivatives need not exceed $n(N-1)$. Some of the $\gamma\cdot r_i$'s will then be reduced to the scalars $r_i\cdot r_j$ and $r_i\cdot e_k$, where e_k $(k=1, \ldots, m)$ are the polarization vectors of the radiation field. The number ν of the scalars $r_i\cdot r_j$ will be at most $(n+R)/2$ or $(n+R-1)/2$. Each $r_i\cdot r_j$ can be obtained by a linear combination of $i\partial/\partial a_{kl}$ on $\exp[iI]$ with coefficients which are homogeneous functions of a_{kl} of degree zero. Thus the effect of these scalars arising from γ-matrices will be absorbed into the function $\sigma(\alpha_{ij}, M^2/m^2)$ and an extra factor $(i/t)^\nu$.

Appendix II.

We will consider the effect of inserting a nucleon closed loop in a group of n meson lines spanned between the points x_1x_2; x_3x_4, ...; $x_{2n-1}x_{2n}$ (in a Euclid space). These lines contribute

$$\text{(A5)}\quad \begin{cases} I = \frac{1}{2}a_1(x_1 - x_2)^2 + \frac{1}{2}a_2(x_3 - x_4)^2 + \dots + \frac{1}{2}a_n(x_{2n-1} - x_{2n})^2 , \\ M^2 = \frac{1}{2}\mu^2\left(\frac{1}{a_1} + \frac{1}{a_2} + \dots + \frac{1}{a_n}\right). \end{cases}$$

After the insertion of a closed loop, the meson lines starting from $x_1, \dots, x_{2n}$ will end on the loop at $y_1, \dots, y_{2n}$, so that now

$$\text{(A6)}\quad \begin{cases} I' = \frac{b_1}{2}(x_1 - y_1)^2 + \frac{b_2}{2}(x_2 - y_2)^2 + \dots + \frac{b_{2n}}{2}(x_{2n} - y_{2n})^2 + \\ \qquad + \frac{c_1}{2}(y'_1 - y'_2)^2 + \frac{c_2}{2}(y'_2 - y'_3)^2 + \dots + \frac{c_{2n}}{2}(y'_{2n} - y'_1)^2 , \\ M'^2 = \frac{1}{2}\mu^2\left(\frac{1}{b_1} + \frac{1}{b_2} + \dots + \frac{1}{b_2}\right) + \frac{1}{2}m^2\left(\frac{1}{c_1} + \frac{1}{c_2} + \dots + \frac{1}{c_{2n}}\right), \end{cases}$$

where $(y'_1, \dots, y'_{2n})$, being a certain permutation of $(y_1, \dots, y_{2n})$, is numbered by counting along the loop.

Let us subdivide c_l as

$$\text{(A7)}\quad \begin{cases} c = \sum_{i=1}^{n} c_{i} , & l = 1, \dots, 2n;\ c_{\cdot i} \geqslant 0, \\ \frac{1}{2}\sum_{l=1}^{2n} c_{\cdot}(y'_l - y'_{l+1})^2 = \frac{1}{2}\sum_{i=1}^{n}\sum_{l=1}^{2n} c_{li}(y'_l - y'_{l+1})^2 , & (y'_{2n+1} \equiv y'_1) \end{cases}$$

and observe that for any pair y'_a and y'_b $(a > b)$ and fixed i

$$\tfrac{1}{2}c_{ai}(y'_a - y'_{a+1})^2 + \tfrac{1}{2}c_{a+1,i}(y'_{a+1} - y'_{a+2})^2 + \dots + \tfrac{1}{2}c_{\ +1,i}(y'_{b-1} - y'_b)^2 \geqslant$$

$$\geqslant \frac{1}{2}\frac{(|y'_a - y'_{a+1}| + \dots + |y'_{b+1} - y'_b|)^2}{(1/c_{ai}) + \dots + (1/c_{;-1,i})} \geqslant \frac{1}{2}\frac{(y'_a - y'_b)^2}{1/c_{ai} + \dots + 1/c_{b-1,i}}$$

$$\tfrac{1}{2}c_{\ i}(y'_b - y'_{b+1})^2 + \tfrac{1}{2}c_{\ +1,i}(y'_{b+1} - y'_{b+2})^2 + \dots + \tfrac{1}{2}c_{a-1,i}(y'_{a-1} - y'_a)^2 \geqslant$$

$$\geqslant \frac{1}{2}\frac{(|y'_b - y'_{b+1}) + \dots + |y'_{a\cdot 1} + y'_a|)^2}{(1/c_{\ i}) + \dots + (1/c_{a-1,i})} \geqslant \frac{1}{2}\frac{(y'_b - y'_a)^2}{(1/c_{\ i}) + \dots + (1/c_{a-1,i})} .$$

Thus

$$\text{(A8)}\quad \frac{1}{2}\sum_l c_{\ i}(y'_l - y'_{l+1})^2 \geqslant \frac{1}{2}(y'_a - y'_b)^2\left(\frac{1}{(1/c_{ai}) + (1/c_{a+1,i}) + \dots + (1/c_{\ -1,i})}\right) +$$

$$+ \left(\frac{1}{(1/c_{bi}) + (1/c_{\ +1,i}) + \dots + (1/c_{a-1,i})}\right) \geqslant \frac{1}{2}(y'_a - y'_b)^2 \frac{4}{\sum_{l=1}^{2n}(1/c_{li})} .$$

Identifying (y'_a, y'_b) with (y_{2i-1}, y_{2i}), we obtain from Eq. (A8)

$$\text{(A9)}\qquad \frac{1}{2}\, b_{2i-1}(x_{2i-1}-y_{2i-1})^2+\frac{1}{2}\, b_{2i}(x_{2i}-y_{2i})^2+\frac{1}{2}\sum c_{li}(y'_l-y'_{l+1})^2 \geqslant$$

$$\geqslant \frac{1}{2}\,\frac{(|x_{2i-1}-y_{2i}|+|x_{2i}-y_{2i}|+|y_{2i-1}+y_{2i}|)^2}{(1/b_{2i-1})+(1/b_{2i})+\frac{1}{4}\sum_{l=1}^{2n}(1/c_{li})} \geqslant$$

$$\geqslant \frac{1}{2}\,(x_{2i-1}-x_{2i})^2\Big/\left(\frac{1}{b_{2i-1}}+\frac{1}{b_{2i}}+\frac{1}{4}\sum_{l=1}^{2n}\frac{1}{c_{li}}\right).$$

Summing this over i, the left-hand side becomes I':

$$\text{(A10)}\qquad I' \geqslant \frac{1}{2}\sum_{i=1}^{n}\left[(x_{2i-1}-x_{2i})^2\Big/\left(\frac{1}{b_{2i-1}}+\frac{1}{b_{2i}}+\frac{1}{4}\sum_{l=1}^{2n}\frac{1}{c_{li}}\right)\right].$$

In order to compare I, M and I', M', we choose a_i in such a way that

$$\text{(A11)}\qquad \frac{1}{a_i}=\frac{1}{b_{2i-1}}+\frac{1}{b_{2i}}+\frac{1}{4}\sum_{l=1}^{2n}\frac{1}{c_{li}}.$$

Then clearly

$$\text{(A12)}\qquad I' \geqslant I.$$

Also

$$\text{(A13)}\qquad M^2=\frac{1}{2}\,\mu^2\sum_{i=1}^{n}\left(\frac{1}{b_{2i-1}}+\frac{1}{b_{2i}}+\frac{1}{4}\sum_{l=1}^{2n}\frac{1}{c_{li}}\right)=$$

$$=\frac{1}{2}\,\mu^2\left(\frac{1}{b_1}+\frac{1}{b_2}+\ldots+\frac{1}{b_{2n}}\right)+\frac{1}{2}\,\frac{1}{4}\,\mu^2\sum_{l=1}^{2n}\sum_{i=1}^{n}\frac{1}{c_{li}}\geqslant$$

$$\geqslant \frac{1}{2}\,\mu^2\left(\frac{1}{b_1}+\frac{1}{b_2}+\ldots+\frac{1}{b_{2n}}\right)+\frac{1}{2}\,\frac{n^2}{4}\,\mu^2\sum_{l=1}^{2n}\frac{1}{c_l},$$

where the inequality

$$\sum_{i=1}^{n}\frac{1}{c_{li}}\geqslant n^2\Big/\sum_{i=1}^{n}c_{li}=n^2/c_l$$

was used. Comparing this with M'^2,

$$\text{(A14)}\qquad M^2-M'^2\geqslant \frac{1}{2}\left(\frac{n^2}{4}\,\mu^2-m^2\right)\sum_{l}\frac{1}{c_l}.$$

The equality holds if the particular choice

$$c_{l1} = c_{l2} = \ldots = c_{ln} = c_l/n \tag{A15}$$

is made for a given c_l.

Now if $n\mu > 2m$, we have the situation $I' \geqslant I$, $M^2 \geqslant M'^2$, which is not necessarily compatible with

$$I'M'^2 \geqslant IM^2 . \tag{A16}$$

On the other hand, if $n\mu \leqslant 2m$, then with the choice Eq. (A15), $M^2 \leqslant M'^2$ so that the inequality Eq. (A16) follows.

RIASSUNTO (*)

Si derivano, nella teoria delle perturbazioni, delle rappresentazioni parametriche per la funzione di Green della teoria dei campi. Tali rappresentazioni sono valide per ogni termine della serie perturbativa corrispondente a un diagramma di Feynman e rispecchiano la sua proprietà analitica e le sue caratteristiche di soglia. Come esempio si dimostra che quando siano fissati due dei tre impulsi la funzione di tre corpi (di vertice) soddisfa una relazione di dispersione con la corretta localizzazione delle singolarità previste in base alle soglie dei processi reali in competizione.

(*) *Traduzione a cura della Redazione.*

Y. NAMBU
1958, Agosto
Il Nuovo Cimento
Serie X, Vol. 9, pag. 610-623

Dispersion Relations for Form Factors (*).

Y. NAMBU

The Enrico Fermi Institute for Nuclear Studies
The University of Chicago - Chicago, Ill.

(ricevuto il 22 Aprile 1958)

Summary. — The dispersion relations for form factors are studied in perturbation theory. In particular, the relations are proved for a realistic nucleon in the form usually assumed in literature. It is demonstrated, however, that in more general cases the mass spectrum in the dispersion relations depends on the masses of the contributing particles in a peculiar way, which is different from the simple relation hitherto believed. Such a situation arises, for example, in the case of some hyperon models and bound systems such as the deuteron. A formula concerning the final state interaction to be used in conjunction with the dispersion relations is also derived from the invariance of the theory under space-time reversal.

1. - Introduction.

In this paper we are concerned with the properties of the form factor F of a particle

$$\langle q^0|F(p-q)|p^0\rangle = \langle q|j(x)|p\rangle \exp[-i(p-q)\cdot x]\,. \tag{1.1}$$

Here $j(x)$ is a local Heisenberg operator such as the charge-current density; $|p\rangle$ and $|q\rangle$ are true eigenstates of a particle characterized by four-momenta p and q, and other quantum numbers which are suppressed; $|p^0\rangle$ and $|q^0\rangle$ are the corresponding eigenfunctions for an interaction-free particle.

It has been generally believed that $F(p-q) \equiv F(k)$ should be representable in a form analogous to the dispersion relations for scattering matrices. For

(*) This work was supported by the United States Atomic Energy Commission.

example, if the particle has no spin, $j(x)$ is a scalar density, and $F(k) \to 0$ sufficiently rapidly as $k^2 \to \infty$, we should have (1)

$$F(k) = \int_{m_0^2}^{\infty} \frac{\varrho(m^2)}{k^2 + m^2}\, d(m^2)\,. \tag{1.2}$$

$\pi\varrho(-k^2)$ is the so-called absorptive part of $F(k)$, which should exist only above a threshold $-k^2 \geqslant m_0^2$ for material creation out of vacuum by the external probing field of frequency k_0 and wave number $\boldsymbol{k}$.

In fact, such « dispersion relations » for form factors have been proved under some restrictive conditions from the general properties of field theory, including notably the local commutativity of field operators (2). For the form factors of a nucleon (mass M) interacting with pion field (mass μ), for example, equations of the type Eq. (1.2) are valid with

$$m_0 = 2\mu \tag{1.3}$$

provided that

$$\mu > (\sqrt{2} - 1)M\,. \tag{1.3'}$$

On the other hand, the relation was proved by the author in every order of perturbation in an example of quantum electrodynamics where $m_0 = 0$ (3). The present paper will deal with an extension of the proof in perturbation theory so as to include the case of nucleon form factors. In particular, it will be shown that the relation (1.3) is valid for actual nucleons without the restriction (1.3'), which is not satisfied by the real pion.

For practical applications, it is also important to relate the absorptive part of F to other observable quantities, thereby making the dispersion relations, at least in principle, equations to be solved for F. The main idea behind this is very similar to the so-called final state interaction in nuclear reactions formulated by WATSON (4).

These problems will be treated in the following sections, specifically with the nucleon form factors in mind. In the course of the development, it will also become clear that some modifications are necessary with regard to the

---

(1) G. F. CHEW, R. KARPLUS, S. GASIOROWICZ and F. ZACHARIASEN: *Phys. Rev.* **110** 265 (1958); M. GOLDBERGER and S. TREIMAN: preprints on form factors for weak interactions. Also a lecture by GOLDBERGER on nucleon form factors at Stanford Conference, 1957.

(2) H. J. BREMERMANN, R. OEHME and J. G. TAYLOR: *Phys. Rev.* **109** 2178 (1958).

(3) Y. NAMBU: *Nuovo Cimento*, **6**, 1064 (1957). This paper will be quoted as I.

(4) K. WATSON: *Phys. Rev.*, **95**, 228 (1954).

usually held statement about the lower limit m_0 of the dispersion integral when more general classes of problems are considered. The discussion of this situation will be made in the last two sections.

2. – Representoation of the nucleon form factors.

We consider here the realistic case of the baryon (nucleon and hyperon) field interacting with π and K meson fields (spin zero). The interaction is to be renormalizable. Among the particles, the nucleon and the pion are characterized by the fact that they are respectively the lowest states of the baryon and the meson families, which will turn out to be an important point in the following. Other details are not essential.

We are interested in the electromagnetic form factor of a nucleon

$$\langle q^0 | F_\mu(k) | p^0 \rangle = \langle q | j_\mu(x) | p \rangle \exp\left[-i(p-q)\cdot x\right], \tag{2.1}$$

$j_\mu(x)$ is the total charge-current density of the participating fields. From Lorentz and gauge invariance it follows immediately that

$$F_\mu^{\text{p,n}} = eF_1^{\text{p,n}}(k^2)i\gamma_\mu + \frac{e}{2M}F_2^{\text{P,N}}(k^2)i\sigma_{\mu\nu}k_\nu\,, \tag{2.2}$$

$$(\text{n} = \text{proton}, \quad \text{p} = \text{neutron})\,.$$

The above decomposition can be carried out in perturbation theory for each Feynman diagram. Let such a contribution from a single diagram $\mathcal{G}$ be simply called F_μ. Then we want to show that

$$\left\{\begin{aligned} F_1^{\text{p n}}(k^2) &= \left(\frac{1+\tau_3}{2}\right)^{\text{p,n}} - k^2\int_{(2\mu)^2}^{\infty}\frac{\varrho_1^{\text{p,n}}(m^2)}{k^2+m^2-i\varepsilon}\,\mathrm{d}m^2\,,\\ F_2^{\text{p,n}}(k^2) &= \int_{(2\mu)^2}^{\infty}\frac{\varrho_2^{\text{p,n}}(m^2)}{k^2+m^2-i\varepsilon}\,\mathrm{d}m^2\,.\end{aligned}\right. \tag{2.3}$$

The different forms of the representation for F_1 and F_2 are due to their different behavior for large k^2. It is also a reflection of the fact that F_1 requires an infinite charge renormalization whereas F_2, the anomalous moment part, converges without a corresponding renormalization. This is well known, and will not be discussed here. Another point to make is that the same function F_μ can represent creation or annihilation of a nucleon pair by an external electromagnetic field. In such cases k^2 will be $\leqslant -4M^2$, whereas for a form factor $k^2 \geqslant 0$.

2·1. – According to the results of I, F_1 and F_2 should have a representation

$$(2.4)\quad \left\{\begin{aligned} &F_1(k^2) = F_1(0) - k^2 \sum_{n=1}^{N_1} \iiint_0^\infty \frac{f_1^{(n)}(\alpha\beta\gamma)}{(J + M^2 - i\varepsilon)^n}\,\mathrm{d}\alpha\,\mathrm{d}\beta\,\mathrm{d}\gamma\,, \\ &F_2(k^2) = \sum_{n=1}^{N_2} \iiint_0^\infty \frac{f_2^{(n)}(\alpha\beta\gamma)}{(J + M^2 - i\varepsilon)^n}\,\mathrm{d}\alpha\,\mathrm{d}\beta\,\mathrm{d}\gamma\,, \\ &J = \alpha p^2 + \beta q^2 + \gamma k^2 = -(\alpha + \beta)M^2 + \gamma k^2\,. \end{aligned}\right.$$

These equations establish the analyticity of $F(k^2)$ for $\mathrm{Im}\,(-k^2) > 0$. In general, there will be a singularity where

$$J + M^2 = 0 \qquad \text{or} \qquad -k^2 = (1 - \alpha - \beta)M^2/\gamma\,.$$

In comparing Eq. (2.4) with Eq. (2.3), the different powers of the denominator will not matter since they do not affect the region of singularity.

2·2. – In order to determine the singularities of Eq. (2.4), we must use Theorem (5) of I, which says that under certain conditions one can find a reduced (lower order) diagram $\mathcal{G}_0$ and a corresponding $F_0(k^2)$ with $J_0(p, q, k) = \alpha_0 p^2 + \beta_0 q^2 + \gamma_0 k^2$ such that

$$(2.5)\qquad J_0(x, y, z) \geqslant J(x, y, z)\,,$$

if x, y, z are real and $x+y+z=0$. and if the «mass term» M^2 is always normalized to the same value.

Let us consider $J(p, q, k)$ as a function of two independent vectors $k = p - q$ and $l = p + q$. Write

$$J(p, q, k) = J(k, l) = J_{11}k^2 + 2J_{12}k\cdot l + J_{22}l^2\,.$$

In our case $k\cdot l = p^2 - q^2 = 0$, and Eq. (2.5) means

$$(2.6)\qquad 0 \leqslant J_{11} \leqslant J_{11}^0\,, \qquad 0 \leqslant J_{22} \leqslant J_{22}^0\,.$$

Now both k and l are time-like if $0 < -k^2 < 4M^2$, so that with $|k| \equiv \sqrt{-k^2}$, $|l| \equiv \sqrt{-l^2}$ we get

$$(2.7)\quad \left\{\begin{aligned} &J(k, l) = -J_{11}|k|^2 - J_{22}|l|^2\,, \\ &\qquad \geqslant -J_{11}^0|k|^2 - J_{22}^0|l|^2 = J_0(k, l)\,. \end{aligned}\right.$$

Suppose $J_0 + M^2 > 0$ at the same time for $0 < -k^2 < m_0^2 \leqslant 4M^2$. Then *a fortiori*

$$J + M^2 > 0 .$$

Once this is satisfied, $J + M^2$ and $J_0 + M^2$ will have no zeros for $-k^2 < m_0^2$ since they are increasing functions of k^2, and Eq. (2.3) will have been proven.

2·3. – Let us classify the Feynman diagrams in the following way. Since the nucleon number is conserved, a diagram consists of

a) a baryon line starting with the initial nucleon line and ending with the final nucleon line, carrying a mass $m \geqslant M$;

b) any number of baryon closed loops;

b') meson (π or K) lines connecting the baryon lines, carrying a mass $m \geqslant \mu$.

The external field (« γ ray ») may act on any of the charge carrying lines. So classify F into F_a and F_b according as the γ-ray acts on the baryon line *a*) or the others. Then we carry out the following operations:

1) Reduce the masses appearing in the Δ_F functions on the baryon line *a*) from their original values to M. (We may leave untouched the factor $\gamma \cdot \partial - m$ of S_F).

2) Reduce the masses appearing in the Δ_F functions in *b*) and *b'*) to the pion mass $\mu < M$. We will now refer to all these lines simply as « meson » lines.

Under these operations, the mass term called M^2 in I obviously decreases (or at least does not increase).

3) If the γ-acts on the baryon line (case F_a), remove all the « meson » lines except one, so that we get a typical lowest order diagram Fig. 1*a*.

4) If the γ-ray acts on one of the meson lines (case F_b), remove as many « meson » lines as possible. We will end up with another typical diagram Fig. 1*b* (*).

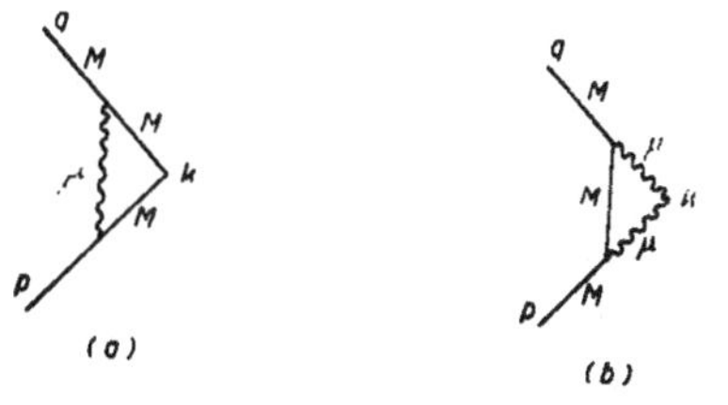

Fig. 1. – Comparison diagrams for the nucleon form factors.

The final diagrams thus obtained we now call $\mathcal{G}_0$. According to Lemma (4)

(*) In case there exists a vector meson, the situation should be different because of different selection rules.

of I, J will increase (or at least not decrease) to J^0 during the operation 3 or 4. Thus after all the operations 1 to 4, the inequality (28) of I; $J/M^2 \leqslant J_0/M_0$ or

$$J \leqslant J_0 \,,$$

for our choice of normalization, is certainly valid.

For the diagrams 1*a* and 1*b*, we can easily verify the dispersion relations. We find (*)

$$(2.8)\qquad \begin{cases} m_0 = 2M(> 2\mu) & \text{for } F_a\,, \\ = 2\mu & \text{for } F_b\,. \end{cases}$$

This is a more precise statement than is indicated in Eq. (2.3).

In the problem of nucleon form factors, a most puzzling point is the apparent big difference of proton and neutron charge radii, which implies near equality of isotopic vector and scalar parts of the charge form factor. Using a similar technique as above, we can prove that the scalar part has, as has been usually claimed, $m_0 = 3\mu$ in contrast to $m_0 = 2\mu$ for the vector part.

3. – A theorem about the final state interaction [5].

The results of the previous sections show that the absorptive part of F exists only if k is time-like and $-k^2 \geqslant 4\mu^2$. As was noticed before, the function F for $-k^2 > 0$ does not correspond to a form factor, but rather to a matrix element for the transformation of a virtual γ-ray into a nucleon pair if $-k^2 \geqslant 4M^2$. In this region the absorptive part can be interpreted as arising from the interaction of the produced pair. Below the threshold, $F(k^2)$ must be considered as an analytic continuation, which can still have an absorptive part because real mesons may be produced in the intermediate states.

The exact mathematical statement about the final state interaction will be derived as follows: Let us assume that the dispersion formulas obtained in perturbation theory are valid for the entire form factors, and consider the matrix element

$$\langle -p, q^{\text{out}} | j_\mu(k) | 0 \rangle \,, \qquad -k^2 \geqslant 4M^2 \,,$$

(*) In deriving the result (2.8), it may seem that a difficulty arises from vanishing of the denominators corresponding to the initial and final nucleon lines in Fig. 1 when we set $p^2 = q^2 = -M^2$. This can be avoided by first setting $p^2 = q^2 > -M^2$ and taking the limit.

([5]) This is an adaptation of a theorem formulated for the case of scattering dispersion relations by S. Fubini, Y. Nambu and V. Wataghin; to be published.

where $\langle -p, q^{\text{out}}|$ is an outgoing state of a nucleon pair. Because of the invariance under time reversal and space reflection, we have

$$(3.1)\qquad \langle -\bar{p}, \bar{q}^{\text{out}}|Uj_\mu U^{-1}|0\rangle^* = \langle 0|Uj_\mu U^{-1}|-\bar{p}, \bar{q}^{\text{out}}\rangle = \\ = \langle -p, q^{\text{in}}|j_\mu|0\rangle = \sum_n \langle -p, q^{\text{in}}|n^{\text{out}}\rangle\langle n^{\text{out}}|j_\mu|0\rangle ,$$

$\langle -\bar{p}, \bar{q}^{\text{out}}|$ here is the space-time reversed state ($p, q \to p, q$, $\sigma \to -\sigma$), and U the associated transformation operator. n runs over a complete set of states, which includes the nucleon pair state. $\langle -p, q^{\text{in}}|n^{\text{out}}\rangle$ is an element of the S-matrix $S^+ = S^{-1}$. By the nature of the operator j_μ, n will consist of states of nucleon number zero, angular momentum one, and odd spatial and charge parities. For a nucleon pair, in particular, it involves $^3S_1$ and $^3D_1$ states, reflecting the fact that there are two invariants $i\gamma_\mu$ and $i\sigma_{\mu\nu}k_\nu$.

Let us now introduce the generalized form factor $\mathcal{F}_\mu$ by

$$(3.2)\qquad \langle n^0|\mathcal{F}_\mu|0\rangle = \langle n^{\text{out}}|j_\mu|0\rangle ,$$

for an arbitrary $|n\rangle$. As in the case of F_μ, $\mathcal{F}_\mu$ may be decomposed into a sum of different invariants

$$(3.3)\qquad \mathcal{F}_\mu = \sum_i O_\mu^{(i)} \mathcal{F}_i(k^2, \alpha_r) ,$$

where O_μ is a hermitian operator. α_r are scalar numbers, other than k^2, which are necessary to fix the state. For a given k^2 there will be only a finite number of invariants since only a finite number of particles can be present in $|n\rangle$.

The relation (3.1) then becomes in this case.

$$(3.4)\qquad \begin{cases} \langle n^0|U\mathcal{F}_\mu U^{-1}|0\rangle^* = \sum\limits_m \langle n^0|S^+|m^0\rangle\langle m^0|\mathcal{F}_\mu|0\rangle \\ \text{or} \\ \sum\limits_i \langle n^0|O_\mu^{(i)}\mathcal{F}_i^*(k^2, \alpha_r)|0\rangle = \sum\limits_i \sum\limits_{m,\alpha_r'} \langle n^0|S^+|m^0\rangle\langle m^0|O_\mu^{(i)}\mathcal{F}_i(k^2, \alpha_r'|0\rangle . \end{cases}$$

We may write it symbolically

$$(3.5)\qquad \sum_i O_\mu^{(i)}\mathcal{F}_i^* = \sum_i S^+ O_\mu^{(i)}\mathcal{F}_i .$$

Alternatively, with $S \equiv 1 + iR$,

$$(3.5')\qquad i\sum_i O_\mu^{(i)}(\mathcal{F}_i^* - \mathcal{F}_i) = \sum_i R^+ O_\mu^{(i)}\mathcal{F}_i .$$

On the left-hand side, $i(\mathcal{F}_i^* - \mathcal{F}_i)$ is twice the absorptive part of $\mathcal{F}$. Assuming S to be given, this expresses a set of linear homogeneous equations to be satisfied by the absorptive and dispersive parts of $\mathcal{F}_i$. It must be understood that they are meaningful even below reaction thresholds through analytic continuation.

Eq. (3.5) or (3.5′) and the dispersion relation (2.3) together constitute coupled integral equations for the absorptive and dispersive parts of F_1 and F_2 as functions of k^2. Since other generalized form factors also appear in Eqs. (3.5) and (3.5′), it may be suggested that all $\mathcal{F}_i$ should have their own dispersion relations like

$$\mathcal{F}_i(k^2, \alpha) = \mathcal{F}_i(0, \alpha) - \frac{k^2}{\pi} \int_{m_i^2}^{\infty} \frac{\operatorname{Im} \mathcal{F}_i(m^2, \alpha_r)}{k^2 + m^2 - i\varepsilon} \frac{dm^2}{m^2}, \tag{3.6}$$

with an appropriate choice of the parameters α_r. Then we would have a set of integral equations for all $\mathcal{F}_i$. In fact, Eq. (3.6), may be easily derived from the representations of general Green's functions. This will be shown in the Appendix I.

We will not attempt to solve the integral equations for $\mathcal{F}_i$. Except for the single channel case where only one $\mathcal{F}_i$ has an absorptive part, the only practical way seems to be perturbation expansion starting from $\operatorname{Re} \mathcal{F}_i = \mathcal{F}_i(0)$. But the solution may not be unique. If there is more than one solution, the difference of two such solutions must satisfy homogeneous dispersion relations.

4. – Form factors in a more general case [6].

As is clear from the proof in Sect. 2, the derivation of the mass limit Eq. (2.8) depends critically on the fact that nucleon and pion are respectively the lightest particles of baryon and meson families. In case, for example, the form factor of a hyperon is considered, the proof will not go through since a hyperon can change to a lighter particle, and destroys the necessary inequalities in the proof.

Such a situation can be handled as follows: Since a hyperon form factor is not of immediate practical interest, we will adopt a simplified model, consisting of a hyperon, a nucleon, and a K-meson field, with the respective masses $M_1 > M > \varkappa$. Then a form factor diagram will contain:

a) A baryon line (or baryonic charge carrying line) connecting the initial and final hyperon lines.

b) A strangeness carrying line, which may overlap with *a*), also connecting the initial and final lines.

([6]) A very similar observation to the one in this section was recently done by KARPLUS, SOMMERFELD and WICHMANN: preprint.

c) Baryonic charge loops and strangeness loops.

These lines will overlap each other to form a connected diagram. The external γ-ray may act on any of the lines.

In order to facilitate the comparison with standard diagrams, we will first observe the following theorem.

Theorem. In order that the inequalities (27) of I

$$
\left\{
\begin{array}{l}
I(z_i, \alpha_r)\, M^2(\alpha_r) \geqslant I_0(z_i, \alpha_r^0)\, M_0^2(\alpha_r^0)\,, \\
J(q_i, \beta_r)/M^2(\beta_r) \leqslant J_0(q_i, \beta_r^0)/M_0^2(\beta_r^0)\,,
\end{array}
\right.
\tag{4.1}
$$

be realized for two diagrams $\mathcal{G}$ and $\mathcal{G}_0$, and for any particular sets z, α or q, β, it is necessary and sufficient that the following inequality be satisfied:

$$
\mathcal{J}(z_i) \geqslant \mathcal{J}_0(z_i)\,. \tag{4.2}
$$

The function $\mathcal{J}(z_i)$ is defined for a diagram by

$$
\mathcal{J}(z_i) = \mathrm{Min}\,(\tfrac{1}{2} \sum_i m_s r_s)^2\,, \tag{4.2$'$}
$$

$r_s \geqslant 0$ is the length of a propagation line, m_s the mass of the propagating particle, and the summation is over all the lines that occur in the diagram. The minimum is taken with respect to the co-ordinates of the internal vertex points. The proof of the theorem rests on the property

$$
I(z_i, \alpha_r)\, M^2(\alpha_r) \geqslant \mathcal{J}(z_i)\,,
$$

which will be derived in the Appendix.

Now proceeding more or less in the same way as before, and making use of the above theorem, we find the following comparison diagrams (with the corresponding I_0)

1) If the γ-ray acts on the baryon line *a*), take Fig. 2*a*.

2) If the γ-ray acts on the strangeness line *b*), take Fig. 2*b*.

3) If the γ-ray acts on a (baryon or strangeness) loop, take either Fig. 2*c* or 2*c*$'$, according as it is connected to the baryon line *a*) or the strangeness line *b*). The masses on the connecting lines have all been reduced to $\varkappa$.

It is not difficult to see that Fig. 2*c* and 2*c*$'$ may further be replaced by Fig. 2*d* and 2*d*$'$ respectively, which are combinations of the vertex diagram 2*a* or 2*b*, and a self-energy diagram.

The main interest is in the vertex diagrams. They are essentially the same as Fig. 1, except that we now put $-p^2 = -q^2 = M_1^2 > M^2$. By direct calculation, the lower limit m_0^2 of the dispersion relation (2.3) now turns out to be different for the following three cases.

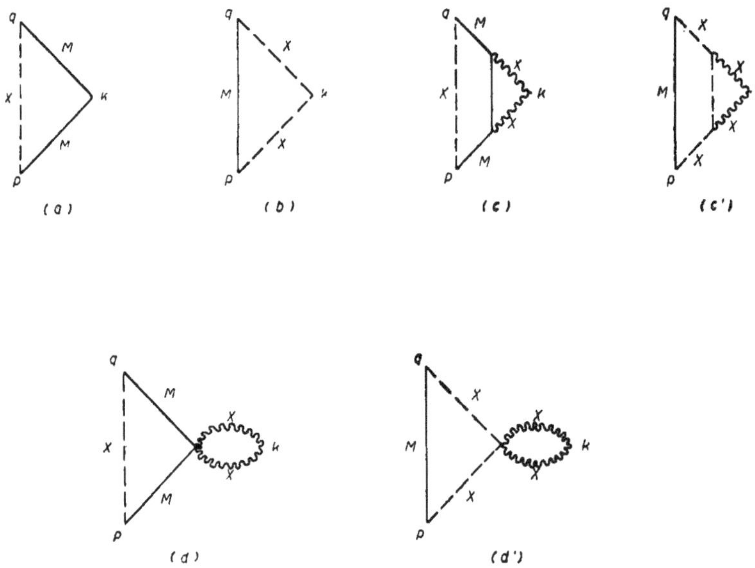

Fig. 2. – Comparison diagrams for the hyperon form factors in our model. Solid line: baryonic charge line. Broken line: strangeness line. These are not necessarily actual diagrams.

Case 1. If $M_1^2 \leqslant M^2 + \varkappa^2$, then

$$m_0 = 2M \qquad \text{Fig. } 2a\,, \tag{4.3}$$
$$\quad = 2\varkappa \qquad \text{Fig. } 2b \text{ and } 2d.$$

The overall lower limit is thus $m_0 = 2\varkappa$.

Case 2. If $M^2 + \varkappa^2 \leqslant M_1^2 < (M+\varkappa)^2$, then

$$m_0^2 = (2M)^2 - \frac{\Delta^2}{\varkappa^2} \quad (\geqslant 0) \qquad \text{Fig. } 2a\,, \tag{4.3'}$$
$$= (2\varkappa)^2 - \frac{\Delta^2}{M^2} \quad (\geqslant 0) \qquad \text{Fig. } 2b \text{ and } 2c,$$
$$= \text{Min}\left(2\varkappa,\ (2m)^2 - \frac{\Delta^2}{\varkappa^2}\right) \qquad \text{Fig. } 2d'.$$
$$\Delta = M_1^2 - M^2 - \varkappa^2\,.$$

The overall lower limit is $m_0^2 = (2\varkappa)^2 - \Delta^2/M^2$.

Case 3. If $M_1 > M + \varkappa$ (the hyperon unstable), then

$$m_0^2 = -\infty \quad (*). \tag{4.3''}$$

This result indicates that the usual argument for the lower limit m_0 as given in Sect. 2 does not necessarily hold for each Feynman diagram, nor for each order of perturbation expansion.

In the general derivation of the dispersion relations from properties of local field operators, the spectrum of the total energy-momentum of intermediate states is specified, but not the masses of individual particles. Thus in the present example we would regard $M_2 = M + \varkappa$ and $2\varkappa$ (or $2M$) as the lowest total masses of the intermediate states. In order to have the Case 1 above, it is then necessary that

$$M_1^2 \leqslant M^2 + \varkappa^2 ,$$

for fixed M_2 and any $\varkappa \leqslant M_2$. Since the right-hand side takes a minimum value $M_2^2/2$ for $\varkappa = M = M_2/2$, that means

$$M_2^2 \geqslant 2M_1^2 \quad \text{or} \quad M_2 - M_1 \geqslant (\sqrt{2} - 1)M_1 . \tag{4.4}$$

This coincides with the condition (1.3′) obtained by BREMERMANN, OEHME and TAYLOR (2) for the derivability of the dispersion relations on the basis of local field theory. It would seem, therefore, that their condition (1.3) is, at least partly, caused by their inability to specify the detailed nature of the intermediate states, rather than their inability to take account of the unitarity condition in general.

5. – Concluding remarks.

We have examined in some generality the validity of the « dispersion relations » for form factors in perturbation theory. In particular, it has been shown that the nucleon form factors should satisfy the usually assumed relations. Thus the large difference of the observed proton and neutron charge radii still remains to be understood. On the other hand, it has become clear that the mass spectrum in the dispersion integral does depend on the masses of the individual particles in a peculiar way.

The last point is of practical significance since any qualitative as well as quantitative appreciation of dispersion relations depends on the mass spectrum. It would be necessary to understand the mass spectrum conditions from a

(*) The actual values of M_1, M, $\varkappa$ correspond to the case 2.

more general point of view. The relation between causality and the validity of the dispersion relations with the modified spectral conditions is not yet clear.

Another interesting point turns up when we consider bound states. It has been shown recently ([7]) that one can describe bound systems in field theory in much the same way as the so-called elementary particles. One may expect then that the form factors of a deuteron, for example, should satisfy the same kind of dispersion relations as the nucleon. On the other hand, we know that the deuteron form factor should be determined largely by the binding energy according to the uncertainty principle.

It would be absurd if in this case the mass spectrum of the dispersion relation started from $m_0 = 2\mu$, as would be expected if the original argument about m_0 was correct. As a matter of fact, if we calculate the deuteron form factor from the virtual process ($M_p = M_n = M$ assumed for simplicity)

$$\mathrm{D} \rightleftarrows \mathrm{n} + \mathrm{p},$$

with *local* interaction, we do fall in the Case 2 of Sect. **4**, with

$$m_0^2 = 4M_{\mathrm{D}}^2(1 - M_{\mathrm{D}}^2/4M^2),$$

$$\sim 16\varepsilon M,$$

where ε is the binding energy. Thus the radius of the deuteron should be $\sim 4\sqrt{\varepsilon M}$. This is just what one gets from the deuteron wave function which falls off as $\exp[-\sqrt{\varepsilon M}r]/r$. ($r$ is the proton-neutron distance, whereas the radius is measured from the center of gravity).

Appendix I

Representation for generalized form factors.

A generalized form factor $\mathcal{F}$, as defined in Eq. (3.2), is essentially a certain n-point Green's function $G_n(k, p_2, ..., p_n)$ $(n \geqslant 2)$, considered as a function of one momentum k, the other $n-1$ momenta p_i being those of free particles. In other words

$$\mathcal{F}(k, p_i) = L_1(k) \prod_{i=2}^{n} L_i(p_i) G_n(k, p_i), \tag{A.1}$$

where the L_i's are the free-field differential operators, and $p_i^2 = -m_i^2$.

According to the results of I, each invariant $\mathcal{F}_i$, Eq. (3.3), will therefore

([7]) K. Nishijima: preprint; W. Zimmermann: preprint.

have a typical representation like

$$\mathcal{F}_i(k, p) = C_i - k^2 \sum_{m=1}^{N_i} \int \dots \int \frac{g_{im}(\beta_r)}{(J + M^2 - i\varepsilon)^m} \Pi \, d\beta_r ,$$

$$J = \beta_1 k^2 + \sum_{i=2}^{n} \beta_i p_i^2 + \sum_{i>j}^{2,n} \beta_{ij}(p_i + p_j)^2 + \sum_{i>j>k}^{2,n} \beta_{ijk}(p_i + p_j + p_k)^2 + \dots \tag{A.2}$$

The summation continues until all combinations of $n-1$ momenta p ; Note that we have deliberately avoided the explicit appearance of combinations like $k+p_i$, $k+p_i+p_j$, ..., by replacing them by the complementary sums. Thus

$$J = \beta_1 k^2 + \sum_{i,j}^{2,n} \gamma_{ij} p_i \cdot p_j = \beta_1 k^2 - \sum_i \gamma_{ii} m_i^2 + 2 \sum_{i>j} \gamma_{ij} p_i \cdot p_j , \tag{A.3}$$

where $\beta_1 \geqslant 0$, $\gamma_{ij} \geqslant 0$ and the matrix γ_{ij} forms a non-negative quadratic form.

Eq. (A.3) contains $n + (n-1)(n-2)/2 = 1 + n(n-1)/2$ scalar products, whereas only $n(n-1)/2$ of them should be lineraly independent since

$$k^2 = (-\sum p)^2 = \sum_{i,j} p_i \cdot p_j . \tag{A.4}$$

So eliminate, for example, $p_2 \cdot p_3$ by the above relation, or

$$2 p_2 \cdot p_3 = k^2 - \sum_{i,j}{}' p_i \cdot p_j . \tag{A.4'}$$

Then J takes the form

$$\begin{cases} J = c k^2 + J'(\alpha_r) , \\ c = \beta_1 + \gamma_{23} \geqslant 0 , \\ \alpha_r = p_i \cdot p_j , \qquad i \geqslant 2,\ j \geqslant 4,\ i < j. \end{cases} \tag{A.5}$$

Thus for fixed parameters α_r, $\mathcal{F}_i(k^2, \alpha)$ is analytic for $\text{Im}(-k^2) > 0$ as was expected. The problem remains as to the lower limit m_0 of the mass spectrum in this case. We will not try to determine it since at any rate such generalized form factors will be of little pratical significance.

APPENDIX II

Proof of the Theorem in Sect. 4.

By the results of I, the quantities $I(z_i)$ and M^2 are

$$\begin{cases} I(z) = \text{Min} \dfrac{1}{2} \sum_s a_s z_s^2 , \\ M^2 = \dfrac{1}{2} \sum \dfrac{m_s^2}{a_s} , \end{cases} \tag{A.6}$$

$a_s \geqslant 0$ is the parameter attached to a propagation line z_s in the diagram. The minimum is taken with respect to the variation of the internal vertex co-ordinates.

Now we want to find the double minimum $\mathcal{J}(z_i)$ of $I(z_i, a)M^2(a_s)$ with respect to both the co-ordinates and the parameters a_s. By taking variation with respect to a_s, we immediately find

$$a_s = \frac{m_s}{r_s}\sqrt{\frac{I}{M^2}}\,, \qquad (r_s = \sqrt{z_s^2}) \tag{A.7}$$

and

$$IM^2 = (\tfrac{1}{2}\sum m\, r_s)^2\,. \tag{A.7'}$$

Thus

$$\mathcal{J}(z) = \text{Min}\,(\tfrac{1}{2}\sum m_s r_s)^2\,, \tag{A.8}$$

the minimum now being with respect to the internal co-ordinates in the diagram.

For any two diagrams, for which one can realize

$$I(z, a)\,M^2(a) \geqslant I_0(z, a_0)\,M_0^2(a_0)\,, \tag{A.9}$$

for ony given set of parameters a, it follows that

$$\left\{\begin{array}{l} IM^2 \geqslant I_0 M_0^2 \geqslant \text{Min}_{a_0}\, I_0 M_0^2 = \mathcal{J}_0(z)\,, \\ \text{Min}_a\, IM^2 = \mathcal{J}(z) \geqslant \mathcal{L}_0(z)\,. \end{array}\right. \tag{A.10}$$

Conversely, if $\mathcal{J}(z) \leqslant \mathcal{J}_0(z)$, this does not necessarily mean the inequality (A.9), to be valid for fixed a, a_0 and all z. However, for any fixed a and z, there **must be some set $a_0(a, z)$ such that (A.9) is satisfied since otherwise $\mathcal{J}(z) \geqslant \mathcal{J}_0(z)$ would not hold. This means in turn that for any fixed a_r and q_i there is some set $a_0\,(a, q)$ such that**

$$J(a, q)/M^2(a) \geqslant J_0(a_0, q)/M_0^2(a_0)\,,$$

is satisfied. For our present purposes in Sect. 4, this should be sufficient.

RIASSUNTO (*)

Colla teoria delle perturbazioni si studiano le relazioni di dispersione pei fattori di forma. Si dimostrano in particolare tali relazioni per un nucleone reale nella forma usualmente assunta in letteratura. Si dimostra, tuttavia, che nei casi più generali lo spettro di massa comparente nelle relazioni di dispersione dipende dalle masse delle particelle che danno contributo in un modo particolare, differente dalla semplice relazione finora ritenuta esatta. Tale situazione si presenta, per esempio, nel caso di qualche modello d'iperone e di sistemi legati quali il deuterone. Dall'invarianza della teoria rispetto all'inversione dello spazio-tempo si deriva anche una formula riguardante l'interazione degli stati finali da usare in unione alle relazioni di dispersione.

(*) *Traduzione a cura della Redazione.*

PHYSICAL REVIEW VOLUME 117, NUMBER 3 FEBRUARY 1, 1960

Quasi-Particles and Gauge Invariance in the Theory of Superconductivity*

YOICHIRO NAMBU

The Enrico Fermi Institute for Nuclear Studies and the Department of Physics, The University of Chicago, Chicago, Illinois

(Received July 23, 1959)

Ideas and techniques known in quantum electrodynamics have been applied to the Bardeen-Cooper-Schrieffer theory of superconductivity. In an approximation which corresponds to a generalization of the Hartree-Fock fields, one can write down an integral equation defining the self-energy of an electron in an electron gas with phonon and Coulomb interaction. The form of the equation implies the existence of a particular solution which does not follow from perturbation theory, and which leads to the energy gap equation and the quasi-particle picture analogous to Bogoliubov's.

The gauge invariance, to the first order in the external electromagnetic field, can be maintained in the quasi-particle picture by taking into account a certain class of corrections to the charge-current operator due to the phonon and Coulomb interaction. In fact, generalized forms of the Ward identity are obtained between certain vertex parts and the self-energy. The Meissner effect calculation is thus rendered strictly gauge invariant, but essentially keeping the BCS result unaltered for transverse fields.

It is shown also that the integral equation for vertex parts allows homogeneous solutions which describe collective excitations of quasi-particle pairs, and the nature and effects of such collective states are discussed.

1. INTRODUCTION

A NUMBER of papers have appeared on various aspects of the Bardeen-Cooper-Schrieffer[1] theory of superconductivity. On the whole, the BCS theory, which leads to the existence of an energy gap, presents us with a remarkably good understanding of the general features of superconducivity. A mathematical formulation based on the BCS theory has been developed in a very elegant way by Bogoliubov,[2] who introduced coherent mixtures of particles and holes to describe a superconductor. Such "quasi-particles" are not eigenstates of charge and particle number, and reveal a very bold departure, inherent in the BCS theory, from the conventional approach to many-fermion problems. This, however, creates at the same time certain theoretical difficulties which are matters of principle. Thus the derivation of the Meissner effect in the original BCS theory is not gauge-invariant, as is obvious from the viewpoint of the quasi-particle picture, and poses a serious problem as to the correctness of the results obtained in such a theory.

This question of gauge invariance has been taken up by many people.[3] In the Meissner effect one deals with a linear relation between the Fourier components of the external vector potential A and the induced current J, which is given by the expression

$$J_i(q)=\sum_{j=1}^{3} K_{ij}(q)A_j(q),$$

with

$$K_{ij}(q)=-\frac{e^2}{m}\langle 0|\rho|0\rangle\delta_{ij}+\sum_n\left(\frac{\langle 0|j_i(q)|n\rangle\langle n|j_j(-q)|0\rangle}{E_n}+\frac{\langle 0|j_j(-q)|n\rangle\langle n|j_i(q)|0\rangle}{E_n}\right). \quad (1.1)$$

ρ and j are the charge-current density, and $|0\rangle$ refers to the superconductive ground state. In the BCS model, the second term vanishes in the limit $q\rightarrow 0$, leaving the first term alone to give a nongauge invariant result. It has been pointed out, however, that there is a significant difference between the transversal and longitudinal current operators in their matrix elements. Namely, there exist collective excited states of quasi-particle pairs, as was first derived by Bogoliubov,[2] which can be excited only by the longitudinal current.

As a result, the second term does not vanish for a longitudinal current, but cancels the first term (the longitudinal sum rule) to produce no physical effect; whereas for a transversal field, the original result will remain essentially correct.

If such collective states are essential to the gauge-invariant character of the theory, then one might argue that the former is a necessary consequence of the latter. But this point has not been clear so far.

Another way to understand the BCS theory and its problems is to recognize it as a generalized Hartree-Fock approximation.[4] We will develop this point a little further here since it is the starting point of what follows later as the main part of the paper.

* This work was supported by the U. S. Atomic Energy Commission.

1 Bardeen, Cooper, and Schrieffer, Phys. Rev. **106**, 162 (1957); **108**, 1175 (1957).

2 N. N. Bogoliubov, J. Exptl. Theoret. Phys. U.S.S.R. **34**, 58, 73 (1958) [translation: Soviet Phys. **34**, 41, 51 (1958)]; Bogoliubov, Tolmachev, and Shirkov, *A New Method in the Theory of Superconductivity* (Academy of Sciences of U.S.S.R., Moscow, 1958). See also J. G. Valatin, Nuovo cimento **7**, 843 (1958).

3 M. J. Buckingam, Nuovo cimento **5**, 1763 (1957). J. Bardeen, Nuovo cimento **5**, 1765 (1957). M. R. Schafroth, Phys. Rev. **111**, 72 (1958). P. W. Anderson, Phys. Rev. **110**, 827 (1958); **112**, 1900 (1958). G. Rickayzen, Phys. Rev. **111**, 817 (1958); Phys. Rev. Letters **2**, 91 (1959). D. Pines and R. Schrieffer, Nuovo cimento **10**, 496 (1958); Phys. Rev. Letters **2**, 407 (1958). G. Wentzel, Phys. Rev. **111**, 1488 (1958); Phys. Rev. Letters **2**, 33 (1959). J. M. Blatt and T. Matsubara, Progr. Theoret. Phys. (Kyoto) **20**, 781 (1958). Blatt, Matsubara, and May, Progr. Theoret. Phys. (Kyoto) **21**, 745 (1959). K. Yosida, *ibid.* **731**.

4 Recently N. N. Bogoliubov, Uspekhi Fiz. Nauk **67**, 549 (1959) [translation: Soviet Phys.—Uspekhi **67**, 236 (1959)], has also reformulated his theory as a Hartree-Fock approximation, and discussed the gauge invariance collective excitations from this viewpoint. The author is indebted to Prof. Bogoliubov for sending him a preprint.

Take the Hamiltonian in the second quantization form for electrons interacting through a potential V:

$$H=\int\sum_{i=1}^{2}\psi_i^+(x)K_i\psi_i(x)d^3x+\tfrac{1}{2}\int\int\sum_{i,k}\psi_i^+(x)\times\psi_k^+(y)V(x,y)\psi_k(y)\psi_i(x)d^3xd^3y$$
$$\equiv H_0+H_{\text{int}}. \tag{1.2}$$

K is the kinetic energy plus any external field. $i=1, 2$ refers to the two spin states (e.g., spin up and down along the z axis).

The Hartree-Fock method is equivalent to linearizing the interaction H_{int} by replacing bilinear products like $\psi_i^+(x)\psi_k(y)$ with their expectation values with respect to an approximate wave function which, in turn, is determined by the linearized Hamiltonian. We may consider also expectation values $\langle\psi_i(x)\psi_k(y)\rangle$ and $\langle\psi_i^+(x)\psi_k^+(y)\rangle$ although they would certainly be zero if the trial wave function were to represent an eigenstate of the number of particles, as is the case for the true wave function.

We write thus a linearized Hamiltonian

$$H_0'=\int\sum_i\psi_i^+K_i\psi_id^3x+\int\int\sum_{i,k}[\psi_i^+(x)\chi_{ik}(xy)\psi_k(y)+\psi_i^+(x)\phi_{ik}(xy)\psi_k^+(y)+\psi_k(x)\phi_{ki}^+(xy)\psi_i(y)]d^3xd^3y$$
$$\equiv H_0+H_s, \tag{1.3}$$

where

$$\chi_{ik}(xy)=\delta_{ik}\delta^3(x-y)\int V(xz)\sum_j\langle\psi_j^+(z)\psi_j(z)\rangle d^3z-V(xy)\langle\psi_k^+(y)\psi_i(x)\rangle, \tag{1.4}$$

$$\phi_{ik}(xy)=\tfrac{1}{2}V(xy)\langle\psi_k(y)\psi_i(x)\rangle,$$

$$\phi_{ik}^+(xy)=\tfrac{1}{2}V(xy)\langle\psi_k^+(y)\psi_i^+(x)\rangle.$$

We diagonalize H_0' and take, for example, the ground-state eigenfunction which will be a Slater-Fock product of individual particle eigenfunctions. The defining equations (1.4) then represent just generalized forms of Hartree-Fock equations to be solved for the self-consistent fields χ and ϕ.

The justification of such a procedure may be given by writing the original Hamiltonian as

$$H=(H_0+H_s)+(H_{\text{int}}-H_s)\equiv H_0'+H_{\text{int}}',$$

and demanding that H_{int}' shall have no matrix elements which would cause single-particle transitions; i.e., no matrix elements which would effectively modify the starting H_0': to put it more precisely, we demand our approximate eigenstates to be such that

$$\langle n|H_{\text{int}}'|0\rangle=\langle n|H|0\rangle=0, \tag{1.5}$$

if in $|n\rangle$ more than one particle change their states from those in $|0\rangle$. This condition is contained in Eq. (1.4).[5] Since in many-body problems, as in relativistic field theory, we often take a picture in which particles and holes can be created and annihilated, the condition (1.5) should also be interpreted to include the case where $|n\rangle$ and $|0\rangle$ differ only by such pairs. The significance of the BCS theory lies in the recognition that with an essentially attractive interaction V, a nonvanishing ϕ is indeed a possible solution, and the corresponding ground state has a lower energy than the normal state. It is also separated from the excited states by an energy gap $\sim 2\phi$.

The condition (1.5) was first invoked by Bogoliubov[2] in order to determine the transformation from the ordinary electron to the quasi-particle representation. He derived this requirement from the observation that H_{int}' contains matrix elements which spontaneously create virtual pairs of particles with opposite momenta, and cause the breakdown of the perturbation theory as the energy denominators can become arbitrarily small. Equation (1.5), as applied to such pair creation processes, determines only the nondiagonal part (in quasi-particle energy) of H_s in the representation in which H_0+H_s is diagonal. The diagonal part of H_s is still arbitrary. We can fix it by requiring that

$$\langle 1'|H_{\text{int}}'|1\rangle=0, \tag{1.6}$$

namely, the vanishing of the diagonal part of H_{int} for the states where one more particle (or hole) having a Hamiltonian H_0' is added to the ground state. In this way we can interpret H_0' as describing single particles (or excitations) moving in the "vacuum," and the diagonal part of H_s represents the self-energy (or the Hartree potential) for such particles arising from its interaction with the vacuum.

The distinction between Eqs. (1.5) and (1.6) is not so clear when applied to normal states. On the one hand, particles and holes (negative energy particles) are not separated by an energy gap; on the other hand, there is little difference when one particle is added just above the ground state.

In the above formulation of the generalized Hartree fields, χ and ϕ will in general depend on the external field as well as the interaction between particles. There is a complication due to the fact that they are gauge dependent. This is because a phase transformation $\psi_i(x)\rightarrow e^{i\lambda(x)}\psi_i(x)$ applied on Eq. (1.3) will change χ and ϕ according to

$$\chi(xy)\rightarrow e^{-i\lambda(x)+i\lambda(y)}\chi(xy),$$
$$\phi(xy)\rightarrow e^{i\lambda(x)+i\lambda(y)}\phi(xy), \tag{1.6}$$
$$\phi^+(xy)\rightarrow e^{-i\lambda(x)-i\lambda(y)}\phi^+(xy).$$

[5] Equation (1.5) refers only to the transitions from occupied states to unoccupied states. Transitions between occupied states or unoccupied states are given by Eq. (1.6). These two together then are equivalent to Eq. (1.4). For the analysis of the Hartree approximation in terms of diagrams, see J. Goldstone, Proc.

It is especially serious for ϕ (and ϕ^+) since, even if $\phi(xy)=\delta^3(x-y)$ times a constant in some gauge, it is not so in other gauges. Therefore, unless we can show explicitly that physical quantities do not depend on the gauge, any calculation based on a particular ϕ is open to question. It would not be enough to say that a longitudinal electromagnetic potential produces no effect because it can be transformed away before making the Hartree approximation. A natural way to reconcile the existence of ϕ, which we want to keep, with gauge invariance would be to find the dependence of ϕ on the external field explicitly. If the gauge invariance can be maintained, the dependence must be such that for a longitudinal potential $A=-\text{grad}\lambda$, it reduces to Eq. (1.6). This should not be done in an arbitrary manner, but by studying the actual influence of H_{int} on the primary electromagnetic interaction when ϕ is first determined without the external field.

After these preliminaries, we are going to study the points raised here by means of the techniques developed in quantum electrodynamics. We will first develop the Feynman-Dyson formulation adapted to our problem, and write down an integral eauation for the self-energy part which corresponds to the Hartree approximation. It is observed that it can possess a nonperturbational solution, and the existence of an energy gap is immediately recognized.

Next we will introduce external fields. Guided by the well-known theorems about gauge invariance, we are led to consider the so-called vertex parts, which include the "radiative corrections" to the primary charge-current operator. When an integral equation for the general vertex part is written down, certain exact solutions are obtained in terms of the assumed self-energy part, leading to analogs of the Ward identity.[6] They are intimately related to inherent invariance properties of the theory. Among other things, the gauge invariance is thus strictly established insofar as effects linear in the external field are concerned, including the Meissner effect.

Later we look into the collective excitations. A very interesting result emerges when we observe that one of the exact solutions to the vertex part equations becomes a homogeneous solution if the external energy-momentum is zero, and expresses a bound state of a pair with zero energy-momentum. Then by perturbation, other bound states with nonzero energy-momentum are obtained, and their dispersion law determined. Thus the existence of the bound state is a logical consequence of the existence of the special self-energy ϕ and the gauge invariance, which are seemingly contradictory to each other.

When the Coulomb interaction is taken into account, the bound pair states are drastically modified, turning into the plasma modes due to the same mechanism as in the normal case. This situation will also be studied.

Roy. Soc. (London) A239, 267 (1957). Compare also T. Kinoshita and Y. Nambu, Phys. Rev. 94, 598 (1953).

[6] J. C. Ward, Phys. Rev. 78, 182 (1950).

2. FEYNMAN-DYSON FORMULATION

We start from the Lagrangian for the electron-phonon system, which is supposed to be uniform and isotropic.[7]

$$\mathcal{L}=\sum_p\sum_i[i\dot{\psi}_i^+(p)\psi_i(p)-\psi_i^+(p)\epsilon_p\psi_i(p)]+\sum_k\tfrac{1}{2}[\dot{\varphi}(k)\dot{\varphi}(-k)-c^2\varphi(k)\varphi(-k)]-g\frac{1}{\sqrt{\mathcal{V}}}\sum_{p,k}\psi_i^+(p+k)\psi_i(p)h(k)\varphi(k). \quad (2.1)$$

p is the phonon field, with the momentum k (energy $\omega_k=ck$) running up to a cutoff value $k_m(\omega_m)$; c is the phonon velocity. ϵ_p is the electron kinetic energy relative to the Fermi energy; $gh(k)$ represents the strength of coupling.[8] ($\mathcal{V}$ is the volume of the system.)

The Coulomb interaction between the electrons is not included for the moment in order to avoid complication. Later we will make remarks whenever necessary about the modifications when the Coulomb interaction is taken into account.

It will turn out to be convenient to introduce a two-component notation[9] for the electrons

$$\Psi(x)=\begin{pmatrix}\psi_1(x)\\ \psi_2^+(x)\end{pmatrix} \quad \text{or} \quad \Psi(p)=\begin{pmatrix}\psi_1(p)\\ \psi_2^+(-p)\end{pmatrix}, \quad (2.2)$$

and the corresponding 2×2 Pauli matrices

$$\tau_1=\begin{pmatrix}0&1\\1&0\end{pmatrix},\quad \tau_2=\begin{pmatrix}0&-i\\i&0\end{pmatrix},\quad \tau_3=\begin{pmatrix}1&0\\0&-1\end{pmatrix}. \quad (2.3)$$

The Lagrangian then becomes:

$$\begin{aligned}\mathcal{L}&=\sum_p\Psi^+(p)\left(i\frac{\partial}{\partial t}-\epsilon_p\tau_3\right)\Psi(p)+\sum_k\tfrac{1}{2}[\dot{\varphi}(k)\dot{\varphi}(-k)-c^2\varphi(k)\varphi(-k)]\\&\quad-g\frac{1}{\sqrt{\mathcal{V}}}\sum_{p,k}\Psi^+(p+k)\tau_3\Psi(p)h(k)\varphi(k)+\sum_p\epsilon_p\\&=\mathcal{L}_0+\mathcal{L}_{\text{int}}+\text{const.}\end{aligned}$$

The last infinite c-number term comes from the rearrangement of the kinetic energy term. This is certainly uncomfortable, but will not be important except for the calculation of the total energy.

The fields obey the standard commutation relations.

[7] We use the units $\hbar=1$.

[8] For convenience, we have included in $h(k)$ the frequency factor: $h(k)=h_1(k)k_0$.

[9] P. W. Anderson [Phys. Rev. 112, 1900 (1958)], has also introduced this two-component wave function.

Especially for Ψ, we have

$$\{\Psi_i(x),\Psi_j^+(y)\}\equiv\Psi_i(x)\Psi_j^+(y)+\Psi_j^+(y)\Psi_i(x)=\delta_{ij}\delta^3(x-y), \tag{2.5}$$
$$\{\Psi_i(p),\Psi_j^+(p')\}=\delta_{ij}\delta_{pp'}.$$

We may now formally treat $H_{\rm int}$ as perturbation, using the formulation of Feynman and Dyson.[10] The unperturbed ground state (vacuum) is then the state where all individual electron states $\epsilon_p<0(>0)$ are occupied (unoccupied) in the representation where $\psi_i^+(p)\psi_i(p)$ is the occupation number.

Having defined the vacuum, the time-ordered Green's functions for free electrons and phonons

$$\langle T(\Psi_i(xt),\Psi_j^+(x't'))\rangle=[G_0(x-x',\,t-t')]_{ij},$$
$$\langle T(\varphi(xt),\varphi(x't'))\rangle=\Delta_0(x-x',\,t-t') \tag{2.6}$$

are easily determined. We get for their Fourier representation (in the limit $\mathcal{V}\to\infty$)[10a]

$$G_0(xt)=(1/(2\pi)^4)\int G_0(pp_0)e^{ip\cdot x-ip_0t}d^3pdp_0,$$

$$\Delta_0(xt)=\frac{1}{(2\pi)^4}\int_{|k|<k_m}\Delta_0(kk_0)e^{ik\cdot x-ik_0t}d^3kdk_0,$$

$$G_0(pp_0)=i\left[P\frac{1}{p_0-\epsilon_p\tau_3}-i\pi\,\mathrm{sgn}(\tau_3\epsilon_p)\delta(p_0-\tau_3\epsilon_p)\right] \tag{2.7}$$
$$=i(p_0+\epsilon_p\tau_3)/(p_0^2-\epsilon_p^2+i\epsilon),$$
$$\Delta_0(kk_0)=i\left[P\frac{1}{k_0^2-c^2k^2}-i\pi\delta(k_0^2-c^2k^2)\right]$$
$$=i/(k_0^2-c^2k^2+i\epsilon).$$

With the aid of these Green's functions, we are able to calculate the S matrix and other quantities according to a well-defined set of rules in perturbation theory.

We will analyze in particular the self-energies of the electron and the phonon. In the many-particle system, these energies express (apart from the self-interaction of the electron) the average interaction of a single particle or phonon placed in the medium. Because the phonon spectrum is limited, there will be no ultraviolet divergences, unlike the case of quantum electrodynamics.

These self-energies may be obtained in a perturbation expansion with respect to $H_{\rm int}$. We are, however, interested in the Hartree method which proposes to take account of them in an approximate but nonperturbational way. It is true that the self-energies are in general complex due to the instability of single particles. But to the extent that the single-particle picture makes physical sense, we will ignore the small imaginary part of the self-energies in the following considerations.

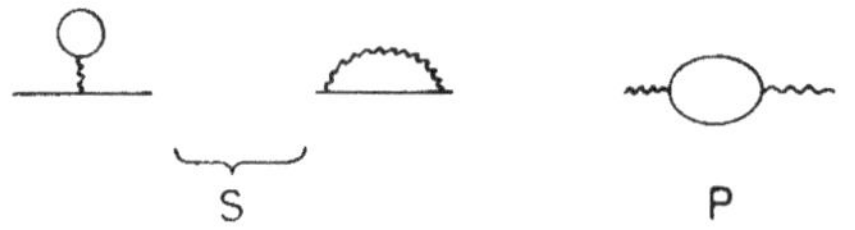

FIG. 1. Second order self-energy diagrams. Solid and curly lines represent electron and phonons, respectively, themselves being under the influence of the self-energies Σ and Π. All diagrams are to be interpreted in the sense of Feynman, lumping together all topologically equivalent processes.

Let us thus introduce the approximate self-energy Lagrangian $\mathcal{L}_s$, and write

$$\mathcal{L}=(\mathcal{L}_0+\mathcal{L}_s)+(\mathcal{L}_{\rm int}-\mathcal{L}_s)\equiv\mathcal{L}_0'+\mathcal{L}_{\rm int}',$$
$$\mathcal{L}_0=\sum_p\Psi_p^+L_0\Psi_p+\sum_k\tfrac{1}{2}\varphi_kM_0\varphi_{-k}, \tag{2.8}$$
$$\mathcal{L}_s=-\sum_p\Psi_p^+\Sigma\Psi_p-\sum_k\tfrac{1}{2}\varphi_k\Pi\varphi_{-k},$$
$$L_0-\Sigma\equiv L,\quad M_0-\Pi\equiv M.$$

The free electrons with "spin" functions u and phonons obey the dispersion law

$$L_0(\mathbf{p},\,p_0=\epsilon_p)u_p=0,\quad M_0(\mathbf{k},\,k_0=\omega_k)=0, \tag{2.9}$$

whereas they obey in the medium

$$L(\mathbf{p},\,p_0=E_p)u_p=0,\quad M(\mathbf{k},\,k_0=\Omega_k)=0. \tag{2.9'}$$

Σ will be a function of momentum p and "spin." Π will consist of two parts: $\Pi(k_0k)=\Pi_1(k)k_0^2+\Pi_2(k)$ in conformity with the second order character (in time) of the phonon wave equation.[11]

The propagators corresponding to these modified electrons and phonons are

$$G(pp_0)=i/(L(pp_0)+i\,\mathrm{sgn}(p_0)\epsilon),$$
$$\Delta(kk_0)=i/(M(kk_0)+i\epsilon). \tag{2.10}$$

We now determine Σ and Π self-consistently to the second order in the coupling g. Namely the second order self-energies coming from the phonon-electron interaction have to be cancelled by the first order effect of $\mathcal{L}_{\rm int}$.

These second order self-energies are represented by the nonlocal operators[12] (Fig. 1)

$$\mathcal{S}((t+t')/2)=\int\int\int\Psi^+(xt)S(x-x',\,t-t')\times\Psi(x't')d^3xd^3x'd(t-t'),$$
$$\mathcal{P}((t+t')/2)=\tfrac{1}{2}\int\int\int\varphi(xt)P(x-x',\,t-t')\times\varphi(x't')d^3xd^3x'd(t-t'), \tag{2.11}$$

[10] F. J. Dyson, Phys. Rev. 75, 486, 1736 (1949); R. P. Feynman, Phys. Rev. 76, 769 (1949); J. Schwinger, Phys. Rev. 74, 1439 (1948). Although we followed here the perturbation theory of Dyson, there is no doubt that the relations obtained in this paper can be derived by a nonperturbational formulation such as J. Schwinger's: Proc. Natl. Acad. Sci. U. S. 37, 452, 455 (1951).

[10a] P stands for the principal value; $i\epsilon$ in the denominator is a small positive imaginary quantity.

[11] In the same spirit Σ should actually be in the form $\Sigma_1(p)p_0+\Sigma_2(p)$. Here we neglect the renormalization term Σ_1 since the two conditions (2.13) can be met without it.

[12] We use the word nonlocal here for nonlocality in time.

where S and P have the Fourier representation

$$S(pp_0)=-ig^2\tau_3\delta^3(p)\delta(p_0)h^2(0)\Delta(0)\times\int \mathrm{Tr}[\tau_3 G(p'p_0)]d^3p\,dp_0$$
$$-ig^2\int \tau_3 G(p-k,\,p_0-k_0)\tau_3 h^2(kk_0)\times\Delta(kk_0)d^3k\,dk_0, \qquad (2.12)$$
$$P(kk_0)=ig^2h(kk_0)^2\int \mathrm{Tr}[\tau_3 G(pp_0)\times G(p+k,\,p_0+k_0)]d^3p\,dp_0.$$

In Eq. (2.11) we have chosen more or less arbitrarily $(t+t')/2$ as the fixed time to which we refer the nonlocal operators $\mathcal{S}$ and $\mathcal{P}$. The self-consistency requirements (1.5) and (1.6) mean in the present case that Σ, Π must be identical with S, P (a): for the diagonal elements [on the energy shell, Eq. (2.9)], and (b): for the nondiagonal matrix elements for creating a pair out of the vacuum.

The pair creation of electrons is possible because Ψ, being a two-component wave function, can have in general two eigenfunctions u_{ps} $(s=1, 2)$ with different energies E_{ps} for a fixed momentum, p, only one of which is occupied in the ground state.

Thus taking particular plane waves $u_{ps}{}^*e^{-ip\cdot x+ip_0t}$, $u_{p's'}e^{ip'\cdot x'-ip_0't'}$ for Ψ^+ and Ψ in (2.11), we easily find that the diagonal matrix element of Σ corresponds to $u_{ps}{}^*S(p,E_{ps})u_{ps}$, while the nondiagonal part corresponds to $u_{ps}{}^*S(p,0)u_{ps}{}'$, $s\neq s'$ $(p_0'=-p_0)$.

A similar situation holds also for the photon self-energy Π. Since Π consists of two parts, the diagonal and off-diagonal conditions will fix these.

With this understanding, the self-consistency relations may be written

$$\Sigma(\mathbf{p}E_p)_{\mathrm{D}}=S(\mathbf{p},E_p)_{\mathrm{D}},\quad \Sigma(\mathbf{p}0)_{\mathrm{ND}}=S(\mathbf{p}0)_{\mathrm{ND}},$$
$$\Pi(\mathbf{k}\Omega_k)=P(\mathbf{k}\Omega_k),\quad \Pi(\mathbf{k}0)=P(\mathbf{k}0), \qquad (2.13)$$

where D, ND signify the diagonal and nondiagonal parts in the "spin" space. As stated before, we have agreed to omit possible imaginary parts in S and P. (The nondiagonal components, however, will turn out to be real.)

Before discussing the general solutions, let us consider the meaning of Eq. (2.13) in terms of perturbation theory. Suppose we expand G occurring in Eq. (2.12), with respect to Σ:

$$G=G_0-iG_0\Sigma G_0-G_0\Sigma G_0\Sigma G_0+\cdots,$$

and expand Σ itself with respect to g^2, then we easily realize that Eq. (2.13) defines an infinite sum of a particular class of diagrams, which are illustrated in Fig. 2. The first term in S of Eq. (2.12) corresponds to the ordinary Hartree potential which is just a constant,

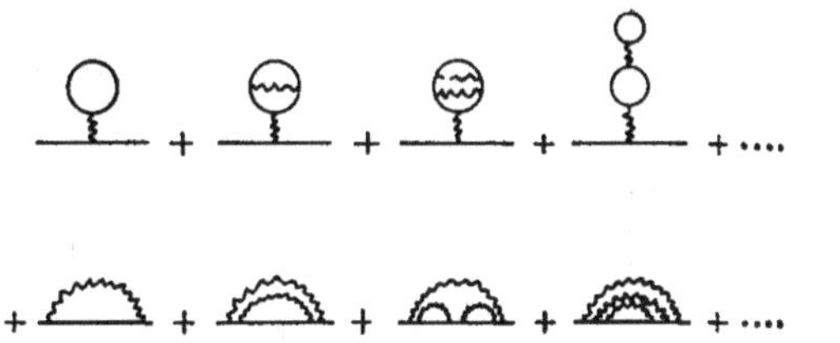

FIG. 2. Expansion of the self-consistent self-energy $\Sigma\sim S$ in terms of bare electron diagrams.

whereas the second term gives an exchange effect. In the latter, the approximation is characterized by the fact that no phonon lines cross each other.

It must be said that the Hartree approximation does not really sum the series of Fig. 2 completely since we equate in Eq. (2.13) only special matrix elements of both sides. For in the perturbation series the Σ obtained to any order is a function of p_0, whereas in Eq. (2.13) it is replaced by a p_0-independent quantity. Hence there will be a correction left out in each order (analogous to the radiative correction after mass renormalization in quantum electrodynamics).

In this perturbation expansion, S in Eq. (2.13) is always proportional to τ_3 on the energy shell since $H_0\propto\tau_3$. Accordingly Σ will be $\propto\tau_3$ and commute with H_0, so that no off-diagonal part exists.[11]

It is important, however, to note the possibility of a nonperturbational solution by assuming that Σ contains also a term proportional to τ_1 or τ_2. Thus, take

$$\Sigma(p)=\chi(p)\tau_3+\phi(p)\tau_1,$$
$$H_0'=(\epsilon+\chi)\tau_3+\phi\tau_1 \qquad (2.14)$$
$$\equiv\bar{\epsilon}\tau_3+\phi\tau_1.$$

This form bears a resemblance to the Dirac equation. Its eigenvalues are

$$E=\pm E_p\equiv\pm(\bar{\epsilon}_p{}^2+\phi_p{}^2)^{\frac{1}{2}}. \qquad (2.15)$$

Since H_0' describes by definition excited states, we have to adopt the hole picture and conclude that the ground state (vacuum) is the state where all negative energy "quasi-particles" ($E<0$) are occupied and no positive energy particles exist. If ϕ remains finite on the Fermi surface, the positive and negative states are separated by a gap $\sim2|\phi|$. The corresponding Green's function G now has the representation

$$G(pp_0)=i\frac{p_0+\bar{\epsilon}_p\tau_3+\phi_p\tau_1}{p_0{}^2-E_p{}^2+i\epsilon}. \qquad (2.16)$$

In order to extract the diagonal and nondiagonal parts in spin space, we will use the trick

$$O_{\mathrm{D}}=\tfrac{1}{2}\,\mathrm{Tr}\,(\Lambda O),$$
$$O_{\mathrm{ND}}=-(i/2)\,\mathrm{Tr}\,(\Lambda O\tau_2), \qquad (2.17)$$
$$\Lambda=[E_p+H_0'(p)]/2E_p.$$

Applying this to Eq. (2.13a) with Eqs. (2.12), (2.14), and (2.15), we finally obtain the following equations for χ and ϕ

$$\frac{\epsilon_p\chi_p+\phi_p^2}{E_p}=\frac{g^2\pi}{(2\pi)^4}P\int\left[\frac{E_p}{\Omega_k}+\frac{E_{p-k}+\Omega_k}{E_pE_{p-k}\Omega_k}\right.$$
$$\left.\times(\bar{\epsilon}_p\bar{\epsilon}_{p-k}-\phi_p\phi_{p-k})\right]\frac{h(k)^2}{E_p^2-(E_{p-k}-\Omega_k)^2}d^3k,$$
$$\epsilon_p\phi_p=\frac{g^2\pi}{(2\pi)^4}\int(\bar{\epsilon}_{p-k}\phi_p+\bar{\epsilon}_p\phi_{p-k})$$
$$\times\frac{h(k)^2d^3k}{E_{p-k}\Omega_k(E_{p-k}+\Omega_k)}. \quad (2.18)$$

The second equation, coming from the nondiagonal condition, has a trivial solution $\phi=0$. If a finite solution ϕ exists, it cannot follow from perturbation treatment since there is no inhomogeneous term to start with.

Equation (2.18) is equivalent to, but slightly different from, the corresponding conditions of Bogoliubov because of a slightly different definition of the nondiagonal part of the self-energy operator, which is actually due to an inherent ambiguity in approximating nonlocal operators by local ones. (This is the same kind of ambiguity as one encounters in the derivation of a potential from field theory. The difference between the local operator Σ and the nonlocal one S shows up in a situation like that in Fig. 3, and the compensation between Σ and S is not complete.) We may avoid this unpleasant situation, by extending the Hartree self-consistency conditions to all virtual matrix elements, but this would mean that ϕ (and χ) must be treated as nonlocal. We will discuss this situation in a separate section since such a generalization brings simplification in dealing with the problem of gauge invariance and collective excitations.

For the moment we consider the second equation of (2.18) and rewrite it

$$\phi_p=A_p\frac{g^2\pi}{(2\pi)^4}\int\frac{\phi_{p-k}}{E_{p-k}}\frac{h(k)d^3k}{\Omega_k(E_{p-k}+\Omega_k)},$$
$$A_p=\bar{\epsilon}_p\bigg/\left(\epsilon_p-\frac{g^2\pi}{(2\pi)^4}\int\frac{\bar{\epsilon}_ph(k)^2d^3k}{E_{p-k}\Omega_k(E_{p-k}+\Omega_k)}\right). \quad (2.19)$$

This is essentially the energy gap equation of BCS if $g^2A_ph(k)^2/\Omega_k(E_{p-k}+\Omega_k)$ is identified with the effective interaction potential V, and if $\bar{\epsilon}_p\sim\epsilon_p(\chi_p\sim 0)$. It has a solution

$$\phi\sim\Omega_m\exp(-1/VN),$$

if $VN \ll 1$, N being the density of states: $N=dn/d\epsilon_p$ on the Fermi surface.

The phonon self-energy Π may be studied similarly from Eq. (2.13), which should determine the renormalization of the phonon field. It does not play an essential role in superconductivity, though it gives rise to an important correction when the Coulomb effect is taken into account. (See the following section.)

FIG. 3. An example of the situation where the cancellation of $\Sigma_{\rm ND}$ *versus* $S_{\rm ND}$ is not complete. The two self-energy parts overlap in time, and their centers of time t_1 and t_2 are such that $t_1>t_2$. If calculated according to the usual perturbation theory, this process will not be eliminated by the condition $\Sigma_{\rm ND}=(S_1)_{\rm ND}$.

From the nature of Eq. (2.12), it is clear that $\tau_1\phi$ can actually be pointed in any direction in the 1–2 plane of the τ space: $\tau_1\phi_1+\tau_2\phi_2$. It was thus sufficient to take $\phi_1\neq 0$, $\phi_2=0$. Any other solution is obtained by a transformation

$$\Psi\rightarrow\exp(i\alpha\tau_3/2)\Psi,$$
$$(\phi,0)\rightarrow(\phi\cos\alpha,\phi\sin\alpha). \quad (2.20)$$

In view of the definition of Ψ, Eq. (2.20) is a gauge transformation with a constant phase. Thus the arbitrariness in the direction of ϕ is the 1–2 plane is a reflection of the gauge invariance.

For later use, we also mention here the particle-antiparticle conjugation C of the quasi-particle field Ψ. This is defined by

$$C:\quad \Psi\rightarrow\Psi^C=C\Psi^+=\tau_2\Psi^+,$$

or

$$\begin{pmatrix}\psi_1{}^C\\ \psi_2{}^{+C}\end{pmatrix}=\begin{pmatrix}-i\psi_2\\ i\psi_1{}^+\end{pmatrix}, \quad (2.21)$$

and changes quasi-particles of energy-momentum (p_0,p) into holes of energy-momentum $(-p_0,-p)$, or interchanges up-spin and down-spin electrons. Under C, the τ operators transform as

$$C:\quad \tau_i\rightarrow C^{-1}\tau_iC=-\tau_i{}^T,\quad i=1,2,3 \quad (2.22)$$

where T means transposition.

As a consequence, we have also

$$C:\quad L(p)\rightarrow L^C(-p)=-L(-p)^T. \quad (2.23)$$

Finally we make a remark about the Coulomb interaction. When this is taken into account, the phonon interaction factor $g^2h(k)^2\Delta(k,k_0)$ in Eq. (2.12a) has to be replaced by

$$[g^2h(k)^2\Delta(kk_0)+ie^2/k^2]/$$
$$\{1-i\Pi(kk_0)[\Delta(kk_0)+ie^2/g^2h(k)^2k^2]\}.$$

As is well known, the denominator represents the screening of the Coulomb interaction. Discussion about this point will be made later in connection with the plasma oscillations.

3. NONLOCAL (ENERGY-DEPENDENT) SELF-CONSISTENCY CONDITIONS

In the last section we remarked that the self-consistency conditions Eq. (2.13) may be extended to all virtual matrix elements, namely, not only on the energy shell (diagonal) and for the virtual pair creation out of the vacuum, but also for the self-energy effects which appear in intermediate states of any process.

This simply means that ϕ and π are now nonlocal; i.e., depend both on energy and momentum arbitrarily, and are to be completely equated with S and P, respectively,

$$\Sigma(pp_0)=S(pp_0),\quad \Pi(kk_0)=P(kk_0). \tag{3.1}$$

Actually, these self-energies can no more be incorporated in H_0' as the zeroth order Lagrangian since they contain infinite orders of time derivatives.[13] Nevertheless, Eq. (3.1) has a precise meaning in the bare particle perturbation theory. It defines the (proper) self-energy parts (in the sense of Dyson) as an infinite sum of the special class of diagrams illustrated in Fig. 2.

The earlier condition of Eq. (2.13) represented, as was noted there, only an approximation to this sum. In other words, Eqs. (2.13) and (3.1) are not exactly identical even on the energy shell.

The Hartree-Fock approximation based on Eq. (3.1) could be interpreted as a nonperturbation approximation to determine the "dressed" single particles (together with the "dressed vacuum") or the Green's function $\langle 0|T(\Psi(xt),\Psi^+(x't'))|0\rangle$ for the true interacting system. Such single particles will satisfy

$$L(p,p_0)u\cong 0,\quad M(k,k_0)\cong 0. \tag{3.2}$$

We use the approximate equality since a really stable single particle may not exist.

Let us assume that these determine the approximate renormalized dispersion law

$$p_0^2=E_r(p)^2,\quad k_0^2=\Omega_r(k)^2. \tag{3.3}$$

If we write for Σ

$$\Sigma(pp_0)=p_0\zeta(pp_0)+\chi(pp_0)\tau_3+\phi(pp_0)\tau_1, \tag{3.4}$$

where ζ, χ, ϕ are even functions of p_0, then

$$\begin{aligned}E_r^2(p)&=[\bar{\epsilon}(pp_0)^2+\phi(pp_0)^2]/[1-\zeta(pp_0)]^2|_{p_0^2=E_r(p)^2}\\&\equiv E(pp_0)^2/Z(pp_0)^2|_{p_0^2=E_r(p)^2}.\end{aligned} \tag{3.5}$$

The Green's functions G and Δ will be given by

$$\begin{aligned}G(pp_0)&=i/L(pp_0)\\&=i\int_0^\infty \frac{dx}{p_0^2-x+i\epsilon}\\&\quad\times\mathrm{Im}\frac{p_0Z(px)+\bar{\epsilon}(px)\tau_3+\phi(px)\tau_1}{x^2Z(px)^2-E(px)^2},\end{aligned} \tag{3.6}$$

$$\begin{aligned}\Delta(kk_0)&=i/M(kk_0)\\&=i\int_0^\infty\frac{dx}{k_0^2-x+i\epsilon}\,\mathrm{Im}\frac{1}{M(kx)}.\end{aligned}$$

This representation assumes that $G(p_0)[\Delta(k_0)]$ is analytic except for a branch cut on the real axis. The imaginary part in the integrand is expected to have a delta function or a sharp peak at $x=E_r^2(p)$ $[\Omega_r(k^2)]$. These properties are necessary in order that the vacuum is stable and the quasi-particles and phonons have a valid physical meaning as excitations.[14] In the following, we will generally consider this quasi-particle peak only, and write

$$G(pp_0)=i\frac{p_0Z(pp_0)+\bar{\epsilon}(pp_0)\tau_3+\phi(pp_0)\tau_1}{p_0Z(pp_0)^2-E(pp_0)^2+i\epsilon},\quad\text{etc.}$$

The Hartree equations now take the form

$$\Sigma(pp_0)=-i\frac{g^2}{(2\pi)^4}\int\tau_3G(p-k,\,p_0-k_0)\tau_3h(kk_0)^2d^3kdk_0,$$

$$\begin{aligned}\Pi(kk_0)=i\frac{g^2}{(2\pi)^4}\int\mathrm{Tr}\,[&\tau_3G(k-p,\,k_0-p_0)\\&\times\tau_3G(pp_0)]d^3pdp_0.\end{aligned} \tag{3.7}$$

This equation for Σ is much simpler than the previous one (2.18) since we may just equate the coefficients of 1, τ_3, τ_1 on both sides. In particular, we get the energy gap equation

$$\begin{aligned}\phi(pp_0)=-\frac{ig^2}{(2\pi)^4}\int&\frac{\phi(p'p_0')}{p_0^2Z(p'p_0')^2-E(p'p_0')^2+i\epsilon}\\&\times h(\mathbf{p}-\mathbf{p}',\,p_0-p_0')^2\Delta(\mathbf{p}-\mathbf{p}',\,p_0-p_0')d^3p'dp_0',\end{aligned} \tag{3.8}$$

which is to be compared with Eq. (2.19).

Although the existence of a solution to Eq. (3.6) may be difficult to establish, the solution, if it exists, should not be much different from the older solution to Eq. (2.19). At any rate, our assumption about the analyiticity of G and Π is consistent with Eq. (3.6) or (3.7) which implies that Σ and Π are also analytic except for a cut on the real axis.

In later calculations we shall encounter various integrals which we may classify into three types regarding their sensitivity to the energy gap. First, a normal self-energy part, for example, represents the effect of the bulk of the surrounding electrons on a particular electron, and is insensitive to the change of the small fraction $\sim\phi/E_F$ of the electrons near the Fermi surface in a superconductor. Such a quantity is

[13] It would seem then that we lose the advantage of the generalization since we cannot find the Bogoliubov transformation. However, we could still start from the older solution (2.13) as the zeroth approximation to Eq. (3.1), and then calculate the correction; namely, the "radiative" correction to the Bogoliubov vacuum and the Bogoliubov quasi-particle. These corrections would take account of the single-particle transitions which remain after the Bogoliubov condition (2.13) is imposed.

[14] This is a representation of the Lehmann type [H. Lehmann, Nuovo cimento 11, 342 (1954)] which can be derived by defining the Green's functions in terms of Heisenberg operators. See also V. M. Galizkii and A. B. Migdal, J. Exptl. Theoret. Phys. U.S.S.R. 34, 139 (1958) [translation: Soviet Phys. JETP 7, 96 (1958)].

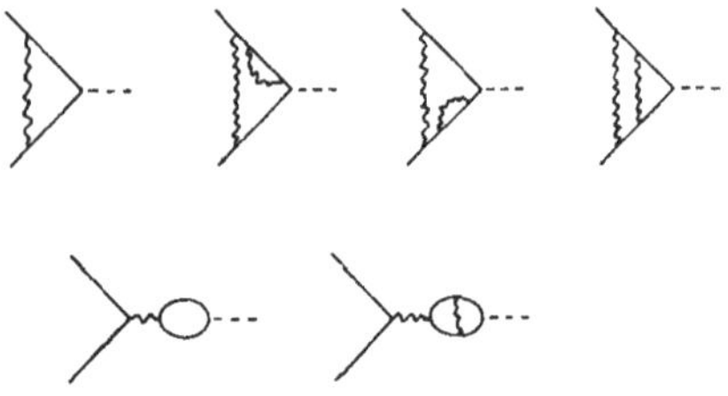

FIG. 4. Construction of the vertex part Γ in bare particle picture. The second line represents the polarization diagrams.

given by an integral like

$$g^2\int \frac{\epsilon_k}{E_k} f(\mathbf{p}-\mathbf{k})d^3k, \tag{3.9}$$

where the region $\epsilon_k \lesssim E_k = (\epsilon_k^2+\phi^2)^{\frac{1}{2}}$ makes little contribution if $f(p-k)$ is a smooth function.

Second, the energy gap itself is determined from an equation of the form

$$g^2\int \frac{d^3k}{E_k} f(\mathbf{p}-\mathbf{k}) \sim g^2 \int_{E_k \lesssim \omega_m} \frac{d^3k}{E_k} f(\mathbf{p}-\mathbf{k}) \sim 1, \tag{3.10}$$

which means that even if g^2 is small, such an expression is always of the order 1.

Finally we meet with integrals like

$$g^2\int \frac{\epsilon_k\phi}{E_k^3} f(\mathbf{p}-\mathbf{k})d^3k, \quad g^2\int \frac{\phi^2}{E_k^3} f(\mathbf{p}-\mathbf{k})d^3k, \quad \text{etc.} \tag{3.11}$$

They have an extra cutoff factor $\sim 1/E$, $1/E^2$, etc., in the integrand which restricts the contribution to an energy interval $\sim 2\phi$ near the Fermi surface. The integrals are thus of the order

$$g^2N\phi/\omega_m, \quad g^2N, \quad \text{etc.}$$

In the following, we will not be primarily concerned with the ordinary self-energy effects. We will assume that proper renormalization has been carried out, or else simply disregard it unless essential. When we carry out perturbation type calculations, we will arrange things so that quantities of the second type are taken into account rigorously, and treat quantities of the third type as small, and hence negligible ($g^2N \ll 1$).

4. INTEGRAL EQUATIONS FOR VERTEX PARTS[15]

In the presence of an electromagnetic potential, the original Lagrangian $\mathfrak{L}$ has to be modified according to the rule

$$i\frac{\partial}{\partial t} \to i\frac{\partial}{\partial t} + eA_0, \quad \mathbf{p} \to \mathbf{p} - \frac{e}{c}\mathbf{A}$$

for the electron. Going to the two-component representation, this corresponds to the prescription

$$i\frac{\partial}{\partial t} \to i\frac{\partial}{\partial t} + e\tau_3 A_0, \quad \mathbf{p} \to \mathbf{p} - \frac{e}{c}\tau_3\mathbf{A} \tag{4.1}$$

acting on Ψ. It can also be inferred from the gauge transformation $\Psi \to \exp(i\alpha\tau_3)\Psi$ as was observed previously. So the ordinary charge-current operator turns out to be in our form given by

$$\rho(x) = \frac{e}{2}([\Psi^+(x), \tau_3\Psi(x)] + \{\Psi^+(x), \Psi(x)\}),$$

$$\begin{aligned} \mathbf{j}(x) = \frac{-ie}{4m}([\Psi^+(x), (\nabla - ie\tau_3\mathbf{A})\Psi(x)] \\ +[(-\nabla - ie\tau_3 A)\Psi^+(x), \Psi(x)] \\ +\{\Psi^+(x), (\nabla\tau_3 - ieA)\Psi(x)\} \\ -\{(\tau_3\nabla - ie\mathbf{A})\Psi^+(x), \Psi(x)\}). \end{aligned} \tag{4.2}$$

The second terms on the right-hand side, being infinite C numbers, arise from the rearrangement of ψ and ψ^+, and will actually be compensated for by the first terms.

This expression, however, has to be modified when we go to the quasi-particle picture.

For we have seen that the self-energy ϕ of a quasi-particle is a gauge-dependent quantity. If we want to have the quasi-particle picture and gauge invariance at the same time, then it is clear that the electromagnetic current of a quasi-particle must contain, in addition to the normal terms given by Eq. (4.2), terms which would cause a physically unobservable transformation of ϕ if the electromagnetic potential is replaced by the gradient of a scalar. In other words, the complete charge current of a quasi-particle has to satisfy the continuity equation, which Eq. (4.2) does not, since

$$\partial\rho/\partial t + \nabla\cdot\mathbf{j} = 2\Psi^+\phi\tau_2\Psi.$$

In order to find such a conserving expression for charge current, it is instructive to go back to the bare electron picture, in which the self-energy is represented by a particular class of diagrams discussed in the previous sections.

It is well known[16] in quantum electrodynamics that, in any process involving electromagnetic interaction, perturbation diagrams can be grouped into gauge-invariant subsets, such that the invariance is maintained by each subset taken as a whole. Such a subset can be constructed by letting each photon line in a diagram interact with a charge of all possible places along a chain of charge-carrying particle lines. The gauge-invariant interaction of a quasi-particle with an electromagnetic potential should then be obtained by attaching a photon line at all possible places in the diagrams of Fig. 2. The result is illustrated in Fig. 4,

[15]Hereafter we will often use the four-dimensional notation $x = (\mathbf{x}, t)$, $p = (\mathbf{p}, p_0)$, $d^4p = d^3p\,dp_0$.

[16] Z. Koba, Progr. Theoret. Phys. (Kyoto) 6, 322 (1951).

which consists of the "vertex" part Γ and the self-energy part Σ.

In this way we are led to consider the modification of the vertex due to the phonon interaction in the same approximation as the self-energy effect is included in the quasi-particle. It is not difficult to see that it corresponds to a "ladder approximation" for the vertex part, and we get an integral equation[17]

$$\Gamma_i(p',p)=\gamma_i(p',p)-g^2\int \tau_3 G(p'-k)\Gamma_i(p'-k,\,p-k)\times G(p-k)\tau_3 h(k)^2\Delta(k)d^4k, \quad (4.3)$$

where γ_i, $i=0, 1, 2, 3$ stand for the free particle charge current $[\tau_3, (1/2m)(\mathbf{p}+\mathbf{p}')]$ which follows from Eq. (4.2). Similar equations may be set up for any type of vertex interactions.

Equation (4.3) is the basis of the rest of this paper. It expresses a clear-cut approximation procedure in which the "free" charge-current operator γ_i of a quasi-particle is modified by a special class of "radiative corrections" due to H_{int}'.

As the next important step, we observe that there exist exact solutions to Eq. (4.3) for the following four types of vertex interactions

$$\begin{aligned}
(a)\quad & \gamma^{(a)}(p',p)=L_0(p')-L_0(p)\\
&\qquad=(p_0'-p_0)-\tau_3(\epsilon_{p'}-\epsilon_p),\\
&\Gamma^{(a)}(p',p)=L(p')-L(p)\\
&\qquad=\gamma_a(p')-[\Sigma(p')-\Sigma(p)],\\
(b)\quad & \gamma^{(b)}(p',p)=L_0(p')\tau_3-\tau_3L_0(p)\\
&\qquad=(p_0'-p_0)\tau_3-(\epsilon_{p'}-\epsilon_p), \qquad (4.4)\\
&\Gamma^{(b)}(p',p)=L(p')\tau_3-\tau_3L(p),\\
(c)\quad & \gamma^{(c)}(p',p)=L_0(p')\tau_1+\tau_1L_0(p'),\\
&\Gamma^{(c)}(p',p)=L(p')\tau_1+\tau_1L(p),\\
(d)\quad & \gamma^{(d)}(p',p)=L_0(p')\tau_2+\tau_2L(p),\\
&\Gamma^{(d)}(p',p)=L(p')\tau_2+\tau_2L(p).
\end{aligned}$$

The verification is straightforward by noting that $G(p)=i/L(p)$, and making use of Eq. (3.7).

The fact that there are simple solutions is not accidental. These solutions express continuity equations and other relations following from the four types of operations, which do not depend on the presence or absence of the interaction:

$$\begin{aligned}
(a)\quad & \Psi(x)\to e^{i\alpha(x)}\Psi(x), & \Psi^+(x)\to\Psi^+(x)e^{-i\alpha(x)},\\
(b)\quad & \Psi\to e^{i\tau_3\alpha}\Psi, & \Psi^+\to\Psi^+e^{-i\tau_3\alpha},\\
(c)\quad & \Psi\to e^{\tau_1\alpha}\Psi, & \Psi^+\to\Psi^+e^{\tau_1\alpha},\\
(d)\quad & \Psi\to e^{\tau_2\alpha}\Psi, & \Psi^+\to\Psi^+e^{\tau_2\alpha},
\end{aligned}\qquad (4.5)$$

where $\alpha(x)$ is an arbitrary real function.

(a) and (b) correspond, respectively, to the spin rotation around the z axis, and the gauge transformation. The entire Lagrangian is invariant under them, so that we obtain continuity equations for the z component of spin and charge, respectively:

$$\begin{aligned}
&\frac{\partial}{\partial t}\boldsymbol{\Psi}^+\boldsymbol{\Psi}+\nabla\cdot\boldsymbol{\Psi}^+\frac{\mathbf{p}}{m}\tau_3\boldsymbol{\Psi}=0,\\
&\frac{\partial}{\partial t}\boldsymbol{\Psi}^+\tau_3\boldsymbol{\Psi}+\nabla\cdot\boldsymbol{\Psi}^+\frac{\mathbf{p}}{m}\boldsymbol{\Psi}=0,
\end{aligned}\qquad (4.6)$$

where $\boldsymbol{\Psi}$ is the true Heisenberg operator.

These equations are identical with

$$\boldsymbol{\Psi}^+\gamma^{(a)}\boldsymbol{\Psi}=0,\quad \boldsymbol{\Psi}^+\gamma^{(b)}\boldsymbol{\Psi}=0. \qquad (4.7)$$

Taken between two "dressed" quasi-particle states, the left-hand side of Eq. (4.7) will become

$$\begin{aligned}
e^{-i(p'-p)\cdot x}\langle|p'|\,\boldsymbol{\Psi}^+(x)\gamma^{(n)}\boldsymbol{\Psi}(x)\,|p\rangle
&=u_{p'}^*\Gamma^{(n)}(p',p)u_p\\
&=0,\quad (n=a,\,b) \qquad (4.7')
\end{aligned}$$

where u_p, $u_{p'}$ are single-particle wave functions satisfying $L(p)u_p=u_{p'}^*L(p')=0$.

In this way we have shown the existence of spin and charge currents $\Gamma_i^{(a)}(p',p)$ and $\Gamma_i^{(b)}(p',p)$ for a quasi-particle, for which the continuity equations

$$(p_0'-p_0)\cdot\Gamma_0^{(n)}-\sum_{i=1}^{3}(p'-p)_i\cdot\Gamma_i^{(n)}=\Gamma^{(n)}(p',p)=0$$

will hold.

The last two transformations of Eq. (4.4) are not unitary, but mix ψ_1 and ψ_2^+ in such a way as to keep $\Psi^+\tau_3\Psi$ invariant. From infinitesimal transformations of these kinds we get

$$\begin{aligned}
&i\boldsymbol{\Psi}^+\tau_1\left(\frac{\overrightarrow{\partial}}{\partial t}-\frac{\overleftarrow{\partial}}{\partial t}\right)\boldsymbol{\Psi}+\nabla\cdot\boldsymbol{\Psi}^+\tau_2\left(\frac{\overrightarrow{\mathbf{p}}}{m}+\frac{\overleftarrow{\mathbf{p}}}{m}\right)\boldsymbol{\Psi}=0,\\
&-i\boldsymbol{\Psi}^+\tau_2\left(\frac{\overrightarrow{\partial}}{\partial t}-\frac{\overleftarrow{\partial}}{\partial t}\right)\boldsymbol{\Psi}+\nabla\cdot\boldsymbol{\Psi}^+\tau_1\left(\frac{\overrightarrow{\mathbf{p}}}{m}+\frac{\overleftarrow{\mathbf{p}}}{m}\right)\boldsymbol{\Psi}=0,
\end{aligned}\qquad (4.8)$$

which bear the same relations to $\gamma^{(c),(d)}$ and $\Gamma^{(c),(d)}$ as Eq. (4.6) did to $\gamma^{(a),(b)}$ and $\Gamma^{(a),(b)}$. Note that the above equations are unaffected by the presence of the phonon interaction.

The fact that we can find a conserved charge-current

---

[17] This equation may also be derived simply by considering the self-energy equation (3.7) in the presence of an external field, and expanding Σ in A. Σ should be now a function of initial and final momenta, and we define

$$\Sigma^{(A)}(p',p)\equiv\Sigma(p)\delta^4(p'-p)+\sum_{i=0}^{3}(\Gamma_i(p',p)-\gamma_i(p',p))\times A^i(p'-p)+O(A^2).$$

In the limit $p'-p=0$, $\Gamma_i-\gamma_i=\partial\Sigma/\partial A^i$, which is the content of the Ward identity.[8] Investigation of the higher order terms in A is beyond the scope of this paper.

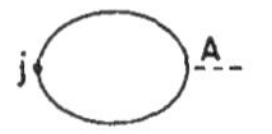

FIG. 5. The diagram for the kernel $K^{(2)}$.

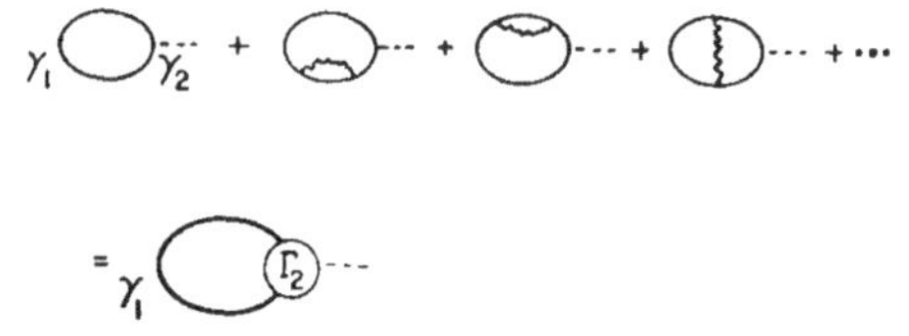

FIG. 6. Graphical derivation of Eq. (5.5). The thick lines represent quasi-particles.

for a quasi-particle is rather surprising. A quasi-particle cannot be an eigenstate of charge since it is a linear combination of an electron and a hole, tending to an electron well above the Fermi surface, and to a hole well below. We must conclude than that an accelerated wave packet of quasi-particles, whose energy is confined to a finite region of space, continuously picks up charge from, or deposits it with, the surrounding medium which extends to infinity. This situation will be studied in Sec. 7, where we will derive the charge current operators Γ_i explicitly.

5. GAUGE INVARIANCE IN THE MEISSNER EFFECT

We will next discuss how the gauge invariance is maintained in the problem of the Meissner effect when the external magnetic field is static. We calculate the Fourier component of the current $J(q)$ induced in the superconducting ground state by an external vector potential $A(q)$:

$$J_i(q)=\sum_{j=1}^{3} K_{ij}(q)A_j(q), \tag{5.1}$$

where q is kept finite.

For free electrons, K is represented by

$$K_{ij}=K_{ij}^{(1)}+K_{ij}^{(2)},\quad K_{ij}^{(1)}=-\delta_{ij}ne^2/m, \tag{5.2}$$

where n is the number of electrons inside the Fermi sphere. $K^{(1)}$ comes from the expectation value of the current operator Eq. (4.2), whereas $K^{(2)}$ corresponds to the diagram in Fig. 5. [Compare also Eq. (1.1).] It is well known that in this case K_{ij} is of the form

$$K_{ij}(q)=(\delta_{ij}q^2-q_iq_j)K(q^2), \tag{5.3}$$

so that for a longitudinal vector potential $A_i(q)\sim q_i\lambda(q)$, we have

$$J_i(q)=K_{ij}q_j\lambda(q)=0, \tag{5.4}$$

establishing the unphysical nature of such a potential.

In the case of a superconducting state, the free electron lines in Fig. 5 will be replaced by quasi-particle lines. But then we have $K^{(2)}(q)\to 0$ as $q\to 0$ since the intermediate pair formation is suppressed due to the finite energy gap, whereas $K^{(1)}$ is essentially unaltered. Thus Eq. (5.2) takes the form of the London equation, except that even a longitudinal field creates a current.

According to our previous argument, this lack of gauge invariance should be remedied by taking account of the vertex corrections. Starting again from the free electron picture, and inserting the phonon interaction effects, as indicated in Fig. 6, we arrive at the conclusion that either one of the vertices γ in Fig. 5 has to be replaced by the full Γ^{10}. In addition, there is the polarization correction represented by a string of bubbles. Let us, however, first neglect this correction. $K_{ij}^{(2)}$ is then

$$K_{ij}^{(2)}(q)=\frac{-ie^2}{(2\pi)^4}\int \mathrm{Tr}[\gamma_i(p-q/2,\,p+q/2)G(p+q/2)$$
$$\times\Gamma_j(p+q/2,\,p-q/2)G(p-q/2)]d^4p. \tag{5.5}$$

Although we do not know $\Gamma_j(p,p)$ explicitly, we can establish Eq. (5.4) easily. For

$$-\sum_{j=1}^{3}\Gamma_j(p+q,\,q)q_j$$

is exactly the solution $\Gamma^{(b)}(p+q,\,p)$ of Eq. (4.4) where q_0 is equal to zero. Substituting this solution in Eq. (4.5) we find

$$K_{ij}^{(2)}(q)q_j$$
$$=\frac{-1}{(2\pi)^4}\int \mathrm{Tr}\{\gamma_i(p-q/2,\,p+q/2)$$
$$\times[\tau_3G(p+q/2)-G(p-q/2)\tau_3]\}d^4p$$
$$=\frac{-1}{(2\pi)^4}\int \mathrm{Tr}\{[\gamma_i(p-q,\,p)-\gamma_i(p,\,p+q)]$$
$$\times\tau_3G(p)\}d^4p$$
$$=\frac{1}{(2\pi)^4}\frac{q_i}{m}\int \mathrm{Tr}[\tau_3G(p)]d^4p, \tag{5.6}$$

where the properties of γ_i and G under particle conjugation and a translation in p space were utilized in going from the first to the second line.

On the other hand, the part $K^{(1)}$ is, according to Eq. (4.2) given by

$$K_{ij}^{(1)}=-\delta_{ij}\frac{e^2}{2m}(\langle 0|[\Psi^+(x),\tau_3\Psi(x)]|0\rangle+\{\Psi^+(x),\Psi(x)\})$$
$$=K_{ij}^{(1a)}+K_{ij}^{(1b)}. \tag{5.7}$$

The first term becomes further

$$-\delta_{ij}\frac{e^2}{2m}\langle 0|[\Psi^+(x),\tau_3\Psi(x)]|0\rangle$$

$$=-\delta_{ij}\frac{e^2}{m}\tfrac{1}{2}\lim_{\epsilon\to 0}\sum_{\pm}\langle 0|T(\Psi^+(x,t\pm\epsilon)\tau_3\Psi(x,t))|0\rangle$$

$$=-\delta_{ij}\frac{e}{m}\mathrm{Tr}[\tau_3 G(xt=0)]$$

$$=-\delta_{ij}\frac{e}{m}\frac{1}{(2\pi)^4}\int\mathrm{Tr}[\tau_3 G(p)]d^4p. \quad (5.8)$$

Thus

$$[K_{ij}^{(1a)}(q)+K_{ij}^{(2)}(q)]q_j=0. \quad (5.9)$$

The second term $K^{(1b)}$ comes from the c-number term of the current operator (4.2), and is just the anticommutator of the electron field, which does not depend on the quasi-particle picture, nor on the presence of interaction. Therefore we may write for this contribution

$$K_{ij}^{(1b)}(q)A_j(q)=\frac{-ie}{2m}\frac{1}{(2\pi)^3}\int e^{-iqx}d^3x$$

$$\times\{\Psi^+(x),(\tau_3\nabla-ieA(x))_i\Psi(x)\} \quad (5.10)$$

to show its formal gauge invariance since $\tau_3\nabla-ieA(x)$ is certainly a gauge-invariant combination for free electron field.

As for the polarization correction, we can easily show in a similar way that it vanishes for the static case ($q_0=0$) because

$$\int\mathrm{Tr}\,\Gamma_i(p-q/2,p+q/2)G(p+q/2)$$

$$\times\gamma_0(p+q/2,p-q/2)G(p-q/2)d^4p=0.$$

Thus the above proof is complete and independent of the Coulomb interaction which profoundly influences the polarization effect. Although the proof is thus rigorous, it is still somewhat disturbing since $K^{(1a)}$, $K^{(1b)}$ and $K^{(2)}$ are all infinite. Actually there is a certain ambiguity in the evaluation of $K^{(2)}$, Eq. (5.6), which is again similar to the one encountered in quantum electrodynamics.[18] An alternative way would be to expand quantities in q without making translations in p space. In this case we may write

$$-\Gamma^{(b)}(p+q/2,p-q/2)=\bar{\epsilon}(p+q/2)-\bar{\epsilon}(p-q/2)$$

$$-i\tau_2[\phi(p+q/2)+\phi(p-q/2)]$$

$$\approx p\cdot q/m-2i\tau_2\phi. \quad (5.11)$$

The first term then gives

$$\frac{e^2}{(2\pi)^3}\int\frac{\phi^2}{4E_p{}^5}\left(\frac{\mathbf{p}\cdot\mathbf{q}}{m}\right)^2 p_i\mathbf{p}\cdot\mathbf{q}d^3p\propto q^2q_i, \quad (5.12)$$

which is convergent and the same as the one obtained from Eq. (1.1) using the bare quasi-particle states. The second term also is finite and equal to

$$\frac{e^2}{(2\pi)^3}\int\frac{\phi^2}{E_p{}^3}\frac{p^2}{3m^2}q_id^3p+O(q^2)q_i\approx N\alpha^2q_i=n(e^2/m)q_i. \quad (5.13)$$

The last line follows from Eqs. (6.11) and (6.11′) below.

The calculation of $K^{(1)}$ from Eqs. (5.7) and (5.10), gives, on the other hand, the same value as Eq. (5.2), so that we get $(K_{ij}^{(1)}+K_{ij}^{(2)})q_j=0$ in the limit of small q. (The polarization correction is again zero.)

Since Eq. (5.13) is a contribution from the collective intermediate state (see Secs. 6 and 7), we may say that the collective state saves gauge invariance, as has been claimed by several people.[3,19]

It goes without saying that the effect of the vertex correction on K_{ij} will be felt also for real magnetic field. But as we shall see later, it is a small correction of order g^2N (except for the renormalization effects), and not as drastic as for the longitudinal case.

6. THE COLLECTIVE EXCITATIONS

In order to understand the mechanism by which gauge invariance was restored in the calculation of the Meissner effect, and also to solve the integral equations for general vertex interactions, it is necessary to examine the collective excitations of the quasi-particles. In fact, people[3] have shown already that the essential difference between the transversal and longitudinal vector potentials in inducing a current is due to the fact that the latter can excite collective motions of quasi-particle pairs.

We see that the existence of such collective excitations follows naturally from our vertex solutions Eq. (4.4). For taking $p=p'$, the second solution $\Gamma^{(b)}(p',p)$ becomes

$$\Gamma^{(b)}(p,p)=L(p)\tau_3-\tau_3L(p)$$

$$=2i\tau_2\phi, \quad (6.1)$$

$$\gamma^{(b)}=0.$$

In other words $\tau_2\phi(p)\equiv\Phi_0(p)$ satisfies a homogeneous integral equation:

$$\Phi_0(p)=-\frac{g^2}{(2\pi)^4}\int\tau_3G(p')\Phi(p')G(p')$$

$$\times\tau_3h(p-p')^2\Delta(p-p')d^4p. \quad (6.2)$$

We interpret this as describing a pair of a particle and an antiparticle interacting with each other to form a bound state with zero energy and momentum $q=p'-p=0$.

[18] H. Fukuda and T. Kinoshita, Progr. Theoret. Phys. (Kyoto) 5, 1024 (1950).

[19] On the other hand, the way in which the collective mode accomplishes this end seems to differ from one paper to another. We will not attempt to analyze this situation here.

In fact, by defining

$$F(p,-p)\equiv -G(p)\Phi_0(p)G(p), \tag{6.3}$$

Eq. (6.2) becomes

$$L(p)F(p,-p)L(p)=-\frac{g^2}{(2\pi)^4}\int \tau_3 F(p',-p')\times\tau_3 h(p-p')^2\Delta(p-p')d^4p',$$

or

$$\sum_{j,l=1}^{2} L(p)_{ij}L^C(-p)_{kl}F(p,-p)_{jl}=\frac{-g^2}{(2\pi)^4}\int\sum_{j,l}(\tau_3)_{ij}(\tau_3)_{kl}F(p',-p')_{jl}\times h(p-p')^2\Delta(p-p')d^4p. \tag{6.4}$$

The particle-conjugate quantity L^C was defined in Eq. (2.23).

Equations (6.2) and (6.4) are the analog of the so-called Bethe-Salpeter equation[20] for the bound pair of quasi-particles with zero total energy-momentum. $F_{ij}(p,-p)$ is the four-dimensional wave function with the spin variables i, j and the relative energy-momentum $(p_0,\mathbf{p})$.

Since there, thus, exists a bound pair of zero momentum, there will also be pairs moving with finite momentum and kinetic energy. In other words, there will be a continuum of pair states with energies going up from zero. We have to determine their dispersion law.

For a finite total energy-momentum q, the homogeneous integral equation takes the form

$$\Phi_q(p)\equiv L(\tfrac{1}{2}q+p)F(\tfrac{1}{2}q+p,\tfrac{1}{2}q-p)L(p-\tfrac{1}{2}q)=-g^2\frac{1}{(2\pi)^4}\int\tau_3F(\tfrac{1}{2}q+p',\tfrac{1}{2}q-p')\times\tau_3h(p-p')^2\Delta(p-p')d^4p. \tag{6.5}$$

From here on we carry out perturbation calculation. Let us expand F and L in terms of the small change $L(p\pm q/2)-L(p)$, thus

$$F(\tfrac{1}{2}q+p,\tfrac{1}{2}q-p)=F^{(0)}(p)+F^{(1)}(p,q/2)+\cdots,\quad L(p\pm q/2)=L(p)+\Delta L(p,\pm q/2). \tag{6.6}$$

Collecting terms of the first order, we get

$$L(p)F^{(1)}(p,q/2)L(p)+U^{(1)}(p,q/2)=-g^2\frac{1}{(2\pi)^4}\int\tau_3F^{(1)}(p',q/2)\times\tau_3h(p-p')^2\Delta(p-p')d^4p', \tag{6.7}$$

$$U^{(1)}(p,q/2)=\Delta L(p,q/2)F^{(0)}(p)L(p)+L(p)F^{(0)}(p)\Delta L(p,-q/2).$$

This is an inhomogeneous integral equation for $F^{(1)}$. In order that it has a solution, the inhomogeneous term $U(p)$ must be orthogonal to the solution $\Phi_0(p)$ of the homogeneous equation. This condition can be derived as follows:

We multiply Eq. (6.7) by $F^{(0)}(p)=-G(p)\Phi_0(p)G(p)$, and integrate thus:

$$\int\mathrm{Tr}\,F^{(0)}(p)L(p)F^{(1)}(p,q/2)L(p)d^4p+\int\mathrm{Tr}\,F^{(0)}(p)U^{(1)}(p,q/2)d^4p=-g^2\frac{1}{(2\pi)^4}\int\int\mathrm{Tr}\,F^{(0)}(p)\tau_3F^{(1)}(p',q/2)\times\tau_3h(p-p')^2\Delta(p-p')d^4pd^4p'.$$

In view of Eq. (6.5) the last line is

$$=\int\mathrm{Tr}\,L(p')F^{(0)}(p')L(p')F^{(1)}(p',q/2)d^4p',$$

so that

$$(F^{(0)},U^{(1)})\equiv\int\mathrm{Tr}\,F^{(0)}(p)U^{(1)}(p,q/2)d^4p=0. \tag{6.8}$$

This is the desired condition.

For the evaluation of Eq. (6.8), we will neglect the p dependence of the self-energy terms. Thus

$$F^{(0)}(p)=\tau_2\phi/(p_0^2-E_p^2+i\epsilon),\quad E_p^2=\epsilon_p^2+\phi^2,\quad \Delta L(p,q/2)=q_0/2-\tau_3(\mathbf{p}\cdot\mathbf{q}/2m+(q/2)^2/2m). \tag{6.9}$$

We then obtain

$$(F^{(0)},U^{(1)})=2\pi i\int\frac{\phi^2}{E_p^3}\left[\left(\frac{q_0}{2}\right)^2-\left(\frac{\mathbf{p}\cdot\mathbf{q}}{2m}\right)^2-\frac{\epsilon_p}{m^2}\left(\frac{q}{2}\right)^2\right]d^3p=0,$$

or

$$\left(\frac{q_0}{2}\right)^2-\left(\frac{q}{2}\right)^2\left[\frac{1}{3}\frac{\bar{p}^2}{m^2}-\frac{\bar{\epsilon}_p}{m}\right]=0, \tag{6.10}$$

where the average $\bar{f}$ is defined

$$\bar{f}=\int f(p)\frac{\phi^2}{E^3}d^3p\bigg/\int\frac{\phi^2}{E^3}d^3p. \tag{6.10'}$$

The weight function $\phi^2/E_p^3=\phi^2/(\epsilon_p^2+\phi^2)^{\frac{3}{2}}$ peaks around the Fermi momentum, so that $p^2\sim p_F^2$, $\epsilon_p\sim0$. Thus

$$q_0^2\approx q^2\frac{1}{3}\frac{\bar{p}^2}{m^2}\equiv\alpha^2q^2,\quad \alpha^2\approx p_F^2/3m^2, \tag{6.11}$$

[20] E. E. Salpeter and H. A. Bethe, Phys. Rev. 84, 1232 (1951).

which is the dispersion law for the collective excitations.[2,3] We also note, incidentally, that

$$\frac{1}{(2\pi)^3}\int \frac{\phi^2}{E^3}d^3p \approx N = mp_F/\pi^2, \tag{6.11'}$$
$$\alpha^2 N \approx p_F^3/3\pi^2 m = n.$$

We would like to emphasize here that these collective excitations are based on Eq. (6.2), which takes account of the phonon-Coulomb scattering of the quasi-particle pairs, but does not take into account the annihilation-creation process of the pair due to the same interaction.

It is well known that this annihilation-creation process is very important in the case of the Coulomb interaction, and plays the role of creating the plasma mode of collective oscillations. We will consider it in a later section.

As for the wave function $F^{(1)}$ itself, we have still to solve the integral equation (6.7). But this can be done by perturbation because on substituting $U^{(1)}$ in the integrand, we find that all the terms are of the type (3.11). In other words, to the zeroth order we may neglect the integral entirely and so

$$F^{(1)}(p,q/2) = -G(p)U^{(1)}(p,q/2)G(p). \tag{6.12}$$

The original function

$$\Phi_q(p) = -L(p+q/2)F(p,q/2)L(p-q/2)$$

is even simpler. We get

$$\Phi_q(p) \approx \Phi_0(p) \tag{6.13}$$

to this order.

7. CALCULATION OF THE CHARGE-CURRENT VERTEX FUNCTIONS

In this section we determine explicitly the charge-current vertex functions Γ_i, $(i=0,\,1,\,2,\,3)$ from their integral equations. Only the particular combination $\Gamma^{(b)}$ of these was given before.

Let us first go back to the integral equation for Γ_0 generated by τ_3:

$$\Gamma_0(p+q/2,\, p-q/2)$$
$$= \tau_3 - g^2\int \tau_3 G(p'+q/2)\Gamma_0(p'+q/2,\, p'-q/2) \times G(p'-q/2)\tau_3 h(p-p')^2\Delta(p-p')d^4p',$$

or

$$L(p+q/2)F_0(p+q/2,\, p-q/2)L(p-q/2)$$
$$= \tau_3 + g^2\int \tau_3 F_0(p'+q/2,\, p'-q/2) \times \tau_3 h(p-p')^2\Delta(p-p')d^4p'. \tag{7.1}$$

For small g^2, the standard approach to solve the equation would be the perturbation expansion in powers of g^2. We know, however, that there are low-lying collective excitations, discussed before, to which τ_3 can be coupled, and these excitations do not follow from perturbation.[21]

Fortunately, if we assume $q=0$, $q_0\neq 0$, then we have an exact solution to Eq. (7.1) in terms of $\Gamma^{(b)}$ of Eq. (4.4). Namely,

$$\begin{aligned}\Gamma_0(p+q/2,\, p-q/2) &= \Gamma^{(b)}(p+q/2,\, p-q/2)/q_0\\ &= \tau_3\{[Z(p+q/2)+Z(p-q/2)\}/2\\ &\quad + (p_0/q_0)[Z(p+q/2)-Z(p-q/2)]\}\\ &\quad - [\chi(p+q/2)-\chi(p-q/2)]/q_0\\ &\quad + i\tau_2[\phi(p+q/2)+\phi(p-q/2)]/q_0,\end{aligned} \tag{7.2}$$

which can readily be verified.

The second term is the result of the coupling of τ_3 to the collective mode. This can be understood in the following way. Γ_0 contains matrix elements for creation or annihilation of a pair out of the vacuum. These processes can go through the collective intermediate state with the dispersion law (6.11), so that Γ will contain terms of the form

$$R_\pm/(q_0\pm\alpha q).$$

The residues $R_\pm$ can be obtained by taking the limit

$$R_\pm = \lim_{q_0\pm\alpha q\to 0} \Gamma_0(p+q/2,\, p-q/2)(q_0\pm\alpha q). \tag{7.3}$$

Applying this procedure to the integral equation (7.1) for Γ_0, we find that $R_\pm$ must be a solution of the homogeneous equation; namely,

$$R_\pm = C_\pm\Phi_q(p), \tag{7.4}$$

under the condition $q_0\pm\alpha q=0$.

For the particular case $q=0$, $\Phi_q(p)$ reduces to $\tau_2\phi(p)$, which in fact agrees with Eq. (7.2) if

$$C_\pm = -2i. \tag{7.5}$$

This observation enables us to write down Γ_0 for $q\neq 0$. According to the results of Sec. 6, $\Phi_q(p)=\Phi_0(p)$ in the zeroth order in g^2N. Since corrections to the non-collective part of Γ_0 also turn out to be calculable by perturbation, we may now put

$$\Gamma_0(p+q/2,\, p-q/2) \approx \tau_3\bar{Z} + 2i\tau_2\bar{\phi}q_0/(q_0^2-\alpha^2q^2),$$
$$\bar{\phi} \equiv [\phi(p+q/2)+\phi(p-q/2)]/2,$$
$$\bar{Z} \equiv [Z(p+q/2)+Z(p-q/2)\}/2$$

to the extent that terms of order g^2N and/or the p-dependence of the renormalization constants are neglected.

In quite a similar way the current vertex $\mathbf{\Gamma}$ may be constructed. This time we start from the longitudinal

[21] If we proceeded by perturbation theory, we would find in each order terms of order 1.

component for $q_0=0$, $q\neq 0$, which has the exact solution

$$
\begin{aligned}
\boldsymbol{\Gamma}(p+q/2,\, p-q/2)\cdot\mathbf{q}/q &= -\Gamma^{(b)}(p+q/2,\, p+q/2)/q \\
&= \frac{\mathbf{p}\cdot\mathbf{q}}{mq}\left\{1+\frac{\chi(p+q/2)-\chi(p-q/2)}{\mathbf{p}\cdot\mathbf{q}/m}\right\} \\
&\quad -\tau_3 p_0[\zeta(p+q/2)-\zeta(p-q/2)]/q \\
&\quad -2i\tau_2\frac{\phi(p+q/2)+\phi(p-q/2)}{2q}. \qquad (7.7)
\end{aligned}
$$

For $q_0\neq 0$, then, we get

$$
\begin{aligned}
\boldsymbol{\Gamma}(p+q/2,\, p-q/2)\cdot\mathbf{q}/q &\approx (\mathbf{p}\cdot\mathbf{q}/q)\bar{Y}+2i\tau_2\phi\alpha^2 q/(q_0^2-\alpha^2q^2), \qquad (7.8) \\
\bar{Y} &\equiv 1+[\chi(p+q/2)-\chi(p-q/2)]/(\mathbf{p}\cdot\mathbf{q}/m).
\end{aligned}
$$

Combining (7.6) and (7.8), the continuity equation takes the form

$$
\begin{aligned}
q_0\Gamma_0-\mathbf{q}\cdot\boldsymbol{\Gamma} &= q_0\tau_3\bar{Z}+(\mathbf{p}\cdot\mathbf{q}/m)\bar{Y}+2i\tau_2\bar{\phi} \\
&\approx \Gamma^{(b)},
\end{aligned}
$$

which is indeed zero on the energy shell.

The transversal part of $\boldsymbol{\Gamma}$, on the other hand, is not coupled with the collective mode because the latter is a scalar wave.[22] We may, therefore, write instead of Eq. (7.8)

$$
\boldsymbol{\Gamma}(p+q/2,\, p-q/2)\approx(\mathbf{p}/m)\bar{Y} +2i\tau_2\bar{\phi}\alpha^2 q/(q_0^2-\alpha^2q^2). \qquad (7.10)
$$

Equations (7.6) and (7.10) for Γ_i have a very interesting structure. The noncollective part is essentially the same as the charge current for a free quasi-particle except for the renormalization $\bar{Z}$ and $\bar{Y}$, whereas the collective part is spread out both in space and time. Neglecting the momentum dependence of $\bar{Z}$, $\bar{Y}$, and ϕ, we may thus write the charge-current density (ρ,j) as

$$
\begin{aligned}
\rho(x,t) &\cong e\Psi^{+}\tau_3 Z\Psi(x,t)+\frac{1}{\alpha^2}\frac{\partial f(x,t)}{\partial t}\equiv\rho_0+\frac{1}{\alpha^2}\frac{\partial f}{\partial t}, \qquad (7.11)\\
\mathbf{j}(x,t) &\cong e\Psi^{+}(\mathbf{p}/m)Y\Psi(x,t)-\nabla f(x,t)\equiv\mathbf{j}_0-\nabla f,
\end{aligned}
$$

where f satisfies the wave equation

$$
\left(\Delta-\frac{1}{\alpha^2}\frac{\partial^2}{\partial t^2}\right)f\approx-2e\Psi^{+}\tau_2\phi\Psi. \qquad (7.12)
$$

$(\rho_0,\, \mathbf{j}_0)$ is the charge-current residing in the "core" of a quasi-particle. The latter is surrounded by a cloud of the excitation field f. In a static situation, for example, f will fall off like $1/r$ from the core. When the particle is accelerated, a fraction of the charge is exchanged between the core and the cloud.

The total charge residing in a finite volume around a core is not constant because the current $-\nabla f$ reaches out to infinity.

[22] There may be transverse collective excitations (Bogoliubov, reference 2), but they do not automatically follow from the self-energy equation nor affect the energy gap structure.

8. THE PLASMA OSCILLATIONS

The inclusion of the annihilation-creation processes in the equations of the previous sections means that the vertex parts get multiplied by a string of closed loops, which represent the polarization (or shielding effect) of the surrounding medium. We will call the new quantities Λ, which now satisfy the following type of integral equations

$$
\begin{aligned}
\Lambda(p',p) &= \gamma - i\int\tau_3 G(p'-k)\Lambda(p'-k,\, p-k) \\
&\quad\times G(p-k)\tau_3 D(k)d^4k \\
&\quad + iD(p'-p)\tau_3\int \mathrm{Tr}[\tau_3 G(p'-k) \qquad (8.1)\\
&\quad\times\Lambda(p'-k,\, p-k)G(p-k)]d^4k, \\
D(q) &\equiv -ig^2h(q)^2\Delta(q)+e^2/q^2.
\end{aligned}
$$

$\tilde{D}(q)$ includes the effect of the Coulomb interaction [see Eq. (2.24)]. Putting

$$
\begin{aligned}
\bar{X}(p'-p) &\equiv i\int \mathrm{Tr}[\tau_3 G(p'-k)\Lambda(p'-k,\, p'-k) \\
&\quad\times G(p-k)]d^4k, \qquad (8.2)
\end{aligned}
$$

Eq. (8.1) takes the same form as Eq. (4.3) for Γ with the inhomogeneous term replaced by $\gamma+\tau_3 D\bar{X}$, so that Λ is a linear combination of the Γ corresponding to γ and Γ_0:

$$
\Lambda=\Gamma+\Gamma_0 D\bar{X}. \qquad (8.3)
$$

Substitution in Eq. (8.2) then yields

$$
\begin{aligned}
\bar{X}(p'-p) &= i\int \mathrm{Tr}[\tau_3 G(p'-k) \\
&\quad\times\Gamma(p'-k,\, p-k)G(p-k)]d^4k \\
&\quad + iD(p'-p)\bar{X}(p'-p)\int \mathrm{Tr}[\tau_3 G(p'-k) \\
&\quad\times\Gamma_0(p'-k,\, p-k)G(p-k)]d^4k,
\end{aligned}
$$

or

$$
\begin{aligned}
\bar{X}(p'-p) &= i\int \mathrm{Tr}[\tau_3 G(p'-k) \\
&\quad\times\Gamma(p'-k,\, p-k)G(p-k)]d^4k \\
&\quad\times\left\{1-iD(p'-p)\int \mathrm{Tr}[\tau_3 G(p'-k)\right. \\
&\quad\left.\times\Gamma_0(p'-k,\, p-k)G(p-k)]d^4k\right\}^{-1} \\
&\equiv X(p'-p)/[1-D(p'-p)X_0(p'-p)]. \qquad (8.4)
\end{aligned}
$$

Especially for $\gamma=\tau_3$, we get

$$\bar{X}_0(p'-p)=X_0(p'-p)/[1-D(p'-p)X_0(p'-p)],$$
$$\Lambda_0(p',p)=\Gamma_0(p',p)/[1-D(p'-p)X_0(p',p)]. \qquad (8.5)$$

To obtain the collective excitations, let us next write down the homogeneous integral equation:

$$\Theta_q(p)=-i\int \tau_3 G(p'+q/2)\Theta_q(p')$$
$$\times G(p'-q/2)\tau_3 D(p-p')d^4p'$$
$$+i\tau_3 D(q)\int \mathrm{Tr}\,[\tau_3 G(p'+q/2)\Theta_q(p')$$
$$\times G(p'-q/2)]d^4p', \qquad (8.6)$$

which means

$$\Theta_q(p)=\Gamma_0(p+q/2,\,p-q/2)D(q)\chi(q),$$
$$\chi(q)\equiv i\int \mathrm{Tr}[\tau_3 G(p'+q/2)$$
$$\times\Theta_q(p')G(p'-q/2)\tau_3]d^4p'. \qquad (8.7)$$

Substituting Θ_q in the second equation from the first, we get

$$1=D(q)X_0(q), \qquad (8.8)$$

where $X_0(q)$ is defined in Eq. (8.4).

The solutions to Eq. (8.8) determine the new dispersion law $q_0=f(q)$ for the collective excitations.

With the solution (7.6), the quantity X_0 in Eq. (8.8) can be calculated. After some simplifications using Eq. (6.11), we obtain

$$X_0=\frac{1}{(2\pi)^3}\left[\frac{\alpha^2q^2}{q_0^2-\alpha^2q^2}\int\frac{\phi^2d^3p}{E_p(E_p^2-\alpha^2q^2/4)}\right.$$
$$+\frac{q^2}{4}\int\frac{\phi^2d^3p}{E_p(q_0^2/4-E_p^2)}\left(\frac{p^2}{3m^2E_p^2}\right.$$
$$\left.\left.-\frac{\alpha^2}{E_p^2-\alpha^2q^2/4}\right)\right]+O(q^4). \qquad (8.9)$$

For $\alpha q\ll\phi$, and $q_0\gg\phi$ or $\ll\phi$, the second integral may be dropped and

$$X_0\cong\alpha^2q^2N/(q_0^2-\alpha^2q^2). \qquad (8.10)$$

For small q^2, the dominant part of $D(q)$ in Eq. (8.8) is the Coulomb interaction e^2/q^2. Equation (8.8) then becomes

$$q_0^2=e^2\alpha^2N=e^2n \quad (q^2\rightarrow 0), \qquad (8.11)$$

where n is the number of electrons per unit volume. This agrees with the ordinary plasma frequency for free electron gas.

We see thus that the previous collective state with $q_0^2=\alpha^2q^2$ has shifted its energy to the plasma energy as a result of the Coulomb interaction.

On the other hand, if Coulomb interaction is neglected, Eq. (8.8) leads to[23]

$$q_0^2=\alpha^2q^2[1-ig^2\Delta(q,q_0)h(q,q_0)^2N]. \qquad (8.12)$$

The correction term, however, is of the order g^2N, hence should be neglected to be consistent with our approximation.

We can also study the behavior of X_0 in the limit $q_0\rightarrow 0$ for small but finite q^2:

$$X_0\approx\frac{1}{(2\pi)^3}\int\frac{\phi^2}{E^3}d^3p\approx N, \qquad (8.13)$$

which comes entirely from the noncollective part of Γ_0, but again agrees with the free electron value.

Another observation we can make regarding $\bar{X}_0(q,q_0)$ is the following. $\bar{X}_0$ represents the charge density correlation in the ground state:

$$\bar{X}_0(q,q_0)=\int\langle 0|T(\rho(xt),\rho(0))|0\rangle e^{-i\mathbf{q}\cdot\mathbf{x}+iq_0t}d^3xdt.$$

If $|0\rangle$ is an eigenstate of charge, $\bar{X}_0$ should vanish for $q\rightarrow 0$, $q_0\neq 0$ since the right-hand side then consists of the nondiagonal matrix elements of the total charge operator Q:

$$\bar{X}_0(0,q_0)\propto\sum_n\left(\frac{1}{q_0-E_n}-\frac{1}{q_0+E_n}\right)|\langle n|Q|0\rangle|^2.$$

The converse is also true if $E_n>|q_0|$, $n\neq 0$ for some $q_0\neq 0$. Our result for $\bar{X}_0$, as is clear from Eqs. (8.5) and (8.9), has indeed the correct property in spite of the fact that the "bare" vacuum, from which we started, is not an eigenstate of charge.

9. CONCLUDING REMARKS

We have discussed here formal mathematical structure of the BCS-Bogoliubov theory. The nature of the approximation is characterized essentially as the Hartree-Fock method, and can be given a simple interpretation in terms of perturbation expansion. In the presence of external fields, the corresponding approximation insures, if treated properly, that the gauge invariance is maintained. It is interesting that the quasi-particle picture and charge conservation (or gauge invariance) can be reconciled at all. This is possible because we are taking account of the "radiative corrections" to the bare quasi-particles which are not eigenstates of charge. These corrections manifest themselves primarily through the existence of collective excitations.

There are some questions which have been left out. We would like to know, for one thing, what will happen if we seek corrections to our Hartree-Fock approximation by including processes (or diagrams) which have not been considered here. Even within our ap-

[23] Compare Anderson, reference 7.

proximation, there is an additional assumption of the weak coupling ($g^2N \ll 1$), and the importance of the neglected terms (of order g^2N and higher) is not known.

Experimentally, there has been some evidence[24] regarding the presence of spin paramagnetism in superconductors. This effect has to do with the spin density induced by a magnetic field and can be derived by means of an appropriate vertex solution. However, this does not seem to give a finite spin paramagnetism at 0°K.[25]

[24] Knight, Androes, and Hammond, Phys. Rev. **104**, 852 (1956); F. Reif, Phys. Rev. **106**, 208 (1957); G. M. Androes and W. D. Knight, Phys. Rev. Letters **2**, 386 (1959).
[25] K. Yosida, Phys. Rev. **110**, 769 (1958).

The collective excitations do not play an important role here as they are not excited by spin density. [$\Gamma^{(a)}$, Eq. (4.4), does not have the characteristic pole.]

It is desirable that both experiment and theory about spin paramagnetism be developed further since this may be a crucial test of the fundamental ideas underlying the BCS theory.

ACKNOWLEDGMENT

We wish to thank Dr. R. Schrieffer for extremely helpful discussions throughout the entire course of the work.

VOLUME 4, NUMBER 7 PHYSICAL REVIEW LETTERS APRIL 1, 1960

AXIAL VECTOR CURRENT CONSERVATION IN WEAK INTERACTIONS*

Yoichiro Nambu
Enrico Fermi Institute for Nuclear Studies and Department of Physics
University of Chicago, Chicago, Illinois
(Received February 23, 1960)

In analogy to the conserved vector current interaction in the beta decay suggested by Feynman and Gell-Mann, some speculations have been made about a possible conserved axial vector current.[1-3] One can formally construct an axial vector nucleon current, which satisfies a continuity equation,

$$\Gamma_\mu^A(p',p) = i\gamma_5\gamma_\mu - 2M\gamma_5 q_\mu/q^2, \quad q = p'-p, \qquad (1)$$

where p and p' are the initial and final nucleon momenta. Such an attempt has some appeal in view of the apparently modest renormalization effect on the axial vector beta decay constant ($g_A/g_V \approx 1.25$), although the second appealing point,[1] namely, the possible forbidding of $\pi \to e + \nu$, has now lost its relevance.

The expression (1), unfortunately, can be easily ruled out experimentally, as was pointed out by Goldberger and Treiman,[3] since it introduces a large admixture of pseudoscalar interaction.

On the other hand, Eq. (1) arouses theoretical curiosity as to the origin of the second term if it really exists; according to our conventional field theory, we would have to interpret the denominator q^2 as implying a massless, pseudoscalar, and charged quantum bridging the nucleon and lepton currents.

We would like to suggest that there may not be a strict pseudovector current conservation, but that we may have an approximate conservation which becomes rigorous in the limit $q^2 >> m_\pi^2$, m_π being the pion mass. Specifically, we propose that the axial vector part of the nucleon beta decay vertex has the following form and properties:

$$g_A \Gamma_\mu^A(p',p)$$

$$= g_V\left[i\gamma_5\gamma_\mu F_1(q^2) - \frac{2M\gamma_5 q_\mu}{q^2+m_\pi^2} F_2(q^2)\right],$$

$$F_1(0) = g_A/g_V \approx F_2(0),$$

$$F_1(q^2) \sim F_2(q^2) \quad \text{for} \quad q^2 >> m_\pi^2. \qquad (2)$$

The pion is then the analog of the massless quantum mentioned above. This is consistent with the dispersion relations expected for Γ_μ^A. Namely, F_1 and F_2 should have in general the form

$$F_i(q^2) = F_i(-m_\pi^2)$$

$$-(q^2+m_\pi^2)\int_{m_0^2}^{\infty} \frac{\rho_i(m^2)dm^2}{(q^2+m^2)(m^2-m_\pi^2)}$$

$$(i=1,2), \qquad (3)$$

where $m_0 = 3m_\pi$ unless there are new particles of low mass. Thus the F's will be slowly varying for $|q^2| << m_0^2$. The conditions in Eq. (2) imply that $F_1/F_2 \approx 1$ for all q^2. If $m_\pi = 0$ and $F_1/F_2 \equiv 1$, then we restore exact current conservation,[2] and we also expect $F_1(0) = g_A/g_V = 1$.

If we adopt Eq. (2), the second term of Γ_μ^A immediately gives a relation between g_A, the pion decay (pseudovector) constant g_π, and the pion-nucleon (pseudoscalar) coupling G_π:

$$2Mg_A \approx 2Mg_V F_2(-m_\pi^2) = \sqrt{2}\, G_\pi g_\pi. \qquad (4)$$

With $g_A = 1.25\, g_V = 1.75\times10^{-49}$ erg cm^3,[4] $G_\pi^2/4\pi = 13.5$, this gives a π-μ decay life of 2.7×10^{-8} sec as compared with the observed value 2.56×10^{-8} sec.

Goldberger and Treiman[5] have arrived at the same relation Eq. (4) (in the limit of their self-energy integral $J\rightarrow\infty$) from an entirely different approach. In our opinion, this is not a coincidence, as will be explained elsewhere.

We are tempted to extend this approximate conservation of the axial vector (and naturally also the vector current) to the strangeness-nonconserving beta decays. We take, for example, the ΛN axial vector in the form

$$\Gamma_\mu^A(p_N', P_\Lambda) \approx i\gamma_5\gamma_\mu - \frac{(M_\Lambda + M_N)\gamma_5 q_\mu}{q^2+m_K^2}, \qquad (5)$$

and attribute the second term to the pseudoscalar K meson.[6] The degree of accuracy of the relation (5) will be poorer than in the previous case in view of the Λ-N mass difference (which destroys vector conservation) and the large K-meson mass. At any rate, we obtain an analog of Eq. (4):

$$(M_\Lambda + M_N) g_A' \approx G_K g_K, \qquad (6)$$

which relates the Λ beta decay axial vector coupling g_A', the ΛNK coupling G_K, and the K_μ decay coupling g_K.

With the observed $K_{\mu 2}$ lifetime 2.1×10^{-8} sec and a tentative value $G_K^2/4\pi = \frac{1}{4}G_\pi^2/4\pi$, we get

$$g_A'/g_A \approx 1/10. \qquad (7)$$

This is not inconsistent with the observed beta decay of Λ which seems an order of magnitude less than predicted from a universal coupling scheme $g_V' = g_A' = g_V$.[7]

We can still go further, though the argument becomes more arbitrary. Let us assume that a fundamental weak coupling $(\bar{N}N\bar{N}\Lambda)$ gives rise to an effective V-A interaction (or at least part of it) of the form

$$g''(\Gamma_\mu^V - \Gamma_\mu^A)_{\bar{N}N}(\Gamma_\mu^V - \Gamma_\mu^A)_{\bar{N}\Lambda}. \qquad (8)$$

Here $\Gamma_\mu^V = i\gamma_\mu$ which is approximately conserved by itself, and Γ_μ^A stands for Eq. (2) or (5). We see easily that Eq. (8) contains information about the $\Lambda\rightarrow N+\pi$ decay matrix element:

$$(2M_N g''/\sqrt{2}\, G_\pi) q_\mu (\Gamma_\mu^V - \Gamma_\mu^A)_{\bar{N}\Lambda}. \qquad (9)$$

Combined with the assumption of $\Delta T = \frac{1}{2}$ selection rule, this gives a lifetime of 2.5×10^{-10} sec

VOLUME 4, NUMBER 7 PHYSICAL REVIEW LETTERS APRIL 1, 1960

for $g'' = g_V$ as compared with the observed value 2.8×10^{-10} sec.

It is possible to apply this kind of consideration to other hyperons. Moreover, if the Feynman–Gell-Mann coupling scheme such as $(\pi\pi e\nu)$ is formally extended to $(K\pi e\nu)$, etc. as has been tried by some people, all the observed decay processes may be covered. Here we would like to point out that if all baryons should satisfy Eqs. (4) and (6), the ratios g_A/G_π and g_A'/G_K must be approximately common constants.

Our final remark concerns the theoretical basis for the assumptions made here. If the baryons are derived from some fundamental field ψ which possesses an invariance under a transformation of the type $\psi \to \exp(i\vec{\alpha}\cdot\vec{\tau}\gamma_5)\psi$,[8] then there will be a conservation of the pseudovector charge-current. A finite observed mass can be compatible with the conservation if the particle is coupled with a boson as was noted in Eq. (1).

This situation may be understood by making an analogy to the theory of superconductivity originated by Bardeen, Cooper, and Schrieffer,[9] and refined by Bogoliubov.[10] There gauge invariance, the energy gap, and the collective excitations are logically related to each other as was shown by the author.[11] In the present case we have only to replace them by γ_5 invariance, baryon mass, and the mesons. In fact, the mathematical method used in superconductivity may be taken over to study the self-energy problem of elementary particles. It is interesting that pseudoscalar mesons automatically emerge in this theory as bound states of baryon pairs. The nonzero meson masses and baryon mass splitting would indicate that the γ_5 invariance of the bare baryon field is not rigorous, possibly because of a small bare mass of the order of the pion mass.

The above-mentioned model of elementary particles will be studied in a separate paper.

---

*This work was supported by the U. S. Atomic Energy Commission.

[1] J. C. Taylor, Phys. Rev. 110, 1216 (1958).

[2] J. C. Polkinghorne, Nuovo cimento 8, 179 and 781 (1958).

[3] M. L. Goldberger and S. B. Treiman, Phys. Rev. 110, 1478 (1958).

[4] A. I. Alikhanov, Ninth Annual International Conference on High-Energy Physics, Kiev, 1959 (unpublished).

[5] M. L. Goldberger and S. B. Treiman, Phys. Rev. 110, 1178 (1958); M. L. Goldberger, Revs. Modern Phys. 31, 797 (1959).

[6] It is also possible to associate a scalar K meson with the ΛN vector current conservation, while leaving the axial vector unaccounted for.

[7] Again Eq. (5) and the subsequent conclusions are essentially the same as those of C. H. Albright, Phys. Rev. 114, 1648 (1959) and B. Sakita, Phys. Rev. 114, 1650 (1959), which are based on the Goldberger-Treiman method. For the Λ-decay case below, see L. Tenaglia, Nuovo cimento 14, 499 (1959).

[8] F. Gürsey (private communication) has recently obtained similar results on the π decay based on this γ_5 invariance. We do not here specify the interaction of the ψ field, which may be of the nonlinear Heisenberg type, or due to an intermediate boson (different from π or K).

[9] J. Bardeen, L. N. Cooper, and J. R. Schrieffer, Phys. Rev. 106, 162 (1957).

[10] N. N. Bogoliubov, V. V. Tolmachev, and D. V. Shirkov, A New Method in the Theory of Superconductivity (Academy of Sciences of USSR, Moscow, 1958).

[11] Y. Nambu, Phys. Rev. 117, 648 (1960).

A 'SUPERCONDUCTOR' MODEL OF ELEMENTARY PARTICLES AND ITS CONSEQUENCES by Y. Nambu (University of Chicago)†

(In absence of the author the paper was presented by G. Jona-Lasinio.)

1

In recent years it has become fashionable to apply field-theoretical techniques to the many-body problems one encounters in solid state physics and nuclear physics. This is not surprising because in a quantized field theory there is always the possibility of pair creation (real or virtual), which is essentially a many-body problem. We are familiar with a number of close analogies between ideas and problems in elementary particle theory and the corresponding ones in solid state physics. For example, the Fermi sea of electrons in a metal is analogous to the Dirac sea of electrons in the vacuum, and we speak about electrons and holes in both cases. Some people must have thought of the meson field as something like the shielded Coulomb field. Of course, in elementary particles we have more symmetries and invariance properties than in the other, and blind analogies are often dangerous.

At any rate, we should expect a close interaction of the two branches of physics in terms of concepts and mathematical techniques, which make up the content of quantum field theory. In this talk we are going to show another possibility of such an interaction, but this time in the opposite direction to what has been the general trend. Namely, the model of elementary particles we are going to talk about is motivated by the mathematical theory of superconductivity which was first worked out with great success by Bardeen, Cooper and Schrieffer[1]. The characteristic feature of the theory is that the ground state of a superconductor is found to be separated by a gap from the excited states, which, of course, has been confirmed experimentally. The gap is caused by the fact that the attractive phonon interaction between electrons produce correlated pairs of electrons with opposite momenta near the Fermi surface, and it takes a finite amount of energy to break the correlation.

The BCS theory was given an elegant mathematical basis by Bogoliubov[2], who introduced a coherent mixture of electrons and holes to discuss the el-

† Retypeset by M. Okai, Nov. 1993
Courtesy of : Purdue University and the Purdue Research Foundation

ementary excitations (quasi-particles) in a superconductor. It is easy to see that such a particle has a finite "rest energy," which corresponds to the finite energy gap. Let us assume the following equations for electrons near the top of the Fermi surface:

$$E\psi_{p+} = \epsilon_p \psi_{p+} + \phi \psi^{\dagger}_{-p-} \ ,$$
$$E\psi^{\dagger}_{-p-} = -\epsilon_p \psi^{\dagger}_{-p-} + \phi \psi_{p+} \ . \qquad (1)$$

ψ_{p+} is the wave function for an electron of momentum p and spin + (up), and $\psi^{\dagger}_{-p-}$ is one for a hole of momentum p and spin +, which means the absence of an electron of momentum $-p$ and spin $-$ (down). ϵ_p is the kinetic energy measured from the Fermi surface; ϕ is a constant.

Eq.(1) gives the eigenvalues

$$E_p = \pm\sqrt{\epsilon_p^2 + \phi^2}. \qquad (2)$$

So it takes an amount of energy $2|E_p| \geq 2\phi$ to excite such a quasi-electron from the lower to the upper state. The quantity ϕ is actually obtained as a self-consistent, self-energy (Hartree-Fock field) from the phonon-electron interaction,

$$\mid \phi \mid \approx \hbar\omega e^{-1/\rho} \qquad (3)$$

where $\hbar\omega$ is the mean phonon frequency, and ρ the effective electron-electron interaction energy density on the Fermi surface.

Eqs.(1) and (2) bear a striking resemblance to the Dirac equation and its eigenvalues. In the Weyl representation, The Dirac equation reads

$$E\psi_1 = \vec{\sigma} \cdot \vec{p}\psi_1 + m\psi_2$$
$$E\psi_2 = -\vec{\sigma} \cdot \vec{p}\psi_2 + m\psi_1$$
$$E_p = \pm\sqrt{p^2 + m^2} \qquad (4)$$

where ψ_1 and ψ_2 are the two eigenstates of the chirality operator γ_5.

This analogy may be a superficial one and devoid of physical significance. But it would also be interesting to see what would happen if we took the analogy seriously and pursued its consequences. The interpretation of Eq.(4) would be then first of all that the mass of a Dirac particle is a self-energy built up by some interaction, a statement which surprises nobody. Indeed we shall find that even though the starting point looks novel, there is nothing unconventional in our model. Nevertheless, we shall also see that the analogy casts a new light on old problems, and reveals some new things which have been overlooked in the usual discussion of the self-energy problem and the symmetry properties of elementary particles.

To give an idea about our program, we draw up a list of correspondences between superconductivity and the elementary particle theory.

Superconductivity	Elementary particles
free electrons	bare fermion (zero or small mass)
phonon interaction	some unknown interaction
energy gap	observed mass (nucleon)
collective excitation	meson bound nucleon pair
charge	chirality
gauge invariance	γ_5–invariance (rigorous or approximate)

As we can see from the table, our problem will be to account for the nucleons (and hyperons) and the mesons in a unified way from some basic field. There is no strong reason why we should not also consider the leptons, but for the time being we would like to exclude them. The reason will become clear later on.

As for the exact nature of the basic interaction which would produce the baryons and mesons, our model does not say what it should be. Some other guiding principles are needed for this purpose, but we do not seem to possess any convincing ones yet. So looking around for some clues, we find two possibilities rather attractive for reasons of simplicity and elegance. One is the Heisenberg type theory[3] where we consider nonlinear spinor interactions. The other one is to use an analogy with the electromagnetic field. The electromagnetic field is inherently related to the conservation of charge, and the dynamics of interaction is uniquely determined by the gauge group. Attempts to generalize this idea to baryon problems have been made by Yang and Mills[4], Yang and Lee[5], Fujii[6], and recently by Sakurai[7].

Both types of theories have attractive points as well as difficulties. The most serious obstacle in any theory dealing with self-energies is the divergence problem, which is more pronounced in the Heisenberg type theory than in the other. The intermediate boson theory runs into trouble because gauge invariance requires such a field to be massless, yet massless boson fields other than electromagnetic and gravitational do not seem to exist. We do not know whether a finite observed mass can be compatible with the invariance assumption.

2

We will consider here the Heisenberg type theory because of its greater practical simplicity. The divergence will be disposed of by simple cut-off, as we do not claim to have found a way to resolve this difficulty.

Thus we adopt the following model Lagrangian for the nucleon. Isotopic spin is ignored.

$$L = -\bar{\psi}\gamma_\mu \partial_\mu \psi - g\left[\bar{\psi}\psi\bar{\psi}\psi - \bar{\psi}\gamma_5\psi\bar{\psi}\gamma_5\psi\right] \tag{5}$$

This Lagrangian is invariant under the transformations

$$\begin{array}{ll} \text{(a)} & \psi \longrightarrow \exp[i\alpha]\psi \;\; ; \;\; \bar{\psi} \longrightarrow \bar{\psi}\exp[-i\alpha] \\ \text{(b)} & \psi \longrightarrow \exp[i\alpha\gamma_5]\psi \;\; ; \;\; \bar{\psi} \longrightarrow \bar{\psi}\exp[+i\alpha\gamma_5] \end{array} \tag{6}$$

where α is a constant c number. (Local gauge transformation is not possible here.) a) implies the nucleon number conservation; b) will be called γ_5 invariance hereafter, which implies the conservation of chirality: the number of right-handed (bare)particles minus the number of left-handed particles is conserved. We get accordingly two conserved currents

$$\frac{\partial}{\partial x_\mu}\bar{\psi}\gamma_\mu\psi = 0 \;\; ; \;\; \frac{\partial}{\partial x_\mu}\bar{\psi}\gamma_5\gamma_\mu\psi = 0 \tag{7}$$

which can be directly verified.

Now we want to derive the observed nucleon mass in the Hartree-Fock approximation. Namely, we determine the mass by linearizing the interaction in which process the assumed mass is used in taking expectation values. We then have the relation

$$\begin{aligned} m &= 2g\left[\langle\bar{\psi}\psi\rangle - \gamma_5\langle\bar{\psi}\gamma_5\psi\rangle\right] \\ &= -2g\left[\mathrm{Tr}S^{(m)}(0) - \gamma_5\mathrm{Tr}\gamma_5 S^{(m)}(0)\right] \end{aligned} \tag{8}$$

where $S^{(m)}(x)$ is the nucleon Green's function having a mass m. In momentum space this becomes

$$m = -\frac{g}{(2\pi)^3}\int \frac{md^3p}{\sqrt{p^2+m^2}} \ . \tag{9}$$

A trivial solution is of course $m = 0$. But with a cut-off we find also a non-trivial one

$$\frac{\pi^2}{|g|K^2} = \sqrt{1+m^2/K^2} - (m^2/K^2)\sinh^{-1}|K/m| \tag{10}$$

provided that $g < 0$ and $\pi^2 < |g|K^2$. For $|m/K| \equiv x \ll 1$,

$$1 - \pi^2/|g|K^2 \approx x^2 \log(2/x), \tag{9'}$$

Eq.(9) is of the same form as the "energy gap equation" in the BCS theory. The non-analytic character of the solution with respect to the coupling constant is easily recognizable.

Our approximation scheme for the self-energy is illustrated by the following Feynman diagrams (Fig.1).

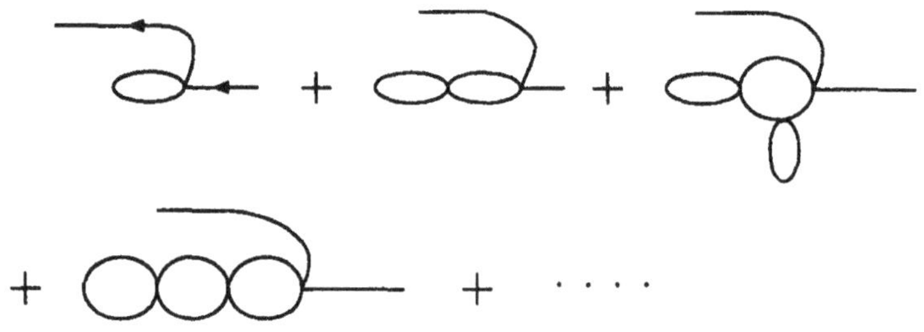

Fig. 1

Because of the attractive interaction ($g < 0$) between the virtual pair, the bubble diagrams give rise to a catastrophic change not obtained by perturbation expansion.

Thus we have created a mass out of nothing. But there are two solutions corresponding to $\neq m$, not to speak of the trivial solution $m = 0$. Presumably the vacuum corresponding to the latter solution is unstable like the normal state below the critical temperature for a superconductor. But would the two non-trivial solutions correspond to two different particles with equal mass? Heisenberg has found a similar situation in his theory. He wants to identify the two massive particles with the proton and the neutron.

Before discussing this problem, let us first worry about the conservation laws. Eq.(7) represents operator equations, so they should hold, among other things, for the matrix element between real one particle states. If we use the Dirac equation with a mass for ψ and $\bar{\psi}$, the ordinary current is all right, but the γ_5 current conservation breaks down:

$$\langle p_2|\frac{\partial}{\partial x_\mu}\bar{\psi}\gamma_5\gamma_\mu\psi|p_1\rangle = -2m\langle p_2|\bar{\psi}\gamma_5\psi|p_1\rangle \neq 0. \tag{11}$$

This means that $\gamma_5\gamma_\mu$ is not the correct vertex operator for the "dressed" particle where the mass is entirely due to the interaction. We would have to take into account the "radiative corrections" also for the vertex. The general form of the γ_5 current vertex operator $\Gamma_{5\mu}$ can be determined from Lorentz invariance and the continuity equation as follows:

$$\Gamma_{5\mu}(p_2, p_1) = \left(\gamma_5\gamma_\mu + \frac{2im\gamma_5 q_\mu}{q^2}\right) F(q^2), \quad q = p_2 - p_1 \tag{12}$$

where $F(q^2)$ is a form factor.

On the other hand, the local field theory requires that the coefficients of $\gamma_5\gamma_\mu$ and γ_5 obey dispersion relations of the type

$$f(q^2) = f(0) - \frac{q^2}{\pi}\int_{k_1^2}^{\infty} \frac{\mathrm{Im} f(-k^2)}{(q^2+k^2)k^2} dk^2. \tag{13}$$

If $F(0) \neq 0$, then Eq.(12) has a pole at $q^2 = 0$, which means, in terms of Eq.(13), the existence of a massless, pseudoscalar, boson contributing to the form factor. A Feynman diagram showing this situation is given in Fig.2.

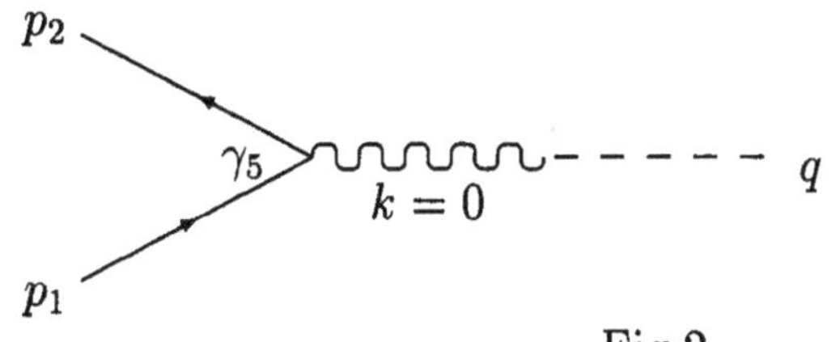

Fig.2

Since no such boson was assumed in our theory in the beginning, we have to manufacture it somehow out of the original fermion field. A natural way would be to interpret the boson as a bound state of a fermion pair—bound because of the attractive interaction. Thus we are forced to the conclusion that if a finite mass can arise from a γ_5 invariant theory there must also be zero-mass bound states of pairs. We would not have this situation if $F(0) = 0$, but then the total γ_5 charge

$$\langle p|\bar{\psi}\gamma_5\gamma_4\psi|p\rangle$$

would be zero, which leads to a contradiction (see the statement after Eq.(19)). We would not have the trouble either if the interaction were not γ_5 invariant, e.g. if the pseudoscalar term were missing in Eq.(5), which would not change Eq.(9). But we have invited the trouble deliberately because we need pseudoscalar bosons in the nucleon problem.

We can easily verify our conclusion within the present Hartree-Fock approximation. For this purpose let us set up the Bethe-Salpeter equation for the nucleon-antinucleon pair in the lowest order:

$$\begin{aligned}\Phi\left(p+\frac{q}{2},p-\frac{q}{2}\right) = ig\Bigg[\int \mathrm{Tr}S\left(p'+\frac{q}{2}\right)\Phi\left(p'+\frac{q}{2},p'-\frac{q}{2}\right)S\left(p'-\frac{q}{2}\right)d^4p' \\ -\gamma_5\int \mathrm{Tr}\gamma_5 S\left(p'+\frac{q}{2}\right)\Phi\left(p'+\frac{q}{2},p'-\frac{q}{2}\right)S\left(p'-\frac{q}{2}\right)d^4p'\Bigg] \quad (14)\end{aligned}$$

$\Phi\left(p+\frac{q}{2},p-\frac{q}{2}\right)$ is the wave function for a nucleon with momentum $p+\frac{q}{2}$ and an antinucleon with momentum $\frac{q}{2}-p$, the total momentum being q.

We also note that if we add an inhomogeneous term $\Gamma_0\left(p+\frac{q}{2},p-\frac{q}{2}\right)$ on the right-hand side of Eq.(14) and write Γ instead of Φ, then it would represent an integral equation for a vertex part Γ generated by Γ_0.

Now it is easy to see that the following functions satisfy the integral equation,

$$\begin{aligned}&\Gamma\left(p+\frac{q}{2},p-\frac{q}{2}\right) = L\left(p+\frac{q}{2}\right)\gamma_5+\gamma_5 L\left(p-\frac{q}{2}\right) = -i\gamma\cdot q\gamma_5 - 2m\gamma_5, \\ &\Gamma_0\left(p+\frac{q}{2},p-\frac{q}{2}\right) = L_0\left(p+\frac{q}{2}\right)\gamma_5+\gamma_5 L_0\left(p-\frac{q}{2}\right) = -i\gamma\cdot q\gamma_5, \\ &\left(L(p) = -i\gamma\cdot p - m = -i\left[S^{(m)}(p)\right]^{-1}; \quad L_0(p) = -i\gamma\cdot p\right). \qquad (15)\end{aligned}$$

Taking $q = 0$, we get $\Gamma = -2m\gamma_5$, $\Gamma_0 = 0$. In other words $\Phi = 2m\gamma_5$ is a solution of the B-S equation for zero energy and momentum, which is a limiting case of a particle with zero rest mass. In fact we can construct bound state wave functions for q_0, $\vec{q} \neq 0$, $q^2 = 0$ starting from the above solution as the zeroth approximation.

For arbitrary q, Eq.(15) is a generalized Ward identity for the divergence of the γ_5 current $\Gamma_{5\mu}$ (Eq.(12 ($F(q^2) = 1$ in our approximation.)) A similar relation is known for the ordinary current[8], and can be derived in our case too. Since $L(p)|p\rangle = 0$ and $\langle p|L(p) = 0$, the continuity equation (11) is obviously satisfied.

Now the question of the mass degeneracy. By the γ_5 transformation $\psi \rightarrow e^{i\alpha\gamma_5}\psi$, the mass m changes into $m\exp[2i\alpha] = m(\cos 2\alpha + i\gamma_5 \sin 2\alpha)$. In fact we could have assigned this form to the mass operator from the beginning. There would be no change of physical content. This arbitrariness is due to the fact that we can add any number of the zero-mass, zero-energy "mesons" to the system. The vacuum itself is degenerate in this sense.

To see this, consider the vacuum state Ω defined by

$$\psi^{(+)}\Omega = \bar{\psi}^{(+)}\Omega = 0.$$

The infinitesimal γ_5 transformation is generated by

$$R = -\int \bar{\psi}\gamma_5\gamma_4\psi d^3x. \tag{16}$$

The zero-mass, zero-energy mesons are similar to the longitudinal photons encountered in quantum electrodynamics. The γ_5 and gauge transformations change the distribution of the respective quanta, but cause no physical effects. What is, then, the γ_5 quantum number for the vacuum or the one particle state? This is the eigenvalue of R, and may be written as

$$R = n_+ - n_- + \bar{n}_+ - \bar{n}_- \tag{17}$$

where $n_\pm(\bar{n}_\pm)$ is the number of bare ferminons (anti-fermions) with $\gamma_5 = \pm 1$. The nucleon number N is given by

$$N = n_+ + n_- - \bar{n}_+ - \bar{n}_- \tag{18}$$

so that $N \pm R = 2(n_\pm - \bar{n}_\mp)$ is an even number.

Real nucleons with finite mass certainly are not eigenstates of R if they transform in the conventional way under R. Such a situation would be possible only if we had a complete degeneracy with respect to R. This seems to be the present case. Our world is conveniently described as a superposition of states with different R's, but there are no realizable processes which change R (a superselection rule), and hence the degeneracy does not manifest itself as physical effects.

Such an interpretation may not be quite satisfactory, but we would like to point out that a similar situation appears in the BCS theory too with respect to charge conservation. At any rate we have not found pseudoscalar photons in nature,but rather we are inclined to identify them with the mesons, which have mass in reality. Thus we will have to admit that after all nature is not γ_5 invariant. We do not know whether there is any other way out. But if the violation of the invariance were small, the foregoing results would guarantee automatically the existence of meson states, this time with a finite mass.

There are two ways to achieve our goal. One is to assume a finite bare mass m_0, which should be small compared to the observed mass m. The other is to destroy the nice symmetry of the interaction a little bit. The former looks simpler, and esthetically less objectionable. In this case, we can confirm by calculation the following results. The nucleon self-energy has now lost the freedom of γ_5 rotation, but we still get two different masses of opposite sign, in addition to the trivial perturbation solution. As m_0 increases, the mass splitting grows larger, until at a certain point one of them merges with the trivial branch and disappears thereafter, leaving only the larger of the non-trivial solution.

On the other hand, the meson mass μ is proportional to $(m_0/m)^{1/2}$ so that only the solution with $m_0/m > 0$ (the largest of the three solutions), can give rise to a stable bound pair, while the other two give rise to "ghost" mesons.

Apart from the ghost trouble, this result raises the interesting question of the possibility of producing mass multiplets of nucleons and mesons since the superselection rule does not operate any more. But the question has to be left for future study.

In our model, of course, we have obtained only neutral mesons because isospin is neglected. But the generalization is easy. For example, the interaction

$$L_{int} = -g\left[(\bar{\psi}\psi) - \sum_{s=1}^{3} \bar{\psi}\gamma_5\tau_s\psi\bar{\psi}\gamma_5\tau_s\psi\right] \tag{19}$$

(where the τ's are the isotopic spin matrices) immediately leads to pseudoscalar mesons of isospin 1. The gauge group in this case consists of

$$\begin{aligned} \psi &\longrightarrow \exp[i\alpha]\psi \\ \psi &\longrightarrow \exp[i\vec{\alpha}\cdot\vec{\tau}]\psi \\ \psi &\longrightarrow \exp[i\vec{\alpha}\cdot\vec{\tau}\gamma_5]\psi \end{aligned} \tag{20}$$

which correspond respectively to nucleonic charge, isotopic spin, and the $\gamma_5\times$isotopic spin conservation[9]. The last two form a four-dimensional rotation group.

3

Because of some interesting features in their own right, we will discuss briefly the intermediate boson theory of primary interaction. As before, we would like to take a γ_5 invariant theory, which means that the boson is either vector or pseudovector[10]. We can immediately write down a self-energy equation in our Hartree approximation. Namely we equate the observed mass with the familiar lowest order self-energy. Actually the self-energy consists of two parts: $\Sigma(p) = i\gamma\cdot p\Sigma_1(p) + \Sigma_2(p)$, Σ_1 being the wave function renormalization. Thus we get equations for Σ_1 and Σ_2 separately. These are direct analogs of the equations we encounter in superconductivity, where the boson means the phonon. It turns out that the vector interaction can give a non-trivial solution, whereas the pseudovector does not, because of the wrong sign of the self-energy. Physically speaking, the vector case causes an attractive interaction between virtual pairs, and hence a catastrophic change. Fig.3 shows the diagrams corresponding to our approximation and the above mentioned effect.

It should be interesting that this is the approximation considered by Landau[11] in his discussion of the Green's functions.

The qualitative feature of the solution is similar to the one obtained in the Heisenberg type theory, except for the nature of divergences. We also

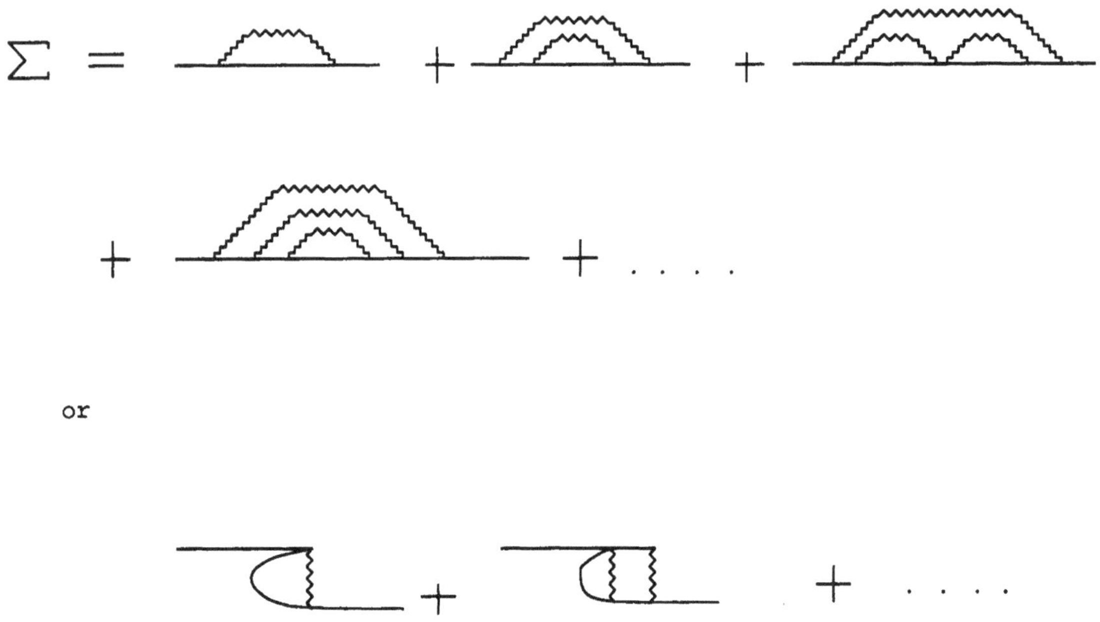

Fig. 3

obtain the zero mass bound states by solving the B-S equation in the ladder approximation. This solution was first discovered by Goldstein[12], but was considered as an abnormal object. Its raison d'etre has now become clear. According to Goldstein, however, the bound state wave function is not always normalizable. For small coupling constants, which is not really strong enough to cause such a binding, the norm of the wave function becomes negative (a ghost state).

Another intriguing question in this type of theory is whether one can produce a finite effective mass for the boson in a gauge invariant theory. Since we have done this for the fermion, it may not be impossible if some extra freedom is given to the field. Should the answer come out to be yes, the Yang-Mills-Yang-Lee-Sakurai theory of vector bosons would become very interesting indeed.

4

We finally come to the predictions or applications of our theory. The theory essentially boils down to the compound particle model of mesons, which has found some advocates in the past (Fermi-Yang, Sakata, Okun, Heisenberg,....). The new feature in our theory is that pseudoscalar mesons arise naturally and necessarily together with the nucleon mass as a consequence of the symmetry properties of the theory. Their type depends on the symmetry we assume. For example, it is possible to produce an isospin 1 meson but not an isospin 0 meson according to Eq.(20). There may, of course, be ordinary (or rather "accidental") bound states, but they will appear as excited or compound states of these basic mesons, like the now fashionable 2π and 3π resonances.

Being serious as we are about the compound particle model of the mesons, we should also try to calculate the meson-nucleon coupling constants from the basic constants. As is well known, there is no difference in the formal description of the particles whether they are elementary or compound. But if we know the wave function of a compound system, then the coupling constant is determined by it. In the case of a loosely bound system, the answer is simple. The pseudovector coupling constant is given by

$$\frac{f^2}{4\pi} = \frac{4a\mu}{m} \tag{21}$$

where m is the nucleon mass, μ the "meson" mass$= 2m - \epsilon$, and $a = \sqrt{m\epsilon}$ is the range of the wave function.

For strongly bound systems, however, the result should depend sensitively on the detailed dynamics since there are no "anomalous thresholds" in the dispersion relations for the form factors which reflect the structure of the wave function. Nevertheless let us extrapolate the above formula and see what happens. We find for the pion

$$\frac{f^2}{4\pi} = \frac{1}{n}\frac{4\mu}{m} \tag{22}$$

where we have also taken into account that different fermion pairs (proton, neutron, Λ, Σ, Ξ) may contribute to the pion state and n is the number of such pairs, assuming equal amount of contribution and neglecting mass fine structure. If we include all possible baryon combinations, then $n = 8$, and $f^2/4\pi = 0.075$!

We have to admit that we do not yet know what type of Lagrangian can possibly give rise to the observed baryon and meson spectrum. How much internal degree of freedom do we need to start with in order to account for the observed particles (and not to account for non-existing ones)? Why do the baryon masses split? Can we assume a high degree of symmetry properties in the beginning and yet come out with a smaller amount of apparent symmetries? The last question seems particularly relevant since we have found an example in the conflict between γ_5 invariance and finite mass. Here it should be helpful to seek analogies in solid state physics for better physical understanding.

For example, it is not hard to foresee how nature can manifest itself in an unsymmetric way while keeping the basic laws symmetric if we compare the situation with ferromagnetism. In an ordinary material, the ground state of a macroscopic body has spin zero practically, so that there is no preferred axis in space. In the ferromagnetic case, on the other hand, all the spins are parallel in the ground state, and they must point in <u>some</u> direction, thereby creating an asymmetry in reality. Such spontaneous polarizations may be happening in the world of elementary particles too[13].

We close this section with an application to weak interactions. So far this seems the most interesting and useful result coming out of our model.

The first question is , what is the renormalization of the vector and axial vector currents of the nucleon due to "strong" interactions, assuming there was a universal Fermi coupling in the beginning? Adopting our Heisenberg model, it can be shown that there will be no renormalization effect, and $g_V = g_A$ as long as we keep strict γ_5 invariance. The fact that g_A/g_V is only approximately unity implies then that there is a small violation of the invariance in agreement with the previous conclusion.

But this is not the whole story. We have already derived the nucleon axial vector vertex $\Gamma_5\mu$, which will also appear in the weak processes. The small violation of the invariance gives the meson mass μ, but will affect the form factor F of Eq.(12) relatively little. Thus we may be able to write

$$\Gamma_{5\mu}(p_2, p_1) = \gamma_5\gamma_\mu F_1(q^2) + \frac{2im\gamma_5 q_\mu}{q^2 + \mu^2} F_2(q^2)$$

$$\approx \left(\gamma_5\gamma_\mu + \frac{2im\gamma_5 q_\mu}{q^2+\mu^2}\right) F(q^2) \tag{23}$$

where now $F(0) = g_A/g_V = 1.25$. The second term of Eq.(23) is small compared to the first for the actual beta decay since $q^2 \ll \mu^2$. For large $q^2 \gg \mu^2$ we expect to recover the strict conservation: $F_1/F_2 \longrightarrow 1$ as $q^2 \longrightarrow \infty$.

Now we see that this second term enables one to determine the pion decay constant since, according to the dispersion theory, it represents the process going through the pion channel. Denoting the pion-nucleon (ps) coupling and the pion-lepton(pv) coupling as G_π and g_π respectively, we find

$$\sqrt{2}G_\pi g_\pi = 2mg_V F_2(-\mu^2) \approx 2mg_A. \tag{24}$$

Using $g_V = 10^{-5}/m^2$, $G^2/4\pi = 13.5$, we get $2.7 \times 10^{-8} sec$ for the pion life time, as compared to the observed $2.56 \times 10^{-8} sec$.

Eq.(24) is exactly the same as Goldberger-Treiman[14] formula derived by an entirely different approach and rather special assumptions. But we do not think that the agreement is a coincidence. There is a certain class of models which can more or less predict this relation[15]. The essential point seems to be that the pion is effectively treated as a bound state in the G-T theory. This manifests itself through the pion renormalization constant being zero or practically zero[16].

We can blindly generalize Eq.(24) to the strangeness changing axial vector current, where the pseudoscalar $K-$meson replaces the pion. Taking the ΛN vertex, for example, we again get a relation between the weak coupling g'_A, ΛNK coupling G_K and the $K-$lepton coupling g_K:

$$G_K g_K \approx (m_N + m_\Lambda)g'_A. \tag{25}$$

This relation does not contradict our present rather meager knowledge about these constants.

Since we are based on the compound particle model, all the considerations that have been made by various people in the past can be adopted in essence. The Gershtein-Zeldovich and Feynman-Gell-Mann idea of $\pi-$lepton vector coupling is, of course, a natural consequence of the model, though it has yet to be tested experimentally.

We feel that a systematic and quantitative calculation of the renormalization effects can be undertaken in our theory with more confidence than in the past because we have better understanding of the interrelation of different phenomena. So far we have tried to estimate the decay life times for most of the decay models of strange particles under very crude assumptions. The result is in general satisfactory, but it is not clear as to what it really means. We have not yet understood such basic questions as the $\Delta T = \frac{1}{2}$ rule and the smallness of the hyperon beta decay rate in any fundamental way.

References

[1] Bardeen, Cooper, and Schrieffer, Phys. Rev. 106, 162 (1957)

[2] N. N. Bogoliubov, J. Exptl. Theoret. Phys. (USSR) 34, 58, 73 (1958) (Soviet Phys. -JETP 34, 41, 51); Bogoliubov, Tolmachev, and Shirkov, "A New Method in the Theory of Superconductivity" (Academy of Sciences of USSR, Moscow, 1958). Also J. G. Valatin, Nuovo cimento 7, 843 (1958).

[3] W. Heisenberg, et al., Zeit.f. Naturf. 14, 441 (1959) Earlier papers are quoted there.

[4] C. N. Yang and R. L. Mills, Phys. Rev. 96, 191 (1954)

[5] T. D. Lee and C. N. Yang, Phys. Rev. 98, 1501 (1955)

[6] Y. Fujii, Progr. Theoret. Phys. (Kyoto) 21, 232 (1959)

[7] J. J. Sakurai, Annal. Phys., to be published.

[8] For example, Y. Takahashi, Nuovo cimento 6, 371 (1957)

[9] This type of theory has been discussed by F. Gursey, Nuovo cimento, to be published.

[10] Derivative interactions of scalar or pseudoscalar fields are also admissible.

[11] Landau, Abrikosov, and Khalatnikov, Dok. Akad. Nauk USSR 95, 497, 773 (1954); 96, 261 (1954): L. D. Landau, "Niels Bohr and the Development of Physics" (McGraw Hill Book Co., New York, 1955) p. 52

[12] J. Goldstein, Phys. Rev. 91, 1516 (1953).

[13] We find similar observations in Heisenberg's paper [3].

[14] M. L. Goldberger and S. B. Treiman, Phys. Rev. 110, 1178 (1958)

[15] Feynman, Gell-Mann, and Levy, to be published.

[16] K. Symanzik, Nuovo cimento 11, 269 (1958).
R. F. Sawyer, Phys. Rev. 116, 236 (1959)

DISCUSSION

Wightman: Is it true that the zero mass bound state moves up to become the $\pi-$meson when the γ_5 invariance is broken?

Jona-Lasinio: When you break the γ_5-invariance, for example by introducing a bare nucleon mass m_0, you dispose of an additional parameter which can be adjusted to give the desired mass.

Guth: How can you say that the model is a Heisenberg type theory if you do not specify the dynamics?

Jona-Lasinio: This theory is a Heisenberg type theory only in the sense that a nonlinear spinor equation is taken as starting point. The model is then linearized self-consistently and the cut-off which has to be introduced to eliminate divergences should be interpreted as a dynamical effect. The mechanism responsible for it is actually unspecified.

J. Sakurai: It is important to emphasize that much of what has been reported is independent of particular models. The only things that are relevant are: a) axial-vector conservation holds; and b) the force between a nucleon and an antinucleon is attractive.

Primakoff: When you break the γ_5 invariance to introduce the $\pi-$meson mass, how do you know that $\lim_{q^2 \to 0} F(q^2) \approx g_A/g_V$?

Jona-Lasinio: It is an assumption. In order to obtain both a finite $\pi-$meson mass and a renormalization of the axial vector coupling, we have to break the γ_5 invariance. So it is assumed the same γ_5 invariance violation is responsible for both effects.

PHYSICAL REVIEW VOLUME 122, NUMBER 1 APRIL 1, 1961

Dynamical Model of Elementary Particles Based on an Analogy with Superconductivity. I*

Y. NAMBU AND G. JONA-LASINIO†
The Enrico Fermi Institute for Nuclear Studies and the Department of Physics, The University of Chicago, Chicago, Illinois

(Received October 27, 1960)

It is suggested that the nucleon mass arises largely as a self-energy of some primary fermion field through the same mechanism as the appearance of energy gap in the theory of superconductivity. The idea can be put into a mathematical formulation utilizing a generalized Hartree-Fock approximation which regards real nucleons as quasi-particle excitations. We consider a simplified model of nonlinear four-fermion interaction which allows a γ_5-gauge group. An interesting consequence of the symmetry is that there arise automatically pseudoscalar zero-mass bound states of nucleon-antinucleon pair which may be regarded as an idealized pion. In addition, massive bound states of nucleon number zero and two are predicted in a simple approximation.

The theory contains two parameters which can be explicitly related to observed nucleon mass and the pion-nucleon coupling constant. Some paradoxical aspects of the theory in connection with the γ_5 transformation are discussed in detail.

I. INTRODUCTION

IN this paper we are going to develop a dynamical theory of elementary particles in which nucleons and mesons are derived in a unified way from a fundamental spinor field.[1] In basic physical ideas, it has thus the characteristic features of a compound-particle model, but unlike most of the existing theories, dynamical treatment of the interaction makes up an essential part of the theory. Strange particles are not yet considered.

The scheme is motivated by the observation of an interesting analogy between the properties of Dirac particles and the quasi-particle excitations that appear in the theory of superconductivity, which was originated with great success by Bardeen, Cooper, and Schrieffer,[2] and subsequently given an elegant mathematical formulation by Bogoliubov.[3] The characteristic feature of the BCS theory is that it produces an energy gap between the ground state and the excited states of a superconductor, a fact which has been confirmed experimentally. The gap is caused due to the fact that the attractive phonon-mediated interaction between electrons produces correlated pairs of electrons with opposite momenta and spin near the Fermi surface, and it takes a finite amount of energy to break this correlation.

Elementary excitations in a superconductor can be conveniently described by means of a coherent mixture of electrons and holes, which obeys the following equations[3,4]:

$$E\psi_{p+}=\epsilon_p\psi_{p+}+\phi\psi_{-p-}^*,\qquad E\psi_{-p-}^*=-\epsilon_p\psi_{-p-}^*+\phi\psi_{p+}, \tag{1.1}$$

near the Fermi surface. ψ_{p+} is the component of the excitation corresponding to an electron state of momentum p and spin $+$ (up), and ψ_{-p-}^* corresponding to a hole state of momentum p and spin $+$, which means an absence of an electron of momentum $-p$ and spin $-$ (down). ϵ_p is the kinetic energy measured from the Fermi surface; ϕ is a constant. There will also be an equation complex conjugate to Eq. (1), describing another type of excitation.

Equation (1) gives the eigenvalues

$$E_p=\pm(\epsilon_p^2+\phi^2)^{\frac{1}{2}}. \tag{1.2}$$

The two states of this quasi-particle are separated in energy by $2|E_p|$. In the ground state of the system all the quasi-particles should be in the lower (negative) energy states of Eq. (2), and it would take a finite energy $2|E_p|\geqslant 2|\phi|$ to excite a particle to the upper state. The situation bears a remarkable resemblance to the case of a Dirac particle. The four-component Dirac equation can be split into two sets to read

$$E\psi_1=\sigma\cdot p\psi_1+m\psi_2,\qquad E\psi_2=-\sigma\cdot p\psi_2+m\psi_1,\qquad E_p=\pm(p^2+m^2)^{\frac{1}{2}}, \tag{1.3}$$

where ψ_1 and ψ_2 are the two eigenstates of the chirality operator $\gamma_5=\gamma_1\gamma_2\gamma_3\gamma_4$.

According to Dirac's original interpretation, the ground state (vacuum) of the world has all the electrons in the negative energy states, and to create excited states (with zero particle number) we have to supply an energy $\geqslant 2m$.

In the BCS-Bogoliubov theory, the gap parameter ϕ, which is absent for free electrons, is determined essentially as a self-consistent (Hartree-Fock) representation of the electron-electron interaction effect.

* Supported by the U. S. Atomic Energy Commission.

† Fulbright Fellow, on leave of absence from Instituto di Fisica dell' Universita, Roma, Italy and Istituto Nazionale di Fisica Nucleare, Sezione di Roma, Italy.

[1] A preliminary version of the work was presented at the Midwestern Conference on Theoretical Physics, April, 1960 (unpublished). See also Y. Nambu, Phys. Rev. Letters 4, 380 (1960); and Proceedings of the Tenth Annual Rochester Conference on High-Energy Nuclear Physics, 1960 (to be published).

[2] J. Bardeen, L. N. Cooper, and J. R. Schrieffer, Phys. Rev. 106, 162 (1957).

[3] N. N. Bogoliubov, J. Exptl. Theoret. Phys. (U.S.S.R.) 34, 58, 73 (1958) [translation: Soviet Phys.-JETP 34, 41, 51 (1958)]; N. N. Bogoliubov, V. V. Tolmachev, and D. V. Shirkov, *A New Method in the Theory of Superconductivity* (Academy of Sciences of U.S.S.R., Moscow, 1958).

[4] J. G. Valatin, Nuovo cimento 7, 843 (1958).

One finds that

$$\phi \approx \omega \exp[-1/\rho], \tag{1.4}$$

where ω is the energy bandwidth ($\approx$ the Debye frequency) around the Fermi surface within which the interaction is important; ρ is the average interaction energy of an electron interacting with unit energy shell of electrons on the Fermi surface. It is significant that ϕ depends on the strength of the interaction (coupling constant) in a nonanalytic way.

We would like to pursue this analogy mathematically. As the energy gap ϕ in a superconductor is created by the interaction, let us assume that the mass of a Dirac particle is also due to some interaction between massless bare fermions. A quasi-particle in a superconductor is a mixture of bare electrons with opposite electric charges (a particle and a hole) but with the same spin; correspondingly a massive Dirac particle is a mixture of bare fermions with opposite chiralities, but with the same charge or fermion number. Without the gap ϕ or the mass m, the respective particle would become an eigenstate of electric charge or chirality.

Once we make this analogy, we immediately notice further consequences of special interest. It has been pointed out by several people[3,5–8] that in a refined theory of superconductivity there emerge, in addition to the individual quasi-particle excitations, collective excitations of quasi-particle pairs. (These can alternatively be interpreted as moving states of bare electron pairs which are originally precipitated into the ground state of the system.) In the absence of Coulomb interaction, these excitations are phonon-like, filling the gap of the quasi-particle spectrum.

In general, they are excited when a quasi-particle is accelerated in the medium, and play the role of a backflow around the particle, compensating the change of charge localized on the quasi-particle wave packet. Thus these excitations are necessary consequences of the fact that individual quasi-particles are not eigenstates of electric charge, and hence their equations are not gauge invariant; whereas a complete description of the system must be gauge invariant. The logical connection between gauge invariance and the existence of collective states has been particularly emphasized by one of the authors.[8]

This observation leads to the conclusion that if a Dirac particle is actually a quasi-particle, which is only an approximate description of an entire system where chirality is conserved, then there must also exist collective excitations of bound quasi-particle pairs. The chirality conservation implies the invariance of the theory under the so-called γ_5 gauge group, and from its nature one can show that the collective state must be a pseudoscalar quantity.

[5] D. Pines and J. R. Schrieffer, Nuovo cimento **10**, 496 (1958).
[6] P. W. Anderson, Phys. Rev. **110**, 827, 1900 (1958); **114**, 1002 (1959).
[7] G. Rickayzen, Phys. Rev. **115**, 795 (1959).
[8] Y. Nambu, Phys. Rev. **117**, 648 (1960).

It is perhaps not a coincidence that there exists such an entity in the form of the pion. For this reason, we would like to regard our theory as dealing with nucleons and mesons. The implication would be that the nucleon mass is a manifestation of some unknown primary interaction between originally massless fermions, the same interaction also being responsible for the binding of nucleon pairs into pions.

An additional support of the idea can be found in the weak decay processes of nucleons and pions which indicate that the γ_5 invariance is at least approximately conserved, as will be discussed in Part II. There are some difficulties, however, that naturally arise on further examination.

Comparison between a relativistic theory and a nonrelativistic, intuitive picture is often dangerous, because the former is severely restricted by the requirement of relativistic invariance. In our case, the energy-gap equation (4) depends on the energy density on the Fermi surface; for zero Fermi radius, the gap vanishes. The Fermi sphere, however, is not a relativistically invariant object, so that in the theory of nucleons it is not clear whether a formula like Eq. (4) could be obtained for the mass. This is not surprising, since there is a well known counterpart in classical electron theory that a finite electron radius is incompatible with relativistic invariance.

We avoid this difficulty by simply introducing a relativistic cutoff which takes the place of the Fermi sphere. Our framework does not yet resolve the divergence difficulty of self-energy, and the origin of such an effective cutoff has to be left as an open question.

The second difficulty concerns the mass of the pion. If pion is to be identified with the phonon-like excitations associated with a gauge group, its mass must necessarily be zero. It is true that in real superconductors the collective charge fluctuation is screened by Coulomb interaction to turn into the plasma mode, which has a finite "rest mass." A similar mechanism may be operating in the meson case too. It is possible, however, that the finite meson mass means that chirality conservation is only approximate in a real theory. From the evidence in weak interactions, we are inclined toward the second view.

The observation made so far does not yet give us a clue as to the exact mechanism of the primary interaction. Neither do we have a fundamental understanding of the isospin and strangeness quantum numbers, although it is easy to incorporate at least the isospin degree of freedom into the theory from the beginning. The best we can do here is to examine the various existing models for their logical simplicity and experimental support, if any. We will do this in Sec. 2, and settle for the moment on a nonlinear four-fermion interaction of the Heisenberg type. For reasons of simplicity in presentation, we adopt a model without isospin and strangeness degrees of freedom, and possessing complete γ_5 invariance. Once the choice is made, we can explore the whole idea mathematically, using essentially the formulation developed in reference 8. It is gratifying that the various field-theoretical techniques can be fully utilized. Section 3 will be devoted to introduction of the Hartree-Fock equation for nucleon self-energy, which will make the starting point of the theory. Then we go on to discuss in Sec. 4 the collective modes. In addition to the expected pseudoscalar "pion" states, we find other massive mesons of scalar and vector variety, as well as a scalar "deuteron." The coupling constants of these mesons can be easily determined. The relation of the pion to the γ_5 gauge group will be discussed in Secs. 5 and 6.

The theory promises many practical consequences. For this purpose, however, it is necessary to make our model more realistic by incorporating the isospin, and allowing for a violation of γ_5 invariance. But in doing so, there arise at the same time new problems concerning the mass splitting and instability. This refined model will be elaborated in Part II of this work, where we shall also find predictions about strong and weak interactions. Thus the general structure of the weak interaction currents modified by strong interactions can be treated to some degree, enabling one to derive the decay processes of various particles under simple assumptions. The calculation of the pion decay rate gives perhaps one of the most interesting supports of the theory. Results about strong interactions themselves are equally interesting. We shall find specific predictions about heavier mesons, which are in line with the recent theoretical expectations.

II. THE PRIMARY INTERACTION

We briefly discuss the possible nature of the primary interaction between fermions. Lacking any radically new concepts, the interaction could be either mediated by some fundamental Bose field or due to an inherent nonlinearity in the fermion field. According to our postulate, these interactions must allow chirality conservation in addition to the conservation of nucleon number. The chirality X here is defined as the eigenvalue of γ_5, or in terms of quantized fields,

$$X=\int \bar{\psi}\gamma_4\gamma_5\psi d^3x. \tag{2.1}$$

The nucleon number is, on the other hand

$$N=\int \bar{\psi}\gamma_4\psi d^3x. \tag{2.2}$$

These are, respectively, generators of the γ_5- and ordinary-gauge groups

$$\psi\rightarrow\exp[i\alpha\gamma_5]\psi,\quad \bar{\psi}\rightarrow\bar{\psi}\exp[i\alpha\gamma_5], \tag{2.3}$$

$$\psi\rightarrow\exp[i\alpha]\psi,\quad \bar{\psi}\rightarrow\bar{\psi}\exp[-i\alpha], \tag{2.4}$$

where α is an arbitrary constant phase.

Furthermore, the dynamics of our theory would require that the interaction be attractive between particle and antiparticle in order to make bound-state formation possible. Under the transformation (2.3), various tensors transform as follows:

$$\begin{aligned}
&\text{Vector:} && i\bar{\psi}\gamma_\mu\psi\rightarrow i\bar{\psi}\gamma_\mu\psi,\\
&\text{Axial vector:} && i\bar{\psi}\gamma_\mu\gamma_5\psi\rightarrow i\bar{\psi}\gamma_\mu\gamma_5\psi,\\
&\text{Scalar:} && \bar{\psi}\psi\rightarrow\bar{\psi}\psi\cos2\alpha+i\bar{\psi}\gamma_5\psi\sin2\alpha,\\
&\text{Pseudoscalar:} && i\bar{\psi}\gamma_5\psi\rightarrow i\bar{\psi}\gamma_5\psi\cos2\alpha-\bar{\psi}\psi\sin2\alpha,\\
&\text{Tensor:} && \bar{\psi}\sigma_{\mu\nu}\psi\rightarrow\bar{\psi}\sigma_{\mu\nu}\psi\cos2\alpha+i\bar{\psi}\gamma_5\sigma_{\mu\nu}\psi\sin2\alpha.
\end{aligned} \tag{2.5}$$

It is obvious that a vector or pseudovector Bose field coupled to the fermion field satisfies the invariance. The vector case would also satisfy the dynamical requirement since, as in the electromagnetic interaction, the forces would be attractive between opposite nucleon charges. The pseudovector field, on the other hand, does not meet the requirement as can be seen by studying the self-consistent mass equation discussed later.

The vector field looks particularly attractive since it can be associated with the nucleon number gauge group. This idea has been explored by Lee and Yang,[9] and recently by Sakurai.[10] But since we are dealing with strong interactions, such a field would have to have a finite observed mass in a realistic theory. Whether this is compatible with the invariance requirement is not yet clear. (Besides, if the bare mass of both spinor and vector field were zero, the theory would not contain any parameter with the dimensions of mass.)

The nonlinear fermion interaction seems to offer another possibility. Heisenberg and his co-workers[11] have been developing a comprehensive theory of elementary particles along this line. It is not easy, however, to gain a clear physical insight into their results obtained by means of highly complicated mathematical machinery.

We would like to choose the nonlinear interaction in this paper. Although this looks similar to Heisenberg's theory, the dynamical treatment will be quite different and more amenable to qualitative understanding.

The following Lagrangian density will be assumed ($\hbar=c=1$):

$$L=-\bar{\psi}\gamma_\mu\partial_\mu\psi+g_0[(\bar{\psi}\psi)^2-(\bar{\psi}\gamma_5\psi)^2]. \tag{2.6}$$

The coupling parameter g_0 is positive, and has dimensions $[\text{mass}]^{-2}$. The γ_5 invariance property of the interaction is evident from Eq. (2.5). According to the Fierz theorem, it is also equivalent to

$$-\tfrac{1}{2}g_0[(\bar{\psi}\gamma_\mu\psi)^2-(\bar{\psi}\gamma_\mu\gamma_5\psi)^2]. \tag{2.7}$$

This particular choice of γ_5-invariant form was taken without a compelling reason, but has the advantage

[9] T. D. Lee and C. N. Yang, Phys. Rev. 98, 1501 (1955).
[10] J. J. Sakurai, Ann. Phys. 11, 1 (1960).
[11] W. Heisenberg, Z. Naturforsch. 14, 441 (1959). Earlier papers are quoted there.

that it can be naturally extended to incorporate isotopic spin.[12]

Unlike Heisenberg's case, we do not have any theory about the handling of the highly divergent singularities inherent in nonlinear interactions. So we will introduce, as an additional and independent assumption, an *ad hoc* relativistic cutoff or form factor in actual calculations. Thus the theory may also be regarded as an approximate treatment of the intermediate-boson model with a large effective mass.

As will be seen in subsequent sections, the nonlinear model makes mathematics particularly easy, at least in the lowest approximation, enabling one to derive many interesting quantitative results.

III. THE SELF-CONSISTENT EQUATION FOR NUCLEON MASS

We will assume that all quantities we calculate here are somewhow convergent, without asking the reason behind it. This will be done actually by introducing a suitable phenomenological cutoff.

Without specifying the interaction, let Σ be the unrenormalized proper self-energy part of the fermion, expressed in terms of observed mass m, coupling constant g, and cutoff Λ. A real Dirac particle will satisfy the equation

$$i\gamma\cdot p+m_0+\Sigma(p,m,g,\Lambda)=0 \tag{3.1}$$

for $i\gamma\cdot p+m=0$. Namely

$$m-m_0=\Sigma(p,m,g,\Lambda)\,|_{i\gamma\cdot p+m=0}. \tag{3.2}$$

The g will also be related to the bare coupling g_0 by an equation of the type

$$g/g_0=\Gamma(m,g,\Lambda). \tag{3.3}$$

Equations (3.1) and (3.2) may be solved by successive approximation starting from m_0 and g_0. It is possible, however, that there are also solutions which cannot thus be obtained. In fact, there can be a solution $m\neq 0$ even in the case where $m_0=0$, and moreover the symmetry seems to forbid a finite m.

This kind of situation can be most easily examined by means of the generalized Hartree-Fock procedure[8,13] which was developed before in connection with the theory of superconductivity. The basic idea is not new in field theory, and in fact in its simplest form the method is identical with the renormalization procedure of Dyson, considered only in a somewhat different context.

Suppose a Lagrangian is composed of the free and interaction part: $L=L_0+L_i$. Instead of diagonalizing L_0 and treating L_i as perturbation, we introduce the self-energy Lagrangian L_s, and split L thus

$$L=(L_0+L_s)+(L_i-L_s)$$
$$=L_0'+L_i'.$$

For L_s we assume quite general form (quadratic or bilinear in the fields) such that L_0' leads to linear field equations. This will enable one to define a vacuum and a complete set of "quasi-particle" states, each particle being an eigenmode of L_0'. Now we treat L_i' as perturbation, and determine L_s from the requirement that L_i' shall not yield additional self-energy effects. This procedure then leads to Eq. (3.2). The self-consistent nature of such a procedure is evident since the self-energy is calculated by perturbation theory with fields which are already subject to the self-energy effect.

In order to apply the method to our problem, let us assume that $L_s=-m\bar{\psi}\psi$, and introduce the propagator $S_F^{(m)}(x)$ for the corresponding Dirac particle with mass m. In the lowest order, and using the two alternative forms Eqs. (2.6) and (2.7), we get for Eq. (3.2)

$$\Sigma=2g_0[\mathrm{Tr}S_F^{(m)}(0)-\gamma_5\,\mathrm{Tr}S_F^{(m)}(0)\gamma_5 -\tfrac{1}{2}\gamma_\mu\,\mathrm{Tr}\gamma_\mu S_F^{(m)}(0)+\tfrac{1}{2}\gamma_\mu\gamma_5\,\mathrm{Tr}\gamma_\mu\gamma_5 S_F^{(m)}(0)] \tag{3.4}$$

in coordinate space.

This is quadratically divergent, but with a cutoff can be made finite. In momentum space we have

$$\Sigma=-\frac{8g_0i}{(2\pi)^4}\int\frac{m}{p^2+m^2-i\epsilon}d^4p\,F(p,\Lambda), \tag{3.5}$$

where $F(p,\Lambda)$ is a cutoff factor. In this case the self-energy operator is a constant. Substituting Σ from Eq. (3.5), Eq. (3.2) gives ($m_0=0$)

$$m=-\frac{g_0mi}{2\pi^4}\int\frac{d^4p}{p^2+m^2-i\epsilon}F(p,\Lambda). \tag{3.6}$$

This has two solutions: either $m=0$, or

$$1=-\frac{g_0i}{2\pi^4}\int\frac{d^4p}{p^2+m^2-i\epsilon}F(p,\Lambda). \tag{3.7}$$

The first trivial one corresponds to the ordinary perturbative result. The second, nontrivial solution will determine m in terms of g_0 and Λ.

If we evaluate Eq. (3.7) with a straight noninvariant cutoff at $|\mathbf{p}|=\Lambda$, we get

$$\frac{\pi^2}{g_0\Lambda^2}=\left(\frac{m^2}{\Lambda^2}+1\right)^{\frac{1}{2}}-\frac{m^2}{\Lambda^2}\ln\left[\left(\frac{\Lambda^2}{m^2}+1\right)^{\frac{1}{2}}+\frac{\Lambda}{m}\right]. \tag{3.8}$$

If we use Eq. (3.5) with an invariant cutoff at $p^2=\Lambda^2$ after the change of path: $p_0\to ip_0$, we get

$$\frac{2\pi^2}{g_0\Lambda^2}=1-\frac{m^2}{\Lambda^2}\ln\left(\frac{\Lambda^2}{m^2}+1\right). \tag{3.9}$$

---

[12] This will be done in Part II.

[13] N. N. Bogoliubov, Uspekhi Fiz. Nauk 67, 549 (1959) [translation: Soviet Phys.-Uspekhi 67, 236 (1959)].

Since the right-hand side of Eq. (3.8) or (3.9) is positive and $\leqslant 1$ for real Λ/m, the nontrivial solution exists only if

$$0<2\pi^2/g_0\Lambda^2<1. \tag{3.10}$$

Equation (3.9) is plotted in Fig. 1 as a function of m^2/Λ^2. As $g_0\Lambda^2$ increases over the critical value $2\pi^2$, m starts rising from 0. The nonanalytic nature of the solution is evident as m cannot be expanded in powers of g_0.

In the following we will assume that Eq. (3.10) is satisfied, so that the nontrivial solution exists. As we shall see later, physically this means that the nucleon-antinucleon interaction must be attractive ($g_0>0$) and strong enough to cause a bound pair of zero total mass. In the BCS theory, the nontrivial solution corresponds to a superconductive state, whereas the trivial one corresponds to a normal state, which is not the true ground state of the superconductor. We may expect a similar situation to hold in the present case.

In this connection, it must be kept in mind that our solutions are only approximate ones. We are operating under the assumption that the corrections to them are not catastrophic, and can be appropriately calculated when necessary. If this does not turn out to be so for some solution, such a solution must be discarded. Later we shall indeed find this possibility for the trivial solution, but for the moment we will ignore such considerations.

Let us define then the vacuum corresponding to the two solutions. Let $\psi^{(0)}$ and $\psi^{(m)}$ be quantized fields satisfying the equations

$$\gamma_\mu\partial_\mu\psi^{(0)}(x)=0, \tag{3.11a}$$

$$(\gamma_\mu\partial_\mu+m)\psi^{(m)}(x)=0, \tag{3.11b}$$

$$\psi^{(0)}(x)=\psi^{(m)}(x) \quad \text{for} \quad x_0=0. \tag{3.11c}$$

According to the standard procedure, we decompose the ψ's into Fourier components:

$$\begin{aligned}\psi_\alpha{}^{(i)}(x)=&\frac{1}{V^{\frac12}}\sum_{\substack{\mathbf{p},s\\ p_0=(p^2+m^2)^{\frac12}}}[u_\alpha{}^{(i)}(\mathbf{p},s)a^{(i)}(\mathbf{p},s)e^{ip\cdot x}\\&+v_\alpha{}^{*(i)}(\mathbf{p},s)b^{(i)\dagger}(\mathbf{p},s)e^{-ip\cdot x}],\\ \psi_\alpha{}^{\dagger(i)}(x)=&\frac{1}{V^{\frac12}}\sum_{\substack{\mathbf{p},s\\ p_0=(p^2+m^2)^{\frac12}}}[u_\alpha{}^{(i)*}(\mathbf{p}\cdot s)a^{(i)\dagger}(\mathbf{p},s)\\&\times e^{-ip\cdot x}+v_\alpha{}^{(i)}(\mathbf{p},s)b^{(i)}(\mathbf{p},s)e^{ip\cdot x}],\\ &i=0 \text{ or } m,\end{aligned} \tag{3.12}$$

where $u_\alpha{}^{(i)}(\mathbf{p},s)$, $v_\alpha{}^{(i)}(\mathbf{p},s)$ are the normalized spinor eigenfunctions for particles and antiparticles, with momentum p and helicity $s=\pm1$, and

$$\begin{aligned}&\{a^{(i)}(\mathbf{p},s),a^{(i)\dagger}(\mathbf{p}',s')\}\\&=\{b^{(i)}(\mathbf{p},s),b^{(i)\dagger}(\mathbf{p}',s')\}=\delta_{\mathbf{pp}'}\delta_{ss'}, \text{ etc.}\end{aligned} \tag{3.13}$$

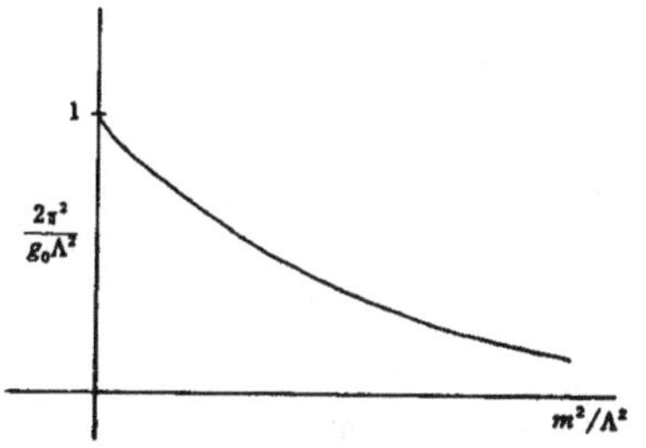

FIG. 1. Plot of the self-consistent mass equation (3.9).

The operator sets $(a^{(0)},b^{(0)})$ and $(a^{(m)},b^{(m)})$ are related by a canonical transformation because of Eq. (3.11c):

$$\begin{aligned}a^{(m)}(\mathbf{p},s)=&\sum_{\alpha,s'}[u_\alpha{}^{(m)*}(\mathbf{p},s)u_\alpha{}^{(0)}(\mathbf{p},s')a^{(0)}(\mathbf{p},s')\\&+u_\alpha{}^{(m)*}(\mathbf{p},s)v_\alpha{}^{(0)*}(-\mathbf{p},s')b^{(0)\dagger}(-\mathbf{p},s')],\\ b^{(m)}(\mathbf{p},s)=&\sum_{\alpha,s'}[v_\alpha{}^{(m)*}(\mathbf{p},s)v_\alpha{}^{(0)}(\mathbf{p},s')b^{(0)}(\mathbf{p},s')\\&+v_\alpha{}^{(m)*}(\mathbf{p},s)u_\alpha{}^{(0)*}(-\mathbf{p},s')a^{(0)\dagger}(-\mathbf{p},s')].\end{aligned} \tag{3.14}$$

Using Eq. (1.3), this is evaluated to give

$$\begin{aligned}a^{(m)}(\mathbf{p},s)=&[\tfrac12(1+\beta_p)]^{\frac12}a^{(0)}(\mathbf{p},s)\\&+[\tfrac12(1-\beta_p)]^{\frac12}b^{(0)\dagger}(-\mathbf{p},s),\\ b^{(m)}(\mathbf{p},s)=&[\tfrac12(1+\beta_p)]^{\frac12}b^{(0)}(\mathbf{p},s)\\&-[\tfrac12(1-\beta_p)]^{\frac12}a^{(0)\dagger}(-\mathbf{p},s),\\ \beta_p=&|\mathbf{p}|/(\mathbf{p}^2+m^2)^{\frac12}.\end{aligned} \tag{3.15}$$

The vacuum $\Omega^{(0)}$ or $\Omega^{(m)}$ with respect to the field $\psi^{(0)}$ or $\psi^{(m)}$ is now defined as

$$a^{(0)}(\mathbf{p},s)\Omega^{(0)}=b^{(0)}(\mathbf{p},s)\Omega^{(0)}=0, \tag{3.16}$$

$$a^{(m)}(\mathbf{p},s)\Omega^{(m)}=b^{(m)}(\mathbf{p},s)\Omega^{(m)}=0. \tag{3.16'}$$

Both $\psi^{(0)}$, $\bar\psi^{(0)}$ and $\psi^{(m)}$, $\bar\psi^{(m)}$ applied to $\Omega^{(0)}$ always create particles of mass zero, whereas the same applied to $\Omega^{(m)}$ create particles of mass m.

From Eqs. (3.15) and (3.16) we obtain

$$\begin{aligned}\Omega^{(m)}=\prod_{\mathbf{p},s}\{&[\tfrac12(1+\beta_p)]^{\frac12}\\&-[\tfrac12(1-\beta_p)]^{\frac12}a^{(0)\dagger}(\mathbf{p},s)b^{(0)\dagger}(-\mathbf{p},s)\}\Omega^{(0)}.\end{aligned} \tag{3.17}$$

Thus $\Omega^{(m)}$ is, in terms of zero-mass particles, a superposition of pair states. Each pair has zero momentum, spin and nucleon number, and carries ±2 units of chirality, since chirality equals minus the helicity s for massless particles.

Let us calculate the scalar product $(\Omega^{(0)},\Omega^{(m)})$ from Eq. (3.15):

$$\begin{aligned}(\Omega^{(0)},\Omega^{(m)})&=\prod_{\mathbf{p},s}[\tfrac12(1+\beta_p)]^{\frac12}\\&=\exp\{\sum_{\mathbf{p},s}\tfrac12\ln[\tfrac12(1+\beta_p)]\}.\end{aligned} \tag{3.18}$$

For large p, $\beta_p\sim1-m^2/2\mathbf{p}^2$, so that the exponent diverges as $V\pi m^2\int d\mathbf{p}/(2\pi)^3$ (V = normalization volume). Hence

$$(\Omega^{(0)},\Omega^{(m)})=0. \tag{3.19}$$

It is easy to see that any two states $\Psi^{(0)}$ and $\Psi^{(m)}$, obtained by applying a finite number of creation operators on $\Omega^{(0)}$ and $\Omega^{(m)}$ respectively, are also orthogonal.

Thus the two "worlds" based on $\Omega^{(0)}$ and $\Omega^{(m)}$ are physically distinct and outside of each other. No interaction or measurement, in the usual sense, can bridge them in finite steps.

What is the energy difference of the two vacua? Since both are Lorentz invariant states, the difference can only be either zero or infinity. Using the expression

$$\begin{aligned} H^{(m)}&=\sum_{\mathbf{p},s}(\mathbf{p}^2+m^2)^{\frac{1}{2}}\{a^{(m)\dagger}(\mathbf{p},s)a^{(m)}(\mathbf{p},s)\\ &\qquad -b^{(m)}(\mathbf{p},s)b^{(m)\dagger}(\mathbf{p},s)\},\\ H^{(0)}&=\sum_{\mathbf{p},s}|\mathbf{p}|\{a^{(0)\dagger}(\mathbf{p},s)a^{(0)}(\mathbf{p},s)\\ &\qquad -b^{(0)}(\mathbf{p},s)b^{(0)\dagger}(\mathbf{p},s)\}, \end{aligned} \tag{3.20}$$

we get for the respective energies

$$E^{(m)}-E^{(0)}=-2\sum_{\mathbf{p}}[(\mathbf{p}^2+m^2)^{\frac{1}{2}}-|\mathbf{p}|], \tag{3.21}$$

which is negative and quadratically divergent. So $\Omega^{(m)}$ may be called the "true" ground state, as was expected.

There remains finally the question of γ_5 invariance. The original Hamiltonian allowed two conservations X and N, Eqs. (2.1) and (2.2). Both $\Omega^{(0)}$ and $\Omega^{(m)}$ belong to $N=0$, and their elementary excitations carry $N=\pm1$. In the case of X, the same is true for the space $\Omega^{(0)}$, but $\Omega^{(m)}$ as well as its elementary excitations are not eigenstates of X, as is clear from the foregoing results. If the latter solution is to be a possibility, there must be an infinite degeneracy with respect to the quantum number X. A ground state will be in general a linear combination of degenerate states with different $X=0$, ±2, $\cdots$:

$$\Omega^{(m)}=\sum_{n=-\infty}^{\infty}C_{2n}\Omega_{2n}^{(m)}. \tag{3.22}$$

Equation (3.17) is in fact a particular case of this. The γ_5-gauge transformation Eq. (2.3) induces the change

$$\begin{aligned} a^{(0)}(\mathbf{p},\pm1)&\rightarrow e^{\mp i\alpha}a^{(0)}(\mathbf{p},\pm1),\\ b^{(0)}(\mathbf{p},\pm1)&\rightarrow e^{\mp i\alpha}b^{(0)}(\mathbf{p},\pm1),\\ a^{(0)\dagger}(\mathbf{p},\pm1)&\rightarrow e^{\pm i\alpha}a^{(0)\dagger}(\mathbf{p},\pm1),\\ b^{(0)\dagger}(\mathbf{p},\pm1)&\rightarrow e^{\pm i\alpha}b^{(0)\dagger}(\mathbf{p},\pm1), \end{aligned} \tag{3.23}$$

and the coefficients of Eq. (3.22) become

$$C_{2n}\rightarrow e^{-2ni\alpha}C_{2n}. \tag{3.24}$$

In particular

$$\begin{aligned} \Omega^{(m)}&\rightarrow\Omega_\alpha^{(m)}\\ &=\exp[-i\alpha X]\Omega^{(m)}\\ &=\prod_{\mathbf{p},\pm}\{[\tfrac{1}{2}(1+\beta_p)]^{\frac{1}{2}}-[\tfrac{1}{2}(1-\beta_p)]^{\frac{1}{2}}\\ &\qquad\times e^{\pm2i\alpha}a^{(0)\dagger}(p,\pm)b^{(0)\dagger}(-p,\pm)\}\Omega^{(0)}. \end{aligned} \tag{3.25}$$

The Dirac equation (3.11b), at the same time, is transformed into

$$[\gamma_\mu\partial_\mu+m\cos2\alpha+im\gamma_5\sin2\alpha]\psi=0. \tag{3.26}$$

The moral of this is that the self-consistent self-energy Σ is determined only up to a γ_5 transformation. This can be easily verified from Eq. (3.4), in which the second term on the right-hand side is nonvanishing when a propagator corresponding to Eq. (3.26) is used. Although Eq. (3.26) seems to violate parity conservation, it is only superficially so since $\Omega_\alpha^{(m)}$ is now not an eigenstate of parity. We could alternatively say that the parity operator undergoes transformation together with the mass operator. Despite the odd form of the equation (3.26), there is no change in the physical predictions of the theory. We shall see more of this later.

Let us calculate, as before, the scalar product of $\Omega_\alpha^{(m)}$ and $\Omega_{\alpha'}^{(m)}$. From Eqs. (3.17) and (3.25) we get

$$\begin{aligned} (\Omega_\alpha^{(m)},\Omega_{\alpha'}^{(m)})&=\prod_{p,\pm}[\tfrac{1}{2}(1+\beta_p)-e^{\pm2i(\alpha'-\alpha)}\tfrac{1}{2}(1-\beta_p)]\\ &=\prod_{p,\pm}[1+(e^{\pm2i(\alpha'-\alpha)}-1)\tfrac{1}{2}(1-\beta_p)]\\ &=\exp\{\sum_{p,\pm}\ln[1+(e^{\pm2i(\alpha'-\alpha)}-1)\tfrac{1}{2}(1-\beta_p)]\}. \end{aligned} \tag{3.27}$$

For large $|\mathbf{p}|$, the exponent goes like

$$\frac{V}{(2\pi)^3}\sum_{\pm}(e^{\pm2i(\alpha'-\alpha)}-1)\int\frac{m^2}{4\mathbf{p}^2}d^3p.$$

The integral is again divergent. Hence

$$\begin{aligned} (\Omega_\alpha^{(m)},\Omega_{\alpha'}^{(m)})&=(\Omega^{(m)},\exp[-i(\alpha'-\alpha)X]\Omega^{(m)})\\ &=0,\quad \alpha'\neq\alpha(\mathrm{mod}\,2\pi), \end{aligned} \tag{3.28}$$

and, of course

$$(\Omega^{(0)},\Omega_\alpha^{(m)})=0. \tag{3.28'}$$

We can evaluate $(\Omega_\alpha^{(m)},\Omega_{\alpha'}^{(m)})$ alternatively from Eqs. (3.22) and (3.24). Then

$$\sum_{m=-\infty}^{\infty}|C_{2n}|^2e^{2ni(\alpha-\alpha')}=0,\quad \alpha\neq\alpha'(\mathrm{mod}\,2\pi), \tag{3.29}$$

implying that

$$|C_0|=|C_{\pm2}|=|C_{\pm4}|=\cdots=C. \tag{3.30}$$

Thus there is an infinity of equivalent worlds described by $\Omega_\alpha^{(m)}$, $0\leqslant\alpha<2\pi$. The states Ω_{2n} of Eq. (3.22) are then expressed in terms of $\Omega_\alpha^{(m)}$ as

$$C_{2n}\Omega_{2n}^{(m)}=\frac{1}{2\pi}\int_0^{2\pi}e^{2ni\alpha}\Omega_\alpha^{(m)}d\alpha, \tag{3.31}$$

which form another orthogonal set. Since the original total H commutes with X, it will have no matrix elements connecting different "worlds." Moreover, as was the case with $\Omega^{(m)}$ and $\Omega^{(0)}$, no finite measurement can induce similar transitions. This is a kind of superselection rule, which effectively avoids the apparent degeneracy to show up as physical effects.[14] The usual description of the world by means of $\Omega^{(m)}$ and ordinary Dirac particles must be regarded as only the most convenient one.

We still are left with some paradoxes. The X conservation implies the existence of a conserved X current:

$$j_{\mu 5}=i\bar{\psi}\gamma_\mu\gamma_5\psi, \tag{3.32}$$

$$\partial_\mu j_{\mu 5}=0, \tag{3.32'}$$

which can readily be verified from Eq. (2.6). On the other hand, for a massive Dirac particle the continuity equation is not satisfied:

$$\partial_\mu\bar{\psi}^{(m)}\gamma_\mu\gamma_5\psi^{(m)}=2m\bar{\psi}^{(m)}\gamma_5\psi^{(m)}. \tag{3.33}$$

If a massive Dirac particle has to be a real eigenstate of the system, how can this be reconciled? The answer would be that the X-current operator taken between real one-nucleon states should not be given simply by $i\gamma_\mu\gamma_5$ because of the "radiative corrections." We expect instead

$$\langle p'|j_{\mu 5}|p\rangle=\bar{u}(p')X_\mu(p',p)u(p), \tag{3.34}$$

where the renormalized quantity $X_{\mu 5}$ should be, from relativistic invariance grounds, of the form

$$X_\mu(p',p)=F_1(q^2)i\gamma_\mu\gamma_5+F_2(q^2)\gamma_5q_\mu,$$
$$q=p'-p,\quad p^2=p'^2=-m^2. \tag{3.35}$$

The continuity equation (3.32′), together with Eq. (3.33), further reduces this to

$$F_1=F_2q^2/2m\equiv F,$$
$$X_\mu(p',p)=F(q^2)\left(i\gamma_\mu\gamma_5+\frac{2m\gamma_5q_\mu}{q^2}\right). \tag{3.36}$$

The real nucleon is not a point particle. Its X-current (3.36) is provided with the dramatic "anomalous" term.

To understand the physical meaning of the anomalous term, we have to make use of the dispersion relations. The form factors F_1 and F_2 will, in general, satisfy dispersion relations of the form

$$F_i(q^2)=F_i(0)-\frac{q^2}{\pi}\int\frac{\mathrm{Im}F_i(-\kappa^2)}{(q^2+\kappa^2-i\epsilon)\kappa^2}d\kappa^2, \tag{3.37}$$

assuming one subtraction. Each singularity at κ^2 corresponds to some physical intermediate state. Thus if $F(0)\neq 0$, Eq. (3.36) indicates that there is a pole at $q^2=0$ for F_2 (and no subtraction), which means in turn that there is an isolated intermediate state of zero mass.

[14] This was discussed by R. Haag, Kgl. Danske Videnskab. Selskab, Mat.-fys. Medd. 29, No. 12 (1955). See also L. van Hove, Physica 18, 145 (1952).

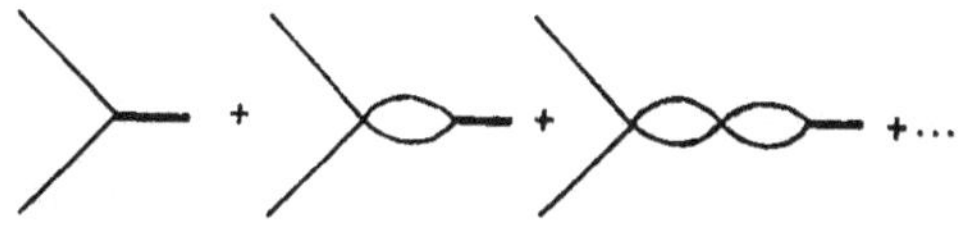

FIG. 2. Graphs corresponding to the Bethe-Salpeter equation in "ladder" approximation. The thick line is a bound state.

To see its nature, we take a time-like q in its own rest frame and go to the limit $q^2\to 0$. The anomalous term has then only the time component, and is proportional to the amplitude for creation of a nucleon pair in a $J=0^-$ state. Hence the zero mass state must have the same property as this pair. It belongs to nucleon number zero, so that we may call it a zero-mass pseudoscalar meson. *In order for a γ_5-invariant Hamiltonian such as Eq. (2.6) to allow massive nucleon states and a nonvanishing X current for $q=0$, it is therefore necessary to have at the same time pseudoscalar zero-mass mesons coupled with the nucleons.* Since we did not have such mesons in the theory, they must be regarded as secondary products, i.e., bound states of nucleon pairs. This conclusion would not hold if in Eq. (3.36) $F(q^2)=O(q^2)$ near $q^2=0$. A nucleon then would have always $X=0$. Such a possibility cannot be excluded. We will show, however, that the pseudoscalar zero-mass bound states do follow explicitly, once we assume the nontrivial solution of the self-energy equation.

IV. THE COLLECTIVE STATES

From the general discussion of Secs. 2 and 3, we may expect the existence of collective states of the fundamental field which would manifest themselves as stable or unstable particles. In particular we have argued that, as a consequence of the γ_5 invariance, a pseudoscalar zero-mass state must exist. We want now to discuss the problem in detail, trying to determine the mass spectrum of the collective excitations (at least its general features) and the strength of their coupling with the nucleons. These states must be considered as a direct effect of the same primary interaction which produces the mass of the nucleon, which itself is a collective effect. We will study the bound-state problem through the use of the Bethe-Salpeter equation, taking into account explicitly the self-consistency conditions. We first verify in the following the existence of the zero-mass pseudoscalar state.

The Bethe-Salpeter equation for a bound pair B deals with the amplitude

$$\Phi(x,y)=\langle 0|T(\psi(x)\bar{\psi}(y))|B\rangle. \tag{4.1}$$

As is well known, the equation is relatively easy to handle in the ladder approximation. In our case we have a four-spinor point interaction and the analog of the "ladder" approximation would be the iteration of the simplest closed loop (see Fig. 2) in which all lines represent dressed particles. We introduce the vertex function

Γ related to Φ by

$$\Phi(p)=S_F^{(m)}(p+\tfrac{1}{2}q)\Gamma(p+\tfrac{1}{2}q,\, p-\tfrac{1}{2}q)S_F^{(m)}(p-\tfrac{1}{2}q). \quad (4.2)$$

All we have to do then is to set up the integral equation generated by the chain of diagrams, looking for solutions having the symmetry properties of a pseudoscalar state. This means that our solutions must be proportional to γ_5. This requirement makes only the pseudoscalar and axial vector part of the interaction contribute to the integral equation. We have

$$\begin{aligned}\Gamma(p+\tfrac{1}{2}q,\, p-\tfrac{1}{2}q) &= \frac{2ig_0}{(2\pi)^4}\gamma_5\int \mathrm{Tr}[\gamma_5 S_F^{(m)}(p'+\tfrac{1}{2}q)\\ &\quad\times\Gamma(p'+\tfrac{1}{2}q,\, p'-\tfrac{1}{2}q)S_F^{(m)}(p'-\tfrac{1}{2}q)]d^4p'\\ &\quad-\frac{ig_0}{(2\pi)^4}\gamma_5\gamma_\mu\int \mathrm{Tr}[\gamma_5\gamma_\mu S^{(m)}(p'+\tfrac{1}{2}q)\\ &\quad\times\Gamma(p'+\tfrac{1}{2}q,\, p'-\tfrac{1}{2}q)S_F^{(m)}(p'-\tfrac{1}{2}q)]d^4p'. \quad (4.3)\end{aligned}$$

For the moment let us ignore the pseudovector term on the right-hand side. It then follows that the equation has a constant solution $\Gamma=C\gamma_5$ if $q^2=0$. To see this, first observe that for the special case $q=0$, Eq. (4.3) reduces to

$$1=-\frac{8ig_0}{(2\pi)^4}\int\frac{d^4p}{p^2+m^2-i\epsilon}, \quad (4.4)$$

which is nothing but the self-consistency condition (3.7), provided that the same cutoff is applied. Since the pseudoscalar term of Eq. (4.3) gives a function of q^2 only, the same condition remains true as long as $q^2=0$.

When the pseudovector term is included, we have still the same eigenvalue $q^2=0$ with a solution of the form $\Gamma=C\gamma_5+iD\gamma_5\gamma\cdot q$, which is not difficult to verify (see Appendix).

We now add some remarks. First, the bound state amplitude for this solution spreads in space over a region of the order of the fermion Compton wavelength $1/m$ because of Eq. (4.2), making the zero-mass particle only partially localizable. We want also to stress the role played by the γ_5 invariance in the argument. We had in fact already inferred the existence of the pseudoscalar particle from relativistic and γ_5 invariance alone, and at first sight the same result seems to follow now essentially from the self-consistency equation. However, we must notice that only the scalar term of the Lagrangian appears in this equation while only the pseudoscalar part contributes in the Bethe-Salpeter equation. It is because of the γ_5-invariant Lagrangian that the Bethe-Salpeter equation can be reduced to the self-consistency condition.

Along the same line we could try to see whether other bound states exist in the "ladder" approximation. However, besides calculating the spectrum, it is also important to determine the interaction properties of these collective states with the fermions. For this purpose the study of the two-"nucleon" scattering amplitude appears much more suitable, as we shall realize after the following remark. Once we have recognized that in the ladder approximation the collective states would appear as real stable particles, we must expect to the same degree of approximation poles in the scattering matrix of two nucleons corresponding to the possibility of the virtual exchange of these particles. For definiteness we shall refer again as an example to the pseudoscalar zero-mass particle. Let us indicate by $J_p(q)$ the analytical expression corresponding to the graph whose iteration produces the bound state [Fig. 3(a)]. We construct next the scattering matrix generated by the exchange of all possible simple chains built with this element. This means that we consider the set of diagrams in Fig. 3(b). The series is easily evaluated and we obtain

$$2g_0 i\gamma_5\frac{1}{1-J_p(q)}i\gamma_5, \quad (4.5)$$

where the γ_5's refer to the pairs (1,1′) and (2,2′), respectively. The meaning of this result is clear: because of the self-consistent equation $J_p(0)=1$, Eq. (4.5) is equivalent to a phenomenological exchange term where the intermediate particle is our pseudoscalar massless boson (Fig. 4). The coupling constant G can now be evaluated by straightforward comparison. Before doing this calculation we need the explicit expression of $J_P(q)$. Using the ordinary rules for diagrams, we have

$$J_P(q)=-\frac{2ig_0}{(2\pi)^4}\times\int\frac{4(m^2+p^2)-q^2}{[(p+\tfrac{1}{2}q)^2+m^2][(p-\tfrac{1}{2}q)^2+m^2]}d^4p. \quad (4.6)$$

It is however more convenient to rewrite J_P in the form of a dispersive integral, and if we forget for a moment that it is a divergent expression, a simple manipulation gives

$$J_P(q)=\frac{g_0}{4\pi^2}\int_{4m^2}^{\Lambda^2}\frac{\kappa^2(1-4m^2/\kappa^2)^{\frac{1}{2}}}{q^2+\kappa^2}d\kappa^2. \quad (4.6')$$

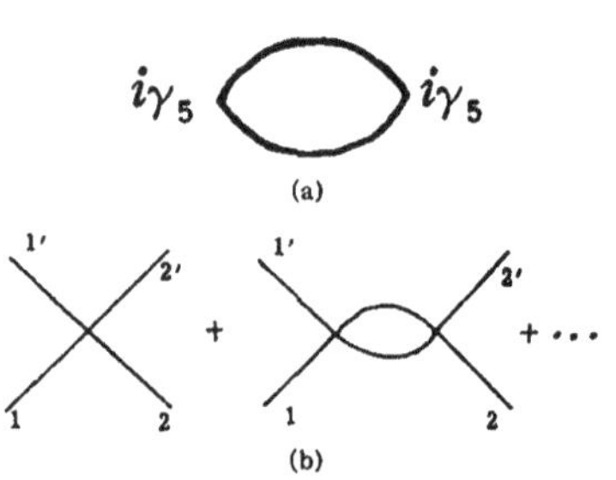

FIG. 3. The bubble graph for J_P and the scattering matrix generated by it.

In order for this expression to be meaningful, a new cutoff Λ must be introduced. There is no simple relation between this and the previous cutoffs. The dispersive form is more comfortable to handle and accordingly we shall reformulate the self-consistent condition $J_P(0)=1$, or

$$1=\frac{g_0}{4\pi^2}\int_{4m^2}^{\Lambda^2}(1-4m^2/\kappa^2)^{\frac{1}{2}}d\kappa^2. \tag{4.7}$$

It may be of interest to remark at this point that Eq. (4.7) can be obtained also if we think of our theory as a theory with intermediate pseudoscalar boson in the limit of infinite boson mass. We are now in a position to evaluate the phenomenological coupling constant G. From Eqs. (4.6′) and (4.7) we have

$$J_P(q^2)=1-q^2\frac{g_0}{4\pi^2}\int_{4m^2}^{\Lambda^2}\frac{(1-4m^2/\kappa^2)^{\frac{1}{2}}}{q^2+\kappa^2}d\kappa^2, \tag{4.8}$$

which leads immediately to the result

$$\frac{G_P^2}{4\pi}=2\pi\left[\int_{4m^2}^{\Lambda^2}\frac{(1-4m^2/\kappa^2)^{\frac{1}{2}}}{\kappa^2}d\kappa^2\right]^{-1}. \tag{4.9}$$

This equation is interesting since it establishes a connection between the phenomenological constant G_P and the cutoff independently of the value of the fundamental coupling g_0. This fact exhibits the purely dynamical origin of the phenomenological coupling G_P. Actually g_0 is buried in the value of the mass m.

So far we have exploited only the γ_5 vertex. What happens then if the scalar part is iterated to form chains of bubbles similar to those we have already discussed? The procedure just explained can be followed again, and a quantity $J_S(q)$ can be defined similarly with the result

$$J_S(q)=\frac{g_0}{4\pi^2}\int_{4m^2}^{\Lambda^2}\frac{(\kappa^2-4m^2)(1-4m^2/\kappa^2)^{\frac{1}{2}}}{q^2+\kappa^2}d\kappa^2. \tag{4.10}$$

It is immediately seen that because of Eq. (4.7)

$$J_S(-4m^2)=1, \tag{4.11}$$

which causes a new pole to appear in the S matrix for $q^2=-4m^2$. This means that we have another collective state of mass $2m$, parity $+$ and spin 0! We observe that it is necessary to assume the same cutoff as in the pseudoscalar case in order that this result may be obtained. The choice of the same cutoff in both cases seems to be suggested by the γ_5 invariance as will be seen later. We also notice the peculiar symmetry existing between the pseudoscalar and the scalar state: the first has zero mass and binding energy $2m$, while the opposite is true for the scalar particle. So in the bound-state picture the scalar particle would not be a true bound state and should be, rather, interpreted as a

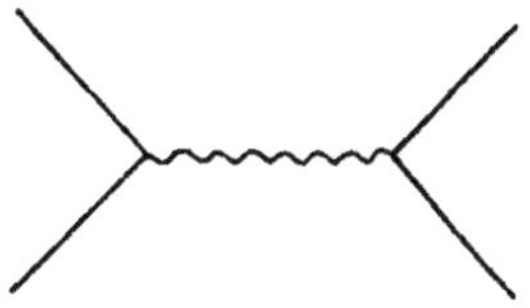

FIG. 4. The equivalent phenomenological one-meson exchange graph.

correlated exchange of pairs in the scattering process.[15] The "nucleon-nucleon" forces induced by the exchange of the scalar particle are, of course, of rather short range. The general physical implications of these results will be discussed more thoroughly later.

The phenomenological coupling constant G_S for the scalar meson is given by

$$\frac{G_S^2}{4\pi}=2\pi\left[\int_{4m^2}^{\Lambda^2}\frac{(1-4m^2/\kappa^2)^{\frac{1}{2}}}{(\kappa^2-4m^2)}d\kappa^2\right]^{-1}. \tag{4.12}$$

Let us next turn to the vector state generated by iteration of the vector interaction. In this case we obtain for each "bubble" a tensor

$$J_{V\mu\nu}=(\delta_{\mu\nu}-q_\mu q_\nu/q^2)J_V,$$
$$J_V=-\frac{g_0}{4\pi^2}\frac{q^2}{3}\int_{4m^2}^{\Lambda^2}\frac{d\kappa^2}{q^2+\kappa^2}\times\left(1+\frac{2m^2}{\kappa^2}\right)(1-4m^2/\kappa^2)^{\frac{1}{2}}. \tag{4.13}$$

Perhaps a remark is in order here regarding the evaluation of J_V. It suffers from an ambiguity of subtraction well known in connection with the photon self-energy problem. The above result is of the conventional gauge invariant form, which we take to be the proper choice.

Equation (4.13) leads to the scattering matrix

$$g_0\left[\gamma_\mu\frac{1}{1-J_V}\gamma_\mu-\gamma\cdot q\frac{J_V}{(1-J_V)q^2}\gamma\cdot q\right], \tag{4.14}$$

where the second term is, of course, effectively zero. It can be easily seen that the denominator can produce a pole below $4m^2$ for sufficiently small Λ^2. In fact, from Eqs. (4.7) and (4.13), we find

$$(8/3)m^2<\mu_V^2. \tag{4.15}$$

The coupling constant is given by

$$\frac{G_V^2}{4\pi}=3\pi\left[\int_{4m^2}^{\Lambda^2}d\kappa^2\frac{\kappa^2+2m^2}{(\kappa^2-\mu_V^2)^2}(1-4m^2/\kappa^2)^{\frac{1}{2}}\right]^{-1}. \tag{4.16}$$

It must be noted that the mass of the vector meson now depends on the cutoff, unlike the previous two cases.

Finally we are left with the pseudovector state. We

[15] Of course this and other heavy mesons will in general become unstable in higher order approximation, which is beyond the scope of the present paper.

find for the bubble[16]

$$J_{A\mu\nu}=-J_{V\mu\nu}+J_A'\delta_{\mu\nu},$$
$$J_A'=\frac{g_0m^2}{2\pi^2}\int_{4m^2}^{\Lambda^2}\frac{d\kappa^2}{q^2+\kappa^2}(1-4m^2/\kappa^2)^{\frac{1}{2}}. \tag{4.17}$$

In view of the self-consistency condition (4.7), it can be seen that this does not produce a pole of the scattering matrix for $-q^2<4m^2$, corresponding to a pseudovector meson.

So far we have considered only iterations of the same kind of interactions. In the ladder approximation there is actually a coupling between pseudoscalar and pseudovector interactions as was explicitly considered in Eq. (4.3). However, the coupling between scalar and vector interactions vanish because of the Furry's theorem.

This coupling of pseudoscalar and pseudovector interactions does not change the pion pole of the scattering matrix, but it affects the coupling of the pion to the nucleon since a chain of the pseudoscalar can join the external nucleon with an axial vector interaction. In other words, the pion-nucleon coupling is in general a mixture of pseudoscalar and derivative pseudovector types (Appendix).

We would like to inject here a remark concerning the trivial solution of the self-energy equation, against which we had no decisive argument. So let us also try to apply our scattering formula to this solution. For the pseudoscalar state we now find $J_P(q=0)>1$, provided that the cutoff Λ is kept fixed and m is set equal to zero in Eq. (4.6′). (The pseudovector interference vanishes.) In other words, there will be a pole for some $q^2>0$ ($\mu^2<0$). This is again a supporting evidence that the trivial solution could be unstable, capable of decaying by emitting such mesons. The final answer, however, depends on the exact nature of the cutoff.

Finally we would like to discuss the nucleon-nucleon scattering in the same spirit and approximation as for the nucleon-antinucleon scattering. In order to make a correspondence with the previous cases, it is convenient to rewrite the Hamiltonian in the following way:

$$\begin{aligned}H_1&=-g_0[\bar\psi\psi\bar\psi^c\psi^c-\bar\psi\gamma_5\psi\bar\psi^c\gamma_5\psi^c]\\&=\tfrac{1}{2}g_0[\bar\psi\gamma_\mu\psi^c\bar\psi^c\gamma_\mu\psi-\bar\psi\gamma_\mu\gamma_5\psi^c\bar\psi^c\gamma_\mu\gamma_5\psi]\\&=-\tfrac{1}{2}g_0[\bar\psi\gamma_\mu C\bar\psi\psi C^{-1}\gamma_\mu\psi-\bar\psi\gamma_\mu\gamma_5C\bar\psi\psi C^{-1}\gamma_\mu\gamma_5\psi],\end{aligned} \tag{4.18}$$

where ψ^c, $\bar\psi^c$ are the charge-conjugate fields.

The last form of Eq. (4.18) is suitable for our purpose. We note first that the vector part of the interaction is identically zero because of the anticommutativity of ψ. Thus only the pseudovector part survives. A "bubble" made of this interaction then is seen to give rise to the same integral J_A, Eq. (4.17). Since the interfering pseudoscalar interaction is missing in the present case,

[16] We meet here again the problem of subtraction. Our choice follows naturally from comparison with the vector case, and is consistent with Eq. (3.33).

TABLE I. Mass spectrum.

Nucleon number	Mass μ	Spin-parity	Spectroscopic notation
0	0	0^-	$^1S_0$
0	$2m$	0^+	$^3P_0$
0	$(8/3)m^2<\mu^2$	1^-	$^3P_1$
±2	$2m^2<\mu^2$	0^+	$^1S_0$

we get the complete scattering matrix by iterating J_A:

$$\begin{aligned}&-\gamma_\mu\gamma_5C\left[\frac{\delta_{\mu\nu}-q_\mu q_\nu/q^2}{1-J_A}+\frac{q_\mu q_\nu/q^2}{1-J_A'}\right]C^{-1}\gamma\gamma_5\\&=\gamma_\mu\gamma_5C\frac{1}{1-J_A}C^{-1}\gamma_\mu\gamma_5\\&\qquad+\gamma\cdot q\gamma_5C\frac{J_V/q^2}{(1-J_A')(1-J_A)}C^{-1}\gamma\cdot q\gamma_5,\end{aligned} \tag{4.19}$$

$$J_A\equiv J_A'-J_V.$$

The first term, corresponding to a scattering in the $J=1^-$ state, does not have a pole. The second term can have one below $4m^2$ for $1=J_A'$. With Eqs. (4.7) and (4.17), this determines the mass μ_D:

$$2m^2<\mu_D^2. \tag{4.20}$$

In this second term of the scattering matrix, the wave function is proportional to $C\gamma\cdot q\gamma_5$, so that the bound state behaves like a scalar "deuteron" (a singlet S state). The residue of the pole determines the nucleon-"deuteron" coupling constant (derivative) G_D^2, which is positive as it should be.

Table I summarizes the main results of this section. Although our approximation is a very crude one, we believe that it reflects the real situation at least qualitatively, because all the results are understandable in simple physical terms. Thus in the nonrelativistic sense, our Hamiltonian contains spin-independent attractive scalar and vector interactions plus a spin-dependent axial vector interaction between a particle and an antiparticle. Between particles, the vector part turns into a repulsion. Table I is just what we expect for the level ordering from this consideration.

V. PHENOMENOLOGICAL THEORY AND γ_5 INVARIANCE

In the previous section special subsets of diagrams were taken into account, and the existence of various boson states was established, together with their couplings with the nucleons. As was discussed there, we can reasonably expect that these results are essentially correct in spite of the very simple approximations. Because the bosons have in general small masses (compared to the unbound nucleon states), they will play important roles in the dynamics of strong interactions at least at energies comparable to these masses.

Thus if we are willing to accept the conclusions of our lowest order approximation, what we should do then is to study the dynamics of systems consisting of nucleons and the different kinds of bosons which all together represent the primary manifestation of the fundamental interaction. These particles will be now assumed to interact via their phenomenological couplings. So we may describe our purpose as an attempt to construct a theory in the conventional sense in which a separate field is introduced for each kind of particle. However, this is not a simple and unambiguous problem because our fundamental theory is completely γ_5 invariant and we must make sure that this invariance is preserved at any stage of our calculations in order that the results be meaningful. For a better understanding of the problem, let us consider our Lagrangian in the lowest self-consistent approximation. We have

$$L'=L_0'+L_I',$$

where

$$\begin{aligned} L_0' &= -(\bar{\psi}\gamma_\mu\partial_\mu\psi+m\bar{\psi}\psi), \\ L_I' &= g_0[(\bar{\psi}\psi)^2-(\bar{\psi}\gamma_5\psi)^2]+m\bar{\psi}\psi. \end{aligned} \tag{5.1}$$

L' is obviously γ_5 invariant. In order to preserve this invariance we must study the S matrix generated by L_I'. Some subsets of diagrams have been considered in the previous section and it will be shown now how those calculations comply with γ_5 invariance. This point must be understood clearly so that we shall discuss it in a rather systematic way. Let us recall first how we constructed the scattering matrix in the "ladder" approximation. The lowest-order contribution is certainly invariant as no internal massive line appears. But what will happen to the next-order terms [Fig. 3(b)]? To these diagrams corresponds the expression

$$J_S(q^2)-\gamma_5J_P(q^2)\gamma_5+iJ_{SP}(q^2)\gamma_5+i\gamma_5J_{PS}(q^2). \tag{5.2}$$

In the gauge in which our calculations were performed, the last two terms happened to be zero. We write down next the transformation properties of the quantities appearing above. By straightforward calculation we find

$$\begin{aligned} \gamma_5 &\to \gamma_5\cos2\alpha+i\sin2\alpha, \\ 1 &\to \cos2\alpha+i\gamma_5\sin2\alpha, \\ J_P &\to J_P\cos^22\alpha+J_S\sin^22\alpha, \\ J_S &\to J_S\cos^22\alpha+J_P\sin^22\alpha, \\ J_{SP} &\to (J_P-J_S)\sin2\alpha\cos2\alpha, \\ J_{PS} &\to (J_P-J_S)\sin2\alpha\cos2\alpha. \end{aligned} \tag{5.3}$$

By simple substitution the invariance follows easily. The argument can now be extended to all orders, provided at each order all the possible combinations of S and P are included. The invariance of the scattering in the "ladder" approximation is thus established. It may look surprising that the SP and PS contributions do not vanish identically. This can be understood by considering the fact that the γ_5 transformation changes the parity of the vacuum which will be in general a superposition of states of opposite parities. In this way products of fields of different parities (as the SP propagator) may have a nonvanishing average value in the vacuum state.

We may now attempt the construction of the phenomenological coupling by introducing two local fields Φ_P and Φ_S describing the pseudoscalar and the scalar particles, respectively. We start by observing that, in the same gauge in which the previous calculations were made, we can write the meson-nucleon interaction as

$$L_I=G_Pi\bar{\psi}\gamma_5\psi\Phi_P+G_S\bar{\psi}\psi\Phi_S. \tag{5.4}$$

In order to find the general expression valid in any gauge, it is convenient to introduce the following two-dimensional notation

$$\varphi\equiv\begin{pmatrix} i\bar{\psi}\gamma_5\psi \\ \bar{\psi}\psi \end{pmatrix},\quad \Phi\equiv\begin{pmatrix} \Phi_P \\ \Phi_S \end{pmatrix},\quad G\equiv\begin{pmatrix} G_P & 0 \\ 0 & G_S \end{pmatrix}. \tag{5.5}$$

The interaction Lagrangian Eq. (5.4) can be written in this notation in a compact form,

$$L_I=\varphi G\Phi. \tag{5.6}$$

The effect of the γ_5 transformation on φ is described with the aid of the matrix

$$U\equiv\begin{pmatrix} \cos2\alpha & -\sin2\alpha \\ \sin2\alpha & \cos2\alpha \end{pmatrix}, \tag{5.7}$$

which satisfies $UU^+=UU^{-1}=UU^T=1$. In other words, the γ_5 transformation induces a unitary transformation in the two-dimensional space, and Eq. (5.6) remains invariant if

$$G\to UGU^{-1},\quad \Phi\to U\Phi. \tag{5.8}$$

To complete the construction of the theory, the free Lagrangian for the fields Φ_P and Φ_S must be added. If we work again in the special gauge $\alpha=0$, we may write

$$L_0=-\tfrac{1}{2}\partial_\mu\Phi_P\partial_\mu\Phi_P-\tfrac{1}{2}\partial_\mu\Phi_S\partial_\mu\Phi_S-\tfrac{1}{2}\mu^2\Phi_S^2, \tag{5.9}$$

where $\mu^2=4m^2$. We use again the two-dimensional notation, and defining the mass operator

$$M^2\equiv\begin{pmatrix} 0 & 0 \\ 0 & \mu^2 \end{pmatrix}, \tag{5.10}$$

we write Eq. (5.9) in the invariant form

$$L_0=-\tfrac{1}{2}\partial_\mu\Phi\partial_\mu\Phi-\tfrac{1}{2}\Phi M^2\Phi. \tag{5.11}$$

In this way we have given a formal prescription for the γ_5 transformation in the phenomenological treatment. We have to emphasize here that the Lagrangians (5.9) and (5.11) are *not* γ_5 invariant in the ordinary sense of the word. In our theory, where the mesons are only phenomenological substitutes which partially represent the dynamical contents of the theory, they may be, however, called γ_5 covariant. In other words, *the masses and the coupling constants are not fixed parameters, but rather dynamical quantities which are subject to transformations when the representation is changed.* It will be legitimate to ask whether this situation corresponds to the one obtained in the framework of the fundamental theory and discussed in the "ladder" approximation in the previous section. We shall examine the transformation rule for the mass operator M^2, since this illustrates the case in point. Let us calculate explicitly M^2 in an arbitrary gauge α. We have

$$M^2 \rightarrow UM^2U^{-1}$$

$$=\mu^2\begin{pmatrix} \sin^2 2\alpha & -\sin 2\alpha\cos 2\alpha \\ -\sin 2\alpha \cos 2\alpha & \cos^2 2\alpha \end{pmatrix}. \quad (5.12)$$

The meaning of this equation is that the pseudoscalar and the scalar particle will have generally different masses in different gauges. In particular we see that the pseudoscalar particle has in the gauge α a mass $\sin 2\alpha\mu$. If this is the case we must expect that after the transformation the pole in the corresponding propagator will move from $q^2=0$ to $q^2=-(\sin^2 2\alpha)\mu^2$. This actually may be verified directly in the "ladder" approximation which shows that the pion propagator changes according to

$$iG_P{}^2\Delta_{PP}=\frac{2g_0}{1-J_P}\rightarrow\frac{2g_0}{1-J_P\cos^2 2\alpha-J_S\sin^2 2\alpha}. \quad (5.13)$$

Using the results of the previous section, it is seen that the denominator of the right-hand side vanishes for $q^2=-(\sin^2 2\alpha)4m^2$. In this way we have seen how our γ_5-invariant theory can be approximated by a phenomenological description in terms of pseudoscalar and scalar mesons. Of course one may add the vector meson as well. Such a description does not look γ_5 invariant. It is only γ_5 covariant, and the masses and coupling constants must be understood to be matrices which, however, can be simultaneously diagonalized.

The reason for this situation is the degeneracy of the vacuum and the world built upon it. Only after combining all the equivalent but nonintersecting worlds labeled with different α do we recover complete γ_5 invariance. Nevertheless, even in a particular world we can find manifestations of the invariance, such as the zero-mass pseudoscalar meson and the conserved γ_5 current.

VI. THE CONSERVATION OF AXIAL VECTOR CURRENT

In this section we will discuss another paradoxical aspect of the theory regarding the γ_5 invariance. In Sec. 3 we argued that the X current should really be conserved, and that this is possible if a nucleon X current possesses a peculiar anomalous term. We now verify the statement explicitly in our approximation.

First we have to realize that the problem is again how to keep the γ_5-invariant nature of the theory at every stage of approximation. It is well known in quantum electrodynamics that, in order to observe the ordinary gauge invariance, a certain set of graphs have to be combined together in a given approximation. The necessity for this is based on a general proof which makes use of the so-called Ward identity. In our present case there also exists an analog of the Ward identity. In order to derive it, let us first consider the proper self-energy part of our fermion in the presence of an external axial vector field B_μ with the interaction $L_B=-j_{\mu5}B_\mu$. The self-energy operator is now a matrix $\Sigma^{(B)}(p',p)$ depending on initial and final momenta. Expanding Σ in powers of B, we have

$$\Sigma^{(B)}(p',p)=\Sigma(p)+\Lambda_{\mu5}(p',p)B_\mu(p'-p)+\cdots. \quad (6.1)$$

We readily realize that the coefficient of the second term gives the desired X-current vertex correction.

On the other hand, the entire Lagrangian remains invariant under a *local* γ_5 transformation if Eq. (2.3) is accompanied by

$$B_\mu\rightarrow B_\mu-\partial_\mu\alpha, \quad (6.2)$$

where α is now an arbitrary function. In other words,

$$e^{i\alpha\gamma_5}\Sigma^{(B-\partial\alpha)}e^{i\alpha\gamma_5}=\Sigma^{(B)} \quad (6.3)$$

in a symbolical way of writing.[17]

Expanding (6.3) after putting $B=0$, we get

$$i\alpha(p'-p)[\gamma_5\Sigma(p)+\Sigma(p')\gamma_5]$$
$$=i\alpha(p'-p)(p'-p)_\mu\Lambda_{\mu5}(p',p),$$

or

$$\gamma_5\Sigma(p)+\Sigma(p')\gamma_5=(p'-p)_\mu\Lambda_{\mu5}(p',p). \quad (6.4)$$

The entire vertex $\Gamma_{\mu5}=i\gamma_\mu\gamma_5+\Lambda_{\mu5}$ then satisfies

$$\gamma_5L'(p)+L'(p')\gamma_5=-(p'-p)_\mu\Gamma_{\mu5}(p',p),$$
$$L'(p)\equiv-i\gamma\cdot p-\Sigma(p), \quad (6.5)$$

which is the desired generalized Ward identity.[18] The right-hand side of Eq. (6.5) is the divergence of the X current, while the left-hand side vanishes when p and p' are on the mass shell of the actual particle. The X-current conservation is thus established. Moreover, the way the anomalous term arises is now clear. For if we assume $\Sigma(p)=m$, Eq. (6.4) gives

$$2m\gamma_5=(p'-p)_\mu\Lambda_{\mu5}(p'-p), \quad (6.6)$$

so that we may write the longitudinal part of Λ as

$$\Lambda_{\mu5}{}^{(l)}(p',p)=2m\gamma_5q_\mu/q^2,\quad q=p'-p, \quad (6.7)$$

which is of the desired form.

Next we have to determine what types of graphs

[17] We assume here that $\alpha(x)$ is different from zero only over a finite space-time region, so that the gauge of the nontrivial vacuum, which we may fix at remote past, is not affected by the transformation. The limiting process of going over to constant α is then ill-defined as we can see from the fact that the anomalous term in $\Gamma_{\mu5}$ has no limit as $q\rightarrow 0$.

[18] See also J. Bernstein, M. Gell-Mann, and L. Michel, Nuovo cimento **16**, 560 (1960).

should be considered for Γ_μ in our particular approximation of the self-energy. Examining the way in which the relation (6.3) is maintained in a perturbation expansion, we are led to the conclusion that our self-energy represented by Fig. 5(a) gives rise to the series of vertex graphs [Fig. 5(b)]. The summation of the graphs is easily carried out to give

$$\Lambda_{\mu 5}=i\gamma_5\frac{1}{1-J_P}J_{PA}, \tag{6.8}$$

where J_P was obtained before [Eq. (4.8)], and

$$J_{PA}=\frac{2ig_0}{(2\pi)^4}\int \mathrm{Tr}\gamma_5 S(p+q/2)\gamma_\mu\gamma_5 S(p-q/2)d^4p$$

$$=-\frac{g_0}{2\pi^2}imq_\mu\int_{4m^2}^{\Lambda^2}\frac{d\kappa^2}{q^2+\kappa^2}(1-4m^2/\kappa^2)^{\frac{1}{2}}. \tag{6.9}$$

Thus

$$\Gamma_{\mu 5}=i\gamma_\mu\gamma_5+\Lambda_{\mu 5}$$
$$=i\gamma_\mu\gamma_5+2m\gamma^5 q_\mu/q^2, \tag{6.10}$$

in agreement with the general formula. We see also that there is no form factor in this approximation.

This example will suffice to show the general procedure necessary for keeping γ_5 invariance. When we consider further corrections, the procedure becomes more involved, but we can always find a set of graphs which are sufficient to maintain the X-current conservation. We shall come across this problem in connection with the axial vector weak interactions.

VII. SUMMARY AND DISCUSSION

We briefly summarize the results so far obtained. Our model Hamiltonian, though very simple, has been found to produce results which strongly simulate the general characteristics of real nucleons and mesons. It is quite appealing that both the nucleon mass and the pseudoscalar "pion" are of the same dynamical origin, and the reason behind this can be easily understood in terms of (1) classical concepts such as attraction or repulsion between particles, and (2) the γ_5 symmetry.

According to our model, the pion is not the primary agent of strong interactions, but only a secondary effect. The primary interaction is unknown. At the present stage of the model the latter is only required to have appropriate dynamical and symmetry properties, although the nonlinear four-fermion interaction, which we actually adopted, has certain practical advantages.

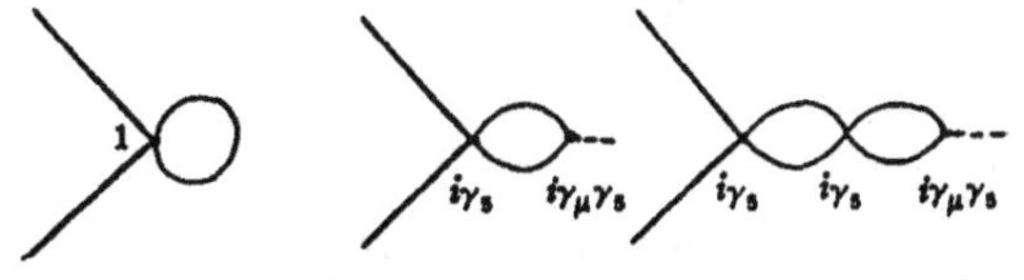

FIG. 5. Graphs for self-energy and matching radiative corrections to an axial vector vertex

FIG. 6. A class of higher order self-energy graphs.

In our model the idealized "pion" occupies a special position in connection with the γ_5-gauge transformation. But there are also other massive bound states which may be called heavy mesons and deuterons. The conventional meson field theory must be regarded, from our point of view, as only a phenomenological description of events which are actually dynamic processes on a higher level of understanding, in the same sense that the phonon field is a phenomenological description of interatomic dynamics.

Our theory contains two parameters, the primary coupling constant and the cutoff, which can be translated into observed quantities: nucleon mass and the pion-nucleon coupling constant. It is interesting that the pion coupling depends only on the cutoff in our approximation. In order to make the pion coupling as big as the observed one (≈ 15) the cutoff has to be rather small, being of the same order as the nucleon mass.

We would like to make some remarks about the higher order approximations. If the higher order corrections are small, the usual perturbation calculation will be sufficient. If they are large compared to the lowest order estimation, the self-consistent procedure must be set up, including these effects from the beginning. This is complicated by the fact that the pions and other mesons have to be properly taken into account.

To get an idea about the importance of the corrections, let us take the next order self-energy graph (Fig. 6). This is only the first term of a class of corrections shown in Fig. 6, the sum of which we know already to give rise to an important collective effect, i.e., the mesons. It would be proper, therefore, to consider the entire class put together. The correction is then equivalent to the ordinary second order self-energy due to mesons, plus modifications arising at high momenta. Thus strict perturbation with respect to the bare coupling g_0 will not be an adequate procedure. Evaluating, for example, the pion contribution in a phenomenological way, we get

$$\frac{\delta m}{m}=\frac{G_P^2}{32\pi^2}\int_{m^2}^{\Lambda'^2}\frac{d\kappa^2}{\kappa^2}\left(1-\frac{m^2}{\kappa^2}\right), \tag{7.1}$$

where Λ' is an effective cutoff. Substituting G_P^2 from Eq. (4.9), this becomes

$$\frac{\delta m}{m}=\frac{1}{4}\int_{4m^2}^{4\Lambda'^2}\frac{d\kappa^2}{\kappa^2}\left(1-\frac{4m^2}{\kappa^2}\right)\Big/\int_{4m^2}^{\Lambda^2}\frac{d\kappa^2}{\kappa^2}\left(1-\frac{4m^2}{\kappa^2}\right)^{\frac{1}{2}}. \tag{7.2}$$

As Λ and Λ' should be of the same order of magnitude, the higher order corrections are in general not negligible. We may point out, on the other hand, that there is a tendency for partial cancellation between contributions from different mesons or nucleon pairs.

We already remarked before that the model treated here is not realistic enough to be compared with the actual nucleon problem. Our purpose was to show that a new possibility exists for field theory to be richer and more complex than has been hitherto envisaged, even though the mathematics is marred by the unresolved divergence problem.

In the subsequent paper we will attempt to generalize the model to allow for isospin and finite pion mass, and draw various consequences regarding strong as well as weak interactions.

APPENDIX

We treat here, for completeness, the problem created by the coupling of pseudoscalar and pseudovector terms encountered in the text. As we have seen, such an effect is not essential for the discussion of γ_5 invariance, but rather adds to complication, which however naturally appears in the ladder approximation.

First let us write down the integral equation for a vertex part Γ:

$$\begin{aligned}\Gamma(p+\tfrac{1}{2}q,\, p-\tfrac{1}{2}q) &= \gamma(p+\tfrac{1}{2}q,\, p-\tfrac{1}{2}q)+\frac{2ig_0}{(2\pi)^4}\gamma_5\int \mathrm{Tr}[\gamma_5 S(p'+\tfrac{1}{2}q) \\ &\quad\times\Gamma(p'+\tfrac{1}{2}q,\, p-\tfrac{1}{2}q)S_F(p-\tfrac{1}{2}q)]d^4p' \\ &\quad -\frac{ig_0}{(2\pi)^4}\gamma_5\gamma_\mu\int \mathrm{Tr}[\gamma_5\gamma_\mu S(p'+\tfrac{1}{2}q) \\ &\quad\times\Gamma(p'+\tfrac{1}{2}q,\, p-\tfrac{1}{2}q)S_F(p-\tfrac{1}{2}q)]d^4p'. \qquad \text{(A.1)}\end{aligned}$$

This embraces three special cases depending on the inhomogeneous term γ:

(a) $\gamma=0$ for the Bethe-Salpeter equation for the pseudoscalar meson;
(b) $\gamma=i\gamma_\mu\gamma_5$ for the pseudovector vertex function $\Gamma_{\mu5}$;
(c) $\gamma=2g_0(\gamma_5)_f(\gamma_5)_i-g_0(\gamma_\mu\gamma_5)_f(\gamma_\mu\gamma_5)_i$ for the nucleon-antinucleon scattering through these interactions.

Here i and f refer to initial and final states, and the integral kernel of Eq. (A.1) operates on the f part.

We will consider them successively.

(a) We make the ansatz $\Gamma=C\gamma_5+iD\gamma_5\gamma\cdot q$. The integrals in Eq. (A.1) then reduce to the standard forms considered in the text. Making use of Eqs. (4.9), (4.17), and (6.9), we get[16]

$$\begin{aligned} C&=C-(C+2mD)q^2I, \\ D&=(C+2mD)mI, \\ I(q^2)&=\frac{g_0}{4\pi^2}\int\frac{d\kappa^2}{q^2+\kappa^2}\left(1-\frac{4m^2}{\kappa^2}\right)^{\frac{1}{2}}, \end{aligned} \qquad \text{(A.2)}$$

which lead to $q^2=0$, and $C:D=1-2m^2I(0):mI(0)$. From Eq. (4.8), we have $0<2m^2I(0)<\frac{1}{2}$.

(b) Put
$$\begin{aligned}\Gamma_{\mu5}=(i\gamma_\mu\gamma_5+2m\gamma_5q_\mu/q^2)F_1(q^2) \\ +(i\gamma_\mu\gamma_5-i\gamma\cdot q\gamma_5q_\mu/q^2)F_2(q^2). \qquad \text{(A.3)}\end{aligned}$$

This is seen to satisfy the integral equation if

$$\begin{aligned} F_1&=1, \\ F_2&=J_A(q^2)/[1-J_A(q^2)], \\ J_A(q^2)&=2m^2I(q^2)-J_V(q^2), \end{aligned} \qquad \text{(A.4)}$$

where $J(q^2)$ was defined in Eq. (4.13).

On the mass shell, $\Gamma_{\mu5}$ reduces to

$$\begin{aligned} &(i\gamma_\mu\gamma_5+2m\gamma_5q_\mu/q^2)F(q^2), \\ &F(q^2)=1+F_2(q^2)=1/[1-J_A(q^2)]. \end{aligned} \qquad \text{(A.5)}$$

For $q^2=0$, we have $J(q^2)=0$ so that $1<F(0)=1/[1-2m^2I(0)]<2$.

(c) From the structure of the inhomogeneous term, it is clear that the scattering matrix is given by

$$M=2g_0(\Gamma_5)_f(\gamma_5)_i+g_0(\Gamma_{\mu5})_f(i\gamma_\mu\gamma_5)_i,$$

where Γ_5 is the pseudoscalar vertex function.

Again, from Eq. (A.1), Γ_5 is determined as

$$\Gamma_5=\gamma_5[1-2m^2I(q^2)]/q^2I(q^2)-mi\gamma\cdot q\gamma_5/q^2, \qquad \text{(A.6)}$$

which has an entirely different behavior from the bare γ_5 for small q^2. The scattering matrix is then

$$\begin{aligned} M=&(\gamma_5)_f(\gamma_5)_i2g_0[1-2m^2I(q^2)]/q^2I(q^2) \\ &-[(i\gamma\cdot q\gamma_5)_f(\gamma_5)_i-(\gamma_5)_f(i\gamma\cdot q\gamma_5)_i]2mg_0/q^2 \\ &-(i\gamma\cdot q\gamma_5)_f(i\gamma\cdot q\gamma_5)_ig_0J_A(q^2)/q^2[1-J_A(q^2)] \\ &+(i\gamma_\mu\gamma_5)_f(i\gamma_\mu\gamma_5)_ig_0/[1-J_A(q^2)]. \qquad \text{(A.7)}\end{aligned}$$

The first three terms have a pole at $q^2=0$. The coupling constants of the pseudoscalar meson are then

pseudoscalar coupling:

$$G_p{}^2=2g_0[1-2m^2I(0)]/I(0),$$

pseudovector coupling:

$$\begin{aligned} G_{pv}{}^2&=g_0J_A(0)/[1-J_A(0)] \\ &=g_02m^2I(0)/[1-2m^2I(0)]. \end{aligned} \qquad \text{(A.8)}$$

Their relative sign is such that the equivalent pseudoscalar coupling on the mass shell is

$$G_p'^2=4m^2g_0\left\{\left[\frac{1-2m^2I(0)}{2m^2I(0)}\right]^{\frac{1}{2}}+\left[\frac{2m^2I(0)}{1-2m^2I(0)}\right]^{\frac{1}{2}}\right\}^2. \qquad \text{(A.9)}$$

Reprinted from THE PHYSICAL REVIEW, Vol. 124, No. 1, 246–254, October 1, 1961

Dynamical Model of Elementary Particles Based on an Analogy with Superconductivity. II*

Y. NAMBU AND G. JONA-LASINIO†
Enrico Fermi Institute for Nuclear Studies and Department of Physics, University of Chicago, Chicago, Illinois
(Received May 10, 1961)

Continuing the program developed in a previous paper, a "superconductive" solution describing the proton-neutron doublet is obtained from a nonlinear spinor field Lagrangian. We find the pions of finite mass as nucleon-antinucleon bound states by introducing a small bare mass into the Lagrangian which otherwise possesses a certain type of the γ_5 invariance. In addition, heavier mesons and two-nucleon bound states are obtained in the same approximation. On the basis of numerical mass relations, it is suggested that the bare nucleon field is similar to the electron-neutrino field, and further speculations are made concerning the complete description of the baryons and leptons.

I. INTRODUCTION

IN Part I of this paper[1] we have proposed a model of strong interactions based on an analogy with the BCS-Bogoliubov theory of superconductivity. It is characterized by a nonlinear spinor field possessing γ_5 invariance, and simulates some important features of the meson-nucleon system. The basic principle underlying the model is the idea that field theory may admit, as a result of dynamical instability, extraordinary (nontrivial) solutions that have less symmetries than are built into the Lagrangian.[2] In fact we have obtained as an extraordinary solution a massive fermion and a massless pseudoscalar boson as idealized proton and pion, together with other heavy mesons.

If we now try to make our model more realistic, a number of problems spring up naturally. First of all, we would have to account for the isospin and strangeness quantum numbers. It seems rather obvious that these degrees of freedom have to be built into the theory from the beginning, although there may be some possibility of utilizing both the ordinary and extraordinary solutions to enlarge the Hilbert space, as will be discussed later.

These quantum numbers will not yet be enough to determine our theory satisfactorily, as we expect to have more additional symmetries which are at least approximately satisfied. Among other things, we have postulated the γ_5 invariance as a cornerstone of our previous model. What would be the proper generalization of the γ_5 invariance? Then there also arises the inevitable question of any possible symmetry among baryons of different strangenesses. Since such a symmetry is at any rate only approximate, the test of the theory will depend on its ability to account for the violation of the symmetry as well.

Finally, we face the problem of the baryon versus the lepton, the electromagnetism, and the weak processes. Here our theory creates a particular incentive for speculation concerning the baryon-lepton problem, since the ordinary and extraordinary solutions immediately remind us of these two families of particles.

We do not profess to have any clear-cut answers to these problems. In the present paper we shall again content ourselves with a rather modest task. We will first discuss a generalization of our model which incorporates the isospin for the nucleon and guarantees the existence of the pion. This can be done by demanding a $\gamma_5\times$isospin gauge group with a slight violation so as to give the pion its finite mass. We find that the bare mass necessary to achieve the latter end is at most several Mev. On this basis a suggestion is made that the bare nucleon field is essentially the same as the electron-neutrino field.

The complete picture of the baryon symmetries and the baryon-lepton problem is largely beyond the scope of the present paper, but some relevant discussions on this subject will also be presented, especially those concerned with the Sakata model and the general γ_5 symmetry

II. MODEL LAGRANGIAN FOR THE NUCLEON

First we would like to observe that the nonlinear spinor field adopted in I is not an essential element of our theory, as is the case with the Heisenberg theory[3] but is rather a model adopted to study our dynamical principles. At least in the present stage of the game, the controlling factors are the symmetry properties and qualitative dynamical characteristics of the basic fermion-fermion interaction, and whether the interaction is due to some fundamental boson, or fundamental nonlinearity (or something entirely new) is of secondary importance. Nevertheless, we have to choose

* This work was supported by the U. S. Atomic Energy Commission.
† Present address: Istituto di Fisica dell'Universita, Roma, and Istituto Nazionale di Fisica Nucleare, Sezione di Roma, Italy.
[1] Y. Nambu and G. Jona-Lasinio, Phys. Rev. **122**, 345 (1961); referred to hereafter as I. Y. Nambu, *Proceedings of the 1960 Annual International Conference on High-Energy Physics at Rochester* (Interscience Publishers, Inc., New York, 1960), p. 858.
[2] See also J. Goldstone, Nuovo cimento **19**, 154 (1961), N. N. Bogoliubov (to be published), V. G. Vaks and A. I. Larkin, *Proceedings of the 1960 Annual International Conference on High-Energy Physics at Rochester* (Interscience Publishers, Inc., New York, 1960), p. 871.
[3] H. P. Duerr, W. Heisenberg, H. Mitter, S. Schlieder, and K. Yamazaki, Z. Naturforsch. **14**, 441 (1959); W. Heisenberg, *Proceedings of the 1960 Annual International Conference on High-Energy Physics at Rochester* (Interscience Publishers, Inc., 1960), p. 851.

some model, and naturally there will arise certain predictions specific to the particular model. We take notice of the fact that the pion, the lightest of the meson family, is pseudoscalar and isovector, whereas its isoscalar counterpart of comparable mass does not seem to exist.[4] If the pion is to be intimately related to a symmetry property as in our previous model, this would imply that the model of nucleons should allow an (approximate) invariance under the $\gamma_5\times$isospin gauge group of Gürsey,[5] but not under the simple (Touschek) γ_5 gauge group, at least not so well as in the former case. For this reason, we would altogether consider the following gauge groups:

$$\psi\rightarrow e^{i\alpha}\psi,\qquad \bar\psi\rightarrow\bar\psi e^{-i\alpha},\tag{2.1a}$$

$$\psi\rightarrow\exp(i\boldsymbol{\tau}\cdot\boldsymbol{\alpha}')\psi,\qquad \bar\psi\rightarrow\bar\psi\exp(-i\boldsymbol{\tau}\cdot\boldsymbol{\alpha}'),\tag{2.1b}$$

$$\psi\rightarrow\exp(i\gamma_5\boldsymbol{\tau}\cdot\boldsymbol{\alpha}'')\psi,\qquad \bar\psi\rightarrow\bar\psi\exp(i\gamma_5\boldsymbol{\tau}\cdot\boldsymbol{\alpha}''),\tag{2.1c}$$

where τ denotes the nucleon isospin matrices.

Obviously, the first two are generators of the nucleon number gauge and the isospin transformation, respectively. The second and third transformations combined form a four-dimensional rotation group on the four components composed by the proton and neutron of both handednesses.[5] Thus we may also replace Eqs. (2.1a) and (2.1b) by the following transformations

$$\psi_R\rightarrow\exp(i\boldsymbol{\tau}\cdot\boldsymbol{\alpha}_R)\psi_R,\qquad \psi_R^\dagger\rightarrow\psi_R^\dagger\exp(-i\boldsymbol{\tau}\cdot\boldsymbol{\alpha}_R),$$
$$\psi_L\rightarrow\exp(i\boldsymbol{\tau}1\boldsymbol{\alpha}_L)\psi_L,\qquad \psi_L^\dagger\rightarrow\psi_L^\dagger\exp(-i\boldsymbol{\tau}\cdot\boldsymbol{\alpha}_L),\tag{2.2}$$

where ψ_R and ψ_L are the right- and left-handed components.

As the simplest Lagrangian that meets our requirements, we adopt the form

$$L=-\bar\psi\gamma_\mu\partial_\mu\psi-\bar\psi M^0\psi+g_0[\bar\psi\psi\bar\psi\psi-\sum_{i=1}^{3}\bar\psi\gamma_5\tau_i\psi\bar\psi\gamma_5\tau_i\psi].\tag{2.3}$$

If the bare mass operator $M^0=0$, this Lagrangian possesses, in addition to Eq. (2.1), an invariance under the discrete "mass reversal" group:

$$\psi\rightarrow\gamma_5\psi,\quad \bar\psi\rightarrow-\bar\psi\gamma_5.\tag{2.4}$$

The bare mass operator M^0 is a possible agent for the breakdown of the Gürsey group, and will be related to the finite pion mass.[6] For the moment, we will assume $M^0=0$. Before going to solve the self-consistent equation for the mass, we give the result of the Fierz transformation on Eq. (2.3): The interaction becomes

$$\begin{aligned}L_{\rm int}=&\tfrac{1}{4}g_0[\bar\psi\psi\bar\psi\psi-\bar\psi\gamma_5\tau_i\psi\bar\psi\gamma_5\tau_i\psi]\\&+\tfrac{1}{4}g_0[\bar\psi\gamma_5\psi\bar\psi\gamma_5\psi-\bar\psi\tau_i\psi\bar\psi\tau_i\psi]\\&-\tfrac{1}{2}g_0[\bar\psi\gamma_\mu\psi\bar\psi\gamma_\mu\psi-\bar\psi\gamma_\mu\gamma_5\psi\bar\psi\gamma_\mu\gamma_5\psi]\\&+\tfrac{1}{8}g_0[\bar\psi\sigma_{\mu\nu}\psi\bar\psi\sigma_{\mu\nu}\psi-\bar\psi\sigma_{\mu\nu}\tau_i\psi\bar\psi\sigma_{\mu\nu}\tau_i\psi],\end{aligned}\tag{2.5}$$

which is a rather complicated combination of all kinds of terms.

We now apply the linearization procedure of I to Eqs. (2.3) and (2.4), and obtain the self-energy

$$m=(1+\tfrac{1}{4})g_0\,{\rm Tr}S_F^{(m)}(0)=-i\frac{10g_0}{(2\pi)^4}\int\frac{d^4p\,m}{p^2+m^2}F(p,\Lambda).\tag{2.6}$$

Note that the trace refers to both spin and isospin variables. This differs from Eq. (3.6) of I only by the change of the effective coupling $g_0\rightarrow 5g_0/2\equiv g_0'$. So we can simply take over the previous formulas, namely,

$$1=\frac{g_0'}{4\pi^2}\int_{4m^2}^{\Lambda^2}d\kappa^2\left(1-\frac{4m^2}{\kappa^2}\right)^{\frac{1}{2}},\tag{2.7}$$

for the nontrivial solution if the dispersion integral (4.7) of I is used.

III. DETERMINATION OF MESON STATES

Since the interaction Lagrangian in Eqs. (1.1) and (1.3) contains a number of different couplings, we expect to get various kinds of "mesons" as bound nucleon-antinucleon pairs in our simple ladder approximation. As was explained in I, this is the proper approximation to match our self-energy equation at least for the pseudoscalar meson which is expected to have zero mass; moreover, even for other types of bound states we may reasonably trust its qualitative validity in predicting the existence and level ordering of possible bound states to the extent that our interaction is regarded basically as a short-range potential between spinor particles.

For general discussion, it is convenient to follow the procedure given in the Appendix of I. The basic equation to be considered is of the type

$$\Gamma(p+\tfrac{1}{2}q,p-\tfrac{1}{2}q)=\gamma+i\sum_n g_nO_n\times\int{\rm Tr}[O_nS_F(p'+\tfrac{1}{2}q)\Gamma(p'+\tfrac{1}{2}q,p'-\tfrac{1}{2}q)\times S_F(p'-\tfrac{1}{2}q)]d^4p',\tag{3.1}$$

where the summation on the right-hand side is over the various tensor forms in the interaction Lagrangian. The "vertex function" $\Gamma(p+\frac{1}{2}q,p-\frac{1}{2}q)$ reduces to a bound state wave function when it becomes a homogeneous solution ($\gamma=0$) for a particular value of $q^2=-\mu^2$. We will briefly discuss those two-nucleon states for which there is a possibility of binding.

A. Pseudoscalar, Isovector Meson

Unlike the case in I, only the pseudoscalar interaction $\sim\bar\psi\gamma_5\tau_i\psi\bar\psi\gamma_5\tau_i\psi$ contributes to this state. Assuming

---

[4] It may not be impossible that the ordinary γ_5 invariance is violated more strongly than the Gürsey γ_5 invariance so that the mass of the $\pi_0{}^0$ meson may come sufficiently high. But to achieve this end by means of a bare mass does not seem to be feasible.

[5] F. Gürsey, Nuovo cimento 16, 230 (1960).

[6] For its possible origin, see Sec. V.

$\Gamma_i^P=\gamma_5\tau_i\Gamma^P$, we obtain

$$\Gamma^P=\gamma^P+\Gamma^P[1-q^2I^P(q^2)],$$

$$I^P(q^2)=\frac{g_0'}{4\pi^2}\int_{4m^2}^{\Lambda^2}\frac{d\kappa^2}{q^2+\kappa^2}\left(1-\frac{4m^2}{\kappa^2}\right)^{\frac{1}{2}}, \qquad (3.2)$$

where, of course, Eq. (2.7) was utilized. This has a homogeneous solution for $q^2=0$, corresponding to the zero-mass "pion." This pion-nucleon coupling is of pure pseudoscalar type, which can be calculated from the inhomogeneous equation with $\gamma^P=g_0'\gamma_5\tau_i$, as was done in the Appendix of I. We get, namely,[7]

$$G_P^2/4\pi=g_0'[4\pi I^P(0)]^{-1}$$

$$=\pi\left[\int_{4m^2}^{\Lambda^2}\frac{d\kappa^2}{\kappa^2}\left(1-\frac{4m^2}{\kappa^2}\right)^{\frac{1}{2}}\right]^{-1}. \qquad (3.3)$$

B. Scalar, Isoscalar Meson

With the ansatz $\Gamma=\Gamma^S$ we have

$$\Gamma^S=\gamma^S+\Gamma^S I^S(q^2),$$

$$I^S(q^2)=\frac{g_0'}{4\pi^2}\int_{4m^2}^{\Lambda^2}\frac{\kappa^2-4m^2}{q^2+\kappa^2}d\kappa^2\left(1-\frac{4m^2}{\kappa^2}\right)^{\frac{1}{2}} \qquad (3.4)$$

$$=1-(q^2+4m^2)I^P(q^2).$$

This leads to a zero-binding state: $q^2=-4m^2$ with the scalar nucleon coupling constant

$$G_S^2/4\pi=g_0'[4\pi I^P(-4m^2)]^{-1}. \qquad (3.5)$$

C. Vector Mesons

There are two vector mesons, with isospin 1 and 0. The isovector meson arises from the tensor interaction $\sim\bar\psi\sigma_{\mu\nu}\tau_i\psi\bar\psi\sigma_{\mu\nu}\tau_i\psi$ with the wave function of the type

$$\Gamma_{\mu i}{}^V=\sigma_{\mu\nu}q_\nu\tau_i\Gamma^V.$$

The mass is determined from[8]

$$1=-\frac{g_0'}{60\pi^2}\int_{4m^2}^{\Lambda^2}\frac{d\kappa^2}{\kappa^2-\mu^2}\left(1-\frac{4m^2}{\kappa^2}\right)^{\frac{1}{2}}$$

$$\times\left[\kappa^2-4m^2-\mu^2\left(2+\frac{4m^2}{\kappa^2}\right)\right],$$

which has a solution (for sufficiently small Λ^2)

$$\mu^2\geqslant 10m^2/3.$$

The coupling of this meson to the nucleon is necessarily of the derivative type.

To the isoscalar meson, both vector and tensor interactions contribute, the former being attractive and the latter repulsive. The wave function will have the form

$$\Gamma_\mu{}^V=\gamma_\mu\Gamma_1{}^V+\sigma_{\mu\nu}q_\nu\Gamma_2{}^V,$$

which yields a coupled equation for Γ_1 and Γ_2. This coupling, however, is rather small, so that we get a solution by neglecting Γ_2:

$$1=\frac{g_0'}{15\pi^2}\mu^2\int_{4m^2}^{\Lambda^2}\frac{d\kappa^2}{\kappa^2-\mu^2}\left(1-\frac{4m^2}{\kappa^2}\right)^{\frac{1}{2}}\left(1+\frac{2m^2}{\kappa^2}\right),$$

$$\mu^2\geqslant 20m^2/7.$$

The nuclear coupling will be predominantly nonderivative.

D. The "Deuteron" States

As in I, we can discuss the nucleon-nucleon states in parallel with the meson states. The interaction may be written conveniently in the form

$$L_{\rm int}=\tfrac{1}{4}g_0[\bar\psi\gamma_\mu\psi^c\bar\psi^c\gamma_\mu\psi-\bar\psi\sigma_{\mu\nu}\psi^c\bar\psi^c\sigma_{\mu\nu}\psi$$
$$+\bar\psi\gamma_\mu\gamma_5\tau_i\psi^c\bar\psi^c\gamma_\mu\gamma_5\tau_i\psi].$$

This is seen to lead to two bound states: a pseudovector, isoscalar ($J=1^+$, $T=0$) coming from the first two interaction terms, and a scalar, isovector ($J+0^+$, $T=1$), coming from the last term. For the $J=1^+$, $T=0$ state (deuteron) the main contribution comes from the attractive tensor interaction, and we get

$$\Gamma_\mu=\sigma_{\mu\nu}q_\nu\Gamma^{A'},$$

$$1=\frac{g_0'}{4\pi^2}\int_{4m^2}^{\Lambda^2}\frac{d\kappa^2}{\kappa^2-\mu^2}\left(1-\frac{4m^2}{\kappa^2}\right)^{\frac{1}{2}}\left[\mu^2+\frac{2}{15}(\kappa^2-4m^2)\right],$$

$$\mu^2\geqslant 17m^2/5.$$

For the $J=0^+$, $T=1$ case we have

$$\Gamma=\gamma_5\gamma\cdot q\Gamma^{S'},$$

$$1=\tfrac{4}{5}m^2I^P(-\mu^2),$$

$$\mu^2\geqslant 16m^2/5.$$

IV. VIOLATION OF γ_5 INVARIANCE

Let us now discuss the violation of the γ_5 invariance as indicated by the finite mass of the real pion. It would be senseless, of course, to talk about the invariance if the observed pion mass implied a large departure from our original Lagrangian, for example, due to a bare nucleon mass as large as the observed mass. So we need to estimate the amount of violation in the Lagrangian.

In general, the bare mass operator, which does not

[7] Note that this is half the value of I because a pion (e.g., π_0) consists of two substates $\bar pp$ and $\bar nn$, which changes the normalization of the pion wave function.

[8] The ambiguity about the subtraction of the most divergent part was discussed in I, section 4. The gross qualitative feature is not altered even if we do not make a subtraction.

violate nucleon number conservation,[9] can have the following form

$$M^0 = m_1^0 + m_2^0 \boldsymbol{\tau}\cdot\mathbf{n} + m_3^0 i\gamma_5 + m_4^0 i\gamma_5 \boldsymbol{\tau}\cdot\mathbf{n}', \qquad (4.1)$$

where $\mathbf{n}$ and $\mathbf{n}'$ are arbitrary unit vectors in the isospin space. The observed mass generated by Eq. (4.1) will also have a similar form. Because of the invariance of the rest of the Lagrangian under the transformations (2.1), we can choose it to be

$$M = m_1 + m_2\tau_3 + m_3 i\gamma_5, \qquad (4.2)$$

which gives two eigenmasses

$$m_p \equiv [(m_1+m_2)^2 + m_3^2]^{\frac{1}{2}},$$
$$m_n \equiv [(m_1-m_2)^2 + m_3^2]^{\frac{1}{2}}. \qquad (4.3)$$

The self-consistent self-energy equation to be solved is now

$$M = M^0 + g_0\{\mathrm{Tr} S_F^{(M)}(0) - \gamma_5\tau_i \,\mathrm{Tr}[\gamma_5\tau_i S_F^{(M)}(0)] + \tfrac{1}{5}\gamma_5 \,\mathrm{Tr}[\gamma_5 S_F^{(M)}(0)] - \tfrac{1}{5}\tau_i \,\mathrm{Tr}[\tau_i S_F^{(M)}(0)]\}. \qquad (4.4)$$

Equating the respective coefficients of both sides, we get

$$\begin{aligned} m_1 &= m_1^0 + \bar{I} m_1 + \tilde{I} m_2, \\ m_2 &= m_2^0 - \tfrac{1}{5}\bar{I} m_2 - \tfrac{1}{5}\tilde{I} m_1, \\ m_3 &= m_3^0 - \tfrac{1}{5}\bar{I} m_3, \\ 0 &= m_4^0 + \tilde{I} m_3, \end{aligned} \qquad (4.5)$$

where

$$\begin{aligned} \bar{I} &= \tfrac{1}{2}[I(m_p) + I(m_n)], \\ \tilde{I} &= \tfrac{1}{2}[I(m_p) - I(m_n)], \\ I(m) &= -\frac{8ig_0'}{(2\pi)^4}\int \frac{d^4p}{p^2+m^2} F(p,\Lambda). \end{aligned} \qquad (4.5')$$

We are interested in a small change of the non-trivial solution due to M^0. From Eq. (4.5) it is clear that $m_3 = 0$ unless

$$m_3^0 = -(\tfrac{1}{5} + \tilde{I}^{-1}) m_4^0 \neq 0.$$

The term m_3 implies a violation of time and space reflections. Since we are not interested in such a violation, we will assume $m_3 = m_3^0 = m_4^0 = 0$ from now on. We further note that

$$\bar{I} = I(m_1) + O[(m_2/m_1)^2], \quad \tilde{I} = O[m_2/m_1].$$

In fact, up to the first order in m_2/m_1, we may put

$$m_1 = m_1^0 + I(m_1) m_1, \qquad (4.6a)$$

$$m_2 = m_2^0 - \tfrac{1}{5}[I(m_1) + 2I'(m_1) m_1^2] m_2, \qquad (4.6b)$$

where

$$I'(m) = dI(m)/d(m^2) < 0.$$

Equation (4.6a) determines m_1 in terms of m_1^0.

The self-consistency condition required, for $m_1^0 = 0$, is that $I(m) = 1$. We may thus expand $I(m)$:

$$I(m_1) = 1 + I'(m)(m_1^2 - m^2),$$

and obtain

$$\Delta m^2 = m_1^2 - m^2 = -m_1^0/[mI'(m)]. \qquad (4.7)$$

Since $I'(m)$ is of the order of $-I(m)/m^2$ (see below), this means

$$\Delta m = m_1 - m \approx m_1^0. \qquad (4.8)$$

From (4.6b), then

$$\begin{aligned} m_2 &\approx m_2^0\{1 - \tfrac{1}{5}[1 + I'(m) m^2]\}^{-1} \\ &\approx m_2^0. \end{aligned} \qquad (4.8')$$

We note that originally there were two solutions $\pm|m|$, which now split into opposite directions according to Eq. (4.7) or (4.8). The meaning of this is as follows. Under the strict γ_5 invariance, there is a complete degeneracy with respect to the transformation (2.1c). The perturbation m_1^0 removes this degeneracy, so that the energy of the vacuum will depend on the orientation of the "γ_5 spin" of the negative energy fermions present in the "vacuum" with respect to this preferred direction. Obviously, the self-consistent procedure, which is similar to the variational method, gives the two extremum configurations corresponding to parallel ($m_0/m > 0$) or antiparallel ($m_0/m < 0$) γ_5-spin lineup. The parallel case has the larger "gap parameter" $|m|$ than the antiparallel case, so that the former will correspond to the stable ground state. The latter, on the other hand, should correspond to a metastable world.

It is perhaps interesting to see the general behavior of the self-consistency equation for arbitrary magnitude of m_1^0, assuming $m_2^0 = 0$ for simplicity. The relevant equation,

$$m[1 - I(m)] = m^0,$$

is plotted schematically in Fig. 1.

Note that the trivial branch of the solution, which goes through the origin, has $m_0/m < 0$. In other words, even in this case the self-consistent solution is qualitatively different from the simple perturbation result. As m^0 increases, it approaches the metastable nontrivial solution, and finally both go into the complex plane.

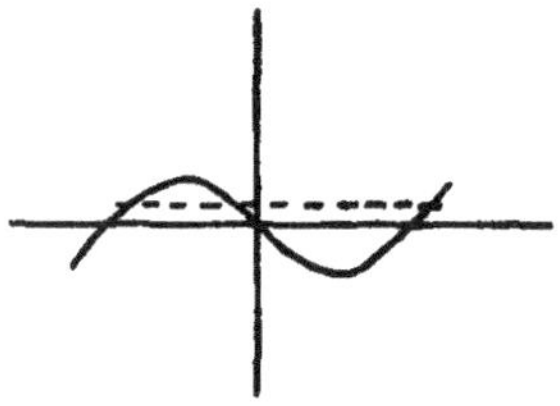

FIG. 1. The three self-consistent mass solutions m (ordinate) as a function of the bare mass m^0 (abscissa).

[9] The most general form of the self-energy Lagrangian (neglecting isospin dependence) is

$$\bar{\psi}[i\gamma\cdot p\,\Sigma_1(p^2) + \Sigma_2(p^2) + i\gamma\cdot p\gamma_5\,\Sigma_3(p^2) + i\gamma_5\,\Sigma_4(p^2)]\psi + \bar{\psi}^c[i\gamma\cdot p\,\Sigma_5(p^2) + \Sigma_6(p^2) + i\gamma\cdot p\gamma_5\,\Sigma_7(p^2) + i\gamma_5\,\Sigma_8(p^2)]\psi + \mathrm{H.c.}$$

We do not attempt to study such a problem at this place.

We now come to the meson problem. The pion mass will be determined from

$$\Gamma_j = ig_0'\tau_i \int \mathrm{Tr}[\tau_i S_F(p'+\tfrac{1}{2}q)\Gamma_j S_F(p'-\tfrac{1}{2}q)]d^4p', \quad (4.9)$$

but

$$\underset{\text{isospin}}{\mathrm{Tr}}[\tau_i S_F{}^{(M)}\tau_j S_F{}^{(M)}]$$

$$= \underset{\text{isospin}}{\mathrm{Tr}}\left[\tau_i\left(S_F{}^{(m_p)}\frac{1+\tau_3}{2}+S_F{}^{(m_n)}\frac{1-\tau_3}{2}\right)\times\tau_j\left(S_F{}^{(m_p)}\frac{1+\tau_3}{2}+S_F{}^{(m_n)}\frac{1-\tau_3}{2}\right)\right]$$

$$=2\delta_{ij}\frac{S_F{}^{(m_p)}+S_F{}^{(m_n)}}{2}\frac{S_F{}^{(m_p)}+S_F{}^{(m_n)}}{2} +2(2\delta_{i3}\delta_{j3}-\delta_{ij})\frac{S_F{}^{(m_p)}-S_F{}^{(m_n)}}{2}\frac{S_F{}^{(m_p)}-S_F{}^{(m_n)}}{2}.$$

The second term yields convergent results, and is $O[(\Delta m/m)^2]$. To the order $\Delta m/m$, therefore, only the first term is important; moreover,

$$(S_F{}^{(m_p)}+S_F{}^{(m_n)})/2 \approx S_F{}^{(m_1)}.$$

In other words, there will be no first-order mass splitting of the pion. The mass is then determined from

$$1 = J_{p1}(-\mu^2) = \frac{g_0'}{4\pi^2}\int_{4m_1^2}^{\Lambda^2}\frac{\kappa^2 d\kappa^2}{\kappa^2-\mu^2}\left(1-\frac{4m_1^2}{\kappa^2}\right)^{\frac{1}{2}}. \quad (4.10)$$

For $m_1^0=0$, we had originally

$$1 = J_p(0) = \frac{g_0'}{4\pi^2}\int_{4m^2}^{\Lambda^2} d\kappa^2\left(1-\frac{4m^2}{\kappa^2}\right)^{\frac{1}{2}},$$

which should now be replaced by

$$1=\frac{m_1^0}{m_1}+\frac{g_0'}{4\pi^2}\int_{4m_1^2}^{\Lambda^2} d\kappa^2\left(1-\frac{4m_1^2}{\kappa^2}\right)^{\frac{1}{2}}, \quad (4.11)$$

according to Eq. (4.6a).

From Eqs. (4.10) and (4.11) follows

$$\frac{m_1^0}{m_1} = \mu^2\frac{g_0'}{4\pi^2}\int_{4m_1^2}^{\Lambda^2}\frac{d\kappa^2}{\kappa^2-\mu^2}\left(1-\frac{4m_1^2}{\kappa^2}\right)^{\frac{1}{2}} \approx \mu^2\frac{g_0'}{4\pi^2}\int_{4m_1^2}^{\Lambda^2}\frac{d\kappa^2}{\kappa^2}\left(1-\frac{4m_1^2}{\kappa^2}\right)^{\frac{1}{2}} \lesssim \frac{\mu^2}{4m_1^2}\left(1-\frac{m_1^0}{m_1}\right). \quad (4.12)$$

For the observed value of $\mu^2/4m_1^2 \approx 1/200$ we then have, for the stable solution

$$m_1^0 \lesssim m_1/200 \approx 5\text{ Mev}. \quad (4.13)$$

The amount of bare mass needed to produce the pion mass is thus surprisingly small.

On the other hand, the metastable solution $(m_1^0/m<0)$ produces an imaginary pion mass, indicating the unphysical nature of the solution.

The pion-nucleon coupling constant at the pion pole becomes [see Eq. (2.3)]

$$G_P^2/4\pi \approx g_0'[4\pi I_P(-\mu^2)]^{-1},$$

which is changed from the old one only by an order $\mu^2/m_1^2 \sim m_1^0/m_1$.

The other heavy meson states can be treated similarly. We see easily that the changes induced by m_1^0 are quite small: In general $\Delta\mu^2/m_1^2 = O(m_1^0/m_1)$ and $\Delta G^2/G^2 = O(m_1^0/m_1)$. Thus the effect of M^0 shows up dramatically only in the pion mass because it was originally zero.

Finally we remark that instead of a bare mass, we could assume slightly different coupling constants g_s and g_p $(<g_s)$ for the scalar and pseudoscalar interaction terms in the Lagrangian (2.3). The nature of the solution is somewhat different from the previous case because the Lagrangian still retains the mass reversal invariance $\psi \to \gamma_5\psi$, and the solution is twofold degenerate $(\pm m)$. The fractional change of the coupling necessary to produce the pion mass is again small: $|\Delta g/g| \approx \mu^2/4m_1^2$.

V. IMPLICATIONS OF THE MODEL

Let us now discuss the relevance of our present model to the physical realities of the nucleons and mesons.

1. We have seen that our Lagrangian (2.3) leads to the nucleon of isospin $\frac{1}{2}$ and the pion of isospin 1. The pion-nucleon coupling constant (pseudoscalar) depends on the cutoff parameter. For the observed large value (≈ 15) of $G_P^2/4\pi$, we see from Eqs. (1.5) and (2.3) that Λ must be of the same order of magnitude as the nucleon mass itself. This is not unreasonable, since the effective nucleon-nucleon interaction in higher approximations would proceed with the exchange of nucleon pairs.

A third parameter, the bare mass, enters our picture in order to make the meson mass finite. It would seem rather unsatisfactory and embarrassing that after all one has to break the postulated symmetry in an *ad hoc* manner. In order to clear up this point, the origin of the effective bare mass then becomes an interesting and important question. Since the required bare mass [Eq. (4.13)] seems to be quite small, a tempting possibility suggests itself that the bare nucleon field is the same as the electron-neutrino field. The electron mass itself could be either intrinsic or of electromagnetic origin.[10] Under this assumption, the bare mass operator would have the form $M^0 = m_e^0(1+\tau_3)/2$, where the word "bare" is used relative to the interaction under consideration. According to the results of the previous section, it is only the isoscalar part of M^0 that produces the large shift of the pion mass, and the amount of violation of the isospin invariance will remain small.

2. Besides the pion, we have also derived vector mesons of both isoscalar ($T=0$) and isovector ($T=1$) types, and a scalar isoscalar meson which is actually unbound. No state corresponding to the isoscalar pion (π_0^0) is found. Of course, these results should depend sensitively on the choice of the interaction in the first place, and to a lesser extent also on the degree of approximation. At any rate, it seems to be a rather interesting and satisfactory feature of the model that these same vector mesons have been anticipated theoretically from various grounds,[11] even though there do not seem to be convincing experimental indications of their existence as yet.[12]

The mass values obtained here are rather high, and these mesons should actually decay into pions very quickly. The coupling constants are generally of the same order as the pion coupling constant, which means a very strong interaction for the vector and scalar mesons. These results, however, may be considerably altered in a better approximation. For one thing, the heavy mesons are coupled strongly to many-pion states which would make the former mere resonances of the latter. Moreover, the nucleon-nucleon and meson-meson interactions can go through long-range forces due to the exchange of these same mesons, which would in turn change the meson states themselves. These processes (the so-called left-hand cuts in the language of the dispersion theory) have not been taken into account in our ladder approximation.

This is a highly cooperative mechanism, and if one wants to handle it in a systematic way, one may be led to the same dispersion theoretical approach that is now widely pursued in pion physics. As a result of such effects, it is then conceivable that the masses of the vector mesons, for example, may come down.[13] Alternatively, it is also conceivable that we have more than one resonance having the same quantum numbers, of which we have obtained the higher ones. These high-energy poles may in turn determine the low-energy resonances.

In addition to the vector mesons, we expect a $T=0$, $J=0^+$ resonance, which has also been postulated by some people.[14] We should try to check these predictions against experimental evidence, such as the characteristic Q-value distributions and angular correlations in meson production processes.

Turning to the nucleon number 2 states, we expect two bound states ($T=0$, $J=1^+$ and $T=1$, $J=0^+$) with comparable masses to those for the vector mesons. This is a qualitatively satisfactory feature in view of the observed deuteron and the singlet virtual states, even though the actual binding is considerably weaker.[15]

3. As was already mentioned in I, our particular model was motivated by the approximate axial vector conservation observed in the nuclear β decay and the role of the pion in it.[5,16] The only difference from I is that (a) we now have the conservation of the isovector axial vector current $i\bar{\psi}\gamma_\mu\gamma_5\tau_i\psi$ instead of the simple axial vector current $i\bar{\psi}\gamma_\mu\gamma_5\psi$, and (b) a small violation of conservation is explicitly introduced. The general treatment of the problem will be completely analogous to the previous case.

Assuming that the β decay occurs through an additional term in the Lagrangian

$$L_\beta = g_\beta\bar{\psi}\gamma_\mu(1+\gamma_5)\tau_+\psi l_\mu + \text{H.c.} \quad [\tau_+ = \tfrac{1}{2}(\tau_1 + i\tau_2)],$$

where l_μ refers to the lepton current, the nuclear β-

[10] The electromagnetic interaction is invariant under the simple γ_5 transformation, but not under the Gürsey transformation since it fundamentally distinguishes between the charged and neutral components. Thus there is a built-in violation which can eventually produce the pion mass.

[11] W. R. Frazer and J. R. Fulco, Phys. Rev. **117**, 1609 (1960); Y. Nambu, *ibid.* **106**, 1366 (1957); G. Chew, Phys. Rev. Letters **4**, 142 (1960); J. J. Sakurai, Ann. Phys. **11**, 1 (1960).

[12] J. A. Anderson, Vo X. Bang, P. G. Burke, D. D. Carmony, and N. Schmitz, Phys. Rev. Letters **6**, 365 (1961); A. Abashian, N. Booth, and K. M. Crowe, *ibid.* **5**, 258 (1960).

[13] A crude way to see the general tendency will be to argue as follows: The $T=1$ vector meson is coupled to the nucleon mainly through tensor coupling, so that it will cause a nucleon-antinucleon interaction of the type $-g^2(\boldsymbol{\sigma}_1\cdot\boldsymbol{\sigma}_2\boldsymbol{\tau}_1\cdot\boldsymbol{\tau}_2)e^{-\mu r}/r$. This tends to raise $T=1$, $J=0^+$ and $T=0$, $J=1^-$ meson states, and lower $T=0$, $J=0^+$ and $T=1$, $J=1^-$ states. Any change in the binding force, however, will be offset by the corresponding change in the nucleon mass, which automatically adjusts the pion mass to lie where it should be. The exchange of the $T=0$, $J=1$, and $J=0$ mesons, therefore, would not be so important in determining the relative shift of the meson levels.

[14] J. Schwinger, Ann. Phys. **2**, 407 (1957); M. Gell-Mann and M. Lévy, Nuovo cimento **16**, 705 (1960); S. Gupta, Phys. Rev. **111**, 1436 (1958), Phys. Rev. Letters **2**, 124 (1959); M. H. Johnson and E. Teller, Phys. Rev. **98**, 783 (1955); H. P. Duerr and E. Teller, *ibid.* **103**, 469 (1956). The σ meson mass obtained here is independent of the cutoff Λ, so that there may be some point in arguing that it is more reliable than for the vector mesons. If so, we may expect a nucleon-antinucleon resonance near zero kinetic energy (taking account of the mass shift due to M^0). The width may be quite broad.

[15] In fact, both $T=0$ and $T=1$ vector meson exchanges work in the direction to reduce the binding relative to the nucleon-antinucleon case.

[16] S. Bludman, Nuovo cimento **9**, 433 (1958); F. Gürsey, Ann. Phys. **12**, 91 (1961); Y. Nambu, reference 1; M. Gell-Mann and M. Levy, reference 14; J. Bernstein, N. Gell-Mann, and L. Michel, Nuovo cimento **16**, 560 (1960); J. Bernstein, S. Fubini, M. Gell-Mann, and W. Thirring, *ibid.* **17**, 757 (1960); Chou Kuang-Chao, J. Exptl. Theoret. Phys. (U.S.S.R.) **39**, 703 (1960) [Soviet Phys.—JETP **12**, 492 (1961)].

decay vertex becomes

$$\Gamma_\mu = g_\beta[i\gamma_\mu\tau_+ F_{V1}(q^2) - i\sigma_{\mu\nu}q_\nu\tau_+ F_{V2}(q^2) + \{i\gamma_\mu\gamma_5\tau_+ + [2m_1\gamma_5 q_\mu\tau_+/(q^2+\mu_\pi^2)]f(q^2)\}F_A(q^2)],$$

where q is the momentum change. In the ladder approximation, $F_{V1}(q^2)$ arises from the vector-type nucleon pairs, and $F_{V1}(0)=1$ (in accordance with the Ward identity, applicable to the isospin current, which shows that $F_{V1}(0)=1$ in general.[17])

In the axial vector part, $F_A(q^2)=1$ in our approximation. $f(q^2)$ arises because of the violation of the γ_5 invariance, but it deviates from 1 only to the order $m_1^0/m_1 \sim \mu^2/m_1^2$, as was already seen in the previous section. For practical purposes, therefore, the axial vector current has the desired form which would lead to the Goldberger-Treiman relation[18]

$$2m_1 g_A \approx \sqrt{2}G_\pi g_\pi,$$

where $g_A = g_\beta F_A(0)$ and G_π, g_π are, respectively, pion-nucleon and pion-lepton couplings.

In higher orders, however, $F_A(q^2)$ will be present, and in general $F_A(0)\neq 1$ even under the strict γ_5 invariance. People have conjectured in the past that $F_A(0)=g_A/g_V=1$ as $\mu_\pi \to 0$, but this does not seem to be easily guaranteed. The generalized Ward identity for the axial vector current[19] suffices to prove the Goldberger-Treiman relation, but is not enough to make $F_A(0)=1$. In order that the latter should come out rigorously, we would need a more subtle mechanism. Nevertheless, we can try a working hypothesis that $g_A/g_V=1$ under the strict invariance, and then estimate the deviation due to the violation. This scheme is carried out in the Appendix.

VI. FURTHER PROBLEM

We will consider here some of the general problems which have not been explored, but which seem to be important in a more comprehensive understanding of the elementary particles.

1. The hyperons. In order to incorporate the strange particles into our picture we would have to increase the dimensions of the fundamental field unless we do further unconventional things (see below). The simplest possibility from the point of view of quantum numbers would be to add a bare Λ-particle field as was originally proposed by Sakata.[20] We would then postulate, in addition, the generalized γ_5 symmetry, which would mean the invariance of the left-handed and right-handed components separately under the unitary transformation among the three fields or some subgroups of it. The mass splitting of the three baryons will be obtained from bare masses of similar magnitude, which destroys the otherwise rigorous symmetry.

This approach will produce easily the pions and K mesons and probably more, and their masses can again be related to the baryon bare masses. But we do not yet have a comparable dynamical method to predict Σ and Ξ particles. Consequently, we shall not be able to say whether or not the present model is dynamically satisfactory in this respect.

2. The leptons. In connection with the above model we are naturally led to the lepton problem. Gamba, Marshak, and Okubo[21] have pointed out an interesting parallelism between the $pn\Lambda$ and $\nu e\mu$ triplets. As was remarked in the beginning, our theory gives a special incentive for speculation about this relation because we have obtained two solutions: one ordinary and one extraordinary, differing in masses. Could they both be realized in nature simultaneously? According to our results in I, the answer is no because they belong to different Hilbert spaces. Moreover, the trivial solution gives rise to unphysical mesons at least under the assumption of fixed cutoff, with a large mass $(-\mu^2 \gtrsim \Lambda^2)$ but not necessarily a weak coupling $(G^2 \lesssim \Lambda^4/\mu^4)$. Nevertheless, it would seem too bad if Nature did not take advantage of the two solutions. A straightforward way to make the two solutions co-exist in the same world is obviously to postulate that the world is represented by the direct product of two Hilbert spaces[22]:

$$\mathcal{H} = \mathcal{H}^{(0)} \otimes \mathcal{H}^{(m)}, \tag{6.1}$$

built upon the vacuum state

$$\Omega = \Omega^{(0)} \otimes \Omega^{(m)}. \tag{6.1'}$$

It is true that this is effectively the same as doubling the fields, but here the choice of the two solutions (particles) is dictated by the dynamics of the original nonlinear theory. In order to describe this situation, we may adopt an *effective* Lagrangian

$$L = L^{(1)} + L^{(2)}, \tag{6.2}$$

where each of the $L^{(i)}$ has the same form, only differing in the charge assignments of the respective triplet fields. The Lagrangian obviously yields four subspaces

$$\mathcal{H}_1^{(0)} \otimes \mathcal{H}_2^{(m)},\quad \mathcal{H}_1^{(m)} \otimes \mathcal{H}_2^{(0)},\quad \mathcal{H}_1^{(0)} \otimes \mathcal{H}_2^{(0)}$$

and

$$\mathcal{H}_1^{(m)} \otimes \mathcal{H}_2^{(m)}.$$

According to our plan, we must say that we happen to live in the first subspace. [In the second space, the masses of $\nu e\mu$ and $pn\Lambda$ are interchanged, whereas in the third (fourth) case we have two kinds of leptons (baryons).]

[17] R. P. Feynman and M. Gell-Mann, Phys. Rev. **109**, 193 (1958).
[18] M. L. Goldberger and S. B. Treiman, Phys. Rev. **111**, 356 (1958).
[19] J. Bernstein *et al.*, reference 16.
[20] S. Sakata, Progr. Theoret. Phys. (Kyoto) **16**, 686 (1956).
[21] A. Gamba, R. E. Marshak, and S. Okubo, Proc. Natl. Acad. Sci. U. S. **45**, 881 (1959); Z. Maki, M. Nakagawa, Y. Ohnuki, and S. Sakata, Progr. Theoret. Phys. **23**, 1174 (1960).
[22] S. Okubo and R. E. Marshak, Nuovo cimento **19**, 1226 (1961), have independently proposed a similar idea. We thank the authors for valuable communications.

So far there is no interaction between leptons and baryons (except the electromagnetic, which is trivial). To introduce the weak interactions, we may, for example, add to Eq. (5.2) a third nonlinear term involving all the (left-handed) fields. This would complete our program of dealing with the strong and weak interactions.

But, of course, it is not yet a truly unified theory; the weak interaction is introduced only as an *ad hoc* additional process. Moreover, we do not know the mathematical consistency of such a procedure, because the additional interaction, if taken seriously, may qualitatively affect the baryon and lepton solutions we already have.

There is an alternative, but less drastic scheme; namely, to assume six different fields from the beginning, of which three (becoming eventually the baryon fields) have additional strong interactions in the Lagrangian. This may not be devoid of elegance if the interaction is mediated by a vector Bose field coupled to the baryon charge. The intermediate bosons, including the photons and possibly also the weak bosons, could then be interpreted as the agents that distinguish between different components of the bare fermions, which otherwise would enjoy a high degree of symmetry.

We would like to throw in another remark here that there may be also a possibility of utilizing the ordinary and extraordinary solutions in distinguishing between electron and muon, or baryons of different strangenesses.

3. The γ_5 invariance for general systems. In our theory the γ_5 invariance is a very essential element. It is a particular symmetry which exists in the Lagrangian, but is masked in reality because of the (approximate) degeneracy of the vacuum with respect to that symmetry. We have used the pion and the β decay in support of the assumption. In order to firmly establish its validity, however, we must try to find more evidences. For one thing, the induced pseudoscalar terms in nucleon β decay and μ capture should be examined more closely.

Furthermore, if such a symmetry is to have a general meaning, we must be able to consider partially conserved currents for processes such as

$$\begin{aligned} &H^3 \rightarrow He^3,\\ &He^6 \rightarrow Li^6,\\ &C^{14} \rightarrow N^{14},\\ &\Sigma^- \rightarrow \Sigma^0,\\ &\Sigma^- \rightarrow \Lambda. \end{aligned} \tag{6.3}$$

An elementary definition of the γ_5 transformation for the general system is obvious: When the wave function of a system is expressed in terms of the fundamental (bare) spinors obeying the rules Eq. (2.1), the transformation is unambiguously defined for each component, and thereby the total axial vector current is determined.

For superallowed transitions with spin $\frac{1}{2}$, the problem is particularly simple, since it is the same as for the neutron case. Thus for $H^3 \rightarrow He^3$ we have the same Goldberger-Treiman relation

$$(M_H + M_{He}) g_A(\mathrm{H,He})/\sqrt{2} G_\pi(\mathrm{H,He}) \approx g_\pi, \tag{6.4}$$

where g_A, G_π now characterize the β decay and the (unknown) pion coupling for the transition under consideration.

Similar relations hold for the Σ decays.[23] For the Σ-Σ case, we have

$$2m_\Sigma g_A(\Sigma\Sigma)/G_\pi(\Sigma\Sigma) \approx g_\pi. \tag{6.5}$$

For the Σ-Λ case, the axial vector vertex becomes[24]

$$\begin{aligned} \Gamma_A \approx [i\gamma_\mu\gamma_5 + (m_\Sigma + m_\Lambda)\gamma_5 q_\mu/(q^2+\mu_\pi^2)]F_{1A}(q^2) \\ + i\gamma_5\sigma_{\mu\nu}q_\nu F_{2A}(q^2) \\ (m_\Sigma + m_\Lambda)F_{1A}(0)/G_\pi(\Sigma\Lambda) \approx g_\pi, \end{aligned} \tag{6.6}$$

if the relative Σ-Λ parity is even. The vector current conservation is also violated because of the Σ-Λ mass difference, and it looks as though this would predict a corresponding scalar meson term. However, the analogy is rather superficial. Firstly, the violation disappears if $m_\Sigma = m_\Lambda$, in which case there would be no need for a scalar meson. The Σ-Λ mass difference itself might be due to the breakdown of the γ_5 symmetry. Secondly, it is an "unfavored" transition ($\Delta T = 1$), so that the vector part, corresponding to the off-diagonal element of the isospin current, should vanish in the ideal limit of strict isospin invariance and $q \rightarrow 0$. In other words, we expect

$$\Gamma_V \approx [q^2 i\gamma_\mu - (m_\Sigma - m_\Lambda)q_\mu]F_{1V}(q^2) + \sigma_{\mu\nu}q_\nu F_{2V}(q^2). \tag{6.7}$$

In case the Σ-Λ parity is odd,[25] the vector and axial vector parts will interchange their roles. The vector part, which now looks like the axial vector current, would have the form

$$\begin{aligned} \Gamma_V \approx [q^2 i\gamma_\mu\gamma_5 + (m_\Sigma + m_\Lambda)\gamma_5 q_\mu]F_{1V}(q^2) \\ + i\gamma_5\sigma_{\mu\nu}q_\nu F_{2V}(q^2). \end{aligned} \tag{6.8}$$

The axial vector part can similarly be put in the form

$$\begin{aligned} [i\gamma_\mu - q_\mu(m_\Sigma - m_\Lambda)/(q^2+\mu_\pi^2)f(q^2)]F_{1A}(q^2) \\ + \sigma_{\mu\nu}q_\nu F_{2A}(q^2). \end{aligned} \tag{6.9}$$

But $f(q^2)$ need not be ≈ 1 if the Σ-Λ mass difference is also due to the violation of the γ_5 symmetry.

There are other processes for which the chirality conservation can be tested in a direct way. Although

[23] L. B. Okun', *Ann. Rev. Nuclear Sci.* **9**, 61 (1959); M. Gell-Mann, *Proceedings of the 1960 Annual International Conference on High-Energy Physics at Rochester* (Interscience Publishers, Inc., New York, 1960), p. 522.

[24] We have in this case three independent terms.

[25] See S. Barshay, Phys. Rev. Letters **1**, 97 (1958); Y. Nambu and J. J. Sakurai, *ibid.* **6**, 377 (1961).

extraordinary solutions are in general not eigenstates of chirality (even under strict γ_5 invariance), the conservation law should still apply to the expectation values of chirality. In fact, we can express the chirality conservation law $\langle X_i \rangle = \langle X_f \rangle$ for any reaction $i \rightarrow f$; for example

$$p+\pi \rightarrow p+\pi, \quad p+\pi+\pi', \text{ etc.},$$
$$p+p \rightarrow p+p, \quad p+p+\pi, \text{ etc.},$$

as a relation between the change of nucleon chirality and the magnitude of the pion production amplitude.

The ideas outlined in this section will be taken up in more detail elsewhere.

APPENDIX

We calculate here the renormalization of the axial vector (Gamow-Teller) coupling constant g_A for nuclear β decay under the following assumptions:

(1) Under strict γ_5 invariance (Gürsey type), there is no renormalization, namely $g_A = g_{A0}$ $(= g_{V0} = g_V)$, where g_{A0} is the bare coupling constant.

(2) The violation of the invariance gives rise to the finite pion mass as well as the deviation of the ratio $R = g_A/g_{A0} = g_A/g_V$ from unity, so that there is a functional relation between the two quantities.

Let us first consider the isovector axial vector vertex Γ_A in the usual perturbation theory. In our model, it consists of various graphs, some of which are shown in Fig. 2(a) and (b). The "ladder" graphs 2(a) have been considered in I as well as in the present paper, since they are intimately related to the γ_5 gauge transformation. In I (Appendix) we found that $R > 1$ when both pseudoscalar and pseudovector type interactions are present.[26] The graphs 2(b) have not been considered yet. These will come into our consideration as soon as we take corresponding higher-order approximations for the self-energy, which was briefly discussed in I. The chain of bubbles in these graphs will act like a meson when there is such a dynamical pole [Fig. 2(c)].

The (divergent) renormalization effect due to intermediate mesons is always negative,[27] irrespective of the type of the meson, so that the effect of these meson-like bubble graphs is also expected to be similar. When the chain does not produce a pole, however, the effect can be opposite.

Combining all these effects, we have no way to predict the resultant magnitude and sign of the renormalization correction. So we simply assume these contributions to cancel out under strict γ_5 invariance.

Next let us suppose that the invariance is slightly violated. This will cause changes in the propagators

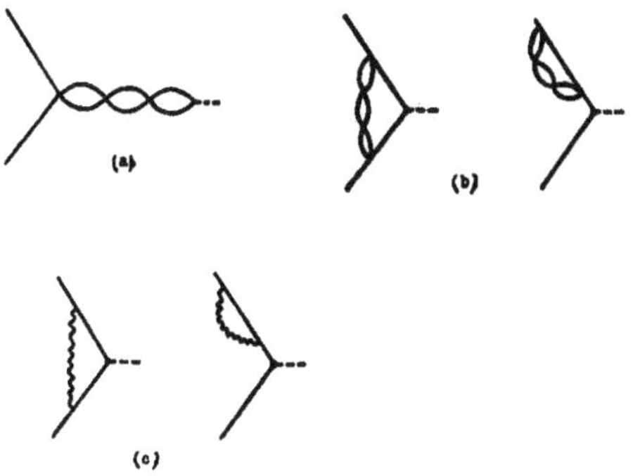

FIG. 2. Typical graphs considered in the evaluation of the axial vector vertex.

in all these graphs. Most of these changes are, however, quite small, being of the order of $m^0/m \approx \mu^2/4m^2$, as will be clear from the results of Sec. IV. The largest effect is naturally expected to come from the "pion" contribution in Fig. 2(b), as this is a change from zero mass (infinite range) to a finite one.

Let us accordingly take the effective pion graph from Fig. 2(c) with an arbitrary pion mass μ. Call its contribution to the vertex renormalization (for zero momentum transfer) $\Lambda(\mu)$. Then according to the above assumption

$$R = \Gamma_A(\mu_\pi)/g_A = \Gamma_A(\mu_\pi)/\Gamma_A(0)$$
$$\approx 1 + \Lambda(\mu_\pi) - \Lambda(0). \qquad \text{(A1)}$$

The difference $\Lambda(\mu) - \Lambda(0)$ is convergent, which turns out to be

$$\Lambda(\mu) - \Lambda(0) = \frac{G^2}{16\pi^2}\frac{\mu^2}{m^2}\left\{\left(3 - \frac{5\mu^2}{2m^2}\right)\ln\frac{m^2}{\mu^2} - 5 + \frac{16\mu}{\sqrt{3}m}\left[\tan^{-1}\left(\frac{4m^2}{3\mu^2} - \frac{2}{3}\right) + \tan^{-1}\left(\frac{2}{3}\right)\right]\right\}, \qquad \text{(A2)}$$

where G is the phenomenological pion coupling constant.

As was expected, this goes like $(\mu^2/m^2)\ln(m^2/\mu^2)$ for small μ, which is more important than the contributions from the neglected processes behaving like μ^2/m^2.

With $(G^2/4\pi)(\mu^2/4m^2) = f^2/4\pi = 0.08$, Eq. (A2) gives

$$R - 1 = \begin{cases} 0.18 \\ 0.24. \end{cases} \qquad \text{(A3)}$$

The first figure is the entire contribution from Eq. (A2), while the second is the contribution from the leading logarithmic term alone. Experimentally, R is estimated to be ≈ 1.25.[28]

[26] See also Z. Maki, Progr. Theoret. Phys. (Kyoto) **22**, 62 (1959).

[27] We owe Dr. J. de Swart the mathematical check on this point.

[28] M. T. Burgy, V. E. Krohn, T. B. Novey, G. R. Ringo, and V. L. Telegdi, Phys. Rev. **120**, 1829 (1961).

PHYSICAL REVIEW VOLUME 125, NUMBER 4 FEBRUARY 15, 1962

Chirality Conservation and Soft Pion Production*

Y. NAMBU AND D. LURIÉ
The Enrico Fermi Institute for Nuclear Studies and the Department of Physics, University of Chicago, Chicago, Illinois

(Received September 25, 1961)

A formally γ_5-invariant system consisting of a Dirac field and a massless pseudoscalar field allows chirality conservation in the sense that its expectation value is a constant of motion. This leads to the consequence that in any reaction a change in the fermion chirality ($\sim$helicity$\times$velocity) is compensated for by the emission of a massless boson at zero energy, which can be expressed by a simple formula relating the radiative amplitude to the elastic amplitude. Assuming the pion-nucleon system to be γ_5-invariant when the pion mass can be neglected, the formula is applied to the processes $N+\pi\rightarrow N+\pi$ and $N+2\pi$. A reasonable agreement with experiment is obtained in a case dominated by the 3–3 resonance.

I. INTRODUCTION

THE observed equality of the vector coupling constant G_V in nuclear and muon β decays has led to the conserved current theory of Feynman and Gell-Mann,[1] according to which the vector part of the nuclear β-decay interaction is proportional to the total isotopic spin current. The nonrenormalization of G_V due to strong interactions is then guaranteed by the isospin conservation in strong interactions.

One expects further a close proportionality of the nucleon electromagnetic and β-decay matrix elements, including the Pauli magnetic term, and the pion β decay $\pi^\pm\rightarrow\pi^0$ would also proceed at the "universal" rate. Although these predictions are yet to be confirmed experimentally, there seems to be enough theoretical motivation for such speculations. For we would then be able to regard the weak interactions, like the electromagnetic interaction, as an agent which reveals the basic symmetries that might exist beneath the confusing effects of strong interactions.

As the nucleon β decay interaction also contains an axial vector part with a comparable strength ($G_A\sim 1.2G_V$), several authors[2] have naturally tried to extend the principle by postulating axial vector current conservation or invariance under the so-called γ_5 transformation. In this case, however, it has not been found possible to guarantee the nonrenormalization of G_A by the axial vector conservation, and besides the conservation law seems to be only approximate under strong interactions. Nevertheless, it has led to one interesting result, namely the Goldberger-Treiman[3] relation between the Gamow-Teller constant G_A, the pion-nucleon constant, and the pion decay constant, which is known to be satisfied.

Recently one of us[4] has proposed a composite model of elementary particles based on an analogy with superconductivity. This model is built on the essential assumption that the Lagrangian describing the nucleons is invariant under the γ_5 transformation but that the physical vacuum state need not be so. As an interesting consequence, we observe that there exist nucleon-antinucleon bound states which behave as massless pseudoscalar mesons. The Goldberger-Treiman relation follows immediately from this if we identify them with the pions.

In the present paper we shall study another consequence of the axial vector conservation which can be used as an experimental test of the assumption. The main point is that for any reaction involving nucleons and mesons, the axial vector current conservation implies a close relation between the elastic amplitude and the "radiative" amplitude where an extra massless pion is emitted at zero energy.

Since the real pion has a finite mass, such a relation cannot actually be satisfied, but one may expect it to be approximately true at sufficiently high energies where the pion mass is negligible.

We shall first consider a simple model due to Nishijima and show how the above-mentioned relation is expected for a process involving the scattering of a nucleon by an external γ_5-invariant potential (Sec. II). We shall verify the relation explicitly in the lowest order perturbation (Sec. III), and then suggest a possible way of proving it in general (Sec. IV). Finally we shall discuss some experiments, in particular those involving pion-nucleon scattering, which would test the above-mentioned relation and hence the assumption of the (approximate) γ_5 invariance of strong interactions.

II. THE NISHIJIMA MODEL

In the actual β decay problem, the relevant symmetry to be associated with the axial vector part is the invariance under the $\gamma_5\times$isospin gauge transformation which acts on the nucleon (proton-neutron) field ψ as

* This work was supported by the U. S. Atomic Energy Commission.

[1] R. P. Feynman and M. Gell-Mann, Phys. Rev. 109, 193 (1958).

[2] S. Bludman, Nuovo cimento 9, 433 (1958); F. Gürsey, Nuovo cimento 16, 230 (1960); Ann. Phys. 12, 91 (1961); M. Gell-Mann and M. Lévy, Nuovo cimento 16, 705 (1960); J. Bernstein, M. Gell-Mann and L. Michel, Nuovo cimento 16, 560 (1960); J. Bernstein, S. Fubini, M. Gell-Mann, and W. Thirring, Nuovo cimento 17, 757 (1960); Chou Kang-Chao, J. Exptl. and Theor. Phys. (U.S.S.R.) 39, 703 (1960) [Soviet Phys. JETP 12, 492 (1961)]; Y. Nambu, Phys. Rev. Letters 4, 380 (1960).

[3] M. L. Goldberger and S. M. Treiman, Phys. Rev. 111, 354 (1958).

[4] Y. Nambu and G. Jona-Lasinio, Phys. Rev. 122, 345 (1961); 124, 246 (1961).

follows:

$$\psi(x)\rightarrow\exp(i\boldsymbol{\alpha}\cdot\boldsymbol{\tau}\gamma_5)\psi(x), \quad (2.1)$$

where $\boldsymbol{\alpha}$ is an arbitrary vector in isospace.

In this and following sections we shall study a simplified model proposed by Nishijima[5] because there would be no loss of the essential features.

The system consists of a single nucleon field ψ and a massless neutral pseudoscalar field ϕ ("pion") coupled through

$$L_{\rm int}=-m\bar{\psi}\exp[(ig/m)\gamma_5\phi]\psi, \quad (2.2)$$

which also incorporates the (bare) mass term for the nucleon. By expanding the exponential, we recognize that the first two terms are the mass and the ordinary meson-nucleon coupling terms. (It is also clear that with the transformation $\psi\rightarrow\exp[-(ig/2m)\gamma_5\phi]\psi$, the theory is equivalent to the simple derivative coupling model. We prefer the Nishijima representation because it brings in the nucleon γ_5 transformation explicitly.) In addition, we shall introduce an external vector (or axial vector) potential $V_\mu(A_\mu)$,

$$L_{\rm ext}=i\bar{\psi}\gamma_\mu\psi V_\mu \quad (\text{or } i\bar{\psi}\gamma_\mu\gamma_5\psi A_\mu). \quad (2.2')$$

The entire Lagrangian is invariant under the γ_5 transformation,

$$\psi\rightarrow e^{i\alpha\gamma_5}\psi, \quad (2.3a)$$

and

$$\phi\rightarrow\phi-(2m/g)\alpha, \quad (2.3b)$$

for constant α. Of course it is essential for the invariance that the pion is massless. One easily verifies the axial vector current conservation which follows from the aforementioned invariance:

$$\partial_\mu j_\mu(x)=0,$$
$$j_\mu=i\bar{\psi}\gamma_\mu\gamma_5\psi-(2m/g)\partial_\mu\phi. \quad (2.4)$$

An important feature of the theory is that the existence of the "pion" field is necessary to preserve the γ_5 invariance in the presence of finite nucleon mass. In this respect it is similar to the model proposed by Nambu and Jona-Lasinio.[4] There the pion field and finite bare nucleon mass are not assumed, but, in the end, we find that, if the nucleon has a finite observed mass, it must also be accompanied by a massless pion field which is to be interpreted as nucleon-antinucleon bound states, or a collective excitation of such pairs. In this case, the conserved current j_μ is simply

$$j_\mu=i\bar{\psi}\gamma_\mu\gamma_5\psi, \quad (2.5)$$

which, however, implicitly contains the pion contribution.

Let us now turn to the meaning of the conservation law associated with Eq. (2.4). We call this conserved quantity the chirality χ, defined by

$$\chi=-i\int j_4d^3x=\int\bar{\psi}\gamma_4\gamma_5\psi d^3x+(2im/g)\int\partial_4\phi d^3x$$
$$=\chi_N+\chi_\pi. \quad (2.6)$$

Equation (2.4) shows that χ is a constant of motion, i.e., a conserved quantity. However, we can easily see that a one-nucleon state or a one-meson state is not an eigenstate of χ. As was analyzed in the model of Nambu and Jona-Lasinio, this seemingly paradoxical situation is related to the fact that the γ_5 transformation generated by $\exp[i\chi\alpha]$ is not a proper operation in the Hilbert space of real particles, but it carries a Hilbert space into another which is orthogonal to it.

Since χ is thus not diagonal, we shall, in the following, work with the expectation value $\langle\chi\rangle$, which should be conserved in any reaction, i.e.,

$$\langle\alpha^{\rm in}|\chi|\alpha^{\rm in}\rangle=\langle\alpha^{\rm out}|\chi|\alpha^{\rm out}\rangle.$$

For a one-nucleon state of momentum p, in particular, we have

$$\langle\chi\rangle=Z\bar{u}_p\gamma_4\gamma_5u_p=-Zhv_p, \quad (2.7)$$

where u_p is the Dirac spinor normalized to $\bar{u}u=m/E$, h the helicity $\langle\boldsymbol{\sigma}\cdot\mathbf{p}/|\mathbf{p}|\rangle$, v_p the velocity $p/(p^2+m^2)^{\frac{1}{2}}$, and Z is a renormalization constant. For convenience, we can replace χ by $Z^{-1}\chi$ and call it chirality. Then Z will drop out in Eq. (2.7). For a one-pion state, we have, naturally, $\langle\chi\rangle=0$.

Let us next consider the scattering of a nucleon by the external potentials. For a static potential, the velocity will not change, so that one might conclude that the helicity remains unchanged: $h_i=h_f$. This, however, is clearly not the case as one can easily check by perturbation calculation. The contradiction is resolved by noting that any scattering can always be accompanied by emission of pions which are massless. The final asymptotic state is then a linear combination.

$$|f\rangle=C_0|N\rangle+C_1|N+\pi\rangle+C_2|N+2\pi\rangle+\cdots, \quad (2.8)$$

and the expectation value $\langle\chi\rangle_f$ must be taken with respect to this complete amplitude. From Eq. (2.6) we further recognize that $\langle\chi\rangle_f$ will have contributions both diagonal and nondiagonal in the decomposition (2.8), the latter being between states $|N+n\pi\rangle$ and $|N+(n\pm1)\pi\rangle$ differing by one zero-energy meson (spurion). We are led to the conclusion that taking the expectation value is essential in interpreting the conservation of chirality. In this sense, we can call it a weak conservation in contrast to the usual "strong" conservation such as charge conservation.

It is clear, furthermore, that one can distinguish various degrees of "weakness" depending on the number of final state degrees of freedom over which the expectation value $\langle\chi\rangle_f$ must be taken in order to satisfy

[5] K. Nishijima, Nuovo cimento 11, 698 (1959).

$\langle\chi\rangle_f=\langle\chi\rangle_i$. The weakest case would involve a summation over all spins, scattering angles, and many-pion production processes. A less weak case might involve only a spin summation and up to one-spurion emission processes. In the following section we shall verify the chirality conservation by perturbation calculation and show that the stronger type of weak conservation (which we might call detailed weak conservation) can sometimes be valid.

III. PERTURBATION CALCULATION

We shall consider the scattering of a nucleon by an external potential in the Born approximation and check the chirality conservation in the lowest order in the meson coupling constant g. For simplicity, the potential will be assumed to be of the vector type

$$L_V=i\bar{\psi}\gamma_\mu\psi V_\mu,$$

where V_μ is static and has an inversion symmetry: $V(\mathbf{x})=V(-\mathbf{x})$. (Of course one can work with an axial vector potential equally well.) We first note that the initial nucleon state can be specified by means of the covariant projection operator

$$\begin{aligned} &P(p,n)=\Lambda(p)S(n)=S(n)\Lambda(p),\\ &\Lambda(p)=(m-i\gamma\cdot p)/2E_p,\quad [\Lambda^2=(m/E_p)\Lambda],\\ &S(n)=(1+i\gamma\cdot n\gamma_5)/2. \end{aligned}\tag{3.1}$$

Here n is the covariant polarization vector which reduces to $(\mathbf{n}',0)$ in the nucleon rest system. In the laboratory frame one has $n=(\mathbf{n},n_0)$ where

$$\mathbf{n}=\mathbf{n}'+\mathbf{p}(\mathbf{p}\cdot\mathbf{n}')/m(E+m),\quad n_0=\mathbf{p}\cdot\mathbf{n}'/m.\tag{3.2}$$

Naturally

$$n\cdot p=0,\quad n^2=1.\tag{3.3}$$

The initial chirality of the nucleon is then given by

$$\begin{aligned}\chi_i&=\bar{u}_i\gamma_4\gamma_5 u_i=\mathrm{Tr}[\gamma_4\gamma_5P(p,n)]\\ &=n_0m/E_p=\mathbf{n}'\cdot\mathbf{v}.\end{aligned}\tag{3.4}$$

It is clear from Eq. (2.6) that in our present approximation (which is the zeroth order in g) only the lowest order elastic and one-pion inelastic processes need be considered. The corresponding diagrams are shown in Fig. 1. These amplitudes are given by

$$\begin{aligned} &M_{\mathrm{el}}(p',p)=i\gamma\cdot V(q),\quad q=p'-p\\ &M_{\mathrm{rad}}(p',p;k)=i\gamma\cdot V(q+k)S(p-k)ig\gamma_5\\ &\qquad\qquad+ig\gamma_5S(p'-k)i\gamma\cdot V(q-k). \end{aligned}\tag{3.5}$$

The chirality of the scattered wave at a given angle may be evaluated with the aid of the formula

$$\begin{aligned}\langle O\rangle_f&=[\bar{u}_i\bar{M}\Lambda(p')O\Lambda(p')Mu_i]/[\bar{u}_i\bar{M}\Lambda(p')Mu_i]\\ &=\mathrm{Tr}[\bar{M}\Lambda(p')OMP(p,n)]/\mathrm{Tr}[\bar{M}\Lambda(p')MP(p,n)],\end{aligned}$$

where O is a Dirac matrix and $\bar{M}=\gamma_4M^\dagger\gamma_4$. Adapting

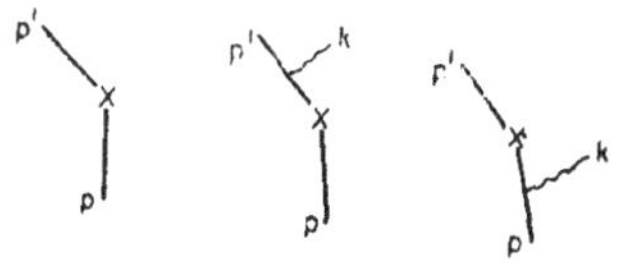

FIG. 1. Lowest order diagrams to be considered for chirality conservation in potential scattering. The point $\times$ is where the potential acts.

this to the present case, we get[6]

$$\begin{aligned}\langle\chi\rangle_f=&\{\mathrm{Tr}[\bar{M}_{\mathrm{el}}\Lambda(p')\chi_N'\Lambda(p')M_{\mathrm{el}}P(p,n)]\\ &+\mathrm{Tr}[\bar{M}_{\mathrm{el}}\Lambda(p')\chi_\pi'\Lambda(p')M_{\mathrm{rad}}P(p,n)]\\ &+\mathrm{Tr}[\bar{M}_{\mathrm{rad}}\Lambda(p')\chi_\pi'^\dagger\Lambda(p')M_{\mathrm{el}}P(p,n)]\}\\ &\times\{\mathrm{Tr}[\bar{M}_{\mathrm{el}}\Lambda(p')M_{\mathrm{el}}P(p,n)]\}^{-1},\end{aligned}$$

where

$$\begin{aligned}&\chi_N'=\gamma_4\gamma_5,\\ &\chi_\pi'=(2m/g)\gamma_4\left\langle 0\left|\int\dot{\phi}d^3x\,\phi(y)\right|0\right\rangle=-(im/g)\gamma_4.\end{aligned}\tag{3.6}$$

M_{rad} does not contribute to the denominator because of its vanishing weight. The elastic contribution can be easily evaluated. We find

$$\begin{aligned}&\mathrm{Tr}[\bar{M}_{\mathrm{el}}\Lambda(p')\chi_N'\Lambda(p')M_{\mathrm{el}}P(p,n)]\\ &\quad=(-im/2E^3)[V\cdot V(m^2+p\cdot p')-2p\cdot V p'\cdot V]n_4\\ &\qquad+(im/E^3)[V\cdot n(m^2+p\cdot p')-p\cdot Vp'\cdot n]V_4,\\ &\mathrm{Tr}[\bar{M}_{\mathrm{el}}\Lambda(p')M_{\mathrm{el}}P(p,n)]\\ &\quad=(1/2E^2)[V\cdot V(m^2+p\cdot p')-2p\cdot Vp'\cdot V],\ V_\mu\equiv V_\mu(q),\end{aligned}$$

so that

$$\begin{aligned}\langle\chi_N\rangle_f=&(m/E)n_0\\ &-(2m/E)V_0\frac{V\cdot n(m^2+p\cdot p')-p\cdot Vp'\cdot n}{V\cdot V(m^2+p\cdot p')-2p\cdot Vp'\cdot V}.\end{aligned}\tag{3.7}$$

Here we have made use of the previous assumptions about V, which means that $E_f=E_i=E$, and the Fourier transform $V_\mu(\mathbf{q})$ is a real vector

$$\begin{aligned}&V_\mu(\mathbf{q})=V_\mu(-\mathbf{q})=\eta_\mu V_\mu^*(\mathbf{q}),\\ &\eta_\mu=1,\ \mu=1,2,3;\quad \eta_\mu=-1,\ \mu=4.\end{aligned}$$

For a general V_μ, the result is more complicated.

As for the inelastic contribution, we first note that although the matrix element of $\chi_\pi\propto\int\partial_4\phi d^3x$ vanishes like $k_0/(2k_0)^{\frac{1}{2}}$ as $k_0\to 0$, the inelastic amplitude contains a factor $1/(2k_0)^{\frac{1}{2}}$ which will cancel the former. Thus there is no trouble in this respect. A more serious difficulty is that the results of the calculation depend on the way the limit $k=0$ is defined. In fact we get different answers depending on whether (a) we approach the limit $k=0$ staying on the zero-mass shell ($k_0=|\mathbf{k}|$), or (b) first we allow a finite mass μ_0, and go to the limit $\mathbf{k}\to 0$, $k_0=\mu_0\to 0$ successively.

We shall show below that the second procedure gives the correct result. Note also that this is the more

[6] Note that χ_π acts as unity operator for the nucleon, which means $O=\gamma_4$. Hence the form of χ_π' below.

appropriate one for application to the real pion problem where μ_0 is indeed finite.

Accordingly we calculate the χ_π part thus:

$$\mathrm{Tr}[\bar{M}_{\rm el}\Lambda(p')\chi_\pi'\Lambda(p')M_{\rm rad}P(p,n)]$$

$$=\lim_{\mathbf{k}\to 0,\mu_0\to 0} m\left\{\int \mathrm{Tr}\left[\gamma\cdot V\frac{m-i\gamma\cdot p'}{2E}\gamma_5\frac{m-i\gamma\cdot(p'+k)}{(p'+k)^2+m^2}\gamma\right.\right.$$

$$\left.\cdot V\frac{m-i\gamma\cdot p}{2E}\frac{1+i\gamma\cdot n\gamma_5}{2}\right]k_0\delta(k^2+\mu_0^2)\theta(k_0)dk_0$$

$$+\int \mathrm{Tr}\left[\gamma\cdot V\frac{m-i\gamma\cdot p'}{2E}\gamma\cdot V\frac{m-i\gamma\cdot(p-k)}{(p-k)^2+m^2}\gamma_5\frac{m-i\gamma\cdot p}{2E}\right.$$

$$\left.\left.\times\frac{1+i\gamma\cdot n\gamma_5}{2}\right]k_0\delta(k^2+\mu_0^2)\theta(k_0)dk_0\right\},$$

which becomes

$$-(m/2E^3)[V\cdot n(m^2+p\cdot p')-p\cdot Vp'\cdot n]V_0.$$

The other interference term in Eq. (3.6) gives an equal contribution so that

$$\langle\chi_\pi\rangle_f=\frac{2m}{E}V_0\frac{V\cdot n(m^2+p\cdot p')-p\cdot Vp'\cdot n}{V\cdot V(m^2+p\cdot p')-2p\cdot Vp'\cdot V}. \tag{3.8}$$

Combining Eqs. (3.7) and (3.8) we get

$$\langle\chi\rangle_f=\langle\chi_N\rangle_f+\langle\chi_\pi\rangle_f=(m/E)n_0=\langle\chi\rangle_i. \tag{3.9}$$

It is interesting to see how the two contributions to $\langle\chi\rangle_f$ add up in the nonrelativistic approximation. The chirality of the nucleon is in this case

$$\langle\chi_N\rangle=-\langle\sigma\rangle_i\cdot\mathbf{p}/m. \tag{3.10}$$

The scattering takes place through V_4 which does not flip the spin but changes the direction of motion so that, after the scattering, we have

$$\langle\chi_N\rangle_f=-\langle\sigma\rangle_i\cdot\mathbf{p}'/m. \tag{3.11}$$

The difference is

$$\Delta\langle\chi_N\rangle=-\langle\sigma\rangle_i\cdot\mathbf{q}/m. \tag{3.12}$$

On the other hand, the above-mentioned limiting procedure for $\langle\chi_\pi\rangle$ corresponds to the S-wave pion production process which goes through the negative energy states, and this is easily seen to cancel $\Delta\langle\chi_N\rangle$.

IV. GENERAL PROOF

The previous particular example shows that the weak conservation of chirality holds in detail, i.e., if summed only over the final nucleon spins, and over elastic and one-pion bremsstrahlung processes at a fixed scattering angle. This is a stronger result than the general statement $\langle\chi\rangle_i=\langle\chi\rangle_f$, and is due to the Born approximation and the special assumption made about V_μ. (In fact, for a completely general static potential V_μ, we do not obtain conservation unless we sum over all the scattered and unscattered waves.)

We expect that the chirality conservation in our sense will hold true to all orders in the pion coupling g, where many-pion processes will also come in. But the calculation is made difficult because of the self-energy effects. In the following, we will therefore try to derive a general relation which follows from the γ_5 invariance and certain other assumptions and is expressed in terms of directly observable (renormalized) quantities.

We assume that there exists a conserved quantity called chirality $\chi=-i\int\chi_4d^3x$ $(\partial_\mu\chi_\mu=0)$. This means that

$$\chi^{\rm in}=\chi^{\rm out}=S^{-1}\chi^{\rm in}S,$$

where S is the S matrix. We rewrite it as

$$S\chi^{\rm in}-\chi^{\rm in}S=0. \tag{4.1}$$

We will further assume, in accordance with the previous discussion, that χ consists of the nucleon part χ_N and the pion part $\chi_\pi=\lambda\int\dot{\phi}d^3x$. Asymptotically, χ_N will be an operator that does not change the number of pions, whereas χ_π, being linear in the pion field, will lead to the absorption or emission of a zero-energy pion. Accordingly Eq. (4.1) becomes

$$S\chi_N^{\rm in}-\chi_N^{\rm in}S=-S\chi_\pi^{\rm in}+\chi_\pi^{\rm in}S.$$

Let us apply this to the case of the potential scattering. Taking the matrix element between elastic states, we get

$$\begin{aligned}i\langle p'|M_{\rm el}\chi_N^{\rm in}-\chi_N^{\rm in}M_{\rm el}|p\rangle&=-\langle p'|S\chi_\pi^{\rm in}-\chi_\pi^{\rm in}S|p\rangle\\&=\textstyle\sum_k[-\langle p'|S|pk\rangle\langle k|\chi_\pi|0\rangle\\&\qquad+\langle 0|\chi_\pi|k\rangle\langle p'k|S|p\rangle],\end{aligned} \tag{4.2}$$

because $\chi_\pi^{\rm in}$ results only in the creation or absorption of a pion. Furthermore,

$$\begin{aligned}\langle p'|S|p,0\rangle=\langle p',0|S|p\rangle&=[i/(2k_0)^{\frac12}]\langle p'|k^2\phi(k)|p\rangle|_{k=0}\\&\equiv[i/(2k_0)^{\frac12}]M_{\rm rad}(p',p;k)|_{k=0},\end{aligned}$$

and

$$\langle k|\chi_\pi|0\rangle=-\langle 0|\chi_\pi|k\rangle=\lambda i(k_0/2)^{\frac12}\delta(\mathbf{k}). \tag{4.3}$$

Equation (4.2) thus may be written

$$\chi_NM_{\rm el}(p',p)-M_{\rm el}(p',p)\chi_N=i\lambda M_{\rm rad}(p',p;0). \tag{4.4}$$

If χ is so normalized that $\chi_N=\gamma_4\gamma_5$ for a free nucleon, then the continuity equation

$$\partial_\mu\chi_{N\mu}-\lambda\Box\phi=0,$$

which follows from χ conservation, means that $1/\lambda$ is more or less the conventional pion coupling constant

$$1/\lambda=f=g/2m. \tag{4.5}$$

It is not proven, however, that this agrees with the coupling constant defined in the dispersion theory. For the time being, we assume it to be the case.

Equation (4.4) represents a general relation between $M_{\rm el}$ and the one-pion emission amplitude $M_{\rm rad}$. The origin of such a simple relation is easily traced back to our starting assumption about the asymptotic behavior of χ. From the above derivation, it is also clear that a similar relation will exist for any reaction amplitude M and the accompanying inelastic amplitude resulting in the emission of an extra zero-energy meson.

We shall next bring the relation (4.4) into a more explicit form. In the covariant form, M is usually defined without the projection operator Λ, so that the left-hand side of Eq. (4.4) should be written

$$\gamma_4\gamma_5\Lambda(p')M_{el}(p',p)-M_{el}(p',p)\Lambda(p)\gamma_4\gamma_5.$$

Observing that

$$\gamma_4\gamma_5\Lambda(p)-\Lambda(p)\gamma_4\gamma_5=-\gamma_5,$$

Eq. (4.4) becomes

$$iM_{rad}(p',p;0)=(g/2m)[(\gamma_4\gamma_5 m/E-\gamma_5)M_{el}(p',p)+M_{el}(p',p)(\gamma_5\gamma_4 m/E-\gamma_5)]. \quad (4.6)$$

If we use the positive energy two-component spinors, the above equation simplifies to

$$iM_{rad}(p',p;0)=-\frac{g}{2m}\left[\frac{\sigma\cdot\mathbf{p}'}{E}M_{el}(p',p)-M_{el}(p',p)\frac{\sigma\cdot\mathbf{p}}{E}\right]. \quad (4.7)$$

In this form the meaning of the relation is easy to understand. Suppose the initial nucleon is in an eigenstate of helicity: $(\sigma\cdot p/E)u_p=\pm vu_p$. The helicity change in the elastically scattered state p' is

$$\left(\bar{u}_pM_{el}^{\dagger}\frac{\sigma\cdot\mathbf{p}'}{E}M_{el}u_p\right)\Big/(\bar{u}_pM_{el}^{\dagger}M_{el}u_p)-\left(\bar{u}_p\frac{\sigma\cdot\mathbf{p}}{E}u_p\right)$$

$$=\left(\bar{u}_pM_{el}^{\dagger}\left[\frac{\sigma\cdot\mathbf{p}'}{E}M_{el}-M_{el}\frac{\sigma\cdot\mathbf{p}}{E}\right]u_p\right)\Big/(\bar{u}_pM_{el}^{\dagger}M_{el}u_p)$$

$$=-i\lambda(\bar{u}_pM_{el}^{\dagger}M_{rad}u_p)/(\bar{u}_pM_{el}^{\dagger}M_{el}u_p).$$

The last expression just corresponds to the interference term $\langle\chi_\pi\rangle$ in the final state with a particular momentum p'. This shows that the weak chirality conservation holds *in detail*, i.e., after summing only over the final nucleon spins, and over elastic and associated inelastic amplitudes *provided that* the initial nucleon is in a helicity eigenstate. Otherwise a more general weak conservation will prevail in general.

V. $\gamma_5\times$ ISOSPIN INVARIANCE AND THE EXPERIMENTAL TEST

The results obtained in previous sections can be generalized to the $\gamma_5\times$ isospin gauge transformation [Eq. (2.1)]. Models that possess this invariance have been considered by Gürsey, Gell-Mann, *et al.*[2] and Nambu and Jona-Lasinio.[4] In this case there are three conserved quantities χ^i, $i=1, 2, 3$, corresponding to the three components of the isospin. The simplest example having such conservation is the conventional meson theory with derivative coupling. The results of Sec. IV can be directly taken over if we replace χ_N by $\chi_N{}^i=\chi_N\tau^i$, and χ_π by $\chi_\pi{}^i=\lambda\int\dot{\phi}^i d^3x$.

In some of the models, the expression for χ is actually more complicated. In general it contains nonlinear terms in the pion field, and terms involving a neutral scalar meson field can also occur. We may assume, however, that these additional terms do not contribute to the asymptotic values $\langle\chi^{in}\rangle$ and $\langle\chi^{out}\rangle$ for the following reason. First, the contribution from non-linear terms $\sim\phi(x)^n$, $n>2$, which involves creation and annihilation of n mesons in the neighborhood of point x, will tend to zero when, before or after the scattering, the particles are well separated from each other, just as the interaction energy is supposed to vanish in this asymptotic region. As for the neutral scalar field, a meson of this kind (σ meson) may exist in nature, but it would be, in any case, quite massive and unstable. We can therefore exclude σ from the fields that contribute to χ^{in} and χ^{out}.

With this observation, Eqs. (4.6) and (4.7) are replaced in the present case by

$$iM_{rad}{}^i=(g/2m)[(\gamma_4\gamma_5 m/E-\gamma_5)\tau^iM_{el}+M_{el}\tau^i(\gamma_5\gamma_4 m/E-\gamma_5)], \quad (5.1)$$

$$iM_{rad}{}^i=-(g/2m)[(\sigma\cdot\mathbf{p}'/E)\tau^iM_{el}-M_{el}\tau^i(\sigma\cdot\mathbf{p}/E)]. \quad (5.2)$$

It is convenient to quantize the spins of the initial and final nucleons along their own direction of motion. Equation (5.2) then reduces simply to

$$iM_{rad}{}^{h'h,i}=-(gv_p/2m)[h'\tau^iM_{el}{}^{h'h}-M_{el}{}^{h'h}\tau^ih], \quad (5.3)$$

where h and h' are the initial and final helicities, and $M^{h'h}$ are the corresponding amplitudes introduced by Jacob and Wick.[7]

We shall apply the above relation to the pion nucleon scattering, which seems to be the simplest case of physical interest. Naturally M_{el} has to be identified with the elastic scattering amplitude $M(p'q',pq)$ for the process $N_p+\pi_q\to N_{p'}+\pi_{q'}$, whereas $M_{rad}{}^i(p'q',pq;k)$ corresponds to the production process $N_p+\pi_q\to N_{p'}+\pi_{q'}+\pi_k{}^i$. The fundamental assumption here is that the pion-nucleon system is γ_5 invariant to the extent that the pion mass can be neglected.

To exhibit the isotopic dependence of M_{el}, we write

$$iM_{el\beta\alpha}=M^{(+)}\delta_{\beta\alpha}+M^{(-)}\tfrac{1}{2}[\tau_\beta,\tau_\alpha], \quad (5.4)$$

so that $M_{rad}{}^{i=3}$ for the process $\pi^\pm+p\to\pi^\pm+p+\pi^0$ takes the form

$$iM_{rad}{}^{h'h}(\pi^0)=-(gv_p/2m)(h'-h)M^{h'h}(\pi^\pm p\to\pi^\pm p)$$
$$=-(gv_p/2m)(h'-h)\times[M^{(+)h'h}\mp M^{(-)h'h}]. \quad (5.5)$$

[7] M. Jacob and G. C. Wick, Ann. Phys. 7, 404 (1959).

The relation becomes somewhat more complicated when $\pi^{\pm}$ are produced. For $\pi^-+p\to\pi^-+n+\pi^+$, we have, for example

$$iM_{\rm rad}{}^{h'h}(\pi^+)=-\frac{\sqrt{2}gv_p}{2m}[h'M^{h'h}(\pi^-p\to\pi^-p)-hM^{h'h}(\pi^-n\to\pi^-n)]$$
$$=-\frac{\sqrt{2}gv_p}{2m}[h'(M^{(+)h'h}+M^{(-)h'h})-h(M^{(+)h'h}-M^{(-)h'h})].\quad(5.6)$$

Let us now compare the elastic and inelastic cross sections. The simplest way to apply our formula will be to take the case where the meson k is produced with small energy ($|\mathbf{k}|\approx0$). We easily find

$$\left(\frac{d^2\sigma_{\rm rad}}{d\omega_k d\Omega_{q'}}\right)\bigg/\left(\frac{d\sigma_{\rm el}}{d\Omega_{q'}}\right)=\frac{|\mathbf{k}|}{(2\pi)^2}|\bar{u}_{p'}M_{\rm rad}u_p|^2/|\bar{u}_{p'}M_{\rm el}u_p|^2.\quad(5.7)$$

For the process $\pi^{\pm}+p\to\pi^{\pm}+p+\pi^0$ [Eq. (5.5)] this reduces to

$$\left(\frac{d^2\sigma_{\rm rad}}{d\omega_k d\Omega_{q'}}\right)\bigg/\left(\frac{d\sigma_{\rm el}}{d\Omega_{q'}}\right)=\frac{1}{\pi}\frac{g^2}{4\pi}\frac{|\mathbf{k}|}{4m^2}v_p{}^2(h'-h)^2.\quad(5.8)$$

Averaging over the nucleon spin, we get

$$\langle(h'-h)^2\rangle_{\rm av}=2[1-A'(\Omega_{q'})],\quad(5.9)$$

where $A'\equiv\langle h'h\rangle_{\rm av}$ is one of the polarization parameters introduced by Wolfenstein.[8] Equations (5.7) and (5.8) are the relations that should hold at each scattering angle. If we integrate Eq. (5.8) over the angles, we get

$$\left(\frac{d^2\sigma_{\rm rad}}{d\omega_k}\right)\bigg/\sigma_{\rm el}=\frac{2}{\pi}\frac{g^2}{4\pi}\frac{|\mathbf{k}|}{4m^2}v_p{}^2(1-\bar{A}'),\quad(5.10)$$

where $\bar{A}'$ means an angular average.

In the actual case of finite meson mass, the low-energy limit $\mathbf{k}=0$ does not have a Lorentz invariant meaning, so that the analysis will depend on the choice of the coordinate system. For example, a meson at rest in the c.m. system will have a momentum $|\mathbf{k}_{\rm L}|=v_{\rm L}\mu/(1-v_{\rm L}{}^2)^{\frac{1}{2}}$ in the laboratory system. In order to keep such an ambiguity reasonably small, e.g., $|\mathbf{k}_{\rm L}|\lesssim\mu$, we must demand that $1/(1-v_{\rm L}{}^2)^{\frac{1}{2}}$ is not too large. In other words, the energy at which we carry out the experiment should not be large compared to the nucleon rest energy.

An alternative way to test our relation is to consider the energy distribution of the inelastically scattered meson near its maximum energy since the produced pion would come out with low energy. In this case, the ratio of cross sections is calculated to be

$$\left(\frac{d^2\sigma_{\rm rad}}{d\omega_{q'}d\Omega_{q'}}\right)\bigg/\left(\frac{d\sigma_{\rm el}}{d\Omega_{q'}}\right)=\frac{1}{(2\pi)^2}\left(\frac{E}{m}\right)^{\frac{1}{2}}[2\mu(\omega_{\rm m}-\omega_{q'})]^{\frac{1}{2}}\times|\bar{u}_{p'}M_{\rm rad}u_p|^2/|\bar{u}_{p'}M_{\rm el}u_p|^2,\quad(5.11)$$

where $\omega_{\rm m}$ is the maximum meson energy at the particular angle, and E is the total c.m. energy of the system. Again for the process $\pi^{\pm}+p\to\pi^{\pm}+p+\pi^0$, this yields

$$\left(\frac{d^2\sigma_{\rm rad}}{d\omega_{q'}d\Omega_{q'}}\right)\bigg/\left(\frac{d\sigma_{\rm el}}{d\Omega_{q'}}\right)=\frac{1}{\pi}\frac{g^2}{4\pi}\left(\frac{E}{m}\right)^{\frac{1}{2}}\frac{[2\mu(\omega_{\rm m}-\omega_{q'})]^{\frac{1}{2}}}{4m^2}v_p{}^2(h'-h)^2.\quad(5.12)$$

Equations (5.8) and (5.11) or the specific forms (5.8) and (5.12) give a relation between elastic and inelastic cross sections in terms of directly measurable quantities alone. When the polarization of the nucleon is not measured, they still can give an inequality since

$$(h'-h)^2\leq4.$$

Unfortunately, these formulas are supposed to apply only at the extreme ends of the meson energy spectrum, for which experimental data are scarce and difficult to obtain.

On the other hand, if the elastic matrix elements are precisely known, we can directly calculate the inelastic amplitudes from Eq. (5.3) [or Eqs. (5.5) and (5.6)]. When, for example, the elastic scattering is dominated

TABLE I. Angular dependence and magnitude of the radiative cross section $\pi+N\to\pi+N+\pi'$, where the last pion is produced nearly at rest, for the case of $T=\frac{3}{2}$, $J=\frac{3}{2}^{\pm}$ elastic channel, $Z=\cos\theta_\pi$.

Process	$\frac{d^2\sigma_{\rm rad}}{d\Omega_{q'}d\omega_k}$	$\left(\frac{d\sigma_{\rm rad}}{d\omega_k}\right)\bigg/\left(\frac{g^2}{4\pi^2}\frac{v_p{}^2}{4m^2}k\sigma_{\rm el}\right)$
$\pi^{\pm}+p\to\pi^{\pm}+p+\pi^0$	$(1+3Z)^2(1-Z)$	2
$\pi^++p\to\pi^++n+\pi^+$	$1+3Z+3Z^2-(27/5)Z^3$	20/9
$\pi^-+p\to\pi^-+n+\pi^+$		20

[8] L. Wolfenstein, Phys. Rev. 96, 1654 (1954).

TABLE II. Angular dependence and magnitude of the radiative cross section for the case of $T=\frac{1}{2}$, $J=\frac{3}{2}^{\pm}$ and $\frac{5}{2}^{\pm}$ channels.

Process	$\frac{d^2\sigma_{\text{rad}}}{d\Omega_q \cdot d\omega_k}$	$\left(\frac{d\sigma_{\text{rad}}}{d\omega_k}\right)\Big/\left(\frac{g^2}{4\pi^2}\frac{v_p^2}{4m^2}k\sigma_{\text{el}}\right)$	
$\pi^-+p\to\pi^-+p+\pi^0$	$(1+3Z)^2(1-Z)$	2	$J=\frac{3}{2}$
	$(1-2Z-5Z^2)^2(1-Z)$		$J=\frac{5}{2}$
$\pi^-+p\to\pi^-+n+\pi^+$	$1+3Z^2$	2	$J=\frac{3}{2}$
	$1-2Z^2+5Z^4$		$J=\frac{5}{2}$

by a resonance, the associated production amplitude is easily obtained.

In Table I we give the results for the $\frac{3}{2}-\frac{3}{2}$ resonance. This is not an ideal case because the resonance energy is only twice the pion mass, which makes the neglecting of the pion mass somewhat dubious. But perhaps one could argue that we should compare the elastic scattering at energy T with the inelastic scattering at incident energy $T+\mu$. For the 3–3 resonance, the latter comes out to be ≈ 300 Mev.

In Fig. 2 we show the measured angular distribution of the inelastically scattered π^- from the reaction[9] $\pi^-+p\to\pi^-+n+\pi^+$ at 290 Mev laboratory energy which is to be compared with the theoretical curve. The agreement is quite reasonable. We note here that, although the experimental points are integral distributions over pion energy, the $\pi^- - n$ resonance in the final system would tend to produce low energy π^+ as is wanted in the comparison with theory.[9a]

As for the magnitude of the cross section, we can make only a crude comparison. We estimate the theoretical production cross section $\sigma(\pi^-p\to\pi^-n\pi^+)$ by extrapolating Eq. (5.7) over the entire energy range. This gives the ratio (with $g^2/4\pi\approx 15$)

$$r=\frac{\sigma(\pi^-p\to\pi^-n\pi^+)}{\sigma(\pi^-p\to\pi^-p)}\approx 1/100$$

at 290 Mev. Since the theoretical peak elastic cross section due to the 3–3 resonance is 20 mb, the above ratio means

$$\sigma(\pi^-p\to\pi^-n\pi^+)\approx 0.2\text{ mb}.$$

The corresponding experimental value is 0.61 ± 0.13 mb at[9] 290 Mev and 0.71 ± 0.10 mb at 317 Mev.[10] These are quite compatible with the prediction considering the crudeness of the estimation. (We also note from Table I that the ratio r for the other production processes would be only 1/10 of the present case. In such an event there may be more contaminations from other partial waves.)

Similar considerations can be made for the higher resonances. We list in Table II the corresponding predictions for resonances with $T=\frac{1}{2}$, $J=\frac{3}{2}^{\pm}$ and $\frac{5}{2}^{\pm}$. Finally it is interesting to compare the preceding results with those of the statistical model. We may identify the Fermi interaction volume Ω with the ratio[11]

$$R=\frac{1}{2\mu}\frac{|M_{\text{rad}}|^2}{|M_{\text{el}}|^2}\approx\frac{g^2}{4\mu m^2}v_p^2\approx\frac{v_p^2}{\mu^3},$$

as far as the soft meson emission is concerned. The corresponding interaction radius R is then

$$R\approx 0.6v_p^{\frac{2}{3}}\mu^{-1},$$

which is energy dependent, and approaches $0.6\mu^{-1}$ at high energies.

VI. FURTHER REMARKS

The main theme of this paper is to show that, under the assumption of γ_5 invariance in strong interactions, the change of nucleon helicity in any reaction will result in the bremsstrahlung of soft pions. This relation is characterized specifically by Eq. (5.1) or (5.2), which expresses a kind of low energy limit theorem. A comparison with the pion-nucleon scattering data around the 3–3 resonance shows an agreement with the formula.

Other interesting applications or tests of our relations may be found in nucleon-nucleon and nucleon-antinucleon scattering and general high energy multiple production processes. Our formulas would give us a way to analyze these events by observing low-energy

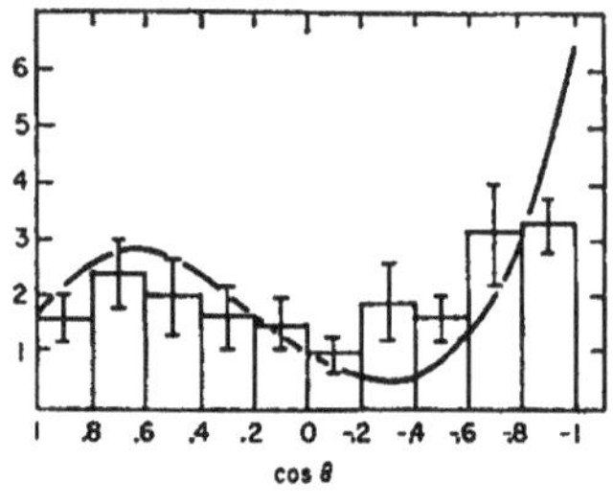

FIG. 2. Angular distribution of π^- from the reaction $\pi^-+p\to n+\pi^+ +\pi^-$ at 290 Mev.[9] The curve is calculated from the last line in Table I.

[9] Ya. A. Batusov, S. A. Bunyatov, V. M. Sidorov, and V. A. Yarba, *Proceedings of the 1960 International Conference on High-Energy Physics at Rochester* (Interscience Publishers, New York, 1960), p. 77; Ya. A. Batusov *et al.*, Doklady Nauk U.S.S.R., **133**, 52 (1960) [Soviet Phys.-Doklady **5**, 731 (1961)].

[9a] *Note added in proof.* However, the agreement may be fortuitous since there are various angular momentum channels that are neglected in this simplified approach. Our formula shows a characteristic dip in the forward direction. But the corresponding behavior of the experimental data may or may not be real.

[10] W. A. Perkins, III, J. C. Caris, R. W. Kenney, and V. Perez-Mendez, Phys. Rev. **118**, 1364 (1960).

[11] E. Fermi, Progr. Theoret. Phys. (Kyoto) **5**, 570 (1950).

mesons emitted. For example, in the case of nucleon-antinucleon annihilation at rest, the initial chirality is zero, and, since the final state contains no nucleons, we would have $\langle \chi_\pi \rangle_f = 0$, i.e., the amplitude for the emission of very low energy pions would be unusually small.

The notion of γ_5 invariance or chirality conservation can be extended to composite systems and strange particles.[4] There is also a possibility that the *K* meson plays a role similar to the pion in the conservation of strangeness-changing chirality current. It is likely, however, that even if such a symmetry existed in essence, the large mass of the *K* meson would tend to make it more approximate in nature than for the case involving pions, except perhaps at sufficiently high energies.

ACKNOWLEDGMENTS

One of the authors (Y.N.) thanks Dr. J. W. Calkin for his hospitality at the Brookhaven National Laboratory where part of the work was done.

Reprinted from The Physical Review, Vol. 128, No. 2, 862–868, October 15, 1962

Soft Pion Emission Induced by Electromagnetic and Weak Interactions*

Y. Nambu and E. Shrauner
The Enrico Fermi Institute for Nuclear Studies and the Department of Physics, The University of Chicago, Chicago, Illinois

(Received May 25, 1962)

We consider some extensions of the relation which has been obtained by Nambu and Lurié between an elastic process and the accompanying soft pion emission process under the assumption of approximate γ_5 invariance. In particular, a generalized formula is found which enables one to describe the electropion production $e+N \rightarrow N+e+\pi$ and the neutrino-pion production $\nu+N \rightarrow N+e+\pi$ at the threshold of $N\pi$ system in terms of the vector (Hofstadter) and the axial-vector form factors. Explicit forms of the cross sections are given.

1. INTRODUCTION

THE possibility of approximate chirality conservation in the pion-nucleon system was first observed in connection with weak interactions, where the lepton seems to be coupled to nearly conserved axial vector as well as vector currents. The physical aspects of the conserved axial vector currents, however, are very much different from those of the vector currents associated with the ordinary conservation laws. One may regard the former as a case of hidden symmetry which does not

* Work supported in part by the U. S. Atomic Energy Commission.

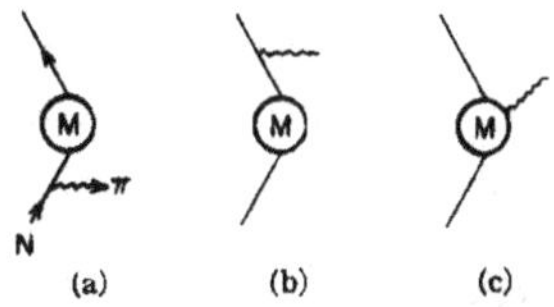

FIG. 1. Three diagrams contributing to M_{rad}.

manifest itself as a quantum number, but only implies that a certain expectation value remains constant in time.

This point was analyzed and elucidated in some detail in a previous paper.[1] It was realized that a change in nucleon isotopic chirality, which we will define to be minus twice the helicity times velocity times isospin, in any reaction results in the emission of a soft pion, so that one obtains a specific relation between the elastic and inelastic reaction amplitudes much resembling the one for ordinary bremsstrahlung. The formula was applied to the case of pion-nucleon elastic scattering and associated pion production, with a reasonable agreement with experimental results in a case where the elastic scattering is dominated by the 3–3 resonance. Since the theoretical formula is actually supposed to be valid only at energies large compared to the pion mass, the agreement may be a fortuitous one.

The purpose of the present paper is twofold. First, as a supplement to the previous paper, NL, we would like to make some more comments about the formula, in particular with regard to its relation to dispersion theory and Feynman diagrams. This enables one to prove the conjectured equality of the pion coupling appearing in our formula and the coupling defined in dispersion theory. Also, it will allow a natural generalization which incorporates meson emission with small but nonzero momentum.

The second and main purpose of the paper is to consider soft pion production induced by a small external perturbation such as a photon (real or virtual) or a neutrino, which does not commute with isotopic chirality. Namely, we have in mind processes of the type

$$e+N\rightarrow N+\pi+e,$$
$$\nu+N\rightarrow N+\pi+e.$$

The violation of chirality conservation due to the external perturbation necessitates a modification of the previous formula. This modification turns out to be very simple to the first order in the external perturbation. Moreover, it is of such a nature that both of the above mentioned processes are controlled by common form factors, namely, the vector (Hofstadter) form factors and the axial vector form factor, which appear intermingled with each other.

Thus, the group theoretical similarity between electromagnetic and weak processes becomes indeed a close one. In practice, this will also enable one to measure these form factors by using either process, and in turn check the validity of our underlying assumptions. Some calculations for this purpose will be presented in later sections.

[1] Y. Nambu and D. Lurié, Phys. Rev. **125**, 1429 (1962), hereafter referred to as NL. References to relevant literature are given there.

2. DIAGRAMMATIC INTERPRETATION OF THE FORMULA

Recapitulating the derivation of the previous paper, let M be a reaction amplitude involving a single nucleon, and $M_{\text{rad}}{}^{\alpha}$ the amplitude for an associated soft pion emission with zero momentum and isospin α. These two are related by

$$iM_{\text{rad}}{}^{\alpha}=f[\chi_N{}^{\alpha,\text{in}},M], \tag{2.1}$$

where $\chi_N{}^{\alpha}$ is the nucleon chirality operator

$$\chi_N{}^{\alpha,\text{in}}=-\tau^{\alpha}\boldsymbol{\sigma}\cdot\mathbf{p}/E_p, \tag{2.1$'$}$$

and $f=g/2m$ is the pion-nucleon coupling constant. This result was obtained from the assumption $\chi^{\text{in}}=\chi^{\text{out}}$, where $\chi=\chi_N+\chi_\pi$, with $\chi_\pi{}^{\alpha,\text{in}}=(1/f)\int\dot{\phi}^{\alpha,\text{in}}d^3x$ being the pion contribution to χ. In case M is the invariant amplitude without the positive energy projection operator, the alternative form of Eq. (2.1) becomes

$$\begin{aligned}iM_{\text{rad}}\alpha=&f\{[(m/E)\gamma_4\gamma_5-\gamma_5]\tau^{\alpha}M\\&\qquad+M\tau^{\alpha}[(m/E)\gamma_5\gamma_4-\gamma_5]\}\\=&f[(m/E)\gamma_4\gamma_5\tau^{\alpha},M]-f\{\gamma_5\tau^{\alpha},M\}.\end{aligned} \tag{2.2}$$

A very simple interpretation of Eq. (2.2), which bears resemblance to a corresponding theorem for bremsstrahlung,[2] can be made in the following way. From the viewpoint of dispersion theory, the pion emission amplitude consists of the nucleon pole contribution corresponding to the diagrams (a) and (b) of Fig. 1, and the rest which is indicated by the diagram (c). The pole diagrams give

$$\begin{aligned}&g\tau^{\alpha}\gamma_5\frac{1}{i\gamma\cdot(p'+k)+m}M+M\frac{1}{i\gamma\cdot(p-k)+m}\gamma_5\tau^{\alpha}g\\&\quad=g\tau^{\alpha}\gamma_5\frac{-i\gamma\cdot k}{2p'\cdot k+k^2}M+M\frac{-i\gamma\cdot k}{2p\cdot k-k^2}\gamma_5\tau^{\alpha}g.\end{aligned} \tag{2.3}$$

In the limit $\mathbf{k}=0$, and then $k_0=\mu\rightarrow 0$, this reduces to the commutator term in Eq. (2.2). The second (anticommutator) term must then correspond to the diagram (c). Note that M above is the mass shell amplitude corresponding to the real elastic scattering. In the Feynman-Dyson picture, on the other hand, diagrams (a) and (b) will contain off-the-mass shell corrections which must be lumped with contributions from the other irreducible diagrams.

It is interesting that the nonpole contribution in Eq. (2.2) is exactly in the form of an infinitesimal change in M generated by a γ_5 transformation. This is not surprising since, for example, in the Nishijima model'

[2] F. Low, Phys. Rev. **110**, 974 (1958).

the pion coupling was generated in this fashion. When M itself happens to be γ_5 invariant, the nonpole contribution will vanish, as was the case considered in Sec. 2 of NL.

The fact that the pole contribution is separated from the rest is due to the singular behavior of the nucleon propagator in the limit $k\rightarrow 0$. As was discussed in NL, this limit is not unique, depending on the direction from which k approaches zero. Our procedure was to let $\mathbf{k}$ go to zero first in the coordinate system considered. Physically, it means that we extrapolate to mesons produced at rest (with mass μ small but finite) in that particular frame of reference. The extrapolated value will change appreciably as soon as we go to a second coordinate system in which the meson at rest in the first coordinate system appears with a momentum $|\mathbf{k}|$ comparable to μ.

The nonpole contribution, on the other hand, consists of a dispersion integral over a continuous mass spectrum starting at $m+\mu$. Its limit as $\mathbf{k}\rightarrow 0$ and $\mu\rightarrow 0$ will not become singular unlike the pole term since the phase space for many pion emission processes vanishes at the threshold. There is no infrared catastrophe in the case of pion bremsstrahlung, but only a mild nonunique behavior of one-pion emission amplitude as $k\rightarrow 0$.

The above observation shows that the interpretation of the two terms of Eq. (2.2) is unambiguous. This, in turn, proves that the coupling constant f in Eq. (2.2) is the same as the dispersion theoretical one since both appear as a residue of the same pole contribution.

Finally, a slight but important generalization of Eq. (2.2) naturally suggests itself. Namely, instead of taking the limit $k\rightarrow 0$, we may keep k small but finite as far as the pole contribution is concerned, since it is most sensitive to k. This brings two advantages. First, the entire formula becomes manifestly Lorentz invariant in contrast to Eq. (2.2). Second, the formula agrees with the exact expression for arbitrary k as far as the pole term is concerned. Our modified formula will thus read

$$iM_{\text{rad}}{}^{\alpha}(k)=-f\left[\tau^{\alpha}\gamma_5\frac{2mi\gamma\cdot k}{2p'\cdot k+k^2}M+M\frac{2mi\gamma\cdot k}{2p\cdot k-k^2}\gamma_5\tau^{\alpha}\right]$$
$$-f(\tau^{\alpha}\gamma_5 M+M\gamma_5\tau^{\alpha}).\quad(2.4)$$

In practical applications, this modification will be an essential improvement since it includes at least part of the p-wave pion contribution (and higher waves) which, combined with the s wave, is necessary to keep M_{rad} Lorentz invariant. Usually such a p-wave contribution will be present in the measurement, and can even be the dominant one unless we select pions with very small momenta.

The second term of Eq. (2.4) is specific to our assumption of chiral invariance, and becomes rigorous in the limit $k\rightarrow 0$. In dispersion dynamics, one may regard it as a new boundary condition to be imposed on M_{rad} as a function of k.

3. REACTIONS INDUCED BY CHIRALITY NONCONSERVING PERTURBATIONS

We now turn to soft pion production initiated by a small perturbation which does not commute with isotopic chirality. As we shall consider later, the electromagnetic interaction and the leptonic interaction of the nucleon belong to this category, since they behave like some particular components of isospin.

Let us denote this perturbing Hamiltonian by $\bar{H}'$. The chirality operator χ^{α} will change with time according to

$$i\dot{\chi}^{\alpha}=[\chi^{\alpha},\bar{H}'],$$

so that

$$\chi^{\alpha,\text{out}}=\chi^{\alpha,\text{in}}-i\int_{-\infty}^{\infty}[\chi^{\alpha}(t),\bar{H}'(t)]dt$$
$$=S^{-1}\chi^{\alpha,\text{in}}S,$$
$$\chi^{\alpha,\text{in}}S-S\chi^{\alpha,\text{in}}=-iS\int_{-\infty}^{\infty}[\chi^{\alpha}(t),\bar{H}'(t)]dt.\quad(3.1)$$

Here S is the S matrix generated by the entire Hamiltonian. By taking an appropriate matrix element, the left-hand side leads to the same combination of M and M_{rad} as before, while the right-hand side gives a correction due to the symmetry breaking effect of $\bar{H}'$. To determine the latter, however, we have to know the complete dynamics of the system.

The problem becomes much simplified if we consider the effect of $\bar{H}'$ only to the first order. Suppose a reaction is generated by $\bar{H}'$ itself. Then the part of S we are interested in on the left-hand side will be of the order $\bar{H}'$, whereas the S on the right-hand side may be replaced by S_0 due to strong interactions alone. We thus obtain

$$(i/f)M_{\text{rad}}{}^{\alpha}=[\chi_N{}^{\alpha,\text{in}},M]+S_0[\chi^{\alpha},H'],\quad(3.2)$$

where we have removed the energy-momentum conservation factor from the amplitude, so that the time integral of $\bar{H}'$ is replaced by the Hamiltonian density H'.

Since χ^{α} is the generator of an infinitesimal γ_5 transformation, the second term on the right involves the infinitesimal change induced on H'. We may therefore state the result in the following way: When a reaction M is induced by $\bar{H}'$, the radiative process M_{rad} consists of two parts, one due to the change in nucleon chirality, and the other directly generated by an equivalent perturbation Hamiltonian $-i[\chi^{\alpha},\bar{H}']$. This latter, when taken between real Heisenberg states, must naturally include all the renormalizations and radiative corrections. The factor S_0 entering in Eq. (3.2) expresses the final-state interaction of the reaction products excluding the soft pion. It will drop out when the final state consists of a single particle.

4. ELECTROPION PRODUCTION

In this section, we apply our previous formula (3.2) to pion production by virtual photon. The electromagnetic interaction of the nucleon and the pion is given by the Lagrangian density

$$L'=j_\mu A_\mu=ei\bar{\psi}\gamma_\mu[(1+\tau_3)/2]\psi A_\mu$$
$$-ei\tfrac{1}{2}(\phi T_3\partial_\mu\phi-\partial_\mu\phi T_3\phi)A_\mu, \quad (4.1)$$

where T is the pion isospin matrix. The chirality operator χ^α is

$$\chi^\alpha=\int[\bar{\psi}\gamma_4\gamma_5\tau^\alpha\psi+(i/f)\partial_4\phi^\alpha]d^3x. \quad (4.2)$$

It is not necessary to use ∂_4-ieA_4 above, as the field may be turned on after $t=-\infty$. We may also use $-L'$ instead of H' without caution to the first order in e. So let us evaluate $[\chi^\alpha,L']$. With the standard commutation relations, we find

$$i[\chi^\alpha,L']=\epsilon_{3\alpha\beta}J_\mu{}^\beta A_\mu,$$
$$J_\mu{}^\beta=e[i\bar{\psi}\gamma_\mu\gamma_5\tau^\beta\psi-(1/f)\partial_\mu\phi^\beta]=e\chi_\mu{}^\beta. \quad (4.3)$$

This part of pion production then is generated by the isotopic axial vector (chirality) current rather than the vector current! In the above calculation, explicit forms (4.1) and (4.2) were used, but actually it is not necessary nor general. Only the group properties of vector and axial vector isotopic currents are essential. Note also that the isoscalar part of L' commutes with χ^α, and that J_μ is twice what should be more properly called isotopic axial vector current.

Now if the electropion production is described by $M_{\rm rad}$,the corresponding "elastic" process is obviously the elastic scattering of the nucleon by a virtual photon described by the Hofstadter form factors; namely,

$$\bar{u}_{p'}Mu_p=\langle p'|j_\mu A_\mu|p\rangle$$
$$=e\bar{u}_{p'}F_\mu(p',p)A_\mu(q)u_p,\quad q=p'-p,$$
$$F_\mu(p',p)=i\gamma_\mu[\tfrac{1}{2}F_1{}^S(q^2)+(\tau_3/2)F_1{}^V(q^2)]$$
$$+(i\sigma_{\mu\nu}q_\nu/2m)[\tfrac{1}{2}F_2{}^S(q^2)+(\tau_3/2)F_2{}^V(q^2)]. \quad (4.3)$$

Similarly, we may write

$$\langle p'|J_\mu{}^\alpha A_\mu|p\rangle=e\bar{u}_{p'}G_\mu{}^\alpha(p',p)A_\mu(q^2)u_p,$$
$$G_\mu{}^\alpha(p',p)=i\gamma_\mu\gamma_5\tau^\alpha G_1(q^2)+\gamma_5\tau^\alpha(q_\mu/q^2)G_2(q^2)$$
$$=[i\gamma_\mu\gamma_5+2m\gamma_5(q_\mu/q^2)]\tau^\alpha G_1(q^2). \quad (4.4)$$

The last form follows, of course, from chirality conservation (neglecting pion mass and proton-neutron mass difference). Combining Eqs. (3.2), (4.3), and (4.4), we obtain

$$(i/f)M_{\rm rad}{}^\alpha=e[\chi_N{}^{\alpha,\rm in},F_\mu(p',p)A_\mu(q)]$$
$$+ie\epsilon_{3\alpha\beta}G_\mu{}^\beta(p',p)A_\mu(q), \quad (4.6)$$

where, for the commutator term, one has to use Eq. (2.2) or (2.4). A few remarks are in order.

(1) It is instructive to interpret the formula in terms of diagrams. The commutator term corresponds to the diagrams in Fig. 1, where M is replaced by the electromagnetic vertex. The additional axial vector term, on the other hand, may be identified with the "photoelectric" term and an associated "direct production" term illustrated in Fig. 2. Again, the identification of the "photoelectric" term is unambiguous because it is the only term with a pion pole. Remember in this connection that the meson pole will have, in general, a denominator $(q-k)^2+\mu^2$, which in the limit $k=0$, $\mu=0$, becomes q^2. The form factor $G_1(q^2)$ must be interpreted as the combined effect of meson electromagnetic form factor, meson propagator correction, and the meson-nucleon vertex form factor. It is then clear that $G_1(0)=1$ since the residue of the pion pole in $M_{\rm rad}$ should equal fe. In other words, the axial vector form factor, like the vector form factor, should have no renormalization in order to be consistent with dispersion theory under strict γ_5 invariance. [In actual β decay, $G_1(0)\approx1.2$.] This property has long been conjectured, but no direct proof seems to exist.

(2) Equation (4.6) is compatible with gauge invariance and the Kroll-Ruderman theorem for photoproduction at threshold. Gauge invariance is obvious because both terms in Eq. (4.6) satisfy the continuity equation. As for the Kroll-Ruderman theorem, we observe that in the case of a real incident photon, the reaction M cannot go as a real process except in the limit $q=0$, which in turn can only lead to $M_{\rm rad}$ for emission of a pion with $k=0$. In the nucleon rest system, which is now equal to the center-of-mass system, the term with χ_N vanishes since $v_N=0$, whereas the axial vector term with $G(0)=1$ agrees with the content of the Kroll-Ruderman theorem. In this sense Eq. (4.6) may be regarded as a generalized Kroll-Ruderman theorem for a virtual photon.

(3) In the more realistic case of finite k (and μ), we may follow the prescription of Sec. 2, and construct $M_{\rm rad}$ as a sum of pole terms and a specific correction characteristic of the γ_5 invariance. There are now two nucleon and one pion pole terms, plus three nonpole additions. However, this is not yet satisfactory, as gauge invariance is violated. In order to restore gauge invariance, another correction has to be added. We propose thus the following form:

$$(1/ef)M_{\rm rad}{}^\alpha=i\tau^\alpha\gamma_5\frac{2mi\gamma\cdot k}{2p'\cdot k+k^2}F_\mu(p'+k,\,p)A_\mu(q)$$
$$+F_\mu(p',\,p-k)A_\mu(q)\frac{2mi\gamma\cdot k}{2p\cdot k+k^2}i\gamma_5\tau^\alpha$$
$$+\frac{1}{2i}[\tau^\alpha,\tau^3]\frac{2m\gamma_5(2k_\mu-q_\mu)A_\mu(q)}{q^2-2k\cdot q}G(q^2)$$
$$+i\tau^\alpha\gamma_5F_\mu(p',p)A_\mu(q)$$
$$+F_\mu(p',p)A_\mu(q)i\gamma_5\tau^\alpha$$
$$-(1/2i)[\tau^\alpha,\tau^3]i\gamma_\mu\gamma_5A_\mu(q)G(q^2)$$
$$-(1/2i)[\tau^\alpha,\tau^3]i\gamma\cdot k\gamma_5q_\mu A_\mu(q)$$
$$\times\{F_1{}^V(q^2)-G(q^2)\}/q^2, \quad (4.7)$$

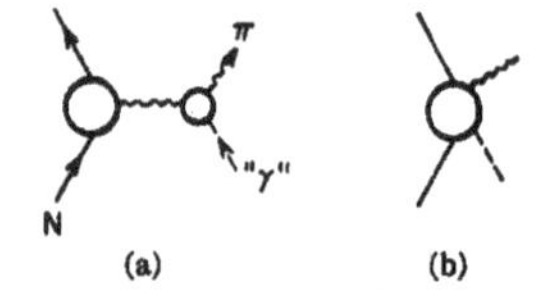

FIG. 2. Diagrams responsible for the axial vector term in electropion production.

where $G(q^2)=G_1(q^2)/G_1(0)$. The last term is the new addition designed to satisfy gauge invariance.[3] Since this is purely longitudinal, however, it does not make any real contribution to the physical amplitude. The assignment of the normalized form factor G to the pion pole term is rather arbitrary unlike the nucleon pole case, but it seems to be the simplest assumption insofar as we do not have any theory about the deviation from strict γ_5 invariance. [Note that this correction term is free of an unphysical singularity at $q^2=0$.]

5. CALCULATION OF CROSS SECTION

We will proceed to make some calculations on the electropion production cross section based on the previous results. Since the amplitude (4.7) is pretty complicated, however, we consider only the easiest problem. This is the total cross section for producing a pion at threshold of the final-nucleon pion system, for a given momentum transfer, q, large compared to pion mass. In this case, the produced pion will be at rest relative to the outgoing nucleon, so that we consider our formula (4.6) in the rest system of the outgoing nucleon. But then the final chirality of the nucleon becomes zero as it is proportional to velocity. Moreover, the pion pole term may be dropped since, after putting $k=0$, it is purely longitudinal ($\sim q_\mu A_\mu$) as shown in Eq. (4.4). We get therefore the simplified form

$$(i/fe)M_{\text{rad}}{}^{\alpha}=\{-F_\mu(p',p)\chi_N{}^{\alpha}(p)-\tfrac{1}{2}[\tau^{\alpha},\tau^{3}]i\gamma_\mu\gamma_5G(q^2)\}A_\mu(q), \quad (5.1)$$
$$A_\mu(q)=j_\mu{}^{(e)}/q^2,$$

$j_\mu{}^{(e)}$ being the electron current. The chirality $\chi_N{}^{\alpha}$ above refers to that of the incident nucleon, which must be measured in the rest frame of the outgoing nucleon. In the helicity representation, $\chi_N{}^{\alpha}$ then turns out to be

$$\chi_N{}^{\alpha}(p)=-\tau^{\alpha}hv=-\tau^{\alpha}h|q|(m^2+q^2/4)^{1/2}\times(m^2+q^2/2)^{-1}. \quad (5.2)$$

In order to calculate the square of M_{rad}, it is now convenient to go to the system where the initial and final momenta of the nucleon are equal and opposite with magnitude $|q|/2$ (the "Breit" system, see Fig. 3). The helicity used in Eq. (5.2) remains unchanged under this change of reference frame. As was shown by Yennie *et al.*,[4] the electromagnetic form factor F_μ contains two incoherent matrix elements in this coordinate system. They are the time component F_0 which flips helicity, and the transverse spatial component $F^{\pm}(\perp q)$ which conserves helicity. Omitting the isotopic spin, these matrix elements are given by

$$\langle p',h'|F_0|p,h\rangle=\delta_{h,-h'}F_c(q^2)(1+q^2/4m^2)^{-1/2},$$
$$\langle p',h'|F_{\pm}|p,h\rangle=[(h\pm1)/2]\delta_{h,h'}F_m(q^2)\times(1+q^2/4m^2)^{-1/2}|q|/2m,$$
$$F_c=F_1-(q^2/4m^2)F_2,$$
$$F_m=F_1+F_2,$$
$$F_{\pm}=(F_x\pm iF_y)/2, \quad (z\|q) \quad (5.3)$$

where the new charge and magnetic form factors F_c and F_m have been introduced.[5] In our problem, there is also the axial vector form factor G_μ, which turns out to have a nonzero matrix element only for the transverse component:

$$\langle p',h'|G_{\pm}|p,h\rangle=-\tfrac{1}{2}(1\pm h)\delta_{h,h'}G(q^2). \quad (5.3')$$

In a similar fashion, the helicity matrix elements of the electron current $j_\mu{}^{(e)}$ in the same reference frame are found to be

$$\langle l',h'|j_0|l,h\rangle=-e\delta_{h,h'}\cos(\theta_B/2),$$
$$\langle l',h'|j_{\pm}|l,h\rangle=e\delta_{h,h'}[h\mp\sin(\theta_B/2)]/2, \quad (5.4)$$

in the relativistic limit. Here θ_B is the electron scattering angle in this frame, being related to the laboratory angle θ by

$$\tan(\theta_B/2)=(1+q^2/4m^2)^{1/2}\tan(\theta/2). \quad (5.4')$$

From Eqs. (5.1)–(5.4), the cross section for soft pion production per unit energy range of the final electron at a fixed laboratory angle is obtained without much difficulty (see reference 4). The result is

$$\frac{d\sigma}{d\epsilon' d\Omega}=\alpha^2\frac{f^2}{4\pi^2}\frac{[2\mu(\epsilon_m'-\epsilon')]^{1/2}}{[1+\mu/m]^2}\frac{\cos^2(\theta/2)}{4\epsilon^2\sin^4(\theta/2)}\times\left[1+\frac{2\epsilon}{m}\sin^2(\theta/2)\right]^{1/2}S,$$

$p\to p+\pi^0$:
$$S=F_{pc}{}^2m^2q^2/(m^2+q^2/2)^2+[1+2\tan^2(\theta_B/2)]\times[F_{pm}q^2/(2m^2+q^2)]^2,$$
$n\to n+\pi^0$:
$$S=F_{nc}{}^2m^2q^2/(m^2+q^2/2)^2+[1+2\tan^2(\theta_B/2)]\times[F_{nm}q^2/(2m^2+q^2)]^2,$$
$p\to n+\pi^+$:
$$S=2F_{nc}{}^2m^2q^2/(m^2+q^2/2)^2+2[1+2\tan^2(\theta_B/2)]\times[G+F_{nm}q^2/(2m^2+q^2)]^2,$$
$n\to p+\pi^-$:
$$S=2F_{pc}{}^2m^2q^2/(m^2+q^2/2)^2+2[1+2\tan^2(\theta_B/2)]\times[G-F_{pm}q^2/(2m^2+q^2)]^2, \quad (5.5)$$

[3] A similar term is proposed in S. Fubini, Y. Nambu, and V. Wataghin, Phys. Rev. **111**, 329 (1958).

[4] D. R. Yennie, M. M. Lévy, and D. G. Ravenhall, Revs. Modern Phys. **29**, 144 (1957), Appendix.

[5] F. J. Ernst, R. G. Sachs, and K. C. Wali, Phys. Rev. **119**, 1105 (1960). See also W. R. Theis, Phys. Rev. Letters **8**, 45 (1962); L. N. Hand, D. G. Miller, and R. Wilson, *ibid.* **8**, 110 (1962).

F_p and F_n refer to the proton and neutron form factors, respectively. The invariant momentum transfer q^2, the incident electron energy ϵ, the maximum energy of inelastic electron ϵ_m', and the scattering angle θ are related by the following relations:

$$q^2=4\epsilon\epsilon_m' \sin^2(\theta/2)=2m(\epsilon-\epsilon_m'-\mu),$$
$$\epsilon_m'=(\epsilon-\mu-\mu^2/2m)/[1+(2\epsilon/m)\sin^2(\theta/2)]. \quad (5.5')$$

In Eqs. (5.5) and (5.5′), pion and electron masses are neglected compared with m, ϵ, and q except for purely kinematical corrections, so that they should be valid only under these conditions.

Equation (5.5) allows one to determine the form factors by measuring the soft pion inelastic cross section at a fixed angle and varying incident energy. In particular, the electric form factors predominate near the forward direction, while the magnetic and axial vector factors are the sole contributions at 180°. These features are common with the elastic cross section given by the Rosenbluth formula. In comparison with Eq. (5.5), the latter reads[4,5]

$$\frac{d\sigma}{d\Omega}=\alpha^2\frac{\cos^2(\theta/2)}{4\epsilon^2\sin^4(\theta/2)}\frac{1}{1+(2\epsilon/m)\sin^2(\theta/2)}\frac{1}{1+q^2/4m^2}$$
$$\times\{F_e^2+[1+2\tan^2(\theta_B/2)](q^2/4m^2)F_m^2\}. \quad (5.6)$$

The electropion production was utilized by Panofsky and Allton,[6] and by Ohlsen[7] in order to determine the neutron magnetic form factor F_{n2}. They measured the cross section for pion production at the 3–3 resonance for varying q^2. Our formula requires a similar experiment to be done near threshold, which would be more difficult because of the smallness of the cross section. On the other hand, it will enable one to determine the axial-vector form factor if the vector form factors are known. The axial-vector form factor, according to our theory, reflects the various effects associated with a virtual pion, and therefore did not appear in the earlier theories.

6. NEUTRINO-PION PRODUCTION[8]

With the standard assumption of the bare V-A lepton-nucleon interaction, H' and M responsible for the elastic process are given by

$$H'=-ig/\sqrt{2}\sum_\pm \bar{\psi}\gamma_\mu(1+\gamma_5)\tau^\pm\psi j_\mu{}^{(L)\mp},$$
$$\bar{u}_{p'}Mu_p=g/\sqrt{2}\sum_\pm \bar{u}_{p'}[F_\mu{}^V+G_\mu]\tau^\pm u_p j_\mu{}^{(L)\mp}, \quad (6.1)$$

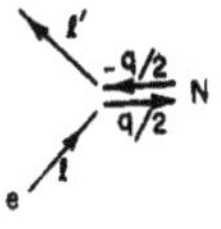

FIG. 3. Nucleon and electron momenta in the "Breit" system.

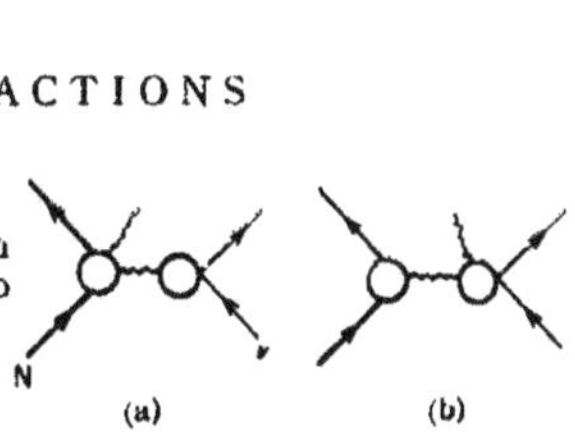

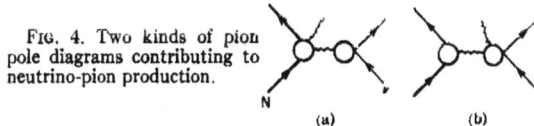

FIG. 4. Two kinds of pion pole diagrams contributing to neutrino-pion production.

where $j^{(L)\pm}$ is the lepton current, and $g=10^{-5}m^{-2}$ the β-decay coupling constant. If the weak interactions are mediated by an intermediate boson, there will arise a characteristic form factor multiplying $j^{(L)}$. Since the necessary modification is obvious, we will in the following ignore this possibility. The commutator $[\chi^\alpha,H']$ in Eq. (3.2) becomes

$$[\chi^\alpha,H']=-(ig/\sqrt{2})\sum_\pm \bar{\psi}\gamma_\mu(1+\gamma_5)\times[\tau^\alpha,\tau^\pm]\psi j_\mu{}^{(L)\mp},$$
$$\langle p'|[\chi^\alpha,H']|p\rangle=-(g/\sqrt{2})\sum_\pm \bar{u}_{p'}[F_\mu{}^V+G_\mu]\times[\tau_\alpha,\tau^\pm]u_p j_\mu{}^{(L)\mp}. \quad (6.2)$$

The radiation amplitude thus takes the form

$$(i/f)M_{\rm rad}{}^\alpha=(g/\sqrt{2})\sum_\pm\{[\chi_N{}^{\alpha,\rm in},(F_\mu{}^V+G_\mu)\tau^\pm]-(F_\mu{}^V+G_\mu)[\tau^\alpha,\tau^\pm]\}j_\mu{}^{(L)\mp}. \quad (6.3)$$

The interpretation of these terms is similar to the electroproduction case, except that we have here also an intrinsic axial vector interaction. The G_μ contains a pion pole, but in the first term the produced pion is emitted from the nucleon-pion vertex, whereas in the second term it is emitted from the pion-lepton vertex like the "photoelectric" term in electropion production (Fig. 4). According to this interpretation, it is again possible to construct a generalized expression corresponding to Eq. (4.7) for finite k, but we will not attempt to write down the explicit formula.

Turning to the easier problem, the inelastic cross section at threshold for a given (and large) momentum transfer q can be calculated by the method in Sec. 5. The formula corresponding to Eq. (5.1) becomes

$$(i/f)M_{\rm rad}{}^\alpha=-(g/\sqrt{2})\sum_\pm\{\tau^\pm(F_\mu{}^V+G_\mu)\chi_N{}^{\alpha,\rm in}+[\tau^\alpha,\tau^\pm](F_\mu{}^V+G_\mu)\}j_\mu{}^{(L)\mp}. \quad (6.4)$$

The matrix elements of F_μ, G_μ, and $j_\mu{}^{(L)}$ in the Breit system are given in Eqs. (5.3)–(5.4′), but in the present case the two-component nature of the neutrino has to be taken into account. We quote the results for the various reactions[9]:

(a) $\nu+p\rightarrow p+e^-+\pi^+$,

(b) $\nu+n\rightarrow p+e^-+\pi^0$,

(c) $\nu+n\rightarrow n+e^-+\pi^+$,

(d) $\bar{\nu}+n\rightarrow n+e^++\pi^-$,

(e) $\bar{\nu}+p\rightarrow n+e^++\pi^0$,

(f) $\bar{\nu}+p\rightarrow p+e^++\pi^-$,

[6] W. K. Panofsky and E. A. Allton, Phys. Rev. **110**, 1155 (1958).

[7] G. G. Ohlsen, Phys. Rev. **120**, 584 (1960).

[8] On this general subject see, for example, Y. Yamaguchi, Progr. Theoret. Phys. (Kyoto) **23**, 1117 (1960); T. D. Lee and C. N. Yang, Phys. Rev. Letters **4**, 307 (1960); S. M. Berman, Proceedings of the International Conference on Theoretical Aspects of Very High-Energy Physics [CERN Report No. 61-22, (unpublished), p. 7.]

[9] For a dispersion-theoretic treatment of the problem, see Ya. N. Azimov, Zhur. Eksp. i. Teoret. Fiz. **41**, 1879 (1961); N. Dombey, Phys. Rev. **127**, 653 (1962); P. Dennery, *ibid.* **127**, 664 (1962).

$$\frac{d^2\sigma}{d\epsilon' d\Omega}=\left(\frac{g}{4\pi}\right)^2\frac{2f^2}{\pi^2}\frac{[2\mu(\epsilon_m'-\epsilon')]^{1/2}}{[1+\mu/m]^2}[1+(2\epsilon/m)\sin^2(\theta/2)]^{1/2}\frac{\epsilon_m'^2\cos^2(\theta/2)}{1+q^2/4m^2}S,$$

$$S_a=2(F_c^V)^2(1+v^2)+\sum_{h=\pm1}[F_m^V(qh/2m)-\lambda G_1]^2(1+hv)^2[1-h\sin(\theta_B/2)]^2/\cos^2(\theta_B/2),$$

$$S_b=(F_c^V)^2(4+v^2)+\tfrac{1}{2}\sum_{h=\pm1}[F_m^V(qh/2m)-\lambda G_1]^2(2+hv)^2[1-h\sin(\theta_B/2)]^2/\cos^2(\theta_B/2),$$

$$S_c=2(F_c^V)^2+\sum_{h=\pm1}[F_m^V(qh/2m)-\lambda G_1]^2[1-h\sin(\theta_B/2)]^2/\cos^2(\theta_B/2),$$

$$S_d=2(F_c^V)^2(1+v^2)+\sum_{h=\pm1}[F_m^V(qh/2m)-\lambda G_1]^2(1+hv)^2[1+h\sin(\theta_B/2)]^2/\cos^2(\theta_B/2),$$

$$S_e=(F_c^V)^2(4+v^2)+\tfrac{1}{2}\sum_{h=\pm1}[F_m^V(qh/2m)-\lambda G_1]^2(2+hv)^2[1+h\sin(\theta_B/2)]^2/\cos^2(\theta_B/2),$$

$$S_f=2(F_c^V)^2+\sum_{h=\pm1}[F_m^V(qh/2m)-\lambda G_1]^2[1+h\sin(\theta_B/2)]^2/\cos^2(\theta_B/2),\qquad(6.5)$$

with the abbreviations

$$v=(q/m)(1+q^2/4m^2)^{1/2}(1+q^2/2m^2)^{-1},\qquad(6.5')$$
$$\lambda=(1+q^2/4m^2)^{1/2}.$$

Other quantities are defined in Sec. 5. As before, these formulas are supposed to be valid when pion and electron masses may be neglected compared with ϵ, q, and m. They should also apply to processes involving the muon under similar conditions.

For the sake of comparison, the elastic cross sections[8] are given below in our notation.

(b′) $\nu+n\rightarrow p+e^-$,

(e′) $\bar{\nu}+p\rightarrow n+e^+$,

$$\frac{d\sigma}{d\Omega}=\left(\frac{g}{4\pi}\right)^2[1+(2\epsilon/m)\sin^2(\theta/2)]^{-3}\frac{8\epsilon^2\cos^2(\theta/2)}{1+q^2/4m^2}S,$$

$$S_{b'}=(F_c^V)^2+\tfrac{1}{2}\sum_{h=\pm1}[F_m^V(qh/2m)-\lambda G_1]^2\times[1-h\sin(\theta_B/2)]^2/\cos^2(\theta_B/2),$$

$$S_{e'}=(F_c^V)^2+\tfrac{1}{2}\sum_{h=\pm1}[F_m^V(qh/2m)-\lambda G_1]^2\times[1+h\sin(\theta_B/2)]^2/\cos^2(\theta_B/2).\qquad(6.6)$$

Reprinted from THE PHYSICAL REVIEW, Vol. 128, No. 6, 2622–2629, December 15, 1962

Magnetic Field Dependence of the Energy Gap in Superconductors*

YOICHIRO NAMBU
The Enrico Fermi Institute for Nuclear Studies, and The Department of Physics, University of Chicago, Chicago, Illinois

AND

SAN FU TUAN†
Department of Physics, Brown University, Providence, Rhode Island
(Received May 17, 1962)

We calculate the reduction of the energy gap at zero temperature for a bulk superconductor in the presence of a static external magnetic field, to the second order in the field strength. Simple formulas are obtained for the long ($\lambda \gg \xi_0$) and short ($\lambda \ll \xi_0$) wavelength limits, where ξ_0 is the coherence length.

We use the general gauge-invariant formulation of the Bardeen-Cooper-Schrieffer (BCS)-Bogoliubov theory developed by one of the authors, but the result is also shown to agree with that of the BCS variational procedure applied in the presence of the field. The gauge invariance is maintained by virtue of the collective excitations as in the Meissner effect. The mathematical proof of gauge invariance is carried out in a completely general manner using Ward identities.

1. INTRODUCTION

WE shall be concerned in this paper with the theoretical derivation of the magnetic field dependence of the energy gap in a superconductor at absolute zero temperature.

Recently, Douglass[1] has performed experiments on superconducting films which show a characteristic dependence of the energy gap on the magnetic field. He has been able to account for the results by means of the Ginzburg-Landau[2] equations which, according to Gor'kov,[3] follow from the Bardeen-Cooper-Schrieffer (BCS) theory[4] and describe just such an effect. However, the validity of Gor'kov's derivation is restricted to the London limit which occurs near the critical temperature. The calculation at zero temperature, which belongs to the Pippard case, has been done by Gupta and Mathur[5] using the theory of Wentzel.[6] Our calculation will also be performed at zero temperature, but in the framework of the gauge-invariant theory developed by one of the authors.[7] Our result will show that, to the second order in the field strength,

$$\delta(\phi^2) \sim -e^2 v_F^2 (\xi_0 q)^2 |\mathbf{A}(\mathbf{q})|^2, \quad (q\xi_0 = q v_F/\pi\phi \ll 1)$$

where ϕ is the gap, v_F is the Fermi velocity, ξ_0 is the coherence length, and $\mathbf{A}(\mathbf{q})$ is the Fourier component of the vector potential present in the medium.[8] The same result is obtained by a simple variational procedure similar to the original BCS work. But the central problem in this kind of derivation is the proof of gauge invariance. The result of Gupta and Mathur, though gauge invariant, gives $\delta(\phi^2) \sim -e^2 v_F^2 |\mathbf{A}(\mathbf{q})|^2$.

In Secs. 2 and 3 we shall present the general mathematical formulation and the actual derivation of the results for both small-q ($\ll \xi_0^{-1}$) and large-q ($\gtrsim \xi_0^{-1}$) regions.

The proof of gauge invariance of our procedure is carried out on a completely general basis in Sec. 4, by means of the Ward identities, whereas the equivalence of our result with the BCS variational procedure in the present problem is demonstrated in Sec. 5.

2. GENERAL FORMULATION

In a previous paper[7] the BCS-Bogoliubov theory was formulated on a general basis using the language and techniques of field theory. According to this, the problem reduces to finding the self-energy of a quasi-particle in the generalized Hartree-Fock approximation. We will briefly recapitulate the main points.

With the two-component notation combining the up-spin and down-spin electron wave functions,

$$\Psi(x) = \begin{pmatrix} \psi_\uparrow(x) \\ \psi_\downarrow^\dagger(x) \end{pmatrix} \quad \text{or} \quad \Psi(p) = \begin{pmatrix} \psi_\uparrow(p) \\ \psi_\downarrow^\dagger(-p) \end{pmatrix}, \tag{2.1}$$

the Hartree-Fock self-consistent equation for the self-energy Σ takes the form

$$\Sigma(x,y) = -\tau_3 G(x,y) \tau_3 V(x,y). \tag{2.2}$$

Here x and y in general refer to space-time coordinates; $G(x,y)$ is the Green's function for the quasi-particle satisfying

$$\begin{aligned} [i(\partial/\partial t) - H_0 - \Sigma] G(x,y) &= i\delta^4(x-y), \\ H_0 = \tau_3[(p^2/2m) - \mu] &= \tau_3 \epsilon_p, \end{aligned} \tag{2.3}$$

* This work is supported by the U. S. Atomic Energy Commission and Advanced Research Projects Agency (ARPA).

† Present address: Purdue University, Lafayette, Indiana.

1 D. H. Douglass, Jr., Phys. Rev. Letters **6**, 346 (1961); *ibid.* **7**, 14 (1961); Phys. Rev. **124**, 735 (1961).

2 V. L. Ginzburg and L. D. Landau, J. Exptl. Theoret. Phys. (U.S.S.R.) **20**, 1064 (1950).

3 L. P. Gor'kov, J. Exptl. Theoret. Phys. (U.S.S.R.) **36**, 1918 (1959) [translation: Soviet Phys.—JETP **9**, 1364 (1959)].

4 J. Bardeen, L. N. Cooper, and J. R. Schrieffer, Phys. Rev. **108**, 1175 (1957).

5 K. K. Gupta and V. S. Mathur, Phys. Rev. **121**, 107 (1961).

6 G. Wentzel, Phys. Rev. **111**, 1488 (1958); Phys. Rev. Letters **2**, 33 (1959). See also K. K. Gupta and V. S. Mathur, Phys. Rev. **115**, 75 (1959).

7 Y. Nambu, Phys. Rev. **117**, 648 (1960), hereafter denoted I for reference purposes.

8 Throughout the paper, natural units $\hbar = c = 1$ are taken.

and $-V(x,y)$ is the interaction potential which is usually assumed to be effective only for states near the Fermi surface, $p\sim p_F$; μ is the chemical potential, being equal to $p_F^2/2m$ for practical purposes. Equation (2.2) has a solution which contains a term $\tau_1\phi$, where $\phi(p)\approx$const is the energy gap parameter. We get

$$\phi=\phi\int\frac{d^3l}{2E_l}V(p-l)\approx\phi\rho\sinh^{-1}\frac{\omega}{\phi};\qquad \rho=\bar{V}N,\quad N=mp_F/2\pi^2,\quad (\bar{V}=\langle V\rangle_{\mathrm{av}}),\tag{2.4}$$

where ω is the Debye cutoff frequency. For small $\rho\ll1$, ϕ becomes

$$\phi\approx 2\omega e^{-1/\rho}.\tag{2.5}$$

Now let us assume that a magnetic field represented by a vector potential $\mathbf{A}(x)$ is present in the medium. $\mathbf{A}(x)$ is the sum of the external $\mathbf{A}^{\mathrm{ex}}$ and the induced field, and should be zero in the case of infinite medium with uniform external field. The relation between the Fourier components of $\mathbf{A}$ and $\mathbf{A}^{\mathrm{ex}}$ is given by

$$A_i(\mathbf{q})=[1-K(q)/q^2]^{-1}A_i^{\mathrm{ex}}(\mathbf{q}),\tag{2.6}$$

where $K(q)$ is the kernel for the induced current

$$j_i(\mathbf{q})=K(q)A_i(\mathbf{q}).\tag{2.7}$$

Usually the kernel K is calculated with the solution of the free superconductive state, and is given by

$$-K(q)=(e^2/m)n+O(q^2)\equiv(1/\Lambda)+O(q^2),\qquad n=p_F^3/3\pi^2.\tag{2.8}$$

This will lead to the Meissner effect, but we do not yet have the field dependence of the energy gap.

In order to obtain the field dependence of the gap, it is necessary to set up the self-energy equation in the presence of the field. Thus, we will modify Eqs. (2.2), (2.3) as

$$\Sigma^{(A)}(x,y)=-\tau_3G^{(A)}(x,y)\tau_3V(x,y),\tag{2.9}$$

$$[i(\partial/\partial t)-H_0^{(A)}-\Sigma^{(A)}]G^{(A)}(x,y)=i\delta^4(x-y),$$
$$H_0^{(A)}=\tau_3\{[(\mathbf{p}-e\tau_3\mathbf{A})^2/2m]-\mu^{(A)}\}=\tau_3\epsilon_p-(e/m)\mathbf{p}\cdot\mathbf{A}+(e^2/2m)\tau_3A^2-\delta\mu\tau_3.\tag{2.9'}$$

The new solution can then be determined by expanding everything in powers of $\mathbf{A}$. Let us define

$$\Sigma^{(A)}(x,y)=\Sigma(x-y)+\int\left[\frac{\delta\Sigma^{(A)}(x,y)}{\delta A_i(z)}\right]_{A=0}A_i(z)dz+\tfrac{1}{2}\int\int\left[\frac{\delta^2\Sigma^{(A)}(x,y)}{\delta A_i(z)\delta A_j(z')}\right]_{A=0}A_i(z)A_j(z')dzdz'+\cdots,$$
$$[\delta\Sigma^{(A)}(x,y)/\delta A_i(z)]_{A=0}\equiv-e\Lambda_i(x,y;z),\quad [\delta^2\Sigma^{(A)}/\delta A_i\delta A_j]_{A=0}\equiv e^2\Lambda_{ij}(x,y;z,z').\tag{2.10}$$

From Eq. (2.9′) we get accordingly

$$G^{(A)}(x,y)=G(x-y)+\int\left[\frac{\delta G^{(A)}(x,y)}{\delta A_i(z)}\right]_{A=0}A_i(z)dz+\tfrac{1}{2}\int\int\left[\frac{\delta^2G^{(A)}(x,y)}{\delta A_i(z)\delta A_j(z')}\right]_{A=0}A_i(z)A_j(z')dzdz',$$
$$[\delta G^{(A)}(x,y)/\delta A_i(z)]_{A=0}=eiG\Gamma_iG,\tag{2.10'}$$

$$\Gamma_i(x,y;z)=i\delta G^{-1}/\delta A_i\equiv e(\gamma_i+\Lambda_i)=-(ei/m)[(\partial/\partial x_i-\partial/\partial y_i)\delta^3(x-y)]\delta^3(x-z)+e\Lambda_i(x,y;z),$$

$$[\delta^2G^{(A)}(x,y)/\delta A_i(z)\delta A_j(z')]_{A=0}\equiv-e^2L_{ij}(x,y;z,z')=\left[ie\frac{\delta}{\delta A_j(z')}(G\Gamma_iG)\right]_{A=0}$$
$$=\left[ie\left(\frac{\delta G}{\delta A_j}\Gamma_iG+G\Gamma_i\frac{\delta G}{\delta A_j}+G\frac{\delta\Gamma_i}{\delta A_j}G\right)\right]_{A=0}=-e^2(G\Gamma_jG\Gamma_iG+G\Gamma_iG\Gamma_jG-iG\Gamma_{ij}G),\tag{2.11}$$

$$e^2\Gamma_{ij}(x,y;z,z')=e[\delta\Gamma_i(x,y;z)/\delta A_j(z')]_{A=0}=-(e^2/m)\delta_{ij}\delta^3(x-y)\delta^3(x-z)\delta^3(z-z')+e^2\Lambda_{ij}(x,y;z,z').$$

Here the obvious shorthand writing such as

$$G\Gamma_iG=\int\int G(x,x')\Gamma_i(x',y';z)G(y',y)d^4x'd^4y'$$

has been used. Substituting Eqs. (2.10) and (2.11) into Eq. (2.9), we obtain the relations

$$\Sigma(x,y)=-\tau_3G(x,y)\tau_3V(x,y),\tag{2.12a}$$

$$\Lambda_i(x,y;z)=i\int\int\tau_3G(x,x')\Gamma_i(x',y';z)G(y',y)\tau_3V(x,y)d^4x'd^4y',\tag{2.12b}$$

$$\Lambda_{ij}(x,y;z,z')=-\tau_3L_{ij}(x,y;z,z')\tau_3V(x,y)=-\tau_3(G\Gamma_jG\Gamma_iG+G\Gamma_iG\Gamma_jG+G\Gamma_{ij}G)\tau_3V.\tag{2.12c}$$

These will determine Σ, Λ_i, $\Lambda_{ij}\cdots$ successively. In general, they are operators containing 1, τ_3, τ_1, τ_2 in the two-component notation. We are, however, interested in that part of Λ_{ij} which is proportional to τ_1, since this gives the correction to the energy gap term in Σ which is also proportional to τ_1.

In the following we will assume that $\Sigma=\tau_1\phi$, $\Gamma_i=\gamma_i$, and with this calculate Λ_{ij} using Eq. (2.12c). The reason is that other terms are either renormalization terms (τ_3 in Σ) or are important only for unphysical longitudinal potentials (τ_2 in Γ_i), so that they will not cause any physically significant changes in our results. For the discussion of gauge invariance of the results, of course, we have to examine those terms which are neglected. We will do this in a later section.

3. CALCULATIONS

The compensation equation (2.9) written in momentum space depends on the initial and final momenta of the quasi-particle. For practical purposes, however, we restrict ourselves to the diagonal elements ($p=p'$). We get

$$\Sigma^{(A)}(p)=\frac{-1}{(2\pi)^4}\int\tau_3G^{(A)}(\mathbf{l},l_0)\tau_3V(\mathbf{p}-\mathbf{l},\,p_0-l_0)d^4l. \tag{3.1}$$

Here

$$\begin{aligned}G^{(A)}(\mathbf{l},l_0)&=i/[l_0-H^{(A)}(\mathbf{l})]\\&=i\bigg[\frac{1}{l_0-H(\mathbf{l})}-\frac{1}{l_0-H(\mathbf{l})}\frac{e\mathbf{l}\cdot\mathbf{A}}{m}\frac{1}{l_0-H(\mathbf{l})}\\&\quad+\frac{1}{l_0-H(\mathbf{l})}\frac{e\mathbf{l}\cdot\mathbf{A}}{m}\frac{1}{l_0-H(\mathbf{l})}\frac{e\mathbf{l}\cdot\mathbf{A}}{m}\frac{1}{l_0-H(\mathbf{l})}\\&\quad+\frac{1}{l_0-H(\mathbf{l})}\tau_3\frac{e^2|\mathbf{A}|^2}{2m}\frac{1}{l_0-H(\mathbf{l})}\\&\quad+\frac{1}{l_0-H(\mathbf{l})}\Sigma^{(2)}\frac{1}{l_0-H(\mathbf{l})}\bigg]+O(A^3),\end{aligned}$$

$$H(\mathbf{l})=H_0+\Sigma=\epsilon\tau_3+\tau_1\phi.$$

The zeroth order term $\Sigma^{(0)}(p)$ is given by

$$\begin{aligned}\Sigma^{(0)}(p)&=\frac{-i}{(2\pi)^4}\int\tau_3\frac{d^4l}{l_0-H(\mathbf{l})}\tau_3V(p-l)\\&=-\frac{1}{2(2\pi)^3}\tau_3\int\frac{\epsilon(\mathbf{l})\bar{V}d^3l}{E(\mathbf{l})}+\frac{1}{2(2\pi)^3}\tau_1\int\frac{\phi}{E(\mathbf{l})}\bar{V}d^3l,\end{aligned}\tag{3.2}$$

where the effective potential is restricted to near the Fermi surface. The first term on the right-hand side makes little contribution, and the second yields the energy gap equation

$$\frac{\tau_1}{2(2\pi)^3}\int\frac{\phi\bar{V}}{E(\mathbf{l})}d^3l=\frac{\tau_1}{2}\int_{-\infty}^{\infty}\frac{\phi\bar{V}N(\mathbf{l})d\epsilon(\mathbf{l})}{[\epsilon^2(\mathbf{l})+\phi^2]^{1/2}}=\tau_1\phi.$$

Hence

$$\Sigma^{(0)}(p)=\tau_1\phi. \tag{3.3}$$

The first-order term $\Sigma^{(1)}$ is the vertex correction which has been investigated in I (cf. reference 7) and does not concern us now. The second-order contribution $\Sigma^{(2)}$ contains three terms,

$$\Sigma^{(2)}=\Sigma_1^{(2)}+\Sigma_2^{(2)}+\Sigma_3^{(2)}.$$

Here $\Sigma_1^{(2)}$ is the contribution from the diagrams of Figs. 1(a) and 1(b), while $\Sigma_2^{(2)}$ and $\Sigma_3^{(2)}$ correspond to Figs. 1(c) and 1(d), respectively. We will consider one particular momentum $\mathbf{q}$ at a time. We have

$$\begin{aligned}\Sigma_1^{(2)}(p)&=\frac{-i}{(2\pi)^4}\int\tau_3\frac{V}{l_0-H(\mathbf{l})}\frac{e\mathbf{l}\cdot\mathbf{A}(-\mathbf{q})}{m}\\&\quad\times\frac{1}{l_0-H(\mathbf{l}+\mathbf{q})}\frac{e\mathbf{l}\cdot\mathbf{A}(\mathbf{q})}{m}\frac{1}{l_0-H(\mathbf{l})}\tau_3d^4l\\&\quad+\frac{-i}{(2\pi)^4}\int\tau_3\frac{V}{l_0-H(\mathbf{l})}\frac{e\mathbf{l}\cdot\mathbf{A}(\mathbf{q})}{m}\frac{1}{l_0-H(\mathbf{l}-\mathbf{q})}\\&\quad\times\frac{e\mathbf{l}\cdot\mathbf{A}(-\mathbf{q})}{m}\frac{1}{l_0-H(\mathbf{l})}\tau_3d^4l\equiv M_1+M_2.\end{aligned}\tag{3.4}$$

M_2 is obtained from M_1 by substitution $\mathbf{q}\to-\mathbf{q}$. M_1 may be written after integrating out l_0 as

$$M_1=-\frac{1}{2(2\pi)^3}\int\frac{e^2}{m^2}\mathbf{l}\cdot\mathbf{A}(-\mathbf{q})\mathbf{l}\cdot\mathbf{A}(\mathbf{q})\bar{V}Xd^3l, \tag{3.5}$$

with

$$\begin{aligned}X&=-A\frac{1}{2E(\mathbf{l})[E(\mathbf{l})+E(\mathbf{l}+\mathbf{q})]^2}\\&\quad+B\frac{2E(\mathbf{l})+E(\mathbf{l}+\mathbf{q})}{2E^3(\mathbf{l})E(\mathbf{l}+\mathbf{q})[E(\mathbf{l})+E(\mathbf{l}+\mathbf{q})]^2},\\A&=[2\epsilon(\mathbf{l})+\epsilon(\mathbf{l}+\mathbf{q})]\tau_3-3\phi\tau_1,\\B&=[\epsilon(\mathbf{l})\tau_3-\tau_1\phi][\epsilon(\mathbf{l}+\mathbf{q})\tau_3-\tau_1\phi][\epsilon(\mathbf{l})\tau_3-\tau_1\phi],\\&=\{\epsilon^2(\mathbf{l})\epsilon(\mathbf{l}+\mathbf{q})+\phi^2[2\epsilon(\mathbf{l})-\epsilon(\mathbf{l}+\mathbf{q})]\}\tau_3\\&\qquad+\phi[\epsilon^2(\mathbf{l})-2\epsilon(\mathbf{l})\epsilon(\mathbf{l}+\mathbf{q})-\phi^2]\tau_1.\end{aligned}\tag{3.6}$$

For $\mathbf{q}=0$, we have $X=0$.

To calculate the London limit $q\ll\xi_0^{-1}$, we may conveniently expand $\epsilon(\mathbf{l}+\mathbf{q})$ in powers of $\mathbf{q}$, retaining for the integrand X terms of order q^2. Making use of the reality condition $A(-\mathbf{q})=A^*(\mathbf{q})$, we get for the τ_1 part of M_1

$$\begin{aligned}(M_1)_{\tau_1}&=-\frac{\tau_1}{2(2\pi)^3}\int\frac{e^2}{m^2}\bar{V}|\mathbf{l}\cdot\mathbf{A}(\mathbf{q})|^2d^3l\bigg[-\frac{\phi\epsilon(\mathbf{l})}{4E^5(\mathbf{l})}\frac{q^2}{m^2}\\&\quad+\frac{\phi}{8E^5(\mathbf{l})}\frac{(\mathbf{q}\cdot\mathbf{l})^2}{m^2}\frac{\phi^2}{E^2}+\frac{3}{4}\frac{\phi\epsilon^2(\mathbf{l})}{E^7(\mathbf{l})}\Big(\frac{\mathbf{q}\cdot\mathbf{l}}{m}\Big)^2\bigg]\\&=-(1/90)(\tau_1/\phi)(e^2/m^2)\\&\qquad\times\rho q^2|\mathbf{A}(\mathbf{q})|^2(p_F^4/m^2\phi^2).\end{aligned}\tag{3.7}$$

We have restricted the integration to the range $-\omega < \epsilon(\mathbf{l}) < \omega$, and neglected terms of the order $(\phi/\omega)^2$. Obviously M_1 is symmetric under $\mathbf{q} \to -\mathbf{q}$, hence $(M_2)_{\tau_1} = (M_1)_{\tau_1}$, and

$$[\Sigma_1^{(2)}(p)]_{\tau_1} = \delta\phi\tau_1 = -\frac{1}{45}\frac{\tau_1}{\phi}\left(\frac{e^2}{m^2}\right)\rho q^2|\mathbf{A}(\mathbf{q})|^2\frac{p_F{}^4}{m^2\phi^2}. \quad (3.8)$$

The next term $(\Sigma_2^{(2)})_{\tau_1}$ is readily shown to vanish independently of the magnitude of $\mathbf{q}$. In fact, for the more general situation $\mathbf{p} \neq \mathbf{p}'$ $(\mathbf{q} \neq \mathbf{q}')$, we have

$$\begin{aligned}
\Sigma_2^{(2)}(\mathbf{p}',\mathbf{p}) &= \frac{-i}{(2\pi)^4}\int \tau_3 \frac{1}{l_0 - H(\mathbf{l}+\boldsymbol{\Delta})}\tau_3\frac{e^2}{2m} \\
&\quad \times \mathbf{A}(\mathbf{q})\cdot\mathbf{A}(-\mathbf{q}')\frac{1}{l_0 - H(\mathbf{l})}\tau_3 V(p-l)d^4l \\
&= \frac{-i}{(2\pi)^4}\frac{e^2}{2m}\mathbf{A}(\mathbf{q})\cdot\mathbf{A}(-\mathbf{q}') \\
&\quad \times \int \frac{[l_0^2 + \epsilon(\mathbf{l})\epsilon(\mathbf{l}+\boldsymbol{\Delta}) - \phi^2]\tau_3 - \phi[\epsilon(\mathbf{l}+\boldsymbol{\Delta}) + \epsilon(\mathbf{l})]\tau_1}{[l_0^2 - E^2(\mathbf{l}+\boldsymbol{\Delta})][l_0^2 - E^2(\mathbf{l})]} \\
&\qquad \times V(p-l)d^4l,
\end{aligned}$$

where

$$\boldsymbol{\Delta} = \mathbf{q} - \mathbf{q}' = \mathbf{p}' - \mathbf{p}.$$

Hence

$$\begin{aligned}
[\Sigma_2^{(2)}(\mathbf{p}',\mathbf{p})]_{\tau_1} &= \frac{i}{(2\pi)^4}\tau_1\frac{e^2}{2m}\frac{\phi}{2m}\mathbf{A}(\mathbf{q})\cdot\mathbf{A}(-\mathbf{q}') \\
&\quad \times \int \frac{\boldsymbol{\Delta}^2 + 2(l^2 - p_F{}^2)}{[l_0^2 - E^2(\mathbf{l})]^2}V(p-l)d^4l \\
&= -(e^2/8m^2)(\rho/\phi)(\mathbf{p}'-\mathbf{p})^2\mathbf{A}(\mathbf{q})\cdot\mathbf{A}(-\mathbf{q}')\tau_1, \quad (3.9)
\end{aligned}$$

which vanishes for $\mathbf{p} = \mathbf{p}'$.

The third term $\Sigma_3^{(2)}$ is the radiative correction to $\Sigma^{(2)}$ itself which must be added in order to obtain self-consistency to this order. Its contribution to the τ_1 part can be calculated from the BCS energy gap Eq. (3.2) as

$$\begin{aligned}
[\Sigma_2^{(2)}(\mathbf{p})]_{\tau_1} &= \tau_1\delta\int\frac{\phi}{2E(\mathbf{l})}\bar{V}d^3l = \tau_1\delta\phi\left[\int\frac{N\bar{V}d\epsilon}{2E}\right. \\
&\quad \left. - \int\frac{\phi^2}{2E^3}N\bar{V}d\epsilon\right] = \tau_1(1-\rho)\delta\phi. \quad (3.10)
\end{aligned}$$

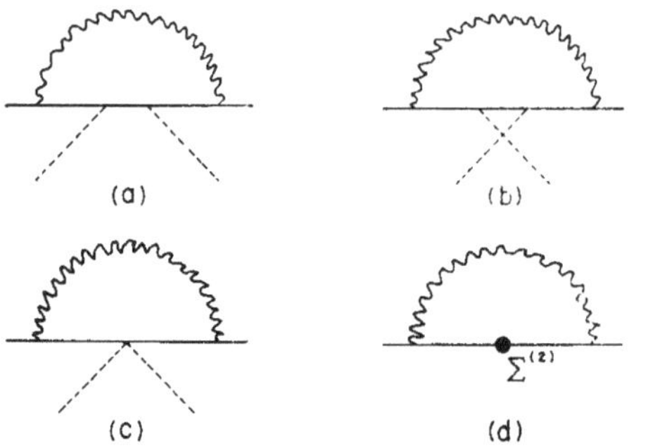

FIG. 1. Compensation equation diagrams to second order in the external electromagnetic field A_μ.

Combining Eqs. (3.8) and (3.10), the compensation equation becomes

$$\delta\phi = (1-\rho)\delta\phi - (1/45\phi)(e^2/m^2)\rho q^2|\mathbf{A}(\mathbf{q})|^2(p_F{}^4/m^2\phi^2), \quad (3.11)$$

or

$$\delta\phi = -(1/45\phi)(e^2/m^2)q^2|\mathbf{A}(\mathbf{q})|^2(p_F{}^4/m^2\phi^2).$$

Introducing the coherence length $\xi_0 = v_F/\pi\phi$ and summing over all different Fourier components $\mathbf{A}(\mathbf{q})$, we obtain

$$\delta(\phi^2) = -(\pi^2/45)e^2v_F{}^2\sum_{\mathbf{q}}(\xi_0\mathbf{q})^2|\mathbf{A}(\mathbf{q})|^2, \quad (3.12)$$

where $\mathbf{q}$ and $-\mathbf{q}$ are to be counted independently.

We now briefly comment on the τ_3 (kinetic energy) part of $\Sigma^{(2)}$. From Eqs. (3.4)–(3.6), we have

$$\begin{aligned}
[\Sigma_1^{(2)}(\mathbf{p})]_{\tau_3} &= -\frac{1}{(2\pi)^3}\int\frac{e^2}{m^2}\mathbf{l}\cdot\mathbf{A}(-\mathbf{q})\mathbf{l}\cdot\mathbf{A}(\mathbf{q})[X]_{\tau_3}\bar{V}d^3l; \\
[X]_{\tau_3} &= \tau_3\{2E^2(\mathbf{l})[E(\mathbf{l})\epsilon(\mathbf{l}+\mathbf{q}) - E(\mathbf{l}+\mathbf{q})\epsilon(\mathbf{l})] \\
&\quad + 2\phi^2[\epsilon(\mathbf{l}) - \epsilon(\mathbf{l}+\mathbf{q})][2E(\mathbf{l}) + E(\mathbf{l}+\mathbf{q})]\}/2E^3(\mathbf{l})E(\mathbf{l}+\mathbf{q})[E(\mathbf{l}) + E(\mathbf{l}+\mathbf{q})]^2.
\end{aligned} \quad (3.13)$$

Figures 1(c) and 1(d) give, respectively,

$$\begin{aligned}
[\Sigma_2^{(2)}(\mathbf{p})]_{\tau_3} &= \frac{-i}{(2\pi)^4}\frac{2e^2|\mathbf{A}(\mathbf{q})|^2\tau_3}{2m} \\
&\quad \times \int\frac{[l_0^2 + \epsilon^2(\mathbf{l}) - \phi^2]V(p-l)d^4l}{[l_0^2 - E^2(\mathbf{l})]^2} \\
&= (e^2/m)|\mathbf{A}(\mathbf{q})|^2\rho\tau_3, \quad (3.14)
\end{aligned}$$

$$\begin{aligned}
[\Sigma_3^{(2)}(\mathbf{p})]_{\tau_3} &= \tau_3\delta\epsilon(\mathbf{p}) = \tau_3\delta\int\frac{\epsilon}{2E}N\bar{V}d\epsilon \\
&= \tau_3\left[\int\frac{\phi^2}{2E^3}N\bar{V}d\epsilon\right]\delta\epsilon = \tau_3\rho\delta\epsilon. \quad (3.15)
\end{aligned}$$

Combining Eqs. (3.13)–(3.15), we obtain

$$\tau_3\delta\epsilon(\mathbf{p}) = \tau_3\rho\delta\epsilon(\mathbf{p}) + \tau_3\frac{e^2}{m}|\mathbf{A}(\mathbf{q})|^2\rho + \Sigma_1^{(2)}(\mathbf{p}),$$

or

$$(1-\rho)\delta\epsilon(\mathbf{p})=\frac{e^2}{m}|\mathbf{A}(\mathbf{q})|^2\rho+[\Sigma_1^{(2)}(\mathbf{p})]_{\tau_3}/\tau_3. \quad (3.16)$$

As is seen from Eq. (3.13), $\Sigma_1^{(2)}(\mathbf{p})$ is independent of p near the Fermi surface, and so is $\delta\epsilon(\mathbf{p})$. This constant shift in ϵ is equivalent to a shift in the chemical potential. Since the particle number n has to be kept fixed, and this depends essentially on the chemical potential, we have to subtract away $\delta\epsilon=\delta\mu$ as the renormalization of the chemical potential.

It must be noted that the "Compton diagrams," which are second order in $\mathbf{A}$, do not contribute to the compensation Eq. (3.1), and hence to the energy gap $\phi(\mathbf{A})$. This is so because the self-energy Σ in the compensation equation is the proper self-energy in the terminology of quantum electrodynamics, whereas the Compton diagrams belong to the improper self-energy. The Compton diagrams are relevant if we are interested in the solution of the quasi-particle Schrödinger equation $H^{(A)}\Psi=E\Psi$. These diagrams then contribute to the second-order term $\delta E^{(2)}$, and will affect the p dependence of E, but not the gap itself. We also emphasize that the Compton diagrams are needed in the discussion of the over-all gauge invariance of the theory as will be shown in Sec. 4.

In Sec. 2 we observed that the field $\mathbf{A}$ is related to the external field $\mathbf{A}^{\text{ex}}$ via Eq. (2.6) which depends on the polarization kernel K. For later purposes we will calculate here $K(q)$ to the second order in q. Following I, K is given by

$$K_{ij}(q)=-(ne^2/m)\delta_{ij}+K_{ij}^{(2)}(q),$$

$$K_{ij}^{(2)}(q)=\frac{-ie^2}{(2\pi)^4}\int \text{Tr}[\gamma_i(p-\tfrac{1}{2}q,\,p+\tfrac{1}{2}q)G(p+\tfrac{1}{2}q) \quad (3.17)$$
$$\times\Gamma_j(p+\tfrac{1}{2}q,\,p-\tfrac{1}{2}q)G(p-\tfrac{1}{2}q)]d^4p.$$

Γ_j may be replaced by γ_j for transversal waves, so that

$$K_{ij}^{(2)}(q)=\frac{-ie^2}{(2\pi)^4}\int \text{Tr}\left[\frac{p_i}{m}G(p+\tfrac{1}{2}q)\frac{p_j}{m}G(p-\tfrac{1}{2}q)\right]d^4p$$
$$=\frac{-ie^2}{(2\pi)^4}i\pi\int\frac{p_ip_j}{m^2}(\mathbf{p}\cdot\mathbf{q})^2\frac{\phi^2}{m^2}\frac{d^3p}{2E^5(\mathbf{p})}. \quad (3.18)$$

Making use of the relation

$$\langle p_ip_jp_kp_m\rangle_{\text{av}}=\langle p^4\rangle_{\text{av}}(\delta_{ik}\delta_{jm}+\delta_{ij}\delta_{km}+\delta_{im}\delta_{kj})/15,$$

we have

$$K_{ij}^{(2)}(q)=(1/45)e^2N[q^2\delta_{ij}+2q_iq_j]p_F^4/m^4\phi^2,$$

or

$$K_{ij}(q)=-(ne^2/m)\delta_{ij}+(e^2/45)N(v_F^4/\phi^2)[q^2\delta_{ij}+2q_iq_j].$$

As was demonstrated in I, the effect of the collective excitations on Γ in Eq. (3.17) is to make the above form vanish for a longitudinal potential. Thus, the correct gauge-invariant result is

$$K_{ij}(q)=-(ne^2/m)(\delta_{ij}-q_iq_j/q^2)$$
$$+(e^2/45)N(v_F^4/\phi^2)[q^2\delta_{ij}-q_iq_j]$$
$$=-(ne^2/m)(\delta_{ij}-q_iq_j/q^2)[1-(\pi^2/30)(\xi_0q)^2]. \quad (3.19)$$

It is appropriate to discuss briefly the type of energy gap behavior expected for large q. For this purpose, we have to evaluate the integral (3.5) without expansion in $\mathbf{q}$. The τ_1 part of $\Sigma_1^{(2)}$ reads

$$[\Sigma_1^{(2)}(\mathbf{p})]_{\tau_1}$$
$$=-\frac{1}{2}\frac{1}{(2\pi)^3}\sum_{\pm\mathbf{q}}\int\frac{e^2}{m^2}\mathbf{l}\cdot\mathbf{A}(-\mathbf{q})\mathbf{l}\cdot\mathbf{A}(\mathbf{q})\bar{V}(X)_{\tau_1}d^3l,$$

$$(X)_{\tau_1}=\frac{3\phi\tau_1}{2E(\mathbf{l})[E(\mathbf{l})+E(\mathbf{l}+\mathbf{q})]^2} \quad (3.20)$$
$$+\frac{\phi\tau_1[\epsilon^2(\mathbf{l})-2\epsilon(\mathbf{l})\epsilon(\mathbf{l}+\mathbf{q})-\phi^2][2E(\mathbf{l})+E(\mathbf{l}+\mathbf{q})]}{2E^3(\mathbf{l})E(\mathbf{l}+\mathbf{q})[E(\mathbf{l})+E(\mathbf{l}+\mathbf{q})]^2}.$$

Although this expression is quite complicated, the integral can be expressed, upon summation over $\pm\mathbf{q}$ (and change of variable $\mathbf{l}-\mathbf{q}\rightarrow\mathbf{l}$ for the case $-\mathbf{q}$), in the simpler form

$$[\Sigma_1^{(2)}(\mathbf{p})]_{\tau_1}=-\frac{1}{4}\frac{1}{(2\pi)^3}\tau_1\frac{\partial}{\partial\phi}\int\frac{e^2}{m^2}\mathbf{l}\cdot\mathbf{A}(-\mathbf{q})\mathbf{l}\cdot\mathbf{A}(\mathbf{q})\bar{V}d^3l$$
$$\times\frac{1}{E(\mathbf{l})+E(\mathbf{l}+\mathbf{q})}\left[1-\frac{\epsilon(\mathbf{l})\epsilon(\mathbf{l}+\mathbf{q})+\phi^2}{E(\mathbf{l})E(\mathbf{l}+\mathbf{q})}\right]. \quad (3.21)$$

Before discussing this integral, we will next consider the polarization kernel $K(q)$ for large q. Without expanding Eq. (3.18) in q, we have (for the transverse part)

$$K_{ij}^{(2)}(q)=\frac{1}{(2\pi)^3}\frac{e^2}{m^2}\int l_il_j\frac{1}{E(\mathbf{l})+E(\mathbf{l}+\mathbf{q})}$$
$$\times\left[1-\frac{\epsilon(\mathbf{l})\epsilon(\mathbf{l}+\mathbf{q})+\phi^2}{E(\mathbf{l})E(\mathbf{l}+\mathbf{q})}\right]d^3l, \quad (3.22)$$

which depends essentially on the same integral that appeared in Eq. (3.21). This is not an accident, as will be shown in a more general way in Sec. 5, where we discuss the equivalence of our method to the variational procedure.

Equation (3.22) has been evaluated by Bardeen, Cooper, and Schrieffer.[4] Their result is

$$K_{ij}(q)=K_{ij}^{(1)}(q)+K_{ij}^{(2)}(q)=-\frac{ne^2}{m}\delta_{ij}+K_{ij}^{(2)}(q)$$
$$=-[\delta_{ij}-q_iq_j/q^2]\frac{3\pi^2}{4}\frac{ne^2\phi}{qmv_F}\left[1-\frac{16\phi}{\pi^2qv_F}\ln(\pi q\xi_0)\right]$$
$$+[K_{ij}(q)]_{\phi=0}. \quad (q\xi_0\gg1) \quad (3.23)$$

Making use of Eq. (3.23), the quantity $[\Sigma_1^{(2)}(\mathbf{p})]_{\tau_1}$ can be easily evaluated. We find

$$[\Sigma_1^{(2)}(p)]_{\tau_1}=-\tau_1\frac{1}{4}\left(\frac{\rho}{N}\right)\frac{3\pi^2}{4}\frac{ne^2}{qmv_F}\times\left[1-\frac{32\phi}{\pi^2 qv_F}\ln(\pi q\xi_0)+\frac{16\phi}{\pi^2 qv_F}\right]. \quad (3.24)$$

This, together with Eqs. (3.8) and (3.10), yields finally

$$\delta\phi=-\frac{\pi^2}{8}e^2v_F\sum_q\frac{1}{q}\left[1-\frac{16\phi}{\pi^2 qv_F}\{2\ln(\pi q\xi_0)-1\}\right]|\mathbf{A}(\mathbf{q})|^2. \quad (3.25)$$

4. PROOF OF GAUGE INVARIANCE

The proof of gauge invariance of our theory is completely general as well as simple. Our procedure will be to show that (a) there exist Ward identities[9] as rigorous manifestations of the gauge invariance, and (b) our Hartree-Fock approximation satisfies these identities.

The (generalized) Ward identities are relations between successive terms of the expansion of the Green's function $G^{(A)}$ in the field A as was carried out in Sec. 2. We start from the gauge invariance condition

$$e^{ie\lambda(x)\tau_3}G(x,y)e^{-ie\lambda(y)\tau_3}\equiv e^{ie\lambda\tau_3}Ge^{-ie\lambda\tau_3}=G^{(A')},\quad A_\mu'=A_\mu+\partial_\mu\lambda. \quad (4.1)$$

By expanding both sides in λ after putting $A_\mu=0$, and using the results of Sec. 2,[10] we first obtain

$$\begin{aligned}ie[\lambda\tau_3,G]&=\int\frac{\delta G}{\delta A_\mu(z)}\partial_\mu\lambda(z)d^4z\\&=ie\int G\Gamma_\mu G\partial_\mu\lambda(z)d^4z\\&=-ie\int G\partial_\mu\Gamma_\mu G\lambda(z)d^4z. \quad (4.2)\end{aligned}$$

In the written-out form,

$$\tau_3\lambda(x)G(x,y)-G(x,y)\tau_3\lambda(y)=-\int\int\int G(x,x')\times\frac{\partial\Gamma_\mu}{\partial z_\mu}(x',y';z)G(y',y)\lambda(z)d^4x'd^4y'd^4z. \quad (4.2')$$

This is the equivalent of the Ward-Takahashi identity for the vertex function Γ_μ. Multiplying (4.2) by G^{-1} from the left and from the right, we obtain also

$$[G^{-1},\tau_3\lambda]=-\int\frac{\partial\Gamma_\mu}{\partial z_\mu}\lambda(z)d^4z. \quad (4.3)$$

This shows that the current is divergenceless when taken between real one-particle states for which we have $G^{-1}=0$.

The above relation has already been utilized in the previous paper I for the proof of gauge invariance of the Meissner effect. For the present problem we need an analogous relation for the two-photon vertex $\Gamma_{\mu\nu}$. This can be done by considering $\delta G/\delta A_\mu$ instead of G in Eqs. (4.2) and (4.3):

$$ie\left[\lambda\tau_3,\frac{\delta G}{\delta A_\mu(z)}\right]=\int\frac{\delta}{\delta A_\nu(z')}\left[\frac{\delta G}{\delta A_\mu(z)}\right]\partial_\nu\lambda(z')d^4z',$$

or

$$[\lambda\tau_3,G\Gamma_\mu G]=-\int\partial_\nu'L_{\mu\nu}\lambda'd^4z'. \quad (4.4)$$

Now

$$\begin{aligned}\int\partial_\nu'L_{\mu\nu}\lambda'dz'&=\int\{G\partial_\nu'\Gamma_\nu G\Gamma_\mu G+G\Gamma_\mu G\partial_\nu'\Gamma_\nu G+G\partial_\nu'\Gamma_{\mu\nu}G\}\lambda'dz'\\&=-[\lambda\tau_3,G]\Gamma_\mu G-G\Gamma_\mu[\lambda\tau_3,G]+i\int G\partial_\nu'\Gamma_{\mu\nu}G\lambda'd^4z'\\&=-[\lambda\tau_3,G\Gamma_\mu G]+G[\lambda\tau_3,\Gamma_\mu]G+i\int G\partial_\nu'\Gamma_{\mu\nu}G\lambda'd^4z',\end{aligned}$$

where Eq. (4.2) is utilized. Comparing this with the left-hand side of Eq. (4.4), we get

$$[\lambda\tau_3,\Gamma_\mu]=-i\int\partial_\nu'\Gamma_{\mu\nu}\lambda'd^4z', \quad (4.5)$$

which is the desired relation for $\Gamma_{\mu\nu}$. The second-order scattering amplitude of the quasi-particle by $A(q)$ is $G^{-1}L_{\mu\nu}G^{-1}A_\mu A_\nu'$. Replacing, say, A_ν' by a longitudinal potential $\partial_\nu'\lambda'$ results in a zero matrix element on the mass shell in view of Eq. (4.4), as is physically required. This is achieved by the cancellation of the three terms making up $L_{\mu\nu}$.

Our next task is to show the consistency of the Hartree-Fock solution with gauge invariance. We expect the self-energy equation (2.9) to satisfy the gauge condition. In other words, a solution $G^{(A)}$ (or $\Sigma^{(A)}$) of Eq. (2.9) must satisfy Eq. (4.1), and consequently solutions Γ_μ, $\Gamma_{\mu\nu}$ of Eq. (2.12) must satisfy the corresponding Ward identities (4.3) and (4.5). This can easily be seen to be the case by taking the latter identities as an ansatz for $\partial_\mu\Gamma_\mu$, $\partial_\nu'\Gamma_{\mu\nu}$. In fact, assume Eqs. (4.2), (4.3) and put this into the right-hand side

[9] J. C. Ward, Phys. Rev. 78, 182 (1950); Proc. Phys. Soc. (London) A64, 54 (1951). Y. Takahashi, Nuovo cimento 6, 370 (1957).

[10] The formulation in this section will be written in four-vector notation; $\mu=1, 2, 3, 4$. It is evident that Eqs. (2.10)–(2.12) of Sec. 2 can be easily recast in the four-verctor form.

of the integral equation for $\partial_\mu\Gamma_\mu$:

$$\partial_\mu\Lambda_\mu = i(\tau_3 G\partial_\mu\Gamma_\mu G\tau_3)V. \tag{4.6}$$

We get

$$\int (\tau_3 G\partial_\mu\Gamma_\mu G\tau_3)V\lambda d^4z = -(\tau_3[\tau_3\lambda,G]\tau_3)V$$
$$= -[\tau_3\lambda,(\tau_3 G\tau_3 V)] = [\tau_3\lambda,\Sigma].$$

But this is exactly the relation (4.3) assumed for the right-hand side of Eq. (4.6). Although we took an integral of Eq. (4.6) with $\lambda(z)$, it is equivalent to the unintegrated form since $\lambda(z)$ is arbitrary. Namely, by taking $\lambda(z')=\delta(z'-z)$, we find that

$$(\partial\Gamma_\mu/\partial z_\mu)(x,y;z)$$
$$= i[G^{-1}(x,y)\tau_3\delta^4(y-z) - \tau_3\delta^4(x-z)G^{-1}(x,y)]$$

is a solution of Eq. (4.6).

Similarly, we assume Eq. (4.5) for $\partial_\nu'\Gamma_{\mu\nu}$, which is equivalent to Eq. (4.4), and substitute it into the equation for $\partial_\nu'\Gamma_{\mu\nu}$:

$$\partial_\nu'\Lambda_{\mu\nu} = -(\tau_3\partial_\nu' L_{\mu\nu}\tau_3)V. \tag{4.7}$$

The verification of consistency is equally straightforward.

In this way we have shown that the self-energy equation, in particular its expanded form (2.12), has Ward identities as solutions. So we can be sure that there are gauge-invariant solutions for Γ_μ and $\Gamma_{\mu\nu}$. To the extent that the integral equations for Γ_μ and $\Gamma_{\mu\nu}$ are assumed to have unique solutions, gauge invariance is then guaranteed.

The fact that the superconductive solution does actually lead to gauge-invariant results was shown in the previous paper I explicitly for the vertex Γ_μ. It was found that there exist collective excitations of the sound-wave type which are strongly coupled to the longitudinal part of the vertex, and thereby forces the Ward identity (or current conservation) to be satisfied. Essentially the same argument will go through for $\Gamma_{\mu\nu}$.

According to the earlier work, I, $\Gamma_\mu(p',p)$ has the form (assuming no Coulomb effect which eliminates the low-lying collective modes)

$$\Gamma_i(p',p) = [(p+p')_i/2m] + [2i\tau_2\alpha^2\phi q_i/(q_0^2-\alpha^2\mathbf{q}^2)],$$
$$\Gamma_0(p',p) = \tau_3 + [2i\tau_2\phi q_0/(q_0^2-\alpha^2\mathbf{q}^2)],$$

where the second term is coupled to the collective states with the dispersion law $q_0^2=\alpha^2\mathbf{q}^2$. Substituting this into Eq. (4.5) we find that there must be in $\partial_\nu'\Gamma_{\mu\nu}$ a term proportional to τ_1 and coupled to the collective states. Since the collective states must contribute to each of the vertices in $\Gamma_{\mu\nu}$, we may assume that $\Gamma_{\mu\nu}(p,p;q,-q)$ contains a term $-4\tau_1\phi h_\mu h_\nu/(q_0^2-\alpha^2q^2)^2$ where $h_\mu=(\alpha^2 q_i,q_4)$, so that this will match the contribution from the collective term in Γ_μ when inserted in the Ward identity (4.5). Actually, the τ_2 term in Γ_μ should contain also contributions from the noncollective part which does not become singular as q_0, $q\to 0$. This implies that $\Gamma_{\mu\nu}$ will, in general, have terms $\propto \tau_1 h_\mu h_\nu/(q_0^2-\alpha^2q^2)$ and $\tau_1 h_\mu h_\nu$. In fact, we have found that Γ_{ij} has a term $\propto \tau_1\delta_{ij}q^2$ which would produce the field dependence of the energy gap. Such a term will be combined with the longitudinal term into $\tau_1(\delta_{ij}q^2-q_iq_j)$ to satisfy gauge invariance, although we have not verified this by explicit calculation.

5. RELATION TO THE SIMPLE VARIATIONAL METHOD

The results obtained in Sec. 3 about the magnetic field dependence of the energy gap can also be derived, in fact in a simpler way, from the variational method originally used in the BCS paper. What we have to do is to minimize the total energy of the system in a magnetic field with respect to the energy gap parameter.[11] The energy change due to the magnetic field is

$$E_A = -\frac{1}{2}\sum_q \frac{K(q)}{1-K(q)/q^2}|A_i^{\mathrm{ex}}(q)|^2, \tag{5.1}$$

to the second order in $\mathbf{A}^{\mathrm{ex}}$. Here $K(q)$ is the kernel in the London-Pippard relation

$$j_i(q) = +K(q)A_i(q). \tag{5.2}$$

The total field $\mathbf{A}$ is related to the external field $\mathbf{A}^{\mathrm{ex}}$ by

$$A_i(q) = A_i^{\mathrm{ex}}(q)/[1-K(q)/q^2]. \tag{5.3}$$

The kernel K has the form

$$K(q) = -(1/\Lambda)f(q^2), \quad 1/\Lambda = ne^2/m. \tag{5.4}$$

The form factor $f(q)$ describes the nonlocal nature of the relation between j and A with a characteristic coherence length $\xi_0 = v_F/\pi\phi$. In other words,

$$f(q) = 1 - C(\xi_0 q)^2 + O((q\xi_0)^4). \tag{5.5}$$

The dependence of E_A on the gap parameter ϕ enters then through the coherence length ξ_0 only. There will be no field dependence of the energy gap in the London limit.

The total energy $E=E_0+E_A$ of the system is then minimized with respect to ϕ. Let ϕ_0 be the solution which minimizes E_0 so that $\phi=\phi_0+\delta\phi$. Then

$$E = E_0(\phi_0) + E_A(\phi_0) + \frac{1}{2}\frac{d^2E_0(\phi_0)}{d\phi_0^2}(\delta\phi)^2 + \frac{dE_A(\phi_0)}{d\phi_0}\delta\phi, \tag{5.6}$$

since E_0 has a minimum at ϕ_0. The new minimum now is displaced by

$$\delta\phi = -\frac{dE_A}{d\phi_0}\Big/\frac{d^2E_0}{d\phi_0^2}. \tag{5.7}$$

With the Hartree wave function of BCS, $d^2E_0/d\phi_0^2$ is

[11] This has been carrried out by J. Bardeen (private communication).

easily calculated to be

$$\frac{d^2E_0}{d\phi_0^2}=\tfrac{1}{2}\sum_p\frac{\epsilon_p^2}{E_p^3}\sum_p\frac{\phi^2}{E_p^3}\bar{V}$$
$$=2N(1-\rho)\approx 2N;$$
$$\frac{dE_A}{d\phi_0}=\tfrac{1}{2}\sum_q\left[\frac{1}{1-K/q^2}\right]^2\frac{2C\Lambda^{-1}(q\xi_0)^2}{\phi_0}|\mathbf{A}^{ex}(\mathbf{q})|^2 \qquad (5.8)$$
$$=\tfrac{1}{2}\sum_q 2C\Lambda^{-1}(q\xi_0)^2|\mathbf{A}(\mathbf{q})|^2;$$
$$N=mp_F/2\pi^2,\quad \Lambda^{-1}=e^2n/m=e^2p_F^3/3\pi^2m,\quad C=\pi^2/30.$$

Here the value of C is taken from Eq. (3.19). This yields

$$\delta(\phi^2)=-\sum_q[C(q\xi_0)^2/\Lambda N]|\mathbf{A}(\mathbf{q})|^2$$
$$=-\sum_q(\pi^2/45)e^2v_F^2(q\xi_0)^2|\mathbf{A}(\mathbf{q})|^2 \qquad (5.9)$$

which agrees with Eq. (3.12).

That this agreement is not accidental can be seen from the following observation. In Sec. 3 we have seen that the quantity $\delta\phi$ is the τ_1-proportional term in $\Sigma^{(2)}(p)/\rho$, where

$$\Sigma^{(2)}(p)=-\frac{1}{(2\pi)^4}\int V(p-l)\tau_3S(l)\tau_3d^4l,$$
$$S(l)=\tfrac{1}{2}\sum_q[G(l)\Gamma_\mu(-q)G(l+q)\Gamma_\nu(q)G(l)$$
$$+G(l)\Gamma_\nu(q)G(l-q)\Gamma_\mu(-q)G(l)]A_\mu(-q)A_\nu(q). \qquad (5.10)$$

Now take

$$\frac{1}{2}\frac{1}{(2\pi)^3}\int \mathrm{Tr}\frac{1}{2E_p}\tau_1\Sigma^{(2)}(\mathbf{p})d^3p$$
$$=\frac{1}{(2\pi)^3}(\rho\delta\phi)\int\frac{d^3p}{2E_p}=N\delta\phi, \qquad (5.11)$$

where the constancy of $\delta\phi$ near the Fermi surface is taken into account. On the other hand, the same quantity is equal to

$$-\frac{1}{2}\frac{1}{(2\pi)^7}\int\int \mathrm{Tr}\left[\tau_1\frac{1}{2E_p}V(p-l)\tau_3S(l)\tau_3\right]d^3pd^4l$$
$$=-\frac{1}{2}\frac{1}{(2\pi)^4}\int \mathrm{Tr}[\tau_1S(l)]d^4l. \qquad (5.12)$$

In view of the definition of $S(l)$ and

$$G(l)\tau_1G(l)=i(d/d\phi)G(l),$$

this in turn is

$$=-\frac{i}{4}\frac{1}{(2\pi)^4}\sum_q\left\{\frac{d}{d\phi}\int \mathrm{Tr}[\Gamma_\mu G(l+q)\Gamma_\nu G(l)]d^4l\right\}$$
$$\times A_\mu(-q)A_\nu(q),$$
$$=+\frac{1}{4}\sum_q\left\{\frac{d}{d\phi}K_{\mu\nu}(q)\right\}A_\mu(-q)A_\nu(q)\to$$
$$\frac{1}{4}\sum_q\left\{\frac{d}{d\phi}K(q)\right\}|A(q)|^2=-\frac{1}{2}\frac{d}{d\phi}E_A\Big|_{A^{ex}}, \qquad (5.13)$$

according to the definition of the kernel K. Note that the last line follows from Eqs. (5.1) and (5.3). Thus combining Eqs. (5.8), (5.11), and (5.13) we get

$$\delta\phi=-\frac{1}{2N}\frac{dE_A}{d\phi_0}=-\frac{dE_A}{d\phi_0}\Big/\frac{d^2E_0}{d\phi_0^2},$$

in agreement with Eq. (5.7).

CONCLUDING REMARKS

Our results on the magnetic field dependence $\delta\phi$ of the energy gap, to the second order in the field, are given by Eqs. (3.12) and (3.25). They are in disagreement with the results obtained by Gupta and Mathur. A characteristic point of our formula is that $\delta\phi$ depends only on the field strength H for small q. According to Gupta and Mathur, on the other hand, $\delta\phi$ depends on the potential A. For large q, both results agree qualitatively, apart from different numerical coefficients.

Since these calculations have been done for an infinite medium, there remains the question about how to apply them to practical cases where thin films or colloidal particles have been used. It would seem appropriate to treat this more interesting problem in a separate paper.

ACKNOWLEDGMENTS

One of us (S. F. T.) acknowledges the partial support of the Advanced Research Projects Agency (ARPA) during this work and wishes to thank Professor R. W. Morse for his interest and courtesy.

Reprinted from THE PHYSICAL REVIEW, Vol. 133, No. 1A, A1–A14, 6 January 1964

Considerations on the Magnetic Field Problem in Superconducting Thin Films*

YOICHIRO NAMBU
The Enrico Fermi Institute for Nuclear Studies and Department of Physics, The University of Chicago, Chicago, Illinois

AND

SAN FU TUAN
Department of Physics, Purdue University, Lafayette, Indiana
(Received 22 July 1963)

We adapt the results obtained in a previous paper on the magnetic field dependence of the energy gap in superconductivity for bulk specimen to thin-film superconductors, using the model of discrete quantization in momentum space. Only the case of parallel and constant external magnetic field along the film surfaces is considered. A series of elementary theorems and some specific calculations lead to the conclusions: (1) A second-order phase transition should be observed at all temperatures for thin film thicknesses. (2) A simple scaling rule exists concerning the field and temperature dependence on the energy gap. (3) The critical field H_c depends on thickness L and reduced temperature t like $H_c \sim L^{-3/2}[\ln(1/t)]^{1/2}$ for not too thin films ($L \gtrsim 0.5 \times 10^{-5}$ cm). The behavior changes as the film becomes very thin or as the temperature becomes moderately low. A crude comparison with available experimental data seems to bear out our conclusions qualitatively.

1. INTRODUCTION

IN a previous paper[1] we derived expressions for the magnetic field dependence of the energy gap in superconductivity for bulk matter. The theoretical assumptions here are that we can somehow introduce a varying magnetic field into the bulk medium and that it makes sense to talk about energy gap depending on the field; of course, for a constant imposed field, the Meissner effects tells us that there is in fact no change in the gap due to the field. Though in some sense the discussion of bulk matter calculations is an academic problem, nevertheless it supplies a useful test model from which we can draw physically realistic conclusions about the case of superconducting thin films—which is directly amenable to experimental corroboration.

It was pointed out earlier[2] how we might expect to apply bulk material results to the actual experimental details of thin-film specimen. We can still introduce momentum pairs, though these are now quantized or discrete at least in one direction, and instead of integrating over momentum variable k (as for the case of infinite or bulk medium)—we have summation over momentum states. We are typically dealing with samples of thickness comparable or smaller than the penetration depth ($\sim 5 \times 10^{-6}$ cm), such that it allows the magnetic field to penetrate the body without much attenuation, yet it is still large compared to the atomic scale in order not to alter drastically the basic dynamics of superconductivity. Thus, we shall not concern ourselves with the intrinsic change of properties due to finite thickness, but only the changes induced by the presence of the magnetic field relative to the field-free case.

Under such circumstances, the bulk material results can be applied by properly observing the discrete

* This work supported by the U. S. Atomic Energy Commission, the U. S. Air Force Office of Scientific Research, the National Science Foundation.

[1] Y. Nambu and S. F. Tuan, Phys. Rev. **128**, 2622 (1962), hereafter denoted I for reference purposes. See also Y. Nambu, Phys. Rev. **117**, 648 (1960). A brief account of the present work is given in Y. Nambu and S. F. Tuan, Phys. Rev. Letters **11**, 119 (1963).

[2] Y. Nambu and S. F. Tuan, in Proceedings of the Eighth International Conference on Low Temperature Physics, London, 1962 (to be published).

momentum for a thin film, but otherwise considering a simulated bulk material formed by a sequence of thin films (of thickness L) placed side by side in which the magnetic field runs parallel to the layers with alternating directions. The Fourier transform $A(q)$ of the vector potential across such superconducting film layers is of the form

$$A(q)\sim 1/q^2, \quad \text{where} \quad q=n\pi/L, \quad n=1,2,3,\cdots.$$

Thus, the lowest q is $=\pi/L$, which becomes large as L becomes small. The need for high q values (Pippard limit) is evident. In Sec. 2, this discrete quantization model is formulated for both the case of thick and thin films and the question of phase transition (energy or magnetic field versus gap ϕ diagrams) at $T=0$°K is briefly discussed.

Section 3 treats the perturbation calculation at $T=0$ for superconducting thin films in the framework of the discrete or lamination model. We shall find that it is essential to take into account the correct gauge and electronic boundary conditions to arrive at a result valid for very small gap or thickness.

In Sec. 4 we derive a formula [Eqs. (4.21) and (4.22)] at zero temperature which is based on the previous perturbation result but can be expected to hold for stronger magnetic fields. It shows the energy gap to decrease steadily with increasing field without ever vanishing until the whole calculation breaks down at extremely high fields. We cannot determine the critical field at zero temperature in our framework.

The above formula can be extended to finite temperatures without carrying out calculations in detail. We shall show in Sec. 5 that there exists an approximate scaling rule with respect to the temperature and magnetic field dependence of the gap, which enables one to treat the finite temperature case more easily than at zero temperature. We find a second-order phase transition at a finite critical field H_c, which depends on thickness L and reduced temperature as $L^{-3/2}(\ln 1/t)^{1/2}$ for not too thin films.

In Secs. 6 and 7 our results are compared with available experiments as well as with other theoretical calculations, notably by Bardeen,[3] by Douglass[4] in the Ginzburg-Landau-Gor'kov (G-L-G) theory,[5] and by Mathur *et al.*[6] The measurements of Tinkham and Morris[7] on thermal conductivity of lead and those of Douglass and Meservey[8] using the more direct tunnel approach on lead, do suggest that for film thicknesses in the range 500 to 1000 Å and reduced temperatures down to 0.12, a second-order phase transition is observed. This agrees with our predictions; moreover, the theoretical curves are in reasonable agreement with the data.

The concluding remarks of Sec. 8 evaluate the problems which confront the present study on thin films, in particular the questions of boundary conditions, a proper and more elaborate extension to finite temperatures, and the case of uneven or nonparallel magnetic fields. Suggestions are made which will test most critically the notions here outlined.

[3] J. Bardeen, Rev. Mod. Phys. **34**, 667 (1962).
[4] D. H. Douglass, Jr., Phys. Rev. Letters **6**, 346 (1961); Phys. Rev. **124**, 735 (1961).
[5] L. P. Gor'kov, Zh. Eksperim. i Teor. Fiz. **36**, 1918 (1959) [translation: Soviet Phys.—JETP **9**, 1364 (1959)].
[6] V. S. Mathur, N. Panchapakesan, and R. P. Saxena, Phys. Rev. Letters **9**, 374 (1962).
[7] D. E. Morris, Ph.D. thesis, 1962 (unpublished).
[8] D. H. Douglass, Jr., and L. M. Falicov, Progress in Low Temperature Physics (to be published); also private communications.

2. THE DISCRETE QUANTIZATION MODEL

We consider a parallel thin film of macroscopically large dimensions along Oy and Oz, and of thickness L along Ox [Fig. 1(a)]. Equal and parallel external magnetic fields H are applied in the plane of film surfaces at $x=0$ and $x=L$. With the boundary conditions $\psi_n(x,y,z)=0$ at $x=0$ and $x=L$, the single electron wave functions $\psi_n(x,y,z)$ are

$$\psi_n(x,y,z)=(2/L)^{1/2}\sin(n\pi x/L)\psi(y)\psi(z), \quad (2.1)$$

with

$$p_{nx}=n\pi/L>0, \quad n=1,2,3,\cdots, \quad (2.2)$$

and so our discrete jump in p_{nx} is $\Delta p_{nx}=\pi/L$, which is still much smaller than the Fermi momentum p_F for typical thin-film values of L (of order 100 to 1000 Å). Thus, we expect that the concept of a quasi-infinite medium should be applicable.

The Meissner effect relation[9] for the current j is

$$j_i(\mathbf{q})=\sum_{j=1}^{3} K_{ij}(\mathbf{q})A_j(\mathbf{q}),$$

with

$$K_{ij}(\mathbf{q})=-(ne^2/m)\delta_{ij}+K_{ij}^{(2)}(\mathbf{q}). \quad (2.3)$$

Here n is the number of conduction electrons, of both spin directions per unit volume. We are interested in the ϕ (energy gap) dependent part of $K(\mathbf{q})$, viz. $K^{(2)}(\mathbf{q})$;

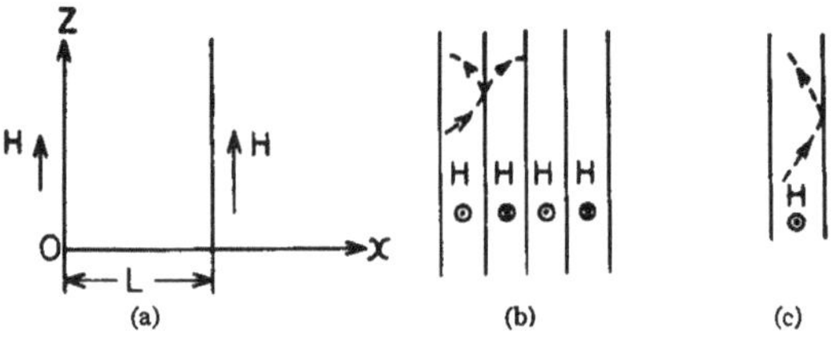

FIG. 1. (a) Thin film of thickness L along Ox, with equal and parallel magnetic fields at film surfaces. (b) Parallel thin films, side by side, with alternating fields H applied in the body of the film. (c) The mathematically equivalent single film, with magnetic field H and specular reflection at film boundaries.

[9] More precisely, Eq. (2.3) must be written

$$j_i(\mathbf{q})=\Sigma_{\mathbf{q}'}\Sigma_{j=1}^{3}K_{ij}(\mathbf{q},\mathbf{q}')A_j(\mathbf{q}').$$

Actually, only $\mathbf{q}'=\mathbf{q}$ gives a dominant contribution as we see from Eqs. (2.6) and (2.7) below. (All unexplained notations are the same as those in Ref. 1 throughout this paper.)

this is given to the second order in field strength by

$$\sum_k K_{ik}^{(2)}(\mathbf{q})A_k(\mathbf{q})=\sum_{\mathbf{p}}\frac{\langle\mathbf{p}|j_i|\mathbf{p}+\mathbf{q}\rangle\langle\mathbf{p}+\mathbf{q}|\mathbf{j}\cdot\mathbf{A}|\mathbf{p}\rangle}{E(\mathbf{p})+E(\mathbf{p}+\mathbf{q})}+\text{crossed terms.} \quad (2.4)$$

$E(p)=[\epsilon(p)^2+\phi^2]^{1/2}$ is the quasiparticle energy. It will be convenient at this stage to work in the London gauge. Since the induced current perpendicular to the plane of film is zero, the external vector potential in this gauge is then just $A(x,y,z)=(0,H(x-L/2),0)$. $A_y(x)$ is chosen to be an odd function with respect to the middle plane of the film. If we want the London equation in the simple form $j\propto A$, this is the proper choice since the induced current should also have the same symmetry when the magnetic field is equal on both sides of the film. However, an additive constant to $A(x)$ does not affect the final result (see below). Equation (2.4) then simplifies to (putting $K_{yy}\equiv K$)

$$K^{(2)}(\mathbf{q})|\mathbf{A}(\mathbf{q})|^2=\sum_{p,\pm}\frac{|\langle\mathbf{p}\pm\mathbf{q}|j_yA_y|\mathbf{p}\rangle|^2}{E(\mathbf{p})+E(\mathbf{p}\pm\mathbf{q})}. \quad (2.5)$$

Here the denominator $E(\mathbf{p}+\mathbf{q})+E(\mathbf{p})$ is just the energy of the quasiparticle pair (hole+particle) created in the intermediate state.

It is evident from Eq. (2.5) that we need to evaluate general matrix elements of the form $\langle\mathbf{p}'|(x-L/2)j_y|\mathbf{p}\rangle$. Using Eq. (2.1) we get

$$\langle\mathbf{p}'|(x-L/2)j_y|\mathbf{p}\rangle=\frac{2p_y}{L}\int_0^L\sin(p_x'x)\sin(p_xx)(x-L/2)\times dx\delta(p_y'-p_y)\delta(p_z'-p_z)$$
$$=\frac{p_y}{L}\left\{\frac{1}{(p_x'-p_x)^2}-\frac{1}{(p_x'+p_x)^2}\right\}\{1-(-1)^{n'-n}\}\times\delta(p_y'-p_y)\delta(p_z'-p_z). \quad (2.6)$$

Comparison of Eqs. (2.5) and (2.6) gives, putting $p'=p+q$, the only nonvanishing $A_y(\mathbf{q})=A_y(q_x,0,0)$:

$$A_y(\mathbf{q})=\frac{-2H}{L}\left\{\frac{1}{q_x^2}-\frac{1}{(2p_x+q_x)^2}\right\}\approx-\frac{2H}{L}\frac{1}{q_x^2}, \quad (2.7)$$
$$q_x=\pm(2r+1)\pi/L,\quad r=0,1,2,3,\cdots.$$

The term $q_x=0$, or $p=p'$ (no scattering) does not appear, and the approximate equality in Eq. (2.7) is generally satisfied for $(p_x+p_x')^2\gg(p_x-p_x')^2$. Note that for the smallest $|q_x|=|p_x'-p_x|=\pi/L$, $(p_x'+p_x)^2>9(p_x'-p_x)^2$.

We have here evaluated the matrix elements in the single electron picture. However, it is completely evident that the transition p to p' does not depend on the details of paired or unpaired states, but rather it is a special nature of the wave function and properties of Fourier transform in going from space variables to momentum variables. Information about the Cooper pairing is inherent in the kernel $K(\mathbf{q})$.

Equation (2.7) is actually an expression for the external vector potential $\mathbf{A}^{ex}(\mathbf{q})$. The physically interesting expression is that of the energy change $\mathcal{E}_A$ due to the magnetic field given by Eq. (5.1) of I

$$\mathcal{E}_A=-\tfrac{1}{2}\sum_{\mathbf{q}}\frac{K(\mathbf{q})}{1-K(\mathbf{q})/q^2}|\mathbf{A}^{ex}(\mathbf{q})|^2 \quad (2.8)$$

to the second order in $A^{ex}(\mathbf{q})$. We see at once that the component $\mathbf{A}(\mathbf{q}=0)$ does not contribute because $K(0)\neq0$ and, hence,

$$K(q)/[1-K(q)/q^2]\to0$$

as $q^2\to0$. This means that the ambiguity of a constant additive term in $A(x)$ is effectively eliminated.

The discussion thus far is restricted to a single thin film. In order to establish connection with the bulk matter case, we prove the following theorem.

Theorem 1. Bulk medium, consisting of parallel layers of thin films placed side by side with alternating magnetic fields (uniform within its periodicity), is *mathematically* equivalent to a single thin film in a uniform magnetic field H, where the electrons confined in the film undergo specular reflection scattering at the film boundaries, except for the difference that in the latter the electron motion has discrete quantization across the film.

Figures 1(b) and 1(c) exhibit the mathematical equivalence in diagrams. An intuitive *physical* interpretation of the theorem is as follows. In the classical sense, an electron follows a curved path inside a film and then reflected at the surface. If we take the mirror image of the reflected motion with respect to the surface, the electron effectively passes into the next film without suffering reflection, but there the curvature is reversed, i.e., the magnetic field within the next film appears to be reversed. We remark that this picture is valid under the adiabatic condition where we neglect scattering of electrons at points of periodicity.

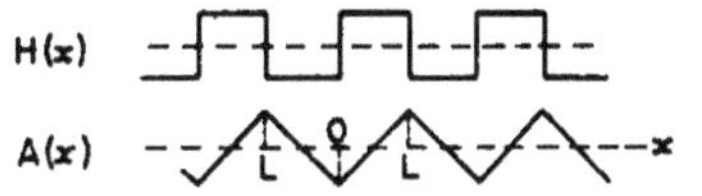

FIG. 2. The distribution of external field and vector potential as a function of distance along Ox for the laminated bulk medium.

Proof. Figure 2 shows the distribution of external field and vector potential as a function of distance along Ox for the laminated bulk medium. It is evident that (writing $A_y\equiv A$)

$$A^{ex}(x)=1/L\sum_{n=0}^{\infty}C_n\cos(n\pi x/L). \quad (2.9)$$

Since $A^{ex}(x)$ is an even function, we have

$$C_n=(2/L)\int_0^L (x-L/2)\cos(n\pi x/L)dx$$
$$=-2/L\left(\frac{L}{n\pi}\right)^2[1-(-1)^n]. \quad (2.10)$$

We see that the even terms vanish identically. The running wave form of the Fourier expansion of $A^{ex}(x)$ is

$$A^{ex}(x)=(1/L)\sum_{n=-\infty}^{\infty}(C_n/2)e^{in\pi x/L}. \quad (2.11)$$

Thus, it is $C_n/2$ which is to be equated with the usual Fourier component $A(q_{nx})$ of $A^{ex}(x)$. Hence,

$$A(q_{nx})=C_n/2=-2/L(L/n\pi)^2H,$$
$$n=2r+1,\ r=0,\ 1,\ 2,\ 3,\ \cdots. \quad (2.12)$$

Equation (2.12) is completely equivalent to Eq. (2.7), provided the term $(2p_x+q_x)^{-2}$ can be neglected, Q.E.D.

We have established the theorem under the assumption that H is constant within its periodicity; in fact the theorem holds for an arbitrary field $H(x)$,[10] as is physically desirable since for films of finite thickness the field inside is generally x-dependent and different from the surface field. The necessary modification requires us to replace $A^{ex}(x)$ by a generalized function $\tilde{A}(x)$, such that

$$\tilde{A}(x)=(2/L)\sum_q A(q)\cos qx.$$

$\tilde{A}(x)$ is an even function of x, with periodicity $2L$, so that the general field $H(x)$ is odd (and hence alternating) but also of period $2L$.

The study of phase transition at $T=0°$K proceeds along the lines of an investigation of the total energy $\mathcal{E}$ versus energy gap parameter ϕ. Here $\mathcal{E}=\mathcal{E}_0+\mathcal{E}_A$, where $\mathcal{E}_0$ is the nonmagnetic BCS ground-state energy relative to the energy of the Fermi sea and $\mathcal{E}_A$ is the magnetic energy calculated from Eq. (2.8). It can be readily shown that $\mathcal{E}_0(\phi)$ in the weak-coupling approximation ($\phi \ll$ the Debye frequency cutoff ω) is

$$\mathcal{E}_0(\phi)=N\phi^2\{\ln(2\omega/\phi)-\tfrac{1}{2}\}-N\rho\phi^2\{\ln(2\omega/\phi)\}^2. \quad (2.13)$$

N is the density of Bloch states of one spin per unit energy at Fermi surface, and $\rho=N\bar{V}$($\bar{V}$ the effective electron-electron interaction). Using Eqs. (2.7) and (2.8), $\mathcal{E}_A$ can be written as

$$\mathcal{E}_A=\tfrac{1}{2}\left(\frac{2H}{L}\right)^2\sum_{q=\pm(2r+1)\pi/L}\left[\frac{1}{q^2}-\frac{1}{q^2+f(q)/\lambda^2}\right]. \quad (2.14)$$

In Eq. (2.14) we have written, following I, $K(\mathbf{q})=(-1/\lambda^2)f(q)$, where λ is the London penetration depth at $T=0°$K: $\lambda=(m/ne^2)^{1/2}$, and $f(q)$ describes the nonlocal nature of the relation between j and A with a characteristic bulk coherence length $\xi_0=v_F/\pi\phi$. In other words, f can be expanded as

$$f(q)=1-C(\xi_0 q)^2+O[(\xi_0 q)^4]. \quad (2.15)$$

The function $f(q)$, in general, decreases with increasing q, and the sum in Eq. (2.14) converges rapidly. For a thick film such that

$$\lambda\pi/L\ll 1 \quad \text{and} \quad \pi\xi_0/L\ll 1,$$

the contribution predominantly comes from the first term in the summand only. Thus,

$$\mathcal{E}_A=\left(\frac{2H}{L}\right)^2\sum_{r=0}^{\infty}\left(\frac{L}{\pi}\right)^2\frac{1}{(2r+1)^2}=\tfrac{1}{2}H^2. \quad (2.16)$$

Typically, $\lambda\sim 5\times 10^{-6}$ cm and $\xi_0\sim 10^{-4}$ cm, so this result applies to thicknesses large compared to 10^{-4} cm.

For such thick films, we see that a first-order phase transition is expected for $\phi=\phi_c$ ($\neq 0$) such that $\mathcal{E}_0(\phi_c)+\mathcal{E}_A(\phi_c)=0$ [since $\mathcal{E}_A$(normal)~ 0], i.e.,

$$\tfrac{1}{2}H_c^2=-\mathcal{E}_0(\phi_c). \quad (2.17)$$

Qualitatively, this type of behavior can be understood from a study of the $\mathcal{E}$ versus ϕ diagram shown in Fig. 3(a). $\mathcal{E}_A(\phi)$ is insensitive to ϕ for a thick film (London limit in bulk matter), so point C is the minimum position (little changed from zero-field value ϕ_0) for $\mathcal{E}$ and point O is the normal state position. As the external field H increases, point C rises above point O, so an abrupt change of phase from superconducting to normal state takes place at $\phi_c\sim\phi_0$.

For thin films such that $L/\lambda\pi\lesssim 1$ or $L\lesssim 10^{-5}$ cm, we have to use the Pippart form of kernel $K(q)$ given by[1]

$$K(q)=(-1/\lambda^2)f(q,\phi),$$
$$f(q,\phi)=\frac{3\pi^2\phi}{4qv_F}\left[1-\frac{16\phi}{\pi^2 qv_F}\ln(qv_F/\phi)\right], \quad (2.18)$$

which is valid for $q\xi_0\gtrsim 1$. For small ϕ we may ignore the logarithm term, and approximate thus

$$\sum_q\frac{1}{q^2+f(q,\phi)/\lambda^2}=\sum_q\frac{1}{q^2}-\sum_q\frac{1}{q^4}\frac{3\pi^2\phi}{4\lambda^2 qv_F}.$$

The general expression for $\mathcal{E}_A$, Eq. (2.14), becomes

$$\mathcal{E}_A\approx\frac{3\phi H^2}{\lambda^2 v_F}\left(\frac{L}{\pi}\right)^3\sum_{r=0}^{\infty}\frac{1}{(2r+1)^5}. \quad (2.19)$$

For small ϕ, the leading term of $\mathcal{E}_0(\phi)$ is of order $\phi^2(\ln\phi)^2$ and hence $|\mathcal{E}_0|<\mathcal{E}_A$ for sufficiently small ϕ. We expect on this basis that the $\mathcal{E}$ versus ϕ diagram will again be of the form that will yield a first-order phase transition [Fig. 3(b)] for a finite $\phi=\phi_c$ at $H=H_c$. This

[10] We wish to thank Professors H. Y. Fan and S. Gartenhaus for an interesting discussion on this point.

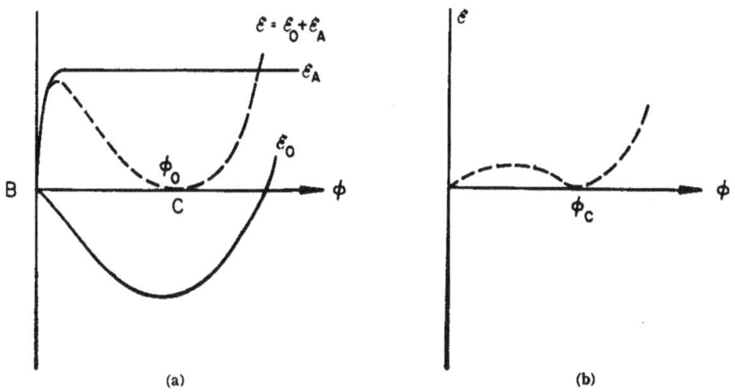

FIG. 3. (a) The energy versus energy gap ϕ diagram for a thick film; ϕ_c is the critical gap for a first-order phase transition. (b) The energy versus energy gap ϕ diagram for a thin film, according to Eq. (2.19).

is the conclusion reached by Bardeen in a recent paper.[3] *It is, however, misleading.* We shall presently prove the following:

Theorem 2. In a thin film which allows complete penetration of the magnetic field, the superconductive ground-state energy is lower than that of the normal state within the framework of diagonal quasiparticle approximation, the meaning of which is specified below. From this follows that no first-order phase transition from superconductive to normal state can be produced at least in this approximation.

The proof is simple. In I, a general approach was developed whereby one sets up the energy gap equation in the presence of an external field. It consists in first considering single electrons as under the influence of the external field and then performing the BCS type pairing in the presence of the field. This alternative was studied using the techniques of field theory in which Green's functions for quasiparticles in a magnetic field was constructed. When an expansion in A was made, it was shown there that the reduction of energy gap to the second order in A was equivalent to the variational result.

We now use this formalism without perturbation expansion. Let the single electron eigenstates and energies in the presence of the field A be labeled by n and ϵ_n. Since the film is thin, A is equal to the external field A^{ex}. We then introduce a pairing between n and its counterpart $\bar{n}$. We may take $\bar{n}$ to be the spin- and space-inversed state of n since the Hamiltonian as well as the boundary conditions remain invariant under spin inversion and $x-L/2\rightarrow -x+L/2$, $y\rightarrow -y$, $z\rightarrow -z$. (It is beyond the scope of this paper to discuss the relation between n and $\bar{n}$ under more general Hamiltonian and geometry.) The only difference from the free-field case is that instead of $\epsilon(p)$ and $\phi(p)$, we use ϵ_n and ϕ_n in setting up the BCS variational procedure.[11] In particular, the energy gap equation becomes

$$\phi_n=\sum_m \frac{\phi_m}{2E_m}V_{nm},\quad E_n=[\epsilon_n{}^2+|\phi_n|^2]^{1/2} \tag{2.20}$$

and the ground-state energy (relative to the normal state) is

$$\mathcal{E}=\sum_n\left(|\epsilon_n|-\frac{\epsilon_n{}^2}{E_n}\right)-\tfrac{1}{4}\sum_{n,m}\frac{\phi_n{}^*}{E_n}V_{nm}\frac{\phi_m}{E_m}$$

$$=\sum_n\left\{|\epsilon_n|-\frac{\epsilon_n{}^2}{E_n}-\frac{1}{2}\frac{|\phi_n|^2}{E_n}\right\}, \tag{2.21}$$

which is <0 unless all $\phi_n=0$.

In the field theoretical treatment of quasiparticles, the energy gap parameter ϕ need not be diagonal in the states n which make the single-particle energy diagonal, but ϕ can be, in general, a matrix. But from the variational point of view, our estimation still gives an upper bound to the total energy $\mathcal{E}$. So as long as this diagonal quasiparticle approximation produces a superconductive state, its true energy is *a fortiori* lower than that of the normal state. This theorem can be extended in a straightforward fashion to finite temperature by replacing $\mathcal{E}$ with thermodynamic free energy.

Bardeen's result is not necessarily contradicted by the above theorem since the above method may not be able to give a superconductive solution when there is actually one in a different treatment such as the perturbation theory. This, however, is unreasonable because the diagonalization with respect to the magnetic eigenstates n will be more justifiable as the field increases.

In the following sections we shall find that a detailed analysis of the implication of our discrete quantization model on the form of the kernel $K(\mathbf{q})$ in perturbation theory actually leads to a different result from

[11] P. W. Anderson, J. Phys. Chem. Solids **11**, 26 (1959); see also *Proceedings of the Seventh International Conference of Low-Temperature Physics, 1960* (University of Toronto Press, Toronto, 1960), p. 298. It might appear that our approach does not allow for a spatially varying energy gap in the presence of the magnetic field, in contrast to the G-L-G theory. This, however, is not true. With our pairing scheme, the pair correlation function in the sense of G-L-G is given by $F(\mathbf{x},\mathbf{x}')=\sum_n(\phi_n/E_n)\psi_n(\mathbf{x})\psi_{\bar{n}}(\mathbf{x}')$, where ψ_n, $\psi_{\bar{n}}$ are the magnetic eigenfunctions. In the G-L-G approximation one defines from this the local gap parameter $\Delta(\mathbf{x})\propto F(\mathbf{x},\mathbf{x})$ which is actually x-dependent.

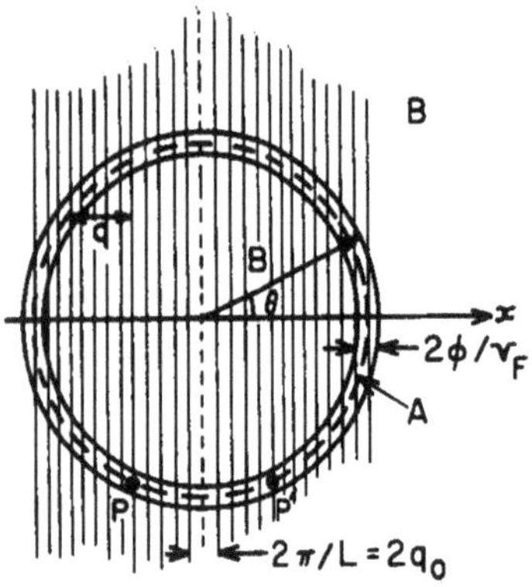

FIG. 4. Cross section of Fermi sphere, with discrete slabs separated by π/L along the x direction. The annulus region represent the BCS band of interaction of width $2\phi/v_F$.

Bardeen's and that the above theorem can be reinforced in the case of a separable potential.

3. PERTURBATION CALCULATION IN THIN FILMS

In this section we shall examine carefully the kernel $K(\mathbf{q})$ under the boundary conditions of a thin film elaborated in Sec. 2. The relation between $K(\mathbf{q})$ and the self-consistent self-energy in the field theoretical method has been established in I. The self-energy equation in a magnetic field reads, to the second order in A,

$$\phi=\Sigma(\phi,A),\quad \Sigma=\Sigma^{(0)}(\phi)+\Sigma^{(2)}(\phi,A). \tag{3.1}$$

Then

$$\Sigma^{(0)}(\phi)\sim\phi\rho\ln(2\omega/\phi),$$
$$\Sigma^{(2)}(\phi,A)\sim\sum_{\mathbf{q}}[\partial K(q,\phi)/\partial\phi]|A(\mathbf{q})|^2. \tag{3.2}$$

According to Eq. (2.18), $K(\mathbf{q})\sim\phi$, so that $\partial K/\partial\phi$ is independent of ϕ. This suggests the possibility that the expansion parameter of the perturbation theory for Σ for small ϕ is something like $|A|^2/\phi$ or $|A|^2/(\phi\ln\phi)$. If this is the case, obviously we cannot terminate our expansion at the second-order terms when ϕ becomes small, but we must include all higher order terms, the sum of which may exhibit a quite different analytic behavior.

The above situation is somewhat paradoxical. For a normal metal state we can stop the expansion at terms proportional to A^2, since in a small sample the electrons do not make circular orbits at such magnitudes of the field as are concerned here. We can start from electrons weakly perturbed by H, then introduce the pairing. It is difficult to imagine how this would introduce higher powers of $|A|^2/\phi$.

If we analyze the form of $K(q,\phi)$ for small ϕ, we find that the dominant contribution comes from the states where both of the intermediate particles are within the energy bandwidth 2ϕ (Fig. 4). This can be seen as follows. We have

$$K(q,\phi)\sim\sum_{\mathbf{p}}\frac{|\langle\mathbf{p}+\mathbf{q}|j|\mathbf{p}\rangle|^2}{E(\mathbf{p})+E(\mathbf{p}+\mathbf{q})}. \tag{3.3}$$

When both intermediate states are in the band, $E(\mathbf{p})+E(\mathbf{p}+\mathbf{q})\approx2\phi$; $|\langle j\rangle|^2$ is essentially constant, while the phase-space volume τ for a q (rather q_x) transition is restricted to

$$|\cos\theta|<2\phi/v_Fq=2/\pi\xi_0q\ll1$$

and $\Delta p<2\phi/v_F$:

$$\tau\approx4\pi p_F^2(2\phi/v_Fq)(2\phi/v_F).$$

Equation (3.3) then gives for the ϕ-dependent part of $K(q)$ a term proportional to

$$16\pi\phi^2p_F^2/2\phi q\sim\phi/q\sim1/\xi_0q,$$

which is precisely the leading term of the BCS kernel (2.18) for small ϕ.

With the discrete electron momenta for a thin film, however, this leading contribution vanishes. We have discrete states separated by $q_0=\pi/L$ along the x direction as shown in Fig. 4, so that for sufficiently thin films the thickness $2\phi/v_F$ of the BCS momentum band becomes small compared to the separation q_0. Since the electronic transition takes place horizontally between different states, there will be a certain critical film thickness below which no transition can take place between two points within the band, except for the special case mentioned below. Simple geometrical considerations show this condition to be

$$(\pi/L)^2=q_0^2\gtrsim m\phi/2=p_F/2\pi\xi_0. \tag{3.4}$$

Taking $\xi_0=10^{-4}$ cm, we have $L\lesssim0.8\times10^{-5}$ cm. (Actually, this thin-film condition will be relaxed later.)

The only situation where both points lie in the bandwidth is when the transition takes place between symmetrical points with respect to the origin O, for instance the p to p' transition shown in Fig. 4. We note, however, that this is forbidden since $q=2p=2r\pi/L$. In fact $K_{ij}^{(2)}(\mathbf{q})$ given by Eq. (3.22) of I gives in our case

$$K_{ij}^{(2)}(\mathbf{q})=\sum_{p_x}\frac{1}{(2\pi)^2}\frac{e^2}{m^2}\int\frac{p_ip_jdp_ydp_z}{E(\mathbf{p})+E(\mathbf{p}+\mathbf{q})}\times\left[1-\frac{\epsilon(\mathbf{p})\epsilon(\mathbf{p}+\mathbf{q})+\phi^2}{E(\mathbf{p})E(\mathbf{p}+\mathbf{q})}\right], \tag{3.5}$$

so for symmetrical transitions where $\epsilon(\mathbf{p})=\epsilon(\mathbf{p}+\mathbf{q})$, we have

$$1-\epsilon(\mathbf{p})\epsilon(\mathbf{p}+\mathbf{q})/[E(\mathbf{p})+E(\mathbf{p}+\mathbf{q})]=0.$$

We see, therefore, that for thin films the leading contribution comes from states where only one of p and p' lies in the band. It is also clear that $K_{ij}=0$ unless $i=j$, and moreover $i=j=y$ is the only physically relevant part in our gauge.

We need now to study and evaluate semiquantitatively the kernel $K_{ij}^{(2)}(\mathbf{q})$ or rather $\partial K_{ij}^{(2)}(\mathbf{q})/\partial\phi$, since ultimately we shall be concerned with a self-energy type compensation equation given by Eqs. (3.1) and

(3.2). For convenience of notation write $E(\mathbf{p})=E$ and $E(\mathbf{p}+\mathbf{q})=E(\mathbf{p}')=E'$, then

$$\partial K_{yy}^{(2)}(\mathbf{q})/\partial\phi=-\sum_{p_x}\frac{\phi}{(2\pi)^2}\frac{e^2}{m^2}\int\int\frac{p_y^2dp_ydp_z}{EE'(E+E')}\times\left[3-(\epsilon\epsilon'+\phi^2)\left(\frac{1}{E^2}+\frac{1}{E'^2}+\frac{1}{EE'}\right)\right]. \quad (3.6)$$

In Fig. 4, we have labeled the band region by A, and the regions outside by B. Since we have shown that contributions from E, E' belonging to A (ϵA) vanish for sufficiently thin-film slabs, it is evident that the leading contribution must come from transitions with $E\epsilon A$ (B) and $E'\epsilon B(A)$. To fix our attention we consider the case where $E\epsilon A$, $E'\epsilon B$; for sufficiently small ϕ we have $E'\approx|\epsilon'|\gg E$. Since $|q|\ll p_F$,

$$|\epsilon'|=|(\mathbf{p}+\mathbf{q})^2/2m-\mathbf{p}^2/2m|\approx|q|v_F\cos\theta, \quad (3.7)$$

where θ is the polar angle shown in Fig. 4. Further,

$$3-(\epsilon\epsilon'+\phi^2)\left(\frac{1}{E^2}+\frac{1}{E'^2}+\frac{1}{EE'}\right)\approx 2$$

effectively since terms proportional to $\epsilon\epsilon'$ vanish upon integration within the narrow band where ϵ runs over both signs while ϵ' remains practically constant. Thus, we get

$$\begin{aligned}\partial K_{yy}^{(2)}(\mathbf{q})/\partial\phi &\approx -\frac{2\phi}{(2\pi)^2}\frac{e^2}{m^2}\sum_{p_x}\int\int\frac{p_y^2}{EE'(E+E')}dp_ydp_z\\ &\approx -\frac{2\phi}{(2\pi)^2}\frac{e^2}{m^2}\sum_{p_x}\int\frac{p_y^2}{E\epsilon'^2}dp_ydp_z\\ &\approx -\frac{4\phi}{(2\pi)^3}\frac{e^2}{m^2}\frac{\pi p_F^2}{q^2v_F^2}\int\int\frac{\tan^2\theta d\cos\theta}{E(l)}l^2dl, \quad (3.8)\end{aligned}$$

where the following crude approximations are understood: (a) The discrete p_x summation is replaced by an integration with cutoff $1\geqslant|\cos\theta|\geqslant\cos\theta_1$ since a large contribution comes from small values of $\cos\theta$, and (b) the l-integration is performed for fixed θ over $0\leqslant|E(l)|\leqslant|\epsilon'|=|qv_F\cos\theta|$ since we have assumed $l=p\epsilon A$, $p'\epsilon B$. A factor of 2 is included to account for the possibility $p\epsilon B$, $p'\epsilon A$. Finally we use Eqs. (3.21), (3.8), and (3.11) of I to obtain

$$\begin{aligned}\Sigma^{(2)}&=\rho\phi\sum_{\mathbf{q}}\int\int\frac{e^2|A(\mathbf{q})|^2\tan^2\theta}{4q^2}\frac{1}{E}d\epsilon d\cos\theta\\ &=\rho\phi\sum_{\mathbf{q}}\int\frac{e^2|A(\mathbf{q})|^2}{2q^2}\sinh^{-1}|qv_F\cos\theta/\phi|\times\tan^2\theta d\cos\theta, \quad (3.9)\end{aligned}$$

and

$$\delta\phi=\Sigma^{(2)}/\rho. \quad (3.10)$$

The self-energy of a quasiparticle in general depends on the state one considers, and in particular it will show some anisotropy due to the quantization of p_x. Such an effect is disregarded here. Furthermore, we see that the summation over q in Eq. (3.9) is heavily weighted in favor of small values of q since $|A(\mathbf{q})|^2/q^2\propto 1/q^6$. Thus, it is practically sufficient to keep only the first term: $q=q_0=\pi/L$; the next term will be smaller by a factor $1/3^6\approx 1/600$. So in the following, we shall limit ourselves to the single term $q=q_0$.

With this in mind, the natural cutoff θ_1 in Eq. (3.9) would appear to be given by

$$\cos\theta_1=\pi/Lp_F, \quad (3.11)$$

since this corresponds to states with the smallest p_x. However, the physical situation requires a closer study of the conditions affecting the problem before deciding upon the appropriate cutoff. We will tentatively integrate Eq. (3.9) with an unspecified cutoff $\cos\theta_c\ll 1$ and obtain

$$\delta\phi=-\phi\frac{2e^2}{q_0^2}\sinh^{-1}(q_0v_F\cos\theta_c/\phi)\frac{1}{\cos\theta_c}|A(q_0)|^2, \quad (3.12)$$

$$q_0=\pi/L.$$

A factor of 2 arises when one sums over $q_x=\pm q_0$. This result is free from the objection raised against the earlier formula (3.2). In fact $\delta\phi$ is proportional to

$$\phi\sinh^{-1}(q_0v_F\cos\theta_c/\phi)\sim\phi\ln(2q_0v_F\cos\theta_c/\phi), \quad (3.13)$$

since the argument of $\sinh^{-1}$ is $\gtrsim 1$ in view of our definitions (3.4) and (3.11). So $\delta\phi$ has the same type of ϕ dependence ($\sim\phi\ln\phi$) as $\Sigma^{(0)}(\phi)$ given in Eq. (3.2).

Although Eqs. (3.9) and (3.12) have been obtained under the thin-film condition (3.4), it appears to be valid also for thicker films. In the latter case, there will be contributions to the kernel $K(q)$ coming from transitions between states within the band A, and one expects the result to agree with the old formula (2.18). This is in fact so except for a minor difference. Observing that $\sinh^{-1}(qv_F\cos\theta_c/\phi)\sim qv_F\cos\theta_c/\phi$ in the "thick" case, we find from Eq. (3.9) that

$$\delta\phi=-\sum_{\mathbf{q}}\frac{e^2}{q}v_F|A(\mathbf{q})|^2, \quad (3.14)$$

which is to be compared with the leading term in Eq. (3.25) of I

$$\delta\phi=-\frac{\pi^2}{8}\sum_{\mathbf{q}}\frac{e^2v_F}{q}\left[1-\frac{16\phi}{\pi^2qv_F}\{2\ln(\pi q\xi_0)-1\}\right]\times|A(\mathbf{q})|^2. \quad (3.15)$$

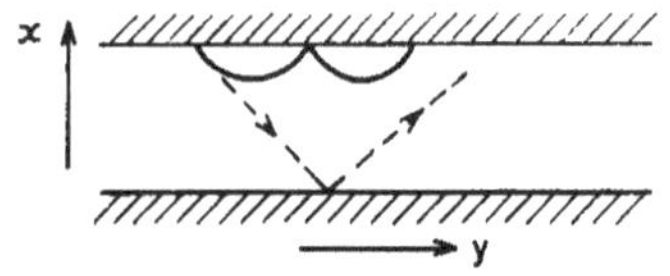

FIG. 5. Situations whereby the electrons are localized to one side of the film.

The correction factor $\pi^2/8=1.24$ may be related to the geometry.

We may therefore regard Eq. (3.12) to be valid irrespective of the restriction (3.4) as long as the thickness is small compared to the penetration depth. In spite of this reasonable behavior of our result we have not yet solved the problem. We need a more critical examination from a nonperturbative point of view before coming to the discussion of phase transition.

4. EXTENSION OF THE PERTURBATION FORMULA

It was remarked earlier that there is in principle no need for perturbation expansion with respect to the magnetic field. Within the framework of diagonal quasiparticle approximation defined there, an exact solution can be obtained by pairing appropriate electron eigenstates in the presence of the magnetic field H.

As was originally assumed in the BCS theory and also adopted in other calculations, let us suppose the interaction to be separable. For its general form take

$$V_{p'p}=\langle p'\bar{p}'|V|p\bar{p}\rangle=S_{p'}S_p^*V_0,$$

where p and $\bar{p}=-p$ are the paired momenta. The energy gap equation takes the form

$$\begin{aligned}1&=\sum_p\frac{V_p}{2E_p}\equiv V_0\sum_p\frac{|S_p|^2}{2E_p},\\ E_p&=[\epsilon_p^2+|\phi_p|^2]^{1/2},\\ \phi_p&=S_p\phi.\end{aligned}\tag{4.1}$$

In particular, $|S_p|=1$ if $V_{pp'}$ is constant within a domain D and zero otherwise. In the presence of the field, we obtain the same equation if we label the states by n instead of p, namely

$$\begin{aligned}1&=\sum_n\frac{V_n}{2E_n}=V_0\sum\frac{|S_n|^2}{2E_n},\\ E_n&=[\epsilon_n^2+|\phi_n|^2]^{1/2},\quad \phi_n=S_n\phi\\ S_n&=\sum_p u_{np}u_{\bar{n}\bar{p}}S_p,\quad u_{np}=\langle n|p\rangle.\end{aligned}\tag{4.2}$$

This may be cast into

$$\begin{aligned}1&=\bar{V}\sum_n\frac{1}{2E_n'},\\ E_n'&=[\epsilon_n^2+\bar{\phi}^2]^{1/2},\end{aligned}\tag{4.2'}$$

by introducing the averages $\bar{V}$ and $\bar{\phi}$. The magnetic field dependence of $\bar{\phi}$ will come from ϵ_n and $\bar{V}$. Let us first consider ϵ_n.

For most of the electrons near the Fermi surface, the magnetic energy is small compared to the kinetic energy, and perturbation theory should be adequate since the classical electron orbits do not deviate much from straight lines. We may still label the states by the unperturbed momenta p, and $\epsilon(H)$ takes the form

$$\epsilon(p,H)=\epsilon(p,0)+e^2H^2L^2(1+p_y^2/p_x^2)/24m,\tag{4.3}$$

as can be seen using the WKB method. This represents the diamagnetic energy increase for electrons confined to within the film. Equation (4.3) is not applicable when $|p_y/p_x|$ becomes very large. This is because these electrons run nearly parallel to the film and get localized to one side of the film by the magnetic field (Fig. 5). The geometrical condition for this to happen is

$$2eHLp_y/p_x^2>1.$$

Those "boundary electrons," however, would not contribute to the pairing energy of superconductivity since we have paired space-reversed states which are now separated to different sides of the film. We shall see this later explicitly.

Equation (4.3) represents an expansion and distortion of the Fermi surface. We must recall here that ϵ must be measured relative to a chemical potential so as to keep the average $\bar{\epsilon}=0$. Hence, only the distortion will affect Eq. (4.1), and this is a fourth-order effect on the effective density of states N at the Fermi surface. The entire picture breaks down only when H is so strong as to make the radius of curvature comparable with the thickness. For $L\sim10^{-5}$ cm, we get $H=p_F/eL\sim10^6$ G.

We see, therefore, that the main effect of the field arises through the change of the matrix element of V, and in going back to Eq. (4.2), we come to the following interesting assertion:

Theorem 3. Under the assumption of separable potential and essentially continuous single-particle energy spectrum, the energy gap in the presence of the field is reduced, but superconductivity is never broken unless $S_n=0$ for all states n near the Fermi surface (except for points of measure zero). In the latter case, the minimum energy gap defined by $\mathrm{Min}_n\{E_n\}$ is zero, but a superconductive solution to Eq. (4.2) may still exist ($\phi_n=S_n\phi$, $\phi\neq0$).

The proof will be rather obvious. When $S_n=0$ for those states with $\epsilon_n=0$ (Fermi surface), Eq. (4.2) does not become singular as $\phi\to0$ since those states do not contribute, so there may or may not be a superconductive solution. On the other hand, if $S_n\neq0$ in some portion of the Fermi surface, one can always find a solution by making ϕ sufficiently small unless the discreteness of ϵ_n becomes important. The minimum energy gap can still be zero. (This observation brings in the necessity of distinguishing the vanishing of energy gap from the vanishing of "superconductive" state and consequent phase transition to the normal state. A "superconductive" state with a vanishing gap will have different physical properties from the ordinary one. In this paper a superconductive state is meant to be a state with $\phi_n \neq 0$ for *some n*.)

Under a condition like our magnetic field problem, it would actually be difficult to realize the special case of theorem 3 since the system is anisotropic, which means that it would be difficult to destroy superconductivity completely by a magnetic field. It is true that the actual interaction does not exactly have the property assumed, so the problem of critical field may depend on the details of interaction. It is also true that even though S_n of theorem 3 may not exactly vanish, it may become so much reduced everywhere that the gap vanishes for practical purposes.

At any rate, it is clear that the effect of the magnetic field, whatever the dependence, may be regarded as a change in the effective coupling parameter $\rho = N\bar{V}$. Namely, ρ will now become a function of H and $\bar{\phi}$. (It depends on $\bar{\phi}$ because $\bar{V}$ depends on the weight function $1/E_n'$ which involves $\bar{\phi}$.) Consequently, we may write the solution to Eq. (4.2′) as

$$\bar{\phi} = 2\omega \exp[-1/\rho(H,\bar{\phi})], \tag{4.4}$$

which is a transcendental equation. If ρ decreases smoothly to zero for some H, $\bar{\phi}$ vanishes at this point and we shall have a second-order transition to the normal state in view of theorem 2.

Setting up Eq. (4.2) and solving it in the magnetic field is an involved task. Furthermore, there is not much sense in doing it since the real potential will be more complicated. We can, however, compare our perturbation formula (3.12) with Eq. (4.4) and thereby identify $\rho(H,\bar{\phi})$. This is legitimate for weak fields, but it seems reasonable to expect that Eq. (4.4) has a larger domain of validity than the simple perturbation formula.

In order to carry out this program, let us take a differential of Eq. (4.2)

$$0 = -\sum_n V_n \frac{\phi_n \delta\phi_n}{2E_n^3} + \sum \frac{1}{2E_n} \delta V_n. \tag{4.5}$$

By the standard technique, the first sum reduces to

$$\begin{aligned}\approx -\bar{\phi}\delta\bar{\phi} \sum_n \frac{\bar{V}}{2E_n'^3} &= -\bar{\phi}\delta\bar{\phi}(\rho/\bar{\phi}^2) \\ &= -\rho\delta\bar{\phi}/\bar{\phi}.\end{aligned} \tag{4.6}$$

Comparing this with Eq. (3.9) we immediately obtain

$$\begin{aligned}\sum_n \frac{1}{2E_n}\delta V_n &= -\rho \sum_q \int\int \frac{1}{2E} \frac{e^2|A(\mathbf{q})|^2}{q^2} \tan^2\theta d\epsilon d\Omega/4\pi \\ &\approx -\rho \int \frac{d\Omega}{4\pi} \int_{-\kappa}^{\kappa} d\epsilon \frac{\tan^2\theta}{2E} \frac{2e^2|A(q_0)|^2}{q_0^2},\end{aligned} \tag{4.7}$$

where $\kappa = q_0 v_F \cos\theta_c$. Hence, from Eq. (4.2) we find

$$\begin{aligned}1-|S_n|^2 = -\delta V_n/V_0 &= \frac{2e^2|A(q_0)|^2}{q_0^2} \tan^2\theta f_\kappa(\epsilon), \\ f_\kappa(\epsilon) &= 1, \quad |\epsilon| < \kappa, \\ &= 0, \quad |\epsilon| > \kappa.\end{aligned} \tag{4.8}$$

Here V_0 is the potential for the field-free case ($V_{p'p} = V_0$ if p, $p' \epsilon D$, and zero otherwise).

Since $|S_n|^2 \geqslant 0$, the perturbation result certainly loses its meaning if the right-hand side of (4.8) exceeds 1. We must then choose a cutoff θ_2 according to

$$\alpha \equiv 2e^2|A(q_0)|^2/q_0^2 = \cot^2\theta_2,$$

or

$$\cos^2\theta_2 = \alpha/(1+\alpha). \tag{4.9}$$

We have ignored here the complication arising from $f_\kappa(\epsilon)$. Equation (4.9) is similar to Eq. (3.11) but more stringent. Beyond this angle up to θ, we shall set simply

$$1-|S_n|^2 = f_\kappa(\epsilon). \tag{4.10}$$

Equation (4.2) now may be written

$$1 = \sum \frac{V_n}{2E_n} = V_0 \sum_n \frac{1}{2E_n} + V_0 \sum_n \frac{|S_n|^2 - 1}{2E_n}. \tag{4.11}$$

Consulting Eqs. (3.12), (4.2), (4.8), and (4.10), we get

$$1 = \rho_0 \sinh^{-1}(\omega/\bar{\phi}) - \rho_0(r_2^{\mathrm{I}} + r_2^{\mathrm{II}})\sinh^{-1}(\kappa_2/\bar{\phi}), \quad \cos\theta_2 > \cos\theta_1. \tag{4.12}$$

Here $\rho_0 = NV_0$; r^{I} and r^{II} come from the two regions corresponding to Eqs. (4.8) and (4.10), respectively:

$$\begin{aligned}r_2^{\mathrm{I}} &\approx \alpha[\cos\theta_2 + (\cos\theta_2)^{-1} - 2], \\ r_2^{\mathrm{II}} &\approx \cos\theta_2 - \cos\theta_1,\end{aligned} \tag{4.13}$$

and $\kappa_2 = \kappa(\cos\theta_2)$. Equations (4.12) and (4.13) are primarily designed for small $\cos\theta_2 \ll 1$ and large $\kappa_2/\bar{\phi}$ (thin film), but should be reasonable for the entire range. Especially r_2^{I} is made to vanish correctly for $\cos\theta_2 = 1$.

In case $\cos\theta_2 < \cos\theta_1$, there is no region II, so that

$$\begin{aligned}1 &= \rho_0 \sinh^{-1}(\omega/\bar{\phi}) - \rho_0 r_1^{\mathrm{I}} \sinh^{-1}(\kappa_1/\bar{\phi}), \\ \kappa_1 &= \kappa(\cos\theta_1), \\ r_1^{\mathrm{I}} &= \alpha/\cos\theta_1.\end{aligned} \tag{4.14}$$

Since

$$1/\rho_0 = \sinh^{-1}\omega/\phi_0 \sim \ln(2\omega/\phi_0)$$

and $\sinh^{-1}(\omega/\bar{\phi}) \sim \ln(2\omega/\bar{\phi})$, Eqs. (4.12) and (4.14) may also be written (dropping the bar)

$$\begin{aligned}\ln(\phi/\phi_0) &= -(r_2^{\mathrm{I}} + r_2^{\mathrm{II}})\sinh^{-1}(\kappa_2/\phi), \\ \ln(\phi/\phi_0) &= -r_1^{\mathrm{I}} \sinh^{-1}(\kappa_1/\phi)\end{aligned}$$

or

$$\phi/\phi_0 = \exp[-(r_2^{\mathrm{I}} + r_2^{\mathrm{II}})\sinh^{-1}(\kappa_2/\phi)], \tag{4.15a}$$

$$\phi/\phi_0 = \exp[-r_1^{\mathrm{I}} \sinh^{-1}(\kappa_1/\phi)], \tag{4.15b}$$

which are transcendental equations for ϕ. For simplicity, we shall write them as a single equation

$$\phi/\phi_0=\exp[-r\sinh^{-1}(\kappa/\phi)]. \tag{4.16}$$

For sufficiently weak fields Eq. (4.15b) applies, but as the field increases and/or the thickness decreases, we go over to Eq. (4.15a). The transition takes place at $\cos\theta_1=\cos\theta_2$, or

$$(\pi/p_FL)^2=\alpha_0/(1+\alpha_0)\approx\alpha_0\ll 1, \tag{4.17}$$

since the left-hand side is $\ll 1$. In terms of H and L, this means

$$\pi/p_FL=8^{1/2}eHL^2/\pi^3$$

or

$$8^{1/2}eHL^2(p_FL)/\pi^4=1. \tag{4.18}$$

For $L=10^{-5}$ cm, it corresponds to $H\approx 25$ G, and for $L=0.5\times 10^{-5}$ cm, $H\approx 200$ G. Fields stronger than these will lead into the domain of Eq. (4.15a). For very strong fields, ϕ becomes small so that $\sinh^{-1}(\kappa/\phi)\approx\ln(2\kappa/\phi)$. Thus, Eq. (4.16) reduces to

$$\begin{aligned}\phi/\phi_0&=(2\kappa/\phi)^{-r}=(2\kappa/\phi_0)^{-r/(1-r)}\\&=\exp[-(r/1-r)\ln(2\kappa/\phi_0)].\end{aligned} \tag{4.19}$$

For large α, then, it decays exponentially like

$$\phi/\phi_0=\exp[-2\alpha\ln(2\kappa_2/\phi_0)]=(2\kappa_2/\phi_0)^{-2\alpha} \tag{4.20}$$

according to Eq. (4.13), but will not vanish at a finite magnetic field.

It is not possible to solve Eq. (4.16) explicitly for ϕ over the entire range, but the following formula

$$\phi/\phi_0=\exp[-(r/1-r)\sinh^{-1}(\kappa/\phi_0)], \tag{4.21}$$

$$r=\alpha/\sqrt{\alpha_0},$$

$$\kappa/\phi_0=\pi q_0\xi_0\sqrt{\alpha_0}\quad\text{for}\quad\alpha<\alpha_0; \tag{4.21a}$$

$$\begin{aligned}r=\alpha\{(\alpha/1+\alpha)^{1/2}+(1+\alpha/\alpha)^{1/2}-2\}\\+(\alpha/1+\alpha)^{1/2}-\sqrt{\alpha_0},\end{aligned}$$

$$\kappa/\phi_0=\pi q_0\xi_0(\alpha/1+\alpha)^{1/2}\quad\text{for}\quad\alpha>\alpha_0; \tag{4.21b}$$

$$\sqrt{\alpha_0}=\pi/p_FL,$$

$$\alpha=8e^2H^2L^4/\pi^6$$

turns out to be a fairly good representation of the solution to Eq. (4.16).

For a relatively thick film and weak field, $\sinh^{-1}\kappa/\phi_0$ may be replaced by κ/ϕ_0, and Eq. (4.21) reduces for $\alpha\ll 1$ to

$$\begin{aligned}\phi/\phi_0&=\exp[-r\kappa/\phi_0]\\&=\exp[-\pi q_0\xi_0\gamma\alpha]\\&=\exp[-(8/\pi^4)\xi_0L^3(eH)^2\gamma],\end{aligned} \tag{4.22}$$

where $\gamma(r,\kappa/\phi_0)=1$ for $\alpha<\alpha_0$, and $\gamma\sim 1$ for $\alpha>\alpha_0$. Since our calculations have been based on thin-film conditions, perhaps the exact value of γ following from Eq. (4.21b) is not to be trusted. But Eq. (4.22) above is a rather convenient way of seeing the qualitative behavior of ϕ/ϕ_0 since $\gamma(r,\kappa/\phi_0)$ defined in this fashion turns out to be not too wildly varying over the entire range of α. In fact, we know from Eq. (4.20) that $\gamma\to 2\ln(2\pi q_0\xi_0)/\pi q_0\xi_0$ as $\alpha\to\infty$, which is ~ 0.1 for $L=10^{-5}$ cm and $\xi_0=10^{-4}$ cm. In the intermediate range $\alpha=0(1)$, γ can be >1.

The fact that our result does not produce a phase transition at a finite field is in accordance with our general considerations, but cannot be taken literally since it is certainly not correct to extrapolate our crude theory to arbitrarily high magnetic fields. It only demonstrates, within our model, the difficulty of completely destroying superconductivity at zero temperature.

We conclude therefore that at zero temperature the gap will decrease steadily with increasing magnetic field, and will undergo a second-order phase transition at a critical field H_c which is probably very high but cannot be estimated within our framework. H_c may be as high as the field necessary to produce complete circular orbits within the film ($H\sim p/eL$). We must, however, also take into account the spin paramagnetic energy[12] which tends to destroy the BCS-type pairing:

$$\frac{eH}{2mc}\Big/kT\approx 6\times 10^{+5}\ \text{G/°K}.$$

Finally, we would like to emphasize that the energy gap will be anisotropic in a magnetic field, and may actually go to zero around the direction perpendicular to the field and parallel to the film since those electron states are most disturbed by the field. The tunnelling and the thermal conductivity experiments will measure different quantities in such a case. If, however, impurity scattering is important ("dirty" superconductor), the anisotropy tends to be smoothed out.[11]

5. PHASE TRANSITION AT FINITE TEMPERATURES

The treatment of the magnetic field problem at finite temperature goes in much the same way as in the previous sections, and qualitatively speaking it is even easier than at zero temperature. The problem of phase transition in these two cases can be treated separately.

The basic procedure is variational calculation where one minimizes the thermodynamic free energy in the presence of the magnetic field. This was carried out in detail by Bardeen[3] in perturbation theory. The objection raised against his calculation of the kernel $K(\mathbf{q})$ still holds, but it does not look as serious as at $T=0$. This is because both the H-independent and dependent parts of thermodynamic free energy F_s behave alike as functions of the variational parameter ϕ [3]:

$$\begin{aligned}F_s^{(0)}&\propto\phi^2\ln t(1-\rho\ln t)\\F_s^{(2)}&\propto\phi_0\phi\tanh(\beta\phi/2)H^2\\&\sim\phi^2H^2,\quad\beta\phi\ll 1,\end{aligned} \tag{5.1}$$

[12] A. M. Clogston, Phys. Rev. Letters 9, 266 (1962).

where $t=T/T_c$, $\beta=1/kT$; ϕ_0 is the gap at $T=0$. This is the reason why he obtains a second-order transition at higher temperatures, $t\gtrsim 0.3$.

From our point of view, we will have to redo all the calculations with our geometrical conditions and using thermal Green's functions. Here let us avoid these troubles and look at the structure of the energy gap equation for finite temperature when the magnetic eigenstates are paired:

$$\phi_n=\sum_m V_{nm}\frac{\phi_m}{2E_m}\tanh(\beta E_m/2). \tag{5.2}$$

E_n and V_{nm} depend on the magnetic field, but not on the temperature. But when we express the solution $\phi_n\approx\phi$ in terms of an effective coupling constant $\rho=N\bar{V}$ after rewriting Eq. (5.2) in the standard form

$$1=\bar{V}\sum_n\frac{1}{2E_n}\tanh(\beta E_n/2), \tag{5.3}$$

a temperature dependence of ρ creeps in through the definition of the average $\bar{V}$ with respect to the weight $\tanh(\beta E_n/2)/E_n$.

We point out, however, that the weight function is rather insensitive to temperature as the actual ϕ changes with temperature to keep the sum in Eq. (5.3) constant. This is also the reason for the well-known fact that the coherence length is nearly independent of temperature. As a result, we may regard ρ as a function of the magnetic field and the gap parameter at $T=0$: $\rho=\rho(H^2,\phi_0)$. This brings in a great simplification of the problem for finite temperatures. Let us plot, in the standard manner, the ϕ versus T curve for the field-free case corresponding to a coupling parameter ρ_0. As the magnetic field is switched on, the effective coupling parameter ρ decreases from ρ_0, but this change is temperature-independent. Consequently, we obtain a family of curves which are only reduced in scale (note that ϕ_0 and the critical temperature T_c are proportional to each other). These curves are shown in Fig. 6 (broken lines).

Now suppose we are operating at a fixed temperature T_1. The energy gap $\phi(H,T_1)$ is given by the intercepts of the family of curves with the vertical plane $T=T_1$. With increasing H, ϕ comes down steadily until it vanishes when T_1 happens to be the critical temperature for a fictitious superconductor with reduced coupling parameter $\rho=\rho(H)$. Hence,

Theorem 4. A thin superconductor in a magnetic field behaves approximately like a fictitious superconductor without the magnetic field, but having a reduced gap and a correspondingly reduced critical temperature. For fixed T and variable H, it undergoes a second-order phase transition when the critical temperature of the fictitious system equals T.

Let us express the above relation quantitatively. For

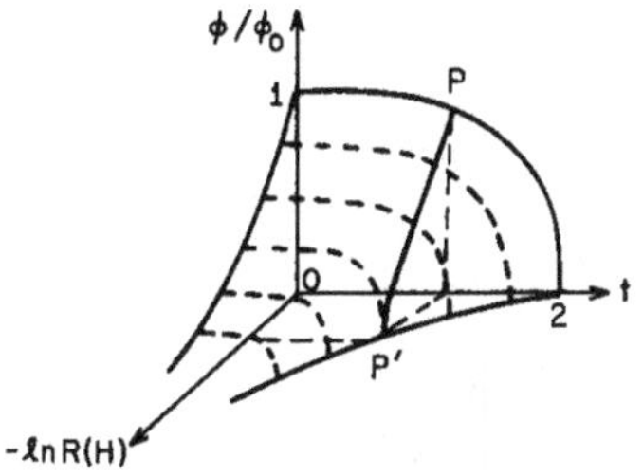

FIG. 6. Three-dimensional plot illustrating the scaling rule. Curve PP' gives the profile of ϕ versus $-\ln R(H)\sim CH^2$ at a fixed t. The letter 2 on the t axis should be read as 1.

$H=0$, the energy gap $\phi(T)$ has the universal form $\phi(t)/\phi(0)=F(t)$,

$$F(0)=1,\quad F(1)=0$$
$$\phi(0)=CkT_c\quad (C=1.75 \text{ in the BCS theory}). \tag{5.4}$$

When $H\neq 0$, $\phi(0)$ and T_c are replaced, respectively, by

$$\phi(0,H)=R(H)\phi(0),$$
$$T_c(H)=R(H)T_c, \tag{5.5}$$

where $R(H)$ is the scaling factor. We get, thus, the general formula

$$\phi(t,H)=R(H)\phi_0F[t/R(H)],$$
$$\phi_0=\phi(T=0,H=0), \tag{5.6}$$
$$t=T/T_c(H=0).$$

Since $F(t)=0$ at $t=1$, it follows immediately that the critical field H_c at which $\phi=0$ is determined by

$$R(H_c)=t. \tag{5.7}$$

Near $t=1$, $F(t)$ behaves like $(1-t)^{1/2}$, so that

$$\phi(t,H)\sim(1-t/R)^{1/2}=[1-R(H_c)/R(H)]^{1/2}. \tag{5.8}$$

$R=\phi(0,H)/\phi_0$ is given by Eqs. (4.21)-(4.22) depending on their applicability. In general, the weak field formula (4.21a) will be valid at high temperatures where $R\approx 1$. At lower temperatures a changeover to Eq. (4.21b) takes place as the field increases. However, this depends on film thickness. We will consider the two cases.

(1) $\pi q_0\xi_0\sqrt{\alpha_0}\lesssim 1$. This happens for a relatively thick film ($L\gtrsim 0.5\times 10^{-5}$ cm for $\xi_0=10^{-4}$ cm). Then Eq. (4.22) (with $\gamma=1$) is certainly valid for temperatures

$$t>t_0=R_0=\exp[-\alpha_0\pi q_0\xi_0]. \tag{5.9}$$

This is very close to 1. For example, with $\xi_0=10^{-4}$ cm, $t_0=1-10^{-3}$ for $L=10^{-5}$ cm and $t_0=0.9$ for $L=0.5\times 10^{-5}$ cm. The ϕ versus H curve following from Eqs. (4.22) and (5.8) becomes then

$$\phi^2(t,H)/\phi^2(t,0)=1-H^2/H_c^2,$$
$$eH_c=(1-t)^{1/2}(\pi^2/\sqrt{8})\xi_0^{-1/2}L^{-3/2}. \tag{5.10}$$

This essentially agrees with the results of Ginzburg-Landau-Gorkov-Douglass theory and of Bardeen near $t=1$.

For temperatures below t_0, there will be a changeover to the region $\alpha>\alpha_0$ at a field determined by Eqs. (4.17)

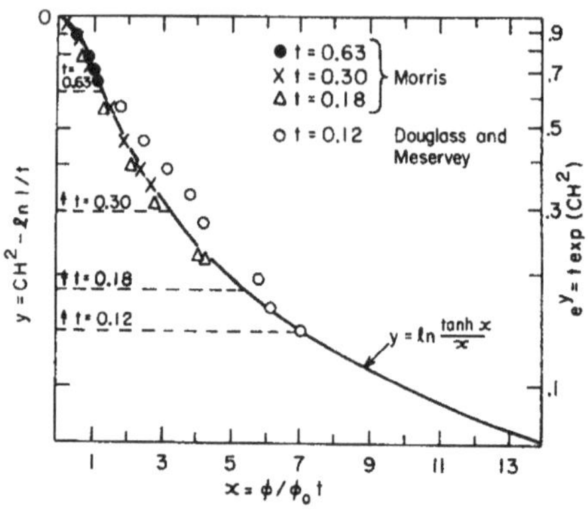

FIG. 7. Universal function $y=\ln(\tanh x/x)$ and experimental points. The experimental points of Morris and those obtained by using the BCS relation $\phi_0=1.75\ kT_c$; $t=T/T_c$ was computed with $T_c=7.2°$K.

and (4.18). As long as $\alpha \ll 1$, $\pi q_0\xi_0\sqrt{\alpha}\lesssim 1$, Eq. (4.22) is still good, but the ϕ versus H behavior will be somewhat more complicated than Eq. (5.10). The critical field is given by

$$eH_c=(\ln 1/t)^{1/2}(\gamma\pi^2/\sqrt{8})\xi_0^{-1/2}L^{-3/2}, \qquad (5.11)$$

which goes over to Eq. (5.10) for $t\approx 1$.

As the temperature is further lowered, we begin to deviate from Eq. (5.11). H_c must now be determined numerically from Eq. (4.21). This happens for the temperature ranges of most experiments.

(2) $\pi q_0\xi_0\sqrt{\alpha_0}\gg 1$. For this very thin-film case, the changeover between $\alpha<\alpha_0$ and $\alpha>\alpha_0$ will take place if

$$t<t_0=R_0=\exp[-\sqrt{\alpha_0}\ln(2\pi q_0\xi_0\sqrt{\alpha_0})]. \qquad (5.12)$$

Above this, we have then

$$\phi^2(t,H)/\phi^2(t,0)=1-H^2/H_c^2,$$

$$eH_c=(\ln 1/t)^{1/2}(\pi^7/8)^{1/2}p_F^{-1/2}L^{-5/2}\times[\ln(2\pi^3\xi_0/p_FL^2)]^{-1/2}. \qquad (5.13)$$

At lower temperatures when Eq. (4.21b) takes over, the behavior again should be determined numerically.

6. COMPARISON BETWEEN THEORY AND EXPERIMENT

In order to express our basic equation (5.6) analytically, it is convenient to use the implicit form[8,13]

$$\phi(t)/\phi_0=\tanh[\phi(t)/\phi_0 t]. \qquad (6.1)$$

Introducing the scaling factor ($\phi_0\to R\phi_0$, $t\to R^{-1}t$), we obtain

$$\phi(t,H)R^{-1}(H)/\phi_0=\tanh[\phi(t,H)/\phi_0 t], \qquad (6.2)$$

which we can write in the form

$$t/R(H)=\tanh x/x$$

or

$$y=\ln[1/R(H)]-\ln 1/t=\ln(\tanh x/x), \qquad (6.3)$$

[13] D. J. Thouless, Phys. Rev. **117**, 1256 (1960).

where

$$x=\phi(t,H)/\phi_0 t.$$

We can plot all data on a single y versus x curve if we measure ϕ in units of $\phi_0 t$ and shift the y coordinate by $\ln 1/t$ for different t. For those relatively thick films and weak fields where Eq. (4.22) applies, $\ln 1/R\sim H^2$, so that Eq. (6.3) becomes of the form

$$CH^2-\ln 1/t=\ln(\tanh x/x). \qquad (6.4)$$

Equation (6.3) is plotted in Fig. 7. Also shown are the experimental points of Morris and those of Douglass and Meservey for lead, based on Eq. (6.4). These are well outside the range of applicability of Eq. (4.22), but numerical calculation has shown that $\ln 1/R\propto H^2$ is still valid to a good approximation. In addition, we find $H_c=3700$ G under Douglass' condition $t=0.12$ and $L=10^{-5}$ cm. Expermentally H_c is ≈ 2300 G. Figure 8 shows the same comparison on more conventional plots. We see that the general trends at lower temperatures are correctly predicted by the theoretical curve.

We have a few remarks to make. (1) Lead is an anomalous superconductor with a large coupling parameter ρ whereas our theory is based essentially on the weak coupling. (2) The effect of the finite penetration depth is neglected in our formulas. (3) The experimental conditions are more complicated than those assumed in our model (parallel surfaces with specular reflection, no impurity scattering, etc.). (4) There are certain uncertainties in the interpretation of experimental data, e.g., the identification of ϕ with measured quantities, and the exact determination of H_c where ϕ vanishes.

Each of these points can be taken into account if necessary, but since all of these possibilities may be present and may cause modifications in different directions, we have not attempted to analyze them. In view of the crudeness of our theory, we therefore conclude that the agreement between theory and experiment is at least qualitatively satisfactory.

7. COMPARISON WITH OTHER THEORIES

Bardeen,[3] in a recent paper on critical fields in superconductors, concluded that the microscopic theory yielded a first-order phase transition for thin films with reduced temperature $T/T_c\lesssim 0.3$. His calculations differ from our approach in two major respects; (a) our use of discrete quantized momentum variables q_z rather than the continuum momentum variable and (b) our choice of the London gauge rather than an arbitrary gauge for purposes of calculations.

Our adoption of the discrete quantization model and the London gauge enables us to conclude that for sufficiently thin films, the leading term in the BCS kernel $K(q)$, ϕ/q, vanishes; this in turn determines in a crucial way in the framework of perturbation theory the question of phase transition at low temperatures. In fact, if we take $\delta\phi$ calculated for the bulk material

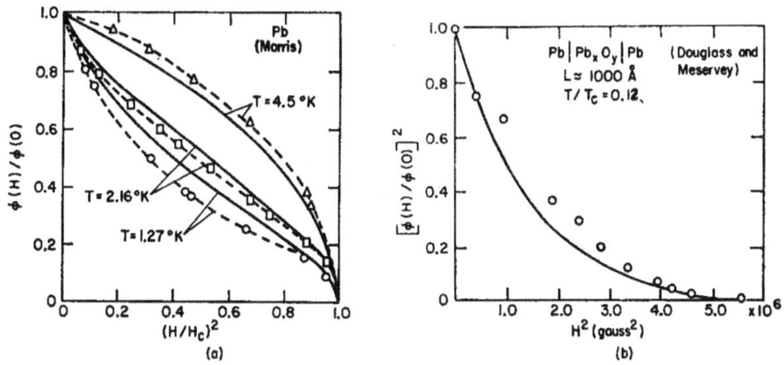

FIG. 8. Comparison of theoretical and experimental curves on conventional plots taken from Ref. 8.

case in I, apply alternating fields H and other specifics appropriate to the thin film case [cf. Eq. (2.19)], we find complete equivalence with Bardeen's calculations for this case if the ϕ/q term in $K(q)$ is naively retained.

It appears that the choice of gauge is important in our case particularly because there is a degeneracy of magnetic eigenstates, i.e., to one energy ϵ belong an infinity of states with different p_y, p_z and the (discrete) quantum number in the x direction. A different gauge would, in general, lead to a different set of eigenstates, which in turn imply a different pairing scheme of electrons. A wrong choice of the pairing would not maximize the pairing energy, and furthermore might not be adiabatically related to the field-free pairing. In the latter case, the perturbation theory would not work.

For cases when there is a large amount of thermal excitation present, or $T\sim T_c$, the Ginzburg-Landau-Gorkov-Douglas theory[4,5] predicts second-order transition for thin films with $H_c\propto L^{-3/2}$. Despite the fact that experimental data[7] seem to suggest that it might be possible to account for the low-temperature properties along the lines of G-L-G theory as well when a sufficiently strong field is applied so that $\phi(H)\sim kT$, we wish to point out that Gorkov's theory (even with the inclusion of a strong magnetic field) cannot reproduce our results for T near $T=0$°K. This is because the Gorkov theory is essentially a London theory, $\xi_0\ll\lambda_0$, where ξ_0 and λ_0 are the bulk coherence distance and penetration depth, respectively. Since $\xi_0=v_F/\pi\phi$, at $T=0$, we have $\xi_0\rightarrow\infty$ as $\phi\rightarrow 0$ (near critical field H_c), and the London limit is not satisfied.

Mathur, Panchapakesan, and Saxena[6] arrived recently at the conclusion that a second-order phase transition is expected at $T=0$°K for thin films based on earlier calculations of Gupta and Mathur[14] which used Wentzel's theory of gauge invariance.[15] It appears that Mathur *et al.* took the London limit for bulk specimen parameters ξ_0 and λ_0 in their study. This is quite evidently not satisfied for thin films where the appropriate limit is the Pippard nonlocal form when expressed in terms of bulk material parameters ξ_0 and λ_0. This is to be contrasted with the work of Douglass[4] on thin films, where the London limit is used appropriately in the form $\xi\ll\lambda$, where ξ and λ are the coherence distance and penetration depth for the thin film itself.

[14] K. K. Gupta and V. S. Mathur, Phys. Rev. **121**, 107 (1961).
[15] G. Wentzel, Phys. Rev. **111**, 1488 (1968); a comparison of Wentzel's formulation of gauge invariance with that of the BCS-Bogoliubov approach is given in Ref. 2.

8. CONCLUDING REMARKS

We have adopted specular reflection boundary conditions for electrons at film surfaces. Though a more exact solution, say for an electron in an image potential[16] is certainly more preferable to any artificial boundary condition on some *ad hoc* "surface," we feel that our conclusions are dependent primarily on the London gauge rather than on the details of boundary conditions or the shape of the boundary. In this connection it will be instructive to study the cases where, for example, the magnetic field is not parallel to the film surfaces, or parallel but unequal fields are applied at the opposite surfaces. This will involve us in the proper choice of current distribution and pairing so as to optimize the balance between magnetic and pairing energies. Significant modifications that would ensue are already suggested in the works of Douglass[4] and Tinkham.[17]

A proper extension of the present work to finite temperatures, which evaluates $K(q,T)$ in terms of thermal Green's functions in the presence of magnetic field,[18] is obviously highly desirable since the temperature dependence here is introduced only qualitatively via theorem 4. Such a comprehensive analysis will allow us to compare reliably our results with those of G-L-G at high temperatures.

Finally, experimental work on thin films in the range 100 to 1000 Å, at reduced temperature $T/T_c<0.3$ for a soft (and weak coupling) superconductor like tin will test most critically the present notions. We note

[16] L. A. MacColl, Bell System Tech. J. **30**, 888 (1951).
[17] M. Tinkham, Phys. Rev. **129**, 2613 (1963).
[18] Such a construction may perhaps be obtained analogously to the Green's functions for the thermal conductivity problem. See, for instance, L. Tewordt, Phys. Rev. **128**, 12 (1962), also private communications.

especially that our formulas are generally valid only in the weak coupling case ($N\bar{V}\ll 1$), thus it may perhaps be improper to infer any definitive conclusions from experimental studies[7,8] on an anomalous (strong coupling) superconductor like lead.

Note added in proof. A recent paper by K. Maki, Progr. Theoret. Phys. (Kyoto) **29**, 603, 945 (1963), starts from the Green's function approach of Gorkov, and obtains results similar to ours.

ACKNOWLEDGMENTS

The authors are greatly indebted to Dr. P. W. Anderson, Dr. J. Bardeen, and Dr. D. H. Douglass, Jr. for illuminating discussions of some of the problems treated herein. Part of the original work for this article was done at the Summer Institute for Theoretical Physics at the University of Washington, Seattle, supported by the National Science Foundation, in the summer of 1962.

Reprinted from THE PHYSICAL REVIEW, Vol. 139, No. 4B, B1006-B1010, 23 August 1965

Three-Triplet Model with Double $SU(3)$ Symmetry*

M. Y. HAN
Department of Physics, Syracuse University, Syracuse, New York

AND

Y. NAMBU
The Enrico Fermi Institute for Nuclear Studies, and the Department of Physics, The University of Chicago, Chicago, Illinois

(Received 12 April 1965)

With a view to avoiding some of the kinematical and dynamical difficulties involved in the single-triplet quark model, a model for the low-lying baryons and mesons based on three triplets with integral charges is proposed, somewhat similar to the two-triplet model introduced earlier by one of us (Y. N.). It is shown that in a $U(3)$ scheme of triplets with integral charges, one is naturally led to three triplets located symmetrically about the origin of I_3-Y diagram under the constraint that the Nishijima–Gell-Mann relation remains intact. A double $SU(3)$ symmetry scheme is proposed in which the large mass splittings between different representations are ascribed to one of the $SU(3)$, while the other $SU(3)$ is the usual one for the mass splittings within a representation of the first $SU(3)$.

I. INTRODUCTION

ALTHOUGH the $SU(6)$ symmetry strongly indicates that the baryon is essentially a three-body system built from some basic triplet field or fields, the quark model[1] is not entirely satisfactory from a realistic point of view, because (a) the electric charges are not integral, (b) three quarks in s states do not form the symmetric $SU(6)$ representation assigned to the baryons, and (c) a simple dynamical mechanism is lacking for realizing only zero-triality states as the low-lying levels.

These difficulties may be avoided if we introduce more than one basic triplet. Recently one of us (Y. N.) has attempted a two-triplet model[2] where the members of the triplets t_1 and t_2 had the charge assignment (1,0,0) and (0, −1, −1), as had been proposed earlier by Bacry *et al.*[3] The baryon would be represented by the combination $t_1t_1t_2$, whereas the mesons would correspond to some combination $\sim at_1\bar{t}_1'+bt_2\bar{t}_2'$. The triplets are assumed to have masses large compared to the baryon mass, which would mean that baryons and mesons have very large binding energies. A dynamical mechanism for this is provided by a neutral field coupled strongly to the "charm number"[4] C, which is 1 for t_1 and −2 for t_2, and therefore $C=0$ for baryons and mesons. In analogy with electrostatic energy, we can argue that the potential energy due to the charm field would be lowest when the system is "neutral," namely, $C=0$. Thus all other unwanted configurations with $C\neq0$, which include among others triplet, sextet, etc. representations, would have high masses, and hence would not be easily observed.

There have been proposed two different ways in which to introduce basic triplet or triplets with integral charges. One approach essentially involves a modification of the Nishijima–Gell-Mann relation by way of introducing an additional quantum number, the triality quantum number,[5] and this has led to considerations of higher symmetry schemes based on rank-three Lie groups.[6] On the other hand, Okubo *et al.*[7] have recently shown that the minimal group required for this purpose is actually the group $U(3)$.[8] It is shown that a triplet scheme may be defined in $U(3)$ such that the triplet always possesses integral values of charge and hypercharge and satisfies the Nishijima–Gell-Mann relation without a modification. The $U(3)$ triplet considered by Okubo *et al.* is of Sakata type; i.e., it consists of an isodoublet and an isosinglet. Actually, the $U(3)$ scheme is much more appealing than those of the rank-three Lie groups on two accounts: firstly, the Nishijima–Gell-Mann relation is satisfied universally by triplets as by octets and decuplets, and secondly as far as the hitherto realized representations are concerned, $U(3)$ is equivalent to $SU(3)$.[9]

In what follows, we show that the $U(3)$ scheme, when fully utilized as described below, naturally and uniquely

---

* Work supported in part by the U. S. Atomic Energy Commission under the Contract No. AT(30-1)-3399 and No. AT(11-1)-264.

[1] M. Gell-Mann, Phys. Letters **8**, 214 (1964); G. Zweig, CERN (to be published).

[2] Y. Nambu, *Proceedings of the Second Coral Gables Conference on Symmetry Principles at High Energy* (W. H. Freeman and Company, San Francisco, 1965).

[3] H. Bacry, J. Nuyts, and L. van Hove, Phys. Letters **9**, 279 (1964).

[4] This name was originally used in connection with the $SU(4)$ symmetry. B. J. Bjørken and S. L. Glashow, Phys. Letters **11**, 255 (1964); A. Salam, Dubna Conference Report, 1964 (unpublished).

[5] G. E. Baird and L. C. Biedenharn, *Proceedings of the First Coral Gables Conference on Symmetry Principles at High Energy* (W. H. Freeman and Company, San Francisco, 1964); C. R. Hagen and A. J. Macfarlane, Phys. Rev. **135**, B432 (1964) and J. Math. Phys. **5**, 1335 (1964).

[6] For example, see I. S. Gerstein and M. L. Whippmann, Phys. Rev. **137**, B1522 (1965). Earlier references are given in this paper.

[7] S. Okubo, C. Ryan, and R. E. Marshak, Nuovo Cimento **34**, 759 (1964).

[8] The use of $U(3)$ in this connection has also been remarked by I. S. Gerstein and K. T. Mahanthappa, Phys. Rev. Letters **12**, 570, 656(E) (1964).

[9] S. Okubo, Phys. Letters **4**, 14 (1963).

leads to a set of three basic triplets with integral charges, namely an I-triplet (isodoublet and isosinglet), a U-triplet (U-spin doublet and U-spin singlet) and a V-triplet (V-spin doublet and V-spin singlet).[10] These triplets arise from three different ways of defining charge Q, hypercharge Y, and a displaced isospin I_3 in the $U(3)$ group as opposed to the $SU(3)$, in such a way that the charge and hypercharge have integral values, while keeping the Nishijima–Gell-Mann relation intact, and they differ from each other in their quantum-number assignments as well as in their transformation properties under the Weyl reflections.[11] This is described in Sec. II. In Sec. III, a double $SU(3)$ symmetry scheme is proposed based on the three-triplet model in which the large mass splittings between different representations are ascribed to one of the $SU(3)$, and the other $SU(3)$ is, as usual, responsible for the mass splittings within a representation. The low-lying baryon and meson states may be taken as singlets with respect to one of the $SU(3)$. The extended symmetry group with respect to the $SU(6)$ symmetry is briefly discussed.

II. THREE TRIPLETS

We shall denote the infinitesimal generators of $U(3)$ by $A_\nu{}^\mu$ which satisfies the following commutation relations:

$$[A_\beta{}^\alpha,A_\nu{}^\mu]=\delta_\beta{}^\mu A_\nu{}^\alpha-\delta_\nu{}^\alpha A_\beta{}^\mu, \tag{1}$$

where all indices take on the values 1, 2, and 3. The corresponding infinitesimal generators $B_\nu{}^\mu$ of $SU(3)$ are then given by

$$B_\nu{}^\mu=A_\nu{}^\mu-\tfrac{1}{3}\delta_\nu{}^\mu A_\lambda{}^\lambda \tag{2}$$

which satisfy the following equations:

$$[B_\beta{}^\alpha,B_\nu{}^\mu]=\delta_\beta{}^\mu B_\nu{}^\alpha-\delta_\nu{}^\alpha B_\beta{}^\mu \tag{3}$$

and

$$B_\lambda{}^\lambda=0. \tag{4}$$

Furthermore, the unitary restriction gives

$$(A_\nu{}^\mu)^\dagger=A_\mu{}^\nu,\quad (B_\nu{}^\mu)^\dagger=B_\mu{}^\nu. \tag{5}$$

Let us now briefly summarize the relevant results of Okubo *et al.* In the $SU(3)$ scheme, the charge Q, the hypercharge Y and the third component of isospin I_3 are identified as follows[12]:

$$Q=-B_1{}^1, \tag{6a}$$

$$Y=B_3{}^3=-B_1{}^1-B_2{}^2\quad[\text{by the relation (4)}], \tag{6b}$$

$$I_3=\tfrac{1}{2}(B_2{}^2-B_1{}^1). \tag{6c}$$

In the $U(3)$ scheme, the corresponding quantities $\tilde{Q}$, $\tilde{Y}$, and $\tilde{I}_3$ are defined as follows:

$$\tilde{Q}=-A_1{}^1=Q-\tfrac{1}{3}\tau, \tag{7a}$$

$$\tilde{Y}=-A_1{}^1-A_2{}^2=Y-\tfrac{2}{3}\tau, \tag{7b}$$

$$\tilde{I}_3=\tfrac{1}{2}(A_2{}^2-A_1{}^1)=I_3, \tag{7c}$$

where

$$\tau=A_1{}^1+A_2{}^2+A_3{}^3. \tag{8}$$

With these definitions, the Nishijima–Gell-Mann relation is seen to be equally satisfied by the $U(3)$ and $SU(3)$ theories, i.e.,

$$Q=I_3+\tfrac{1}{2}Y \tag{9}$$

and

$$\tilde{Q}=\tilde{I}_3+\tfrac{1}{2}\tilde{Y}, \tag{10}$$

respectively. Since the generators $A_1{}^1$, $A_2{}^2$, and $A_3{}^3$ possess integral eigenvalues in any representation,[13] the identifications of $\tilde{Q}$ and $\tilde{Y}$ to be the charge and the hypercharge, respectively, in $U(3)$ theory shall always lead to integral values for the charge and the hypercharge. In particular, in the three-dimensional representation, the $U(3)$ triplet has the eigenvalues

$$\tilde{Q}=\begin{pmatrix}1&0&0\\0&0&0\\0&0&0\end{pmatrix},\quad \tilde{I}_3=\begin{pmatrix}\tfrac{1}{2}&0&0\\0&-\tfrac{1}{2}&0\\0&0&0\end{pmatrix},\quad \tilde{Y}=\begin{pmatrix}1&0&0\\0&1&0\\0&0&0\end{pmatrix}. \tag{11}$$

This triplet corresponds to the Sakata triplet which we call an I triplet for short.

We can now generalize the above constructions of the $U(3)$ triplet in the following way. Comparing (6b) and (7b), we see that a particular choice has been made for $\tilde{Y}$. Had we defined $\tilde{Y}$ to be $A_3{}^3$, it would still have integral eigenvalues but the relation (10) would have been violated. This is because $B_\lambda{}^\lambda=0$ in $SU(3)$ but $A_\lambda{}^\lambda\neq0$ in general in $U(3)$ and thus some care is needed in defining corresponding quantities in $U(3)$. Making use of (4), the definition in (6) can be written more generally as

$$Q=-B_1{}^1=B_2{}^2+B_3{}^3, \tag{12a}$$

$$Y=B_3{}^3=-B_1{}^1-B_2{}^2, \tag{12b}$$

$$I_3=\tfrac{1}{2}(B_2{}^2-B_1{}^1)=\tfrac{1}{2}(2B_2{}^2+B_3{}^3)=-\tfrac{1}{2}(2B_1{}^1+B_3{}^3). \tag{12c}$$

As in (7), replacing $B_\nu{}^\mu$'s in (12) by corresponding $A_\nu{}^\mu$'s, we list all possible candidates for the corresponding quantities in $U(3)$ which are now however not equivalent to each other [they are equivalent, of course, when reduced to $SU(3)$], i.e.,

$$\tilde{Q}:\quad -A_1{}^1,\quad A_2{}^2+A_3{}^3, \tag{13a}$$

$$\tilde{Y}:\quad A_3{}^3,\quad -A_1{}^1-A_2{}^2, \tag{13b}$$

$$\tilde{I}_3:\quad \tfrac{1}{2}(A_2{}^2-A_1{}^1),\quad \tfrac{1}{2}(2A_2{}^2+A_3{}^3),\quad -\tfrac{1}{2}(2A_1{}^1+A_3{}^3). \tag{13c}$$

[10] C. A. Levinson, H. J. Lipkin, and S. Meshkov, Nuovo Cimento 23, 236 (1961); Phys. Letters 1, 44 (1962) and Phys. Rev. Letters 10, 361 (1963).

[11] A. J. Macfarlane, E. C. G. Sudarshan, and C. Dullemond, Nuovo Cimento 30, 845 (1963).

[12] We use the sign convention of S. P. Rosen, J. Math. Phys. 5, 289 (1964).

[13] For a derivation of this result, see Eq. (7) of Ref. 7.

To start with, the alternative choices in (13) provide twelve inequivalent ways in which to choose a set of three quantities $\tilde{Q}$, $\tilde{Y}$ and $\tilde{I}_3$ for the $U(3)$ scheme. In every choice $\tilde{Q}$ and $\tilde{Y}$ will have integral eigenvalues, but as can be easily checked the Nishijima–Gell-Mann relation will not be valid for all of them. In fact, there are only three cases for which it is valid and we are thus naturally led to three inequivalent triplets in the $U(3)$ scheme; they are defined by the following three choices:

$$t_I:\quad \tilde{Q}=-A_1{}^1,\qquad \tilde{Y}=-A_1{}^1-A_2{}^2,\qquad \tilde{I}_3=\tfrac{1}{2}(A_2{}^2-A_1{}^1), \tag{14a}$$

$$t_U:\quad \tilde{Q}=A_2{}^2+A_3{}^3,\qquad \tilde{Y}=A_3{}^3,\qquad \tilde{I}_3=\tfrac{1}{2}(2A_2{}^2+A_3{}^3), \tag{14b}$$

$$t_V:\quad \tilde{Q}=-A_1{}^1,\qquad \tilde{Y}=A_3{}^3,\qquad \tilde{I}_3=-\tfrac{1}{2}(2A_1{}^1+A_3{}^3). \tag{14c}$$

Now the first one, t_I, for which

$$\tilde{Y}=-A_1{}^1-A_2{}^2, \tag{15}$$

$$\tilde{I}_3=\tfrac{1}{2}(A_2{}^2-A_1{}^1)=\tfrac{1}{2}(B_2{}^2-B_1{}^1)=I_3 \tag{16}$$

corresponds to the I triplet mentioned above.

The structure of the remaining triplets t_U and t_V can be brought to much more transparent and symmetric forms in terms of the U-spin and V-spin subalgebras.[10] As in the case of relations (9) and (10) for $SU(3)$ and $U(3)$, we define the U and V spin of $U(3)$ in exactly the same forms as in $SU(3)$ except that all quantities are tilded quantities. From the $SU(3)$ definitions,[12] we then have

$$\tilde{Y}_U=-\tilde{Q}=-A_2{}^2-A_3{}^3, \tag{17}$$

$$\tilde{U}_3=\tilde{Y}-\tfrac{1}{2}\tilde{Q}=\tfrac{1}{2}(A_3{}^3-A_2{}^2)=\tfrac{1}{2}(B_3{}^3-B_2{}^2)=U_3 \tag{18}$$

for (14b), and

$$\tilde{Y}_V=\tilde{Q}-\tilde{Y}=-A_3{}^3-A_1{}^1, \tag{19}$$

$$\tilde{V}_3=-\tfrac{1}{2}(\tilde{Y}+\tilde{Q})=\tfrac{1}{2}(A_1{}^1-A_3{}^3)=\tfrac{1}{2}(B_1{}^1-B_3{}^3)=V_3 \tag{20}$$

for (14c). They correspond, therefore, to a U triplet and a V triplet, respectively, and hence the notations t_I, t_U, and t_V. With respect to the $SU(3)$ triplet (quark), these $U(3)$ triplets have their respective "hypercharges" (i.e., Y, Y_U, and Y_V) shifted by the amount of $\tfrac{2}{3}$ and as such they have quite different transformation properties under the Weyl reflections W_1, W_2, and W_3[11] which are reflections about the axis $I_3=0$, $U_3=0$, and $V_3=0$, respectively. Whereas the $SU(3)$ triplet is invariant under all three Weyl reflections, the $U(3)$ triplets are not. They transform according to

$$W_1:\quad t_I\to t_I,\quad t_U\leftrightarrow t_V; \tag{21a}$$

$$W_2:\quad t_U\to t_U,\quad t_I\leftrightarrow t_V; \tag{21b}$$

$$W_3:\quad t_V\to t_V,\quad t_I\leftrightarrow t_U. \tag{21c}$$

Figure 1 and Table I(a) list the quantum numbers $\tilde{I}_3$ and $\tilde{Y}$ for the single triplet (quark) model; a possible assignment implied by the two-triplet model[2] is shown in Fig. 2 and Table I(b); the corresponding quantum numbers for the three-triplet model are given in Fig. 3 and Table I(c).

TABLE I. Quantum-number assignments for (a) the quark model, (b) the two-triplet model, and (c) the three-triplet model.

(a)

	quark		
$\tilde{I}_3$	$\tfrac{1}{2}$	$-\tfrac{1}{2}$	0
$\tilde{Y}$	$\tfrac{1}{3}$	$\tfrac{1}{3}$	$-\tfrac{2}{3}$
$\tilde{Q}$	$\tfrac{2}{3}$	$-\tfrac{1}{3}$	$-\tfrac{1}{3}$

(b)

	t_1			t_2		
$\tilde{I}_3$	$\tfrac{1}{2}$	$-\tfrac{1}{2}$	0	$\tfrac{1}{2}$	$-\tfrac{1}{2}$	0
$\tilde{Y}$	1	1	0	−1	−1	−2
$\tilde{Q}$	1	0	0	0	−1	−1

(c)

	$t_1(t_I)$			$t_2(t_U)$			$t_3(t_V)$		
$\tilde{I}_3$	$\tfrac{1}{2}$	$-\tfrac{1}{2}$	0	0	−1	$-\tfrac{1}{2}$	1	0	$\tfrac{1}{2}$
$\tilde{Y}$	1	1	0	0	0	−1	0	0	−1
$\tilde{Q}$	1	0	0	0	−1	−1	1	0	0

III. DOUBLE $SU(3)$ SYMMETRY

Let us call the three triplets $t_1(=t_I)$, $t_2(=t_U)$, and $t_3(=t_V)$. Each triplet may be characterized in general by the average values, $\bar{I}_3$ and $\bar{Y}$, of $\tilde{I}_3$ and $\tilde{Y}$ for its three members. This specifies the location of the center of the triplet in the $\tilde{I}_3-\tilde{Y}$ diagram. Since $\bar{A}_1{}^1=\bar{A}_2{}^2=\bar{A}_3{}^3=\bar{\tau}/3=\tau/3$, Eq. (14) gives for the three definitions of

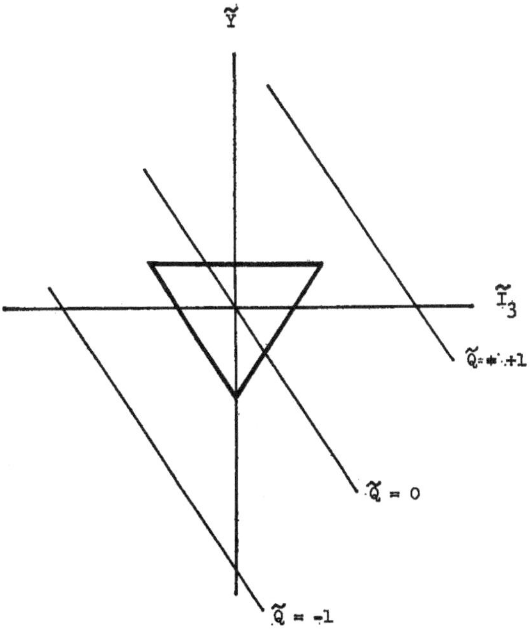

FIG. 1. The single-triplet (quark) model.

$\bar{I}_3$ and $\bar{Y}$,

$$\begin{aligned}\bar{I}_3&=0,\ \tfrac{1}{2}\tau,\ -\tfrac{1}{2}\tau,\\ \bar{Y}&=-\tfrac{2}{3}\tau,\ \tfrac{1}{3}\tau,\ \tfrac{1}{3}\tau,\end{aligned}\tag{22}$$

respectively, where $\tau=-1$ for all the triplets. We may define new quantities I_3, Y and $Q=I_3+\frac{1}{2}Y$ by the relations:

$$\begin{aligned}\tilde{I}_3&=\bar{I}_3+I_3,\\ \tilde{Y}&=\bar{Y}+Y,\\ \tilde{Q}&=\bar{I}_3+\tfrac{1}{2}\bar{Y}+I_3+\tfrac{1}{2}Y=\bar{Q}+Q.\end{aligned}\tag{23}$$

It is clear that I_3 and Y play the role of $SU(3)$ generators within each triplet. The charm number C defined in the two-triplet model[2] is then

$$\tfrac{1}{3}C=\bar{Q}=\bar{I}_3+\tfrac{1}{2}\bar{Y}.\tag{24}$$

Now it is interesting to note that according to Eq. (22) and Fig. 3, the centers of the three triplets form an antitriplet, equivalent to an antiquark, symmetrically located around the origin. Let us suppose that the nine members of the three triplets $t_{1\alpha}$, $t_{2\alpha}$, $t_{3\alpha}$, $\alpha=1, 2, 3$ be combined into a single multiplet $T=\{t_{i\alpha}\}$, $i=1, 2, 3$. We can then imagine two distinct sets of $SU(3)$ operations on T. One is the $SU(3)$ acting on the index α for each triplet, while the other $SU(3)$ acts on the index i, which mixes corresponding members of different triplets. T is then a representation $(3,3^*)$ of this group $G\equiv SU(3)'\times SU(3)''$.[14] The quantum numbers of $SU(3)'$ and $SU(3)''$ are identified as $I_3'=I_3$, $Y'=Y$, $I_3''=\bar{I}_3$ and $Y''=\bar{Y}$ in Eq. (22), so that

$$\begin{aligned}&\tilde{I}_3=I_3'+I_3'',\quad \tilde{Y}=Y'+Y'',\\ &\tilde{Q}=I_3'+I_3''+\tfrac{1}{2}Y'+\tfrac{1}{2}Y'',\\ &\tfrac{1}{3}C=I_3''+\tfrac{1}{2}Y''.\end{aligned}\tag{25}$$

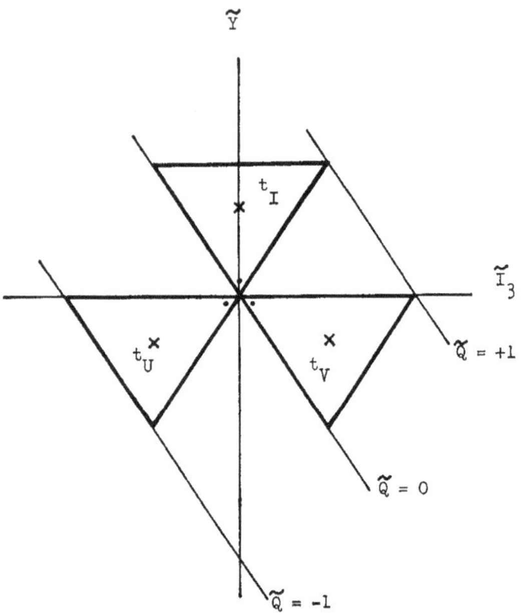

FIG. 3. The three-triplet model.

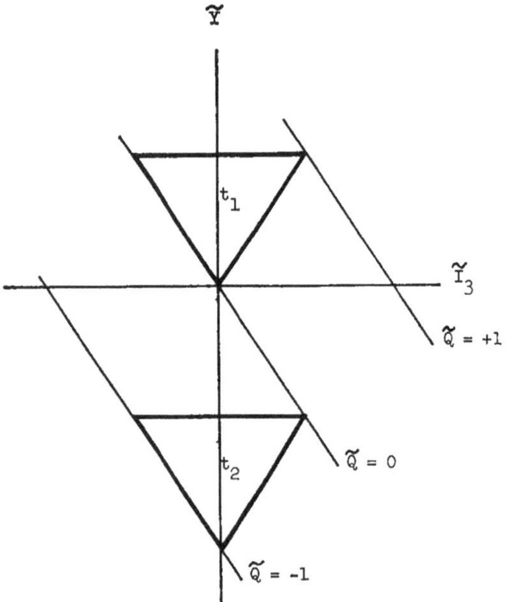

FIG. 2. The two-triplet model.

A general representation of G may be characterized by four numbers p', q', p'', q'' so that $D(p',q',p'',q'')\sim D(p',q')\times D(p'',q'')$, where $D(p,q)$ is a representation of $SU(3)$. However, in our scheme where the nonet T is the fundamental field, we do not get all the possible representations of G. This can be illustrated by means of the triality numbers[5] $t'=p'-q' \bmod(3)$, $t''=p''-q'' \bmod(3)$. The nonet T has $t'=1$, $t''=-1$. All representations constructed out of T and T^* then satisfy $t'=-t''$.

Let us next consider the meson and baryon states $\sim TT^*$ and $\sim TTT$. The $SU(3)'\times SU(3)''$ contents of these 81- and 729-plets are

$$\begin{aligned}&(3,3^*)\times(3^*,3)=(8,1)+(1,1)+(1,8)+(8,8),\\ &(3,3^*)\times(3,3^*)\times(3,3^*)=(1,1)+2(8,1)+2(1,8)\\ &\quad+(1,10^*)+(10,1)+2(8,10^*)+2(10,8)\\ &\quad+4(8,8)+(10,10^*).\end{aligned}\tag{26}$$

It is an attractive possibility to postulate at this point that the energy levels are classified according to $SU(3)''$. The masses will then depend on the Casimir operators of $SU(3)''$. For example, a simple linear form will be

$$m=m_0+m_2C_2''+m_3C_3'',\tag{27}$$

[14] Such a nonet provides a natural basis for the symmetry of $SU(9)$. However, we will not consider it here.

where C_2'', C_3'' are the eigenvalues of quadratic and cubic Casimir operators of $SU(3)''$. In particular, we may assume that the main mass splitting comes from C_2''. Since this increases with the dimensionality of representation, the lowest mass levels will be $SU(3)''$ singlets. This selects the low-lying meson and baryon states to be (8,1), (1,1) and (8,1), (1,1), (10,1), respectively. In general, all low-lying states will have triality zero, $t'=t''=0$.

As for the baryon number assignment to the triplets, the simplest possibility would be to assign an equal baryon number, i.e., $B=\frac{1}{3}$, to them. In this case the triplets themselves would be essentially stable, and their nine members would behave like an octet plus a singlet of "heavy baryons" as may be seen from Fig. 3. Another simple possibility may be $B=\frac{1}{3}+Y''$, namely $B=(1,0,0)$ for (t_1,t_2,t_3). We expect a mass splitting depending on B or Y'', which may be the origin of the Okubo-Gell-Mann mass formula.

The advantage of the three-triplet model is that the $SU(6)$ symmetry can be easily realized with s-state triplets. The extended symmetry group becomes now $SU(6)'\times SU(3)''$. Since an $SU(3)''$ singlet is antisymmetric, the over-all Pauli principle requires the baryon states to be the symmetric $SU(6)$ 56-plet. Other $SU(6)$ representations such as the 70, will be obtained by bringing in either the orbital angular momentum or the "ρ spin" of the Dirac spinor triplets.

As in the two-triplet model mentioned in the Introduction, the mass formula of the type (27) may be derived dynamically. Instead of the charm number field, we introduce now eight gauge vector fields which behave as (1,8), namely as an octet in $SU(3)''$, but as singlets in $SU(3)'$. Since their coupling to the individual triplets is proportional to λ_i'' [the generators of $SU(3)''$], the interaction energy arising from the exchange of these vector fields will yield the first and second terms of Eq. (27). If these mesons obey again a similar type of mass formula, they will be expected to be massive compared to the ordinary mesons. However, it is not clear whether the resulting short-range character of the interaction can be readily reconciled with the postulated largeness of the interaction energy.

We may characterize the hierarchy of interactions and their symmetries implied by the above model as follows. First, the *superstrong* interactions responsible for forming baryons and mesons have the symmetry $SU(3)''$, and causes large mass splittings between different representations. The scale of mass involved would be comparable or large compared to the baryon mass, namely $\gtrsim 1$ BeV. The lowest states, i.e., $SU(3)''$ singlet states, would split according to $SU(3)'$, which would be the $SU(3)$ group observed among the known baryons and mesons, with their *strong* interactions. The scale of mass splitting would then be $\lesssim 1$ BeV.

When we go to the massive $SU(3)''$ nonsinglet states, there may very well be coupling between the two $SU(3)$ groups similar to the $L\cdot S$ coupling. The levels should be classified in terms of the three sets of Casimir operators formed out of λ_i', λ_i'', and $\lambda_i=\lambda_i'+\lambda_i''$, respectively. The splitting due to the coupling would naturally be intermediate between the above two splittings, namely ~ 1 BeV. Because of this coupling, the separate conservation of the two $SU(3)$ spins, I_3' and Y' on the one hand, and I_3'' and Y'' on the other, would be destroyed, and only the sums $I_3=I_3'+I_3''$ and $Y=Y'+Y''$ would be conserved under *strong* interactions. This in turn would mean that all the massive states are in general highly unstable, and decay strongly to the low-lying states. (In the two-triplet model, we considered only weak decays of $C\neq 0$ states. But strong decays are also a possibility as is contemplated here.)

We have discussed here a possible model of baryons and mesons based on three triplets. How can we distinguish this and other different models mentioned already? Certainly different models predict considerably different structure of massive states. These states are characterized by the triality for the quark model, by the charm number for the two-triplet model and by the $SU(3)''$ representation for the present three-triplet model. If we restrict ourselves to the low-lying states only, however, it seems difficult to distinguish them without making more detailed dynamical assumptions.

ACKNOWLEDGMENTS

One of us (M. Y. H.) wishes to thank Professor E. C. G. Sudarshan and Professor A. J. Macfarlane for their encouragement and useful discussions and Professor L. O'Raifeartaigh and J. Kuriyan for helpful comments.

A SYSTEMATICS OF HADRONS IN SUBNUCLEAR PHYSICS

YOICHIRO NAMBU

The Enrico Fermi Institute for Nuclear Studies and the Department of Physics, The University of Chicago, Chicago, Illinois

(Received May 3, 1965)

1.

With the recognition that the SU(3) symmetry is the dominant feature of the strong interactions, the main concern of the elementary particle theory has naturally become directed at the understanding of the internal symmetry of particles at a deeper level. An immediate question that arises in this regard is whether there are fundamental objects (such as triplets or quartets) of which all the known baryons and mesons are composed. These fundamental objects would be to the baryons and mesons what the nucleons are to the nuclei, and the electrons and nuclei are to the atoms. If that was really the case, it would certainly precipitate a new revolution in our conceptual image of the world. At the moment we can only hope that the question will be answered within the next ten to twenty years when the 100 GeV to 1000 GeV range accelerators will have been realized.

Even now, the amusing and rather embarassing success of the SU(6) theory [1] lends support to the existence of those fundamental objects. It is embarassing because this is basically a non-relativistic and static theory, and we do not know exactly how this can cover the realm of high energy relativistic phenomena.

Putting aside those theoretical difficulties mainly associated with relativity, let us make the working hypothesis that there are fundamental objects which are heavy ($\gg 1$ GeV), though not necessarily stable, and that inside each baryon or meson they are combined with a large binding energy, yet moving with non-relativistic velocities. Though this might look like a contradiction, at least it does not violate the uncertainty principle in non-relativistic quantum mechanics since the range of the binding forces ($10^{-14}-10^{-13}$ cm) is large compared

Reprinted from Preludes in Theoretical Physics, eds. A. De-Shalit, H. Feshbach, L. van Hove, © 1966 North-Holland, pp. 133–142.

to the Compton wave lengths of those constituents, and the strength of the forces can be arbitrarily adjusted. In other words, we have a model very similar to the atomic nuclei except for large binding energies. Theoretical justification of such a hypothesis must await future investigation.

In a previous article [2], we have put forward such a model with the following characteristic features.

1) There exist two fundamental fermion triplets t_1 and t_2 with charge assignments (1, 0, 0) and (0, −1, −1) for their three members. The baryons have the structure $\sim t_1 t_1 t_2$, and the mesons $\sim a t_1 \bar{t}_1 + b t_2 \bar{t}_2$.

2) To t_1 and t_2 are assigned "charm" charge $C = +1$ and -2 respectively. Thus the baryons and mesons (zero triality states) have $C = 0$. The primary binding forces acting on them are proportional to C. Let us imagine these forces to be mediated by a field (C-field). The resulting Coulomb-like energy though probably of finite range, then stabilizes the $C = 0$ ("uncharmed") systems against the $C \neq 0$ ("charmed") states, such as the triplets themselves.

3) The SU(6) symmetry can be brought in, with the Pauli principle taken into account, since the constituent particles are non-relativistic. In another paper, we also considered a three-triplet model, in which t_1, t_2 and t_3 have charge assignments (1, 0, 0), (1, 0, 0) and (0, −1, −1) respectively. This has the advantage that the baryon states (the 56-dimensional representation of SU(6)) may be realized with s-state triplets as $\sim t_1 t_2 t_3$.

The reasoning that has gone into the above stability problem is similar to the one used in nuclear physics in deriving the semi-empirical formula of Weizsäcker. The purpose of the present paper is to put this idea into a more precise form, even though the outcome should still be called at best semi-quantitative.

2.

Let us first consider states composed of an arbitrary number of t_1 and t_2, but without antiparticles $\bar{t}_1$ and $\bar{t}_2$. Their masses are M_1 and M_2, respectively, and the "charm" numbers 1 and −2, as was mentioned already. The pairwise interaction energy through the C-field will depend on the spatial configurations of the particles, but we will represent it, in the first approximation, by a constant V_c, as long as the size of the system is comparable with the range of the force. If the number of t_1's and t_2's are n_1 and n_2, respectively, the total energy of the system is

$$\begin{aligned} E(n_1, n_2) &= M_1 n_1 + M_2 n_2 + \\ &\quad + V_c \tfrac{1}{2} n_1(n_1-1) + 4V_c \tfrac{1}{2} n_2(n_2-1) - 2V_c n_1 n_2 \\ &= M_1 n_1 + M_2 n_2 + \tfrac{1}{2}V_c(n_1-2n_2)^2 - \tfrac{1}{2}V_c(n_1+4n_2) \\ &= (M_1 - \tfrac{1}{2}V_c)n_1 + (M_2 - 2V_c)n_2 + \tfrac{1}{2}V_c C^2, \\ C &= n_1 - 2n_2. \end{aligned} \tag{1}$$

As expected, the leading quadratic term depends only on the total charm C. If V_c is sufficiently large, this will favor $C = 0$ as the lowest states, which means $n_1 = 2n_2$. Restricting ourselves to $C = 0$ states now, the remaining terms are linear in n_1 and n_2, implying a saturation property. With $n_1 = 2n_2$, we have

$$E(2n_2, n_2) = (2M_1 + M_2 - 3V_c)n_2. \tag{2}$$

From the physical requirement that this increases with n_2 and that the baryon ($n_2 = 1$) be lighter than the triplets, we further need

$$M_1, M_2 > 2M_1 + M_2 - 3V_c > 0. \tag{3}$$

Thus the energy surface in the $n_1 - n_2$ plane has a valley running along the line $C = n_1 - 2n_2 = 0$, and its level rises linearly with increasing coordinates. However, it will be further necessary to make sure that the $C = 0$ states are actually lower than their neighbors even for small n's. Namely

$$\begin{aligned} E(2n_2 \pm 1, n_2) &> E(2n_2, n_2), \\ E(2n_2, n_2 \pm 1) &> E(2n_2, n_2). \end{aligned} \tag{4}$$

This gives two more conditions

$$V_c - M_1 > 0, \qquad 4V_c - M_2 > 0. \tag{5}$$

Combining Eqs. (3) and (5), we obtain

$$3V_c - 2M_1 > M_2 - M_1 > 3(V_c - M_1) > 0. \tag{6}$$

The second triplet, therefore, must be heavier than the first, but not too much heavier. This is because we have to maintain a balance between the energy due to rest masses and that due to interaction.

Eq. (1) may be expressed in terms of C and the baryon number B if we make an appropriate assignment: $B = x$ for t_1 and $B = y$ for t_2. Since the baryon $\sim t_1 t_1 t_2$ has $B = 1$, we require $2x+y = 1$. Possible choices given in ref. [2] are

$$\begin{aligned} (x, y) &= (\tfrac{1}{3}, \tfrac{1}{3}) \\ \text{or} \quad &(0, 1) \\ \text{or} \quad &(1, -1). \end{aligned} \tag{7}$$

The numbers n_1 and n_2 may be then expressed in terms of C and B as

$$\begin{aligned} n_1 &= 2B+yC \\ n_2 &= B-xC \end{aligned} \tag{8}$$

and thus

$$E(B, C) = \tfrac{1}{2}V_c C^2 + (2M_1+M_2-3V_c)B + [(M_1-\tfrac{1}{2}V_c)-(2M_1+M_2-3V_c)x]C. \tag{9}$$

At this point we should add a reservation that the linear terms in the above mass formula are not as meaningful as the leading quadratic terms since the effects depending on spatial configurations, such as those due to the finite range character of the C-field and the exchange energy, can be of the same order as the former.

3.

In order to consider the meson states, we will next bring in anti-particles as well in the picture. We make the basic assumption that a system consists of definite numbers of $n_1, \bar{n}_1, n_2, \bar{n}_2$ of $t_1, \bar{t}_1, t_2$ and $\bar{t}_2$. This means that *we regard pair creation and annihilation as forbidden processes*, which is consistent with our basic non-relativistic approach.

The formula corresponding to Eq. (1) becomes

$$E(n_1, \bar{n}_1, n_2, \bar{n}_2) = (M_1-\tfrac{1}{2}V_c)(n_1+\bar{n}_1)+(M_2-2V_c)(n_2+\bar{n}_2)+ +\tfrac{1}{2}V_c C^2, \tag{10}$$

$$C = n_1-\bar{n}_1-2(n_2-\bar{n}_2).$$

The requirement that $E > 0$ demands

$$M_1 - \tfrac{1}{2}V_c > 0, \qquad M_2 - 2V_c > 0 \tag{11}$$

in contrast to Eq. (5), which was derived for the special case $\bar{n}_1 = \bar{n}_2 = 0$. We find, together with Eqs. (3) and (5),

$$M_1 > V_c - M_1 > M_2 - 2V_c > 0 \tag{12}$$

which replaces Eq. (6).

We will now relate the constants M_1, M_2 and V_c to the baryon $(t_1 t_1 t_2)$ and meson $(t_1 \bar{t}_1$ and $t_2 \bar{t}_2)$ masses m, μ_1 and μ_2:

$$\begin{aligned} m &= 2M_1 + M_2 - 3V_c, \\ \mu_1 &= 2M_1 - V_c, \\ \mu_2 &= 2M_2 - 4V_c, \end{aligned} \tag{13}$$

from which we obtain an identity

$$2\mu_1 + \mu_2 = 2m. \tag{14}$$

Because of this, we cannot determine the three unknowns M_1, M_2, V_c uniquely. Instead, we can express Eq. (10) in terms of μ_1 and μ_2:

$$E(n_1, \bar{n}_1, n_2, \bar{n}_2) = \tfrac{1}{2}\mu_1(n_1 + \bar{n}_1) + \tfrac{1}{2}\mu_2(n_2 + \bar{n}_2) + \tfrac{1}{2}V_c C^2. \tag{15}$$

Turning to the relation (14), we put $m \sim 1.2$ GeV, $\mu_1 \sim 600$ MeV $= \frac{1}{2}m$ corresponding to the average baryon and meson masses, and *predict* a value

$$\mu_2 \sim m \sim 2\mu_1 \tag{16}$$

for the second meson. This is not an unreasonable value in view of the fact that a large number of unidentified meson resonances seem to exist in this energy range. Eq. (15) reduces then to the simple form

$$E(n_1, \bar{n}_1, n_2, \bar{n}_2) = \tfrac{1}{2}\mu_1[n_1 + \bar{n}_1 + 2(n_2 + \bar{n}_2)] + \tfrac{1}{2}V_c^2. \tag{17}$$

It is rather surprising that such a naive picture as ours can yield non-trivial and qualitatively reasonable results.

By way of a remark, we note from Eq. (13) that

$$\begin{aligned} M_1 &= \tfrac{1}{2}V_c + \tfrac{1}{2}\mu_1 \sim \tfrac{1}{2}V_c, \\ M_2 &= 2V_c + \tfrac{1}{2}\mu_2 \sim 2V_c \sim 4M_1 \end{aligned} \tag{18}$$

since $V_c \gg \mu_1, \mu_2$ by assumption. Interestingly enough, the above relation admits the interpretation that the mass of each triplet is made up of a self-energy due to the C-field plus a small "bare mass" $\frac{1}{2}\mu$.

4.

We will now turn to the three-triplet model [3] proposed as an alternative to the two-triplet model. The three triplets t_1, t_2 and t_3 altogether contain nine fermions $T_{i\alpha}$, $i, \alpha = 1, 2, 3$, where the index i distinguishes different triplets, and α the different members of a triplet. Two different SU(3) operations, called SU(3)′ and SU(3)″, are introduced, acting respectively on α and i, and in these spaces $T_{i\alpha}$ behave as a representation (3, 3*). The electric charge is assigned to each particle according to

$$Q = I_3' + \tfrac{1}{2}Y' + I_3'' + \tfrac{1}{2}Y'' \tag{19}$$

which takes integral values. In fact both $T_{1\alpha}$ and $T_{2\alpha}$ have the assignment (1, 0, 0), and $T_{3\alpha}$ have (0, −1, −1), exactly like t_1 and t_2 of the previous two-triplet model.

An important difference from the two-triplet case is that instead of the charm gauge group $U(1)$, we have the group SU(3)″. The charm gauge field C must then be replaced by an octet of gauge fields G_μ, $\mu = 1, \ldots, 8$, coupled to the infinitesimal SU(3)″ generators (currents) λ_μ'' of the triplets, with a strength g. For a system containing altogether N particles, the exchange of such fields between a pair then results in an interaction energy

$$\begin{aligned} V_G = +g^2 \sum_{n>m} \lambda_\mu''^{(n)} \cdot \lambda_\mu''^{(m)} &= \tfrac{1}{2}g^2 \Big[\sum_{n=1}^{N} \lambda_\mu''^{(n)}\Big]\Big[\sum_{m=1}^{N} \lambda_\mu''^{(m)}\Big] - \tfrac{1}{2}g^2 \sum_{n=1}^{N} \lambda_\mu''^{(n)}\lambda_\mu''^{(n)} \\ &= \tfrac{1}{2}g^2[C_2 - NC_{20}], \end{aligned} \tag{20}$$

where $\lambda_\mu^{(n)}$ refers to the n-th particle, C_2 is the quadratic Casimir operator of SU(3), and $C_{20} = 4/3$ is its value for a triplet representation $D(1, 0)$ or $D(0, 1)$. In general C_2 is given by

$$C_2(l_1, l_2) = \tfrac{1}{3}(l_1^2 + l_1 l_2 + l_2^2) + (l_1 + l_2) \tag{21}$$

for a representation $D(l_1, l_2)$.

Note that the only dependence on the total number N of constituents appears in the second term of Eq. (17).

We add to V_G the rest masses (M = common mass), and obtain the total energy

$$E = (M - \tfrac{1}{2}C_{20}g^2)N + \tfrac{1}{2}g^2C_2. \tag{22}$$

Bound states are characterized by $V_G < 0$, and the low lying states by the smallest value of C_2, namely $C_2 = 0$ for the singlet $D(0, 0)$. For the latter, E is simply proportional to the total number N of constituents, starting with the meson $(N = 2) \sim t_1\bar{t}_1 + t_2\bar{t}_2 + t_3\bar{t}_3$ and the baryon $(N = 3) \sim t_1t_2t_3$ (antisymmetric combination). Their masses are thus related by

$$\mu = 2(M - \tfrac{1}{2}C_{20}g^2) = \tfrac{2}{3}m, \tag{23}$$

and Eq. (22) becomes

$$E = \tfrac{1}{2}\mu N + \tfrac{1}{2}g^2C_2. \tag{24}$$

These are to be compared with Eqs. (14) and (15). Because of the high symmetry among the three triplets, we have found only one set of mesons with $N = 2$. In any case, the energy is simply proportional to the total number of constituents as long as $C_2 = 0$, as if it were made up of non-interacting basic units of mass $\tfrac{1}{2}\mu$.

5.

Having disposed of the gross mass spectrum of many-triplet compound systems, we now turn our attention to the "fine structure" of low lying states, which in our view comprise all the mesons and baryon resonances known so far. In all probability, however, our crude qualitative arguments are not really satisfactory for discussing these details. We will therefore restrict ourselves to general remarks only.

Because of our basic assumptions about the superstrong interactions and the static behaviour of particles, the dynamics we have been dealing with so far does not depend on the spin and the SU(3) spin variables, therefore the system possesses the symmetry of superstrong interactions, the SU(6) symmetry of combined spin and SU(3) spin, and the symmetry of orbital angular momentum. The overall Pauli principle imposes constraints among these symmetries, and thereby single out certain SU(6) and orbital states for the lowest configuration with respect to the superstrong interaction. The general classification of these states can be done as in the case of nuclear and atomic physics, but this will be beyond the scope of the present paper.

In the three-triplet model, however, the problem is relatively simple if we take only s-state triplets. The low lying three particle configuration is a SU(3)″ singlet, so the baryon must go into a complete symmetric SU(6) representation 56. No other states are possible without changing the spatial configuration, but this will cause some change in the superstrong interaction. For the mesons, we obviously obtain $36 = 35+1$ SU(6) states which are degenerate. These results are in accordance with those of the original SU(6) theory, as well as its "relativistic" version.

We must next discuss the two additional effects which do exist and tend to upset the symmetries. One arises from the internal motion of particles, and the other from the presence of virtual mesons. Contrary to the prevalent view, we regard the mesons as perturbing forces rather than the decisive factors in the physics of hadrons. Since the strong interactions are then merely first forbidden processes, so to speak, the meson and baryon resonances are really bound states decaying via violation of superstrong interaction symmetry. Nevertheless, these secondary effects can affect, and may even decide, the "fine structure" of low lying states. Perhaps we may compare the situation to the electronic levels of an atom where the main spectrum is determined by the static Coulomb force, and both the fine structure and the photon emission processes are higher order effects. In this sense, we do not necessarily find a contradiction between the present approach and the conventional strong interaction theory as far as the low lying states are concerned.

The reason we consider the strong interaction as generally symmetry breaking is that the virtual exchange of 36 virtual mesons do not possess an SU(6) symmetric form. An ideal SU(6) symmetric interaction would involve the 35 generators χ_μ as in Eq. (20):

$$V \sim \pm g^2 \sum_{n>m} \chi_\mu^{(n)} \chi_\mu^{(m)} = \pm \frac{g^2}{8} \sum_{n>m} [\tfrac{2}{3}\sigma_i^{(n)}\sigma_i^{(m)} + \lambda_\alpha^{(n)}\lambda_\alpha^{(m)} + \sigma_i^{(n)}\sigma_i^{(m)}\lambda_\alpha^{(n)}\lambda_\alpha^{(m)}]. \tag{25}$$

Viewed as a static force, this requires an exchange of 35 scalar and axial vector mesons.

if the relative signs of the various terms are to be correctly maintained for both particle-particle and particle-antiparticle interactions. [For processes involving meson-baryon scattering, however, Capps [4] and Belinfante and Cutkosky [5] have shown their compatibility with SU(6).]

Next consider the effect of the internal motion. This disturbs the basic symmetry in two senses. It mixes the Dirac spinor components, introducing corrections to the static superstrong forces. Further it simply adds the kinetic energy of orbital motion to the system. As far as the symmetry is concerned, these perturbations act like adding a neutral singlet meson with a suitable spin-parity. Its order of magnitude will depend on the internal velocity v of the particles, which should be of the order $1/MR$ where R is the size of the system. If we take this correction to be of the order $Mv^2 \sim 100$ MeV, and $R \sim 1/M$, we obtain the estimation

$$M \sim 10m, \qquad \frac{v}{c} \sim \frac{1}{10}$$

as we did before [2].

6.

Finally we would like to comment on some obvious difficulties and intriguing problems concerning our model of the subnuclear structure of hadrons.

a) What is the origin of superstrong interactions?

Are these another kind of vector fields or something entirely new? If they are ordinary fields, their range must be at least of the order of the baryon size, and moreover sufficiently smooth and well-behaved in order to keep the kinetic energy small. It is conceivable that no single or a relatively few well defined meson states are responsible for this. A direct confirmation of such interactions would be difficult.

b) The magnetic moments of baryons, for example, agree closely with the SU(6) symmetry, yet obviously the bulk of contributions come from the meson cloud. This means that regardless of whether the meson cloud obeys SU(6) symmetry or not, the baryon should not be considered as composed of three bare triplets without structure. How, then, can we justify our picture that each system, including the mesons, is composed of a definite number of triplets? The answer to this probably should be that the quantities like charm are at any instant well localized at a definite number of centers in space, and these centers are accompanied by large concentrations of energy, moving with slow velocities; whereas the quantities like ordinary charge are more uniformly spread out and carried by faster moving matter. In order to test such a picture experimentally, we would have to use some phenomena which depend on the energy distribution, the correlation functions of charges and energies at different points, the internal velocity of particles, etc.

c) The notion that decays and resonances are actually forbidden processes was first recognized as a surprising paradox in the process of adapting SU(6) to relativity. In our view, this is not only natural, but also simplifies the whole picture. We should be able to discuss the classes of first forbidden, second forbidden, etc. transitions, and they will be accessible to experimental test [6]. For this we should look especially for small, inconspicuous bumps in cross sections, many particle decay modes, and relatively rare events.

d) It has been widely speculated that an axial vector current conservation as relativistic chival symmetry has physical significance. If this is actually the case, it is probably beyond the capacity of our extreme static approach, since we have first to explain away the large masses of triplets, even though we can formally apply group theoretical arguments and the Goldberger-Treiman type relations to individual problems.

REFERENCES

1) F. Gürsey and L. A. Radicati, Phys. Rev. Letters, **13** (1964) 173;
A. Pais, Phys. Rev. Letters, **13** (1964) 175;
B. Sakita, Phys. Rev. **136** (1964) B1765.
2) Y. Nambu, Proc. of the Second Coral Gables Conference on Symmetry Principles at High Energy, University of Miami, January, 1965.
3) M. Y. Han and Y. Nambu, Syracuse University preprint 1206-SU-31.
4) R. Capps, Phys. Rev. Letters, **14** (1965) 31.
5) J. G. Belinfante and R. E. Cutkosky, Phys. Rev. Letters, **14** (1965) 33.
6) A more detailed study of this problem will be done elsewhere.

VOLUME 16, NUMBER 9 PHYSICAL REVIEW LETTERS 28 FEBRUARY 1966

NONLEPTONIC DECAYS OF HYPERONS*

Y. Hara†

Institute for Advanced Study, Princeton, New Jersey

and

Y. Nambu and J. Schechter

The Enrico Fermi Institute for Nuclear Studies and the Department of Physics, The University of Chicago, Chicago, Illinois

(Received 19 January 1966)

Recently Sugawara[1] and Suzuki[2] have shown that the current commutation relations based on the quark model,[3] and the partial conservation of axial-vector currents (PCAC),[4] lead to predictions about the nonleptonic hyperon decays which are compatible with experiment as far as the s-wave amplitudes are concerned. In this paper we show that the above two assumptions plus octet dominance suffice to determine both s- and p-wave amplitudes. At the end we shall also comment on the $K_{2\pi}$ decays.

We shall assume that the nonleptonic weak interaction Hamiltonian H_w is represented either by a product of quark currents in the Cabibbo form,

$$H_w \sim J_\mu{}^\dagger J_\mu + J_\mu J_\mu{}^\dagger, \qquad J_\mu = i\bar{q}\gamma_\mu(1+\gamma_5)(\lambda_1+i\lambda_2)q\cos\theta + i\bar{q}\gamma_\mu(1+\gamma_5)(\lambda_4+i\lambda_5)q\sin\theta, \tag{1}$$

as was done by the above authors, or by a sum of products of charged as well as neutral currents so that the $|\Delta S|=1$ part of H_w behaves as

$$H_w(|\Delta S|=1) \sim d_{bij} J_\mu{}^{(i)} J_\mu{}^{(j)}, \qquad J_\mu{}^{(i)} = i\bar{q}\gamma_\mu(1+\gamma_5)\lambda_i q = V_\mu{}^{(i)} + A_\mu{}^{(i)}. \tag{2}$$

In the former case dynamical enhancement of the octet part will have to be assumed in addition.[5]

The nonleptonic decay amplitude can be written as

$$(2k_0)^{1/2}\langle B^{(a)}\pi^{(b)}(k)|H_w(0)|B^{(c)}\rangle = i\int d^4x\, e^{-ikx}(\mu^2-\Box^2)\langle B^{(a)}|[\varphi^{(b)}(x), H_w(0)]|B^{(c)}\rangle\theta(-x_0), \tag{3}$$

where $\varphi^{(b)}(x)$ is the pion field. From the PCAC relation

$$\partial_\mu A_\mu{}^{(b)}(x) = c\varphi^{(b)}(x) \quad [c = (2m_N\mu^2/g_{\pi NN})(-G_A/G_V)], \tag{4}$$

we obtain

$$\begin{aligned}
&c\int d^4x\langle B^{(a)}|[\varphi^{(b)}(x), H_w(0)]|B^{(c)}\rangle\theta(-x_0)\\
&= \int d^4x\langle B^{(a)}|[\partial_\mu A_\mu{}^{(b)}(x), H_w(0)]|B^{(c)}\rangle\,\theta(-x_0)\\
&= \lim_{k\to 0}\{\int d^4x\langle B^{(a)}|[\partial_\mu(A_\mu{}^{(b)}(x)e^{ikx}), H_w(0)]|B^{(c)}\rangle\theta(-x_0)\\
&\quad -ik_\mu\int d^4x\langle B^{(a)}|[A_\mu{}^{(b)}(x)e^{ikx}, H_w(0)]|B^{(c)}\rangle\theta(-x_0)\}, \quad (\mathrm{Im}k_0<0)\\
&= \langle B^{(a)}|[\int d^3x A_0{}^{(b)}(\vec{x}), H_w(0)]|B^{(c)}\rangle - \sum_{E_n=m(B_8)}\int d^3x[\langle B^{(a)}|A_0{}^{(b)}(\vec{x})|n\rangle\langle n|H_w(0)|B^{(c)}\rangle\\
&\qquad -\langle B^{(a)}|H_w(0)|n\rangle\langle n|A_0{}^{(b)}(\vec{x})|B^{(c)}\rangle].
\end{aligned} \tag{5}$$

From Eqs. (3) and (5) follows

$$(c/\mu^2)\lim_{k\to 0}(2k_0)^{1/2}\langle B^{(a)}\pi^{(b)}(k)|H_w(0)|B^{(c)}\rangle$$
$$=\langle B^{(a)}|[\int d^3xA_0^{(b)}(\vec{x}),H_w(0)]|B^{(c)}\rangle-\sum_{E_n=m(B_8)}\int d^3x[\langle B^{(a)}|A_0^{(b)}(\vec{x})|n\rangle\langle n|H_w(0)|B^{(c)}\rangle$$
$$-\langle B^{(a)}|H_w(0)|n\rangle\langle n|A_0^{(b)}(\vec{x})|B^{(c)}\rangle], \tag{6}$$

if we assume that both sides of Eq. (6) satisfy unsubtracted dispersion relations. The sum must be taken only over the states degenerate in energy and spin with the initial and final states, i.e., the octet baryons B_8.

Equation (6) is identical with the formula for soft pion emission (as induced by H_w) developed by Nambu and Schrauner[6] on the basis of explicit chiral invariance of strong interactions. According to this work, the second term corresponds to "pole diagrams" where the meson is emitted by the initial or final baryon. These diagrams are important in the sense that in the limit of degeneracy $E_n\to m(B_8)$ and zero meson mass $\mu\to 0$, the p-wave meson-emission amplitude remains finite because of a vanishing energy denominator.[7] This justifies the retention of the pole terms even though in the above derivation they would seem to survive only under strict degeneracy.

Working in the above limit, Eq. (6) can now be expressed in terms of the spurion $\langle B^{(a)}|H_w(0)\times|B^{(c)}\rangle=S^{(ac)}$ and the pion-baryon coupling constants alone. The CP invariance implies that $S^{(ac)}$ is a scalar spurion.[2] The first term (Sugawara and Suzuki) gives only S waves since $[\int d^3xA_0^{(b)}(\vec{x}),H_w(0)]$ is effectively a scalar spurion due to CP invariance.[2] The second term, interpreted as pole diagrams, gives only p waves (induced by the scalar spurion).[8] The final result can be cast in the form

$$(c/\mu^2)(2k_0)^{1/2}\langle B^{(a)}\pi^{(b)}|H_w(0)|B^{(c)}\rangle$$
$$=A_{abc}\langle 1\rangle-B_{abc}\langle\vec{\sigma}\rangle\cdot\Delta\vec{p}/\Delta m. \tag{7}$$

Here

$$A_{abc}/\sqrt{2}=(D+F)\operatorname{tr}(\bar{B}_a[M_b,\lambda_6]B_c)-(D-F)\operatorname{tr}(\bar{B}_aB_c[\lambda_6,M_b]),$$
$$B_{abc}/\sqrt{2}=(D+F)(d+f)\operatorname{tr}(\bar{B}_a[M_b,\lambda_6]B_c)+(D-F)(d-f)\operatorname{tr}(\bar{B}_aB_c[\lambda_6,M_b])$$
$$-\tfrac{4}{3}Dd[\operatorname{tr}(\bar{B}_aM_b)\operatorname{tr}(B_c\lambda_6)-\operatorname{tr}(\bar{B}_a\lambda_6)\operatorname{tr}(B_cM_b)]. \tag{8}$$

$\bar{B}$, B, and M are normalized 3×3 baryon and meson matrices, and

$$\langle B'^{(a)}|H_w|B^{(c)}\rangle=(2DD_{6ac}+2FF_{6ac})\bar{u}'u,$$
$$\langle B'^{(a)}|A_\mu^{(b)}|B^{(c)}\rangle$$
$$=(2dD_{bac}+2fF_{bac})i\bar{u}'\gamma_\mu\gamma_5u. \tag{9}$$

In deriving Eq. (7), we have made the simplifying assumption $m_\Lambda=m_\Sigma$, namely that the intermediate states in Eq. (6) have either $m=m^{(a)}$ or $m=m^{(c)}$. There is an ambiguity as to whether (a) we should stick to the degeneracy limit $|\Delta p/\Delta m|=1$; or (b) take the actual values $|\Delta p/\Delta m|\approx q_\pi/190$ MeV $[\Delta m\approx(m_\Xi-m_N)/2]$. The A's and B's are listed in Table I. They satisfy the three $|\Delta I|=\frac{1}{2}$ sum rules as well as the Lee-Sugawara relation[9]

$$\Lambda_-^-+2\Xi_-^-=\sqrt{3}\Sigma_0^+, \tag{10}$$

together with $A(\Sigma_+^+)=0$.

Equation (8) contains only four parameters: D, F, d, and f. We have tried two different fits to 10 (s- and p-wave) amplitudes under the prescriptions (a) and (b) above.[10] Λ_0^0 and Ξ_0^0 are ignored as the $|\Delta I|=\frac{1}{2}$ rule is well satisfied for Λ and Ξ. The solutions are (fit b in

VOLUME 16, NUMBER 9 PHYSICAL REVIEW LETTERS 28 FEBRUARY 1966

Table I. Coefficients A and B of Eq. (8).

	A	B
Λ_-^0	$-(D+3F)/\sqrt{3}$	$-[2(D+F)(d+f)+(D-F)(d-f)]/\sqrt{3}$
Λ_0^0	$(D+3F)/\sqrt{6}$	$[2(D+F)(d+f)+(D-F)(d-f)]/\sqrt{6}$
Ξ_-^-	$(-D+3F)/\sqrt{3}$	$[(D+F)(d+f)+2(D-F)(d-f)]/\sqrt{3}$
Ξ_0^0	$(D-3F)/\sqrt{6}$	$-[(D+F)(d+f)+2(D-F)(d-f)]/\sqrt{6}$
Σ_+^+	0	$\sqrt{2}(\frac{4}{3})Dd$
Σ_0^+	$-(D-F)$	$(D-F)(d-f)$
Σ_-^-	$\sqrt{2}(D-F)$	$\sqrt{2}[(\frac{4}{3})Dd-(D-F)(d-f)]$

brackets)

$$D=0.95(1.1)\times r\times 10^{-7}\mu,$$

$$F=-2.2(-2.2)\times r\times 10^{-7}\mu,$$

$$d=1.02(1.95),\quad f=0.48(1.10), \tag{11}$$

where $r=(-G_A/G_V)$ appears through the coefficient c in Eq. (4). Table II displays the comparison with experiment. The agreement is reasonable except for $B(\Sigma_+^+)$ in fit a. We must bear in mind that the $|\Delta I|=\frac{1}{2}$ rule for the Σ's and the Lee-Sugawara relation are not quite satisfied by experiment in the case of p waves.

Eq. (11) gives $d+f(=-G_A/G_V)=1.5(3.0)$ and $d/f=2.1(1.8)$, which are to be compared with their expected values 1.18 and 1.7 (leptonic decay data[11]), or $\frac{5}{3}$ and $\frac{3}{2}$ [SU(6) theory]. The ratio $D/F=-0.42(-0.5)$ for the spurion is,[12] interestingly enough, comparable to those for the strong and electromagnetic splittings, viz. $D/F\sim -0.33$, suggesting a universal coupling of spurions.

We may thus conclude that the nonleptonic decays can be described reasonably well by means of the four parameters. Their values are theoretically satisfactory except that $d+f$ is uncertain by a factor of 2 and tends to be too large. The deviation from the sum rules in p waves must be attributed to various corrections to our formulas.

Finally we remark that the same procedure may be applied to K decays, relating any two processes which differ by an extra π emission. In this case, however, there are no pole diagrams. Now let us assume the universal spurion coupling, and relate $K\to 2\pi$ decays to the transition $K\to\pi$ due to a spurion D'. D' may be determined by requiring that D'/D (or D'/F) is equal to the corresponding one for the strong (Okubo-Gell-Mann) splitting. The $K_1^0\to 2\pi$ decay rate predicted from this[13] is $\sim 2\times 10^{10}$/sec as compared to the experimental 1.1×10^{10}/sec. The $K_{2\pi}^{\pm}$ decays are forbidden since we have assumed a strict $|\Delta I|=\frac{1}{2}$ rule.

One of the authors (Y.H.) would like to thank Professor Robert Oppenheimer for his hospitality at the Institute for Advanced Study.

*Research sponsored by the U. S. Air Force Office of Scientific Research, Office of Aerospace Research, U. S. Air Force, under AFOSR Nr 42-65, and by the U. S. Atomic Energy Commission under Contract No. AT(11-1)-264.

†On leave of absence from the Physics Department, Tokyo University of Education, Tokyo, Japan.

[1]H. Sugawara, Phys. Rev. Letters 15, 870, 997 (1965).

[2]M. Suzuki, Phys. Rev. Letters 15, 986 (1965).

[3]M. Gell-Mann, Physics 1, 63 (1964).

[4]M. Gell-Mann and M. Lévy, Nuovo Cimento 16, 705 (1960); Y. Nambu, Phys. Rev. Letters 4, 380 (1960).

[5]See, for example, R. F. Dashen and S. Frautschi, Phys. Rev. 137, B1331 (1964); R. F. Dashen, S. Frautschi, and D. Sharp, Phys. Rev. Letters 13, 777 (1964).

[6]Y. Nambu and E. Shrauner, Phys. Rev. 128, 862 (1962). The two terms of Eq. (6) correspond to the second and the first terms of Eq. (3.2) of this reference,

Table II. Comparison with experiment.[10] The alternative theoretical values correspond to the two solutions in Eq. (11).

		Λ_-^0	Ξ_-^-	Σ_+^+	Σ_0^+	Σ_-^-
$(\mu^2/c)A\times 10^7$	Expt.	3.3 ±0.1	−4.4 ±0.1	−0.1 ±0.2	−1.9 ±0.3 −3.8 ±0.3	4.2 ±0.1
	Theory	3.3(3.2)	−4.4(−4.5)	0(0)	−3.2(−3.3)	4.5(4.7)
$(\mu^2/c)\lvert\Delta p/\Delta m\rvert\times B\times 10^7$	Expt.	1.2 ±0.1	0.9 ±0.1	4.2 ±0.2	3.7 ±0.3 1.5 ±0.2	−0.5 ±0.6
	Theory	1.2(1.2)	0.9(0.9)	1.8(4.0)	1.7(2.8)	−0.6(−0.1)
Decay rate,	Expt.	2.6	5.7	6.2	6.4	6.3
$\Gamma\times 10^9$ sec	Theory	2.6	5.7(5.9)	1.0(5.6)	5.0(7.0)	7.3(7.8)
Asymmetry α	Expt.	0.66 ±0.05	−0.41 ±0.05	−0.05 ±0.08	−0.79 ±0.1	−0.16 ±0.21
	Theory	0.66	−0.40(−0.39)	0	−0.83(−0.99)	−0.26(−0.04)

respectively.

[7]We can see this, also, by observing that the invariant amplitude (6) is to be evaluated in a reference frame in which the emitted meson is at rest. As the meson mass decreases, the baryon velocities increase, and hence the matrix elements of A_0 remain finite.

[8]The pion coupling in the pole diagrams of Ref. 6 is of the derivative type. We compute here the pole contributions in the sense of dispersion theory.

[9]H. Sugawara, Progr. Theoret. Phys. (Kyoto) 31, 213 (1964); B. W. Lee, Phys. Rev. Letters 12, 83 (1964). For the s waves, this follows directly from the properties of H_w: M. Gell-Mann, Phys. Rev. Letters 12, 155 (1964).

[10]A. Rosenfeld et al., Rev. Mod. Phys. 37, 633 (1965); R. H. Dalitz, Properties of the Weak Interactions, Lectures at the International School of Physics at Varenna (Oxford University Press, New York, 1965), pp. 1-106.

[11]T. D. Lee and C. S. Wu, Ann. Rev. Nucl. Sci. 15, 381 (1965).

[12]B. W. Lee and A. R. Swift, Phys. Rev. 136, B229 (1964).

[13]There is a peculiar paradox in that the answer depends on to which of the three mesons we apply the PCAC formula. This is due to the fact that the PCAC formula becomes exact only in the limit where the meson represented by the axial current becomes massless and the corresponding chiral SU(2) symmetry is exact. In the SU(3)-symmetry limit (all masses equal), the $K_1^0 \to 2\pi$ amplitude vanishes in agreement with Gell-Mann's general argument (Ref. 9). The finite values quoted here are obtained because we are off the symmetry point. More details about this problem will be reported elsewhere.

368 Supplement of the Progress of Theoretical Physics, Nos. 37 & 38, 1966

Relativistic Wave Equations for Particles with Internal Structure and Mass Spectrum*)

Yoichiro NAMBU

The Enrico Fermi Institute for Nuclear Studies and the Department of Physics The University of Chicago, Illinois, U.S.A.

(Received August 5, 1966)

We consider a class of wave equations which couple an infinite number of tensors or spinors of all ranks. Such a system of equations naturally possesses an infinite number of mass levels, and each eigenfunction implicitly contains a built-in form factor. Two simple examples of first order differential equations are examined. One is based on the set of all finite representations

$$D\left(\frac{n}{2},\ \frac{n}{2}\right) \quad \text{or} \quad D\left(\frac{n}{2}+s,\ \frac{n}{2}\right)+D\left(\frac{n}{2},\ \frac{n}{2}+s\right),$$

$n=0, 1, 2, \cdots$, of the Lorentz group, and gives a mass spectrum resembling the hydrogen atom, but probability densities and form factors are unphysical. The other model is based on a unitary representation of the group $0(4, 2)$, and has an inverted hydrogen-like spectrum. The latter corresponds to a generalization of a model first proposed by Majorana.

§1. Introduction

Relativistic wave equations may be expected to play two rôles, which we will call kinematical and dynamical. In the kinematical sense, a wave equation singles out a particle of definite mass and spin, usually together with its anti-particle counterpart, and specifies how the components of the wave function transform as we change its state of motion or our reference system. This law of transformation must be in accordance with the Poincaré group,**) and therefore a wave equation should characterize an irreducible unitary representation of that group appropriate to quantum mechanical description of a particle.[1]

In the dynamical sense, however, a wave equation ought to serve more purposes. It should be able to describe not only an isolated free particle, but also a particle in interaction, at least with classical fields such as the electromagnetic and gravitational fields, and ideally with all other dynamical fields and associated particles. A wave equation then becomes an equation

---

*) This work is supported by the U. S. Atomic Energy Comission.

**) The Poincaré group, as opposed to the Lorentz group, may be called dynamical, but we use the term kinematical for this purpose.

of motion, embodying the dynamical laws that govern the behavior of the fields and particles. It is generally assumed that such dynamical wave equations are derivable from a local Lagrangian density.

Kinematical wave equations are useful in the S matrix theory, which provides us with a framework into which all dynamics can be fed from the outside. Wave equations serve as boundary conditions for the asymptotic part of the S matrix, but they cannot by themselves describe any dynamical property of a particle, such as the magnetic moment and the form factor. For example, the Bargmann-Wigner equation has turned out to be a convenient tool to describe baryons and mesons of arbitrary spin in the quark model with $SU(6)$ symmetry.[2] One cannot, however, bring in an external field into the equation to make it a dynamical one. The reason for this is that an external field destroys Poincaré invariance so that one cannot in general remain in an irreducible representation of the Poincaré group.

In this paper we explore the possibility of dynamical wave equations which possess an infinite number of mass levels and spins, and exhibit intrinsic form factors in the presence of external fields. Actually, examples of wave equations incorporating a limited number of masses and spins are well known. Even equations with an infinite number of levels have also been studied in the past, including the little known but remarkable work by Majorana in 1932.[3]

In order to show how a form factor of a particle can be described by a "free field", we will take as an example a spin zero particle of mass κ. If simplicity and economy are not the consideration, we may represent a spin zero particle in terms of all possible symmetric tensor fields $\phi(x)$, $\phi_\mu(x)$, $\phi_{\mu\nu}(x)$, $\cdots$ by taking their divergences $\phi(x)$, $\partial_\mu\phi_\mu(x)/\kappa$, $\partial_\mu\partial_\nu\phi_{\mu\nu}(x)/\kappa^2$, etc. So let us assume that the wave function $\psi(x)$ of the particle is expanded in the space of all tensors $\phi^{(n)}$

$$\psi(x)=\sum C_n\phi^{(n)}(x), \tag{1}$$

where the $\phi^{(n)}$ are appropriately normalized. For a plane wave of momentum $\boldsymbol{p}$, we have

$$\psi(\boldsymbol{p})=\sum C_n\phi^{(n)}(\boldsymbol{p}). \tag{2}$$

In particular, a particle at rest $[\boldsymbol{p}=(\boldsymbol{0}, m)]$ will be represented as

$$\psi(0)=\sum C_n\phi^{(n)}_{\underbrace{00\cdots0}_{n}}, \tag{3}$$

and all other components $\phi_{\mu_1\cdots\mu_n}$, $\mu_1\mu_2\cdots\mu_n\neq 00\cdots0$ will be zero. We can construct $\psi(\boldsymbol{p})$ from $\psi(0)$ by carrying out the Lorentz transformation Λ_p, so that $\phi^{(n)}(\boldsymbol{p})=\Lambda_p\phi^{(n)}(0)$.

Now let us define a local scalar product of two wave functions

$$(\psi'(x), \psi(x)) = \psi'^{+}(x)\psi(x) = \sum_n C_n^{*\prime} C_n (\phi^{(n)\prime}(x), \phi^{(n)}(x)),$$

$$(\phi^{(n)\prime}(x), \phi^{(n)}(x)) = \sum_{\mu} g_{\mu'_1\mu_1} \cdots g_{\mu'_n\mu_n} \times \phi^{(n)*\prime}_{\mu'_1\cdots\mu'_n}(x) \phi^{(n)}_{\mu_1\cdots\mu_n}(x),$$

$$g_{11} = g_{22} = g_{33} = -g_{00} = 1\ , \tag{4}$$

which we may regard as a scalar vertex function. Taken between plane wave states $\psi(0)$ and $\psi(\boldsymbol{p})$, this vertex function has a Fourier transform

$$(\psi(\boldsymbol{p}), \psi(0)) = \sum_n |C_n|^2 (\phi(\boldsymbol{p}), \phi(0))$$

$$= \sum_n \epsilon_n |C_n|^2 \gamma^n \phi^{(n)*\prime}_{0\ldots0} \phi^{(0)}_{0\ldots0}$$

$$= \sum_n \epsilon_n |C_n|^2 \gamma^n, \tag{5}$$

if $(\phi^{(n)}(0), \phi^{(n)}(0))$ is normalized to $\epsilon_n = (-1)^n$. Here γ is the Lorentz factor $1/(1-v^2)^{1/2}$ necessary to bring the four-vector $(\boldsymbol{0}, m)$ into $(\boldsymbol{p}, (p^2+\kappa^2)^{1/2})$, and is related to the invariant momentum transfer t by

$$t = 2(\gamma - 1)\kappa^2. \tag{6}$$

Eq. (5) amounts to the expansion of a form factor $F(t)$ into a power series in γ. For example, a simple pole-dominated form $F(t) = \mu^2/(t+\mu^2)$ has a power series

$$\mu^2/(t+\mu^2) = \sum (\alpha/\alpha - 1)^n \gamma^n, \quad \alpha = 2\kappa^2/\mu^2. \tag{7}$$

However, the series converges only if $\gamma < |(\alpha-1)/\alpha|$. Therefore we must have $\alpha < 1/2$ or $\mu^2 > 4\kappa^2$ in order that the series converges at least for $\gamma = 1$.

In the above consideration, we did not assume the tensors $\phi^{(n)}$ to be irreducible, or traceless. From the viewpoint of group representation, it would be more appropriate to use irreducible tensors. Each irreducible tensor of rank n then contains $n+1$ pieces having spin 0, 1, $\cdots$, n, each occurring once, and this decomposition can be done for each plane wave in its respective rest frame. ψ can again be expanded as a sum of spin 0 components of all $\phi^{(n)}$, but the scalar product (5) is now replaced by

$$(\psi(\boldsymbol{p}), \psi(0)) = \sum \epsilon_n |C_n|^2 P_n(\gamma),$$

$$P_n(\gamma) = \sinh[(n+1)\vartheta]/(n+1)\sinh\vartheta, \ \gamma = \cosh\vartheta\ . \tag{8}$$

Here $P_n(\gamma)$ is the four-dimensional Legendre polynomial of order n. Eq. (7) can be written

$$\mu^2/(t+\mu^2) = (1-a)^2 \sum (-a)^n (n+1) P_n(\gamma),$$

$$a=(\alpha-1+\sqrt{1-2\alpha})/\alpha \tag{9}$$

by making use of the generating function

$$1/(1+a^2+2a\gamma)=\sum(-a)^n(n+1)P_n(\gamma).$$

Again, α must be $<1/2$ in order that the expansion exists at least for $\gamma=1$ and can be identified with the scalar product (5).

We shall now pose the following question. Take a set of all possible irreducible tensor fields $\{\phi^{(n)}(x)\}$, or more generally all multi-spinor fields $\{\psi_n^m(x)\}$, where ψ_n^m has n undotted and m dotted spinor indices, is completely symmetric in each set of indices, and therefore corresponds to a representation $D(n/2, m/2)$ of the Lorentz group. A particle of mass κ, momentum p and spin S may be expressed as a superposition of these fields. Is it possible then to set up a system of linear wave equations which couple all the fields in such a way that a plane wave solution will naturally determine, for any spin, the above superposition of fields as an eigenvector, and its mass as an eigenvalue? If this is possible, such a system will give a mass spectrum of an infinite number of excitations, and the form factors associated with such levels. Conversely, given the mass spectrum and form factors we should be able to write down the wave equations.

The use of all finite dimensional non-unitary representations without restriction has the obvious trouble that some of the states will have negative probability density. Besides, as we have seen above, an expansion in terms of irreducible tensors does not converge if the "size" of a particle is larger than its own Compton wave length. To avoid these difficulties, one would have to use infinite-dimensional unitary representations of the Lorentz group instead of non-unitary finite representations. At any rate it can be an interesting exercise to construct possible examples of various types.

In the following we shall examine some simple models which are first order differential equations and utilize either unitary or non-unitary representations. Their mass spectra have a point of accumulation, as given by Eqs. (28) and (49), and "accidental degeneracy" reminiscent of the hydrogen atom. But all these models exhibit some unphysical features one way or another. The problem of finding a more realistic model, or finding wave equations having given spectrum and form factors, will not be attempted here.

§2. General structure of wave equations and the use of operator method

In the last section we were led to consider the set $\{\psi_n^m(x)\}$ of all multi-spinor fields comprising all finite representations $D(n/2, m/2)$ of the

Lorentz group. Since the differential operator $\partial/\partial x_\mu$ behaves as $D(1/2, 1/2)$, $\partial_\mu\psi_n^m$ decomposes into four neighboring representations according to

$$D\left(\frac{1}{2},\frac{1}{2}\right)\otimes D\left(\frac{n}{2},\frac{m}{2}\right)=D\left(\frac{n+1}{2},\frac{m+1}{2}\right)\oplus D\left(\frac{n-1}{2},\frac{m+1}{2}\right)$$
$$\oplus D\left(\frac{n+1}{2},\frac{m-1}{2}\right)\oplus D\left(\frac{n-1}{2},\frac{m-1}{2}\right). \quad (10)$$

The most general set of linear differential equations which couple $\{\psi_n^m\}$ therefore takes the form[4)]

$$\sum_{N'} C^{NN'}\Gamma_\mu^{NN'}\partial_\mu\psi^{N'}=M^N\psi^N. \quad (11)$$

Here ψ^N stands for ψ_n^m; $\Gamma_\mu^{NN'}$ is a Clebsch-Gordan coefficient, appropriately normalized, which couples representations $D(n/2, m/2)$, $D(1/2, 1/2)$ and a neighboring $D(n'/2, m'/2)$; $C^{NN'}=C^{N'N}$ and M^N are arbitrary real scalar factors, and specify the physical contents of the equations. If only a finite number of $C^{NN'}$ are non-zero, we have a finite set of equations, which will possess only a finite number of masses and spins, and their form factors will be polynomials in γ. We are, on the contrary, interested in an infinite set. We will also require in the following that parity should be conserved. This will naturally impose certain conditions on $C^{NN'}$ and M^N.*)

Eq. (10) can be rendered more transparent by use of operational techniques.[5),10)] Let us introduce two sets of spinor operators a_s, a_s^+ and b_s, b_s^+, $s=1, 2$ which will transform like

$$a_s\sim\psi_s\,,\quad a_s^+\sim(\psi^{\dot s})^*,$$
$$b_s\sim\psi^{\dot s}\,,\quad b_s^+\sim(\psi_s)^*, \quad (12)$$

and satisfy the Lorentz invariant commutation relations

$$[a_s, a_{s'}^+]=[b_s, b_{s'}^+]=\delta_{ss'}\,,$$
$$[a_s, b_{s'}]=[a_s, b_{s'}^+]=0\,,\ \text{etc.} \quad (13)$$

We regard a and a^+ (b and b^+) as annihilation and creation operators (but not Hermitian conjugates) for the undotted (dotted) indices so that a spinor

$$\psi_{1\ldots1\,2\ldots2}^{\dot1\ldots\dot1\,\dot2\ldots\dot2}\equiv\psi_{n_1n_2}^{m_1m_2}$$

where 1, 2, $\dot1$, $\dot2$ occur n_1, n_2, m_1, m_2 times respectively, can be generated from a scalar ("vacuum") Ψ_0 as

$$\psi_{n_1n_2}^{m_1m_2}=(n_1!\,n_2!\,m_1!\,m_2!)^{-1/2}(a_1^+)^{n_1}(a_2^+)^{n_2}(b_1^+)^{m_1}(b_2^+)^{m_2}\Psi_0\,. \quad (14)$$

*) The spirit of the present parer is most akin to that of Le Couteur's[4)] except that we are dealing with an infinite system of equations.

The operators a, a^+, b, b^+ thus operate in the space of all irreducible spinors ψ^m_n, into which an arbitrary state vector Ψ may be decomposed.

We can now construct various tensor operators. First of all, the generators of Lorentz transformation $(M_{ij}, M_{i0}) = (i\boldsymbol{L}, \boldsymbol{K})$ are

$$\boldsymbol{L} = a^+\boldsymbol{\sigma}a + b^+\boldsymbol{\sigma}b\,,$$
$$\boldsymbol{K} = a^+\boldsymbol{\sigma}a - b^+\boldsymbol{\sigma}b\,, \tag{15}$$

under which a four-vector (V_i, V_0) will transform according to

$$[V_i, L_j] = 2i\epsilon_{ijk}V_k\,,\ [V_i, K_j] = 2\delta_{ij}V_0\,,$$
$$[V_0, K_i] = -2V_i\,. \tag{16}$$

Other tensors are

Scalar : $S = a^+a + b^+b$

Pseudoscalar : $P = a^+a - b^+b$

Vector or axial vector :

$$V^{(1)}_{I\mu} = A^{(2)}_{I\mu} = (a^+\boldsymbol{\sigma}b - b^+\boldsymbol{\sigma}a,\ a^+b + b^+a)$$
$$A^{(1)\prime}_{I\mu} = V^{(2)\prime}_{I\mu} = (ia^+\boldsymbol{\sigma}b + ib^+\boldsymbol{\sigma}a,\ ia^+b - ib^+a)$$
$$A^{(1)}_{II\mu} = V^{(2)}_{II\mu} = (a^+\boldsymbol{\sigma}Cb^+ + bC\boldsymbol{\sigma}a,\ a^+Cb^+ - bCa)$$
$$A^{(1)\prime}_{II\mu} = V^{(2)\prime}_{II\mu} = (ia^+\boldsymbol{\sigma}Cb^+ - ibC\boldsymbol{\sigma}a,\ ia^+Cb^+ + ibCa)$$
$$\mu = 1, 2, 3, 0;\quad C = i\boldsymbol{\sigma}_2 = \begin{pmatrix} 0 & 1 \\ -1 & 0 \end{pmatrix}. \tag{17}$$

In defining the effect of spatial reflection, we have two alternatives for parity operation:

$$\eta^{(1)}:\quad a \to b\,,\ b \to a\,,$$
$$\eta^{(2)}:\quad a \to b\,,\ b \to -a\,. \tag{18}$$

In Eq. (17), the superscript designates which of the parity operations is to be used.*) These parity operators also serve as metric operators in defining a Lorentz invariant norm (scalar density) of a vector Ψ,

$$(\Psi, \Psi) = \Psi^+\eta^{(1)}\Psi \quad \text{or} \quad \Psi^+\eta^{(2)}\Psi\,. \tag{19}$$

We shall be concerned with a space of vectors Ψ with finite non-zero norms in this sense. The difference between the two definitions is, as can be seen

*) We are concerned with the transformation of the operators like Eqs. (15) and (16). For the state vectors Ψ constructed from Ψ_0 there is an additional freedom of assigning intrinsic parity ± 1 to the "vacuum" Ψ_0.

easily, that $\eta^{(1)}$ corresponds to the metric convention $g_{\mu\nu}=(1, 1, 1, -1)$, and $\eta^{(2)}$ to $g_{\mu\nu}=(-1, -1, -1, 1)$, of the Minkowski space. This means that the spin 0 part of a nth rank tensor has a norm $(-1)^n$ according to $\eta^{(1)}$, or $+1$ according to $\eta^{(2)}$. Thus the metric used in Eq. (5) corresponds to $\eta^{(1)}$.

The general wave equation (11) can now be expressed by means of the above defined operators. For the matrix coefficients Γ_μ we take a set of vector operators V_μ multiplied by functions of operators S, P^2, $V_\mu V^\mu$, etc., whereas the mass operator M will be a function of S, P^2, $V_\mu V^\mu$, etc. In particular, we note that for a representation $D(n/2, m/2)$ we have

$$S=n+m\equiv N\,, \quad P=n-m\equiv N_5\,. \tag{20}$$

These numbers characterize twice the maximum and minimum spin values contained in D. N_5 may also be regarded as the "chirality", being the number of "left-handed" components minus the number of "right-handed" components.

The two types I and II of operators differ in that I keeps N constant $([V_I, N]=0)$ whereas II keeps N_5 constant $([V'_{II}, N_5]=0)$. Thus if V_I is used for Γ, we have equations which link the spinors ψ^m_n for which $n-m$ is fixed: $\psi^0_N, \psi^1_{N-1}, \ldots, \psi^N_0$; whereas if V'_{II} is used, those spinors with $n-m$ constant are linked, e.g., $\psi^0_0, \psi^1_1, \psi^2_2, \ldots$, which can go to infinity. The latter is the sequence that was considered previously. We will study this case below in more detail.

§3. An infinitely coupled equation based on non-unitary representations

For no other reason than the simplicity of mathematical handling, we shall assume that Γ_μ and M of Eq. (11) depend only linearly on the tensor operators defined in the last section. In other words, we assume that they are at most bilinear functions of a, a^+, b, b^+. Thus we write

$$\begin{aligned}
\Gamma_\mu &= (\boldsymbol{V}_{II}, V_{II0}) = (a^+\boldsymbol{\sigma}Cb^+ + bC\boldsymbol{\sigma}a,\ a^+Cb^+ - bCa),\\
\Gamma'_\mu &= (\boldsymbol{V}'_{II}, V'_{II0}) = (ia^+\boldsymbol{\sigma}Cb^+ - ibC\boldsymbol{\sigma}a,\ ia^+Cb^+ + ibCa),\\
M &= \kappa_0(\mathcal{N}-2\alpha),\\
\mathcal{N} &= N+2 = a^+a+b^+b+2\,,
\end{aligned} \tag{21}$$

where κ_0 and α are arbitrary real constants. Since Γ_μ must be a vector, we take $\eta^{(2)}$ of Eq. (18) for parity. There is no essential difference between V_{II} and V'_{II}, however, since they are related by a simple phase transformation on the a's and b's. The Lagrangian is therefore given by

$$\mathcal{L}=\Psi^+\eta^{(2)}(\Gamma_\mu p^\mu - M)\Psi, \quad p^\mu=(-\boldsymbol{p}, E), \tag{22}$$

from which follows the wave equation

$$(\Gamma_\mu p^\mu - M)\Psi = 0\,,$$
$$\Psi^+(\Gamma_\mu \tilde{p}^\mu - M) = 0\,, \quad \tilde{p}^\mu = (\boldsymbol{p}, E)\,. \tag{23}$$

We note that Γ_0 and M are Hermitian, Γ_i anti-Hermitian, but $\eta^{(2)}M = M\eta^{(2)}$, $\eta^{(2)}\Gamma_0 = \Gamma_0\eta^{(2)}$, $\eta^{(2)}\Gamma_i = -\Gamma_i\eta^{(2)}$ so that M and $\eta^{(2)}\Gamma_\mu$ are all Hermitian.

In order to diagonalize Eq. (23) it is expedient to utilize commutation relations among the operators (21) as listed below

$$[\Gamma_0, \Gamma_i] = [\Gamma'_0, \Gamma'_i] = 2K_i\,,$$
$$[\Gamma_i, \Gamma_j] = [\Gamma'_i, \Gamma'_j] = 2i\epsilon_{ijk}L_k$$
$$[\Gamma_0, \Gamma'_i] = [\Gamma'_0, \Gamma_i] = 0$$
$$[\Gamma_0, \Gamma'_0] = 2i\mathcal{N}, \quad [\Gamma_i, \Gamma'_j] = -2i\mathcal{N}\delta_{ij}$$
$$[\Gamma_0, \mathcal{N}] = 2i\Gamma'_0, \quad [\Gamma_i, \mathcal{N}] = 2i\Gamma'_i\,,$$
$$[\Gamma'_0, \mathcal{N}] = -2i\Gamma_0\,, \quad [\Gamma'_i, \mathcal{N}] = -2i\Gamma_i\,. \tag{24}$$

This shows, among other things, that each of the triplets $(i\Gamma_0, i\Gamma'_0, -\mathcal{N})$ and $(\Gamma_i, \Gamma'_i, -\mathcal{N})$ satisfy among themselves commutation relations of the angular momentum operators.

Suppose now we work in the rest frame of a particle of mass κ, assuming p to be time-like, so that

$$[\kappa\Gamma_0 - \kappa_0(\mathcal{N} - 2\alpha)]\Psi = 0\,. \tag{25}$$

We can carry out a rotation (by an imaginary angle) by the operator $i\Gamma'_0$ and bring the vector $(\kappa\Gamma_0, -\kappa_0\mathcal{N})$ to a diagonal form $\sim(0, \mathcal{N})$. This is achieved by the transformation

$$\psi = \exp\left[-i\frac{\theta}{2}\Gamma'_0\right]\Phi, \quad \tanh\theta = \kappa/\kappa_0\,, \tag{26}$$

with the result

$$(\sqrt{\kappa_0^2 - \kappa^2}\,\mathcal{N} - 2|\kappa_0|\alpha)\Phi = 0\,. \tag{27}$$

Hence

$$\kappa = \pm\kappa_0\left(1 - \frac{4\alpha^2}{\mathcal{N}^2}\right)^{1/2} = \pm\kappa_0\left(1 - \frac{4\alpha^2}{(N+2)^2}\right)^{1/2}, \tag{28}$$

provided that $|\alpha| \leq 1$. We have thus obtained a hydrogen-like spectrum! It should be pointed out that the unitary transformation (26) corresponds to an imaginary rotation.*) Since the metric operator $\eta^{(2)}$ commutes with

*) On the other hand, if p is space-like, we have to make a real rotation by a spatial Γ_i'. But this does not lead to real eigenvalues if $|\alpha| \leq 1$. Thus we have only particle-like solutions (27).

Γ'_0, (26) makes the Lagrangian itself diagonal.

In the space of transformed states Φ, N stands for the order of representation, $N=n+m$. The other constant $N_5=n-m$ is arbitrary but can be fixed. Our model is γ_5-invariant in this sense, even though the masses are non-zero. If $N_5=0$, we have a boson sequence $\{\phi_n^m\}$ having even N. For a fixed $n=N/2$, there are $n+1$ different components having spin-parity 0^+, 1^-, $\cdots$, $n^\pm$, all of which are degenerate, with the mass

$$\kappa=\pm\kappa_0[1-\alpha^2/(n+1)^2]^{1/2}.$$

Therefore we have an "accidental degeneracy" (due to $O(3,1)$ symmetry) similar to the hydrogen case, and $n+1$ corresponds to the principal quantum number.

Similarly, a fermion sequence will be obtained if $N_5=1$ or -1, consisting of spinors ϕ_{n+1}^n or ϕ_n^{n+1}. For a given $N=2n+1$, we have $n+1$ degenerate states with spin 1/2, 3/2, $\cdots$, $(n+1)/2$. The two sets $N_5=\pm1$ must be brought in simultaneously to form the Lagrangian since $\eta^{(2)}$ transforms one to the other. Each set, however, has both positive and negative energy solutions. This means essentially that there is a parity doubling of all spin states, formed by two linear combinations $\phi_{n+1}^n\pm\phi_n^{n+1}$. In a similar way we can consider a set $\{\phi_{n+s}^n, \phi_n^{n+s}\}$ of bosons or fermions.

Although the mass spectrum we have found looks very interesting, there are other properties which are unphysical. First of all, the probability density of a state that follows from the Lagrangian is

$$\rho(x)=\Psi^+(x)\eta^{(2)}\Gamma_0\Psi(x). \tag{29}$$

In a given degenerate set of levels characterized by a principal quantum number, this changes sign as the parity (and spin) of the state alternates. So does the energy density

$$\epsilon(x)=i\Psi^+(x)\eta^{(2)}\Gamma_0\Psi(x). \tag{30}$$

In order to avoid the trouble we would have to introduce indefinite metric when $\Psi(x)$ is second quantized, for both boson and fermion cases.

Next we shall compute the form factor (or the Fourier component) of the scalar vertex operator, $F(\boldsymbol{p}'-\boldsymbol{p})=\Psi^+(\boldsymbol{p}')\eta^{(2)}\Psi(\boldsymbol{p})$, between two states of different momenta. For simplicity, take the "$1S$ state" of the boson sequence. The wave function at rest is, according to Eq. (26)

$$\Psi_0=\exp\left[-i\frac{\theta}{2}\Gamma'_0\right]\Phi_0, \tag{31}$$

where Φ_0 is the "vacuum". Now from Eq. (21), we have

$$\Gamma'_0=i(X^+-X),\quad X^+\equiv a^+Cb^+=a_1^+b_2^+-a_2^+b_1^+, \tag{32}$$

and further

$$[X, X^{+}] = \mathcal{N}. \tag{33}$$

Obviously $(X^{+})^{n}\Phi_0 \equiv \Phi_n$ is the spin 0 part of the nth rank tensor ϕ_n^n. It has a norm

$$\Phi_0^{+}\eta^{(2)}\Phi_n = n!\,(n+1)\,!. \tag{34}$$

Using these properties it can be shown that Eq. (31) has the expansion

$$\Phi_0 \equiv \sum_n [\tanh(\theta/2)]^{n}\Phi_n/[n!\cosh^2(\theta/2)]. \tag{35}$$

When the state is set in motion in the z direction with velocity v, each Φ_n in (35) is Lorentz transformed, which amounts to changing X^{+} in Eq. (34) into

$$\begin{aligned} X^{+\prime} &= \exp[-K_z\vartheta/2]\,X^{+}\exp[K_z\vartheta/2] \\ &= e^{-\vartheta}a_1^{+}b_2^{+} - e^{\vartheta}a_2^{+}b_1^{+},\ \tanh\vartheta = v\,, \end{aligned} \tag{36}$$

resulting in the scalar product

$$\begin{aligned} \Phi_n^{+\prime}\eta^{(2)}\Phi_n &= (n!)^2\ (e^{-n\vartheta} + e^{-(n-2)\vartheta} + \cdots + e^{n\vartheta}) \\ &= n!\,(n+1)\,!P_n(\gamma),\ \gamma = \cosh\vartheta\,. \end{aligned} \tag{37}$$

From Eqs. (35) and (37) we obtain the form factor

$$\begin{aligned} F(\boldsymbol{p}'-\boldsymbol{p}) &= \sum\left(\tanh\frac{\theta}{2}\right)^{2n}(n+1)P_n(\gamma)/\cosh^4\frac{\theta}{2} \\ &= \frac{1}{1-\tanh^2\frac{\theta}{2}e^{\vartheta}}\ \frac{1}{1-\tanh^2\frac{\theta}{2}e^{-\vartheta}}\ \frac{1}{\cosh^4\frac{\theta}{2}} \\ &= \frac{1}{1-(\sinh^2\theta)(\gamma-1)/2} = \frac{4(\kappa_0^2-\kappa^2)}{4(\kappa_0^2-\kappa^2)-t}\,. \end{aligned} \tag{38}$$

The denominator has a pole at $t = 4(\kappa_0^2 - \kappa^2)$ which, for small "binding energy" $\kappa_0 - \kappa = B$, reduces to $8\kappa_0 B$. This agrees with the singularity (anomalous threshold) for a real hydrogenic wave function except that we have a wrong sign! Equation (38) corresponds to an unphysical oscillating density distribution $|\psi(x)|^2 \sim \sin\mu r/r$ rather than the exponential form $e^{-\mu r}$ of the hydrogen atom. It is obvious from the deviation that this analytic structure is common to all form factors associated with different tensor operators, the only difference being in the appearance of polynomials in the numerator.

As a further example of the properties of the solution, we compute the magnetic moment assuming minimal electromagnetic interaction. Since our Lagrangian admits a gauge group, the electromagnetic interaction can be brought in by the prescription $\partial_\mu \to \partial_\mu - ieA_\mu$. The magnetic moment operator is then

$$\boldsymbol{\mu} = \frac{e}{2}(\boldsymbol{r} \times \boldsymbol{\Gamma}) = \frac{ei}{2}(\partial/\partial \boldsymbol{p} \times \boldsymbol{\Gamma}). \tag{39}$$

Now, $\partial/\partial \boldsymbol{p} = \partial/\kappa\partial \boldsymbol{v}$, for a particle at rest. This amounts to an infinitesimal Lorentz transformation, the spin part of which is induced by the operator K, Eq. (15). From this argument we arrive at the general formula,

$$\begin{aligned} \langle \boldsymbol{\mu} \rangle &= -ie\langle(\boldsymbol{K} \times \boldsymbol{\Gamma})\rangle/4\kappa \\ &= -e(\Psi^+ \eta^{(2)}(\boldsymbol{K} \times \boldsymbol{\Gamma})\Psi)/4\kappa(\Psi^+ \eta^{(2)} \boldsymbol{\Gamma}_0 \Psi). \end{aligned} \tag{40}$$

If we carry out the transformation (26) the operators become

$$\begin{aligned} &\boldsymbol{K} \to \boldsymbol{K} \cosh\theta - i\boldsymbol{\Gamma}' \sinh\theta\,, \quad \boldsymbol{\Gamma} \to \boldsymbol{\Gamma}\,, \\ &\Gamma_0 \to \Gamma_0 \cosh\theta + \mathfrak{N} \sinh\theta\,. \end{aligned} \tag{41}$$

Since K and $\mathfrak{N}$ are diagonal, and Γ_0, $\boldsymbol{\Gamma}$ off-diagonal for an eigenstate $\boldsymbol{\Phi}$, Eq. (40) reduces to

$$\langle \boldsymbol{\mu} \rangle = \Phi^+ \eta^{(2)}(\Gamma \times \Gamma')\Phi/4\kappa\Phi^+ \eta^{(2)} \mathfrak{N} \Phi\,. \tag{42}$$

Evaluating this explicitly using the representation (21), we find

$$\begin{aligned} \langle \boldsymbol{\mu} \rangle &= \frac{eg}{2\kappa}\langle \boldsymbol{S} \rangle, \\ g &= -1 + \frac{N_5^2}{4S(S+1)}, \end{aligned} \tag{43}$$

where $\boldsymbol{S}$ is the spin angular momentum. For example, for the fermion sequence, the ground spin 1/2 state has $N=1$, $N_5 = \pm 1$, which gives $g = -2/3$ in contrast to the Dirac value $g=2$.

§4. An example based on a unitary representation

We will show in this section that the general formalism developed above can also accommodate wave equations based on infinite dimensional unitary representations, just by reinterpreting various operators which appear in the commutation relations (24). The essential point is that the three Hermitian operators $i\boldsymbol{\Gamma}'$ (or $i\boldsymbol{\Gamma}$) satisfy the same commutation relations as for the non-Hermitian operators $i\boldsymbol{K}$, so that if we identify $(i\boldsymbol{L}, -\boldsymbol{\Gamma}')$ with the Lorentz generators, the space of spinors $\{\psi_n^m\}$ will become a basis for a unitary representation of the Lorentz group with space reflection. The parity operator in this case is either $\eta^{(2)}$ of Eq. (18) or

$$\eta^{(3)}=\eta^{(2)}\eta^{(1)}:\ a\to a,\ a^+\to a^+,\ b\to -b,\ b^+\to -b^+ \tag{44}$$

The operators a, a^+ (b, b^+) are now Hermitian conjugates. Furthermore, we find the following Hermitian tensors

$$\begin{array}{ll} \Gamma_0 & \text{scalar (pseudoscalar)} \\ (i\boldsymbol{\Gamma}, \mathcal{N}) & \text{vector} \\ (\boldsymbol{K}, \Gamma'_0) & \text{vector (axial vector)} \end{array} \tag{45}$$

if the metric operator is 1 (i.e. $\langle\Gamma'_0\rangle$ means $\Psi^+\Gamma'_0\Psi$, etc.), and the parity is $\eta^{(2)}$ (brackets correspond to the case of $\eta^{(3)}$). The reason for this remarkable property is that the set $\{\psi^m_n\}$ actually forms a basis for a unitary representation of 6-dimensional rotation group $O(4,2)$ $(\sim U(2,2))$ which leaves the quadratic form $x_1^2+x_2^2+x_3^2+x_4^2-x_5^2-x_6^2$ invariant. According to Eq. (24), its generators $M_{\mu\nu}$ can be identified with

$$\begin{aligned} &M_{ij}=i\epsilon_{ijk}L_k\,,\quad M_{i4}=iK_i\,,\quad M_{i5}=-\Gamma'_i\,,\\ &M_{i6}=-\Gamma_i\,,\quad M_{45}=i\Gamma'_0\,,\quad M_{46}=i\Gamma_0\,,\\ &M_{56}=i\mathcal{N}\,. \end{aligned} \tag{46}$$

In view of Eq. (45), let us now consider an analog of the Lagrangian (22)

$$\mathcal{L}'=\Psi^+[\mathcal{N}p_0-i\boldsymbol{\Gamma}\cdot\boldsymbol{p}-\kappa_0(\Gamma_0+2\alpha)]\Psi \tag{47}$$

with the parity operator defined by $\eta^{(2)}$.*) From this follows, in the rest system,

$$[\kappa\mathcal{N}-\kappa_0(\Gamma_0+2\alpha)]\psi=0\,. \tag{48}$$

Compared with Eq. (25), the two operators $\mathcal{N}$ and $\boldsymbol{\Gamma}_0$ are switched here. The solution is thus

$$\begin{aligned} &\Psi=\exp\left[-i\frac{\theta}{2}\Gamma'_0\right]\Phi\,,\quad \tanh\theta=\kappa_0/\kappa\,,\\ &\kappa=|\kappa_0|\left(1+\frac{4\alpha^2}{\mathcal{N}^2}\right)^{1/2}\operatorname{sgn}(\kappa_0\alpha)\,, \end{aligned} \tag{49}$$

instead of Eqs. (26) and (28). The spectrum is now inverted, approaching κ_0 from above with increasing principal quantum number. Since m has only one sign, there are no anti-particle states, energy and charge (probability) density being both definite. This is a characteristic feature which is possible only for unitary representations. In order to accommodate anti-

*) There is also a possibility of using the second vector set of Eq. (44). This, however, does not give meaningful solutions since both Γ_0 and Γ_0' are non-diagonalizable in the space of Ψ with finite norm.[12]

particles, therefore, it is necessary to double the Ψ space, as in the previous model.

The wave equation (48) and its mass spectrum (49) are generalizations of results obtained by Majorana[8] and others.[6] Majorana considered an infinite system of wave equations which couple all spinors ψ_n having only one type of indices since such a set does form a basis for a unitary irreducible representation of the Lorentz group.*) In our operator formalism his equation has the following form

$$\mathcal{L}=\Psi^+(\gamma_\mu p^\mu-\kappa_0)\Psi\,,$$
$$\gamma_0=a^+a+1\,,\ \boldsymbol{\gamma}=\frac{1}{2}(a^+\boldsymbol{\sigma}Ca^+-aC\boldsymbol{\sigma}a),\tag{50}$$

whereas the Lorentz generators $M_{\mu\nu}$ and parity operator are given by

$$M_{ij}=ia^+\boldsymbol{\sigma}a\,,\ M_{i0}=-\frac{1}{2}(a^+\boldsymbol{\sigma}Ca^++aC\boldsymbol{\sigma}a),$$
$$\eta:\ a\to ia\,,\ a^+\to -ia^+.\tag{51}$$

In the rest system, Eq. (50) leads to the mass spectrum

$$\kappa=\kappa_0/\gamma_0=\kappa_0/(n+1),\quad n=0,1,2,\cdots,\tag{52}$$

where n is now uniquely related to the spin S and parity P by $n=2S$, $P=(-1)^S$, with no accidental degeneracy. The difference between these two cases is obviously due to the fact that we took the set $\{\psi_n^m\}$ and therefore had a larger space to play with. In Majorana's case, for example, there is no non-trivial mass operator (scalar) which would alter the above mass spectrum.**) This is because Majorana equation is embedded in a single irreducible unitary representation $D_{0,1/2}$ or $D_{1/2,0}$ of the Lorentz group, where D_{S_1,S_2} designates a representation containing minimum spin S_1 and maximum spin (if it is non-negative integer or half-integer) S_2-1.

Our equation, on the other hand, is not irreducible with respect to the Lorentz group. However, if we take the square $\sum M_{\mu\nu}M_{\mu\nu}$ of the generators of the 6-dimensional group, from Eq. (46) we find by explicit computation

$$-\sum M_{\mu\nu}M_{\mu\nu}=\boldsymbol{L}^2+\boldsymbol{K}^2+\boldsymbol{\Gamma}^2+\boldsymbol{\Gamma}'^2-\Gamma_0^2-\Gamma_0'^2+\mathfrak{N}^2$$
$$=3N_5^2-12\,,\tag{53}$$

which is a constant since N_5 commutes with all generators. Thus we are dealing with a manifold belonging to a definite value of the invariant

*) Majorana's equation also is a simple realization of irreducible unitary representations of the Poinceré group. There exist another class of equations which use continuous internal variables to realize unitary representations of the Poincaré or Lorentz group.[7] Such equations are more directly related to the non-local theory of Yukawa.

**) It is a general feature of unitary representations (with constant mass operator) that the mass spectrum tends to zero. See Naimark[4] and Gel'fand et al.[4]

(Casimir operator), Eq. (53), of the group $O(4, 2)$.

It has not been possible to compute explicitly the form factors associated with our solution.[9] We only remark that there is no danger of singularities in the unphysical region. In fact, the scalar form factor $F(p)=(\Psi(p), \Psi(0)) =\Psi^{+}(p)\Psi(0)$ is real and satisfies the Cauchy inequality $|F(p)|\leq|F(0)|$.

The magnetic moment, computed according to the technique developed in the last section, yields the same results as Eqs. (42) and (43). In the Majorana theory, on the other hand, it is given by[3]

$$\langle \mu\rangle=-\frac{e}{2\kappa}\langle S\rangle \tag{54}$$

for all spin states.

Summary and discussion[10),11)]

We have studied two examples of infinitely coupled system of wave equations. Both are based on the set $\{\psi_n^m\}$, $n-m=$constant, of multi-spinors. Depending on the definition of Lorentz generators, we may regard the set either as a sum of finite representations $D(n/2, m/2)$, or of certain infinite unitary representations of the Lorentz group, imbedded in a unitary representation of $O(4, 2)$. The simplicity of the mass spectra, Eqs. (28) and (49), together with their $O(3,1)$ or $O(4)$ degeneracy, appears to be due to the existence of this larger group.

The unphysical nature of the first model is rather unfortunate since it yields a mass spectrum very much like that of the hydrogen atom, although this trouble was to be expected from the beginning since non-unitary representations are used without subsidiary conditions. The second model, on the other hand, does not obviously conflict with quantum mechanical interpretations, but the mass spectrum as well as the magnetic moment is not quite natural. Consequently the relevance of these examples to physical reality is purely of academic nature at the moment. Nevertheless an attack along this line may give us useful insight into possible new ways of describing real particles. For this purpose, the problems of field quantization, causality and unitarity, behavior under C, T and P, etc., would be worth exploring. We also remark that the internal quantum numbers such as $SU(3)$ can be readily incorporated into our scheme. Our approach will then become more closely related to various recent papers dealing with higher hadron symmetries.[5),8)] For example, we can conceive of equations of the type we have considered here, where the algebraic structure of the current and mass operators Γ_μ and M are in accord with the one that follow from the quark model.

References

1) E. P. Wigner, Ann. Math. (Princeton) **40** (1939), 149; *Group Theoretical Concepts and Methods in Elementary Particles*, edited by F. Gürsey (Gordon and Breach, New York, 1963), p. 37.
2) V. Bargmann and E. P. Wigner, Proc. Nat. Acad. Sci. **34** (1948), 211.
A. Salam, R. Delbourgo and J. Strathdee, Proc. Roy. Soc. A**284** (1965), 146.
B. Sakita and K. C. Wali, Phys. Rev. **139** (1965), B1355.
3) E. Majorana, Nuovo Cim. **9** (1932), 335.
See also an interesting review article by D. M. Fradkin, Am. J. Phys. **34** (1966), 314. The author thanks Prof. F. Gürsey and Mr. J. Cronin for first calling his attention to these papers. For related papers see references 4) and 6).
4) General wave equations are discussed, for example, in E. M. Corson, *Introduction to Tensors, Spinors, and Relativistic Wave Equations* (Blackie and Son Ltd., London, 1953). I. M. Gel'fand, R. A. Minlos and Z. Ya. Shapiro, *Representations of the Rotation and Lorentz Group and their Applications*, English translation (Pergamon Press, Oxford, 1963). M. A. Naimark, *Linear Representations of the Lorentz Group*, English translation (Pergamon Press, Oxford, 1964). A. J. Le Couteur, Proc. Roy. Soc. A**202** (1950), 284, 394.
5) Similar operator methods were proposed by
B. Kursunoglu, Phys. Rev. **135** (1964), B761.
Y. Dothan, M. Gell-Mann and Y. Ne'eman, Phys. Letters **17** (1965), 148.
6) Majorana's equation was rediscovered later: I. M. Gel'fand and A. M. Yaglom, Zhur. Exper. i Teoret. Fiz. **18** (1948), 703.
7) P. A. M. Driac, Proc. Roy. Soc. A**183** (1944), 284.
E. P. Wigner, Z. Physik **124** (1947), 665.
E. M. Corson, reference 4).
H. Yukawa, Phys. Rev. **77** (1953), 219; **91** (1953), 415, 416.
V. L. Ginzburg and I. Ye. Tamm, Zhur. Eksper. i Teor. Fiz. **17** (1947), 227.
V. L. Ginzburg, Acta Phys. Polonica **15** (1956), 163, etc.
8) For form factors and mass spectra, see C. Fronsdal, Symmetry Principles at High Energy, *Proceedings of the Third Coral Gables Conference* (W. H. Freeman and Co., New York, 1966); and other preprints from Trieste.
9) We have in the meantime determined the (scalar) form factor. It is the same as Eq. (38), but has now the correct analytic structure since $\kappa_0^2 < \kappa^2$.
10) It has come to our attention that a number of papers have been published by T. Takabayashi who considers internal structure of particles in terms of Bose operators corresponding to a relativistic oscillator, regarded as a model for $SU(3)$. He also discusses wave equations and mass spectra using these operators. The author thanks Drs. Fronsdal and Takabayashi for private communication about their respective works in relation to the present paper.
T. Takabayashi, Prog. Theor. Phys. **36** (1966), 185; H. Kase and T. Takabayashi, Prog. Theor. Phys. **36** (1966), 187; and references therein.
11) Working in the product space of the unitary representation of $O(4,2)$ and a finite (Dirac) representation of the Lorentz group, we have now obtained an equation which properly simulates the hydrogen atom regarding mass spectrum, magnetic moment and form factors.
12) In this paper we considered only discrete spectra corresponding to normalizable internal wave functions. Whether we should also include continuum (ionized) states remains an open question.

Infinite Multiplets

Y. Nambu

University of Chicago

I. Introduction

The main ingredients of the method of infinite multiplets consists of:[1]

1) the use of wave functions with an infinite number of components for describing an infinite tower of discrete states of an isolated system (such as an atom, a nucleus, or a hadron),
2) the use of group theory, instead of dynamical considerations, in determining the properties of the wave functions.

Group theory is used in three ways. It determines the relativistic transformation properties of the wave function as an infinite-dimensional (not necessarily irreducible) representation of the Lorentz group. It determines the internal quantum numbers and degeneracy structure of the mass levels as a consequence of an assumed symmetry group in the rest frame of the particle (i.e., a little group). It is also used to define various observables. The smallest of the symmetry group is the rotation group O(3), which gives rise to the $(2j+1)$-fold degeneracy of states with spin j. In general, we can have a larger degree of degeneracy, including that due to the so-called internal symmetry like $SU(2)$, $SU(3)$, etc. Thus we are led to a hierarchy of groups

$$\begin{array}{ccc} S_0 & \subset & S \sim L \subset \rho \\ \cap & & \cap \\ G_0 & \subset & G\,. \end{array}$$

Here $S_0 = SO(3)$, and G_0 is the degeneracy group (dynamical symmetry group). S (the internal Lorentz group) and G are their

Published in Proceedings of 1967 International Conference on Particles and Fields, ed. C. R. Hagen et al., © 1967 John Wiley & Sons, pp. 347–373.

Lorentz closures obtained by boosting them. Usually G_0 is compact and semi-simple. G_0 may be just a direct product $G_0 = S_0 \otimes A$, where A is an internal symmetry group, and hence $G = S \otimes A$. There are, however, other non-trivial examples, like the hydrogen atom, the isotropic harmonic oscillator, or the $SU(6)$ model of hadrons, in which G is not such a direct product. We take an (infinite dimensional) representation of the non-compact group G and consider it, in addition, as a function of the space-time coordinates, thus making it an infinite component field $\Psi_n(x)$. The discrete index n arises because we reduce Ψ with respect to the compact subgroup G_0. In general we would like to choose G and Ψ in such a way that the set $\{\Psi_n\}$ coincides with the complete set of actual physical levels. However, sometimes it may be more convenient to relax this condition. The energy-momentum 4-vector and the total angular momentum (or the physical Lorentz group L) can be defined on this field, and $\{\Psi_n(x)\}$ becomes also a representation of the Poincaré group (inhomogeneous Lorentz group) P. However, we are not trying to combine P and G into a larger group, so we dismiss herewith all the problems related to the theorems of McGlinn, O'Raifeartaigh, etc.

As was mentioned already, the group G and its Lie algebra serve the purposes of defining the spin and other quantum numbers, ensuring relativistic covariance, and expressing physical observables such as the mass and currents. In this sense G is sometimes called a dynamical group[2] or a non-invariance group[3], and its Lie algebra forms a spectrum generating algebra and a transition operator algebra[4] as well as a symmetry algebra.

The replacement of dynamics by group theory has been one of the recent trends in elementary particle physics. Its motivation comes from the fact that, having seen a general qualitative success of the quark model picture of hadrons, one wants a more quantitative, but simple and unified description of hadron phenomena without assuming detailed dynamical mechanism or the existence of the yet unknown quarks as real physical particles. In this kind of approach, the wave function $\Psi(x_1, x_2)$ for a two-particle bound state, e.g., a meson as a quark-antiquark system, can be brought into a more abstract form (departing from a space-time description of the internal structure) by expanding it with respect to an appropriate discrete basis $\{f_n\}$:

$$\Psi(x_1, x_2) \to \Psi(X, r) \to \sum_n c_n(X) f_n(r) .$$

The set $\{c_n(X)\} = \{\Psi_n(X)\}$ then defines an infinite component field. Thus the emergence of infinite component fields seems entirely natural from this viewpoint.

There are other considerations that have led some people to turn their attention to infinite component fields. For example, the scattering amplitude cannot be made well behaved (i.e., to satisfy superconvergent relations and sum rules) unless one introduces an infinite tower of states; whereas, vertex functions acquire damping form factors if the field transforms as a unitary (and therefore infinite-dimensional) representation of the Lorentz group. I would like to remark, however, that the unitarity requirement in quantum mechanics for the probability density does not, in itself, lead to the necessity of having a unitary representation of G. As is well known, it is the unitary representation of the Poincaré group P that we need, not necessarily of L or G. (In order to secure the unitarity with respect to P, however, it will be necessary to invoke wave equations.)

Starting from more abstract principles, one can also arrive at the infinite component fields. Thus Takabayasi's long series of work[5] has been motivated by the non-local (bilocal, and later, quadrilocal) theory of Yukawa.[6] The pioneering work of Majorana[7] was motivated by the non-existence (at that time) of anti-particles; a fact which is not valid any longer.

Because of the different viewpoint from which different people have started, there seems to be no consensus about the position of the infinite component fields relative to the conventional mathematical tools in elementary particle physics. One may use them, for example, as part of the S-matrix theory in defining an infinite multiplet of asymptotic fields, or simply as a basic for the expansion of an amplitude. Or one may regard them as a new possibility within the framework of axiomatic field theory. One may also consider them as belonging to Lagrangian field theory. In the last circumstance, a field will be subject to a wave equation which will determine the mass spectrum of a system, as well as its interaction with external fields, etc. The question is whether one can make a formulation which is free of internal inconsistencies, and besides, is useful and relevant to the actual physics of hadrons. However, the more we demand out of it, the more problems we must also face. This will be seen in the following.

II. Choice of the dynamical group G and its representation

The first contact between mathematics and physics is made when one decides on the choice of the group G and its representation. For this purpose, the following considerations are in order:

1) The degeneracy structure (exact or approximate) of levels will depend on the symmetry group G_0.
2) The nature of low lying excitation; i.e., whether there exist rotational excitations, radial excitations, etc.
3) The behavior of form factors and transition amplitudes between levels.

For example, if radial excitations in a two-body system are frozen, we have for G_0 the smallest possible one $G_0 = SO(3)$ ignoring intrinsic spin and other internal degrees of the constituents, and $G = SO(3,1)$. We should then consider an irreducible representation of $SO(3,1) \sim SL(2,c)$, or a finite sum of IR's in order to accommodate parity and other additional operators. Since an IR contains, when reduced with respect to the rotation group, different spin values j (either integer or half-integer) only once for each j, we get a single Regge trajectory (combining both signatures).

When radial excitations are included and the degeneracy group G_0 is larger than $SO(3)$, we run into two conspicuous physical examples that have been extensively studied.

1) $G_0 = SO(4)$, $G = SO(4,1)$ or $SO(4,2) \sim SU(2,2)$, (hydrogen atom).

2) $G_0 = SU(3)$, $G = SU(3,1)$ or $U(3,1)$ (harmonic oscillator).

The number of times an IR of the subgroup G_0 occurs in an IR of G depends on the representation, so we have to choose a particular kind (usually a degenerate representation) that corresponds to the actual multiplet structure of levels. When reduced with respect to the rotation group, it gives rise to an infinite recurrence of Regge trajectories, but each trajectory is not simply related to the reduction with respect to the internal Lorentz group $S \subset G$.

III. Models based on $SL(2,c)$

Since $SO(3,1) \sim SL(2,c)$ is the minimal group in our approach, we will take this as a basic example to discuss some characteristic problems. A finite-dimensional IR $D(j_1,j_2)$ of $SL(2,c)$ is given in terms of a symmetric tensor product $\sim \psi^{t_1 \ldots t_n}_{s_1 \ldots s_m}$; $m = 2j_1$, $n = 2j_2$; $s,t = 1,2$, built out of basic spinors $\psi_s \sim D(\frac{1}{2},0)$, $\psi^t \sim D(0,\frac{1}{2})$, with the property $D(j_1,j_2) \approx D(j_2,j_1)^*$. It decomposes into the spin substance as $D(|j_1-j_2|) \oplus D(|j_1-j_2|+1) \oplus \ldots \oplus D(|j_1+j_2|)$. The general IR of $SL(2,c)$ can be obtained from this by analytic continuation, fixing $k = j_1 - j_2$, and letting $j_1 + j_2 + 1 \equiv c$ take an arbitrary complex value. Thus we write an IR as $D(k,c)$, which will decompose in general as $\sum_{n=0}^{\infty} \oplus D(|k|+n)$. It also turns out that $D(k,c) = D(-k,-c)$.

The space reflection (parity) may be defined on $D(j_1,j_2)$ as $j_1 \leftrightarrow j_2$, which means $D(k,c) \leftrightarrow D(-k,c) = D(k,-c)$.

Let us write the six infinitesimal generators as $iL_i = \varepsilon_{ijk}\Lambda_{jk}$ $K_i = \Lambda_{io}$. $\mathbf{L}$ operates within each subspaces $D(j)$; $\mathbf{K}$ has matrix elements between neighboring $D(j)$'s. Different representations differ in the matrix elements of $\mathbf{K}$.

An IR $D(k,c)$ is unitary, i.e., $\mathbf{K}$ is Hermitian, if and only if

a) $c =$ imaginary (principal series), or
b) $k = 0$, $0 < c < 1$ (supplementary series).

Since k is the lowest spin value that occurs in an IR, it is fixed by physical considerations. Thus c is the only variable parameter.

1. *Form factors*

Define a local scalar product $\sum_n \bar{\Psi}_n(x)\Psi_n(x)$ where $\{\Psi_n\}$ transforms as $D(k,c)$ under S. In a unitary representation, $\bar{\Psi} = \Psi^*$. In general we have to use an adjoint field $\{\bar{\Psi}_n\} \sim D(k,-c)$. Now let us consider a plane wave state which is at rest and has definite spin values j and j_z. A moving state can be obtained from this by applying the Lorentz transformation L to both spin and orbital parts (boosting). In this way we can define a scalar vertex function for a transition between two Fourier components (momenta) p_μ and p'_μ having spin and helicity (j,h) and (j',h') respectively. Essentially it is given by the matrix element

$$\langle j', h'| \exp(i\theta K_z)|j,h\rangle \tag{1}$$

between two spin substates of Ψ_n and $\bar{\Psi}_n$, $\tanh\theta = v$ is the relative velocity between the two states. We have the simple relation

$$\gamma \equiv \cosh\theta = p\cdot p'/mm' = 1 + \frac{(m-m')^2 - t}{2mm'}\ . \tag{2}$$

For a finite-dimensional representation, the t-dependence (form factor) of (1) is essentially like a polynomial, but we obtain non-trivial results in the infinite-dimensional case. For example, with

$$j = j' = 0, \quad \langle 0| \exp(i\theta K_z)|0\rangle = \sinh(c\theta)/c\sinh\theta\ . \tag{3}$$

We make the following remarks.

a) The maximum degree of growth of (1) for large $\gamma \sim -t/2m^2$ is like $\lesssim \gamma^{j_1+j_2} = \gamma^{c-1}$ in the sense that there exists at least one matrix element for any fixed j and j' which behaves like $\sim \gamma^{c-1}$. (This follows from elementary considerations about the behavior of a finite-dimensional $D(j_1,j_2)$ which can be analytically continued.) Thus for unitary representations the matrix element $\to 0$ as $\gamma \to \infty$, but it does so only like $\sim \gamma^{-1}$ asymptotically. It should be mentioned that this applies to the overall matrix element, and not to the invariant form factors defined after separating out kinematic factors. Similar arguments can be made for vector (current) and other vertices when these can be defined.

b) The matrix elements (1) have the right threshold behavior. This can be understood from the fact that the Lorentz generator **K** has a selection rule $|j-j'|<1$, so that the matrix element must behave like $\sim \theta^{|j-j'|} \sim v^{|j-j'|}$.

c) Analyticity and crossing. The matrix elements (1) for infinite representations have a branch point at $\gamma=-1$, corresponding to the normal threshold $t_0=(m+m')^2$, and at $\gamma=\infty$. This is due to the fact that the imaginary Lorentz transformation $\gamma=\cosh i\pi=-1$, which is a compact rotation, acts on an infinite basis. For the same reason, the vertices do not have crossing symmetry[8]: the transition element between positive ($p_0>0$) and negative ($p_0'<0$) Fourier components as calculated from Eq. (1) is not the same as the usual analytic continuation of the positive-positive matrix element. In fact, the former should be regular at threshold ($\gamma=-1$) and singular at $\gamma=1$. The two form factors can be related only by analytic continuation around ∞. These points can be explicitly seen from Eq. (3).

2. *Wave equations*

Most wave equations that have been considered are limited to quadratic or linear (Majorana[7]–Gelfand–Yaglom[9] type) differential equations, if only for reasons of mathematical simplicity and nothing else. These equations will be of the form

$$(p_\mu p^\mu + \alpha w_\mu w^\mu - \beta)\Psi = 0 \quad \text{or} \tag{3a}$$

$$(\Gamma_\mu p^\mu - \kappa)\Psi = 0\ , \tag{3b}$$

where w_μ is the Pauli–Lubanski vector, Ψ is in general reducible under $SL(2,c)$, and α, β, κ may be functions of the Casimir operators. In fact, we cannot take a single arbitrary IR for Ψ in general. To define parity, we need a pair (k,c) and $(k,-c)$. In order to define Hermitian operators, we need also their adjoints. Finally, in the case of linear equations, we need a pair of IR's (k,c) and $(k',c')=(k\pm 1,c)$ or $(k,c\pm 1)$ in order to realize current operators Γ_μ.[9,1] Thus we end up with up to eight IR's. The Majorana equation ($(k,c)=(0,\frac{1}{2})$ or $(\frac{1}{2},0)$) is the only case where a single IR suffices.

These equations, as they may look simple and elegant, suffer from well-known deficiencies from the physical point of view:

a) The mass spectrum $m(J) \to 0$ as the spin $J \to \infty$.[9] (Exceptions to this are the cases, with Eq. (3a), i) $\alpha = 0$, $m^2 =$ const., and ii) $\beta = 0$, $J(J+1) = \alpha$, m^2 arbitrary.)
b) There is also a continuous family of lightlike solutions, and even worse, of unphysical spacelike solutions as well.[7,10,11] (There have been proposals[12,13] to interpret the spacelike solutions in physical terms.)
c) In the case of Majorana representation, the massive and massless states come in with only one sign of the energy (or frequency). Thus a particle is not accompanied by an anti-particle. If, however, a pair of IR's are used in Eq. (3b), we can have both signs. So this is not a general feature of infinite-dimensional representations.

The above remarks do not necessarily apply to more complicated (non-Gelfand–Yaglom type) equations; e.g., when κ in Eq. (3b) is a function of p^2. We can get a rising spectrum $m \sim J$ by taking $k \sim p^2$. But the other difficulties still persist.

Another way is to take an infinite (or finite) set of IR's $D(k, c)$, $k = 0, 1, 2,$ (Boson) or $\frac{1}{2}, \frac{3}{2}, \ldots$ (Fermion) with fixed c, and postulate

$$\begin{aligned} [p^2 - f(k)]\Psi &= 0 \\ [w^2 - g(k)]\Psi &= 0 \end{aligned} \tag{3c}$$

in such a way that only certain spin values are selected for each k. In this case the space $\{\Psi\}$ is much larger than is necessary to accommodate physical levels, but it has the advantage of being free of all unphysical solutions. For example, we obtain a rising linear spectrum $m^2 = aj + b$ by choosing $f(k) = ak + b$, $g(k) = -f(k)k(k+1)$. Eq. (3c) can be derived from a Lagrangian by introducing an auxiliary field.

3. *Minimal electromagnetic interaction*

We discuss here briefly the problem of electromagnetic interaction, especially the magnetic moment. It is indeed an interesting feature of our approach that non-trivial electromagnetic properties can result even with the assumption of minimal electromagnetic interaction. This assumption is reasonable at least if we are effectively dealing with a composite system in which only one of the constituents, considered elementary, is charged.

The current is uniquely defined as

$$j_\mu(x) = \bar{\Psi}\omega V_\mu \Psi(x), \quad V_\mu = \frac{\delta L}{\delta p^\mu}$$

$$\partial_\mu j_\mu(x) = 0$$

if the free Lagrangian density is $\bar{\Psi}L\Psi$. In addition to the form factors discussed above, we will also obtain non-trivial static magnetic moment (and higher moments) by taking the matrix element of j_μ. The result is, however, again rather unphysical in the case of the already unphysical equation (3), i.e., the g factor is zero for Eq. (3a), and < 0 for Eq. (3c) if only a pair of IR's are used.[14] For non-Gelfand–Yaglom type equations, the situation can be different.

IV. Models based on $SO(4,2) \sim SU(2,2)$

As has been discovered by various authors,[15–17] the group $SO(4,2) \sim SU(2,2)$, rather than $SO(4,1)$, provides a nice framework in which to formulate the problem of the hydrogen atom. It is also interesting that $SU(2,2)$ is the group of Dirac matrices, and therefore, it serves as prototype of theories like $SU(6,6)$. But this connection is probably accidental.

$SO(4,2)$ contains $SO(4) \otimes SO(2)$ as its compact subgroup, so that there are three discrete quantum numbers to label a state. Since the actual hydrogen atom has only two, corresponding to $SO(4)$ degeneracy, we need to select a special degenerate representation in which the third quantum number is not independent of the other two. In fact, we find a unique representation that serves our purpose.

We label the six fictitious dimensions $0, 1, 2, \ldots, 5$, with metric $+ - - - - +$. The physical space is identified with $1, 2, 3$. The secret of success in reformulating the non-relativistic hydrogen atom lies in the fact that, using the physical space as the carrier space, the 15 generators $M_{\alpha\beta}$ of $SO(4,2)$ may be identified as

$$\begin{aligned}
M_{45} + M_{50} &= r = (\mathbf{r}^2)^{1/2} \\
M_{45} - M_{50} &= r\Delta \\
M_{4i} - M_{0i} &= x_i \\
M_{0i} + M_{4i} &= x_i\Delta + 2x_k\partial^k\partial_i + 2\partial_i \\
M_{5i} &= ir\partial_i \\
M_{4i} &= i(x_i\partial^i + 1) \; .
\end{aligned} \tag{4}$$

These operators are Hermitian, and hence give a unitary representation, if the metric is defined by

$$(\Psi, \Psi) = \int \Psi^*(\mathbf{r})\frac{1}{r}\Psi(\mathbf{r})d^3x \; . \tag{5}$$

Now multiply the Schrödinger equation

$$(H + B)\Psi = 0, \qquad H = -\frac{1}{2}\Delta - \frac{\alpha}{r} \tag{6}$$

by r, and use Eq. (4). We get immediately

$$\left[\left(\frac{1}{2} + B\right)M_{50} + \left(\frac{1}{2} - B\right)M_{54} - \alpha\right]\Psi = 0 \; . \tag{7}$$

Once this is achieved, we can also go to a discrete basis $\Psi(\mathbf{r}) \to \Psi_n$ using the quantum numbers of the compact subgroup.

Eq. (7) can be solved by making a hyperbolic rotation in the (04) plane:

$$\Psi = \exp(i\theta M_{04})\Psi', \qquad \tanh\theta = \frac{(\frac{1}{2} - B)}{(\frac{1}{2} + B)}, \qquad \text{if } B > 0 \; . \tag{8}$$

The transformed equation

$$\left\{\left[\left(\frac{1}{2} + B\right)^2 - \left(\frac{1}{2} - B\right)^2\right]^{1/2} M_{50} - \alpha\right\}\Psi' = 0$$

immediately gives the familiar spectrum

$$\sqrt{2B}\, M_{50} = \alpha, \qquad \text{or} \qquad \beta = \frac{\alpha^2}{2M_{50}^2}$$

where the discrete "principal quantum number" $M_{50} = n$ takes the values $1, 2, 3, \ldots$, in this representation. Similarly, the ionized states $B < 0$ follow by rotating into the direction 4: $B = -\alpha^2/2M_{54}$, where the non-compact generator M_{54} now takes continuous values.

It was shown by Fronsdal[17] that the wave function Ψ_n can be made into a field $\Psi_n(x)$ with a Galilei invariant Hamiltonian. The electromagnetic interaction may be introduced according to the rule $\mathbf{P} \to \mathbf{P} - e\mathbf{A}(\mathbf{x})$. The equation is still linear in the $SO(4,2)$ generators. This corresponds to a description of the hydrogen atom in which

$$M_p \gg M_e, \qquad \mathbf{X} = \mathbf{r}_e, \qquad \mathbf{r} = \mathbf{r}_p - \mathbf{r}_e\,.$$

The peculiar feature of this problem is the rotation (8), first noted by Barut.[15] Since θ is different for different levels, the eigenfunctions Ψ_B are not orthogonal; they are orthogonal only with respect to a metric operator $M_{50} - M_{54}$. This nicely solves the old dilemma that physical transition operators cannot be represented by generators since the latter induce transition only between neighboring orthogonal states in a representation.

So much for the non-relativistic hydrogen atom. By assigning the subspace (0123) to the Minkowski space, one can formally write down linear relativistic wave equation with $SO(4)$ degeneracy. But again we tend to run into the difficulties encountered in the $SL(2,c)$ case. Nevertheless, some model equation without spacelike solutions have been found.[16,17] As an example, we mention

$$\begin{gathered} \left[\Gamma_\mu p^\mu + \frac{1}{k} S p_\mu p^\mu - \alpha\gamma_\mu p^\mu\right] \Psi = 0 \\ \Gamma_\mu = M_{5\mu}, \qquad S = M_{54} \end{gathered} \tag{9}$$

where Ψ is a direct product of Dirac and $SO(4,2)$ representations. This equation has a discrete hydrogen-like mass spectrum

$$M = \pm\kappa\sqrt{1 - \alpha^2/n^2}$$

a continuous spectrum $|M| > \kappa$, and a family of massless solution, but is free of spacelike solutions.

V. Models based on $SU(3,1)$

This group is the minimal Lorentz extension of $SU(3)$, the degeneracy group of the 3-dimensional isotropic harmonic oscillator. Perhaps the harmonic oscillator is an appealing idealization of the orbital motion in the quark model of hadrons as well as the nuclear shell model. At the same time, it has an advantage in the simplicity of its mathematical properties.

In order to construct a representation, we introduce the 4-dimensional oscillator variables a_μ, a_μ^+, $(\mu = 0,1,2,3)$ with $[a_\mu^+, a_\nu] = G_{\mu\nu}$ (metric $+---$). (Annihilation operators are thus a_1, a_2, a_3 and a_0^+.) Then the 16 combinations $a_\mu^+ a_\nu$ form generators of $U(3,1)$. The Lorentz rotation group $O(3,1)$ is generated by the antisymmetric tensor $a_\mu^+ a_\nu - a_\nu^+ a_\mu = iL_{\mu\nu}$ whereas the symmetric tensor $i(a_\mu^+ a_\nu + a_\nu^+ a_\mu) = i\Gamma_{\mu\nu}$ corresponds to deformation and dilatation. The $SU(3)$ subgroup (G_0) is generated by a_i, a_i^+ $(i = 1,2,3)$ alone.

In order to describe the harmonic oscillator, then, we have to suppress the fourth degree of freedom. This can be done by going to $SU(3,1)$.[18–20] We impose the condition

$$-a_\mu^+ a^\mu = \sum_{i=1}^{3} a_i^+ a_i - a_0 a_0^+ = \lambda - 1$$

which relates $a_0^+ a_0$ to $N = \sum a_i^+ a_i$. It can be shown that the substitution $a_0 \to \sqrt{N-\lambda}$, $a_0^+ \to \sqrt{N-\lambda}$ in the $U(3,1)$ generators leaves their commutation relations unchanged except that now $\Gamma_\mu^\mu = \lambda$. We get the following IR's

i) $\lambda \leq 0$, unitary.
ii) $\lambda =$ integer > 0, $N < \lambda$, finite non-unitary.
iii) $\lambda =$ integer ≥ 0, $N > \lambda$, discrete unitary.

The behavior of form factors according to these representations is $\sim \gamma^\lambda$ for i) and ii), and $\sim \gamma^{-1-\lambda}$ for iii). They have an anomalous singularity at $\gamma = 0$ or $t = M^2 + M'^2$, the physical meaning of which is unclear. (It may be a little surprising that we do not obtain the Gaussian form factor of the harmonic oscillator, but the latter follows as a non-relativistic limit in which $|\lambda| \to \infty$.)[19] At any rate, by a suitable choice of λ we can get any asymptotic power behavior.

The wave equations based on $SU(3,1)$ are necessarily quadratic, since $\Gamma_{\mu\nu}$ is the only available tensor. Thus a typical form will be

$$[\Gamma_{\mu\nu}p^\mu p^\nu - \alpha p_\mu p^\mu + \beta]\Psi = 0 \tag{10}$$

with the mass spectrum

$$m^2 = \beta/(\alpha + \lambda - N), \qquad N = 0, 1, 2, \ldots . \tag{11}$$

This is physical only if $\beta > 0$, $\alpha + \lambda - N > 0$. There are no other solutions corresponding to a real four-vector p_μ. If we take a fourth order equation, we can also obtain an infinity of levels with a hydrogen-like accumulation point.

Another form similar to Eq. (10) is based on $SU(3,1)$ and eliminates the unphysical solutions by means of a supplementary condition.[20,21] We take

$$\begin{gathered} [p_\mu p^\mu + \kappa^2(a_\mu^+ a^\mu - c)]\Psi = 0, \\ (p_\mu a^{+\mu})\Psi = 0 \quad [\text{or } (p_\mu a^\mu)(p_\nu a^{\nu+})\Psi = 0]. \end{gathered} \tag{12}$$

The mass spectrum is strictly linear

$$m^2 = \kappa^2(N + c) \tag{13}$$

whereas the supplementary condition suppresses the fourth degree of freedom: $a_0^+\Psi = 0$ for time-like solutions, and eliminates space-like solution completely: $a_i^+\Psi = 0 \to \Psi = 0$. In this scheme, $\lambda = a_\mu^+ a^\mu = a_1^+ a_1 - 1 = N - 1$, so that each level belongs to a different IR $D(\lambda)$ (discrete series) of $SU(3,1)$. Eq. (12) may be derived from a Lagrangian.

VI. Problems of quantized fields

The peculiar unconventional properties of infinite component fields with regard to the general conditions of quantum field theory have been widely noted.[22–27] We will confine ourselves here to very brief remarks. Certainly more work is necessary than has been done so far to clarify these problems.

First we note that the results will, in general, depend on whether: a) we regard $\Psi(x)$ simply as a field, or b) as also satisfying a wave equation explicitly or implicitly to define single particle states. In the case a) we may assume all the conventional axioms of field theory (spectral conditions on the states, local commutativity of fields, etc.) except that the fields transform as an infinite dimensional representation of $SL(2,c)$.

A major difference between finite and infinite representations is that the latter cannot be regarded as a (non-singular) representation of the complex Lorentz group $\sim SL(2,c) \otimes SL(2,c)$ because it is infinite dimensional with respect to the compact imaginary Lorentz transformation. So all the theorems like TCP and spin-statistics, based on such a transformation, are in general expected to break down. (The lack of crossing symmetry in vertex functions was understood in this way earlier.)

This does not mean, however, that these theorems always have to break down. The representations $D(\lambda)$ of $SU(3,1)$ do have the crossing symmetry (up to sign) if $|\lambda|$ is an integer. Perhaps this is another attractive feature of $SU(3,1)$.

In the case b), a great deal depends on the assumed wave equation. Since we have a Lagrangian, it is possible to define the energy-momentum tensor, conserved current vector, etc. and formally introduce canonical quantization.

For a physical theory, the energy density should be positive. Thus, if the wave equation is of a simple Klein–Gordon type (as in the so-called index-invariant theories), we would need Bose statistics, whether the field describes integer or half-integer particles. This requirement also turns out to lead to causal commutation relations. On the other hand, in a Gelfand–Yaglom type equation, Fermi statistics are to be taken for any spin.

There may remain some doubt about the validity of canonical quantization and local commutativity. For if the field $\Psi(x)$ is supposed to describe a composite system, local commutativity with respect to the "center of mass" coordinates x will not hold.

This is probably related to the problem of unphysical spacelike solutions since physical solutions alone would not form a complete set of states, and therefore would not lead to the local commutation relation required in the canonical quantization scheme.

Although the infinite component fields seem to require a relaxation of general theorems, we emphasize that not everything goes overboard. The conventional TCP and spin-statistics can be preserved if a right group and a right representation are chosen. Besides, they offer a new possibility of deliberately violating, for example, the TCP Theorem, micro-causality and crossing relations.

VII. Scattering processes

As a next step in the physical application of infinite component wave equations, we take up the scattering with an external field in Born approximation. We are interested in the intrinsic structure effect such as we have seen in the vertex function. For simplicity, we take a scalar external field $\Phi(x)$ coupled to the local current $g(\Psi(x), \Psi(x))$, so that the wave equation reads $L_0\Psi = g\Phi\Psi$. We assume a perturbation expansion and obtain for the transition $p + k \rightarrow p' + k'$ a formal expression

$$\left(\Psi(p', n'), g^2 \left\{\frac{1}{L_0(p+k)} + \frac{1}{L_0(p-k')}\right\} \Psi(p, n)\right) = M_1 + M_2 \quad (14)$$

corresponding to the usual second order Feynman diagrams. Actually the propagator $\sim 1/L_0$ will involve a summation over all the intermediate states, and this is where interesting properties are expected to emerge.

We emphasize, however, that Feynman perturbation theory and dispersion theory are not necessarily equivalent for infinite component fields, since the usual analyticity assumption do not hold. Thus we have to compute Eq. (14) directly. For this purpose, it is convenient to use the trick

$$\frac{1}{L_0 + i\varepsilon} = -i \int_0^\infty d\tau \exp[iL_0\tau] \tag{15}$$

(assuming that this gives the right boundary condition). In case L_0 is linear in the generators of a group G, with the coefficients being functions of the momentum, the integrand represents an element of G, so that its action on Ψ can be determined within the framework of group theory. Similarly, the initial and final states can be obtained by boosting appropriate rest states. In this way, Eq. (14) is reduced to

$$\begin{aligned} -ig^2 \int_0^\infty d\tau \big(\Psi(\mathbf{0}, n') \exp(-i\theta_{p'}K_{p'}) \exp(-iL_0(p+k)\tau) \\ \times \exp(i\theta_p K_p)\Psi(\mathbf{0}, n)\big) + \dots , \end{aligned} \tag{16}$$

where K_p, K'_p are the boost generators.

We will treat this in the $SU(3,1)$ theory with a unitary representation $\lambda < 0$ since mathematical manipulation is particularly easy.[28,29] For simplicity, we consider the ground state $j = 0$ and equal masses $p^2 = p'^2 = k^2 = k'^2 = 1$, and work in the C.M. system. The wave equation is generally taken as

$$L_0(p) = -\Gamma_{\mu\nu}p^\mu p^\nu f(p^2) + g(p^2), \qquad \Gamma_{00} = 2(N - \lambda) .$$

The result is, for the direct channel (s-channel) amplitude M_1,

$$\begin{aligned} M_1 &= -ig^2 \int_0^\infty d\tau [\cosh^2\theta_p - \cos\theta e^{-2i\tau s f(s)} \sinh^2\theta_p]^\lambda e^{i\tau[g(s)+2\lambda f(s)]} \\ &= -ig^2 \int_0^\infty d\tau \left[\frac{s}{4} - e^{-2i\tau s f(s)} \left(\frac{t}{2} + \frac{s}{4} - 1\right)\right]^\lambda e^{i\tau[g(s)+2\lambda f(s)]} \\ &= \sum_{n=0}^\infty \frac{-g^2}{2sf(s)} \oint_{c_n} dz\, z^{\alpha(s)-1} \left[\frac{s}{4} - z^{-1}\left(\frac{t}{2} + \frac{s}{4} - 1\right)\right]^\lambda , \end{aligned} \tag{17}$$

$$\alpha(s) = \frac{g(s)}{2sf(s)} + \lambda .$$

In the last equation, the contour integral is taken along a unit circle around zero repeatedly, as is indicated by the summation sign. This can be summed up to give

$$M_1 = \frac{-g^2}{2sf(s)} \frac{1}{1-\exp[2\pi i\alpha(s)]} \oint_{c_0} dz\, z^{\alpha(s)-1} \left[\frac{s}{4} - z^{-1}\left(\frac{t}{2} + \frac{s}{4} - 1\right)\right]^{\lambda} \tag{18}$$

which exhibits the Regge-like resonance factor: there is a resonance whenever $\alpha(s) = N = 0, 1, 2, \ldots$, in accordance with the wave equation.

The integrand in Eq. (18) is not singular in z for physical values of s and t, but as $s \to \infty$ the singularity z_0 approaches the point $z = 1$ along the contour, so that an asymptotic expansion may be made. If we assume a straight trajectory $\alpha(s) = s/\mu$, we find

$$M_1 \sim \frac{g^2}{2sf(s)} \frac{\mu}{s} \cot[\pi\alpha(s)/\mu] e^{2t/\mu}, \qquad (s \to \infty) \tag{19}$$

showing a resonance factor and an optical potential type angular distribution.

The crossed amplitude M_2 may be computed in a similar way, but we can easily see the lack of crossing symmetry: it is not the analytic continuation $s \to \mu < 0$ of M_1 (s,t). Moreover, we run into the danger of resonance factors arising from the discrete spacelike solutions unless the wave equation is so chosen that $\alpha(s) < 0$ for $s < 0$.

The lack of crossing symmetry also fails to make the Regge behavior come out right. The contribution from M_2 in backward scattering ($u \sim 0$) looks rather uninteresting as it behaves like $\sim s^{\lambda}$. One may perhaps argue that Regge behavior should not be expected for the present case since it has to do with an external source belonging to a finite multiplet. However, the situation does not seem to change by coupling three infinite component fields through a local vertex.

VIII. Application to hadron physics

Let us now briefly discuss the application of our ideas to real hadron physics. It is obvious that the model has to be more complicated than the ones so far considered. But at any rate, the most interesting and promising approach seems to be the one based on the quark model, assuming the mesons and baryons to belong to definite configurations $q\bar{q}$ and qqq without adding higher configurations. Then the hadronic fields become a product of finite quark spinors and an infinite representation for the orbital motion of the quarks. If only the rotational excitations are considered, the classification of states may be carried out according to the group $G_0 = SU(3) \otimes SU(2) \otimes SO(3)$, or $SU(6) \otimes O(3)$ in the well known manner, and the corresponding G will be $SU(3) \otimes SL(2,c) \otimes SO(3,1)$, or $SL(6,c) \otimes SO(3,1)$. Infinite representations are taken only for the $SO(3,1)$ part. If we include all the internal orbital modes of the quarks, thus making quarks full-fledged quasi-particles, G_0 and G could be made larger. For example, for the mesons $q\bar{q}$, one could use $SO(4,2)$ or $SU(3,1)$. However, the harmonic oscillator type representation $D(\lambda)$ of $SU(3,1)$ is not appropriate in this case since it is not self-conjugate. In order to define self-conjugate systems, we have to take $D(\lambda) \pm D(\lambda)^*$, thus doubling the space. (Using $SL(4,R)$ instead of $SU(3,1)$ will serve the same purpose.) The number of states becomes essentially the same as in the case of $SO(4,2)$.

For the treatment of strong interaction of the hadrons, it is necessary to introduce the coupling of infinite component fields with each other, which we have not discussed yet. If only a (quasi-) local interaction like $C_{nml}\Psi_n^+(x)\Psi_m(x)\Phi_l(x)$ is considered, the problem reduces to that of the Clebsch–Gordan coefficient for infinite representations. It is known, for example, that there is a unique coupling scheme for $SL(2,c)$ IR's. In the case of $SU(3,1)$, a simple coupling scheme holds among $D(\lambda)$, $D(\lambda')$ and $D(\lambda + \lambda')^*$.

I believe, however, that there is no reason to insist on quasi-local coupling. The best way to avoid unphysical properties may be to stick to a realistic quark model. That is to say, we start from a quark-antiquark wave function $\psi(r_1, r_2)$ for the meson, go to the abstract basis $\psi_n(x)$ and assume an equation like (12). For interaction, we can take a form like $\psi(r_1 r_2)\psi(r_2 r_3)\psi(r_3 r_1)$. This will introduce non-local coupling when expressed in terms of the $\Psi_n(x)$.

Finally a word about the current algebra. It has been suggested that infinite dimensional representation of a group may provide non-trivial solutions to the algebra of currents. All indications are, however, that this is not so, and in fact it may be impossible in general.

Let us assume a current of the form $j_\mu(x) = \bar{\Psi}(x)\Gamma_\mu\Psi(x)$, where Γ_μ belongs to the Lie algebra of a group, we know the commutator $[\Gamma_\mu, \Gamma_\nu]$. But this is not the same thing as the commutator $[j_\mu(x), j_\nu(x')]$. For the latter depends on the commutator of the $\Psi(x)$'s, which must be defined as

$$[\Psi_n(x), (\bar{\Psi}(x')\Gamma_0)_m]_\pm = \delta(\mathbf{x} - \mathbf{x}')\delta_{nm} \ .$$

Thus the current commutator amounts to taking

$$\Gamma_\mu\Gamma_0^{-1}\Gamma_\nu - \Gamma_\nu\Gamma_0^{-1}\Gamma_\mu$$

which in general does not have a simple tensor transformation property required by the current algebra. We can verify this by using IR's of $SL(2, c)$ for example. For the Majorana representation, the commutator vanishes.

References

1. The basic ideas can be found in a number of papers, e.g. A. O. Barut and A. Böhm, *Phys. Rev.* **139**, B1107 (1965); P. Budini and C. Fronsdal, *Phys. Rev. Lett.* **14**, 968 (1965); Y. Dothan, M. Gell-Mann and Y. Ne'eman, *Phys. Lett.* **17**, 148 (1965); N. Mukunda, L. O'Raifeartaigh and E. C. G. Sudarshan, *Phys. Rev. Lett.* **15**, 1041 (1965); T. Takabayashi, *Prog. Theor. Phys. Suppl.* (Kyoto), Extra Number, 339 (1965); A. O. Barut, in *Non-Compact Groups in Particle Physics*, edited by Y. Chow (W. A. Benjamin, New York, 1966).
2. A. O. Barut and A. Böhm, Ref. 1.
3. N. Mukunda, L. O'Raifeartaigh and E. C. G. Sudarshan, Ref. 1.
4. Y. Ne'eman, *Algebraic Theory of Particle Physics* (W. A. Benjamin, New York, 1967).
5. T. Takabayashi, Ref. 1.
6. H. Yukawa, *Phys. Rev.* **77**, 219 (1953); **91**, 415 (1953).
7. E. Majorana, *Nuovo Cimento* **9**, 335 (1932).
8. C. Fronsdal (preprint).
9. I. M. Gelfand and A. M. Yaglom, *Zh. Exper. Teoret. Fiz.* **18**, 703 (1948); M. A. Naimark, *Linear Representations of the Lorentz Group* (English transl., Pergamon Press, Inc., London, 1964).
10. V. Bargmann, *Ann. Math.* **48**, 568 (1947).
11. W. Rühl, (preprint).

12. F. Gürsey, *Proc. Coral Gables Conference on Symmetry Principles at High Energy*, 1967.
13. G. Feinberg, preprint.
14. Y. Nambu, Dalhousie Summer School Lectures, Univ. Delhi, 1967.
15. A. O. Barut and H. Kleinert, Proc. NATO Advanced Study Inst., Istanbul, 1966.
16. Y. Nambu, *Prog. Theor. Phys. Suppl.* **37** and **38**, 368 (1966); *Phys. Rev.* **160**, 1171 (1967).
17. C. Fronsdal, *Phys. Rev.* **156**, 1665 (1967).
18. Y. Nambu, *Proc. Coral Gables Conference on Symmetry Principles at High Energy*, 1967 and Ref. 18; S. P. Rosen (unpublished).
19. E. Abers, R. E. Norton and C. Fronsdal, preprint.
20. T. Takabayashi, preprint.
21. M. Gell-Mann and R. P. Feynman, private communication.
22. I. M. Gelfand and A. M. Yaglom, *Zh. Exper. Teoret. Fiz.* **18**, 1096 (1948).
23. B. Zumino, *Proc. Seminars on Unified Theories of Elementary Particles*, (Munich) 1965.
24. G. Feldman and P. T. Matthews, *Phys. Rev.* **151**, 1176 (1966); **154**, 1241 (1967).
25. C. Fronsdal, *Phys. Rev.* **156**, 1653 (1967).
26. E. Abers, I. T. Grodsky and R. E. Norton, *Phys. Rev.* **159**, 1222 (1967).
27. I. T. Todorov, preprint.
28. G. Cocho, preprint.
29. Y. Nambu, to be published.

Discussion

Nauenberg: I wonder if you would explain in a bit more detail your very interesting propagator for the infinite component case. How does one obtain those propagators?

Nambu: Let me go back to an equation like

$$L_0\bar{\Psi} \equiv (\Gamma_{\mu\nu} p^\mu p^\nu - \alpha p^2 - \beta)\bar{\Psi} = 0.$$

Consider for simplicity the case of scattering by equal mass particles. I take the ground state of this tower, namely the lowest level of the harmonic oscillator. Now suppose you

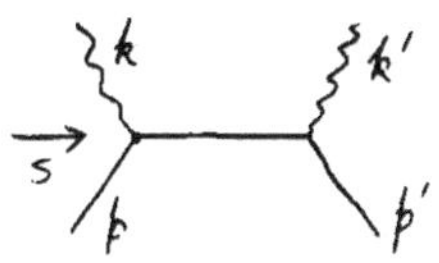

work in the centre of mass system. Then of course P =p + k. Then the term multiplying Γ_{00} is $s = (p+k)^2$. I make use of a formal expression (which may or may not be valid)

$$\frac{1}{L_0 - i\epsilon} \sim \left(\bar{\Psi}_{0,J'}\, e^{-i\theta_{p'}K_3'},\ \int_0^\infty d\tau\, \exp(-iL_0\tau)\, e^{i\theta_p K_3}\, \bar{\Psi}_{0,J}\right).$$

The advantage of writing it in such a fashion is that L_0 contains only the generator of SU(3,1) and hence the above expression becomes a finite element of the group. Consequently one knows the effect of this operator. To prepare the initial state you can start from a particle at rest with spin J and boost this in the direction of p by supplying a suitable boosting operator. The group rotation $\exp(-i\tau L_0)$ operates on this state to produce the final state. When the opposite boosting operator acts on the final state we get a particle at rest with spin J'. So I get the complete matrix-element between two rest states of the product of two rotation operators. This way one gets an integral representation for the scattering amplitude of the type:

$$M \simeq \oint_C dz\, z^{\alpha(s)-1} [\qquad] , \quad z = e^{-i\Gamma_0 \tau s} .$$

The main point is that this integral goes around the unit circle and there is a branch point at Z=0 and another singularity inside the circle. If one expands the denominator in a power series in Z, then the successive terms will give you the contributions from the successive intermediate states that appear in this propagator. But this is a closed expression and is nothing but a hypergeometric function.

When the energy gets large the pole moves very close to the origin so you can make an asymptotic expansion around this point. I believe this is a reflection of the fact that somehow this harmonic oscillator simulates the non-relativistic harmonic oscillator which has a Gaussian shape. This happens in spite of the fact that relativistic form factors do not have a Gaussian (but rather power law) behavior.

Bialynicki-Birula: I want to ask two questions concerning your propagators. Is this propagator causal in the sense of Stuckelberg, Fierz and Feynman? And secondly, is this propagator positive definite?

Nambu: The answer to that question will depend on the exact quantization procedure adopted for this field and I don't think I have time to go into that. It will be covered by later speakers. But I should say that in general you may violate the spin-statistics relation or CPT and also the positive definiteness of energy. There are now two different approaches. The first one is to regard it as just a field without assuming any sort of wave equation but to make the usual assumptions about positivity of energy and mass spectrum on the Hilbert space of states. Then, you look at an experiment involving this field and study the analytic structure of the matrix elements and so on. The other approach would be to impose a certain kind of wave equation which determines also the one particle states that appear here. An equation like this can always be derived from an appropriate Lagrangian and so we can carry out the canonical quantization. However, one has to see whether this cononical quantization is compatible with positiveness of energy and so on. If for example, I take as a simple case the so-called index invariant theories, where all masses are degenerate, the canonical commutation (as opposed to anticommutation) relations are compatible with the positive definiteness of energy and locality, irrespective of whether this refers to integer spin or half-integer spin. On the other hand, if I start from the Majorana type or general

Gelfand-Yaglom type equation, the quantization can be done in such a way that this energy density is positive definite by adopting anti-commutation rules for all spins, i.e.,

$$\{\Psi_n(x), (\bar{\Psi}^\dagger\Gamma_c)_{n'}(x')\}_+ = \delta_{nn'}\delta(\underset{\sim}{x}-\underset{\sim}{x}').$$

It looks formally as if this satisfies local commutativity and also preserves positive definiteness. However, I suspect that there could be some trouble with local commutativity even though formally the canonical quantization seems to imply it. But I am not quite sure.

Yukawa: I'd like to mention that there is a recent paper by S. Tanaka and others dealing with scattering in which the exchanged particle is extended. They took the simplest model of bi-local fields which I proposed in 1950 soon after Fierz showed that the simplest kind of bi-local field is equivalent to a superposition of point particles with different spin. So, as long as this exchanged particle has a momentum in a timelike direction it behaves like several real particles with different spins. But if you go over to spacelike momentum then it behaves like a Regge particle. I don't remember the details but the general feature looks very similar to what Nambu discussed.

Pais: I would like to ask a question about applying these ideas to particle physics, but I can ask my question most easily if I also put it in terms of the hydrogen atom. You start out, say with the ordinary hydrogen atom in the static approximation, you take the Coulomb field. Then you discover that it is convenient to describe the spectrum as a whole in terms of representations of the groups you have mentioned. The hydrogen atom also interacts with the radiation field. And now the question arises, how do you introduce the interaction? The problem which is brought to me is this. The interaction is incorporated into a group which leads you to a certain wave equation. Then you must, of course, (by some rule to be specified) introduce this interaction or what you may call the residual part of the electromagnetic field, and I am a little bit confused about gauge invariance. That is part one of my question, and I can now say it in general terms. Namely, it is an interesting hope that such a noncompact group may be a guide for the spectrum of particles. Now if I push the analogy with the hydrogen atom, then you might conceive that by looking at the representation of an appropriate group, you have incorporated a certain part of the interaction, which effectively generates the spectrum. There is then a similar question when you come to the interaction of the fundamental particles with a baryon spectrum, with say, the meson fields. Because a certain part of it is already in the guts of the wave equation, and then you ask, what is the interaction with the radiating pion field. In particular when one draws graphs I am a little bit confused. Must you then also draw the old fashioned kind of loops which would correspond in some sense to what we call a self energy graph, and if I have to do that, am I then not confusing the bookkeeping?

Nambu: The first question is about the H atom and transverse EM field. I should have mentioned this, though I forgot to do so. I replaced the hydrogen atom wave function $\psi(\underset{\sim}{r})$, where $\underset{\sim}{r}$ is the relative coordinate, by some abstract representation $\psi_n(\underset{\sim}{x})$ where n is a substitute for $\underset{\sim}{r}$. Now, however, I can go from $\psi(\underset{\sim}{x},\underset{\sim}{r})$ to $\psi_n(\underset{\sim}{x})$ where $\underset{\sim}{x}$ stands for the center of mass coordinates of a real two-particle problem of the hydrogen atom, and where $\underset{\sim}{r}$ is the relative coordinate. We can separate out the two coordinates in the Hamiltonian. It can be shown that after the Hamiltonian is expressed in terms of p and the generators of SO(4,2), $M_{\mu\nu}$, one makes the replacement $\underset{\sim}{p} \to \underset{\sim}{p} - e\underset{\sim}{A}$ (just the transverse part and no electric field). This correctly describes the interaction of the transverse field with the electron in the hydrogen atom. We get something like $r\underset{\sim}{p}\cdot\underset{\sim}{A}$. This can again be expressed in terms of the generators of SO(4,2) so everything becomes linear in the generators. This is true only if the mass of the proton is practically infinite so that the kinetic energy of the proton motion can be neglected and thus the effect of the proton current can be ignored. Only under such a condition does this scheme work. Otherwise, you get into a much more complicated situation, namely, the equation is no longer linear in the generator, even though one can formally write down a certain equation in terms of the generators alone. In the case where there is an external field which acts both on the proton and the electron, one cannot treat it in a similar fashion.

With regard to the next question, of course, when one begins to discuss strong interactions, one must be guided by simplicity. In the quark model, for example, for the wave function of the meson one starts with the wave function of two particles of the type $\Psi(x_1,x_2)$. For the interaction, if one takes the quark model seriously, one may wish to take a coupling of the form

$$\Psi(x_1,x_2)\,\Psi(x_2,x_3)\,\Psi(x_3,x_1).$$

If you translate this into group theoretical language, then the coupling would not be local with respect to the center of mass coordinates of the respective fields. It is highly nonlocal and involves operators of the internal space. I don't know which is the most appropriate way to proceed. Perhaps one should start with the simple picture--just with the phenomenological field theory. If it doesn't work, perhaps one has to take the contact model more seriously, and then this approach becomes just some kind of very approximate description of the complicated dynamics by means of group theoretical methods. But this could go in the other direction. This $\Psi_n(x)$ may have a meaning of its own and it would be interesting to pursue the consequences of looking at this as a general field.

Ne'eman: I have just two or three comments with respect to the actual use of these freedoms to classify the states that we observe in the hadrons. In the beginning when we started that way we thought we would suggest two different methods that could be used in order to classify the recurrences of the higher and higher states. One was to think of an internal kind of ℓ, something which was not a spin-like part but the difference between the angular momentum at rest and the axial-vector component of SU(6). Then we could multiply the ℓ. The alternative was like in U(6,6) in which both the spin-like and the other part were growing. Now my comment is that I think we are going in the direction where we shall have to do both things and perhaps the structure will not be simple, and may be very complicated. It will turn out to be something like a U(2,2) of the hydrogen type, multiplying a U(6,6). The reason is the following. If you think of the Lorentz group, the job done by the Lorentz transformation contains an ℓ in it. So you do increase the ℓ and if you want to look at the tower, where the ℓ increase, then you will make either an SL(2,C) or a U(2,2) at the end.

That's one side of it. On the other hand, there seems to be experimental evidence for growth of the internal degrees of freedom. Recently there were five new 2^+ particles reported in the Phys. Rev. Letters found in some photon production experiments. The general experience with baryon resonances has been that they stay after further experiments and if this general experience is realized here then we have evidence of a growth of the system. This cannot be done with three quarks. Perhaps in some models more complicated than just quarks one can still achieve a certain growth. But there seems to be some general evidence for an increase of internal degrees of freedom. So I think that the final answer may have to be an unpleasant and complicated one which is that one will have both towers of ℓ and the internal degrees of freedom. There seems to be some evidence of this when you look at the Regge trajectory picture, because you have the growth in ℓ along the trajectory but you also have some other systematics occurring in this plane. It may very well be connected with the other increase.

The last point I wanted to make is in answer to Pais' question. This is just a very qualitative thing on the point you are making, about whether we were counting the dynamics twice. I think that this is why I must suggest that the infinite component theory is ideal for the bootstrap type of physics, because if no particle is elementary, then the only thing you can treat is a system in which all the phenomenological particles are all treated equivalently when you have this kind of infinite multiplet and there would be some kind of nonlinear relation between multiplets reproducing themselves in this relation.

190 Supplement of the Progress of Theoretical Physics, Extra Number, 1968

Quantum Electrodynamics in Nonlinear Gauge*)

Yoichiro NAMBU

The Enrico Fermi Institute and Department of Physics
The University of Chicago, Chicago, Illinois 60637, U. S. A.

(Received September 27, 1968)

§1.

Can the photons be regarded as massless excitations ("Goldstone bosons") arising as a result of spontaneous breakdown of Lorentz invariance? This is a question of some theoretical interest, and has in fact been studied in the past. Heisenberg,[1] for example, tries to invoke the concept of spontaneous breakdown in order to derive electromagnetism from his nonlinear unified field theory. Bjorken[2] shows more explicitly how ordinary quantum electrodynamics can be reproduced in a conventional cutoff treatment of a nonlinear theory of current-current type interaction. Spontaneous breakdown implies here a non-vanishing current $\langle j_\mu \rangle = \eta_\mu = \text{const}$ in the vacuum. The massless mode of excitation corresponds to a vector potential whose components are proportional to the infinitesimal change $\delta\eta_\mu$ under the Lorentz transformation. (To restore full Lorentz and charge conjugation invariance, η_μ is allowed to vanish in the end as the cutoff$\rightarrow\infty$.)

In the present note, we will give a simple mathematical formulation to this interpretation of electromagnetism, independently of the dynamical mechanism which causes spontaneous breakdown. For this purpose we will apply the technique of nonlinear transformations[3] which have been successful in handling the spontaneous breakdown of chiral symmetry. As we shall see, this formulation corresponds to fixing the gauge of the vector potential in a special manner. The fact that the vector potential is unobservable makes the breakdown of Lorentz invariance only superficial, just as in the case of taking the Coulomb gauge in the ordinary formulation.

§2.

Let us subject a vector field $A_\mu(x)$ to the constraint

$$A_\mu(x)A^\mu(x) = \lambda = \text{const.} \tag{1}$$

Thus there are three components which can be regarded as dynamically independent, the fourth one being a function of them through Eq. (1). Under

*) This work supported in part by U. S. Atomic Energy Commission.

a Lorentz transformation, the four components transform linearly, but the three independent variables will have to transform nonlinearly among themselves. Three distinct cases have to be considered:

$\lambda>0$ (timelike), $\lambda<0$ (spacelike) and $\lambda=0$ (lightlike).

1) $\lambda>0$. We may choose the spatial components to be independent, so that

$$\begin{aligned} A_0 &= \pm(\lambda+A_1^2+A_2^2+A_3^2)^{1/2} \\ &= \pm[\sqrt{\lambda}+(A_1^2+A_2^2+A_3^2)/2\sqrt{\lambda}+\cdots], \end{aligned} \tag{2}$$

if we assume A_i^2 to be small. The components $A_i(x)$ transform linearly under the group $O(3)$, but nonlinearly under boosting. In a standard case, the latter goes, in infinitesimal form, as

$$\begin{aligned} &\delta A_1=\delta A_2=0, \\ &\delta A_3=\varepsilon A_0=\pm\varepsilon(\lambda+A^2)^{1/2}. \end{aligned} \tag{3}$$

One can easily see that the fourth relation $\delta A_0=\varepsilon A_3$ is also satisfied by Eq. (3).

2) $\lambda<0$. We may choose, rather arbitrarily, A_0, A_1, A_2 to be independent. Then

$$\begin{aligned} A_3 &= (|\lambda|+A_0^2-A_1^2-A_2^2)^{1/2} \\ &= \pm[|\lambda|+(A_0^2-A_1^2-A_2^2)/2\sqrt{\lambda}+\cdots]. \end{aligned} \tag{4}$$

The independent components transform linearly under $O(2, 1)$, and nonlinearly otherwise. (Actually they are not completely independent, as the inside of the brackets above must not be negative.) The analog of Eq. (3) can be easily written down.

3) $\lambda=0$. In this case we have, as a limiting case of 1),

$$A_0=\pm(A_1^2+A_2^2+A_3^2)^{1/2}. \tag{5}$$

Unlike the previous cases, however, there is no natural expansion parameter.*)

The next step is to give $A_\mu(x)$ a Lagrangian. Different Lagrangians will lead to different physical contents. According to our schedule, we would like

---

*) An alternative choice may be to take A_1, A_2 and $(A_0+A_3)/2\equiv\xi$ as independent. Then,

$$\eta\equiv(A_0-A_3)/2=(A\ +A\ )/4\xi.$$

Further ansatz $\xi(x)=c+\xi'(x)$ enables one to make an expansion of η, corresponding to a spontaneous breakdown:

$$\langle A_\mu\rangle_0=(c, 0, 0, c).$$

In contrast, the preceding cases correspond to $\langle A_\mu\rangle_0=(\sqrt{\lambda}, 0, 0, 0)$ and $(0, 0, 0, \sqrt{-\lambda})$, respectively.

to identify A_μ with the electromagnetic potential. Let us assume therefore the usual Maxwell Lagrangian

$$L=-\frac{1}{4}F_{\mu\nu}F^{\mu\nu}-j_\mu A^\mu,$$
$$F_{\mu\nu}=\partial_\mu A_\nu-\partial_\nu A_\mu, \tag{6}$$

where $j_\mu(x)$ is a conserved current, say that of the electron. The difference from ordinary electrodynamics lies in the constraint (1) When Euler equations are derived for the independent components, we obtain equations different from the Maxwell form. These need not be, however, written down explicitly, as our standard procedure will be perturbation theory, expanding everything in powers of independent variables if possible.

In the following we will assume $\lambda>0$. When the expansion (2) is substituted for $A_0(x)$, we find

$$L=-\frac{1}{2}(\boldsymbol{\nabla}\times\boldsymbol{A})^2+\frac{1}{2}(\partial_0\boldsymbol{A}+\boldsymbol{\nabla}A_0)^2+\boldsymbol{j}\cdot\boldsymbol{A}-j_0A_0$$
$$=-\frac{1}{2}(\boldsymbol{\nabla}\times\boldsymbol{A})^2+\frac{1}{2}(\partial_0\boldsymbol{A}\pm\boldsymbol{\nabla}\boldsymbol{A}^2/2\sqrt{\lambda}+\cdots)^2$$
$$+\boldsymbol{j}\cdot\boldsymbol{A}\mp j_0(\sqrt{\lambda}+\boldsymbol{A}^2/2\sqrt{\lambda}+\cdots). \tag{7}$$

The "free" part of the Lagrangian gives

$$\boldsymbol{\nabla}\times\boldsymbol{\nabla}\times\boldsymbol{A}+\partial^2\boldsymbol{A}/\partial x_0^2=0. \tag{8}$$

In terms of Fourier components, it means

$$\Box A_t(k)=k^2A_t(k)=0,$$
$$\partial^2A_l/\partial x_0^2=-k_0^2A_l(k)=0, \tag{9}$$

where

$$A_l(k)=\boldsymbol{k}\cdot\boldsymbol{A}(k)/|\boldsymbol{k}|,$$
$$A_t(k)=\boldsymbol{A}(k)-\boldsymbol{k}A_l(k)/|\boldsymbol{k}| \tag{10}$$

are the longitudinal and transverse parts of $\boldsymbol{A}(k)$ Thus the transverse part behaves like the normal light wave, whereas the longitudinal part is essentially static, and hence will not behave as a particle.

The rest of the Lagrangian (7) contains terms describing interaction of A_μ with itself, with the current, and a term $\mp\sqrt{\lambda}j_0$, corresponding to a constant potential acting on the charge. This last term, however, can be eliminated by a gauge transformation on the charged field (or redefinition of its momentum), and would not cause physical effects.

§3.

Although our Lagrangian and the constraint (1) are Lorentz invariant, the choice of independent variables and the separation of free and interaction Lagrangians are not. This is in line with the concept of spontaneous breakdown. At the same time we would expect that the predictions of such a theory in general would not be Lorentz invariant. We shall see, however, that in the present case we get the same S-matrix as in the ordinary quantum electrodynamics, keeping Lorentz invariance intact.

For this purpose we first examine all lowest order scattering processes (without closed loops). In view of Eq. (7), these are electron-electron, electron-photon, and photon-photon scattering. The Feynman propagator for $A_l(x)$ will be, from Eq. (9),*)

$$\begin{aligned} D_{ij}(k) &= -i\left[\delta_{ij} - \frac{k_i k_j}{\boldsymbol{k}^2}\right]\frac{1}{\boldsymbol{k}^2 - k_0^2 - i\varepsilon} + i\frac{k_i k_j}{\boldsymbol{k}^2}\frac{1}{k_0^2} \\ &= -i\left[\delta_{ij} - \frac{k_i k_j}{k_0^2}\right]\frac{1}{\boldsymbol{k}^2 - k_0^2 - i\varepsilon}. \end{aligned} \tag{11}$$

1) Electron-electron scattering. The amplitude for this is

$$ij_i(k)D_{ij}(k)j_j(-k)$$

corresponding to exchange of a virtual "photon". By virtue of charge conservation $k_0 j_0 - \boldsymbol{k}\cdot\boldsymbol{j} = 0$, the second term of D_{ij} in Eq. (11) gives

$$\boldsymbol{k}\cdot\boldsymbol{j}(k)\boldsymbol{k}\cdot\boldsymbol{j}(-k)/k_0^2 = j_0(k)j_0(-k),$$

so that Eq. (12) is equivalent to the Møller formula

$$-j_\mu(k)j^\mu(-k)/(\boldsymbol{k}^2 - k_0^2).$$

2) Electron-photon scattering. In addition to the usual Compton diagrams, we have a contact term $\mp j_0 A^2/2\sqrt{\lambda}$ and a "longitudinal photon" exchange term due to the second brackets in Eq. (7). However, we can easily check that these two extra terms cancel each other when the propagator is inserted and the continuity equation is taken into account.

3) Photon-photon scattering. Contributing to this are direct and longitudinal photon exchange terms, but these are again seen to cancel each other, so there is no photon-photon scattering to this order, as is consistent with quantum electrodynamics.

We have verified that the S-matrix elements in the second order are Lorentz invariant and agree with those of quantum electrodynamics. To show the equivalence to higher orders, however, would be rather complicated. We

*) We need not worry about the $i\varepsilon$ for the longitudinal part. It will not change the results.

will instead proceed by making the general argument that the condition (1) amounts to a special choice of the gauge of the vector potential, and therefore should not change the physical content of the theory. It is certainly true that the freedom of gauge transformation allows one to impose a variety of conditions on A_μ. In the present case this means that, for a given $A_\mu(x)$, we seek a gauge function $\chi(x)$ such that

$$[A_\mu(x)-\partial_\mu\chi(x)]^2=\lambda. \tag{12}$$

$A_\mu-\partial_\mu\chi$ is to be identified with the A_μ used above. Now the relevant question is whether Eq. (12) is admissible, i.e. has solutions for arbitrary A_μ. To answer this, we observe that the classical equation of motion for a charged particle

$$\begin{aligned} m du_\mu/d\tau &= eF_{\mu\nu}u_\nu\,, \\ u_\mu=dx_\mu/d\tau, &\quad u_\mu u^\mu=1, \end{aligned} \tag{13}$$

is equivalent to a Hamilton-Jacobi equation

$$[\partial_\mu S(x)-eA_\mu(x)]\,[\partial^\mu S(x)-eA^\mu(x)]=m^2. \tag{14}$$

Comparison of Eqs. (12) and (14) shows the correspondence $\chi=S/e$, $\lambda=m^2/e^2$. Thus Eq. (12) will have a solution inasmuch as there is a solution to the classical problem (13).*) The above argument is admittedly not rigorous, being more of physical than mathematical nature. But it suggests at the same time that the constant λ has to be taken positive as we did here. We would expect different results for the cases $\lambda\leqslant 0$, but this has not been explicitly checked.

§4.

We have seen that the nonlinear constraint (1) is a possible gauge condition in quantum electrodynamics. The S-matrix seems to remain unaltered under such a gauge convention (at least for $\lambda>0$). This particular gauge allows one to interpret quantum electrodynamics in terms of a spontaneous breakdown of Lorentz invariance, because the perturbation expansion corresponds to the assumption $\langle A_\mu\rangle_0=(\pm\sqrt{\lambda},\boldsymbol{0})^4$. The breakdown, however, is superficial as it affects only the gauge of the vector potential. This would not have been the case had we taken a Lagrangian different from Eq. (6). In fact, adding a term $\sim(\partial_\mu A^\mu)^2$ would lead to an entirely different theory; Lorentz invariance would be actually broken, and there would be three massless modes, two transversal and one longitudinal.

Apart from theoretical curiosity, it is an open question whether our

*) Note that e and m are arbitrary, not necessarily being related to those of the electron with which the photon interacts.

formulation has any practical use, and whether it has really the same physical content as the ordinary theory. (It is known that different gauges are convenient for different processes, and that the divergences in quantum electrodynamics are gauge dependent.) These points need further study.

After writing this article, it was brought to the author's attention that Dirac[4)] had proposed the gauge condition (1) and noted its equivalence to the Hamilton-Jacobi equation (14). His motivation, however, was different from ours, and his considerations were confined mostly to classical theory. We thank Professors S. Deser, Y. Takahashi and G. Wentzel for their advice on Dirac's work.

References

1) W. Heisenberg, Rev. Mod. Phys. **29** (1957), 269.
H. D. Dürr, W. Heisenberg, H. Yamamoto and K. Yamazaki, Nuovo Cim. **38** (1965), 1220.
2) J. B. Bjorken, Ann. of Phys. **24** (1963), 174.
See also G. S. Guralnik, Phys. Rev. **136** (1964), B1404, 1417.
I. Bialynicki-Birula, Phys. Rev. **130** (1963), 465.
3) See, for example, F. Gürsey, Nuovo Cim. **16** (1960), 230.
S. Weinberg, Phys. Rev. **166** (1968), 1568.
Other references are quoted in the latter paper.
4) P. A. M. Dirac, Proc. Roy. Soc. **A209** (1951), 291; **A212** (1952), 330; **A223** (1954), 438; Nature **168** (1951), 906; **169** (1952), 146, 702.
See also E. Schrödinger, Nature **169** (1952), 538.

Note added in proof:

The Euler equation under the condition (2) reads

$$(\nabla\times \boldsymbol{A}-\boldsymbol{E}-\boldsymbol{j})-(\boldsymbol{A}/A_0)(\nabla\cdot\boldsymbol{E}-j_0)=0.$$

In order to recover the Maxwell equations, we have to impose the supplementary condition $\nabla\cdot\boldsymbol{E}-j_0=0$. In perturbation theory, this condition is effectively implied by assuming the initial absence of longitudinal modes.

We have also confirmed that 1) the S-matrix is independent of the choice of the η_μ as long as $\lambda=\eta_\mu\eta^\mu>0$; 2) the cases $\lambda\leq 0$ are not equivalent to quantum electrodynamics.

Volume 26B, number 10 PHYSICS LETTERS 15 April 1968

S-MATRIX IN SEMICLASSICAL APPROXIMATION *

Y. NAMBU
The Enrico Fermi Institute and Department of Physics,
The University of Chicago, Chicago, Illinois, USA

Received 18 March 1968

It is shown that the tree approximation currently utilized in connection with phenomenological chiral Lagrangians corresponds to a semiclassical approximation applied to the S-matrix in quantum field theory. Some invariance properties of the S-matrix are derived in this framework.

Recently the method of effective Lagrangians has been widely used in formulating the dynamical contents of chiral symmetry [1]. One writes down in this approach a non-linear Lagrangian invariant under the chiral $SU(2) \times SU(2)$ (or $SU(3) \times SU(3)$) group, in which the pion field (or its SU(3) extention) transforming non-linearly under the group plays a crucial role [2]. It is characteristic of such a Lagrangian that the invariance does not apply to the free and the interaction part separately; these two will mix under the group transformation. Thus the asymptotic one-particle states defined by (appropriately renormalized) free fiels do not form a representation of the group, in sharp contrast to the ordinary case in which a symmetry implies a multiplet structure and associated quantum numbers for one-particle states. This situation may be interpreted in terms of a "spontaneous breakdown" of chiral symmetry; each time a chiral transformation is performed, one has to redefine the free and interaction parts, corresponding to a change of the vacuum state [3].

For the actual construction of the S-matrix, in chiral dynamics, one carries out a perturbation expansion with respect to a coupling parameter characterizing the degree of nonlinearity, collecting consistently all terms that contribute to a specific n-particle process in the lowest possible order. The corresponding Feynman diagrams turn out to be "tree diagrams", or simply connected diagrams having no closed loops [1]. Obviously it is an approximation which does not take into account unitarity (rescattering), self-energy and other higher order corrections to a given process. This may be reasonable in the low energy (soft pion) limit. On the other hand it is not clear whether the highly singular Lagrangian one uses here is really meaningful beyond such an approximation.

It is the purpose of this note to point out that the above procedure may be characterized as a semiclassical approximation to the S-matrix, which is an analog in quantum field theory of the WKB approximation to the Schrödinger equation. This statement is entirely general, being independent of the Lagrangian chosen. In the following we present the proof, and discuss its implications.

We start from the S-matrix in interaction representation, expressed as a time-ordered (T-) product ‡

$$S = T \exp[(\mathrm{i}/\hbar c) \int L_{\mathrm{int}}(\phi)\, \mathrm{d}^4x] \,. \qquad (1)$$

Its contribution to individual processes can be obtained by converting it to a normal product, for which there is a compact formula [4] †

* Supported in part by the U.S. Atomic Energy Commission.

‡ When L_{int} contains derivatives, one should perform the differentiation after time-ordering.

† Eq. (10') could have been derived directly from Feynman's path integral formulation, or from the matrix representation

$$\langle \phi' | U | \phi'' \rangle \sim \exp[(\mathrm{i}/2\hbar c) \int (\phi' - \phi'') K(\phi' - \phi'')\, \mathrm{d}^4x] \,.$$

Applying this to eq. (2), we get

$$S(\phi) = U \exp[(\mathrm{i}/\hbar c)\, \overline{L}_{\mathrm{int}}(\phi)]$$

$$= \int \langle \phi | U | \widetilde{\phi} \rangle\, D[\phi] \exp[(\mathrm{i}/\hbar c)\, \overline{L}_{\mathrm{int}}(\widetilde{\phi})]$$

which is an exact result. Approximating the functional integral by a stationary "path" determined from the classical Euler equation for $\widetilde{\phi}$, we arrive at the formulas (9') and (10'). However, their relation to the tree approximation would have been obscure.

Volume 26B, number 10 PHYSICS LETTERS 15 April 1968

$$T \exp[(i/\hbar c) \int L_{int}(\phi)\, d^4x] =$$

$$= :U \exp[(i/\hbar c) \int L_{int}(\phi)\, d^4x]: \equiv :S:,$$

$$U \equiv \exp[\tfrac{1}{2}\hbar c \iint \frac{\delta}{\delta\phi(x)} \Delta(x-x') \frac{\delta}{\delta\phi(x')} d^4x\, d^4x'] . \quad (2)$$

Here we have chosen for simplicity a neutral scalar field $\phi(x)$; $\Delta(x-x')$ is the corresponding free Feynman propagator $(1/\hbar c)\langle T(\phi(x), \phi(x'))\rangle_0$ ‡. The functional differentiation is carried out regarding all $\phi(x)$ with different x as if they were independent c-numbers; i.e., not subject to a free field equation. This formula can be readily generalized to all types of field. For a complex field ϕ_α, we replace U with

$$\exp[\hbar c \iint \frac{\delta}{\delta\phi_\alpha} \Delta_{\alpha\beta}(x-x') \frac{\delta}{\delta\phi^*_\beta(x')} d^4x\, d^4x'] .$$

When ϕ_α is a Fermi field, we make the further stipulation that $\phi_\alpha(x)$, $\delta/\delta\phi_\alpha(x)$, etc., at different points all anticommute. In order to built up the operation with U step by step, we may supply a fictitious parameter η in front of $\Delta(x-x')$, which will vary from 0 to 1. Differentiating with respect to η, we obtain

$$\frac{\partial}{\partial\eta} :S_\eta: = \quad (3)$$

$$= :\left(\tfrac{1}{2}\hbar c \iint \frac{\delta}{\delta\phi(x)} \Delta(x-x') \frac{\delta}{\delta\phi(x')} d^4x\, d^4x'\right) S_\eta: .$$

Now let us write

$$:S: = :\exp[-(i/\hbar c)\Theta(\phi)]: . \quad (4)$$

It is not difficult to see that $\Theta(\phi)$ corresponds to a collection of all connected Feynman diagrams; this will in fact become clear below. Substituting in eq. (3), we may drop the normal product symbol and manipulate S as a mere c-number functional. We get thus

$$\frac{\partial\Theta}{\partial\eta} + \tfrac{1}{2}i \iint \frac{\delta\Theta}{\delta\phi(x)} \Delta(x-x') \frac{\delta\Theta}{\delta\phi(x')} d^4x\, d^4x + \quad (5)$$

$$- \tfrac{1}{2}\hbar c \iint \frac{\delta^2\Theta}{\delta\phi(x)\,\delta\phi(x'x)} \Delta(x-x')\, d^4x\, d^4x' = 0 ,$$

‡ The dimensional convention adopted here is such that the field equations are "classical" (i.e., use wave numbers rather than momenta), with the free Lagrangian given by

$$-\tfrac{1}{2}(\partial_\mu\phi\partial_\mu\phi + \kappa^2\phi^2) \quad \text{for a scalar, and}$$

$$-(\bar\psi\gamma_\mu\partial_\mu\psi + \kappa\bar\psi\psi) \quad \text{for a spinor, where}$$

$$x_\mu = (x_1, x_2, x_3, x_4 = ict) , \quad \kappa = mc/\hbar .$$

with the initial condition

$$\Theta = \int L_{int}(\phi)\, d^4x \equiv -\bar{L}_{int} \quad \text{for} \quad \eta = 0 . \quad (6)$$

If Θ is expanded in a power series in η, the second term in eq. (5) means the contraction of two disconnected diagrams by means of a propagator to form a single connected diagram, whereas the third term corresponds to a contraction within a connected diagram, creating an internal loop. If we drop this third term, which is multiplied by $\hbar c$, eq. (5) takes the form of a classical "Hamilton-Jacobi equation" (writing Θ_c for Θ)

$$\frac{\partial\Theta_c}{\partial\eta} + \bar{H}(\pi(x)) = 0 ,$$

$$\bar{H} = \tfrac{1}{2}i \iint \pi(x)\Delta(x-x')\pi(x')\, d^4x\, d^4x' ,$$

$$\pi(x) \equiv \delta\Theta_c/\delta\phi(x) . \quad (7)$$

where $\phi(x)$ and $\pi(x)$ are canonical conjugates, and η plays the role of time. It is obvious that we generate only tree diagrams.

The "Hamiltonian" contains only the "kinetic energy" dependent on $\pi(x)$, so the solution can be immediately obtained by integrating the equations of motion

$$\frac{d\pi(x)}{d\eta} = 0 , \quad \frac{d\phi(x)}{d\eta} = i \int \Delta(x-x')\pi(x')\, d^4x ,$$

$$\phi(x) = \eta i \int \Delta(x-x')\pi(x')\, d^4x' + F(\pi) . \quad (8)$$

$F(\pi)$ is determined from the initial condition

$$\pi(x) = \delta\Theta/\delta\phi(x) = -\delta \bar{L}_{int}/\delta\phi(x) \quad \text{for} \quad \eta = 0 .$$

Thus

$$\pi(x) = -\delta \bar{L}_{int}(\tilde\phi)/\delta\tilde\phi ,$$

$$\tilde\phi(x) \equiv \phi(x) - \eta i \int \Delta(x-x')\pi(x')\, d^4x' , \quad (9)$$

which is an intrinsic equation relating ϕ and π. The Hamilton-Jacobi functional Θ_c is now determining from integration of the relation

$$\delta\Theta_c(\phi,\eta) = \int \pi(x)\delta\phi(x)\, d^4x - \bar{H}\delta\eta$$

We find readily

$$\Theta_c = -\bar{L}_{int}(\tilde\phi) + \eta\bar{H}(\pi) . \quad (10)$$

This, together with eq. (9), gives the solution Θ_c when it is expressed in terms of $\phi(x)$, and η put equal to 1. Eqs. (9) and (10) can be brought to a more transparent form. From eq. (9) follows

$$\pi(x) = K(\tilde\phi(x) - \phi(x)) = -\delta \bar{L}_{int}/\delta\tilde\phi(x) ,$$

Volume 26B, number 10 PHYSICS LETTERS 15 April 1968

or

$$\delta \bar{L}_{tot}(\tilde{\phi})/\delta\tilde{\phi}(x) = K\phi(x) ;$$

$$K \equiv i\Delta^{-1}, \quad \bar{L}_{tot}(\tilde{\phi}) \equiv \bar{L}_{int}(\tilde{\phi}) + \bar{L}_0(\tilde{\phi}) \qquad (9')$$

$$\equiv \bar{L}_{int}(\tilde{\phi}) + \tfrac{1}{2}\int \tilde{\phi} K \tilde{\phi}\, d^4x ,$$

and eq. (10) becomes (with $\eta = 1$)

$$\Theta_c(\phi, \tilde{\phi}) = -\bar{L}_{int}(\tilde{\phi}) - \bar{L}_0(\tilde{\phi} - \phi) \qquad (10')$$

$$= -\bar{L}_{tot}(\tilde{\phi}) + \int \phi K \tilde{\phi}\, d^4x - \tfrac{1}{2}\int \phi K \phi\, d^4x .$$

Here K is the Klein-Gordon operator $\Box - \kappa^2$, so $\bar{L}_{tot}$ is the total Lagrange functional. Eq. (10') is nothing but a Lagrange functional for the field $\tilde{\phi}$, with ϕ regarded as an external source field, and eq. (9') is the resulting Euler equation. When the latter is solved for $\tilde{\phi}$ in terms of ϕ with the boundary condition characterized by the Feynman kernel Δ, and in the final result we let $K\phi = 0$, then we get a solution of the classical field equation which approaches asymptotically the free field ϕ. Θ_c becomes a functional of ϕ as the dummy (internal) field $\tilde{\phi}$ has been eliminated †.

This proves our contention about the nature of the approximation involved. Eqs. (9') and (10'), or (9) and (10), provide a concise formula for the construction of Θ_c. When there is more than one field, we can simply add a respective source term to each field. The whole thing is treated as a quasi-*c* number theory, in the sense that for Fermi fields anticommutatively is to be respected. We will now make use of the formalism to examine invariance properties of the *S*-matrix in our approximation.

1) Suppose

$$\Theta_c(\phi_\alpha, \tilde{\phi}) = -\bar{L}_{int}(\tilde{\phi}_\alpha) - \sum_\alpha \bar{L}_{o\alpha}(\tilde{\phi}_\alpha - \phi_\alpha) \qquad (11)$$

is invariant under an infinitesimal change $d\phi_\alpha(x)$, $d\tilde{\phi}_\alpha(x)$, $\alpha = 1, 2, \dots,$ of the fields. This will be true of the free and interaction parts of $\bar{L}$ are separately invariant. After the $\tilde{\phi}$'s have been eliminated, $\Theta_c(\phi_\alpha)$ and the *S*-matrix remain invariant under the change $d\phi_\alpha$.

2) It is possible that $\Theta_c(\tilde{\phi}_\alpha, \tilde{\phi}_\alpha)$ is invariant under a change of the ϕ_α's alone due to a special feature of the free Langrangian. The *S*-matrix will again be invariant. A notable example is the familiar gauge invariance of the *S*-matrix under $dA_\mu(x) = \partial_\mu \lambda(x)$ (without making a phase transformation on charged fields). This is because $L_0(A_\mu)$ depends only on the antisymmetric tensor $F_{\mu\nu}$ *.

The case of a massless spin - 0 field does not belong to this class, although $L_0(\phi)$ is invariant under a constant displacement $d\phi(x) = c$. The reason is that the Klein-Gordon (or rather the d'Alembertian) operator appearing in the interference term in eq. (10') is spurious, and cannot be counted on to kill a constant, as actually $\tilde{\phi}$ has a singularity $\sim \Delta = iK^{-1}$ according to eq. (9).

3) Next let us assume that $\bar{L}_{tot}(\tilde{\phi}_\alpha)$ as a whole is invariant, but not necessarily for its parts L_{int} and L_0 separately. We have then

$$\delta \bar{L}_{tot}(\tilde{\phi}_\alpha) = \int d\tilde{\phi}_\alpha(x)\, \delta \bar{L}_{tot}/\delta\tilde{\phi}_\alpha(x)\, d^4x = 0 .$$

From the Euler equation for $\tilde{\phi}_\alpha$: $\delta \bar{L}_{tot}/\delta\tilde{\phi}_\alpha(x) = K_\alpha \phi_\alpha(x)$, it follows that

$$\sum_\alpha \int d\tilde{\phi}_\alpha K_\alpha \phi_\alpha\, d^4x = 0 .$$

As a result, the entire $\Theta_c(\phi_\alpha)$ is invariant. This, however, is not a statement about the symmetry of the *S*-matrix, as the asymptotic fields ϕ_α are not varied. It is rather a statement about the independence of the *S*-matrix (as well as its off-shell extrapolation $K_\alpha\phi_\alpha \neq 0$) on a redefinition of the dummy fields, which is a symmetry operation on L_{tot}.

4) We can go even further, and subject the dummy field variables to an arbitrary substitution which is not a symmetry operation:

$$\tilde{\phi}_\alpha(x) = F_\alpha(\tilde{\phi}'_\beta(x)) = \tilde{\phi}'_\alpha(x) + f_\alpha(\tilde{\phi}'_\beta(x)) , \qquad (12)$$

where f contains only nonlinear terms in the fields. We claim that the *S*-matrix (on the mass shell) is again invariant under such a redefinition of the $\tilde{\phi}$'s. To be more precise, we get the same result if we start from

$$\Theta_c(\phi_\alpha, \tilde{\phi}'_\alpha) = -\bar{L}_{tot}(F_\alpha(\tilde{\phi}'_\beta)) - \sum_\alpha \bar{L}_{o\alpha}(\tilde{\phi}'_\alpha - \phi_\alpha) \quad (13)$$

instead of eq. (11), provided that we construct the solution as a power series in the fields. It will suffice to show this for infinitesimal f.

† See footnote † on page 626.

* We know from the Ward-Takahashi identity that gauge invariance in the present sense does not hold off the mass shell of charged particles in general, whereas the present argument would seem to apply also off the mass shell. The latter, however, is not true. In order for the Maxwell equations to be consistent, their currents have to be conserved. This would not be true if the external charged fields were arbitrary.

Volume 26B, number 10 PHYSICS LETTERS 15 April 1968

Comparing eqs. (11) and (13), we find

$$\Theta_c'(\phi,\tilde{\Phi}') - \Theta_c(\phi,\tilde{\Phi}) =$$

$$= \int \sum_\alpha f_\alpha(\tilde{\Phi}')K_\alpha \phi_\alpha = \Theta_c'(\phi) - \Theta_c(\phi)$$

when the respective Euler equations are satisfied (stationary values of $\tilde{\Phi}'$ and $\tilde{\Phi}$). But the f_α's have no mass shell singularities in a perturbative solution because they are non-linear. Thus the difference vanishes as $K_\alpha \phi_\alpha \to 0$. (The condition on the f's may be further relaxed.)

Summarizing, we found that the S-matrix in the "tree" approximation corresponds to a semiclassical (WKB) approximation applied to field theory, and that invariance properties expected of the exact S-matrix are also valid here. In particular, the item 4) above can be interpreted as the irrelevance of the choice of the "interpolating fields" $\tilde{\Phi}_\alpha$, a property implied by the Lehmann-Symanzik-Zimmermann formalism of quantum field theory [5]. This property has been explicitly noted in chiral dynamics [6] where the definition of the pion field is not unique, as there is no obvious principle by which to choose a particular nonlinear transformation.

1. S. Weinberg, Phys. Rev. Letters 18 (1967) 188; J. Schwinger, Phys. Letters 24B (1967) 473; J. A. Cronin, Phys. Rev. 161 (1967) 1483; J. Wess and B. Zumino, Phys. Rev. 163 (1967) 1727; L. S. Brown, Phys. Rev. 163 (1967) 1802; P. Chang and F. Gürsey, Phys. Rev. 164 (1967) 1752; T. Minamikawa and Y. Miyamoto, Prog. Theor. Phys. 38 (1967) 1195; W. A. Bardeen and B. W. Lee, Canadian Summer Institute Lectures, to be published; B. W. Lee, Proc. 1957 Conf. on Particles and fields (John Wiley and Sons, Inc., New York, 1957) p. 546; Y. Ohnuki and Y. Yamaguchi, Inst. Nucl. Study (Tokyo) preprint (1967).
2. F. Gürsey, Nuovo Cimento 16 (1960) 230; Ann. Phys. (N.Y.) 12 (1961) 91; M. Gell-Mann and M. Lévy, Nuovo Cimento 16 (1960) 705.
3. Y. Nambu and G. Jona-Lasinio, Phys. Rev. 122 (1961) 345.
4. S. Hori, Prog. Theor. Phys. (Kyoto) 7 (1952) 578.
5. H. Lehmann, K. Symanzik and W. Zimmermann, Nuovo Cimento 5 (1955) 205; R. Haag, Phys. Rev. 112 (1958) 669; K. Nishijima, Phys. Rev. 111 (1958) 995; W. Zimmermann, Nuovo Cimento 10 (1958) 597.
6. J. Cronin, Ref. 1; B. Zumino, private communication. Actual differences that appear in π-π scattering, for example, are due to different choices of symmetry breaking terms in the Lagrangian, and not due to different choices of field variables in the present context.

* * * * *

Reprinted from Proceedings of the International Conference on Symmetries and Quark Models, ed. R. Chand, © 1970 Gordon and Breach, pp. 269–278.

*Quark model and the factorization of the Veneziano amplitude**

YOICHIRO NAMBU

The Enrico Fermi Institute
The University of Chicago, Chicago, Illinois

IF WE TAKE the quark model in the most realistic sense, a meson in the $q\bar{q}$ configuration, for example, would have a wave function which, beside the spin and the $SU(3)$ indices, depends on the relative coordinates of the quarks. We would therefore expect in general excitations in L as well as in the radial mode, and hence a family of Regge trajectories forming 36-plets of $U(6)$. So far we do not have a clearcut evidence for the radial excitations, but the apparent complexity and structure we see in higher mesonic levels suggest that there may well be excitations other than the orbital L type. This is also supported by the successes of the duality concept[1] and its elegant mathematical realization embodied in the Veneziano[2] amplitude. Although we do not yet know the exact meaning of duality, nor do we know whether it is to be regarded as a new fundamental principle, the Veneziano model seems to contain a lot of implications which have yet to be tested.

A close connection between duality and the quark model has been pointed out by a number of people recently.[3] It is based on a diagrammatic interpretation of hadronic reactions in terms of quark exchanges which satisfy simple topological rules, the most important of which is to first consider only "planar diagrams" with no crossed quark lines. These rules make qualitative but powerful predictions, and it should be highly interesting if we could

* This work supported in part by the U.S. Atomic Energy Commission.

make a more quantitative and precise fusion of the quark model and the analytic structure of the Veneziano representation. Work along this line seems to be going on already at various places. Our results which we would like to discuss here are in fact very similar to those of the Berkeley[4] and the MIT groups,[5] but it would be useful to put the problem in a clear, if not perfect, perspective.

The Veneziano amplitude for a two-body reaction is typically represented by the function (or more generally, a sum of functions)

$$V_{lmn}(s, t) = \Gamma(1 - \alpha_s)\,\Gamma(m - \alpha_t)/\Gamma(n - \alpha_s - \alpha_t) \qquad (1)$$

with $\alpha_s = s + a_s$, $\alpha_t = s + a_t$ and an appropriate set of integers l, m, n. V has the duality in the sense that it can be decomposed into an infinite series of resonances either in the s or in the t channel, which converges in its respective physical range, and exhibits the right Regge behavior but for the unphysical infinitely sharp resonance structure superimposed on it. This can probably be remedied by unitarizing V through iteration, perhaps at the risk of violating strict duality, but we will not worry about this question since we do not hold duality to be a sacred principle. Of course we expect that the Regge cuts and even the Pomeranchukon may be generated by starting from V as the input Born term.[6]

Our primary concern will be to find a self-consistent, factorizable set of Veneziano-type amplitudes for all hadronic processes. By this we mean that we want to find a set of hadronic states (represented e.g. by a master wave function with infinite components) and their basic coupling among themselves in such a way that if we build up "tree diagrams" one by one using the basic vertices we would obtain amplitudes for arbitrary many-particle processes and that the result will automatically exhibit duality. That the Veneziano amplitudes can be generalized to many-particle processes has been shown,[7] and we will use the general formula as the basis for our program. In order to build a realistic theory from the quark model it is of course important to discuss the way the $SU(3)$ and the spinor indices are to be coupled, but this is a straightforward problem once the basic topological rules are given.[4] On the other hand, the factorization of analytic functions of many variables requires introduction of a large number of internal coordinates, many more than would correspond to a naive bound state picture. A remarkable point, however, is that this can be done with a finite degree of degeneracy for each level, although the solution is not necessarily a unique one.

To focus our attention on the essential part, we will ignore the spins and consider an n-point Veneziano function for neutral scalar particles. Such a function can be written down according to the known rules, and it takes a particularly simple form in the multiperipheral configuration:

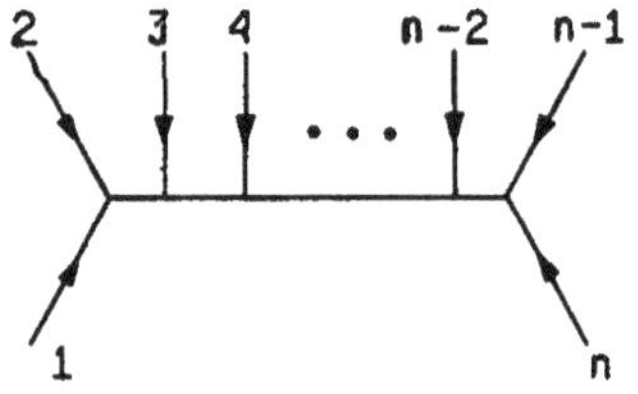

$$
\begin{aligned}
V = \int & x^{(12)} y^{(123)} z^{(1234)} \cdots w^{(n-1,n)} \times \\
& \times \left(\frac{1-x}{1-xy}\right)^{(23)} \left[\frac{(1-y)(1-xyz)}{(1-xy)(1-yz)}\right]^{(34)} \cdots \times \\
& \times \left(\frac{1-xy}{1-xyz}\right)^{(234)} \left[\frac{(1-yz)(1-xyzu)}{(1-xyz)(1-yzu)}\right]^{(345)} \cdots \times \\
& \times \cdots \times \frac{dx\, dy\, dz \cdots dw}{(1-xy)(1-yz)\cdots}
\end{aligned} \tag{2}
$$

where

$$
\begin{aligned}
(12) &\equiv -\alpha_2(s_{12}) - 1 = -s_{12} - a_2 - 1, \qquad s_{12} = (p_1 + p_2)^2, \quad \text{etc.} \\
(123) &\equiv -\alpha_3(s_{123}) - 1 = -s_{123} - a_3 - 1, \quad s_{123} = (p_1 + p_2 + p_3)^2, \quad \text{etc.}
\end{aligned} \tag{3}
$$

This function is invariant under the cyclic permutation $(12 \cdots n) \rightarrow (23 \cdots n1)$ and the inversion $(12 \cdots n) \rightarrow (n \cdots 21)$. The a's are the intercepts for the i-body Regge trajectories, which need not be equal to each other. Regrouping the individual factors and expanding the variables: $s_{12} = (p_1 + p_2)^2 = 2p_1 \cdot p_2 + 2m^2$, etc., we get

$$
\begin{aligned}
V = \int & x^{(12)}\, y^{(123)} \cdots dx\, dy \cdots \times \\
& \times (1-x)^{-2p_2 \cdot p_3 - c_2} (1-y)^{-2p_3 \cdot p_4 - c_2} \cdots \times \\
& \times (1-xy)^{-2p_2 \cdot p_4 - c_3} (1-yz)^{-2p_3 \cdot p_5 - c_3} \cdots \times \\
& \times (1-xyz)^{-2p_2 \cdot p_5 - c_4} (1-yzu)^{-2p_3 \cdot p_5 - c_3} \cdots \times \\
& \times \cdots \times \\
& \times (1 - xyz \cdots w)^{-2p_2 \cdot p_{n-1} - c_{n-2}}
\end{aligned} \tag{4}
$$

where

$$c_2 = a_2 + 2m^2 + 1$$
$$c_3 = a_3 - 2a_2 - m^2 \tag{5}$$
$$c_n = a_n - 2a_{n-1} + a_{n-2}, \quad n \geq 4.$$

The last relations may be written in a uniform fashion for all c_n's by defining

$$a_0 \equiv 1, \quad a_1 \equiv -m^2. \tag{6}$$

Thus the external particles belong to the ground state $\alpha_1(m^2) = 0$ of "one-body" resonances.

When a chain of resonances occur simultaneously for $\alpha_2(s_{12}) = k$, $\alpha_3(s_{123}) = m$, $\alpha_4(s_{1234}) = l$, etc., the corresponding residue is given by the coefficient of $x^k y^m z^l \ldots$ in the product of factors $(1 - x)$, $(1 - xy)$, $(1 - xyz)$, $\ldots$ in the integrand of Eq. (4), and it is this coefficient which we would like to factorize as

$$\sum \langle n| \cdots |l\rangle \langle l| \Gamma |m\rangle \langle m| \Gamma |k\rangle \langle k| \Gamma |1\rangle. \tag{7}$$

Here the summation is over an as yet unspecified set of states which make up each of the levels $k, m, l \ldots$ Note that we are regarding the process as a chain of transitions $1 \to k$, $k \to m$, $m \to l$, etc., induced by the external particles $2, 3, \ldots n - 1$. Thus we are not treating the lines 1 and n on an equal footing with the rest, but this is a natural consequence of the way Eq. (4) is constructed.

As the next step, consider the factor $(1 - x)^{-2p_2 \cdot p_3 - c_2}$. For a 4-point amplitude this would be the only factor to be dealt with. We rewrite it as

$$(1 - x)^{-2p_2 \cdot p_3 - c_2} = \exp\left[-(2p_2 \cdot p_3 + c_2)\ln(1 - x)\right]$$
$$= \exp\left[(2p_2 \cdot p_3 + c_2)\sum_{r=1}^{\infty} \frac{x^r}{r}\right]. \tag{8}$$

The coefficient of x^n is then to be equated with

$$\sum_{-k} \langle 4| \Gamma(p_3) |k\rangle \langle k| \Gamma(p_2) |1\rangle, \tag{9}$$

where p_2 and p_3 are the respective momentum transfers. Let us now introduce a set of Bose-like creation and annihilation operators $a_\alpha^{+(r)}$ $a_\alpha^{(r)}$, ($\alpha = 1, 2, \ldots, 5$) $r = 1, 2, \ldots$; with the property

$$[a_\alpha^{(r)}, a_\beta^{+(s)}] = -\delta_{rs} g_{\alpha\beta} \quad (\text{metric } - - - + -)$$
$$[a_\alpha^{(r)}, a_\beta^{(s)}] = [a_\alpha^{+(r)}, a_\beta^{+(s)}] = 0. \tag{10}$$

For each r, $a_\alpha^{+(r)}$ and $a_\alpha^{(r)}$ correspond to a 5-dimensional harmonic oscillator in which the 5th dimension is purely fictitious, and the timelike (4th) excitations are "ghosts" with a negative metric. Observe then that the well known formula

$$e^A e^B = e^B e^A e^{[A,B]}$$

(valid for a c-number commutator) gives

$$\langle 0|\, e^{ik'\cdot a/\eta}\, e^{ik\cdot a^+\xi}\, |0\rangle = e^{k'\cdot k\xi/\eta}\,, \tag{11}$$

where $k\cdot a^+ = k^\alpha a^+_\alpha$. It is obvious that if a complete set of intermediate states characterized by the occupation numbers $\{n_\alpha\}$ are inserted on the left, the states with $n = \Sigma\, n_\alpha$ contribute to the term $\sim(\xi/\eta)^n$ on the right.

Eq. (11) forms the basis of the factorization program. In fact Eq. (8) can be factorized as

$$\exp\,[(2p_2\cdot p_3 + c_2)\textstyle\sum x^r/r]$$
$$= \left\langle 0\left|\exp\left[i\sum\sqrt{\frac{2}{r}}\,a^{(r)}\cdot k'\eta^{-r}\right]\exp\left[i\sum\sqrt{\frac{2}{r}}\,a^{+(r)}\cdot k\xi^r\right]\right|0\right\rangle, \tag{12}$$

$$k_\alpha = \left(p_2\,,\, i\sqrt{c_2/2}\right),\quad k'_\alpha = \left(p_3\,,\, i\sqrt{c_2/2}\right),\quad \xi/\eta = x\,.$$

The term $\sim x^N$ obviously comes from intermediate states such that

$$\sum_r\sum_\alpha r n_\alpha^{(r)} = -\sum_r r a^{+(r)}\cdot a^{(r)} = N. \tag{13}$$

We may therefore write down the vertex as

$$\langle N|\,\Gamma\,|0\rangle = \left\langle N\left|\exp\left[i\sum\sqrt{\frac{2}{r}}\,a^{+(r)}\cdot k\right|0\right\rangle,$$

$$k_\alpha = \left(p_N - p_0\,,\, i\sqrt{c_2/2}\right),\quad \langle 0|\,\Gamma\,|N\rangle = \langle N|\,\Gamma\,|0\rangle^+. \tag{14}$$

The level $s = s_N$ then consists of all possible excitations satisfying the condition (13). The 5th dimension was introduced for the purpose of handling the term c_2, assumed to be >0. If $c_2 < 0$ we have to change the metric of the 5th dimension. (The factor i in k_5 is not really necessary; its purpose is to bring Γ to the form (18) below.)

How will this scheme work out for the general n-point function? We can see without much difficulty that if all the c's are equal, we can factorize

Eq. (4) by the ansatz

$$\langle N'|\Gamma|N\rangle = \left\langle N' \left| \exp\left[i\sum\sqrt{\frac{2}{r}}\,a^{+(r)}\cdot k\right] \exp\left[i\sum\sqrt{\frac{2}{r}}\,a\cdot k\right]\right| N\right\rangle ,$$

$$k = \left(p_{N'} - p_N,\ \ i\sqrt{c_2/2}\right). \tag{15}$$

This vertex operator is Hermitian except for the timelike excitations, and reduces to Eq. (14) when it operates on $|0\rangle$. In order to see how the ansatz works, we form a product of such Γ's according to Eq. (7), and take its vacuum expectation value. By shifting the annihilation operators step by step to the right we will note that the 5-momenta $k_1, k_2, \ldots$ associated with the external lines get contracted with one another, producing an exponential factor for each contraction. To extract a particular set of intermediate states $N_1, N_2, \ldots$, multiply $a^{(r)}$ and $a^{+(r)}$ in each vertex $\langle N_l|\,\Gamma\,|N_{l-1}\rangle$ by the power-counting variables $(\xi_{l-1})^{-r}$ and $(\xi_l)^r$ respectively, and pick up the coefficient of $\xi_1^{N_1}(\xi_2/\xi_1)^{N_2}(\xi_3/\xi_2)^{N_3}\ldots$ from the final result. We can verify raedily that the variables $\xi_1, \xi_2/\xi_1, \xi_3/\xi_2, \ldots$ correspond to $x, y, z, \ldots$ in Eq. (4).

In actuality, the constants c_n are not all equal. If all intercepts a_n are equal, i.e. all multiparticle resonances belong to the same trajectories, we have $c_m = 0$ for $m \geq 4$, but c_2 and c_3 are different. So the 5th dimension is necessary only for the nearest neighbor and the next nearest neighbor contractions, and we need to devise a scheme which treats these special cases. Another example is the case where we distinguish between even-n and odd-n resonances so that

$$a_0 = a_2 = \cdots = a_{\text{even}},\quad a_1 = a_3 = \cdots = a_{\text{odd}},\quad c_{\text{even}} = -c_{\text{odd}}$$

for all n. According to Eq. (5), however, this requires $a_{\text{even}} = 1$ and the external lines belong to α_{odd}. The alternating signs of c_n can be realized by a suitable definition of Γ. This example is relevant to the many-pion processes which have, however, $a_{\text{even}} = \frac{1}{2}$ rather than 1. We shall not go into the details of how one would handle these problems more precisely.

We have outlined above the basic procedure for factorizing n-point Veneziano amplitudes in a consistent way. For this we had to pay the price of introducing an infinite set of 5-dimensional harmonic oscillators, although this does not affect the leading trajectory. It is clear that each of them generates a tower of non-unitarity representations of $U(4, 1)$, leading to the unpleasant presence of ghosts. However, this does not necessarily mean that the amplitudes themselves are unphysical since there can be cancellations from various intermediate states.[5] For example, the elastic pion–pion scattering amplitude seems to be free of ghosts.[8] But the general answer to this question is not known yet.

The appearance of harmonic oscillators in our problem is intriguing since the simple bound-state picture of quarks with a harmonic oscillator potential would naturally give rise to linear trajectories and the $U(3, 1)$ level scheme. We can bring out this analogy more clearly in the following way. Let us introduce a Bose field $\phi_\alpha(\xi)$ and its canonical conjugate $\pi_\alpha(\xi)$, which are even and periodic with periodicity 2π in ξ. In analogy to the ordinary field theory, we decompose it into plane waves:

$$\phi_\alpha(\xi) = \sum_{r=1}^{\infty} \frac{1}{\sqrt{2r}} (a_\alpha^{(r)} + a_\alpha^{+(r)}) \cos r\xi$$

$$\pi_\alpha(\xi) = \sum_{r=1}^{\infty} i\sqrt{\frac{r}{2}} (a_\alpha^{(r)} - a_\alpha^{+(r)}) \cos r\xi \tag{16}$$

where we have excluded the constant mode $r = 0$. The a's and a^+'s are the operators we have defined above. In view of Eq. (13), the quantum number N which determines the resonance energy can be written

$$N = -\sum_r r a^{+(r)} \cdot a^{(r)}$$

$$= \frac{-1}{\pi} \int_0^{2\pi} :(\partial_\xi \phi(\xi) \cdot \partial_\xi \phi(\xi) + \pi(\xi) \cdot \pi(\xi)): d\xi. \tag{17}$$

Furthermore, the vertex operator Γ can be written

$$\Gamma = \, :\exp[2ik \cdot \phi(0)]: \tag{18}$$

Eq. (17) suggests that the internal energy of a meson is analogous to that of a quantized string of finite length (or a cavity resonator for that matter) whose displacements are described by the field $\phi_\alpha(\xi)$. A master field Ψ representing the mesons will depend on the space-time coordinates x_μ as well as the internal variables $\phi_\alpha(\xi)$, and satisfy the free wave equation

$$(\Box - N - c)\Psi(x; \phi) = 0. \tag{19}$$

Its interaction is described by a Lagrangian $\sim \Psi^+(x)\, \Gamma \Psi(x)\, \varphi(x)$ where φ is the field for the external particles.

The form (18) has a striking resemblance to the form factor in a naive quark model, in which one of the quarks interacts with the external field carrying momentum k. If the $q\bar{q}$ system is described by a wave function $\Psi(x; r)$, where x and r are the center of mass and relative coordinates respectively, the interaction will carry a form factor exp $[ik \cdot r/2]$. An unsatisfactory feature of the result is that the external field φ and the master field Ψ have to be distinguished, whereas clearly the starting program was to include everything in Ψ. To achieve this end one would need to find a trilinear coupling of Ψ's which agrees with the coupling that arises when the n-point Veneziano amplitude is in a configuration in which three resonances meet. The above interpretation of the form factor suggests that one might be able to construct such a coupling geometrically from quark diagrams like

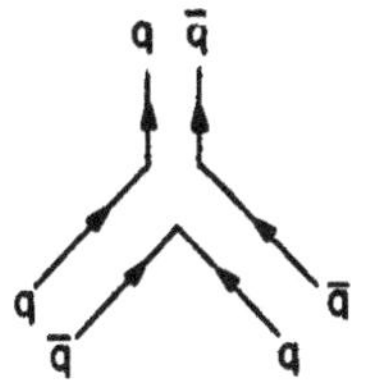

If this can be carried out, we are essentially led to a nonlocal picture of the Veneziano model, though the nonlocality may not manifest itself in a conspicuous way.

Another problem that confronts a field theory of duality is that of the multiple counting of diagrams. With a trilinear coupling of Ψ's one can construct diagrams of both s-channel and t-channel types. Because of duality they will of course give the same contributions, but the amplitude is doubled nevertheless. We can adjust for this by reducing the trilinear coupling parameter, but what about the higher order processes? Even if the factorization may work formally, it is not clear whether or not we should build a field theory in this fashion.[9] That will depend on how one interprets duality, and must eventually be solved through a confrontation with experiment.

Putting aside these problems, we can make some remarks about high energy phenomena based on our results. Since the internal excitations of hadrons resemble those of a cavity resonator, we are able to use statistical concepts like temperature, entropy and the Bose–Einstein distribution in a well defined way. For example, at extreme high energies ($\gg 1$ GeV) where the discreteness of eigenfrequencies becomes unimportant, we find

$$M^2 \sim E = TS + F \sim -F \sim (5\pi^2/6)\,(kT)^2, \tag{20}$$

where M is the rest energy of the system, and the other symbols are the standard thermodynamic ones. Thus the density of the levels goes like exp $[S/k]$ $\sim$ exp $[2M\,(5\pi^2/6)^{1/2}]$.

References

1 R. Dolen, D. Horn and C. Schmid, *Phys. Rev.* **166**, 1768 (1968).
2 G. Veneziano, *Nuovo Cimento* **57A**, 190 (1968).
3 H. Harari, *Phys. Rev. Letters* **22**, 562 (1969); J. Rosner, *Phys. Rev. Letters* **22**, 689 (1969); J. Mandula, C. Rebbi, R. Slansky, J. Weyers and G. Zweig, *Phys. Rev. Letters* **22**, 1147 (1969); K. Matsumoto, *Lettere al Nuovo Cimento* **1**, 620 (1969); T. Matsuoka, K. Ninomiya and S. Sawada, Nagoya Univ. preprint.
4 S. Mandelstam, A Relativistic Quark Model Based on the Veneziano Representation, Berkeley preprint; K. Bardakci and M. B. Helpern, A Possible Born Term for the Hadron Bootstrap, Berkeley perprint; K. Bardakci and S. Mandelstam, Analytic Solution of the Linear-Trajectory Bootstrap, Berkeley preprint.
5 S. Fubini and G. Veneziano, Level Structure of Dual Resonance Models, MIT preprint.
6 P. G. O. Freund, *Phys. Rev. Letters* **22**, 565 (1969); K. Kikkawa, B. Sakita and M. A. Virasoro, University of Wisconsin preprint.
7 C. Goebel and B. Sakita, *Phys. Rev. Letters* **22**, 257 (1969); H. M. Chan, *Phys. Letters* **28B**, 425 (1969); H. M. Chan and T. S. Tsou, *Phys. Letters* **28B**, 485 (1969); J. F. L. Hopkins and E. Platt, *Phys. Letters* **28B**, 489 (1969); Z. Koba and H. B. Nielsen, *Nucl. Phys.* **B10**, 633 (1969).
8 Y. Nambu and P. Frampton, University of Chicago preprint EFI 69-56.
9 I thank P. Freund for a discussion on this point.

DISCUSSION

ROSNER Could you amplify that discussion in which you show by statistical argument the way the states should go considering the square root of N.

NAMBU You can use the standard formulae of statistical mechanics for a cavity resonator. The eigenfrequencies are quantized in units of $\omega_0 \simeq 1$ GeV, but a high temperatures $kT \gg \omega_0$ we may ignore the discreteness of energy, and integrate over the Bose–Einstein distribution. We find

$$E = TS + F \simeq -F \simeq (5\pi^2/6) \quad (kT)^2$$

where F is the Gibbs free energy, S is the entropy, and $E = M^2 \simeq N$, the square of the rest energy of the system since we are dealing with the Mandelstam variable s. This relation shows that $T \simeq S \simeq M \simeq \sqrt{N}$. So the level density goes like $\exp [S/k] \simeq \exp \left[c \sqrt{N}\right]$.

GUNZIK How do you get rid of the ghosts in your approach?

NAMBU I do not know. One thing I have thought about is the following. The vertex Γ contains the operators $a^+ \cdot k$ and $a \cdot k$ which cause excitations or dexcitations in the direction k of the oscillator space. If k is spacelike, $k \cdot k < 0$, you might say the excitation is normal; if it is timelike, $k \cdot k > 0$, you excite ghosts. This will give you a condition of the Regge intercept a since k depends on it. Applied to a_2, we get $1 + a_2 > 0$. But this is not what you get from the positivity of partial wave residues in a 4-point amplitude. Thus the above argument seems to be shaky.

GUNZIK Is k the momentum vector in the intermediate state?

NAMBU The first four components of k is the momentum transfer, the fifth is essentially the square root of the constant c, which depends on the Regge intercept.

BREIT Why do you have factorization?

NAMBU Well, without factorization, it would be hard to understand the Veneziano amplitudes in terms of the usual quantum mechanical description of resonances. I wanted to show that factorization is possible, but at the price of having ghost states.

SYMMETRY BREAKDOWN AND SMALL MASS BOSONS*

Yoichiro Nambu
The University of Chicago

1. Introduction

The past decade has been a decade of symmetries. To be more exact, I should say an era of broken symmetries: the discovery and establishment of symmetry patterns in elementary particle phenomena which are recongized right from the beginning only as approximately valid. As has been once remarked by Salam,[1] one first tries hard to establish a symmetry, but the next moment he is figuring out how to break it. This is an entirely human attitude. The complexities of high energy phenomena convince everybody that there cannot be perfect regularities beyond the classical conservation laws--and even those have been crumbling down since 1956. (I am, of course, referring to the violations of P C and T.) Yet to achieve any meaningful phenomenological description of complex phenomena, one must organize them into regular patterns, and then allow for exceptions and deviations. But these exceptions and deviations themselves in turn must follow certain rules, and so on down the line. In this way we have been accustomed to various approximate symmetries, some of which are by now well established, like SU(3) and probably chiral SU(2) x SU(2) and SU(3) x SU(3), while others are of lesser validity or significance, as in the case of SU(6) and collinear SU(6).

In order to formulate the above problem mathematically, people have been led to consider a hierarchy of symmetries which are embodied in an imaginary (hypothetical) Hamiltonian, which consists of several "pieces" of varying magnitudes. Each piece is characterized by its symmetry properties, and is responsible for certain processes and their regularities (or irregularities depending on the point of view). One usually hastens to mention apologetically that any explicit expression for the Hamiltonian only serves as a model, and should not be taken seriously except for its symmetry properties. (I do not really subscribe to such a philosophy, but this need not be discussed here.)

*This work supported in part by the U. S. Atomic Energy Commission.

Reprinted from Fields and Quanta 1, 33 (1970).

Now there are several ways in which the symmetries (exact or approximate) can manifest themselves.

1) Symmetries (invariances) of H ⟹ conservation laws (selection rules)
⟹ degeneracies
⟹ relations among S-matrix elements

2) Transformation properties and commutation relations of particular pieces of H ⟹ Symmetry breaking patterns
⟹ Current algebra relations

3) Spontaneous breakdown of symmetries of H
⟹ Incomplete multiplets
⟹ Massless bosons.

My assigned task is to address myself to the last problem.[1a] Whether this is to be regarded as a manifestation of symmetries (of a Hamiltonian) or a violation of symmetries (of state vectors) depends on one's point of view, but I assume that no explanation is necessary about what is meant by it.

2. Spontaneous breakdown of symmetries

So far as is known, the only example in elementary particle physics where the notion of spontaneous breakdown definitely seems to make sense is, according to my view, the case of chiral symmetry, especially SU(2) x SU(2). Even in this case there are people who do not see the necessity of invoking such a principle, and I will come back to this point later. What I would like to do first is to give a historical account of the problem as I have seen it from my side.

The nomenclature "spontaneous breakdown", which is generally used now, originated in the paper by Baker and Glashow (1962)[2] bearing the same title. I had been searching for an appropriate term but had not hit on it, although a term like the "spontaneous polarization" of a ferromagnet was in common usage.[4]

The "Goldstone boson" of course owes its origin to Goldstone's paper (1961)[3] in which he quotes, as his starting point, the preprint of my talk on the superconductivity model at the Midwest Theoretical Physics Conference (1960)[4] at Purdue University. At that time I was kicking around the same idea, but was kept busy working out the consequences of the chiral (γ_5) symmetry problem. [Later, Freund and I (1964)[5] tried the name "zeron".]

Curiously, Goldstone's paper seems to be mainly concerned with the non-perturbative nature of asymmetric solutions; he works out two examples of discrete symmetry [the mass reversal $\gamma_5 \to -\gamma_5$ in a neutral pseudoscalar meson theory, and the amplitude reversal $\phi \to -\phi$ (Bronzan and Low's A parity[6]) in a ϕ^4 meson theory], and only casually discusses the breaking of continuous symmetry (gauge transformation of the 1st kind) and massless excitations in a charged ϕ^4 theory, although he conjectures the result to be more general.

The concept of degenerate vacua (or ground states) seems to have been first stated in a paper by Heisenberg (1959)[7] and co-workers on the non-linear spinor theory. Heisenberg tries to identify the Pauli-Gürsey group[8] inherent in his theory with the isospin group, and in doing so is led to consider a vacuum state (or a "world") having very large isotopic spin, because he does not have two distinct "bare" fields to represent proton and neutron. This is like saying, in the usual case of chiral symmetry, that a left-handed nucleon or quark becomes right-handed by picking up a spurion, or that a $1/2^+$ nucleon becomes a $1/2^-$ isobar in a similar fashion. [In the language of nonlinear realizations, $q_r \sim Mq_L$, where $M = 1 + if\underset{\sim}{\tau} \cdot \underset{\sim}{\pi} + \ldots$]. He also refers to Mach's principle, emphasizing the logical independence of the boundary conditions from the intrinsic symmetries of a Hamiltonian. In later publications, Heisenberg's group has been attempting to interpret photons as Goldstone bosons resulting from broken isospin symmetry. I must confess, however, that I have not understood the mathematical handling of the problem. (Some more discussions on the photon problem later.)

Let me now come back to my own story. The idea of linking massless pion with chiral symmetry came as an accident through the study of the BCS theory of superconductivity (1958)[9]. When I first heard Schrieffer's seminar at Chicago on the theory they were just working out, I was struck by the boldness of taking a trial wave function which did not conserve charge. But how can such a wave function represent a physical solution? This started to bother me more and more as time went on, and eventually led me to an investigation (1960)[10] of the gauge invariance requirements and the role of collective excitations which had previously been taken up by Anderson and others.[11,12] This problem arises in the static London relation

$$J_i(q) = K_{ij}(q)A_j(q) \tag{1}$$

4

in an infinite medium. The Meissner effect and gauge invariance require the response function K_{ij} to be of the form

$$(\delta_{ij} - \frac{q_i q_j}{q^2})\ K(q^2), \quad K(0) \neq 0. \qquad (2)$$

There are two possibilities as to the origin of the kinematic pole in the second term: either it is of dynamic origin so that in nonstatic cases q^2 is replaced by $q^2 - \alpha^2\omega^2$, implying acoustic-type collective excitations, or else the pole remains static. Since the response function is a dynamic quantity, the latter could not happen unless a $1/q^2$ singularity were built into the Hamiltonian from the beginning. Indeed it is the Coulomb interaction that actually produces such an effect. In the absence of the Coulomb interaction, the pole would be dynamic. In this case K_{ij} reduces to a non-gauge invariant form $\sim \delta_{ij}$ if we take the limit $q \to 0$, $\omega \neq 0$, and then $\omega \to 0$. One might say it reveals the symmetry breaking aspect of superconductivity.

In relativistic field theories the above mentioned two alternatives still exist, as was first suggested by Schwinger[13] and Anderson.[14]. The fact that second alternative escapes the relativistic Goldstone theorem developed by various people[15] just because of the lack of manifest Lorentz invariance of gauge vector fields was discovered later.[16] In particular, Higgs[17] envisages a realization of massive vector and pseudoscalar mesons from a single dynamical principle. But as far as the actual finite pion mass is concerned, it seems to be due to a non-spontaneous symmetry breaking effect (see below).

The obvious analogy[11] of the Bogoliubov-Valatin quasi-particle in the BCS theory with the Dirac electron naturally led me to consider the latter in light of the former. It is the γ_5 invariance that is broken by the mass (energy gap) in the case of the Dirac electron. If the breaking is spontaneous, we must also have massless pseudoscalar excitations. Very similar ideas seemed to have occurred to Vaks and Larkin,[18] who considered the Heisenberg model and derived massless excitations. At any rate, this reasoning naturally directs one to the nucleon and pion problem rather than to the electron.[19] Taking the matrix element of axial nucleon current $J_{\mu 5}$ for $q_\mu \to 0$, we find a zero-mass pion pole and a Goldberger-Treiman condition[20] just as the two terms in Eq. (2) are uniquely related by gauge invariance. Since the current is an observable, manifest Lorentz covariance requirement closes the escape hatch that is available for gauge fields, and so the finite pion mass has to be attributed to nonspontaneous breaking. More direct physical arguments for this conclusion are: 1) $g_A(0) \neq 0$ for the nucleon β decay, and 2) $\pi^{\pm} \to \ell^{\pm}\nu$ is not forbidden.[20a] If the weak axial current were really divergenceless and yet $M_{\pi} \neq 0$ these could not happen.

The large value of nucleon to pion mass ratio, and the miraculous Goldberger-Treiman relation gave me motivation for building a dynamical theory of strong interactions.[21] The problem was to choose a model which was simple and would not cloud the main theme with side issues. The Heisenberg-type nonlinear model was chosen for this purpose, but without the indefinite metric nor any attempt at reducing the number of independent fermion fields. Heisenberg's suggestion regarding the degenerate vacua was not known to me at first. Rather, the idea was generated by the study of the BCS ground state. This also led me to contemplate other examples such as ferromagnetism, in which obviously the spin waves were the symmetry-restoring collective modes.

I must refer at this point to the series of work developed by Gürsey[22] (1960) independently and from a quite different viewpoint. His elegant mathematical formalism (in terms of the operator M quoted already) was in fact a realization of the spontaneous breakdown. Much later his formalism was revived and developed to a higher degree of sophistication, to which topic I will come back after a while.

3. PCAC, Current Algebra, and Soft Pion

Gell-Mann and Lévy (1960)[23,24] initiated the well-known approach to the Goldberger-Treiman relation. It explicitly exploits the fact that pion mass is nonzero, as one can see from their formula, often called the field theoretical version of the PCAC (partially conserved axial vector current) condition.

$$\partial_{\mu} J^{i}_{\mu 5} = C\phi^{i}, \qquad C \propto m_{\pi}^{2} \tag{3}$$

In this sense it is orthogonal to the spontaneous breakdown approach. As is often remarked, Eq. (3) actually amounts to a mere definition of the pion field ϕ^{i}, and becomes meaningful only with the additional "gentleness" assumption about the variation of matrix elements between $q^2 = m_{\pi}^2$ and $q^2 = q_{\mu} = 0$, which would be reasonable only if m_{π} is small compared to other mass scales and no singularities are crossed in the process of extrapolation. Advantages of Eq. (3) are that one can easily extract various consequences from it, with the help of usual reduction techniques; the mass of external pions may be actually switched off by extrapolation while keeping the internal pion mass fixed, thus leading, for example, to the Adler self-consistency condition.[25] On the other hand, Eq. (3) does not explain why m_π is small, nor predict what would happen if actual pion mass could be turned off.

The spontaneous breakdown approach starts from the situation where $m_\pi = 0$, so that the mathematical handling of the problem becomes more complicated, though the results are the same in most cases. Clearly, however, this is a more restrictive assumption than the formal PCAC condition. What it says is that the nucleon mass should remain finite if the agent responsible for the pion mass were turned off. Most models that have been constructed do possess both features, so there are no practical differences between the two approaches.

The group theoretical structure of chiral symmetry, however, is important in making spontaneous breakdown of chiral symmetry a meaningful statement. For there exists the derivative coupling model[23] for which the symmetry group is an inhomogeneous SU(3) (the inhomogeneous part being realized by the displacement of massless pseudoscalar fields). The fermions do not take part in chiral transformations, so their masses need not be created by a spontaneous breakdown. This may be part of the reasons why some people do not regard spontaneous breakdown as necessary for the success of PCAC. As far as single pion emission is concerned, SU(3) x SU(3) and the inhomogeneous SU(3) are equivalent. But a difference shows up in the commutator of two axial generators, as in π-N scattering. It is here that the significance of current algebra emerges. The success of the Adler-Weisberger relation[26] (1965) indicates that SU(3) x SU(3) [or at least SU(2) x SU(2)] is indeed the relevant group for the strong interaction of hadrons.

The soft pion theorems, including part of the SU(2) x SU(2) current algebra relations, were created for the purpose of testing the basic idea of chiral symmetry.[26a] But these earlier attempts at applications of the soft pion theorems did not yield much because of a lack of useful data. For example, for the electro-pion production process, the data has become available only recently.[27] Applications to semileptonic or nonleptonic processes could have been done more profitably. In fact this was being contemplated in the early days. What prevented its success was the absence of the quark model.

7

Without it the extension of chiral SU(2) to the baryon octet would have too much arbitrariness.[28] So one had to wait until Gell-Mann (1964) introduced the quark model[29] and subsequently laid down the principle of current algebra,[30] followed by Fubini and Furlan's discovery[31] of combining current algebra with dispersion relations, and Adler and Weisberger's remarkable result on g_A/g_V,[26,32] before Suzuki, Sugawara, Callan and Treiman, etc., could start with confidence the rush to weak processes.[33]

We have yet to discuss the choice between mere PCAC and spontaneous (or almost spontaneous) breakdown. Relevant in this respect is the work of Kim and Von Hippel[34] on baryon-pseudoscalar meson scattering lengths, which is an extension of Tomazawa's formulation[35] of pion nucleon scattering as a low energy theorem. The point can be explained in terms of the effective Lagrangian method (see next Section). If the nucleon mass is of spontaneous origin, the mass is part of the Gürsey operator M. The S wave pion-nucleon scattering length $a_{1/2} + a_{3/2}$ (the so-called σ term) computed from this Lagrangian is zero. On the other hand, if the mass is actually a symmetry breaking term, it would be a pure constant, behaving like (2,2*) of SU(2) x SU(2), and would give a large scattering length. The same argument applies to to SU(3) x SU(3) too. The Kim-Von Hippel analysis indicates that the actual breaking of SU(3) x SU(3) is essentially in the bare mass of the strange quark only, and its magnitude is of the order of Gell-Mann - Okubo mass splitting. Thus bulk of the baryon mass appears to have been spontaneously created, and SU(2) x SU(2) symmetry is a better symmetry than broken SU(3) (because $m_\pi^2 \ll m_K^2$). More quantitatively, one can compute the ratios of bare quark masses from the masses of the octet pseudoscalar mesons. This scheme was considered by Gell-Mann, Oakes and Renner.[36] If one assumes the quark masses to be $(m_p, m_n, m_\lambda) = (m_1, m_1, m_2)$, one gets

$$2m_1/(m_1 + m_2) = m_\pi^2/m_K^2 \simeq 1/12. \tag{4}$$

Actually, all quark masses may be different. If the K mass difference is attributed to this cause, the relation

$$m_n - m_p : m_n + m_p : 2m_\lambda + m_n + m_p = m_{K^\circ}^2 - m_{K^\pm}^2 : m_{\pi^\pm}^2 : m_{K^\circ}^2 + m_{K^\pm}^2$$

then yields the result

$$m_p : m_n : m_\lambda \approx 2 : 3 : 60. \tag{5}$$

(There is no pion mass difference in this approximation because it belongs to an I=2 piece.)

The significance of nearly spontaneous breaking of SU(3) x SU(3) as a valid picture of hadrons has been emphasized by Dashen.[37] There are, however, some unsolved puzzles. One is the p wave nonleptonic decays of hyperons for which PCAC is not as successful as for the s wave decays. Another is the ξ parameter in $K_{\ell 3}$ decays, which is ≈ -1 experimentally in contrast to the theoretical estimate ≈ 0. A more serious problem, it seems to me, is the question why chiral SU(3) is a better symmetry than chiral U(3), as is evidenced by the large mass of η' compared to those of the pseudoscalar octet, with little mixing between them. The symmetry breaking scheme in terms of quark masses alone cannot explain it.

Related to this kind of program is the attempt at seeking causal relations among Gell-Mann - Okubo SU(3) breaking, electromagnetic interaction, and the Cabibbo angle in weak interactions. These create three different preferred axes in the SU(3) space, and might correspond to a self-consistent pattern of symmetry breaking similar to the Jahn-Teller effect. The discovery by Brout[38] and Cabibbo[39] that in SU(3) a symmetry breaking driving force and its response need not be in the same direction made such a mechanism an intriguing possibility. This has been further pursued by Gatto and others[40] from somewhat different angles, but it is still an open question whether the picture is fundamentally correct or not.

4. Unified description of hadron symmetries

The physical consequences of spontaneous breakdown can be most elegantly displayed by the method of nonlinear realizations. It originated in the work of Gürsey (1960)[22] and one of the models considered by Gell-Mann and Lévy.[23] A forerunner to Gürsey's nonlinear Lagrangian is the model of Nishijima.[41] Nishijima's idea was to save chirality invariance in leptonic and hadronic weak currents by considering the phase of a chiral gauge transformation as a neutral pseudoscalar massless field. He also discussed a kind of low energy theorem analogous to the soft pion theorem.

Gürsey considered massless pion fields as the phases of a chiral SU(2) gauge transformation operator,[42] and constructed an invariant Lagrangian out of such gauge operators. As it happens, this model is essentially equivalent to Gell-Mann - Lévy's nonlinear σ model in spite of their apparent differences.

9

They can be transformed into each other by a nonlinear redefinition of the pion field without affecting the S-matrix elements. This point has been clarified after Weinberg[43] (1968) and others[44] started to develop a theory of nonlinear realizations, which reached its most general form in the work of Coleman, et al.[45]

A significant new element which emerged as a result is the marriage of nonlinear realization with the gauge theory of massive vector and axial vector fields, as was initiated by Lee, Weinberg and Zumino[46] (1967). It offered an elegant unified approach to PCAC and vector dominance, enabling one to handle the "hard pion" problem.

Eventually, one must come to an understanding of the role of the chiral group in all hadronic states. The chiral group may be realized either linearly (as parity doublets) or nonlinearly (via small mass mesons). It is possible that some baryon resonances belong to the former. But if that is the case, how does a state decide to choose one or the other? And how can one reconcile the chiral symmetry with the conflicting SU(6)-type symmetries? In spite of some suggestive possibilities offered by Weinberg's attempts[47] and Lovelace's discovery[48] of the connection between chiral symmetry and the Veneziano model, we are still far from having a clear cut picture.

5. Further search for spontaneous breakdown

Except in nonrelativistic phenomena, we do not seem to have any genuine cases of spontaneous breakdown. The only known massless particles are gravitons, photons and neutrinos. Each has already a perfectly respectful description without invoking any symmetry breaking. Besides, neutrinos could not fit the picture because they are fermions. Some attempts have been made to reformulate electromagnetism, and perhaps also gravity, as a spontaneous breakdown phenomenon.

In either case, the relevant symmetry must be some space-time symmetry. The trouble is that once we break a symmetry even spontaneously, it is no longer a symmetry in the ordinary sense. Thus one loses, at least, manifest space-time invariance of a theory, and probably the actual invariance as well unless one is lucky. Such a lucky situation, however, seems to prevail in electrodynamics. A preferred direction, characterized by a timelike vector n_μ, generates Goldstone modes $A_\mu(x)$ which are orthogonal to it, and may be identified with the vector potential. A particular choice of η_μ merely corresponds to a particular choice of gauge, and therefore does not destroy the Lorentz invariance of physical predictions. This can be most clearly understood if we work in the Dirac gauge:[49] $A_\mu(x)\, A^\mu(x) = \text{const.}$, and regard this relation as a nonlinear realization of the Lorentz group.[50] So electrodynamics is compatible with spontaneous breakdown, but the latter is not necessary for the understanding of the former.

A more ambitious program in exploiting spontaneous breakdown is the quantum electrodynamics of Baker, Johnson and Willey,[51] who try to create electron mass spontaneously. But the electron mass breaks γ_5 as well as scale invariance, and it is not clear why one can escape the consequences of the Goldstone theorem. Nevertheless, it is perhaps interesting to seek differences, if any, in the predictions of their theory and the conventional quantum electrodynamics with infinite subtractions.

I would like to thank Professors E. C. G. Sudarshan and Jagdish Mehra for giving me an opportunity to stay at the University of Texas and prepare part of this talk.

FOOTNOTES

1. A. Salam, quoted by J. J. Sakurai, Ann. Phys. 11, 1 (1960)(footnote, p. 5).

1a. My earlier attempts along this line are found in: Group Theoretical Concepts and Methods in Elementary Particle Physics (Ed. F. Gürsey), Gordon and Breach, New York, 1964; Lectures at the Instanbul Summer School, 1966 (Univ. Chicago Preprint EFINS 66-107);Talk at the Amer. Phys. Soc. Meeting, Chicago, 1968 (Univ. Chicago Preprint EFINS 68-11).

2. M. Baker and S. L. Glashow, Phys. Rev. 128, 2462 (1962).

3. J. Goldstone, Nuovo Cimento 19, 155 (1961).

4. Proc. Midwest Theoretical Physics Conference at Purdue Univ. (1960).

5. P. G. O. Freund and Y. Nambu, Phys. Rev. Letters 12, 714 (1964).

6. J. B. Bronzan and F. E. Low, Phys. Rev. Letters 12, 522 (1964).

7. H. P. Dürr, W. Heisenberg, H. Mitler, S. Schlieder and K. Yamazaki, Z. Naturf. 14a, 441 (1959).

8. W. Pauli, Nuovo Cimento 6, 204 (1957).

9. J. Bardeen, L. N. Cooper and J. R. Schrieffer, Phys. Rev. 106, 162 (1957).

10. Y. Nambu, Phys. Rev. 117, 648 (1960).

11

11. P. W. Anderson, Phys. Rev. 110, 827 (1958); 112, 1900 (1958). G. Rickayzen, Phys. Rev. 111, 817 (1958); Phys. Rev. Letters 2, 91 (1959).

12. For a historical perspective, see P. W. Anderson in Superconductivity (Ed. by R. D. Parks), Marcell Dekker, Inc., New York, 1969 (p. 1343).

13. J. Schwinger, Phys. Rev. 125, 394 (1962). A massive photon emerges in two-dimensional electrodynamics. Note that the charge has dimensions of mass in this case.

14. P. W. Anderson, Phys. Rev. 130, 439 (1963).

15. J. Goldstone, A. Salam and S. Weinberg, Phys. Rev. 127, 965 (1962). For a detailed review of the mathematical aspects of the Goldstone theorem see articles by T. W. Kibble (p. 277) and D. Kastler (p. 305) in Proc. 1967 International Conference on Particles and Fields (Ed. C. R. Yagen et al.), Interscience, New York, 1967; and by G. S. Guralnik, C. R. Hagen and T. W. Kibble in Advances in Particle Physics, Vol. 2.

16. P. W. Higgs, Phys. Letters 12, 132 (1964); Phys. Rev. Letters 13, 508 (1964). F. Englert and R. Brout, Phys. Rev. Letters 13, (1964).

17. P. W. Higgs, Phys. Rev. 145, 1156 (1966).

18. V. G. Vaks and A. I. Larkin, JETP 40, 282 (1961) [Soviet Phys. 13, 192 (1961)].

19. Y. Nambu, Phys. Rev. Letters 4, 380 (1960).

20. M. L. Goldberger and S. B. Treiman, Phys. Rev. 110, 1178 (1958).

20a. J. C. Taylor, Phys. Rev. 110, 1216 (1958).

21. Y. Nambu and G. Jona-Lasinio, Phys. Rev. 122, 345 (1961); 124, 246 (1961).

22. F. Gürsey, Nuovo Cimento 16, 230 (1960); Ann. Phys. 12, 705 (1960).

23. M. Gell-Mann and M. Lévy, Nuovo Cimento 16, 705 (1960). J. Bernstein, M. Gell-Mann and L. Michel, Nuovo Cimento 16, 560 (1960). J. Bernstein, S. Fubini, M. Gell-Mann and W. Thirring, Nuovo Cim. 17, 757 (1960). Chou Kuang-Chao, JETP 39, 703 (1963) [Soviet Phys. 12, 492 (1961)].

24. For the state of affairs prevailing around 1960, see Proc. 1960 International Conference on High Energy Physics at Rochester (Ed. E. C. G. Sudarshan et al.), Interscience, 1960, especially articles by Gell-Mann (p. 508), Gürsey (p. 572), Heisenberg (851), Goldberger (p. 733), Nambu (p. 85), Okun (p. 743), Vaks and Larkin (p. 873). Landau is also credited with the idea of PCAC (pp. 741, 749).

25. S. L. Adler, Phys. Rev. 137, B1022 (1965); 139, B1638 (1965).

26. S. L. Adler, Phys. Rev. Letters 14, 1051 (1965); Phys. Rev. 140, B736 (1965). W. I. Weisberger, Phys. Rev. Letters 14, 1047 (1965); Phys. Rev. 143, 1306 (1966).

26a Y. Nambu and D. Lurie, Phys. Rev. 125, 1429 (1962). Y. Nambu and E. Shrauner, Phys. Rev. 128, 862 (1962). E. Shrauner, Phys. Rev. 131, 1847 (1963).

27. Y. Nambu and M. Yoshimura, Phys. Rev. Letters 24, 25 (1970).

28. This is also the reason that the Baker-Glashow program[2] based on the Sakata model, and the complicated octet model of S. Glashow [Phys. Rev. 130, 2132 (1962)] and of N. Byrne, C. Iddings and E. Shrauner [Phys. Rev. 139, B918, B933 (1965)] did not yield interesting results.

12

29. M. Gell-Mann, Phys. Letters 8, 214 (1964); G. Zweig, CERN Reports Nos. 8182/TH401, 8419/TH.412, 1964 (unpublished).

30. M. Gell-Mann, Physics 1, 63 (1964). A detailed exposition of the subject, including many of the topics discussed in this section, is found in S. L. Adler and R. F. Dashen, Current Algebras, W. A. Benjamin, Inc., New York, 1968.

31. S. Fubini and G. Furlan, Physics 1, 229 (1965).

32. In early days there was a misguided conjecture that $g_A/g_V \to 1$ in the chiral symmetry limit. There is no general proof for it when spontaneous breakdown sets in.

33. M. Suzuki, Phys. Rev. Letters 15, 986 (1965).
H. Sugawara, Phys. Rev. Letters 15, 870; 997 (1965)
C. G. Callan, Jr., and S. B. Treiman, Phys. Rev. Letters 16, 153 (1966).

34. F. Von Hippel and J. K. Kim, Phys. Rev. Letters 22, 740 (1969).
Yuk-Ming P. Lam and Y. Y. Lee, Phys. Rev. Letters 23, 734 (1969).

35. Y. Tomozawa, Nuovo Cimento 46A, 707 (1966).
S. Weinberg, Phys. Rev. Letters 17, 616 (1966).

36. M. Gell-Mann, R. J. Oakes and B. Renner, Phys. Rev. 175, 2195 (1968).

37. R. Dashen, Phys. Rev. 183, 1245 (1969).
R. Dashen and M. Weinstein, Phys. Rev. 183, 1261 (1969).

38. R. Brout, Nuovo Cimento 47A, 932 (1967).

39. N. Cabibbo, in Hadrons and Their Interactions (Ed. A. Zichichi), Academic Press, New York, 1968.

40. R. Gatto, G. Sartori and M. Tonin, Phys. Letters 28B, 128 (1968).
N. Cabibbo and L. Maiani, Phys. Letters 28B, 131 (1968).

41. K. Nishijima, Nuovo Cimento 11, 698 (1959).

42. A similar idea is found in G. Kramer, H. Rollnik and B. Stech, Z. Phys. 154, 564 (1959).

43. S. Weinberg, Phys. Rev. 166, 1568 (1968).

44. For a survey of nonlinear realizations and related topics see S. Gasiorowicz and D. A. Geffen, Rev. Mod. Phys. 41, 531 (1969).

45. S. Coleman, J. Wess and B. Zumino, Phys. Rev. 177, 2239 (1969).
C. G. Callan, Jr., S. Coleman, J. Wess and B. Zumino, Phys. Rev. 177, 2247 (1969).

46. T. C. Lee, S. Weinberg and B. Zumino, Phys. Rev. Letters 18, 1029 (1967).

47. S. Weinberg, Phys. Rev. 177, 2604 (1969).

48. C. Lovelace, Phys. Letters 28B, 265 (1968).

49. P. A. M. Dirac, Proc. Roy. Soc. A209, 291 (1951). For further references, see Ref. 50.

50. Y. Nambu, Progr. Theoret. Phys. Suppl., Extra Number, 190 (1968).
The vector n_μ may be interpreted, in the context of Dirac's work, as the velocity of "ether". Compare also F. London, Superfluids, John Wiley & Sons, New York, 1950, Vol. I, p. 62.

51. K. Johnson, M. Baker and R. Willey, Phys. Rev. 136, B1111 (1964).

Duality and Hadrodynamics†

Y.Nambu

Enrico Fermi Institute and

Department of Physics

University of Chicago

Notes prepared for the Copenhagen
High Energy Symposium, Aug. 1970
(Unpublished and undelivered)

(Supported in part by the U.S.Atomic Energy
Commission under the Contract No.AT(11-1)-264)

†Retypeset by M. Okai, Nov. 1993

1 Introduction

We would like to explore here possible dynamical properties of hadrons suggested by the duality principle of Dolen, Horn and Scmid, and in particular by its mathematical realization due to Veneziano. Our primary concern will be not in the very interesting details of the mathematical properties that the Veneziano model seems to possess in abundance, but rather in guessing at the internal structure and dynamics of hadrons which underlie the Veneziano model, recognizing that the latter is in all likelihood only an approximate and imperfect representation of the former.

On crucial step we take in interpreting the Veneziano model is factorization. Of course this is a trivial problem, at least in principle if not in practice, if one is dealing with a given four-point amplitude. But it becomes a very restrictive condition if one demands that the same set of resonances saturate all $n-$point amplitudes. Perhaps such a condition is unwarranted; in discussing as revolutionary a concept as duality one may have to give up all the conventional notions about resonances, e.g. that the intrinsic properties of a resonance is independent of how it is prepared, and a multiple resonance can be analyzed in terms of the individual resonances, etc. Nevertheless, the fact that a set of resonances have been found to saturate all the $n-$point Veneziano amplitudes of the standard variety is very significant. It suggests that our conventional notions about resonances still make some sense, and the know techniques of field theory can be applied to them.

The condition of factorizability immediately rules out an ad hoc addition of satellite terms to a scattering amplitude because it would in general destroy factorization unless more and more resonances are introduced. If satellite terms do arise, they must do so in a well defined and self-consistent way.

As it stands now, the Veneziano model is still beset with many difficulties. For all its mathematical elegance, its practical successes are few. It would therefore be appropriate to list here its basic predictions as well as difficulties without going into the details.

Basic predictions.

1) Linearly rising trajectories. This is in accordance with observation. Whether the trajectories really keep rising indefinitely or not is of course an open question, but it seems to be a valid simplification to assume that they do.

2) Regularly spaced daughter trajectories, implying a highly degenerate level structure. Again the actual degeneracy may be only approximate, but these

daughters must exist if the model makes sense at all. So far there is some evidence for ϵ (daughter of ρ), but none for ρ' and ϵ' (daughter of f).
3) The factorized Veneziano model implies an even higher degree of degeneracy. At each level many distinguishable states having the same spin exist. The number of states increases with energy as $\sim \exp(cE)$. This situation bears a striking similarity to Hagedorn's model of high energy reactions.

Unsolved practical problems.
1) Baryons. We do no yet have a quantitatively satisfactory picture of meson-baryon and baryon-baryon scattering based on the Veneziano model, in spite of the fact that duality was first discovered in meson-baryon scattering.
2) A general satisfactory unification of the quark model (or its $SU(3)$, $SU(6)$ and chiral $SU(3)$ aspects) with the Veneziano model does not exist yet.
3) No convincing theory of form factors exists.
4)We do not know what the Pomeron is within the framework of the Veneziano model.

We must emphasize, however, that there are numerous attempts and speculations regarding all these problems.

More fundamental difficulties.
1) Unitarity. The original Veneziano model is a zero width approximation. The amplitude wildly oscillates with energy, and only after averaging over an interval does it reproduce the smooth Regge behavior at high energy. If a basic Hamiltonian for the model is given, unitarization might be formally carried out by taking into account higher order processes, although the highly singular nature of the Hamiltonian casts doubt about its meaningfulness.
2) A more serious problem, however, is the existence of ghosts. Here we mean by ghosts unphysical particles having a) negative probability, or b) spacelike momenta (tachyons), or both. The levels of the factorized Veneziano model contain those of four-dimensional harmonic oscillators, where the timelike excitations have an intrinsic negative norm. This does not necessarily mean that the Breit-Wigner residue of a partial wave projection of a given amplitude is not positive. Contributions from many degenerate states can add up to a positive value, but there is no guarantee that this will always happen. Fubini and Veneziano have found a Ward-like identity which accomplishes this cancellation to a certain extent, and Virasoro has extended their result. They rely, however, on very special properties of the Hamiltonian, and it is not clear whether and how these can be preserved in general. Besides, Virasoro's scheme still leaves us with a tachyon ghost ($m^2 = -1$). The killing of

tachyons is easy in special cases like $\pi - \pi$ scattering (where the ρ trajectory has a potential tachyon), but a general prescription in a factorized model seems very complicated, if not impossible.

We have emphasized the ghost problem because this is not peculiar to the Veneziano model alone, but it is rather a common disease afflicting all attempts at a description of hadron states as infinite multiplets. The program of current algebra saturation, as well as the use of infinite-component wave equations, have floundered on the same difficulty. To a certain extent, the two kinds of ghosts seem to be complementary: The use of finite-dimensional Lorentz tensors (as in the factorized dual model) involves negative probabilities, whereas infinite-dimensional unitary representations in general lead to tachyons.

If these ghosts are so difficult to eliminate, why not accept them and look for them? Maybe ghosts of one kind or the other do exist, which would make either T.D.Lee or G.Feinberg happy (or both, plus Sudarshan and others). But the trouble is that the extent to which these ghosts appear in a particular problem seems to depend on one's cleverness and ability to avoid them. How many of the ghosts are "real"? This is the most serious question of principle that haunts us, especially us the theorists.

2 Factorized Veneziano model

As has been shown by various people, the n−point dual amplitude for scalar external particles can be factorized in terms of a set of harmonic oscillators corresponding to an elastic string of finite intrinsic length. In classical theory, the motion of a free mass point can be derived from the action integral

$$I = -m \int d\tau, \quad d\tau^2 = -dx_\mu dx^\mu \quad (\text{metric } (- + + +)) \tag{1}$$

Alternatively one may take

$$I' = \frac{1}{2} m \int \left(\frac{dx_\mu}{d\tau} \frac{dx^\mu}{d\tau} - 1 \right) d\tau \tag{2}$$

τ being an independent parameter, and $x^\mu(\tau)$ the dynamical variables. (The constant in the integrand is added for convenience.) This form is more suitable for the transition to quantum theory. Because of the translational invariance under $\tau \to \tau + c$ and $x^\mu \to x^\mu + a^\mu$, both the Hamiltonian and the momenta

$$\begin{aligned} H &= (p_\mu p^\mu + m^2)/2m \\ p^\mu &= m dx^\mu/d\tau \end{aligned} \tag{3}$$

are conserved. By imposing the constraint

$$\frac{dx_\mu}{d\tau}\frac{dx^\mu}{d\tau} = -1 \tag{4}$$

we can normalize the parameter τ, which then becomes the proper time of the mass point. This condition (4) amounts to

$$H = 0. \tag{5}$$

In quantum mechanics, one postulates the commutation relations

$$[x^\mu, p^\nu] = i g^{\mu\nu} \tag{6}$$

and the Schrödinger equation

$$i\frac{\partial}{\partial\tau}\Psi = H\Psi \tag{7}$$

Eq.(5) is to be replaced by

$$H\Psi = 0 \tag{8}$$

which is nothing but the Klein-Gordon equation

$$(p^\mu p_\mu + m^2)\Psi = 0. \tag{9}$$

We see thus the usefulness of the proper-time formalism. If we just integrate Eq.(7), we get

$$\Psi(\tau) = \exp[-i(p^2 + m^2)\tau/2m]\Psi(0) \tag{10}$$

or

$$\begin{aligned} (\Psi(x';\tau), \Psi(x;0)) &= \langle x'|\exp[-i\tau(p^2+m^2)/2m]|x\rangle \\ &= (-im^2/4\pi^2)\exp[imx^2/2\tau - i\tau m/2] \\ &= (-im^2/4\pi^2)\exp[im\tau\{(x/\tau)^2 - 1\}/2] \end{aligned} \tag{11}$$

This is the transition amplitude $x \to x'$ after an elapsed "time" τ. Integrating over τ from 0 to ∞ (we might say it does not matter how much time it has elapsed between the events x and x'), we obtain the Feynman propagator

$$\frac{1}{2m}\int_0^\infty e^{-i\tau(p^2+m^2)/2m}d\tau = -i(p^2+m^2-i\epsilon)^{-1}. \tag{12}$$

It is well known that the above results may be interpreted or derived from the path integration method.

After this digression, let us come to the elastic string. We can imagine it to be the limit of a chain of N mass points as $N \to \infty$. Each mass point will trace out a world line, so that in the limit we are dealing with a two-dimensional world sheet. This sheet may be parametrized by two intrinsic coordinates ξ ($0 \leq \xi \leq \pi$, let us say) and τ ($-\infty < \tau < \infty$), corresponding to spacelike and timelike coordinates. We assume the action integral

$$I = \frac{1}{4\pi}\iint\left(\frac{\partial x_\mu}{\partial \tau}\frac{\partial x^\mu}{\partial \tau} - \frac{\partial x_\mu}{\partial \xi}\frac{\partial x^\mu}{\partial \xi}\right)d\xi d\tau \tag{13}$$

from which follows the equation

$$(\partial^2/\partial\tau^2 - \partial^2/\partial\xi^2)x^\mu = 0 \tag{14}$$

with the boundary condition

$$\partial x^\mu/\partial\xi = 0 \qquad \text{at } \xi = 0,\ \pi. \tag{15}$$

Duality is essentially a result of the symmetry between ξ and τ though it is not yet a perfect symmetry because of the differences in domain and metric. We have chosen the hyperbolic form because only then can one formulate the Hamiltonian principle. Actually it turns out that in computing the scattering amplitudes a switch to an elliptic form through the change $\tau \to -i\eta$ brings out duality more explicitly.

Eq.(13) is invariant under the translations $\tau \to \tau + c$ and $x^\mu \to x^\mu + a^\mu$, which imply the conservation laws

$$\begin{aligned} H &= \int[\pi p_\mu p^\mu + \frac{1}{4\pi}\left(\frac{\partial x_\mu}{\partial\xi}\right)\left(\frac{\partial x^\mu}{\partial\xi}\right)]d\xi = \text{const.} \\ p^\mu &= (1/2\pi)(\partial x^\mu/\partial\tau), \\ P^\mu &= (1/2\pi)\int(\partial x^\mu/\partial\tau)d\xi = \int p^\mu d\xi = \text{const.} \end{aligned} \tag{16}$$

in direct analogy with Eq.(3). A normal mode decomposition of Eqs.(14) and (15) yields

$$\begin{aligned} x^\mu &= x_0^\mu + 2\sum_{n=1}^{\infty} x_n^\mu \cos n\xi \\ p_n^\mu &= \partial x_n^\mu/\partial\tau \;\; (n\neq 0), \;\; p_0^\mu = \frac{1}{2}\partial x_0^\mu/\partial\tau \\ H &= p_{0\mu}p_0^\mu + \frac{1}{2}\sum_{n=1}^{\infty}(p_{\mu n}p_n^\mu + n^2 x_{\mu n}x_0^\mu \\ P^\mu &= p_0^\mu \end{aligned} \tag{17}$$

We may interpret x_0^μ and p_0^μ as the center-of-mass coordinates and momenta.

When the system is quantized, we get the familiar expression

$$\begin{aligned} &H = P_\mu P^\mu + H_0, \;\; H_0 = \sum_{n=1}^{\infty} n a_{\mu n}^\dagger a_n^\mu (+\text{c number}) \\ &x_n^\mu = (a_n^\mu + a_n^{\mu\dagger})/\sqrt{2n} \\ &p_n^\mu = -i(a_n^\mu - a_n^{\mu\dagger})\sqrt{n/2} \;\; (n\neq 0) \\ &[a_n^\mu, a_n^{\nu\dagger}] = g^{\mu\nu}\delta_{nm} \end{aligned} \tag{18}$$

By imposing the subsidiary condition

$$(H-\alpha)\Psi = 0 \tag{19}$$

we can single out an infinite tower of states with the mass spectrum

$$M^2 = H_0 + \alpha = \sum_{\mu,n} nN_{\mu n} + \alpha \tag{20}$$

In order to construct dual scattering amplitudes we introduce an external scalar field φ and postulate

$$H = P_\mu P^\mu + H_0 + g :\varphi(x(\xi=0)): \tag{21}$$

We will not discuss how this leads to $n-$point dual amplitudes in the multiperipheral configuration since it is well known.

We now add several remarks.

1) The condition $0\leq\xi\leq\pi$ fixes a fundamental scale of length. ξ is here

measured in $(GeV/c)^{-1}$ to fit the trajectory slope of ~ 1. Whether the system is to be interpreted as a rubber string or a rubber band depends on the boundary condition to be imposed. The rubber band, as twice as many modes as the string, but the difference does not show up in a case like Eq.(21) because those modes which have nodes at $\xi = 0$ cannot be excited. Clearly there will be differences if one tries to extend the model, and this will be an important point in constructing a general theory of hadrons.

2) For a point particle Eq.(1) has a purely geometric meaning (as the length of a world line), but Eq.(2) does not since it depends on the scale (gauge) of the unphysical parameter τ. In the case of the string, on the other hand, Eq.(13) is invariant under the scaling $\tau, \xi \to \lambda\tau, \lambda\xi$. This is one aspect of the conformal invariance of two-dimensional Laplace equations, a property which has widely been utilized to study the Veneziano model.

Nonetheless, Eq.(13) is not a purely geometrical quantity. For curiosity, then, let us try to construct a geometric action integral as one does in general relativity. Obviously a natural candidate for it is the surface area of the two-dimensional world sheet; another would involve its Riemann curvature. The sheet is imbedded in the Minkowskian 4-space, so one can parametrize its points as $y^\mu(\xi^0, \xi^1)$, $(\xi^0 \sim \tau, \xi^1 \sim \xi)$. The surface element is a $\sigma-$tensor

$$\begin{aligned} d\sigma^{\mu\nu} &= G^{\mu\nu} d^2\xi, \\ G^{\mu\nu} &= \partial(y^\mu, y^\nu)/\partial(\xi^0, \xi^1) \end{aligned} \tag{22}$$

whereas its line element is

$$\begin{aligned} ds^2 &= g_{\alpha\beta} d\xi^\alpha d\xi^\beta \qquad (\alpha, \beta = 0, 1) \\ g_{\alpha\beta} &= (\partial y_\mu/\partial\xi^\alpha)(\partial y^\mu/\partial\xi^\beta) \end{aligned} \tag{23}$$

A possible action integral would be

$$I = \int |d\sigma_{\mu\nu} d\sigma^{\mu\nu}|^{1/2} = \iint |2\det g|^{1/2} d^2\xi \tag{24}$$

to be compared with the old one (13) which can be written $(y \to x)$

$$I = -\frac{1}{4\pi} \iint g_{\alpha\beta} \overset{\circ}{g}{}^{\alpha\beta} d^2\xi, \quad \overset{\circ}{g}{}^{\alpha\beta} = \begin{pmatrix} -1 & 0 \\ 0 & 1 \end{pmatrix} \tag{25}$$

It is obvious that Eq.(24) leads to nonlinear equations. More complicated equations involving curvature would be not only nonlinear, but also have

higher derivatives.
3) It is sometimes useful to consider the "energy-momentum tensor" in the (ξ, τ) space:

$$T_{\alpha\beta} = \frac{1}{2\pi}(g_{\alpha\beta} - \frac{1}{2}\overset{\circ}{g}_{\alpha\beta}\, g_{\gamma\delta}\, \overset{\circ}{g}^{\gamma\delta}) \tag{26}$$

In particular, let us take its space integral over a test function $f(\xi)$,

$$\bar{T}_{\alpha\beta}[f] = \int_0^\pi T_{\alpha\beta}(\xi)f(\xi)d\xi. \tag{27}$$

By virtue of the canonical commutation relations, they generate a commutator algebra

$$\begin{aligned}
&\left[\ (\bar{T}_{00} \pm \bar{T}_{01})[f],\quad (\bar{T}_{00} \pm \bar{T}_{01})[g]\ \right] = -2i(\bar{T}_{00} \pm \bar{T}_{01})[h],\\
&\left[\ (\bar{T}_{00} \pm \bar{T}_{01})[f],\quad (\bar{T}_{00} \mp \bar{T}_{01})[g]\ \right] = 0,\\
&h = f'g - fg'
\end{aligned} \tag{28}$$

as an integral form of the Schwinger conditions. These relations, when applied to the set $f_n = 1 - e^{-2in\xi}$, amount to

$$\begin{aligned}
&\left[\ L_n^{\pm},\ L_m^{\pm}\ \right] = 2(n-m)L_{n+m}^{\pm}\\
&\left[\ L_n^{\pm},\ L_m^{\mp}\ \right] = 0\\
&L_n^{\pm} = (\bar{T}_{00} \pm \bar{T}_{01})[e^{2in\xi}],\quad L^{\pm}[f_n] = L_0^{\pm} - L_{-n}^{\pm}
\end{aligned} \tag{29}$$

These operators $L^{\pm}[f_n]$ have been found useful in generating the various gauge operations.
4) As we have mentioned already, the most serous defect of the above formulation is the indefinite metric that appears in defining the covariant commutation relations (18). The mass operator (20), however, acquires as a result the nice property of being positive. The transition from a classical to quantum picture of 4-dimensional harmonic ocsillators is a drastic one. A wave function in coordinate space would behave like $\exp[-c^2(\underline{x}^2 - x_0^2)]$, which explodes in the timelike direction. We could actually insist that it should behave instead like $\exp[-c(\underline{x}^2 + x_0^2)]$. This would amount to interchanging the creation and annihilation operators and using positive metric for the time component. But then the mass operator would not be positive, and moreover

each level would become infinitely degenerate. In group theoretical terms, the former corresponds to non-unitary, and the latter to unitary representations of $U(3,1)$. The former have negative probability ghosts while the latter have negative mass squared ghosts. The Veneziano model seems to prefer the former. Such a choice is necessary to ensure a Regge behavior à la Van Hove, but runs into trouble with form factors. This is a general agony of making the choice, not restricted to Mr.Veneziano alone.

5) We have ignored the problems of the extra scalar excitations which are needed to incorporate the proper trajectory intercepts in the dual channel. This is another rather unphysical aspect of factorization. These extra modes may be taken either as a set of harmonic oscillators in a fifth dimension or as a modification of the propagator. Whether these states have positive metric or not depends on the intercept and the external masses. Actually a sixth dimension would be necessary to take care of two-particle trajectories correctly. We have no illuminating interpretations to offer on this subject.

3 Quarks and the dual model

What we propose here is a program of building a general picture of the structure of hadrons on the basis of the factorized Veneziano model. It has been noted by Harari and Rosner that the duality may be interpreted schematically in terms of quark diagrams, which have a strong predictive power, albeit of qualitative nature. These diagrams are indeed very suggestive. First of all they agree with the foregoing picture that the hadrons form two-dimensional sheets in space-time. Furthermore, they imply that quarks and antiquarks form the boundary lines of the sheets. A meson system, for example, is then a $q-\bar{q}$ molecule bound by an elastic string. We could also imagine a rubber band in which q and $\bar{q}$ are attached to diametrically opposite points. To go on further, we have to make a choice.

On the basis of the duality diagram picture, we will adopt the linear molecule rather than the benzene ring. An advantage of the linear picture is that a linear chain can be broken in two linear chains, thereby accounting for the production mechanism

$$A \longrightarrow B + C \tag{30}$$

In fact the duality diagrams can be interpreted exactly in this way.

To make things a bit more sophisticated, we will present a modified version of the quark model. This is the three-triplet model proposed by Dr. Han and me some time ago. It had the advantage of a) having integral charges, b) naturally accounting for the zero triality of known hadrons, as well as c) for the $SU(6)$ classification of the baryons. In this scheme there are nine fundamental fermions grouped into three $SU(3)$ triplets. We may use the notation T_i^n; $i, n = 1, 2, 3$, where n distinguishes between different triplets. T_i^n behaves like a triplet representation in the ordinary $SU(3)$ space (lower index), and like an antitriplet representation in the new $SU(3)$ space (upper index). These $SU(3)$ spaces are denoted as $SU(3)'$ and $SU(3)''$ respectively. Each space has its own isospin and hypercharge, and the electric charge is the sum of two charge operators $\lambda'_Q + \lambda''_Q$. We call $3\lambda''_Q$ the charm number C. We will also make things a bit more exciting by naming the three different triplets D, N, and A. Their quantum number assignments are given in the table.

	C		Q	
D_i	1	1,	0,	0
N_i	-2	0,	-1,	-1
A_i	1	1,	0,	0

In the lowest approximation, $SU(3)'$ and $SU(3)''$ are separately good symmetries. In particular, all low lying hadron states are assumed to belong to $SU(3)''$ singlets, which turn out to correspond to only zero triality states in $SU(3)'$. (The same is accomplished also by assuming zero charm for hadrons.) Baryons and mesons have the usual pattern TTT and $T\bar{T}$. More precisely,

$$
\begin{aligned}
&B \sim DNA \\
&M \sim D\bar{D} + N\bar{N} + A\bar{A}
\end{aligned}
\tag{31}
$$

where B is completely antisymmetric in the $SU(3)''$ space in order to be a singlet, and this takes care of the Pauli principle.

The preference of zero triality is thus reduced in this model to the $SU(3)''$ symmetry, which one may attribute to a dynamical property of superstrong interactions having a larger scale of masses ($\gtrsim 1GeV$) than for the strong interactions ($\sim 1GeV$).

The combination of duality and the three-triplet model will then produce the following picture. The hadrons are "molecules" bound by superstrong interactions, with the bond structure

$$\begin{aligned} B &\sim T - T - T \\ M &\sim T - \bar{T} \end{aligned} \tag{32}$$

The bonds must have a saturation property for zero triality. In the original scheme demanding perfect $SU(3)''$ symmetry, the baryon would have to have either a ring structure

$$B \sim \begin{array}{c} D \\ \diagup \;\; \diagdown \\ N \text{ — } A \end{array} \tag{33}$$

or a resonating linear structure

$$B \sim D - N - A \quad + \quad \text{permutations.} \tag{34}$$

If $SU(3)''$ symmetry is abandoned and only the neutral charm condition is imposed, we may simply assume

$$B \sim D - N - A \tag{35}$$

In the latter case the charm number is equivalent to valency. But then we lose the distinction between D and A, drifting back to a two-triplet model. Our tentative preference is in Eq.(34) though the other two possibilities should not be ignored. The meson scheme would follow Eq.(31).

What can we do with this model? We have now hadrons endowed with $SU(3)$ and Dirac spin. These are more or less localized at certain points along the string whose function is to carry bulk of the energy and momentum of the system. The triplets (or simply quarks) themselves are massless, or have only small masses. Several remarks are in order.

1) The interaction process (30) is viewed as a creation of pair $T\bar{T}$ at a point where the break occurs. After the cut, each portion subsequently grows into a full grown string, like an earthworm! This would not be possible if the

string is made up of a fixed number of mass points. We must conclude that the number is not only large but also indefinite. Let us examine this situation a little further. Take a string stretched between two fixed points in space with a distance L apart. If the number of mass points is N, the potential energy is

$$V \sim N(L/N)^2 = L^2/N \tag{36}$$

which depends on N. However, if the number of discrete steps in the "time" direction τ also increases with N to sustain duality, the action integral $I \sim NV \sim L^2$ will be independent of N. This is the scale invariance we have discussed. In units of the fictitious time τ, a system lives longer the larger the number N. The actual state of a hadron would be a linear superposition of configurations having different values of N, but their contributions are all proportional to each other.

If this view is accepted, we can define the Hamiltonian (or vertex) responsible for the process (30) as an overlap integral of the three wave functions corresponding to the states A, B, and C. There is a selection rule

$$N_A = N_B + N_C \qquad (N_i > 0) \tag{37}$$

and its two cyclic permutations, either one of which must be satisfied. A more explicit expression satisfying (37) would look like

$$\sum_{0<N'<N} C_{N'N} \int \cdots \int \Psi_B^*(x^{(1)}, \cdots, x^{(N')}) \Psi_C^*(x^{(N'+1)}, \cdots, x^{(N)})$$
$$\Psi_A(x^{(1)}, \cdots, x^{(N)}) F(x^{(1)}, \cdots, x^{(N)}) \prod d^4x^{(i)} \tag{38}$$

where F is some scalar function. This integral can be appropriately rewritten in terms of the wave functions $\Psi[x(\xi)]$ in the limit N, $N' \to \infty$.
In the actual Veneziano model, the interaction is such that the breaking of a string occurs only at one of the ends, which may be interpreted to mean the limit

$$N \to \infty; \;\; N'/N \to 0 \;\; \text{or} \;\; N'/N \to 1$$

In general, however, there is no reason to impose such a condition. We would still get a dual theory of sorts; an amplitude corresponding to one

configuration will have singularities in the crossed channels too, though there may not be a symmetry between dual channels in individual amplitudes.
2) The triplet or the quark fields introduce extra spins to the system in conformity with the $SU(6)$ type theories, so π and ρ mesons belong to the s wave states of the string. But this also brings in the old headaches of relativistic $SU(6)$ theories as well. How can one eliminate half of the Dirac components in order to avoid parity doubling and ghost states? The difficulty is compounded by the fact that we would like to maintain duality too. A possible scheme based on the Carlitz-Kisslinger type cut mechanism has been developed by Freund et al. We will not discuss it here. Instead, we would like to propose a general formalism which tries to accomplish this in a dynamical way. The basic idea is as follows. Instead of regarding the triplets and the strings as separate entities, let us take a unified picture and replace the string with a chain of $T\bar{T}$ pairs, so that it would look like a polarized medium with opposite charges created at its ends. There will be interactions between neighbors which depend on their Dirac, $SU(3)'$, and $SU(3)''$ spins. To be dual, these interactions must occur in the "time" direction too, thus forming a two-dimensional polarizable medium. These interactions would contribute to the action integral in addition to the kinetic term represented by the string Hamiltonian. Roughly speaking, that would produce a spin and $SU(3)$ dependence of various trajectories. It is conceivable that parity doubling and other problems can be reduced in the same way to ones of dynamical stability. An example of $T\bar{T}$ interaction might be

$$I = g \sum_{(n,n')} (\gamma_\mu^{(n)} \gamma^{\mu(n')} + \text{const}) \tag{39}$$

where $\gamma_\mu^{(n)}$ refers to a triplet sitting on a site n of a two-dimensional lattice, and n' refers to one of its neighbors. We must choose the constant g in such a way that a chain $T\bar{T}T\bar{T}\cdots\bar{T}$ will have the lowest energy (which may be adjusted to zero). To be dual, the same patter must be repeated in the time direction too. We thus end up with a pattern

$$\begin{array}{l} T\bar{T}T\bar{T}...........\bar{T} \\ \bar{T}T\bar{T}T...........T \\ T\bar{T}T\bar{T}...........\bar{T} \\ \bar{T}T\bar{T}T...........T \end{array}$$

....

....

In other words, it is a two-dimensional antiferromagnet or ionic crystal! External particles should couple to it like a magnetic field couples to the spin:

$$I' = \sum_n \gamma_\mu^{(n)} \phi^{\mu(n)} \tag{40}$$

The scattering amplitude would then be obtained, following the Feynman principle, from an expression like

$$\mathrm{Tr} \quad \exp[I + I'] \tag{41}$$

the trace being taken with respect to the γ matrices. Eq.(41) is nothing but a partition function! It is not scale invariant, but rather an extensive quantity proportional to the number of constituents (unless $I = 0$). Thus the probability of creating exotic states having many disordered atoms (Large $|I|$) would be severely cut down.

The simplest mathematical model of the above type is the well known Ising model with its glorious Onsager solution. Perhaps we can adopt the Ising model here as a prototype. The transition from a Lagrangian to a Hamiltonian formalism is accomplished by means of the so-called transfer matrix.

3) The world sheet formalism presented here may accommodate Dirac's monopoles, because the monopoles can be described, according to Dirac, in terms of the same kind of world sheets swept out by strings attached to them. If this is the case, the triplets can have magnetic charges. Since all hadrons are magnetically neutral, these charges must add up to zero, which reminds us of the fact that the charm number is also zero for them. Thus we are tempted to identify the charm with the magnetic charge, which would cause strong binding between opposite charges. The two spaces $SU(3)'$ and $SU(3)''$ may be called electric and magnetic $SU(3)$ respectively, corresponding to the 3×3 ways of assigning electric and magnetic quantum numbers.

Such a formalism has been independently proposed by Schwinger from a different motivation. He calls these nine objects dyons. The direct parallelism between dyons and the triplets has been pointed out by Biedenhahn and Han.

Of course there are all sorts of problems associated with monopoles. The most serious and perhaps most intriguing is the large P and T violation one must expect off hand. This could give a natural explanation for CP violation, but the problem is how to suppress it to a degree $\lesssim 10^{-10}$ which is required by neutron electric dipole moment.

At any rate, the close mathematical connection between the dual model and the monopoles lies in the fact that the Maxwell field due to a pair of monopoles is given by

$$F_{\mu\nu}(x) = g\epsilon_{\mu\nu\lambda\rho} \iint \delta^4(x-y)G^{\lambda\rho}(y)d^2\xi \tag{42}$$

where $G_{\mu\nu}d^2\xi = d\sigma_{\mu\nu}$ is the surface element, Eq.(22), of a world sheet spread between the world lines of the pair. Eq.(42) is independent of the choice of the sheet; one gets the correct equations if Eq.(42) is substituted in the Maxwell Lagrangian and the y's are varied, including the end points.

4 Statistical approximation

We will briefly discuss here a high energy approximation to dual amplitudes which was originally based on an intuitive argument, but can also be justified more rigorously. The point is that in a high energy process a very large number of states are available according to the factorized dual theory. In fact the number of states $\rho(s)$ increases like $\exp[c\sqrt{s}]$, which one can easily derive from the Stephan-Boltzmann law for a one-dimensional black body radiation. The only difference is that $s =$ (center-of mass energy)^2 takes the place of the ordinary energy. The constant c is numerically

$$c = 2\pi\sqrt{n/6} \tag{43}$$

where n is the dimensionality (4, or 5, or more) of the oscillator vectors. As has been pointed out by Fubini et al, this happens to give the same number ($c \approx 1/160MeV^{-1}$ for $n = 6$) as the Hagedorn constant.

The absorptive part of a scattering amplitude consists of a sum over intermediate states with fixed s, i.e. over a microcanonical ensemble:

$$A = \sum_{s_n=s} \langle p'|j(-q')|n\rangle\langle n|j(q)|p\rangle \tag{44}$$

If s is large, one is tempted to replace it with a sum over a canonical ensemble. More precisely

$$\begin{aligned} A &\sim c(\beta)\sum_n \langle p'|j(-q')|n\rangle e^{-\beta s_n}\langle n|j(q)|p\rangle \\ &\equiv c(\beta)\sum_n F_n e^{-\beta s_n} \end{aligned} \tag{45}$$

Here the s_n's are the eigenvalues of the mass operator (20), not the actual $s = (p+q)^2$. This amounts to relaxing the subsidiary condition (19). $c(\beta)$ is a normalization factor. If the sum (45) had a sharp peak around $s_n = s$, it would be a good approximation to Eq.(44) as in the usual statistical mechanics. The parameter β should then be the inverse temperature ($\sim 1/\sqrt{s}$) of the string. Actually, things do not work out that way because F_n does not grow like $\rho(s_n)$ but much more slowly like s^α, which is the Regge behavior. If $\alpha > 0$, still there will be a peak around $s_n = s$ where

$$\alpha/s \approx \beta \tag{46}$$

So we can use this as the definition of β.

In the operator formalism, Eq.(45) can be explicitly evaluated from

$$A = c(\beta)\langle 0|\Gamma' e^{-\beta H_0}\Gamma|0\rangle \tag{47}$$

where Γ and Γ' are appropriate interaction vertices. We find

$$A \sim (1-e^{-\beta})^{-1-\alpha(t)} \sim \beta^{-1-\alpha(t)} \tag{48}$$

which give the correct Regge behavior in view of Eq.(46), if $c(\beta) \sim \beta$. The above idea makes physical sense, perhaps better than the Veneziano model itself, because we apply it to the absorptive parti and smear the resonance peaks as in the discussion of finite energy sum rules. But the real part, when smeared, should also show a similar behavior for reasons of analyticity. Justification of the method depends on the assumption that the absorptive part grows with s ($\alpha > 0$). The formula may still be valid for $\alpha \leq 0$, but that does not follow from the above argument. Assuming the general validity of the procedure, we can also handle the high energy behavior of many particle processes. The main point is that any high energy ("hot") propagator $\sim 1/(s - H_0)$ is replaced by a Boltzmann factor $\sim \beta\exp(-\beta H_0)$

where $\beta \sim 1/s$, which would give a correct answer as far as the $s-$dependence is concerned.

At this point let us indulge in some speculations. The problems are unitarity and the Pomeron, both of which are lacking in the Veneziano model. It is generally assumed that the Pomeron (in the $t-$channel) is equivalent to non-resonant background in the $s-$channel. But the background is after all made up of hadrons, so a unitarized dual theory should naturally contain the Pomeron. Now the unitarization means taking account of dissociation and recombination of resonances among themselves. Wouldn't it be reasonable, then, to consider a grand canonical ensemble of resonances? The main problem is of course how to define the vertex operator in a scattering problem. We would like to suggest the ansatz that instead of Eq.(46), β^{-1} should be the thermodynamic temperature, or

$$\beta \approx c/\sqrt{s} \tag{49}$$

because the distribution of M^2 would be decided by the interaction among the strings in the ensemble, and not by the coupling of these states to the external channels. We would then get the result

$$A \sim s^{\alpha(t)/2} \tag{50}$$

suggesting that the Pomeron trajectory has half the universal slope of resonances. Whether this is a pole or a cut, and what the intercept is, cannot be decided in such a crude picture. It is interesting that the same behavior as Eq.(50) has been obtained by explicitly computing certain higher order diagrams of the dual model.

There is still another possibility for the Pomeron which will be discussed later.

5 Electromagnetic interactions and inelastic $e-p$ scattering

We would like to tackle the problem of inelastic $e-p$ scattering on the basis of our model. Before doing this, we have to make some remarks about the electromagnetic interactions in general. If we just have an elastic spring with a charge distribution along the string but without the quark spin, the problem of setting up the gauge principle is very simple. In the Hamiltonian (16) one makes the replacement

$$p^\mu(\xi) \longrightarrow p^\mu(\xi) - \rho(\xi)A^\mu(x(\xi)) \tag{51}$$

where $\rho(\xi)$ is an arbitrary distribution function. But this is not the most general form. Instead of $x(\xi)$ and $p(\xi)$, we should choose the set $\{x_n, p_n\}$, or any other orthogonal basis, and apply the gauge principle. Because $A(x)$ is nonlinear in x, we get inequivalent results.

The trouble with this method is that the form factors are all Gaussian, an undesirable characteristic of harmonic oscillators. For a pointlike charge distribution the Gaussian peak is infinitely sharp (in momentum space) since the string has an infinite zero-point length. If we renormalize away the Gaussian factor, on the other hand, what is left is in general a polynomial which is not good either.

Another popular approach to form factors is the spurion method which allows one to obtain pole dominance form factors. A difficulty here is the gauge invariance, or the conservation law, which cannot be automatically guaranteed. At any rate there is a large amount of arbitrariness in either method.

Another serious problem is dealing with external fields in a factorized dual model. It arises from the fact that one needs fifth-dimension oscillators for factorization of a general dual amplitude, but the parameters of the fifth dimension depends on the individual external masses in a non-factorizable way. Thus one cannot maintain duality and factorizability for arbitrary external fields. Full duality must then be abandoned in our model when dealing with electromagnetic interaction. For example, a virtual Compton amplitude should not be dual;

One might argue that in the pole dominance model the lines p and p', instead of the photons, may be treated as external:

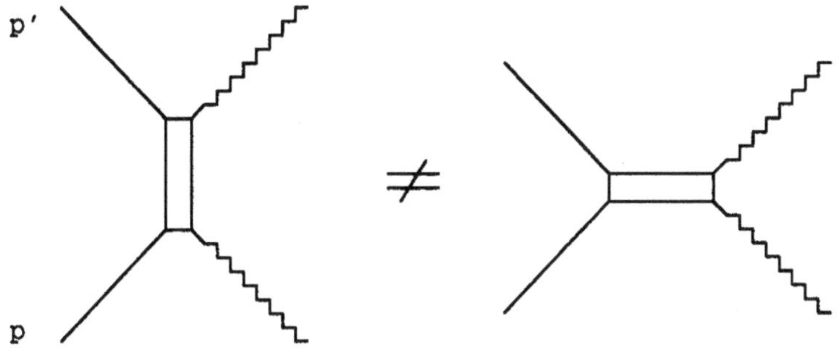

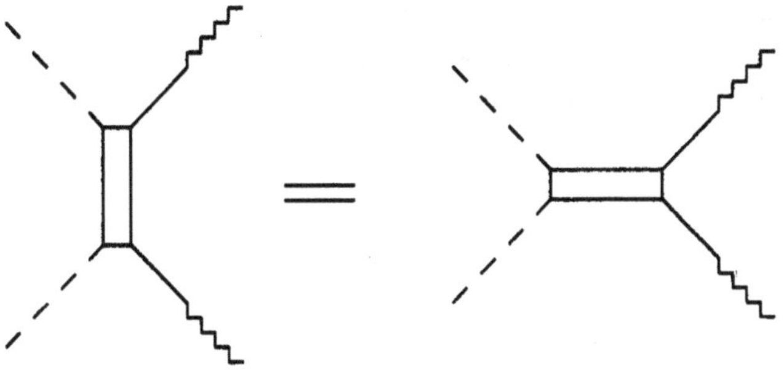

but this does not work for many photon cases.

In spite of the unpleasant Gaussian nature of form factors in the first method described above, let us see what will come out if duality is not demanded. The crossed channel singularities in the usual case arises from the singular nature of the vertex operator

$$\Gamma(q) \sim \exp[iq \cdot x(0)] = \exp\left[iq \cdot x_0 + 2iq \cdot \sum_{n=1}^{\infty} x_n\right] \tag{52}$$

In order to blunt the singularity, we assume that the charge is located at a coordinate

$$\bar{x} = \int x(\xi) f(\xi) d\xi = x_0 + 2\sum_{n=1}^{\infty} x_n f_n \tag{53}$$

such that

$$\sum f_n^2 < \infty \tag{54}$$

According to the gauge principle, the electromagnetic interaction is obtained by the substitution $p_0^\mu \to p_0^\mu - eA^\mu(\bar{x})$, $p_n^\mu \to p_n^\mu - 2ef_n A^\mu(\bar{x})$ in the Hamiltonian. The current operator is then

$$j^\mu = e\left\{ p_0^\mu + \sum_{n=1}^{\infty} p_n^\mu f_n \ , \ e^{iq\cdot\bar{x}} \right\}_+ \tag{55}$$

For the ground state (spin zero), this gives a vertex

$$\Gamma_\mu = e(p+p')_\mu \exp\left[-q^2 \sum_{n=1}^{\infty} f_n^2/n\right] \tag{56}$$

Now let us discuss off-diagonal elements $\langle n|j_\mu|0\rangle$ corresponding to inelastic processes. We would like to compare them with the $e-p$ scattering data ignoring the effect of spin. The familiar structure functions W_1 and W_2 should be obtained from

$$\sum_{s_n=n} \langle 0|j_\nu(-q)|n\rangle\langle n|j_\mu(q)|0\rangle \tag{57}$$

where the statistical method could be applied for large s.

There still remains a problem. In the SLAC data, W_1 and W_2 seem to have an s dependence ($W_1 \sim s$, $W_2 \sim 1/s$) which is consistent with the Pomeron picture. On the other hand, Eq.(55) will not give any Regge (or power) behavior. We propose to fix this by including the fifth dimension:

$$\exp[iq\cdot\bar{x}] \longrightarrow \exp[iq\cdot\bar{x} + iq_5 x_5(0)] \tag{58}$$

where q_5 is an appropriate constant. Since $x_5(0)$ is not smeared out, it gives rise to a fixed power behavior $\sim s^\alpha$ where α is constant, corresponding to a flat trajectory. We do not know what all this means, but the Pomeron might be of this nature. At any rate we can evaluate Eq.(57) with the new ansatz, and obtain the result

$$W_1 \sim (s-m^2)^{\alpha_1} \exp\left[-2q^2 \sum_{n=1}^{\infty} \frac{1-e^{-n\beta}}{n} f_n^2\right] \tag{59}$$

and similarly for W_2. Here m^2 is the initial Hadron mass. Actually we have lost gauge invariance from a) addition of the fifth dimension and b) the statistical approximation, so we cannot get the ratio W_1/W_2 exactly. But the main feature of Eq.(59) is that for large s we have

$$\beta \sim c/(s-m^2) \tag{60}$$

so that

$$\begin{aligned} W_i &\sim (s-m^2)^{\alpha_i} \exp[-2q^2\beta_i \sum f_n^2] \\ &\sim (s-m^2)^{\alpha_i} \exp[-\lambda_i q^2/(s-m^2)] \\ &= (s-m^2)^{\alpha_i} \exp[-\lambda_i/(\omega-1)], \quad i=1,2 \\ \lambda_i &= 2c_i \sum_{n=1}^{\infty} f_n^2, \quad \omega = 1+(s-m^2)/q^2 \ (= 2m\nu/q^2) \end{aligned} \tag{61}$$

Which $\alpha_1 = 1$, $\alpha_2 = -1$, we get the scaling law (for large s)

$$\begin{aligned} (s-m^2)W_2 &\sim \exp[-\lambda_2/(\omega-1)] \\ \text{or} \qquad 2m\nu W_2 &\sim (\omega/\omega-1)\exp[-\lambda_2/(\omega-1)], \\ (s-m^2)^{-1}W_1 &\sim \exp[-\lambda_1/(\omega-1)] \end{aligned} \tag{62}$$

The general behavior of the exponential factor in Eq.(62) is:

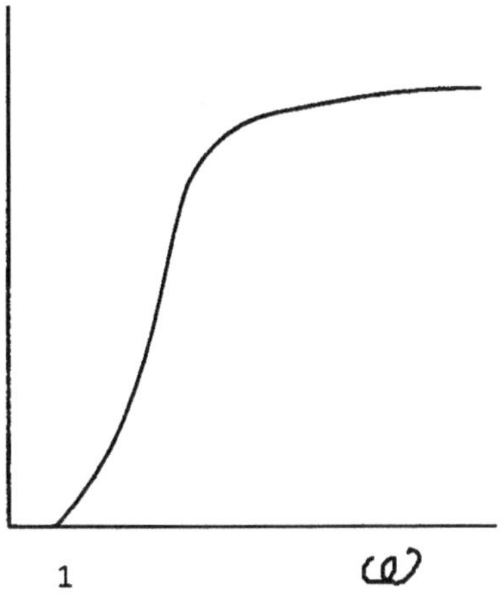

Reprinted from:
PHYSICAL REVIEW D VOLUME 7, NUMBER 8 15 APRIL 1973

Generalized Hamiltonian Dynamics*

Yoichiro Nambu
The Enrico Fermi Institute and the Department of Physics, The University of Chicago, Chicago, Illinois 60637
(Received 26 December 1972)

Taking the Liouville theorem as a guiding principle, we propose a possible generalization of classical Hamiltonian dynamics to a three-dimensional phase space. The equation of motion involves two Hamiltonians and three canonical variables. The fact that the Euler equations for a rotator can be cast into this form suggests the potential usefulness of the formalism. In this article we study its general properties and the problem of quantization.

I. INTRODUCTION

A notable feature of the Hamiltonian description of classical dynamics is the Liouville theorem, which states that the volume of phase space occupied by an ensemble of systems is conserved. The theorem plays, among other things, a fundamental role in statistical mechanics. On the other hand, Hamiltonian dynamics is not the only formalism that makes a statistical mechanics possible. Any set of equations which lead to a Liouville theorem in a suitably defined phase space will do (provided of course that ergodicity may be assumed). With this in mind, let us consider the following scheme.

Let $(x, y, z) \equiv \vec{r}$ be a triplet of dynamical variables (canonical triplet) which span a three-dimensional phase space. This is a formal generalization of the conventional phase space spanned by a canonical pair (p, q). Next introduce two functions, H and G, of (x, y, z), which serve as a pair of "Hamiltonians" to determine the motion of points in phase space. More precisely, we postulate the following "Hamilton equations":

$$\frac{dx}{dt} = \frac{\partial(H, G)}{\partial(y, z)}, \quad \frac{dy}{dt} = \frac{\partial(H, G)}{\partial(z, x)}, \quad \frac{dz}{dt} = \frac{\partial(H, G)}{\partial(x, y)}, \tag{1}$$

or in a vector notation

$$\frac{d\vec{r}}{dt} = \vec{\nabla}H \times \vec{\nabla}G . \tag{1'}$$

For any function $F(x, y, z)$, then, we have

$$\frac{dF}{dt} = \frac{\partial(F,H, G)}{\partial(x, y, z)} = \vec{\nabla}F \cdot (\vec{\nabla}H \times \vec{\nabla}G) . \tag{2}$$

We may call the right-hand side of (2) a generalized Poisson bracket (PB), to be denoted by $[F,H, G]$. Obviously a PB is antisymmetric under interchange of any pair of its components. As a result we have $\dot{H} = \dot{F} = 0$, i.e., both H and G are constants of motion. The orbit of a system in phase space is thus determined as the intersection of two surfaces, $H = \text{const.}$ and $G = \text{const.}$

Equation (1) or (1′) also shows that the velocity field $d\vec{r}/dt$ is divergenceless,

$$\vec{\nabla} \cdot (\vec{\nabla}H \times \vec{\nabla}G) \equiv 0 , \tag{3}$$

and this amounts to a Liouville theorem in our phase space.

The above properties immediately tempt us to construct a statistical mechanics where a canonical ensemble is characterized by a generalized Boltzmann distribution in phase space with a weight factor

$$e^{-\beta H - \gamma G} .$$

Two temperaturelike intensive parameters are thus required to specify the ensemble, much as in a grand canonical ensemble.

It is obvious that this kind of generalization can be extended to a phase space of any dimensionality, n. We would introduce an n-component vector x_i and $n-1$ Hamiltonians H_k, and postulate in lieu of Eqs. (1) and (2)

$$\frac{dx_i}{dt} = \sum_{jk\cdots l} \epsilon_{ijk\cdots l} \frac{\partial H_1}{\partial x_j} \frac{\partial H_2}{\partial x_k} \cdots \frac{\partial H_{n-1}}{\partial x_l} , \quad \frac{dF}{dt} = \frac{\partial(F, H_1, H_2, \ldots, H_{n-1})}{\partial(x_1, x_2, \ldots, x_n)} , \tag{4}$$

where $\epsilon_{ijk\cdots l}$ is the Levi-Civita tensor.

From the standpoint of physics, however, we must first examine the relevance and applicability of such generalizations. Are there real physical systems which may be described in this way? Or else can one think of these generalizations as a possible direction in which classical and quantum mechanics might develop?

In this paper we will limit ourselves to the three-dimensional case only. Then the first question can be answered affirmatively: Equation (1) is nothing but the Euler equation for a rigid rotator, if we identify $\vec{r}$ with the angular momentum $\vec{L}$ in the body-fixed frame, and G and H, respectively, with the total kinetic energy and the square of angular momentum in this frame:

$$G = \frac{1}{2}\left(\frac{L_x^2}{I_x} + \frac{L_y^2}{I_y} + \frac{L_z^2}{I_z}\right),$$
$$H = \tfrac{1}{2}(L_x^2 + L_y^2 + L_z^2). \quad (5)$$

We believe this to be justification enough to explore further the proposed ideas at least for the three-dimensional case. Needless to say, one may in general consider a number of canonical triplets $\vec{r}_n$, $n = 1, \ldots, N$ which form a $3N$-dimensional phase space, and write in place of Eq. (2)

$$\frac{dF}{dt} = \sum_n \frac{\partial(F,H,G)}{\partial(x_n, y_n, z_n)}. \quad (6)$$

For example, this will enable one to handle a model which simulates coupled spin systems.[1] For the moment, however, we are interested in the basic formalism only.

There is another direction in which Eq. (2) can be generalized. It is to assume

$$\frac{dF}{dt} = \sum_i \frac{\partial(F,H_i,G_i)}{\partial(x,y,z)}, \quad (7)$$

where (H_i, G_i) are a given set of functions. Since the Liouville theorem holds with respect to each term separately, it also holds for the sum. But the individual "Hamiltonians" H_i, G_i, are no longer constants of motion in general, though there may nevertheless be some constants of motion which are not directly related to the Hamiltonians. This may sound like an uninteresting and unnecessary digression, but we will see later that it becomes more natural when one tries quantization.

II. CANONICAL TRANSFORMATION

In this section we examine canonical transformations on the triplet (x, y, z), or more generally on the set of triplets (x_n, y_n, z_n).

We may call a mapping $(x,y,z) \to (x',y',z')$ a canonical transformation if

$$[x', y', z'] = \frac{\partial(x',y',z')}{\partial(x,y,z)} = 1. \quad (8)$$

Then

$$\frac{dF}{dt} = \frac{\partial(F,H,G)}{\partial(x,y,z)} = \frac{\partial(F,H,G)}{\partial(x',y',z')}, \quad (9)$$

i.e., the Hamilton equations are form-invariant if we use the new set of variables. In particular, the Hamilton equations themselves generate an infinitesimal canonical transformation in view of the Liouville theorem (3). On the other hand, not all infinitesimal canonical transformations need be generated this way. Moreover, two sets (H, G) and (H', G') generate the same transformation if there is a functional relationship $H' = h(H,G)$, $G' = g(H,G)$ such that

$$\frac{\partial(H',G')}{\partial(H,G)} = 1, \quad (10)$$

because then

$$\frac{\partial(H,G)}{\partial(y,z)} = \frac{\partial(H',G')}{\partial(y,z)}, \quad \text{etc.}$$

Thus the Hamiltonians are defined up to the usual type of canonical transformations (10) where (H, G) are regarded as a canonical pair of variables. In order not to confuse this with the transformations on (x, y, z) we will refer to (10) as "gauge" transformations.

Linear canonical transformations form a special class. We may use the matrix notation

$$\vec{r}' = A\vec{r}. \quad (11)$$

Equation (8) implies that the matrix A must be unimodular: $\det A = 1$. In other words, linear canonical transformations form the group $SL(3, R)$. In order to generate them we can conveniently take H and G to be, respectively, linear and quadratic forms in $\vec{r}$:

$$H = \sum_i a_i r_i, \quad G = \sum_{ij} \tfrac{1}{2} r_i B_{ij} r_j, \quad (12)$$

where B_{ij} is a symmetric 3×3 matrix. Then

$$\dot{\vec{r}} = \vec{a} \times (B\vec{r})$$

or

$$\dot{r}_i = \sum_{jkl} \epsilon_{ijk} a_j B_{kl} r_l.$$

Hence

$$A_{ik} = \sum_{jl} \epsilon_{ijl} a_j B_{lk}. \quad (13)$$

The number of parameters is 3 for H and 6 for G but there is a redundancy under the scaling $H \to \lambda H$, $G \to G/\lambda$. Therefore the degrees of freedom are 8, the correct number for the group $SL(3, R)$.

The Euler equations for a rotator belongs to a slightly more complicated case where both H and G are quadratic forms. Let us assume in general

$$H = (\vec{r}, A\vec{r}), \quad G = (\vec{r}, B\vec{r}), \quad (14)$$

where A and B are symmetric matrices. It is possible, however, to bring them into a standard diagonal form

$$H=\lambda(\vec{r}',1\vec{r}'),\quad G=(\vec{r}',\Lambda\vec{r}') \tag{15}$$

by a linear canonical transformation, provided that A is positive (or negative) definite. For example, first diagonalize A by a rotation, then make it proportional to a unit matrix by a unimodular rescaling of the three coordinates. Finally, make G diagonal by a second rotation. Equation (5) for the Euler top corresponds precisely to such a form. One should note also that there is a freedom of linear gauge transformations between H and G, or equivalently between A and B:

$$\begin{aligned} &A\to\alpha A+\beta B,\\ &B\to\gamma A+\delta B,\quad \alpha\delta-\beta\gamma=1 \end{aligned} \tag{16}$$

which forms a group SL(2, R).

Let us now examine the case of many canonical triplets as represented by Eq. (6). Here one can define a PB to mean

$$[A,B,C]=\sum_{n=1}^{N}\frac{\partial(A,B,C)}{\partial(x_n,y_n,z_n)}\,. \tag{17}$$

The canonical variables then have the properties

$$\begin{aligned} [x_l,y_m,z_n]&=1 \text{ if } l=m=n\\ &=0 \text{ otherwise},\\ [x_l,x_m,z_n]&=[x_l,x_m,x_n]=0,\quad \text{etc.} \end{aligned} \tag{18}$$

A canonical transformation $\{r_n\}\to\{r_n'\}$ will be a transformation which leaves the PB form-invariant. In other words, Eq. (17) must hold true for $\{r_n\}$ when evaluated in terms of the new variables $\{r_n'\}$ and vice versa.

In the case of an infinitesimal canonical transformation $\vec{r}_n\to\vec{r}_n+\delta\vec{r}_n$, Eq. (18) leads to

$$\begin{aligned} \delta[x_l,y_m,z_n]&=[\delta x_l,y_m,z_n]\\ &\quad+[x_l,\delta y_m,z_n]+[x_l,y_m,\delta z_n]\\ &=0,\\ \delta[x_l,x_m,z_n]&=[\delta x_l,x_m,z_n]\\ &\quad+[x_l,\delta x_m,z_n]+[x_l,x_m,\delta z_n]\\ &=0,\\ \delta[x_l,x_m,x_n]&=[\delta x_l,x_m,x_n]\\ &\quad+[x_l,\delta x_m,x_n]+[x_l,x_m,\delta x_n]\\ &=0,\quad\text{etc.} \end{aligned} \tag{19}$$

Most of these conditions are satisfied trivially because the $\{r_n\}$ satisfy Eq. (18). The only nontrivial ones turn out to be

$$\begin{aligned} \delta[x_l,y_l,z_l]&=\frac{\partial}{\partial x_l}\delta x_l+\frac{\partial}{\partial y_l}\delta y_l+\frac{\partial}{\partial z_l}\delta z_l\\ &=0, \end{aligned} \tag{20a}$$

$$\delta[x_l,x_m,z_m]=-\frac{\partial}{\partial y_m}\delta x_l=0\quad(m\neq l),\quad\text{etc.} \tag{20b}$$

The condition Eq. (20b) means that $\delta\vec{r}_n$ does not depend on any r_l, $l\neq n$. If so, then the first condition (20a) is a simple statement for canonical transformations of individual triplets. We are thus led to a rather unexciting result that the only continuous canonical transformations consist of independent transformations of the individual triplets. This contrasts with the case of usual canonical doublets where a general transformation involve all the variables simultaneously.

A corollary to the above result is that a Hamilton equation like Eq. (6) cannot be regarded as generating successive canonical transformations unless H and G are simple sums

$$H=\sum_n H_n(\vec{r}_n),\quad G=\sum_n G_n(\vec{r}_n). \tag{21}$$

More general forms of H and G will satisfy Eq. (20a) but not (20b). On the other hand, Eq. (20a) alone is sufficient to guarantee that the Liouville theorem holds in the $3N$-dimensional phase space.

The problem of an infinitesimal canonical transformation can also be tackled from a different angle. The standard general solution to an equation like (20a) is

$$\delta\vec{r}=\vec{\nabla}\times\vec{A}, \tag{22}$$

where $\vec{A}$ is a vector field, determined up to the gradient of a scalar. Our Hamilton equation (2) corresponds to a special choice

$$\vec{A}=H\vec{\nabla}G\ (\text{or } -G\vec{\nabla}H). \tag{23}$$

As is well known, a transformation on (H,G) satisfying Eq. (10) has the property

$$H\delta G-H'\delta G'=\delta S \tag{24}$$

for some function $S(G,G')$. This induces a transformation on $\vec{A}$,

$$\vec{A}\to\vec{A}+\vec{\nabla}S, \tag{25}$$

which is indeed a gauge transformation in the conventional sense. On the other hand, Eq. (23) does not exhaust all possible fields $\vec{A}$, which have three independent components.

This brings us finally to the type of generalization represented by Eq. (7). Since more than two functions $\{H_n,G_n\}$ are now available, we should be able to represent an arbitrary $\vec{A}$ this way. The Hamiltonians are, however, again subject to gauge transformations $\{H_n,G_n\}\to\{H_n',G_n'\}$ such that

$$[H'_i, G'_j] \equiv \sum_n \frac{\partial(H'_i, G'_j)}{\partial(H_n, G_n)} = \delta_{ij}, \tag{26}$$
$$[H'_i, H'_j] = [G'_i, G'_j] = 0.$$

These are precisely the ordinary canonical transformations with (H_n, G_n) being regarded as canonical pairs.

III. QUANTIZATION

Can one "quantize" our system of equations just as the ordinary canonical formalism can be quantized? This is an intriguing question which we would like to investigate. The first problem is how to define a quantization. One supposes that quantization would be an algebraic mapping of the relationships which characterize the canonical formalism developed so far. As it turns out, this is not an easy task. The PB defined in (2) has two properties:

(a) Alternation law:

$$[A, B, C] = -[B, A, C] = [B, C, A] = \cdots.$$

In particular, $[A, A, C] = 0$, etc. (27a)

(b) Derivation law:

$$[A_1 A_2, B; C] = [A_1, B, C]A_2 + A_1[A_2, B, C], \text{ etc.} \tag{27b}$$

The first property guarantees that the Hamiltonians are constants of motion. The second makes the PB appropriate for a differential equation. Therefore it should be natural to characterize a PB by Eqs. (27a) and (27b). But the main problem lies in satisfying both of them simultaneously. In fact we have not been able to find a solution beside the classical one (2).

In order to gain some more insight into the situation we recall that the correct Eulerian equations for a top are obtained in quantum theory by the Heisenberg equation (with $\hbar = 1$)

$$i\dot{F} = [F, G], \tag{28}$$

where G is the kinetic energy, Eq. (7), and the angular momenta L_i satisfy the commutation relations

$$[L_i, L_j] = -i\epsilon_{ijk} L_k. \tag{29}$$

The derivation law is duly satisfied by Eq. (28). [The anomalous sign on the right-hand side of Eq. (29) is well known,[2] but it is irrelevant to the discussion.] Comparing Eq. (28) with our classical form, Eqs. (2) and (5), we realize that the relations (29) were translated there as

$$[L_i, L_j] \to -i\epsilon_{ijk}\, \partial H/\partial L_k. \tag{30}$$

This suggests a commutator algebra different from (29):

$$[L_i, L_j] = i\epsilon_{ijk} K_k, \quad [L_i, K_j] = i\delta_{ij}, \tag{31}$$

and

$$\partial H/\partial L_k = i[H, K_k].$$

However Eq. (31) is incompatible with the Jacobi identity

$$[[L_1, L_2], L_3] + [[L_2, L_3], L_1] + [[L_3, L_1], L_2] = 0.$$

These considerations seem to indicate that we had better relax the constraints on a PB in order to find other solutions. If we want to keep only the physical consequences of Eqs. (27a) and (27b) the following conditions are also admissible for a PB:

(a′) $[H, H, G] = [G, H, G] = 0,$ (32a)

(b′) $[A_1 A_2, H, G] = A_1[A_2, H, G] + [A_1, H, G]A_2,$ (32b)

where H and G are fixed. Still, we have three options,

(1) (a)+(b′), (2) (a′)+(b), and (3) (a′)+(b′).

In the following we study in some detail the case (1) because the example of Pauli spin matrices falls in this category and suggests a way to define a PB. Case (2) retains the derivation law with respect to A, B, and C. This is the most difficult condition, and probably cannot be met. Case (3) will be discussed later. Let us consider an operator algebra $\mathfrak{A}$ generated by three elements, X, Y, and Z, and define a PB by

$$[A, B, C] \equiv ABC + BCA + CAB - BAC - ACB - CBA \tag{33a}$$
$$= A[B, C] + B[C, A] + C[A, B]$$
$$= [A, B]C + [B, C]A + [C, A]B$$
$$([A, B] \equiv AB - BA), \tag{33b}$$

which clearly satisfies Eq. (27a). The canonical triplet (X, Y, Z) will then have the property

$$[X, Y, Z] = i. \tag{34}$$

The Hamilton equation should read

$$i\frac{dF}{dt} = [F, H, G] \tag{35}$$

for arbitrary F. According to Eq. (33), a PB of Hermitian operators is anti-Hermitian, so that Eq. (35) is consistent if all physical quantities are translated into Hermitian operators.

We will next check the derivation law (27b). First of all it requires

$$[1,H,G]=0 \tag{36}$$

as can be seen by putting $A_2=1$ in Eq. (27b). But this means, from Eq. (33b),

$$[H,G]=0. \tag{37}$$

For general A_1 and A_2, then, Eq. (27b) can be reduced to

$$\begin{aligned}&\dot{A}_1A_2'-A_1'\dot{A}_2=0,\\&\dot{A}\equiv[A,H],\quad A'\equiv[A,G],\end{aligned} \tag{38}$$

or

$$[A_1']^{-1}\dot{A}_1=\dot{A}_2[A_2']^{-1}=\dot{A}_1[A_1']^{-1}=[A_2']^{-1}\dot{A}_2$$

whenever the division is possible. Thus

$$\dot{A}=\alpha A'\quad(\text{or } A'=\alpha\dot{A}) \tag{39}$$

for any A such that A'^{-1} (or $\dot{A}^{-1}$) exists, which in turn implies

$$H=\alpha G+\beta\quad(\text{or } G=\alpha H+\beta). \tag{40}$$

Here α and β commute with all operators and therefore are c numbers. But then

$$[A,H,G]=-[A,\beta G]\quad(\text{or } [A,\beta H]) \tag{41}$$

and by redefining $\beta G\to -G$ (or $\beta H\to H$) we recover the Heisenberg equation.

We conclude thus that if Eq. (33) satisfies the conditions (27a) and (32b), one of the Hamiltonians is a c number, and Eq. (35) is equivalent to a Heisenberg equation. We may wonder at this point whether we can avoid the rather disappointing result if we go over to Eq. (7). Following the same arguments as before, we then arrive at the conditions

$$\begin{aligned}&\sum_i[H_i,G_i]=0,\\&\sum_i[A_1,H_i][A_2,G_i]=\sum_i[A_1,G_i][A_2,H_i]\end{aligned} \tag{42}$$

in place of Eqs. (37) and (38). It is not clear, however, whether or not these more general relations admit nontrivial solutions.

Our next task is then to find realizations of the PB for the canonical triplet. For this purpose let us assume that the canonical variables $\vec{R}=(X,Y,Z)$ generate, under repeated commutator operations, a Lie subalgebra $\mathfrak{a}'$ of $\mathfrak{a}$ which is semisimple. Let the generators of $\mathfrak{a}'$ be $\{R_n\}$, where the first three elements coincide with (X,Y,Z). Further let

$$\begin{aligned}&[R_2,R_3]\equiv iR_1',\\&[R_3,R_1]\equiv iR_2',\\&[R_1,R_2]\equiv iR_3'.\end{aligned} \tag{43}$$

Then (R_1',R_2',R_3') belong to the vector space of $\{R_n\}$. In view of Eq. (33b) we have

$$\begin{aligned}[X,Y,Z]&=i\sum_{i=1}^{3}R_iR_i'\\&=i\sum_{i=1}^{3}R_i'R_i.\end{aligned} \tag{44}$$

Equation (34) demands that this be a c number (in an irreducible Hilbert space generated by $\vec{R}$). Therefore it must be a Casimir operator of the Lie algebra. Then $\vec{R}$ and $\vec{R}'$ must exhaust the generators, and hence $N\leq 6$. There are only two such possibilities, $N=6$ and $N=3$.

(a) If $N=6$, the algebra is either $SO(4)\approx SU(2)\times SU(2)$, $SO(3,1)\approx SL(2,c)$, or $SO(2,2)$. For the first two cases, their generators $(\vec{L},\vec{K})$ satisfy

$$\begin{aligned}&[L_1,L_2]=iL_3,\quad\text{etc.},\\&[L_1,K_2]=[K_1,L_2]=iK_3,\quad\text{etc.},\\&[K_1,K_2]=\pm iL_3,\quad\text{etc.},\quad\begin{cases}SO(4)\\SO(3,1)\end{cases}\end{aligned} \tag{45}$$

and the Casimir operators are

$$\begin{aligned}C_1&=\vec{L}^2\pm\vec{K}^2,\\C_2&=\pm\vec{L}\cdot\vec{K}\\&=-i(K_1[K_2,K_3]+K_2[K_3,K_1]+K_3[K_1,K_2]).\end{aligned} \tag{46}$$

Without loss of generality one may put $R_i=icK_i$. In this way the solution is found to be

$$\begin{aligned}&\vec{R}=C_2^{-1/3}\vec{K},\\&\vec{R}'=\pm C_2^{-2/3}\vec{L}.\end{aligned} \tag{47}$$

The results for the case of SO(2, 2) are obtained from that of SO(4) by a similar change of metric.

(b) If $N=3$, the algebra is either $SO(3)\approx SU(2)$ or $SO(2,1)\approx SU(1,1)$. Their generators $\vec{L}$ satisfy

$$\begin{aligned}&[L_1,L_2]=\pm iL_3,\quad\begin{cases}SO(3)\\SO(2,1)\end{cases}\\&[L_2,L_3]=iL_1,\quad[L_3,L_1]=iL_2,\end{aligned} \tag{48}$$

with the Casimir operator

$$\begin{aligned}C&=L_1^2+L_2^2\pm L_3^2\\&=-i(L_1[L_2,L_3]+L_2[L_3,L_1]+L_3[L_1,L_2]).\end{aligned} \tag{49}$$

Thus the solution is

$$\begin{aligned}&\vec{R}=(X,Y,Z)=C^{-1/3}(L_1,L_2,L_3),\\&\vec{R}'=C^{-2/3}(L_1,L_2,\pm L_3),\quad\begin{cases}SO(3)\\SO(2,1).\end{cases}\end{aligned} \tag{50}$$

We are left with the cases where $\mathcal{A}'$ is not semisimple. We will discuss some representative cases without claiming completeness.

(c) Euclidean algebra E(3) and its noncompact version E(2, 1). These are derived from SO(4) and SO(3, 1) by group contraction. The six generators $(\vec{L}, \vec{P})$ obey the commutation relations

$$\begin{aligned} &[L_1, L_2] = \pm i L_3, \quad [L_2, L_3] = i L_1, \quad [L_3, L_1] = i L_2, \\ &[L_1, P_2] = -[L_2, P_1] = i P_3, \\ &[L_2, P_3] = \pm[L_3, P_2] = \pm i P_1, \\ &[L_3, P_1] = \pm[L_1, P_3] = i P_2, \\ &[P_i, P_j] = 0, \qquad [P_i, L_i] = 0. \end{aligned} \tag{51}$$

The Casimir operators are

$$\begin{aligned} C_1 &= P_1^2 + P_2^2 + P_3^2, \\ C_2 &= L_1 P_1 + L_2 P_2 + L_3 P_3. \end{aligned} \tag{52}$$

The solutions for R and R' are given by

$$\begin{aligned} \vec{R} &= (\pm C_2)^{-1/3}(L_1, L_2, P_3), \\ \vec{R}' &= \pm C_2^{-2/3}(P_1, P_2, L_3) \end{aligned} \tag{53}$$

or

$$\begin{aligned} \vec{R} &= C_2^{-1/3}(P_1, L_2, L_3), \\ \vec{R}' &= C_2^{-2/3}(L_1, P_2, P_3). \end{aligned}$$

[These are equivalent for E(3).]

(d) E(2) and E(1, 1) derived from SO(3) and SO(2, 1) by contraction. The generators are (L, P_1, P_2) with the commutation relations

$$[L, P_1] = i P_2, \quad [L, P_2] = \mp i P_1, \quad [P_1, P_2] = 0, \tag{54}$$

and the Casimir operator is

$$C = \pm P_1^2 + P_2^2. \tag{55}$$

The solutions are

$$\begin{aligned} \vec{R} &= C^{-1/3}(L, P_1, P_2), \\ \vec{R}' &= C^{-2/3}(0, \pm P_1, P_2). \end{aligned} \tag{56}$$

(e) This case is characterized by the commutation relations of the Galilean group in one dimension,

$$[X, Y] = i Z^{-1}, \quad [X, Z] = [Y, Z] = 0, \tag{57}$$

for which we have a representation

$$\vec{R} = \left(i \frac{d}{dq}, \frac{q}{x}, x\right), \quad \vec{R}' = \left(0, 0, \frac{1}{x}\right). \tag{58}$$

(f) By putting $x = \text{const.} = 1$ in Eq. (58) we recover the canonical formalism of quantum mechanics:

$$\vec{R} = (-p, q, 1), \quad [p, q] = -i. \tag{59}$$

IV. USE OF NONASSOCIATIVE ALGEBRAS

In this section we pose the question whether it is possible to realize the PB relations by means of nonassociative algebras.[3] As a matter of fact, we immediately see that there is some hope, because it takes three elements to characterize the nonassociative nature of an algebra, just as we have a PB involving three observables. The analog of a PB in a nonassociative algebra is the associator

$$(a, b, c) \equiv (ab)c - a(bc) \tag{60}$$

which is zero if the multiplication table is actually associative.

Let us now see whether or not one can identify the associator with the PB under some combination of conditions (27) and (32). First take the alternative (1) stated there. Equation (27a) then demands that the associator is alternative:

$$(a, b, c) = -(b, a, c) = (b, c, a) = \cdots. \tag{61}$$

That means that we are dealing with the alternative algebra of Cayley and Dickson.[3] The Cayley-Dickson algebra C is an algebra over real numbers (or any suitable field) with eight basic elements u_i, $i = 0, \ldots, 7$, where u_0 serves as the unit element; it is the only possible algebra of this kind.

Unfortunately, we run into trouble with the derivation law (32b). The associator does not in general satisfy the derivation law, and conversely any derivation algebra over the Cayley numbers is known to be generated instead by operations of the form $\sum_{a,b} D_{a,b}$ on C, where

$$\begin{aligned} D_{a,b}\, x &= a(bx) - b(ax) + (xb)a - (xa)b + (bx)a - b(xa), \\ &\equiv D(a, b; x). \end{aligned} \tag{62}$$

$D(a, b; x)$ is antisymmetric only with respect to a, b, and for associative algebras reduces to

$$D(a, b; x) = [x, [a, b]]. \tag{63}$$

Suppose we now identify the PB $[A, B, C]$ with $D(B, C; A)$, and $B = H$, $C = G$ in the Hamilton equation. We have then switched to the alternative (3), which requires Eq. (32a):

$$D(H, G; H) = D(H, G; G) = 0. \tag{64}$$

This amounts to

$$[H, (HG) - (GH)] = [G, (HG) - (GH)] = 0, \tag{65}$$

but there is no nontrivial solution to it.

On the other hand, we can give up Eq. (32a) if we are considering Eq. (9) as the classical basis. Then the form

$$\frac{dF}{dt}=\sum_i D(H_i G_i ; F) \tag{66}$$

will certainly serve as a possible Hamilton equation. Since the multiplication table is preserved under Eq. (66), it induces an automorphism on C. It is known that automorphisms on C form a Lie group of type G_2. An element of C which is left invariant under (66) will be a constant of motion.

Let us leave the Cayley algebra and next shift our attention to the Jordan algebra J, which is commutative (but not associative).[3,4] The associator has the property

$$(a,b,a)=0, \quad (a,b,a^2)=0. \tag{67}$$

(The first is an identity which follows from the definition.) A derivation operator on J is given by

$$D_{a,b}x=(a,b,x)-(b,a,x). \tag{68}$$

This allows one to define the PB as

$$-i[A,B,C]=D_{B,C}A$$

$$=(B,C,A)-(C,B,A) \tag{69}$$

and the Hamilton equation as

$$i\frac{dF}{dt}=[F,H,G], \tag{70a}$$

or more generally

$$=\sum_i [F,H_i,G_i]. \tag{70b}$$

As before, we may add the condition (32a) in the case of Eq. (70a).

Now it is known that a Jordan algebra can be derived from an associative algebra $\mathfrak{A}$ if we define the multiplication to mean

$$axb=\tfrac{1}{2}(ab+ba), \tag{71}$$

where ab stands for a multiplication in the associative algebra $\mathfrak{A}$. Going back to this realization, we then find the PB (69) is given by

$$[A,B,C]=i[A,[B,C]]. \tag{72}$$

This is the same form as Eq. (63), and Eqs. (70a) and (70b) reduce to

$$i\frac{dF}{dt}=[F,\mathcal{H}], \quad \mathcal{H}=i[H,G] \text{ or } i\sum_i [H_i,G_i]. \tag{73}$$

In other words, it is equivalent to a Heisenberg equation with Hamiltonian $\mathcal{H}$. In addition, an example of canonical triplet $\vec{R}$ satisfying

$$-i[X,Y,Z]=[X,[Y,Z]]=1 \tag{74}$$

can readily be found. In fact

$$\vec{R}=\left(i\frac{\partial}{\partial x}, xy, i\frac{\partial}{\partial y}\right). \tag{75}$$

Finally there exists one Jordan algebra (the exceptional Jordan algebra) which is not isomorphic to an algebra generated by Eq. (71).[3,4] It is an algebra of 3×3 matrices M with elements in a Cayley algebra. M has the typical form

$$M=\begin{pmatrix} \alpha & c & \bar{b} \\ \bar{c} & \beta & a \\ b & \bar{a} & \gamma \end{pmatrix}, \tag{76}$$

where α, β, γ are c numbers; a, b, c are Cayley numbers, where $\bar{a}$ is obtained from a by the involution $u_0\to u_0$, $u_i\to -u_i$ $(i\neq 0)$. This should again offer another possibility of realizing the PB relations.

V. SUMMARY

Taking the Liouville theorem as a guiding principle, we have proposed a possible generalization of classical Hamiltonian dynamics to a three-dimensional phase space. The equation of motion involves two Hamiltonians and three canonical variables. A more general form may have many triplets and many Hamiltonian pairs. Such a formalism does not seem totally irrelevant to physics because the Eulerian top problem can be cast into this form, and it offers a new possibility in statistical mechanics. An attempt to find a "quantized" version of the formalism, however, has been only partially successful. In the process, the correspondence between classical and quantized versions is largely lost. One is repeatedly led to discover that the quantized version is essentially equivalent to the ordinary quantum theory. This may be an indication that quantum theory is pretty much unique, although its classical analog may not be.

On the other hand, there remains some possibility that nonassociative algebras may also be incorporated into the new formalism. Jordan was first led to what is now known as Jordan algebra in an attempt to reformulate and generalize quantum mechanics.[4] Although our starting point and motivation were different from Jordan's we have also found the potential significance of nonassociative algebras.

I would like to express my appreciation to Professor K. Husimi who kindly took an interest in my ideas when the contents of the first section were conceived more than twenty years ago. I would also like to acknowledge the inspirations I derived from recent communications with Professor F. Gürsey[5] and Dr. Pierre Ramond regarding nonassociative algebra.

*Work supported in part by the National Science Foundation (Contract No. NSF GP 32904 X) and the John Simon Guggenheim Foundation.

[1]T. Bell and Y. Nambu (unpublished report).

[2]O. Klein, Z. Physik 58, 730 (1920); J. H. Van Vleck, Rev. Mod. Phys. 23, 213 (1951); L. D. Landau and E. M. Lifshitz, *Quantum Mechanics* (Addison-Wesley, Reading, Mass. 1958), p. 373.

[3]For nonassociative algebras see, for example, R. D. Schafer, *An Introduction to Nonassociative Algebras* (Academic, New York, 1966).

[4]P. Jordan, Z. Physik 80, 285 (1933); Nachr. Ges. Wiss. Göttingen (1933), p. 209; P. Jordan, J. von Neumann, and E. Wigner, Ann. Math. 35, 29 (1934); A. Albert, Ann. Math. 35, 65 (1934).

[5]M. Günaydin and F. Gürsey, Yale University report (unpublished).

PHYSICAL REVIEW D VOLUME 10, NUMBER 12 15 DECEMBER 1974

Strings, monopoles, and gauge fields

Y. Nambu
The Enrico Fermi Institute and The Department of Physics, The University of Chicago, Chicago, Illinois 60637
(Received 9 September 1974)

The Nielsen-Olesen interpretation of dual strings as Abrikosov flux lines is extended to the case of open-ended strings by adapting Dirac's description of magnetic monopoles to a London-type theory. The mathematical formalism turns out to be similar to that of Kalb and Ramond. Translated to hadron physics, it implies that the quarks will act as carriers of magnetic charge, permanently bound in pairs by the string bonds. However, massive axial-vector gluons can be created by hadrons.

I. INTRODUCTION

In a very interesting paper[1] Nielsen and Olesen have pointed out a parallelism between the Higgs model of broken gauge invariance and the Landau-Ginzburg theory of superconductivity on the one hand and the dual string model and the Abrikosov flux lines in type II superconductors on the other. According to their suggestion, a dual string is nothing but a mathematical idealization of a magnetic flux tube in equilibrium against the pressure of the surrounding charged superfluid (Higgs-scalar field) which it displaces. Only strings with no ends (infinite strings or loops) were considered by them. It is known that a closed string could be a candidate for the Pomeron. But what will happen if the string is open-ended? Obviously the magnetic flux will terminate at the end points, thus creating a pair of magnetic charges.[2] In the dual quark model ordinary hadrons are viewed as being made up of quarks bound by dual strings, or, from the string's point of view, as open strings having quarks at their ends. From the Nielsen-Olesen picture it then follows that these quarks will act as a source of magnetic charge. (Here we are using the words electric and magnetic not to refer to actual electromagnetism, but as an analogy in a simplified model of strong interactions.)

At any rate we are led to the following picture. The quarks carry magnetic-type charge g whereas the Higgs field (boson) carries electric-type charge e, although it is a matter of convention to call one magnetic and the other electric. The two charges will be related by the Dirac quantization condition

$$\frac{eg}{4\pi}=\tfrac{1}{2}n .$$

The theory also contains two mass parameters m_V and m_S of the vector and Higgs scalar fields A_μ and ϕ, respectively. These are in turn related to two characteristic lengths $\lambda_V=1/m_V$ and $\lambda_S=1/m_S$ which determine the transverse dimensions of the vector-field concentration and of the scalar-field rarification, respectively, around the string. Thus the string is actually two things, a flux of a magnetic field and a hollow vortex line in the Higgs field.

Once we recognize these basic features, it is possible to idealize the situation and formulate the problem without referring to the particular Higgs model. We will do this in the next section by modifying Dirac's description of magnetic monopoles.

II. MODIFIED DIRAC MONOPOLE THEORY

Dirac's extension[3] of the Maxwell equations reads

$$\partial_\mu F_{\mu\nu}=-j_\nu \,, \tag{1a}$$

$$\partial_\mu F^*_{\mu\nu}=-k_\nu \,, \tag{1b}$$

where $F^*_{\mu\nu}$ is the dual of $F_{\mu\nu}$; j_ν and k_ν are electric and magnetic currents, respectively.[4] The only new step we take now is to go to the London theory of superconductivity by making the ansatz[5]

$$j_\mu=-m_V{}^2A_\mu \tag{2}$$

so that Eq. (1a) is replaced by

$$\partial_\mu F_{\mu\nu}-m_V{}^2A_\nu=0 . \tag{1a$'$}$$

A classical solution to Eqs. (1b) and (1a′) can be obtained with the aid of the Dirac string. Let us consider a pair of point magnetic charges $\pm g$ joined by a string. Then define

$$F_{\mu\nu}=\partial_\mu A_\nu-\partial_\nu A_\mu-G^*_{\mu\nu}\,,$$
$$G_{\mu\nu}(x)=g\iint dv\,\delta^4(x-y)[y_\mu,y_\nu]\,, \tag{3}$$
$$[y_\mu,y_\nu]\equiv\frac{\partial(y_\mu,y_\nu)}{\partial(\tau,\sigma)}\,,\quad dv=d\tau d\sigma .$$

Here $y_\mu(\tau,\sigma)$ represents the position of a point on the world sheet swept out by the string, and the sheet is parametrized by the internal coordinates τ, σ. $G_{\mu\nu}$ is independent of parametrization.[6] However, for definiteness we fix the range of the parameters as $-\infty<\tau<\infty$, $0\leqslant\sigma\leqslant\pi$ so that $y_\mu(\tau,0)=y_\mu^{(1)}$ and $y_\mu(\tau,\pi)=y_\mu^{(2)}$ represent the world lines of the two magnetic charges.

The $F_{\mu\nu}$ as defined in Eq. (3) automatically satisfies Eq. (1b):

$$\begin{aligned}\partial_\mu F^*_{\mu\nu}&=\partial_\mu G_{\mu\nu}\\&=g\iint dv\frac{\partial}{\partial x_\mu}\,\delta^4(x-y)[y_\mu,y_\nu]\\&=-g\int dv\Big(\frac{\partial}{\partial y_\mu}\,\delta^4(x-y)\Big)[y_\mu,y_\nu]\\&=-g\int dv[\,\delta^4(x-y),y_\nu]\\&=g\int\frac{dy_\nu}{d\tau}\delta^4(x-y)\,d\tau\,\Big|_{\sigma=0}^{\sigma=\pi}\\&=-\sum_i k_\nu^{(i)}(x)\,,\end{aligned} \tag{4}$$

$$k_\nu^{(i)}(x)=g^{(i)}\int\frac{dy_\nu^{(i)}}{d\tau}\,\delta^4(x-y)d\tau\,,$$

$$g^{(1)}=g,\quad g^{(2)}=-g .$$

One may therefore regard only Eq. (1a′) as an equation of motion. However, one also needs an equation of motion for the string and the magnetic monopoles at its ends.

For this purpose we take a Lagrangian density in space-time

$$L=-\tfrac{1}{4}F_{\mu\nu}F_{\mu\nu}-\tfrac{1}{2}m_V{}^2A_\mu A_\mu-\chi\,, \tag{5}$$
$$\chi(x)=\sum_{i=1,2}M^{(i)}\int\Big(\frac{\partial y_\mu^{(i)}}{\partial\tau}\,\frac{\partial y_\mu^{(i)}}{\partial\tau}\Big)^{1/2}d\tau\,\delta^4(y-x).$$

Here the last term is a contribution from the monopoles carrying mechanical masses $M^{(i)}$. By varying A_μ in the action integral $\int L\,d^4x$, one gets Eq. (1a′), or

$$(\Box-m_V{}^2)A_\nu=\partial_\mu G^*_{\mu\nu} \tag{6}$$

after substituting Eq. (3) and observing that $\partial_\mu A_\mu=0$.

By varying y_μ at a point (τ,σ) in the interior of the world sheet, we get

$$\frac{\delta}{\delta y_\lambda}\int L\,d^4x = -\frac{1}{2}\int \frac{\delta F_{\mu\nu}}{\delta y_\lambda} F_{\mu\nu}\,d^4x$$

$$= -\frac{1}{2}\int \frac{\delta G_{\mu\nu}}{\delta y_\lambda} F^*_{\mu\nu}\,d^4x$$

$$= -\frac{g}{2}\iint \frac{\delta}{\delta y_\lambda}\{[y_\mu, y_\nu]\delta^4(y-x)\}\,dv \times F^*_{\mu\nu}(x)\,d^4x$$

$$= -\frac{g}{2}\int \frac{\delta}{\delta y_\lambda}\{[y_\mu, y_\nu]\,F^*_{\mu\nu}(y)\}\,dv$$

$$= 0\,, \qquad (7)$$

which leads, in view of the definition of $[y_\mu, y_\mu]$, to

$$\frac{\partial}{\partial\tau}\left(\frac{\partial y_\lambda}{\partial\sigma}F^*_{\mu\nu}(y)\right) - \frac{\partial}{\partial\sigma}\left(\frac{\partial y_\lambda}{\partial\tau}F^*_{\mu\nu}(y)\right) - \tfrac{1}{2}[y_\mu, y_\nu]\frac{\partial F^*_{\mu\nu}}{\partial y_\lambda}$$

$$= [F^*_{\lambda\nu}, y_\nu] - \tfrac{1}{2}[y_\mu, y_\nu]\frac{F^*_{\mu\nu}}{y_\lambda}$$

$$= \tfrac{1}{2}[y_\mu, y_\nu]\left(\frac{\partial F^*_{\lambda\nu}}{\partial y_\mu} + \frac{\partial F^*_{\nu\mu}}{\partial y_\lambda} + \frac{\partial F^*_{\mu\lambda}}{\partial y_\nu}\right)$$

$$= 0\,. \qquad (8)$$

This amounts to

$$[y_\mu, y_\nu]^* \frac{\partial F_{\rho\nu}}{\partial y_\rho} = 0\,, \qquad (9a)$$

or

$$[y_\mu, y_\nu]^* A_\nu(y) = 0 \qquad (9b)$$

because of Eq. (1a′).

By varying $y_\mu^{(i)}$, the coordinates of the magnetic poles, we get

$$M^{(i)}\frac{d}{d\tau}\left[\frac{\dot y_\mu^{(i)}}{(\dot y_\lambda^{(i)}\dot y_\lambda^{(i)})^{1/2}}\right] = g^{(i)}F^*_{\mu\nu}(y^{(i)})\dot y_\nu^{(i)}\,,$$

$$\dot y_\mu^{(i)} \equiv dy_\mu^{(i)}/d\tau \qquad (10)$$

where the right-hand side comes from boundary contributions in Eq. (7).

Equations (6), (9), and (10), then, are our basic equations, the first two of which are equivalent to Eqs. (1a′) and (1b). Equation (6) is a differential equation in the real space-time, whereas Eqs. (9) and (10) are differential equations for y_μ on the two-dimensional world sheet and its boundary. A first observation to make is that Eq. (9) is a linear constraint on A_μ at any point on the sheet. For the existence of a nonzero solution, it is then necessary that the coefficient matrix $\sigma_{\mu\nu} \equiv [y_\mu, y_\nu]$ satisfy

$$\det(\sigma^*_{\mu\nu}) = \tfrac{1}{16}(\sigma_{\mu\nu}\sigma^*_{\mu\nu})^2 = 0\,. \qquad (11)$$

This is indeed true because the six-tensor $\sigma_{\mu\nu}\,dv$ represents the (oriented) surface element of the sheet embedded in the real space-time, and one can always choose, unless the surface intersects itself, a local Lorentz frame ("normal frame") so that only one component $\sigma_{\lambda\rho} = -\sigma_{\rho\lambda}$ is nonvanishing, where λ and ρ are in the tangential plane of the sheet. Equation (9) then implies that the vector A_μ must lie in the tangential plane. In physical terms, the London current can only flow along the string.

For a further study of the motion of the string, let us solve Eq. (6) for A,

$$A_\nu(x) = \int \Delta(x-y)\partial_\mu G^*_{\mu\nu}(y)\,d^4y\,, \qquad (12)$$

and substitute it in Eqs. (5), (9), and (10). Here $\Delta(x)$ is a Green's function for a field of mass m_V. In this way we obtain nonlinear equations for y_μ, and an effective Lagrangian from which these equations will follow. A straightforward calculation shows that the action integral, after discarding total divergences, can be brought to the form

$$\int d^2v\,\mathcal{L}_{\text{eff}} = \tfrac{1}{4}g^2 m_V^2 \iint d^2v\,d^2v'\sigma_{\mu\nu}\Delta(y-y')\sigma'_{\mu\nu}$$

$$+\tfrac{1}{2}\sum_{i,j}\iint g^{(i)}g^{(j)}\,\dot y_\mu^{(i)}\Delta(y^{(i)} - y^{(j)\prime}) \times \dot y_\mu^{(j)\prime}\,d\tau\,d\tau'$$

$$-\sum_i \int M^{(i)}(\dot v_\mu^{(i)}\dot y_\mu^{(i)})^{1/2}\,d\tau\,. \qquad (13)$$

It is now defined entirely in the two-dimensional world, y_μ merely being regarded as fields with four components. The first term represents a short-range Yukawa interaction between two surface elements; the second is another Yukawa interaction between magnetic currents, including self-interaction; the third is, of course, the mechanical mass term. Everything is manifestly independent of the choice of internal coordinates, and therefore can be given a direct geometrical interpretation.

It is interesting to observe that if $m_V = 0$, the first term goes out and we recover the familiar result of magnetic charges interacting via the long-range Maxwell field. The string is unphysical in this case since it does not carry energy-momentum. Once $m_V \neq 0$, however, the string acquires physical reality. From dimensional considerations and the short-range nature of $\Delta(x)$, it is clear that the first term of Eq. (13) is proportional to the surface area of the world sheet, as in the dual string model, with a characteristic coefficient $\sim g^2 m_V^2$.

A somewhat more elaborate derivation of the string Lagrangian from Eq. (13) will be as follows.

At a point (τ,σ) on the sheet we choose, as before, a "normal frame" so that

$$\frac{\partial y_\mu}{\partial_\sigma}=\delta_{\mu 3}\,,\quad \frac{\partial y_\mu}{\partial\tau}=i\delta_{\mu 4}\,,$$

for example. If the curvature of the surface is small over the range $\sim 1/m_V$ of $\Delta(x)$, and the point in question is not within a distance $\sim 1/m_V$ from the boundaries, one can approximate

$$\begin{aligned}\int \Delta(y-y')\sigma_{34}\,dv &= -i\int \Delta(x)\,dx_3\,dx_4\\ &=\frac{-i}{(2\pi)^4}\iint \frac{e^{i(k_3x_3+k_4x_4)}}{k^2+m_V^2}\,d^4k\,dx_3\,dx_4\\ &=\frac{-i}{(2\pi)^2}\int\frac{dk_1\,dk_2}{k_1^2+k_2^2+m_V^2}\\ &=\frac{-i}{4\pi}\ln\left(\frac{K^2}{m_V^2}+1\right),\end{aligned}\tag{14}$$

where a momentum cutoff K has been introduced in directions perpendicular to the sheet, corresponding to the finite thickness of the string. From the work of Nielsen and Olesen[1,5] one may equate it with a mass parameter m_S characteristic of the Higgs field.

With Eq. (14) the first term of Eq. (13) becomes a single integral of surface elements. In covariant notation, then, one gets the string Lagrangian (density)[7] in the (τ,σ) space

$$\mathcal{L}_{\text{string}}=-\frac{1}{2\pi\alpha'}\,|\det\sigma_{\mu\nu}|^{1/2}\,,\tag{15}$$

where

$$\frac{1}{2\pi\alpha'}=\frac{g^2}{8\pi}\,m_V^2\ln\left(\frac{m_S^2}{m_V^2}+1\right).\tag{16}$$

This relates the Regge trajectory slope α' to the parameters of our theory.

We have thus seen that Eq. (13) contains the string Lagrangian in the local limit. Beside that, however, Eq. (13) has the following additional important features.

(a) The end points of the string behave like particles with mass $M^{(i)}$ and charge $g^{(i)}$, coupled to a massive vector field. This leads to a Yukawa interaction between the end points and their own self-energies. In the static picture, this interaction energy is

$$-\frac{g^2}{4\pi}\,\frac{e^{-m_Vl}}{l}\,,\tag{17}$$

where l is the distance between the end points. The string energy, on the other hand, will be

$$\gtrsim l\,\frac{g^2m_V^2}{8\pi}\ln\left(\frac{m_S^2}{m_V^2}+1\right).\tag{18}$$

For a sufficiently long string ($l\gtrsim 1/m_V$), the string energy is dominant; for a short string ($l\lesssim 1/m_V$) the singular Yukawa interaction becomes important if the size of the end-point monopoles is even smaller.

(b) The string is oriented, i.e., has an intrinsic sense of polarization, like a magnet.[8] When two portions of the world sheet nearly overlap, there will be a Yukawa interaction between their surface elements. The interaction is attractive when the two string elements line up antiparallel, and repulsive when they are parallel. Such an effect will become most pronounced if the string folds on itself. An example of normal modes of this type is

$$\begin{aligned}x&=\cos n\sigma\,\cos n\tau\,,\\ y&=\cos n\sigma\,\sin n\tau\,,\\ z&=0\,,\\ t&=n\tau\,.\end{aligned}\tag{19}$$

If we ignore the end-point effects, its energy will be $1/n$ of that in the naive string model if n is odd, and zero if n is even.

(c) It is an easy matter to generalize our considerations to cases with more than one string. The total action integral will now consist of a sum over actions of the type (13) for individual strings, plus similar terms representing interstring contributions to the surface-surface and boundary-boundary interactions. Interestingly, a picture of this kind has been proposed by Kalb and Ramond.[9]

III. QUANTIZATION

Quantization of our theory can be done following Dirac's procedure. Starting from Eq. (7) one first defines the canonical conjugates to A_μ and y_μ, respectively, under displacement of real time t and fictitious time τ. Dirac has shown that the single-valuedness of the wave function for a system containing both electric and magnetic charges requires the quantization condition

$$\frac{eg}{4\pi}=\tfrac{1}{2}n\,,\quad n=0,\pm1,\pm2,\ldots\,.\tag{20}$$

As is well known, this condition also appears as flux quantization in superconductivity. In the Landau-Ginzburg-Higgs model, e is the electric charge of a scalar field ϕ, and m_V is given by

$$m_V=|e\langle\phi\rangle|\,,\tag{21}$$

so that the characteristic coefficient in Eq. (13),

$$g^2m_V^2=g^2e^2\langle\phi\rangle^2\,,$$

is independent of e, because $\langle\phi\rangle$ is determined by other parameters in the model.

Beyond this, the actual passage to a quantum theory of strings is beset with various well-known difficulties. We will nevertheless indicate how it might be carried out.

The Lagrangian (5) contains three dynamical quantities, the vector field A_μ, the string variable y_μ, and the magnetic source variables $y_\mu^{(i)}$. For a clear separation of the first two, we use the definition (3) to write Eq. (5) as

$$L = -\tfrac{1}{4}F^0_{\mu\nu}F^0_{\mu\nu} + \tfrac{1}{2}F^0_{\mu\nu}G^*_{\mu\nu} + \tfrac{1}{4}G_{\mu\nu}G_{\mu\nu} - \tfrac{1}{2}m_V{}^2 A_\mu A_\mu - \chi\,, \quad F^0_{\mu\nu} \equiv \partial_\mu A_\nu - \partial_\nu A_\mu\,. \tag{22}$$

The second term shows that $G_{\mu\nu}(x)$ acts as a source for the field A, whereas the third term has the nature of a free Lagrangian for the string. From the point of view of the string, it is more convenient to go to a Lagrangian (density) $\mathcal{L}$ in the (τ,σ) space, related to L by

$$L_{\text{string}}\, d^4x = \mathcal{L}_{\text{string}}\, dv\,, \quad L_{\text{string}} = -\tfrac{1}{4}G_{\mu\nu}G_{\mu\nu} + \tfrac{1}{2}F^{0*}_{\mu\nu}G_{\mu\nu}\,. \tag{23}$$

We find

$$\mathcal{L}_{\text{string}} = \tfrac{1}{4}[y_\mu, y_\nu]\int \delta^4(y-y')[y'_\mu, y'_\nu]\,dv' + \tfrac{1}{2}[y_\mu, y_\nu]\,F^{0*}_{\mu\nu}(y)\,. \tag{24}$$

The first term of Eq. (24) is very much like the first term of Eq. (13) except that it is quadratically divergent instead of logarithmically. It seems sensible, therefore, to replace this term by $\mathcal{L}_{\text{string}}$ of Eq. (15). The second term is responsible for emission and absorption of vector-field quanta. Thus we get a new Lagrangian

$$\mathcal{L}_{\text{string}} = -\frac{1}{2\pi\alpha'}(\det\sigma_{\mu\nu})^{1/2} + \tfrac{1}{2}\sigma_{\mu\nu}F^0_{\mu\nu}(y)\,, \tag{25}$$

which can further be reduced to

$$\mathcal{L}_{\text{string}} = \frac{1}{4\pi\alpha'}\left[\left(\frac{\partial y_\mu}{\partial\tau}\right)^2 - \left(\frac{\partial y_\mu}{\partial\sigma}\right)^2\right] + \tfrac{1}{2}\sigma_{\mu\nu}F^{0*}_{\mu\nu}(y) \tag{26}$$

under the well-known Virasoro gauge condition[10]

$$\left(\frac{\partial y_\mu}{\partial\tau}\right)^2 + \left(\frac{\partial y_\mu}{\partial\sigma}\right)^2 = \frac{\partial y_\mu}{\partial\tau}\frac{\partial y_\mu}{\partial\sigma} = 0\,. \tag{27}$$

The canonical conjugate to y_μ following from Eq. (26) is

$$\pi_\mu = \frac{1}{2\pi\alpha'}\frac{\partial y_\mu}{\partial\tau} + F^{0*}_{\mu\nu}\frac{\partial y_\nu}{\partial\sigma}\,, \tag{28}$$

and the Hamiltonian is

$$\bar{\mathcal{H}}_{\text{string}} = \int \mathcal{H}_{\text{string}}\, d\sigma\,, \quad \mathcal{H}_{\text{string}} = \pi\alpha'\left(\pi_\mu - F^0_{\mu\nu}\frac{\partial y_\nu}{\partial\sigma}\right)^2 + \frac{1}{4\pi\alpha'}\left(\frac{\partial y_\mu}{\partial\sigma}\right)^2. \tag{29}$$

The gauge constraint (27) now reads

$$\mathcal{H}_{\text{string}} = 0\,, \quad \mathcal{P}_{\text{string}} \equiv \pi_\mu\frac{\partial y_\mu}{\partial\sigma} = 0\,. \tag{30}$$

$\mathcal{H}_{\text{string}}$ and $\mathcal{P}_{\text{string}}$ are nothing but the components $T_{\tau\tau} = T_{\sigma\sigma}$ and $T_{\tau\sigma} = T_{\sigma\tau}$ of the energy-momentum tensor in the (τ,σ) space. Since $\mathcal{L}_{\text{string}}$ does not explicitly depend on (τ,σ), they satisfy the continuity equations

$$\frac{\partial\mathcal{H}}{\partial\tau} - \frac{\partial\mathcal{P}}{\partial\sigma} = \frac{\partial\mathcal{P}}{\partial\tau} - \frac{\partial\mathcal{H}}{\partial\sigma} = 0 \tag{31}$$

in the interior of the world sheet. Imposing the condition (30) everywhere on the sheet, including the boundaries, is then compatible with the equations of motion. Actually, one must consider the contribution of magnetic poles in Eq. (22). In quantum theory let us assume these poles to be Dirac particles, and postulate the following equations[3]:

$$\gamma_\mu^{(i)}\pi_\mu^{(i)} - iM^{(i)} = 0 \quad (i = 1, 2)\,. \tag{32}$$

Equations (31) and (32) are to be regarded as constraints on the wave function $\Psi_{\alpha,\beta}(y_\mu(\sigma), A_\nu(\vec{x}))$ which depends on the Dirac spin indices α,β of the poles in addition to the string and field variables.

The real question is, of course, the compatibility of these constraints with each other and with the Hamiltonian as operator relations. Unless this can be shown, the present formalism will remain only a superficial one.

IV. IMPLICATIONS FOR HADRON PHYSICS

The foregoing model theory has many interesting features which are relevant to the actual strong interactions, although some important pieces are missing. For a more realistic model, one would have to seek generalizations to non-Abelian gauge fields.[11] Nevertheless, it will still be instructive to take stock of what the present model already has. First of all, isolated magnetic charges cannot exist because, in contrast with the case for Dirac monopoles, an infinite amount of energy is required to infinitely stretch a string. If these charges are carried by quarks, then single quarks cannot exist. Only quark-antiquark pairs ("mesons") would exist, which have zero total magnetic charge. Unfortunately, there are no "baryons." However, massive axial-vector gluons can also be produced by mesons, as we shall see below.

The forces that bind "quarks" and "antiquarks" are of two kinds. One is the tension of a string; the other is a Yukawa force. For long strings (highly excited mesons) the former will be dominant, leading to the usual linear trajectories of the dual resonance model. However, the finite thickness of the string and the short-range interactions between string elements will distort deep-lying daughter trajectories. If we use Eq. (16) with the known value $\alpha' = 1\ \text{GeV}^{-2}$ and the reasonable ansatz $m_V, m_S \gtrsim 1$ GeV, we find that $g^2/4\pi$ cannot be very large ($\lesssim \frac{1}{2}$). It is the electric coupling $e^2/4\pi$ which is large ($\gtrsim \frac{1}{2}$).

For low-lying states of the meson (short strings) the Yukawa interaction as well as the kinetic energy of the quarks becomes important since both go like 1/length (the quark mass is here ignored). The former is attractive, while the latter acts like a repulsive force. Their effect would lead to a shift in the trajectory intercepts.

Another effect of the Yukawa interaction would be to change the short-distance behavior of the wave function from the Gaussian form of the dual resonance model to a power form. This should be highly desirable in view of what we know about elastic and inelastic form factors of hadrons.

Strings interact with each other via exchange of the gluon field, whose source is distributed over the entire length of a string. This results in joining and splitting of strings not only by end-to-end contact of opposite magnetic charges, but also by antiparallel lineup of two string segments. (See Fig. 1.)

Another straightforward consequence of the theory is emission of a gluon by a string through the

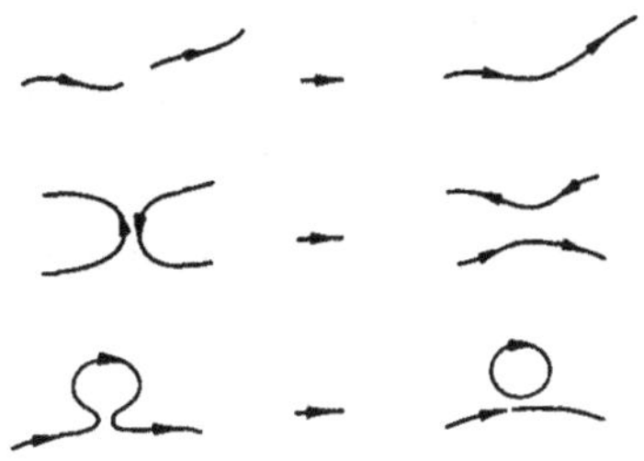

FIG. 1. Examples of joining and breaking of strings.

interaction term in Eq. (29),

$$-8\pi\alpha' \int F^{0*}_{\mu\nu}\pi_\mu \frac{\partial y_\nu}{\partial\sigma}\, d\sigma\,. \tag{33}$$

For a gluon of momentum k_μ and polarization ϵ_μ, this is

$$\sim \int k_\mu \epsilon_\nu \pi_\lambda y'_\rho e^{-ik\cdot y}\epsilon_{\mu\nu\lambda\rho}\, d\sigma\,. \tag{34}$$

Obviously the gluon behaves as an axial vector ($J^P = 1^+$) instead of a vector (1^-) particle.[8] A simple example of such a process would involve the transition of a meson string from 0^- to 1^+, as in a hypothetical hadronic reaction

$$\pi \to A_1 + G\,,$$

where G stands for the gluon. It is a curious fact that the axial-vector field nevertheless gives rise to a vector-type interaction between magnetic charges, according to Eq. (13). In any case it appears that unlike the "quarks" the gluons cannot be permanently contained in our Abelian model.

---

*Work supported in part by the National Science Foundation under Grant No. NSF GP 32904 X2.

[1]H. B. Nielsen and P. Olesen, Nucl. Phys. B61, 45 (1973).

[2]This has been independently suggested by G. Parisi, Columbia University Report No. CO-2271-24, 1974 (unpublished). A preliminary account of the present article is given by Y. Nambu, in *Proceedings of the Johns Hopkins Workshop on Current Problems in High Energy Particle Theory, 1974,* edited by G. Domokos *et al*. (Johns Hopkins Univ., Baltimore, 1974).

[3]P. A. M. Dirac, Phys. Rev. 74, 817 (1948).

[4]We use the pseudo-Euclidean metric convention: $g_{\mu\nu} = \delta_{\mu\nu}$; $\mu, \nu = 1,\ldots,4$. A dual is defined as $F^*_{\mu\nu} = \frac{1}{2} i\,\epsilon_{\mu\nu\lambda\rho}F_{\lambda\rho}$.

[5]As is well known, this is a local approximation to a nonlocal (Pippard) relation $j_\mu(x) = \int K_{\mu\nu}(x-x')A_\nu(x')d^4x'$, where the polarization tensor $K_{\mu\nu}$ has a characteristic range $1/m_S$. With this form the cutoff parameter in Eq. (14) will automatically appear.

[6]It is implicitly assumed that the sheet is essentially Minkowski-like, τ and σ playing the roles of timelike and spacelike coordinates, respectively. The Poisson bracket notation for the Jacobian in Eq. (3) naturally carries an implied suggestion for possible abstractions. However, it is not our intention to pursue it here.

[7]Y. Nambu, lectures prepared for the Copenhagen Summer Symposium, 1970 (unpublished); T. Goto, Prog. Theor. Phys. 46, 1560 (1971).

[8]The theory of Abrikosov flux lines in type II superconductors shows that these lines move only in helical modes of one sign, as determined by the sense of the magnetic flux, whereas the present model does not have such a feature. Herein lies a significant difference between real superconductors and corresponding relativistic models. In the former only electrons exist; in the latter, as in the Higgs model, fluids of both charges respond to magnetic fields. This charge symmetry is responsible for the lack of apparent time-reversal violation in relativistic models. Mathematically, the difference can be traced to different assumptions about static charge density: $\rho \sim \langle\phi^*\phi\rangle \neq 0$ in the

Landau-Ginzburg theory, but $\rho \sim i\,\langle \phi^* \dot{\phi} - \dot{\phi}^* \phi \rangle = 0$ in the Higgs theory.

[9]M. Kalb and P. Ramond, Phys. Rev. D 9, 2273 (1974). See also E. Cremmer and J. Scherk, Nucl. Phys. B72, 117 (1974).

[10]L. N. Chang and F. Mansouri, Phys. Rev. D 5, 2235 (1972); Goto, Ref. 7; G. Goddard, J. Goldstone, C. Rebbi, and C. B. Thorn, Nucl. Phys. B56, 109 (1973).

[11]A clearcut answer to this problem seems to be lacking. See, however, Nielsen and Olesen, Ref. 1; G. 't Hooft, CERN Report No. TH-1873-CERN, 1974 (unpublished); Y. Nambu, Ref. 2.

Strings, Vortices, and Gauge Fields

YOICHIRO NAMBU
University of Chicago
Chicago, Illinois

In this talk I will make a few scattered observations on mathematical formalisms which are aimed at describing stringlike objects and their interactions. The attempt was originally motivated by the string model of hadrons, which may well be a phenomenological manifestation of a gauge theory of strong interactions. Here, however, I will put emphasis on broader aspects of the problem, keeping in mind that there may exist as yet unknown phenomena which are to be described by these mathematical formalisms. Of course it is sheer speculation at the moment.

There are three main topics to be discussed. They are: relativistic hydrodynamics of vortices, its broken symmetry schemes and non-Abelian versions, and a unified algebraic characterization of string and gauge field equations.

I. RELATIVISTIC HYDRODYNAMICS OF VORTICES

At the outset I would like to make clear that this is an elaboration of the ideas developed originally by Kalb and Ramond[1]. The emphasis is here on its physical interpretation and implications. First let us define the following fields.

Velocity field 4-vector (dimension M^2)	V_μ
Velocity potential 6-vector (dimension M)	$W_{\mu\nu}$ $(=-W_{\nu\mu})$
Vorticity 6-vector (dimension M^3)	$\Omega_{\mu\nu}$ $(=-\Omega_{\nu\mu})$
Vorticity source 6-vector (dimension M^2)	$\omega_{\mu\nu}$ $(=-\omega_{\nu\mu})$ (1)

V_μ and $\Omega_{\mu\nu}$ are derived from $W_{\mu\nu}$ according to

Reprinted from Quark Confinement and Field Theory, eds. D. R. Stump and D. H. Weingarten, © 1977 John Wiley & Sons, pp. 1–12.

$$V_\lambda \equiv i\epsilon_{\lambda\mu\nu\rho}\,\partial_\mu W_{\nu\rho} \equiv \partial_\mu \tilde{W}_{\mu\lambda},$$
$$\partial_\lambda V_\lambda \equiv 0$$
$$\Omega_{\mu\nu} \equiv \epsilon_{\mu\nu\lambda\rho}\partial_\lambda V_\rho \quad (\text{or } \tilde{\Omega}_{\mu\nu} = -\partial_\mu V_\nu + \partial_\nu V_\mu)$$
$$\partial_\mu \Omega_{\mu\nu} \equiv 0 \,. \tag{2}$$

The source $\omega_{\mu\nu}$ also is assumed to satisfy the continuity equation

$$\partial_\mu \omega_{\mu\nu} = 0 \,. \tag{3}$$

The velocity field V_μ is a physical observable; actually it should be interpreted as a current density, namely density × 4-velocity. The velocity potential $W_{\mu\nu}$, on the other hand, is not uniquely determined by V_μ, but is subject to a gauge transformation

$$W_{\mu\nu} \rightarrow W_{\mu\nu} + \partial_\mu \Lambda_\nu - \partial_\nu \Lambda_\mu \,, \tag{4}$$

with an arbitrary vector field Λ_μ. $W_{\mu\nu}$, $\Omega_{\mu\nu}$ and $\omega_{\mu\nu}$ are defined in such a way that in the nonrelativistic limit the space-time (or "electric") components are nonzero and correspond to the usual definition in nonrelativistic hydrodynamics. Note also that in comparison with electrodynamics, the roles of a 4-vector and a 6-vector are reversed, but otherwise we are describing rather similar phenomena. This becomes clearer when we look at the Ramond Lagrangian

$$L = \frac{1}{2} V_\mu V_\mu - \frac{k}{2} W_{\mu\nu}\omega_{\mu\nu} \,, \tag{5}$$

which satisfies gauge invariance (4) because of Eq. (3). k is a parameter with the dimensions of a mass. By varying $W_{\mu\nu}$, we get from this

$$\partial_\mu V_\nu - \partial_\nu V_\mu = k\,\tilde{\omega}_{\mu\nu} \,, \tag{6a}$$

or

$$\Box\, W_{\mu\nu} - (\partial_\mu U_\nu - \partial_\nu U_\mu) = k\,\omega_{\mu\nu} \,. \tag{6b}$$

Here we have introduced another auxiliary field

$$U_\mu \equiv \partial_\lambda W_{\lambda\mu} \,. \tag{7}$$

Equations (6a) and (6b) amount to an identity involving W, U and V. From Eq. (6a) follows

$$\Box\, V_\nu = \partial_\mu \tilde{\omega}_{\mu\nu} \,, \tag{8a}$$

and with a special gauge $U_\mu = 0$, $W_{\mu\nu}$ also satisfies

$$\Box\, W_{\mu\nu} = k\, \omega_{\mu\nu} \,. \tag{8b}$$

The content of these equations becomes more tranparent if we write out the components of $W_{\mu\nu}$ and $\omega_{\mu\nu}$ as

$$\begin{aligned} (W_{ij}, W_{io}) &= (\varepsilon_{ijk}\, b_k, e_i) \,, \\ (\tilde{W}_{ij}, \tilde{W}_{io}) &= (-\varepsilon_{ijk}\, e_k, b_i) \,, \\ (\omega_{ij}, \omega_{io}) &= (\varepsilon_{ijk}\, \nu_k, \omega_i) \,, \end{aligned} \tag{9}$$

where $\vec{e}$ and $\vec{b}$ are analogs of the electric and magnetic fields. Then

$$(V_i, V_o) = (-\vec{\nabla} \times \vec{e} + \vec{b}, + \vec{\nabla} \cdot \vec{b})$$

$$L = \frac{1}{2}(\vec{b} + \vec{\nabla} \times \vec{e})^2 - \frac{1}{2}(\vec{\nabla} \cdot \vec{b})^2 - k(\vec{\nu} \cdot \vec{b} - \vec{e} \cdot \vec{\omega}), \tag{10}$$

$$\vec{b} - \vec{\nabla}(\vec{\nabla} \cdot \vec{b}) + \vec{\nabla} \times \vec{e} = -k\vec{\nu}$$

$$\vec{\nabla} \times (\vec{b} + \vec{\nabla} \times \vec{e}) = -k\vec{\omega} \,. \tag{11}$$

From this it is easy to see that 1) $\vec{e}$ is nondynamical (like the Coulomb field in electrodynamics); 2) $\vec{b} + \vec{\nabla} \times \vec{e}$ is longitudinal; and 3) the dynamical field $\vec{b}$ describes massless longitudinal radiation (and quanta).

The emergence of a massless radiation associated with hydrodynamic motion is characteristic of a relativistic field theory, as in the case of electromagnetism. The importance of this observation, however, lies in the implied possibility that such hydrodynamic media with attendant massless, longitudinal radiation (propagating with the speed of light!) might actually exist in nature. What will be the source of such radiation?

The Dirac construction of the source field $\omega_{\mu\nu}$ is well known. The latter represents the surface element of a world sheet swept out by a closed string, which in our interpretation constitutes the core of a vortex ring. Thus

$$\omega_{\mu\nu}(x) = \int \delta^4(x-y)\, d\tau d\sigma\, \sigma_{\mu\nu}(y)$$

$$\sigma_{\mu\nu}(y) = \dot{y}_\mu y_\nu' - y_\mu' \dot{y}_\nu \equiv [y_\mu, y_\nu] (\dot{y}_\mu \equiv \partial y_\mu/\partial\tau, y_\mu' \equiv \partial y_\mu/\partial\sigma). \quad (12)$$

Here the $y_\mu(\tau,\sigma)$ are the coordinates of a point on the sheet, which is parametrized by the internal coordinates τ and σ. The notation is familiar except for the Poisson bracket symbol $[y_\mu, y_\nu]$ which is designed to emphasize the fact that τ and σ act as if they were a pair of canonical variables. They range from $-\infty$ to ∞, but y_μ is assumed to be periodic in σ, so that σ parametrizes a closed string at a given instant τ. If a segment of the string is at rest, and τ is chosen to be proportional to the time coordinate, the only nonzero components $\omega_{oi} = \omega_i$ of $\omega_{\mu\nu}$ represent a tangential vector to the string. Equation (8) then gives rise to a long-range potential created by the string element, which is similar to the vector potential due to an electric current element. The string-string interaction arising from such a potential, however, is of opposite sign to that for the current-current interaction because ω_i is "electric" rather than "magnetic". This is exactly the property of the interaction between hydrodynamic vortex elements.

When a vortex line element parallel to the x axis vibrates in the y direction, it radiates a longitudinal (compressional or "acoustic") wave, with a $\cos^2\theta$ intensity distribution relative to the z axis. This is the new effect mentioned above. But creation and detection of such waves depend on the existence (in vacuo) of vortices. So the next question is: do the quanta of vortex rings exist as elementary particles? A corollary to this question is: does a conserved source field $\omega_{\mu\nu}(x)$ exist in local field theory? And if so, is it associated with any symmetry principle?

There is a trivial answer to the second question. We can put

$$\omega_{\mu\nu} \propto \tilde{F}_{\mu\nu} , \quad (13)$$

where $F_{\mu\nu}$ is an analog of the familiar electromagnetic field, which in turn is generated by a 4-vector current. Or else we can simply write

$$\tilde{\omega}_{\mu\nu} = \partial_\mu j_\nu - \partial_\nu j_\mu , \quad (14)$$

with an arbitrary vector field j_μ. But it is unlikely that any symmetry principle can be associated with these currents. More attractive is the spirit of the original string model, namely to quantize the vortex as a nonlocal object. To each closed vortex, we then associate a wave function $\psi[y(\sigma)]$, which is a functional of the string configuration embedded in the Minkowski space. As was proposed by Ramond, we can define a covariant derivative

$$D_\mu(y) = y'_\nu \left(\frac{\delta}{\delta\sigma_{\mu\nu}(y)} - ig\, W_{\mu\nu}(y) \right) , \qquad (15)$$

where the first term represents the change induced when the vortex line at point y is displaced to create a 2-surface $\delta\sigma_{\mu\nu}$. The second term is designed to absorb the effect of the gauge transformation (4) on $W_{\mu\nu}$ into a phase transformation on Ψ:

$$\Psi[y] \to \exp[i \int_s \Lambda_\mu\, d\sigma_\mu]\, \Psi[y] \; . \qquad (16)$$

Unfortunately, there is a formidable mathematical problem in choosing an action integral and defining a proper measure in the space of vortex configurations. The well known difficulties encountered in the dual string model are reflections of this problem.

It appears that vortices as elementary particles do not exist;if relativistic vortices exist at all, they are probably composite, nonlocal objects with very large internal degrees of freedom. And the question remains whether long-range hydrodynamic waves can be emitted and absorbed by such objects.

A concrete example of relativistic vortex derived from local field theory as a nonlocal object is the Nielsen-Olesen string.[2] However, it is not coupled to a long-range field, and hence not an analog of hydrodynamic vortex line. Rather, it is a magnetic flux line in a superconductive medium.

II. BROKEN SYMMETRY SCHEMES AND OTHER GENERALIZATIONS

We will discuss here various generalizations of the basic Lagrangian we introduced above (Eq. (5)). These seem to be more interesting and relevant as providing models of hadron confinement.

A. Massive 6-Vector Field

We add a mass term to the Lagrangian (5):

$$L = \frac{1}{2} V_\mu V_\mu - \frac{m^2}{4} W_{\mu\nu} W_{\mu\nu} - k\, W_{\mu\nu}\omega_{\mu\nu} , \qquad (17)$$

in analogy to the case of a massive vector field. From Eq. (17) follows

$$\partial_\mu V_\nu - \partial_\nu V_\mu - m^2 \tilde{W}_{\mu\nu} = k \tilde{\omega}_{\mu\nu} , \tag{18}$$

so that

$$(\Box - m^2) V_\nu = k \partial_\mu \tilde{\omega}_{\mu\nu} , \tag{19a}$$

and

$$-m^2 \partial_\mu W_{\mu\nu} = -m^2 U_\nu = k \partial_\mu \omega_{\mu\nu} . \tag{19b}$$

V_ν now becomes a massive spin 1 field. This is nothing but an unorthodox way of describing a vector meson. If $\omega_{\mu\nu}$ is conserved, U_ν must be zero, but the scheme allows for the possibility of nonconserved $\omega_{\mu\nu}$. For example, if the vortex line is open ended, we will have

$$\partial_\mu \omega_{\mu\nu} = j_\nu , \tag{20}$$

where j_ν is the 4-vector current generated by the two end points, one carrying an opposite charge from the other.

The Lagrangian (17) serves as a phenomenological substitute for the Nielsen-Olesen model. The self-energy of the vortex string is proportional to $m^2 \ell$ where ℓ is its length.[4] The end points of the string may be regarded as a pair of quark and antiquark with equal and opposite charges according to Eq. (20). Thus it will serve as a model for mesons, but baryons cannot be formed this way.

B. Confinement with Nonlocal Fields

We bring in an electromagnetic-type field in addition to $W_{\mu\nu}$ and consider in particular

$$L = \frac{1}{2} V_\mu V_\mu - \frac{1}{4} (m W_{\mu\nu} + F_{\mu\nu})(m W_{\mu\nu} + F_{\mu\nu}) - A_\mu j_\mu . \tag{21}$$

Here j_μ is a conserved source current; m is a mass parameter inserted for dimensional reasons. The equations following Eq. (21) are

$$\partial_\mu V_\nu - \partial_\nu V_\mu - m(m \tilde{W}_{\mu\nu} + F_{\mu\nu}) = 0$$

$$\partial_\mu (m W_{\mu\nu} + F_{\mu\nu}) = j_\nu , \tag{22}$$

which lead to

$$(\Box - m^2)V_\mu = 0 \tag{23a}$$

$$j_\nu = 0 \ . \tag{23b}$$

In other words, the equations are not self-consistent unless the current vanishes. The underlying reason is simple: $F_{\mu\nu}$ can be eliminated from Eq. (21) by the gauge transformation (4) with $\Lambda_\mu = A_\mu/m$ and hence Eq. (23b).

It is not unreasonable to interpret Eq. (23b) as a statement of quark confinement. For this, however, we have to regard m as a dynamical parameter. If m is replaced by a field m(x), the current j_μ can exist only where $m(x) \sim 0$. Conceivably the sources $\omega_{\mu\nu}$ and j_μ coupled respectively to $W_{\mu\nu}$ and A_μ develop expectation values $\langle\omega,\omega\rangle \sim m^2$, $\langle\omega,j\rangle \sim m$ in vacuo, but that in the vicinity of charges they are forced to vanish. Such a mechanism would reproduce the Nielsen-Olesen string with quarks attached to its ends. The corresponding Lagrangian would be

$$L = \frac{1}{2} V_\mu V_\mu - \frac{1}{4} F_{\mu\nu} F_{\mu\nu} - k\, W_{\mu\nu}\omega_{\mu\nu} - A_\mu j_\mu \ , \tag{24}$$

where the sources are supposed to satisfy Eq. (20) so that the gauge invariance under Eq. (4) holds with Λ_μ replaced by A_μ/k. The problem here is to find a natural (and renormalizable) candidate for $\omega_{\mu\nu}$ in local field theory. One way to simulate this mechanism would be to regard m in Eq. (21) as a Goldstone scalar field with a nonzero value $\langle m(x)\rangle$ in the absence of j_μ.

C. Non-Abelian Analogs

Most of what has been done above can be formally generalized to non-Abelian cases by the replacement of the derivatives ∂_μ and the gauge field $F_{\mu\nu}$ by their covariant versions. In other words, all fields except the gauge potentials A_μ will be regarded as transforming linearly under local gauge transformations. However, the gauge principle of Ramond cannot be reconciled with the Yang-Mills gauge principle. This is because the transformation (4) cannot be properly generalized. To see this, observe that the ansatz

$$W_{\mu\nu} \to W_{\mu\nu} + D_\mu\Lambda_\nu - D_\nu\Lambda_\mu \ , \tag{24}$$

does not leave $D_\mu\tilde{W}_{\mu\nu}$ invariant unlike the Abelian case, but rather

$$(D_\mu\tilde{W}_{\mu\nu})^a \to (D_\mu\tilde{W}_{\mu\nu})^a - g\, f^{abc}\tilde{F}^b_{\mu\nu}\Lambda^c_\mu \ , \tag{26}$$

(the superscript refers to the internal spin). The second term vanishes only if $\Lambda_\mu^c = 0$, or det $\tilde{F} = 0$ when $\sum_b F_{\mu\nu}{}^b f^{abc}$ is regarded as a transformation matrix acting on Λ_μ^c.

In place of Eq. (25), however, there is a residual invariance under a one-parameter family of transformations

$$W^a_{\mu\nu} \to W^a_{\mu\nu} + \alpha F^a_{\mu\nu} , \tag{27}$$

with constant α. As a result, the non-Abelian analogs of Eqs. (21) and (22) lead to results similar to Eq. (23), namely

$$D_\mu(D_\mu V_\nu - D_\nu V_\mu) - m^2 V_\nu = 0$$

$$j_\nu(\psi) + j_\nu(W) = 0 . \tag{28}$$

Here j_ν is made up of contributions from the quarks and the W.

Apparently related to the above peculiarities of the non-Abelian cases is the fact that the field $W^a_{\mu\nu}$ has very limited degrees of freedom even when it is coupled only to the gauge field. The equation of motion (cf. Eq. (6a))

$$D_\mu V_\nu - D_\nu V_\mu = 0 , \tag{29}$$

entails, upon taking a covariant divergence of its dual,

$$f^{abc}\tilde{F}^b_{\mu\nu} V^c_\nu = 0 , \tag{30}$$

i.e., $V_\nu = 0$ or det $F = 0$. Thus a consistent theory of non-Abelian massless antisymmetric field does not seem to exist, and again this may have something to do with confinement.

III. ALGEBRAIC CHARACTERIZATION OF STRING AND GAUGE FIELD EQUATIONS

As is well known, the string Lagrangian

$$L = \frac{1}{2} g \equiv \frac{1}{2} \sqrt{-\sigma_{\mu\nu}\sigma_{\mu\nu}} = \sqrt{\dot{y}^2 y'^2 - (\dot{y} \cdot y')^2} , \tag{31}$$

can be linearized by going to the Virasoro gauge:

$$\dot{y}^2 + y'^2 = \dot{y} \cdot y' = 0$$

$$L \to \frac{1}{2} (\dot{y}^2 - y'^2) . \tag{32}$$

In the general case, Eq. (31) leads to

$$[\frac{1}{g}\sigma_{\mu\nu}, y_\nu] = [\frac{1}{g}[y_\mu, y_\nu], y_\nu] = 0 . \qquad (33)$$

Recently Schild[6] has proposed another choice of gauge g^2 = const, with a Lagrangian

$$L' = \frac{1}{4} g^2 = -\frac{1}{4}\sigma_{\mu\nu}\sigma_{\mu\nu} . \qquad (34)$$

The resulting equations

$$[[y_\mu, y_\nu], y_\nu] = 0 . \qquad (35)$$

indeed lead to $\dot{g} = g' = 0$, and hence equivalent to Eq. (33). (This can be seen by contracting Eq. (35) with $\dot{y}_\mu$ or y'_μ.) The transition from Eq. (31) to Eq. (34) is very similar to what is often done in point mechanics, which is an essential step in going over to quantum mechanics. Interestingly, the Schild gauge also works in the presence of the field $W_{\mu\nu}$. The Lagrangian

$$L' = \frac{1}{4} g^2 + \frac{1}{2}\sigma_{\mu\nu}W_{\mu\nu}(y) , \qquad (36)$$

leads to

$$[[y_\mu, y_\nu] + W_{\mu\nu}, y_\nu] = \frac{1}{2}\sigma_{\nu\lambda}\partial W_{\nu\lambda}/\partial y_\mu ,$$

or

$$[[y_\mu, y_\nu], y_\nu] + (W_{\mu\nu,\lambda} - \frac{1}{2}W_{\lambda\nu,\mu}) \quad [y_\lambda, y_\nu] = 0, \qquad (37)$$

from which follows g^2 = const.

Although Eq. (33) is nonlinear, a class of solutions can be obtained with the ansatz

$$y_\mu = u_\mu\tau + f_\mu(\tau,\sigma), \quad u^2 = -1, u\cdot f = 0 . \qquad (38)$$

In particular, we can choose $u_\mu=(0,0,0,i)$, $f_\mu=f(\sigma)\times(\cos\tau,\sin\tau,0,0)$ and obtain

$$f'' - f(ff')' = 0 , \qquad (39)$$

which can be easily integrated. The solution represents a spinning rod. It differs from the corresponding solution in the Virasoro gauge by a redefinition of the coordinate σ.

An interesting possibility that suggests itself is to take the Poisson bracket notation in Eq. (35) seriously, and go to its "quantum mechanics" version, by regarding the internal coordinates τ and σ as noncommuting operators. It is totally unclear what this means, but we try it nevertheless. Some of the properties of the classical theory are lost. For example, g^2 will not automatically be a c-number, i.e., a constant. The difference stems from the fact that y_μ is in general a nonlinear function of τ and σ. Thus two exponentials $\exp[im\tau]$ and $\exp[in\sigma]$ commute if $mn/2\pi$ is an integer, but their classical Poisson bracket does not vanish.

An ansatz similar to Eq. (38) will be

$$f_x + if_y = e^{i\tau}f(\sigma), \quad f_x - if_y = f(\sigma)e^{-i\tau} . \tag{40}$$

We then get, in place of Eq. (39),

$$f''(\sigma) - \frac{1}{2} f(\sigma)[f(\sigma+1)^2 + f(\sigma-1)^2 - 2f(\sigma)^2] = 0. \tag{41}$$

An explicit solution of this equation has not been found yet.

We remark also that under the quantum mechanical interpretation, the ordinary action integral is replaced by

$$\mathrm{Tr}[y_\mu, y_\nu][y_\mu, y_\nu] . \tag{42}$$

By taking a variation δy_μ, we recover Eq. (35).

The structure of Eq. (35) reminds us of Maxwell's equations. In fact we can write them in an identical form by defining

$$Y_\mu \equiv p_\mu - e A_\mu(x) \equiv - i D_\mu ,$$
$$[x_\mu, p_\nu] = i . \tag{43}$$

Then

$$[Y_\mu, Y_\nu]/ei = - F_{\mu\nu},$$
$$[Y_\mu,[Y_\mu, Y_\nu]] = e\, \partial_\mu F_{\mu\nu} = 0 . \tag{44}$$

Observe the similarity between Eqs. (38) and (43). $A_\mu(x)$ is the analog of $f_\mu(\tau,\sigma)$. Since A_μ depends only on x, there is no difference between classical and quantum interpretations of $[x_\mu,p_\nu]$. The Maxwell Lagrangian in our language becomes

$$L = \frac{1}{4e^2} [Y_\mu, Y_\nu][Y_\mu, Y_\nu] . \tag{45}$$

Encouraged by this observation, we now try to include the sources. For this purpose, we will let $\hat{e}$ be a charge operator such that

$$[\hat{e},\psi] = -e\, \psi , \tag{46}$$

where ψ_i is a field carrying charge e. Then define

$$\hat{Y}_\mu \equiv p_\mu + \hat{e}\, A_\mu(x) \ , \tag{47}$$

so that

$$[\hat{Y}_\mu, \psi] = (-i\partial_\mu - e\, A_\mu)\psi \, . \tag{48}$$

If ψ is, for example, a scalar field, we add a Lagrangian

$$L_\psi = -[\hat{Y}_\mu, \psi]^+ [\hat{Y}_\mu, \psi] \ , \tag{49}$$

to L, Eq. (45). If $\hat{Y}$ instead of Y is used in the latter, however, we have to divide by $\hat{e}^2$ instead of e^2, or else project $\hat{e}$ onto a unit charge sector.

The non-Abelian generalization is straightforward. We write

$$\hat{Y}_\mu = p_\mu + \hat{e}^i\, A^i_\mu(x),$$

with

$$[\hat{e}^i, \hat{e}^j] = ig\, f^{ijk}\, \hat{e}^k \, . \tag{50}$$

Then

$$[\hat{Y}_\mu, \hat{Y}_\nu] = i\hat{e}^i\, F^i_{\mu\nu} \ , \tag{51}$$

$$L = \frac{1}{4Cg^2}\, [\hat{Y}_\mu, \hat{Y}_\nu][\hat{Y}_\mu, \hat{Y}_\nu]P \ . \tag{52}$$

The $\hat{e}^i$'s are g times the generators of a group. In Eq. (52), P stands for projection onto the space of an appropriate (e.g., fundamental) representation D of $\hat{e}^i$ such that $\mathrm{Tr}_D\, \hat{e}_i \hat{e}_k = \delta_{ik} C$.

It remains to be seen how one can exploit the observations made here, but there arise a few natural questions:

1. The parallelism between string and gauge field equations suggests that perhaps there is more than a formal connection between them. For example, the string may be a special realization of gauge fields in which some dynamical degrees of freedom are frozen while the others have become classical in a sense.
2. A non-Abelian string may be possible taking a cue from non-Abelian gauge fields.
3. Similarly, sources for a string may be introduced by following the way in which sources were introduced to gauge fields.

REFERENCES

1. M. Kalb and P. Ramond, Phys. Rev. D9, 2273 (1974); see also Y. Nambu, Proc. Symposium on Extended Systems in Field Theory, Paris, 1975 (Univ. of Chicago preprint EFI 75-43).
2. H. B. Nielsen and P. Olesen, Nucl. Phys. B61, 45 (1973).
3. E. Kyriakopoulos, Phys. Rev. 183, 1318 (1969).
4. Y. Nambu, Phys. Rev. D10, 4262 (1974) and Ref. 1.
5. See Y. Nambu, Lecture Notes for the Erice Workshops on Quark Models, 1975 (Univ. of Chicago preprint EFI 76-42).
6. A. Schild, Univ. of Texas preprint, 1976.

ADDENDUM

A more precise meaning of the charge operator $\hat{e}$ of Eq. (46) is as follows. It may be regarded as a 2 × 2 matrix which acts on the real and imaginary parts of $\psi(x)$. Then $\hat{e}^2 = e^2 \times 1$ (unit matrix). In rewriting Eq. (45) using $\hat{e}$, a trace average of the matrix should be taken. In the non-Abelian case, the $\hat{e}^i$ will be a set of matrices ($n \times n$ for SU(n)) acting, let us say, on a quark multiplet.

Nuclear Physics B130 (1977) 505–516

STRING-LIKE CONFIGURATIONS IN THE WEINBERG-SALAM THEORY *

Y. NAMBU
The Enrico Fermi Institute and the Department of Physics, University of Chicago, Chicago, Illinois 60637

Received 1 September 1977

The Weinberg-Salam theory of electromagnetic and weak interactions admits classical configurations in which a pair of magnetic monopoles is bound by a flux string of the Z^0 field. They give rise to Regge trajectories of excitations with a mass scale in the TeV range.

1. Introduction

Much of the interest that has been aroused recently in the monopoles, strings and other special solutions of gauge field equations seems to be directed at the general existence and structure of solutions in arbitrary model systems, rather than at the nature of solutions in specific models which are more relevant to the actual physical problems. This is understandable because, first of all, the very possibility and beauty of such solutions is so fascinating, with implications which probably will extend beyond our current concerns, and also because there are still a great deal of uncertainties in pinning down the right theories.

The existence of string-like objects in an Abelian Higgs model has been pointed out by Nielsen and Olesen [1] as a simplified model for hadronic strings. This analogy can be carried a bit closer by attaching monopoles to the ends of a string [2], but the problem of constructing non-Abelian strings, especially those representing baryons, has not been satisfactorily resolved yet. In other words, it is not yet clear whether the Nielsen-Olesen type picture is the correct picture for hadronic strings.

As for monopoles, 't Hooft [3] and Polyakov [4] have shown the existence of finite-energy solutions in an SO(3) Higgs model. However, the more relevant model for electromagnetic and weak interactions is the SU(2) × U(1) model of Weinberg [5] and Salam [6]. We would like to assert in this paper that the Weinberg-Salam theory admits classical solutions which look very much like the Nielsen-Olesen hadronic string with a pair of monopoles attached to its ends. The energy scales are

* Work supported in part by the NSF: contract no. PHY74-08833.

vastly different, however. Both the mass of the monopole and the tension of the string are estimated to be in the TeV range.

The qualitative arguments for the existence of such solutions are as follows. Consider a monopole solution of the Wu-Yang type [7], in which the isospin direction is locked with the radial vector. A Higgs field is necessary to smooth out the singularity at the origin and make the energy finite. In the model of 't Hooft and Polyakov the Higgs field is an isovector

$$\phi^i(r) = f(r)x^i/r\,, \tag{1}$$

where $f(r) \sim$ const asymptotically, and $f(0) = 0$ so that ϕ^i is well defined everywhere.

In an SU(2) $\times$ U(1) model where the Higgs field is an isospinor ϕ^α ($\alpha = 1,2$), the corresponding solution would be (in polar coordinates)

$$\phi \sim \begin{pmatrix} \cos \frac{1}{2}\theta \\ \sin \frac{1}{2}\varphi \, e^{i\varphi} \end{pmatrix}, \tag{2}$$

for $r \neq 0$, but this still leaves an ill-defined phase along the negative z axis ($\theta = \pi$). Therefore ϕ would have to vanish for $\theta = \pi$. That means that one cannot have a simple monopole, but it must be accompanied by a string. Suppose now we go far away from the origin along the negative z direction: Since there is a U(1) gauge field, and the effect of the monopole may be ignored, ϕ is essentially of the form

$$\phi = \begin{pmatrix} 0 \\ f(\rho)\, e^{i\varphi} \end{pmatrix}, \qquad \rho = \sqrt{x^2 + y^2}\,. \tag{3}$$

Thus one is dealing with a semi-infinite Nielsen-Olesen string made up of a flux of a combination of gauge fields. Such a system is unstable, but we can obtain a stable system if we terminate the string by putting another monopole of opposite sign and spin the system. The emerging picture is exactly that of the string model of mesons.

The long-range fields associated with the system behave as follows. Far away from the system, we have a linear combination (the real electromagnetic field) of U(1) and SU(2) gauge fields created by the pair of magnetic poles. Along the string, however, the two components behave differently. The U(1) part has a return flux through the string, whereas the SU(2) part does not. Thus the poles are genuine SU(2) monopoles. Because of this situation we expect such a system to be essentially stable if the two poles are kept sufficiently far apart; the poles will maintain their integrity since each has a topological quantum number.

The rest of the paper is devoted to the mathematical details.

2. General formalism for dealing with monopoles and strings

In this section we review a mathematical formalism developed elsewhere [8] *, which makes transparent the handling of classical solutions of a gauge field in the presence of a non-vanishing Higgs field. We denote the latter by ϕ^α, ($\alpha = 1,2, ...$) or simply ϕ; it corresponds to a minimum of the total energy of the system. Usually this means that both the potential and kinetic energies are separately minimal:

$$V(\phi) = \min , \qquad D_\mu \phi = 0 , \tag{4}$$

where D_μ denotes a covariant derivative. These two conditions are compatible with each other since the second equation states that the ϕ's at two neighboring points are related by an infinitesimal action of a gauge group, under which $V(\phi)$ is invariant. Thus an arbitrary path joining two points in space is mapped into a path in the space of ϕ's with $V(\phi) = \min$. In particular, a closed loop must be mapped into a closed loop since ϕ must be one-valued. This puts a stringent condition on the admissible distribution of the gauge fields. In fact from eq. (4) follows

$$[D_\mu, D_\nu]\phi \propto F_{\mu\nu}\phi = 0 \qquad \text{for all } (\mu\nu) , \tag{5}$$

where $F_{\mu\nu}$ acts on ϕ as a transformation matrix. Eq. (5) means det $F_{\mu\nu} = 0$ for all $(\mu\nu)$, since by assumption $\phi \neq 0$. Thus the gauge field can have a non-vanishing component in a medium (with $\phi \neq 0$) only if the corresponding generator of the gauge group has a zero eigenvalue in the representation ϕ. One may call this the statement of a generalized Meissner effect. In the real Meissner effect, the group is U(1), ϕ is a complex one-component field, and hence $F_{\mu\nu} = 0$. A similar situation holds for an SU(2) gauge field with an isospinor ϕ,: all the components $F_{\mu\nu}{}^i$, $i = 1,2,3$, must vanish because no generators have zero eigenvalues on ϕ.

On the other hand, an isovector ϕ allows one of the generators to vanish, so that the corresponding component of $F_{\mu\nu}{}^i$ survives the Meissner effect. This phenomenon also occurs in general if there are more than one commuting generator (i.e., if the group is not simple, or if its rank is $\geqslant 2$), because we can make a linear combination of such commuting generators annihilate a ϕ.

The above conditions, however, are not sufficient in general for satisfying eqs. (4) and (5) everywhere. There may be points (monopoles) or lines (strings) on which the gauge potential is singular and the distribution of ϕ ill defined. So the true solution must necessarily make ϕ vanish at the singularities and violate these equations in their neighborhood. But the topological properties of the mapping $x \to \phi(x)$ around the singularities are not affected by such deviations.

As was said above, eq. (4) or (5) puts a severe restriction on the gauge potentials

* The line of reasoning that follows is often found implicitly stated in the literature, but it has not been properly emphasized in our opinion. See in particular, ref. [9].

$A_\mu{}^i$. In fact, we can solve eq. (5) for A_μ. For example, for the case of isovector ϕ and A_μ, eq. (4) reads

$$\partial_\mu \boldsymbol{\phi} - g\boldsymbol{A}_\mu \times \boldsymbol{\phi} = 0 \tag{6}$$

from which follows first

$$\boldsymbol{\phi} \cdot \boldsymbol{\phi} = \text{const} , \tag{7}$$

so we will conveniently choose $\boldsymbol{\phi}$ to be a unit vector. Then the solution to eq. (5) is [8,9]

$$g\boldsymbol{A}_\mu = \boldsymbol{\phi} \times \partial_\mu \boldsymbol{\phi} + a_\mu \boldsymbol{\phi} , \tag{8}$$

where a_μ is arbitrary, and

$$\begin{aligned} \boldsymbol{F}_{\mu\nu} &= \partial_\mu \boldsymbol{A}_\nu - \partial_\nu \boldsymbol{A}_\mu + g\boldsymbol{A}_\mu \times \boldsymbol{A}_\nu \\ &= \frac{1}{g}(\partial_\mu \boldsymbol{\phi} \times \partial_\nu \boldsymbol{\phi} + f_{\mu\nu}\boldsymbol{\phi}) , \end{aligned} \tag{9}$$

$$f_{\mu\nu} = \partial_\mu a_\nu - \partial_\nu a_{\mu}{}'$$

$$F_{\mu\nu} \equiv \boldsymbol{\phi} \cdot \boldsymbol{F}_{\mu\nu} = \frac{1}{g}(\boldsymbol{\phi} \cdot \partial_\mu \boldsymbol{\phi} \times \partial_\nu \boldsymbol{\phi} + f_{\mu\nu}) .$$

The gauge field equations

$$D_\mu \boldsymbol{F}_{\mu\nu} = \boldsymbol{j}_\nu , \tag{10}$$

where the source $\boldsymbol{j}_\nu$ comes from fields other than $\boldsymbol{\phi}$ and $\boldsymbol{A}_\mu$, are equivalent to a U(1) equation

$$\partial_\mu F_{\mu\nu} = j_\nu , \qquad j_\nu = \boldsymbol{\phi} \cdot \boldsymbol{j}_\nu \tag{11}$$

because $\boldsymbol{F}_{\mu\nu} // \boldsymbol{\phi}, D_\mu \boldsymbol{\phi} = 0$.

3. String-monopole solution in a SU(2) × U(1) model

We apply the method of sect. 2 to the Weinberg-Salam model. In this case ϕ is an isodoublet, for which we write

$$D_\mu \phi = (\partial_\mu - i\,\tfrac{1}{2} g\tau^i \cdot A_\mu{}^i - i\,\tfrac{1}{2} g' A_\mu{}^0)\phi = 0 . \tag{12}$$

Again ϕ may be normalized to unity:

$$(\phi^\dagger\phi) = \sum \phi^{\dagger\alpha}\phi^\alpha = 1\,. \tag{13}$$

Furthermore, there are a series of Fierz identities

$$\sum_i (\phi^\dagger\tau^i\phi)^2 = (\phi^\dagger\phi)^2\,,$$

$$\sum_i (\phi^\dagger\tau^i\phi)(\phi^\dagger\tau^i\partial_\mu\phi) = (\phi^\dagger\phi)(\phi^\dagger\overleftrightarrow{\partial_\mu}\phi)\,, \tag{14}$$

$$(\phi^\dagger\tau^i\phi)(\phi^\dagger\overleftrightarrow{\partial_\mu}\phi) - (\phi^\dagger\phi)(\phi^\dagger\tau^i\overleftrightarrow{\partial_\mu}\phi) = i(\phi^\dagger\tau^j\phi)\partial_\mu(\phi^\dagger\tau^k\phi)\epsilon^{ijk}\,, \text{ etc.}$$

These useful identities can be derived with the aid of the Dirac exchange operator.

It is not difficult to solve eq. (12) for the potentials. We find

$$gA_\mu{}^i + g'A_\mu{}^0(\phi^\dagger\tau^i\phi) = -i(\phi^\dagger\tau^i\overleftrightarrow{\partial_\mu}\phi)\,, \tag{15}$$

or, in view of eq. (14),

$$= -(\phi^\dagger\tau^j\phi)\partial_\mu(\phi^\dagger\tau^k\phi)\epsilon^{ijk} - i(\phi^\dagger\tau^i\phi)(\phi^\dagger\overleftrightarrow{\partial_\mu}\phi)\,, \tag{16}$$

which can be split into two equations

$$gA^i_{\mu\perp} = -(\phi^\dagger\tau^j\phi)\partial_\mu(\phi^\dagger\tau^k\phi)\epsilon^{ijk}\,,$$

$$gA^i_{\mu\parallel} + g'A_\mu{}^0 = -i(\phi^\dagger\overleftrightarrow{\partial_\mu}\phi)\,, \tag{17}$$

with $A_\perp$ and $A_\parallel$ referring respectively to the components perpendicular and parallel to $\phi^\dagger\tau^i\phi$. We will therefore parametrize the general solution as *

$$gA^i_{\mu\perp} = -(\phi^\dagger\tau^j\phi)\partial_\mu(\phi^\dagger\tau^k\phi)\epsilon^{ijk}\,,$$

$$gA^i_{\mu\parallel} = -\xi i(\phi^\dagger\tau^i\phi)(\phi^\dagger\overleftrightarrow{\partial_\mu}\phi)\,,$$

$$g'A_\mu{}^0 = -\eta i(\phi^\dagger\overleftrightarrow{\partial_\mu}\phi)\,, \qquad \xi + \eta = 1\,. \tag{18}$$

Note the similarity between eqs. (8) and (18); $(\phi^\dagger\tau^i\phi)$ plays the role of the normalized isovector field, and obviously satisfies $D_\mu(\phi^\dagger\tau^i\phi) = 0$.

* Actually, we may add to $gA_{\mu\parallel}^i$ and $g'A_\mu{}^0$, terms proportional to an arbitrary potential a_μ and still satisfy eq. (17). Here, we drop these "external" potentials.

From eq. (18) we compute the fields. After some algebra the result may be cast in alternative forms related by the Fierz identities:

$$g'F_{\mu\nu}{}^0 = \eta f_{\mu\nu}\,,$$

$$gF_{\mu\nu}{}^i = -\partial_\mu(\phi^\dagger\tau^j\phi)\partial_\nu(\phi^\dagger\tau^k\phi)\epsilon^{ijk} + \xi f_{\mu\nu}(\phi^\dagger\tau^i\phi)\,,$$

$$f_{\mu\nu} = -2i(\partial_\mu\phi^\dagger\partial_\nu\phi - \partial_\nu\phi^\dagger\partial_\mu\phi)\,,$$

or

$$gF_{\mu\nu}{}^i = -(1-\xi)f_{\mu\nu}(\phi^\dagger\tau^i\phi) = -\eta f_{\mu\nu}(\phi^\dagger\tau^i\phi)\,, \tag{19}$$

so that we have always

$$gF_{\mu\nu}{}^i + g'F_{\mu\nu}{}^0(\phi^\dagger\tau^i\phi) = 0\,. \tag{20}$$

This in turn means

$$|gF_{\mu\nu}| = |g'F_{\mu\nu}{}^0|\,,$$

$$gF_{\mu\nu}{}^i(\phi^\dagger\tau^i\phi) + g'F_{\mu\nu}{}^0 \equiv \sqrt{g^2+g'^2}\,F_{\mu\nu}{}^Z = 0\,. \tag{21}$$

Here $F_{\mu\nu}{}^Z$ stands for the field associated with the neutral massive boson Z^0, defined in a covariant way. The electromagnetic field $\mathcal{F}_{\mu\nu}$ should then be identified with

$$\begin{aligned}\mathcal{F}_{\mu\nu} &\equiv [g'F_{\mu\nu}{}^i(\phi^\dagger\tau^i\phi) - gF_{\mu\nu}{}^0]/(g^2+g'^2)^{1/2} \\ &= [(g^2+g'^2)^{1/2}/g]F_{\mu\nu}{}^0\,.\end{aligned} \tag{22}$$

Let us now assume for ϕ the form mentioned before,

$$\phi = \begin{pmatrix}\cos\frac{1}{2}\theta \\ \sin\frac{1}{2}\theta\; e^{i\varphi}\end{pmatrix}. \tag{23}$$

Because of the phase factor, ϕ has a line of singularity along $x = y = 0$, $z \leqslant 0$, but $(\phi^\dagger\tau^i\phi)$ is regular except at the origin. This means that $A^i_{\mu\perp}$ is always free of the line singularity. In fact we find

$$(\phi^\dagger\tau^i\phi) = x^i/r\,, \qquad gA_k{}^0 = -\epsilon^{kl3}x_l/r(r+z)\,,$$

$$gA^i_{k\perp} = \epsilon^{ikl}x_l/r^2\,, \qquad f_{kl} = \epsilon^{klm}x_m/r^3\,. \tag{24}$$

The last line shows that $F_{\mu\nu}{}^i$ and $F_{\mu\nu}{}^0$ are precisely monopole fields, and therefore a solution of field equations except for the singularities. The line singularity is manifest only in ϕ, $A_\mu{}^0$ and $A_\mu{}^i$. To grasp the physical significance of this fact, compute the magnetic flux out of the origin, by using the Gauss theorem for $F_{\mu\nu}{}^0$ and $F_{\mu\nu}{}^i(\phi^\dagger\tau^i\phi)$, and also compute the singular flux through the negative z axis by using the Stokes theorem for $A_\mu{}^0$ and $A_\mu{}^i$. The former computation yields

$$4\pi\eta/g' \quad \text{and} \quad -4\pi\eta/g \tag{25}$$

for U(1) and SU(2) parts respectively, and the latter computation yields

$$-4\pi\eta/g' \quad \text{and} \quad -4\pi\xi/g\,. \tag{26}$$

The former has missed the singular contributions, so the total flux out of the origin should be the sum of the two, namely

$$\text{zero for U(1) and } -4\pi/g \text{ for SU(2)}\,. \tag{27}$$

The emerging picture may then be stated as follows. The U(1) flux is sourceless, like that of a solenoid. The flux $4\pi\eta/g'$ that spreads out to infinity is returned surreptitiously through the string or tube along the negative z axis. The SU(2) flux, on the other hand, forms a genuine monopole of the correct quantized value $4\pi/g$, part of which spreads out and the rest goes through the string. There is one parameter $\xi = 1 - \eta$ which determines the division, and it so far remains arbitrary.

Another way to put it would be to say that the spreading field is an electromagnetic (since $F_{\mu\nu}{}^Z = 0$) field due to a magnetic charge

$$4\pi\eta(g^2 + g'^2)^{1/2}/gg' = 4\pi\eta/e \equiv Q\,, \tag{28}$$

whereas the flux tube contains a mixture of both,

$$4\pi\left(-\frac{g'}{g}\xi + \frac{g}{g'}\eta\right)\Big/\sqrt{g^2+g'^2} = (4\pi/e)[-\xi\sin^2\theta + \eta\cos^2\theta] \qquad \text{for } \mathcal{F}_{\mu\nu}\,,$$

$$-4\pi/\sqrt{g^2+g'^2} = -(4\pi/e)\sin\theta\cos\theta \qquad \text{for } F_{\mu\nu}{}^Z\,, \tag{29}$$

where θ now stands for the Weinberg angle.

To proceed further, we must resort to the dynamics of the original field theory. Then the singularities will be smeared out, and the total energy will be minimized. Since the flux of the Z field through the tube is fixed by quantization, this means that the energy of the tube per unit length will be minimum if the electromagnetic flux is zero, or in other words

$$\xi = \cos^2\theta\,, \qquad \eta = \sin^2\theta\,. \tag{30}$$

The monopole will then carry a magnetic charge

$$Q_0 = (4\pi/e)\sin^2\theta\,. \tag{31}$$

The system still has an infinite energy, and will not be stationary: the infinitely long string will keep pulling the monopole with a constant force. But a quasi-stationary finite energy system can be formed by spinning a pair of monopoles joined by a string. A mathematical description of such a dumb bell system [10,11] (monopolium) * is relatively simple only in the classical string picture. We will not discuss in detail how to carry out the transition to such a picture except to remark that the Higgs field ϕ for a pair of monopole and antimonopole placed at two fixed positions on the z axis is given by [8]

$$\phi(x) = \begin{pmatrix} \cos \frac{1}{2}\Theta \\ \sin \frac{1}{2}\Theta \, e^{i\varphi} \end{pmatrix} , \tag{32}$$

$$\cos \Theta = \cos \theta_1 - \cos \theta_2 + 1$$

instead of eq. (23). Here $\theta_1(\theta_2)$ is the polar angle of point x as seen from the monopole (antimonopole). The line singularity along the z axis now exists only between the poles. Far away from the system, $\cos \Theta \sim 1$ in all directions, i.e., ϕ approaches the standard constant solution, as it should.

4. Dynamics of a rotating dumb bell

We will skip any attempt to solve the field equations near the singularities because exact solutions cannot be obtained. Assuming that such solutions exist, however, it is easy to make estimates for the string tension and the monopole mass. The relevant part of the Weinberg-Salam Lagrangian is

$$-L = \tfrac{1}{4}F_{\mu\nu}{}^i F_{\mu\nu}{}^i + \tfrac{1}{4}F_{\mu\nu}{}^0 F_{\mu\nu}{}^0 + \tfrac{1}{2}(D_\mu \phi^\dagger)(D_\mu \phi) + \tfrac{1}{8}\lambda^2(\phi^\dagger\phi - 1)^2 , \tag{33}$$

where the energy is measured in units of $\langle\phi\rangle = 246$ GeV. This unit of energy will be adopted from here on. The W-boson and Higgs boson masses are respectively

$$m_{\mathrm{W}} = \tfrac{1}{2}g , \qquad m_{\mathrm{H}} = \lambda . \tag{34}$$

Consider first the flux tube between two poles. We approximate it by a cylinder of radius ρ and length $l >> \rho$, within which $\phi = 0$ and the fluxes given by eqs. (25), (26) are uniformly distributed. Minimizing the total energy with respect to ρ, we find the tension τ of the tube (string)

$$\tau = 2\lambda\pi^{1/2}\,[\xi^2/g^2 + \eta^2/g'^2]^{1/2} , \tag{35}$$

* This name is due to Maki [12], who has considered similar configurations in superfluid ^{3}He.

with

$$\rho = (8/\lambda)^{1/2}(\xi^2/g^2 + \eta^2/g'^2)^{1/4} .$$

Minimizing this further with respect to ξ, we recover the condition (30) and

$$\tau_0 = (2\lambda\pi/e)\sin\theta\cos\theta = \pi\cos\theta\,(m_{\rm H}/m_{\rm W}) ,$$

$$\rho_0 = ((8/\lambda e)\sin\theta\cos\theta)^{1/2} = 2((m_{\rm W}m_{\rm H})^{-1}\cos\theta)^{1/2} . \qquad (36)$$

To estimate the monopole mass M, we assume a hollow sphere of radius r inside which all the fields are zero. Minimizing with respect to r gives

$$M = (8\pi/3e)(\lambda/2e)\eta^{3/2} , \qquad r = (2\eta/\lambda e)^{1/2} , \qquad (37)$$

and for

$$\eta = \sin^2\theta , \quad M_0 = (4\pi/3e)(\sin\theta)^{5/2}(m_{\rm H}/m_{\rm W})^{1/2} ,$$

$$r_0 = ((m_{\rm H}m_{\rm W})^{-1}\sin\theta)^{1/2} . \qquad (38)$$

The energy of a pair of monopoles separated by a distance l is then $2M - Q^2/4\pi l$, but the second term may be ignored if the poles are well separated from each other.

With τ_0 and M_0 fixed by eqs. (36) and (38), the total energy and angular momentum of a rotating dumb bell are computed to be

$$E \sim \tfrac{1}{2}\pi l\tau_0 , \qquad L \sim \tfrac{1}{8}\pi l^2\tau_0 , \qquad l\tau_0/2M_0 = \beta^2/(1-\beta^2) , \qquad (39)$$

where β is the velocity of the poles. This amounts to a straight Regge trajectory

$$L \sim \alpha_0' E^2 , \qquad \alpha_0' = 1/2\pi\tau_0 . \qquad (40)$$

We have shown here only asymptotic relations which are valid in the relativistic limit $\beta \simeq 1$. In non-asymptotic cases, not only is the trajectory non-linear, but also the magnetic charge (visible from the outside) becomes L dependent, because the parameter ξ has to be chosen to minimize the total energy.

Let us now make rough numerical estimates. From eqs. (36) and (38) we obtain

$$(1/\alpha_0')^{1/2} = (2\pi\tau_0)^{1/2} \simeq \sqrt{2}\pi((m_{\rm H}/m_{\rm W})\cos\theta)^{1/2}$$

$$= 1.1\ \text{TeV} \times ((m_{\rm H}/m_{\rm W})\cos\theta)^{1/2} ,$$

$$M_0 \simeq (4\pi/3e)(\sin\theta)^{5/2}(m_{\rm H}/m_{\rm W})^{1/2}$$

$$= 3.4 \text{ TeV} \times (\sin\theta)^{5/2}(m_H/m_W)^{1/2} . \tag{41}$$

Unless m_H/m_W is unusually small, we are dealing with multi-TeV phenomena. For example, with $\sin^2\theta = \frac{1}{3}$, $m_H/m_W = 1$, eq. (41) gives $(1/\alpha_0')^{1/2} = 1$ TeV, $M = 0.9$ TeV.

On the other hand, the dimensions of the string and monopoles are of the order of $(1/m_W m_H)^{1/2}$, which would be in the $(100 \text{ GeV})^{-1}$ range. So our strings and poles are very fat in comparison with their mass scale. This is due to the smallness of the coupling constants. The object would not look like a rod unless $l \gg \rho$, which translates into

$$L \gg \tfrac{1}{2}\pi \times 137 \sin^2\theta \cos\theta . \tag{42}$$

The right-hand side of the above relation is 70 for $\sin^2\theta = \frac{1}{3}$. Thus by the time the object takes on a well-defined linear shape, it really behaves like a classical relativistic system.

How stable and well defined will such an object be once it is formed? Because the coupling constants are small, we expect that the decay widths as well as the quantum corrections to the classical solutions will be small. The system can decay by emitting γ's, W's and Z's, pairs of leptons and quarks, or by breaking up. The last process becomes more difficult for higher L because of energetics and angular momentum barrier, but it also means that production of such a state would be more difficult.

As an example, we estimate the radiative energy loss of rotating magnetic poles in classical electrodynamics. The process is the familiar synchrotron radiation. The radiated power is given by

$$P \sim \tfrac{8}{3}(Q^2/4\pi)\gamma^4/l^2 = \tfrac{8}{3} \times 137(\tau_0/M_0)^2 \sin^4\theta$$

$$\simeq (m_H/m_W) \text{ TeV}^2 \qquad \text{for} \qquad \sin^2\theta = \tfrac{1}{3} . \tag{43}$$

P is related to the radiative width Γ by $\Gamma = P/E$, and the measure of stability Γ/E takes the simple form

$$\Gamma/E = P/E^2 \simeq 1/L . \tag{44}$$

For the range of L we are considering, the system is indeed rather stable.

5. Discussion

The arguments and calculations presented here are certainly not rigorous, but the existence of massive string-like solutions in the framework of Weinberg-Salam theory seems very plausible. As already remarked, the smallness of the coupling constant makes the mass scale uncomfortably large, but at the same time it keeps the quantum effects small, once a well defined string-like state is formed. For this to be the case, the angular momentum has to be large, $L \gtrsim 100$, and the energy $E \gtrsim 10$ TeV. It is a simple consequence of the fact that the size of a monopole is ~137 times its own Compton wave length.

A pertinent question would be the production mechanism and the magnitude of cross section. In principle, a 10 TeV object may be produced in a proton-proton collision with projectile energy of 10^5 TeV and impact parameter of $10^{-4}\, m_W^{-1}$. At such an impact parameter, the electromagnetic interaction between point-like quarks would be $\simeq \frac{1}{137} \times 10^4\, m_W \simeq 10^2\, m_W$, which is of the order of the mass to be produced.

In addition to the rotating states, there must also exist longitudinal vibrations [10,11] of the monopoles. We did not discuss these states before because of the theoretical uncertainties when the poles overlap. They are probably very unstable and ill-defined. However, the production of a spin-1 rate through one-photon annihilation of an electron and a positron, for example, is forbidden because in the c.m.s. the e^-–e^+ pair converts into a purely electric field, whereas the monopole pair converts into a purely magnetic field.

As a final remark we emphasize the extreme similarity between the present results and the string model of mesons. If we scale down the masses by 1000, we obtain almost exact analogs of the mesons except for an absence of quarks and associated quantum numbers. It remains to be seen whether this is more than an analogy.

References

[1] H.B. Nielsen and P. Olesen, Nucl. Phys. B61 (1973) 45.
[2] Y. Nambu, Phys. Rev. 183 (1974) 4262.
[3] G. 't Hooft, Nucl. Phys. B79 (1974) 276.
[4] A. Polyakov, ZhETF Pisma 20 (1974) 430; (JETP Lett. 20 (1974) 194).
[5] S. Weinberg, Phys. Rev. Lett. 29 (1967) 1264.
[6] A. Salam, Elementary particle physics, ed. N. Swartholm, (Almquist and Wiksells, Stockholm, 1968) p. 367.
[7] T.T. Wu and C.N. Yang, Properties of matter under unusual conditions, ed. Fernbach and Mark (Interscience Publishers, New York 1969) p. 349.
[8] Y. Nambu, U. Chicago preprint EFI 77/17, to be published in Proc. Int. Symp. "Five decades of weak interactions" in honor of the 60th Birthday of R.E. Marshak, 1977.
[9] E. Corrigan, D.B. Fairlie, J. Nuyts and D.I. Olive, Nucl. Phys. B106 (1976) 475.
[10] A. Chodos and C.B. Thorn, Nucl. Phys. B81 (1974) 525.
[11] I. Bars and A.J. Hanson, Phys. Rev. D13 (1976) 1744;
W.A. Bardeen, I. Bars, A. Hanson and R.D. Pecei, Phys. Rev. D13 (1976) 2364.
[12] K. Maki, Physics 90B (1977) 84.

Volume 92B, number 3,4 PHYSICS LETTERS 19 May 1980

HAMILTON–JACOBI FORMALISM FOR STRINGS ☆

Y. NAMBU
The Enrico Fermi Institute and the Department of Physics, The University of Chicago, Chicago, IL 60637, USA

Received 11 February 1980

It is shown that a Hamilton–Jacobi-type formalism can be set up to deal with the classical dynamics of relativistic strings and other one-dimensional extended systems. A special feature is that the formalism involves two evolution parameters which are treated on an equal footing. The corresponding Hamilton–Jacobi functions turn out to be vector potentials or Clebsch potentials, and in this sense we find a link between the string model and gauge field theory.

The relativistic string is an interesting example of extended objects which permit a simple geometrical description. Its relevance, of course, derives from the success of the string model of hadrons as a phenomenological theory, as well as the close physical and mathematical link that seems to exist between the string model and quantum chromodynamics (QCD). It has recently been pointed out [1] that the path ordered phase factor, which is a natural stringlike construct in QCD, can be shown to satisfy, under certain restricted conditions, the equations for a quantized string.

In the present article we will try to reverse the process. We will reanalyze the classical string mechanics by developing a Hamilton–Jacobi formalism appropriate to systems having two evolution parameters. The ensuing results can be interpreted in terms of an abelian gauge field. This program goes smoothly up to the semi-classical level, but logical uncertainties appear when one tries to develop a full quantum theory or to extend it to non-abelian cases.

Our starting point is the Schild form of string lagrangian [2] (in the pseudo-euclidean metric)

$$L = \tfrac{1}{4}\{y_\mu, y_\nu\}\{y_\mu, y_\nu\}\,,$$
$$\{y_\mu, y_\nu\} \equiv \partial(y_\mu, y_\nu)/\partial(\sigma, \tau)\,, \tag{1}$$

☆ This work was supported in part by the NSF, contract No. PHY-78-01224.

where $y_\mu(\sigma, \tau)$ denotes a point lying on the two-dimensional surface swept out by a string. This form is the square of the usual lagrangian representing the surface area element, and hence does not have a purely geometrical meaning. As has been shown by Schild, however, it is nevertheless equivalent to the latter as a consequence of the equation of motion

$$\{y_\mu\{y_\mu, y_\nu\}\} = 0\,, \tag{2}$$

which implies L = const. The equivalence is secured by choosing this constant to have a fixed value, $-C^2 = -1/2(1/2\pi\alpha')^2$, where α' is the Regge slope parameter. Recently, Eguchi [3] has shown how the Schild lagrangian may be used for quantization of the string, and thereby provide an alternative to the usual Virasoro equations. Our aim is somewhat different, being directed at a generalization of the hamiltonian dynamics.

The Hamilton–Jacobi formalism is based on the differential one-form relation

$$\mathrm{d}S = \sum_i p_i\, \mathrm{d}q_i - H\, \mathrm{d}t\,,$$
$$H = H(p_i, q_i)\,, \quad S = S(q_i, t)\,, \tag{3}$$

from which all the rest follows. We replace this with a two-form relation

Volume 92B, number 3,4 PHYSICS LETTERS 19 May 1980

$$dS_1 \wedge dT_1 + dS_2 \wedge dT_2 = \sum_{i>j}^{1,N} p_{ij}\, dq_i \wedge dq_j - H\, d\sigma \wedge d\tau\,, \tag{4}$$

$$S_m = S_m(q_i, \sigma, \tau)\,, \quad T_m = T_m(q_i, \sigma, \tau)\,, \quad m = 1, 2\,,$$

$$H = H(p_{ij}, q_k)\,.$$

There are N coordinates q_i, $N(N-1)/2$ momenta p_{ij}, and two evolution parameters σ, τ which are treated on an equal footing. On the left-hand side, an exact two-form is written in terms of pairs of one-forms, or "Clebsch potentials" (see below) S_m, T_m. Obviously we have

$$\begin{aligned} p_{ij} &= \sum_m \partial(S_m, T_m)/\partial(q_i, q_j)\,, \\ -H &= -H(p_{ij}, q_k) = \sum_m \partial(S_m, T_m)/\partial(\sigma, \tau)\,, \\ 0 &= \sum_m \partial(S_m, T_m)/\partial(\sigma, q_i)\,, \\ 0 &= \sum_m \partial(S_m, T_m)/\partial(q_i, \tau)\,. \end{aligned} \tag{5}$$

When a functional relation $H = H(p_{ij}, q_k)$ is given, these represent Hamilton–Jacobi-type equations for the S_m and T_m. The last two equations are constraints, and because of them we have taken two pairs. For example, the ansatz $S_1(\sigma,\tau)$, $T_1(\sigma,\tau)$, $S_2(q_i)$, $T_2(q_i)$ would do the job.

We now would like to set up the corresponding Hamilton-type equations of motion. For this purpose we will regard all the variables q_i, p_{ij} as independent. Physically, the momenta represent planes tangent to an evolving world sheet, so clearly we have introduced too many degrees of freedom. We will have to make sure that these degrees of freedom are properly suppressed as a result of the equations of motion (i.e., on the "mass shell"), so the equations become integrable. But integrability is guaranteed if the tangential planes are generated by a pair of global functions as in eq. (5). The Hamilton equations should then be derived by noting that the exterior derivatives of eq. (4) are zero:

$$0 = \sum_{i>j} dp_{ij} \wedge dq_i \wedge dq_j - \left(\sum_{i>j} \frac{\partial H}{\partial p_{ij}}\, dp_{ij} + \sum_i \frac{\partial H}{\partial q_i} dq_i\right) \wedge d\sigma \wedge d\tau\,. \tag{6}$$

Equating the coefficients of dp_{ij} and dq_i, respectively, we find

$$\{q_i, q_j\} = \partial H/\partial p_{ij}\,, \quad \sum_j \{p_{ij}, q_j\} = -\partial H/\partial q_i\,. \tag{7}$$

Substituting them back into eq. (4), we also obtain the "on-shell" relation

$$\sum dS_m \wedge dT_m = \left(\sum_{i>j} p_{ij} \frac{\partial H}{\partial p_{ij}} - H\right) d\sigma \wedge d\tau \equiv L\, d\sigma \wedge d\tau\,, \tag{8}$$

which defines the lagrangian L from the hamiltonian H. Furthermore, it follows from eq. (7) that

$$\begin{aligned} \frac{\partial H}{\partial \tau} &= \sum_{i>j} \frac{\partial p_{ij}}{\partial \tau} \{q_i, q_j\} - \sum_{i,j} \frac{\partial q_i}{\partial \tau} \{p_{ij}, q_j\} = 0\,, \\ \frac{\partial H}{\partial \sigma} &= \sum_{i>j} \frac{\partial p_{ij}}{\partial \sigma} \{q_i, q_j\} - \sum_{i,j} \frac{\partial q_i}{\partial \sigma} \{p_{ij}, q_j\} = 0\,, \end{aligned} \tag{9}$$

so the hamiltonian is a constant of motion, being independent of the evolution parameters. Eq. (7) is the set of Hamilton-type equations that we were looking for. To specialize to the Schild equations, we choose H to be

$$H = \frac{1}{2} \sum_{\mu>\nu} p_{\mu\nu}^2\,. \tag{10}$$

Eq. (7) then reduces to eq. (2) (with the identification $q = y$), and we have $H = L = -C^2$.

Let us now go back to the Hamilton–Jacobi system (5). It is a straightforward matter to check that a set of functions S_m, T_m satisfying eq. (5) do actually generate a family of surfaces that evolve according to eq. (7). Suppose, then, that a solution $y_\mu(\sigma,\tau)$ spans in the six-dimensional space $(y_\mu, \sigma, \tau) \equiv y_\alpha$, $\alpha = 1, \ldots, 6$, a two-surface domain D bounded by a closed curve Γ. According to eq. (8), the action is

$$I_D = \int_D L\, d\sigma\, d\tau = \sum \int_D dS_m \wedge dT_m = \int_\Gamma A_\alpha\, dy_\alpha\,, \tag{11}$$

$$A_\alpha \equiv \sum S_m \partial_\alpha T_m \quad \left(\text{or alternatively} - \sum T_m \partial_\alpha S_m\right).$$

If A is regarded as a vector potential, the corresponding field $F_{\alpha\beta}$ is given by

Volume 92B, number 3,4 PHYSICS LETTERS 19 May 1980

$$F_{\alpha\beta} = \partial_\alpha A_\beta - \partial_\beta A_\alpha = \sum_m \frac{\partial(S_m, T_m)}{\partial(y_\alpha, y_\beta)}, \tag{12}$$

and eqs. (5) and (1) may be stated as

$$\{y_\mu, y_\nu\} = p_{\mu\nu} = F_{\mu\nu}, \quad F_{5\mu} = F_{6\mu} = 0,$$
$$H = -F_{56} = \frac{1}{2} \sum_{\mu>\nu} F_{\mu\nu}^2 = -C^2 . \tag{13}$$

We realize at this point that the ansatz for the left-hand side of eq. (4) in terms of scalar functions S_m, T_m could have been replaced with

$$\sum_\alpha \mathrm{d}A_\alpha \wedge \mathrm{d}y_\alpha , \tag{14}$$

from the beginning. However, the fields $F_{\mu\nu}$ do not satisfy the sourceless Maxwell equations; instead, they are constrained by eq. (13c). We also notice that the $F_{\mu\nu}$ can, in general (i.e., except at singular points), be reduced to one nonvanishing component as they represent a tangent plane. This property is characterized by

$$F_{\mu\nu} F_{\lambda\rho} \epsilon_{\mu\nu\lambda\rho} = 0 , \tag{15}$$

together with eq. (13c). Under these circumstances, the "Clebsch potentials" [4] S, T may be more natural than the vector potentials.

Going to quantum theory, we may associate with the action integral (11) an amplitude

$$W = \exp(\mathrm{i} I_D)$$
$$= \exp\left(\mathrm{i} \int_\Gamma A_\mu \, \mathrm{d}x_\mu\right) \exp\left(-\mathrm{i}C^2 \int \mathrm{d}\sigma \, \mathrm{d}\tau\right) . \tag{16}$$

The first factor is the usual Wilson loop factor, whereas the second measures the area in the (σ, τ) space [3], which does not have a direct physical significance. Eq. (16) may be interpreted as a semiclassical approximation to the full quantum theory. However, it is not entirely obvious what the full theory should be. One possibility would be to generate the $F_{\mu\nu}$ by keyboard variations $\delta\sigma_{\mu t}$ of the boundary Γ, and express the constraint as

$$\left[\sum_\mu (\delta/\delta\sigma_{\mu t})_z^2 - C^2\right] W = 0 , \quad z \in \Gamma , \tag{17}$$

where W is now freed from the semiclassical form (16). This is precisely the equation derived before [1]. However, it does not quite correspond to the eq. (13c) as only half of the $F_{\mu\nu}$ are generated this way. Generating the other half by means of twisting variations of Γ is conceivable, but the question is whether such operations can be adequately defined or not.

For now, we will content ourselves with the semiclassical approximation, and treat the spinning string (rod) as an example. The solution to eq. (2) for this case is

$$y_0 = \tau , \quad y_1 + \mathrm{i}y_2 = \mathrm{e}^{\mathrm{i}\omega\tau} \rho(\sigma) ,$$
$$\sigma = \frac{1}{C\omega} \int_0^{\rho\omega} (1 - x^2)^{1/2} \, \mathrm{d}x , \quad y_3 = 0 . \tag{18}$$

The fields $F_{\mu\nu}$ are then

$$F_{\rho 0} = C/(1 - \omega^2\rho^2)^{1/2} , \quad F_{\rho\theta} = C\omega\rho/(1 - \omega^2\rho^2)^{1/2} , \tag{19}$$
$$\text{others} = 0 .$$

The corresponding potentials (naturally not unique) are found to be

$$S_2 = \rho, \quad T_2 = C(t + \omega\theta\rho^2)/(1 - \omega^2\rho^2)^{1/2}$$
$$(-1 < \omega\rho < 1, \theta = \tan^{-1} y_2/y_1) , \tag{20}$$
$$A_\rho = -T_2 \partial_\rho S = -T_2, \quad \text{others} = 0 ,$$

We note here that the potentials are not single valued; in particular, ρ has to be extended to negative values in order to cover the entire rod. This latter feature seems inevitable because of the oriented nature of the string.

Under this choice of gauge (20), a rectangular loop integral visiting the points (t_1, ρ), $(t_1, -\rho)$, $(t_2, -\rho)$, (t_2, ρ) at fixed θ reduces to two spacelike segments at times t_1 and t_2, so the factor

$$\exp\left(\mathrm{i} \int_{-1/\omega}^{1/\omega} A_\rho \, \mathrm{d}\rho\right)\Bigg|_t = \exp\left[-\mathrm{i} \int_{-1/\omega}^{1/\omega} \frac{C \, \mathrm{d}\rho}{(1 - \omega^2\rho^2)^{1/2}} - \mathrm{i}\theta \int_{-1/\omega}^{1/\omega} \frac{C\omega\rho^2}{(1 - \omega^2\rho^2)^{1/2}} \mathrm{d}\rho\right] \tag{21}$$
$$= \exp[-\mathrm{i}\pi C(2\omega t + \theta)/2\omega^2]$$

may be interpreted as the wave function of the rod. Indeed, the two integrals are nothing but the total energy E and angular momentum $l = \pi C/2\omega^2$ of the spinning rod, related by the Regge formula $l = \alpha' E^2$.

We also find that the total "magnetic flux" in the z-direction is

$$2\pi \int_{-1/\omega}^{1/\omega} F_{\rho\theta}\, \rho \mathrm{d}\rho = 2\pi l\,. \tag{22}$$

The condition of single valuedness of the wave function then leads in the usual manner to the quantization of l as well as of the flux, whatever the significance of the latter might be. Although such an interpretation seems natural, it leaves some open questions. For example, what is the meaning of the amplitude when the integral does not extend from end to end?

If one is curious about how special the above example is, it is instructive to try a more general ansatz: $S_2 = \rho$, $T_2 = tf + \theta\rho(1-f^2)^{1/2}$ which satisfies eq. (13c) for arbitrary $f(\rho)$. Minimizing the energy E for fixed angular momentum l, however, leads one back to the previous solution.

Finally, we will briefly comment on two generalizations. First, eq. (4) is a special case of the more general form

$$\begin{aligned}&\sum_{i>j} p_{ij}\, \mathrm{d}q_i \wedge \mathrm{d}q_j + \sum_i P_i\, \mathrm{d}q_i \wedge \mathrm{d}\sigma \\ &\quad + \sum_i Q_i\, \mathrm{d}q_i \wedge \mathrm{d}\tau - H\, \mathrm{d}\sigma \wedge \mathrm{d}\tau\,,\end{aligned} \tag{23}$$

$$H = H(p_{ij}, P_i, Q_i, q_i)\,.$$

Such a form will be relevant to strings weighted with mass points.

The second comment concerns non-abelian extensions. There is no problem in formally generalizing the quantum equations (16) and (17) as has been done before [1], but from the classical string point of view it is not so natural because only one set of $F_{\mu\nu}$ may be associated with the tangential plane at each point. Thus the non-abelian degrees of freedom must be reduced to a locally abelian level.

Discussions with T. Eguchi are duly appreciated.

References

[1] Y. Nambu, Phys. Lett. 80B (1979) 372;
A.M. Polyakov, Phys. Lett. 82B (1979) 247;
S.L. Gervais and A. Neveu, Phys. Lett. 80B (1979) 255.

[2] A. Schild, Phys. Rev. D16 (1977) 1722;
see also Y. Nambu, in: The quark confinement and field theory, eds. D.R. Stump and D.H. Weingarten (Wiley, New York, 1977).

[3] T. Eguchi, Phys. Rev. Lett. 44 (1979) 111.

[4] See H. Lamb, Hydrodynamics (Dover Publ, New York, 1945) p. 248

PHYSICS REPORTS (Review Section of Physics Letters) 104, Nos. 2–4 (1984) 237–258. North-Holland, Amsterdam

Concluding Remarks

Y. NAMBU

1. The paradigms of particle physics

The modern search for the fundamental constituents of matter started in the nineteenth century chemistry and physics but its real progress essentially belongs to the present century. By the early thirties, the basic structure of atoms had been established. One knew that there were electron, proton and neutron, the latter two being the constituents of atomic nuclei. All elements, out of which matter is composed, seemed in turn to be built up of just these three particles. Moreover, one also had a beautiful and successful theoretical framework, namely quantum mechanics, which was able to account for the dynamical properties of atoms in terms of the known electromagnetic forces among the constituents.

For the purposes of the present conference, however, this is a mere prehistory. In discussing what is now called particle physics (or elementary particle physics), I think I am allowed to say that it began around fifty years ago with the achievements of two physicists, E.O. Lawrence and H. Yukawa. These two physicists introduced the basic tools, one experimental and one theoretical, which have since served as the backbone of particle physics research. Today we are building accelerators with energies 10^6 times that of Lawrence's cyclotron, but the underlying principles are still the same. Yukawa postulated one new particle, the pion (as it is known today) to account for the nuclear forces. There are now hundreds of particles discovered since then, but we are trying to invent still more new particles and new forces.

These two methods, one experimental and one theoretical, have worked hand in hand, and have proved to be enormously successful. Will this continue to be so? Surely there is a practical limit to the scale of accelerators one can build, and that limit is already in sight. On the other hand, the power of theory and the imagination of theorists do not have an obvious limit. Whether by coincidence or not, the rapid progress of theory in the past ten years or so is outstripping the pace of experimental development. Fortunately (at least for the theorists) the very theoretical progress has enabled them to explore the realm of cosmology. They feel that they are now equipped to tackle the problems of cosmology, and use cosmological and astrophysical evidence to guide them. The theoretical side of particle physics is changing its character.

At this juncture it seems quite appropriate for us to reassess the basic paradigms of particle physics which have been developed over the past fifty years, and have proved so successful so far. I use the work paradigm in the sense defined by T. Kuhn [1]. According to Kuhn, a scientific achievement (or a body of achievements) becomes a paradigm if it is sufficiently unprecedented to attract an enduring group of adherents away from competing modes of activity, and simultaneously if it is sufficiently open-ended to leave all sorts of problems for the redefined group of practitioners to resolve.

The above description seems particularly relevant to the current state of affairs. I will, however, first look back over the entire fifty years, and identify what I think is the basic set of paradigms. These are:

1. Model building, mainly in terms of new particles, leading to an ever-deepening sequence of substructures of matter.
2. Quantum field theory with the addition of the concept of renormalization.

3. Symmetries and symmetry breaking as pervasive phenomena in nature.
4. Gauge principle and unification of forces.

Below I will elaborate on them.

1. Model building in terms of particles

In the present context, I am speaking of the line of thought represented by the works of Yukawa [2], Sakata [3], Gell-Mann [4] and Zweig [5], leading to the more recent activities in the grand unified theories as well as in the models at the subquark level.

The tradition of such model building itself is certainly not new. In fact it is the very basis of the search for the basic constituents of matter. But I have always felt that there is something new in Yukawa's approach. Up to that time, physicists had searched for the constituents of matter, matter that was available in ordinary life and was supposed to be permanent. The electron and the proton therefore had to be permanent, too. If one can build all the elements with these two, then that will be all there is to the structure of matter.

In Yukawa's case, on the other hand, he felt a theoretical need to invent a new particle in order to account for the dynamics of nuclear forces. Moreover, the new particle had to be unstable, because it was not found in ordinary matter. One cannot escape some uneasiness in postulating a fundamental constituent which is not permanent. But once this is accepted, Yukawa's method becomes a powerful tool in particle physics, and indeed it has led to an ever-increasing proliferation of new particles. Theory and experiment have gone hand-in-hand in confirming this.

The changed attitude toward the basic constituents – or elementary particles as they came to be called – is not an abstract invention. I think there are two elements that made it happen. One is the relativistic quantum field theory which not only led to a natural association of particles with forces (and vice versa), but also the possibility of pair creation and annihilation of particles. If one takes both quantum theory and relativity seriously, one must also accept these consequences.

The other element is the discovery of weak interactions, which actually goes back to earlier times. Radioactivity, especially the β-decay, darkly hinted that there were obscure forms of interaction which somehow made particles impermanent. Nowadays, we take it for granted that most particles are unstable. Only a few just happen to be stable, and these are the ones that form the ordinary matter. In the fundamental sense, though, we do not understand why this should be so.

It should be pointed out that in trying to resolve the problem of β-decay, Pauli was led to postulate (or invent) the neutrino, and Fermi gave an explicit field theory of β-decay. Certainly they were the forerunners who paved the way for Yukawa.

After Yukawa's success, his ideas were consciously developed by Sakata (who was a collaborator of Yukawa) into a methodology. This aspect has remained little known outside of Japan, and only recently has it begun to be appreciated [6]. As a person who was educated in those early years of particle physics in Japan, I must confess to a considerable amount of influence I (and my contemporaries) received from Sakata's philosophy.

Sakata is credited with the two meson hypothesis [7] which postulated the existence of two particles (today's pion and muon), a typical example of solving a puzzle by means of more new particles. He also expanded the ideas of Fermi and Yang [8], and proposed a composite model of hadrons in which the mesons were no longer elementary. But the decisive step forward in this direction belongs to the quark model of Gell-Mann and Zweig which placed the constituents at a lower level than the hadrons themselves.

The quark of Gell-Mann and Zweig has since undergone a gradual evolution. The color attribute was added out of sheer theoretical needs regarding quantum statistics [8a]. A fourth flavor (charm) was also anticipated for similar reasons, first motivated by the Cabbibo mixing [9] and later by neutral weak current phenomenology [10]. In the three generations of Kobayashi–Maskawa quarks [11] one again finds the predictive power of applying a simple logic to a subtle clue (*CP* violation).

We are now at the stage of three generations of quarks and leptons as fundamental fermions, but without any understanding of the origin of the generations. Moreover, this time evidence does not seem to support as yet, the existence of more generations, nor of the next layer of substructure. Instead, Yukawa's approach has been successfully transferred to the realm of weak interactions. The Weinberg [12]–Salam [13] theory has achieved for weak interactions what Yukawa has done for strong interactions, this time with the help of more sophisticated tools of field theory, to which I will now turn.

2. *Renormalization theory*

The difficulties in quantum field theory are inherently related to the infinite degrees of freedom of the fields, but one of them, the self-energy problem, already shows up in classical theory. Actually Weisskopf [14] discovered that the divergence of electron self-energy is less severe, i.e., only logarithmic, in quantum electrodynamics because of the polarizability of the vacuum. However it takes one more step to recognize the intrinsic necessity for renormalization, whether it is finite or not. The idea of renormalization seems to have developed in the thirties (e.g., Dirac), but for its execution one needed a fully covariant perturbation theory (as developed by Tomonaga [15], Schwinger [16], Feynman [17] and Dyson [18]) so that the identification of divergent but observable quantities such as mass and charge could be done unambiguously.

This is a rather conservative approach to quantum field theory. Contrast it with what seems to have been a widespread expectation in the thirties (as typically expressed by Heisenberg [19]) that quantum mechanics would break down at the scale of nuclear forces, that the latter must be handled by a new mechanics. [Heisenberg invented the *S*-matrix theory [20] with this limitation in mind. It is no coincidence that the *S*-matrix theory proved to be extremely useful in the fifties and sixties when the real origin of strong interactions was not known.] But the real progress was made by those who faithfully pursued the logical consequences of quantum field theory without radically altering it, as in the cases of meson theory, electron shower theory, and quantum electrodynamics. The radical side of progress was achieved not by changing quantum mechanics, but by means of the previous paradigm of particle model building, although the latter does not solve all the problems. An attempt to make electron self-energy finite by cancellation between two fields (Pais [21] and Sakata [22]) could not replace the renormalization theory. The recent supersymmetry, by coincidence, has some logical link to this problem.

In spite of the success of renormalization theory, I do not think anybody around that time was entirely satisfied with it. It was regarded as a temporary solution. [Tomonaga often referred to renormalization as the principle of renunciation.] Certainly this has been my feeling, too. I would like to suggest that it has taken virtually a generation, in real time as well as in the human sense, to shake off the inhibition and accept renormalization without reservation. After quantum electrodynamics, there have been two important events which may be called real progress. One is the application of renormalization ideas to statistical mechanics, developed largely by Wilson [23]. This supplied concrete examples and a sound physical basis to the concept of renormalization. The other is the discovery of asymptotic freedom in non-Abelian gauge theories by 't Hooft [24], Gross and Wilczek [25], and Politzer [26]. I think quantum chromodynamics is a real testing ground for renormalization theory. The emergence of a scale in an intrinsically scaleless theory, and the nonclassical antiscreening behavior leading to quark confinement (or the possibility thereof, at least) are its genuinely new consequences.

Quantum chromodynamics seems to be essentially the correct theory of strong interactions. Its main features, mentioned above, have stood quantitative experimental tests. Even the spectrum of hadrons has been successfully calculated in the lattice gauge theory formulation [27], although an adequate analytic description of the confinement mechanism is still lacking. This is all the more impressive if one recalls that after fifty years from Yukawa, still one does not have an ab initio derivation of nuclear binding energies (of deuteron and triton, for example) based on meson theory. It is an indication that the dynamics of quarks and gluons is simpler and more fundamental.

The essence of renormalization theory lies in the logarithmic dependence of parameters (like charge) on energy scale. In quantum electrodynamics, this dependence is positive, so one faces a potential breakdown of renormalization theory at extremely high energies. These energies happen to be in the realm of the Planck energy and beyond, depending on the number of flavors. This was first recognized by Landau [28]. The modern grand unified theories [29, 30, 31], with their asymptotic freedom, are free of such trouble. Yet they may still be said to follow the same spirit, namely to carry the renormalization theory to its logical conclusion. Moreover, the trouble comes back if the gravity is to be included. Thus those who are working on unified theories implicitly recognize a limit to renormalization theory without emphasizing it. Rather, the emphasis is on the validity of the theory until that limit is reached. We are coming back full circle to Heisenberg's fundamental length after all! But the length has been pushed to the Planck scale. [Heisenberg [19] did mention the Planck length.] Is the history going to repeat itself? That will be an interesting question.

3. Symmetries and symmetry breaking

Symmetry is basically a geometrical concept. Mathematically it can be defined as the invariance of geometrical patterns under certain operations. But when abstracted, the concept applies to all sorts of situations. It is one of the ways by which the human mind recognizes order in nature. In this sense symmetry need not be perfect to be meaningful. Even an approximate symmetry attracts one's attention, and makes one wonder if there is some deep reason behind it.

Symmetries, both exact and approximate, occupy an important place in physics. They have served very useful purposes. In the first place, symmetries imply conservation laws through Noether's theorem. Thus the homogeneous and isotropic nature of space-time guarantees the conservation of energy, momentum and angular momentum, which are among the most fundamental statements of physics. In the second place, physicists have been led to read meanings in approximate symmetries, too, and discover new order by analogy, so to speak. Flavor symmetry is a good example, I think. Whether flavor symmetry is a genuine symmetry or not, various flavors were found basically through such reasonings. The approximate equality of proton and neutron masses strikes one as something suggestive. Of course physicists are free to brush it aside as mere accident, or to take it more seriously, but the symmetry looks more meaningful if one knows that proton and neutron can transform into each other [32].

Are the two kinds of symmetry, one exact and the other approximate, basically of the same nature? Or is it that one is real while the other is illusory? I do not know the answer. It is customary to assume that when symmetry is not exact, it is due to a small perturbation. This procedure works well in general, and not only the symmetry itself, but the way it is broken becomes a meaningful question. In this view, symmetry is only an idealization which is realized if perturbations are turned off, or if only a part of the total Hamiltonian is isolated. So the symmetry is something of an artifact. One would be more satisfied if the perturbation can be externally controlled, but that is usually not the case, or at least it looks so.

Nevertheless, there are situations where the apparent lack of a symmetry actually results from underlying physical laws which are symmetric. This is because the symmetry of a Lagrangian, or an equation of motion, and the symmetry of a state are two entirely different things. The latter may be a member of a multiplet which as a whole represents the symmetry. A seeming paradox arises when the ground state of a system belongs to such a multiplet. An example for a finite system is the Jahn–Teller effect [33] in molecules, but the phenomenon becomes more dramatic for infinite systems, of which examples are plenty, like ferromagnetism and superconductivity. The occurrence of a nonsymmetric ground state (vacuum) of a system is what is usually called spontaneous breakdown of symmetry. It implies that the ground state is actually degenerate. In an infinite system the degeneracy may be infinite. Each ground state generates a tower of excited states, and different towers generated from different ground states are orthogonal. Since an infinite system usually occupies an infinite volume, it is operationally impossible to change the ground state from one to another by local perturbations. Thus one lives in a world built upon a particular vacuum, and cannot see the others. This makes the spontaneous breakdown a conspicuous phenomenon, if one somehow knows the existence of the other worlds, but a very inconspicuous one if one doesn't. In the latter case, the only remnants of the lost symmetry are the Goldstone modes [34, 35]. But the former case also happens very often, as in phase transitions between symmetric and asymmetric phases and in the formation of domain structures.

The spontaneous breakdown can occur in ungauged symmetries as well as in gauged ones. I must postpone a detailed discussion on the latter, but the Goldstone modes appear very differently in these two cases, i.e., in the gauged case the induced zero mass modes can mix with the gauge fields, thereby lifting the mass to a finite value, as in the familiar plasma phenomenon.

The idea of spontaneous breakdown has been extensively applied to particle physics. Mostly it has to do with weak interactions and flavor dynamics, as in the creation of mass for the W and Z boson according to the Weinberg–Salam theory. The fact that the effect appears in flavor dynamics is not surprising, because it is the domain which looks most irregular. It does not seem to respect any symmetry at all, but the ultimate cause for this remains unknown.

One is thus led to the question whether spontaneous breakdown alone is sufficient to explain all the symmetry violations, assuming that the symmetries are real. The answer seems to be no, in my opinion. For the renormalizable field theory contains built-in mechanisms for inducing symmetry breaking which are characteristically quantum effects. They are: the appearance of a scale parameter due to renormalization, breaking of chiral invariance due to chiral anomaly, and the topological effects due to instantons, monopoles, etc., and perhaps some more. Before discussing this, however, I will turn to the next topic.

4. *The gauge principle*

By gauge principle I mean the view that dynamics is derived from certain geometrical principles, modeled after the two prototypes, Maxwell's electromagnetism and Einstein's general relativity. In its narrow sense, it covers the gauge theories of Maxwell and Yang–Mills type (Abelian and non-Abelian), but Einstein's theory of gravity is the most geometrical of all in the conventional sense of geometry. The geometrical view of dynamics actually extends beyond these gauge theories. Supersymmetry, for example, can be formulated in geometrical terms. Unified field theories are based on the view that all forces (or interactions) in nature follow from a single, comprehensive principle, and are subject to a unified description. The developments of the last twenty years or so have given us the hope that the gauge theory will accomplish this.

The idea of unification of course dates back to much earlier times. Inspired by Einstein's gravity theory based on space-time geometry, Weyl [36] and Kaluza [37] respectively made an attempt to incorporate electromagnetism by enlarging the Einstein geometry. Einstein himself devoted his later years to the search for a correct unified theory. All these efforts failed, but we owe them some important concepts like gauge transformation and higher dimensional space.

Actually, the course of events in particle physics since the 1930's evolved in the opposite direction from the ideals of the unified theory. As was already discussed, more particles and more complicated phenomena turned up both in strong and weak interactions. Clearly gravity and electromagnetism were not everything, but the other forces did not show any trace of the characteristics of a gauge theory. Nevertheless, this was the period when more experimental knowledge was accumulated in preparation for the theoretical developments of the 1970's. Also some new theoretical ideas like non-Abelian gauge fields were born.

An important characteristic of a gauge theory is that dynamics is inherently related to certain conserved quantities. As was once remarked by Wigner, there are two operationally different ways to establish a conservation law. One is by means of selection rules in various reactions, and the other is by directly measuring the conserved quantities. The latter is possible only for the gauged symmetries, which are accompanied by long range fields coupled to the conserved charges. When Yang and Mills [38] discovered a generalization of Maxwell's theory to the non-Abelian SU(2) gauge group, its relevance to physics was not clear because an exact non-Abelian symmetry of this kind, or a corresponding long range field, did not seem to exist. The only candidate for an exact conservation law beside energy-momentum and electric charge was that of baryon number, but the associated Abelian gauge field did not exist either, as was pointed out by Lee and Yang [39]. In fact, according to the current viewpoint, the only exact symmetry is a gauged symmetry; hence one does not expect the baryon number to be exactly conserved, in spite of the extremely high stability of matter. The current degree of interest shown by the physicists in proton decay experiments is an indication of how much the gauge principle has taken hold in physicists' minds.

Beside Yang and Mills, the contributions of Utiyama [40] deserve mention. He independently developed the idea of non-Abelian gauge fields, and recognized that Einstein gravity can also be derived from the same principle. Utiyama classifies interactions in field theory into two kinds: the first kind which follows from the gauge principle, and the second kind which does not, like the Yukawa interactions. Logical preference of the first over the second is obviously implied. Equally obviously, the Yukawa type interactions do exist among the hadrons. From the current viewpoint, however, one had better differentiate the two types as primary and secondary. The Yukawa interaction is a secondary and phenomenological interaction among mesons and baryons, which are composite systems formed out of quarks due to the primary interactions with the color gauge fields. Unfortunately, this redefinition is not quite satisfactory at the moment, because the Weinberg–Salam theory of weak interactions still contains both types as elementary interactions. The modern grand unified theories (GUTs), with their Higgs fields and couplings, would not satisfy gauge theory purists (to whom I am now sympathetic).

The non-Abelian gauge theory developed in several stages after its introduction. To be relevant to strong and weak interactions, it was immediately obvious that the gauge fields had to be massive and their symmetry nonexact. One way was to assume them blindly and identify the gauge fields with vector mesons carrying various flavor quantum numbers. This approach certainly served useful and enlightening purposes as in the vector meson dominance model [41] of strong interaction and the early attempts at the $SU(2) \times U(1)$ weak interaction [42]. On a more theoretical side, the plasma phenomenon as an example of mass generation in electromagnetism was readily recognized, but a complete understanding of the mechanism was achieved only when it was combined with spontaneous breakdown. As it turned out, superconductivity served as an ideal model for building a theory of weak interactions. The

Ginzburg–Landau [43] formulation of superconductivity is a direct prototype of the Higgs model [44] for gauge field mass generation, whereas the Weinberg–Salam model uses also a BCS-like mechanism [45, 33] for fermion mass generation, although this part does not amount to a spontaneous breaking of symmetry.

Next important steps in non-Abelian gauge theory had to do with its quantum version. The discovery of the asymptotic freedom opened up the possibility that the color gauge theory of quarks is the correct theory of primary strong interactions. But for this to happen, a sophisticated mathematical machinery [46] for non-Abelian quantum gauge theory had to be developed. This machinery then was also found to render the Weinberg–Salam theory renormalizable even after spontaneous breakdown [47].

From here to the grand unification, the step is a very logical and very significant one. Because of the different behaviors of Abelian and non-Abelian theories regarding the scale dependence of renormalized coupling constant, the apparently different strengths of strong, electromagnetic and weak gauge fields can merge at high energies, and in fact they seem to do so at the same scale point $\sim 10^{15}$ GeV. Thus a unification of the three nongravitational forces that we know has been made possible through a combination of three elements: non-Abelian gauge theory, renormalization theory, and spontaneous symmetry breaking.

Another important development concerns the topology of gauge fields. It started with Dirac's theory of monopoles [48]. He introduced configurations of gauge fields which are topologically nontrivial, and these configurations are characterized by a topological quantum number, i.e., magnetic charge. The gauge theory becomes a theory in the general mathematical framework of fibre bundles, which the mathematicians developed independently. Essentially the same topological problems arise in the real example of superconductors with frozen-in magnetic fluxes, as was noted by London [49], but it was Aharonov and Bohm [50] who illustrated in a profound way the intimate relations between geometry and gauge field in quantum mechanics. Although the Aharonov–Bohm effect may nowadays look self-evident to the theorist, it is gratifying that a clear experimental demonstration of the effect has recently been achieved [51]. The topological phenomena become richer when we include non-Abelian gauge fields. We now know that there are configurations like walls, strings, monopoles and instantons [52–55]. Unlike the Dirac monopole which was artificially introduced, these configurations are natural solutions of gauge field equations (in interaction with suitable matter fields except in the case of instantons), and so one must regard them as predictions of gauge theory.

Of course topological questions are very old and more extensively studied in the case of gravity which deals with the topology of space-time itself. Particle physics is simply rediscovering these phenomena which are characteristic of a geometrical principle.

2. Problems and issues

In this second section I will discuss specific problems and issues we are facing at the moment. My intention is not to make a comprehensive review of the situation, but to give a general perspective and then focus attention on certain individual problems I am interested in. From the nature of our endeavor it is inevitable that the problems cannot be restricted to those specific to particle physics and related subfields, but will extend to a more general domain of physics.

There are numerous problems and questions, open and unresolved, and they may be categorized in various ways. For example, there are experimentally established facts which remain to be explained or understood, and there are problems which are created by a theory itself. From a different viewpoint, one may divide the problems into those of substantive nature and those concerning theoretical framework, although the distinction is not necessarily exclusive. Taking these into account, I will group the problems as follows before taking them up individually.

(a) General substantive questions.

What kinds of particles and fields actually exist in nature?
What are their interactions?
What are the basic symmetries?
Are there new forces, fields or particles that fall outside of current paradigms?

These are very general questions, and essentially summarize the entire goals of particle physics proper. Therefore there is actually not much one can discuss without coming to more specific issues.

(b) Specific questions that are both substantive and theoretical. These include some of the most elusive but urgent questions one has at the moment:

Origin of mass and mass spectra in general.
The problem of generations.
The problem of hierarchy.
The problem of subconstituents.
Primary origin of symmetry breaking.

(c) Questions concerning theoretical framework. These are questions generated by our theoretical endeavors, some of which may turn out to be irrelevant to physics, but others may lead to profound new discoveries.

General questions on quantum mechanics.
Limits of quantum field theory, characterized by renormalizability, locality, etc.
Topological excitations and other nonperturbative effects.
Chiral anomaly and related phenomena.
Quantum mechanics of extended objects.
Higher-dimensional space-time.
Supersymmetry and supergravity.
Grand unification and ultimate unification.

It is not possible to give an orderly discussion of these problems listed above because they are intricately connected with one another. On the whole, however, the task will be easier if one starts from general questions concerning theoretical framework, and proceeds to more specific issues in particle physics, but not necessarily sticking to a logical order.

1. General questions on quantum mechanics

As was mentioned earlier, there was a time when breakdown of quantum mechanics was anticipated by some people at the nuclear length scale of 10^{-13} cm (10^8 eV in energy). However, it is not my intention to question quantum mechanics itself, which is one of the most beautiful and profound manifestations of physical laws. The various attempts at an alternative to quantum mechanics do not seem very relevant or to serve constructive purposes.

My first question concerns the two formulations of quantum mechanics: the canonical (Heisenberg–Schrödinger) formulation and functional integration (Feynman) formulation. Actually a third formulation was recently introduced by the name of stochastic quantization [56], about which I am not prepared to say much. So, restricting myself to the first two, they are generally regarded as equivalent, at least in a naive sense. However, one may assert that the functional formalism is, in many ways, more general than the canonical one. To illustrate the difference between them, it is instructive to appeal to the similarity between quantum action in the functional formalism and the partition function in statistical mechanics. The former is schematically given by

$$Z = \int \exp\left[\frac{\mathrm{i}}{\hbar} I\right] \mathrm{D}[\psi]$$

where I is a classical action integral involving dynamical variables ψ. If the degree of freedom is infinite, as in the case of field theory, the usual quantum mechanics is derived from this by restricting the function space to the vicinity of a classical extremum, which corresponds in statistical mechanics to a particular thermodynamic phase. Clearly the original function space is much larger and richer in general, and can accommodate many different phases and transitions between them. The canonical formalism deals only with a local tangent space of the function space.

This point is well known nowadays, and the actual richness of the entire function space has been uncovered and explored in field theories in examples of spontaneous breakdown, topological excitations, etc., some of which will be further discussed below. At this point, however, I would like to pose questions bearing on the interpretation of quantum mechanics. Traditional discussions on the fundamentals of quantum mechanics, such as the problem of measurement, are mostly based on the canonical formalism. I do not know whether or not the functional formalism can shed more light on them, but it can complicate the issues further. For example, the conventional Hilbert space of physical states applies to a tangent space only, and the questions like completeness and unitarity do not seem to be obvious in the larger functional space. So my proposal is to examine these old and new problems seriously.

Another point I would like to bring up is the above mentioned analogy between quantum action and statistical partition function. In statistical mechanics, one knows that the thermodynamic concepts like entropy and temperature have fundamental and profound significance. Very general and important results are often obtained from thermodynamic principles only. It is possible that one can apply similar reasonings to quantum action not as a mere analogy, but in a more constructive sense. The correspondence here is, roughly speaking: energy $E \to$ classical action I_C, free energy $F \to$ quantum action I_Q, temperature $T \to$ Planck constant $\hbar$, entropy $S \to$ quantum corrections represented by S_Q, with the relation

$$F = E - TS \to I_Q = I_C - \hbar S_Q .$$

In gauge theories, the situation becomes much more interesting because one can also interpret the coupling constant g^2 (κ^2 in the case of gravity) as temperature, and g^2 is actually a variable. The above remarks are not new, and the analogy has been frequently invoked in gauge theory problems. What I am advocating is a matter of more general principle. I will come to a concrete discussion about it later on.

2. *Limits of renormalizable field theory*

In the subsection on renormalizable theory (RT) in section 1 no discussion was made on the quantum field theory as a whole, so it seems appropriate now to start by commenting on this general subject. Relativistic quantum field theory (QFT for short) is a theoretical framework whose ingredients are local fields, with local interactions, satisfying the principles of relativity and quantum mechanics. The axiomatic structure of QFT and its consequences have been worked out over the years. Among them are some fundamental theorems like spin-statistics connection, TCP, and dispersion relations. On the basis of such general theorems, QFT has firmly stood experimental test. Details need not be discussed here, but one notes that many of the successful results came after the RT was established, although the latter deals with dynamical questions, and so is not a proper subject for an axiomatic approach.

QFT is a rather rigidly constrained framework. It is practically impossible to conceive of a more general field theory (for example, a nonlocal theory) which does not run into conflict with some of the axioms and thus with sound physical principles. Renormalizable field theories are those which are manifestly compatible with these axioms at least in a perturbative sense. One sees immediately the difficulty of finding a substitute for RT.

Nevertheless, RT is clearly inadequate in dealing with quantum gravity. Here one sees manifest limitations of RT and perhaps QFT as well. The fact that this limiting scale, the Planck length 10^{-33} cm or the Planck energy 10^{19} GeV, is beyond the domain of grand unified theories is significant. One may trust the latter as an extrapolation from our current knowledge of particle physics, while at the same time one knows its limit. I would like to regard this as a situation that is to be welcomed, comparable to the difficulties encountered in reconciling atomic phenomena with classical physics. Unfortunately, however, this limit is far off our accessible scales, and there is quite a lot of unknown ground to explore before that goal is reached.

Setting aside this ultimate problem, there are still others to be resolved or understood in QFT and RT. First of all, RT must be satisfactorily extended to nonperturbative domains. Nonperturbative effects manifest themselves in strong coupling regimes and nontrivial topological configurations of fields. Lattice theories have been devised to handle the former. Another problem concerns the so-called anomalies, which needs to be better understood. Let me discuss these separately.

(a) *Lattice theories.*

The lattice formulation of QFT is not only a mathematical device, but it also gives one a deeper insight into the underlying principles, especially in the case of gauge fields. At the same time, it sacrifices the continuity of space-time and the associated invariance properties, assuming that these will be recovered in the end by a limiting procedure. The validity of this assumption, however, is not entirely obvious.

A gauge field is a fibre bundle: at each point in space-time, an abstract compact space is associated with it, representing, so to speak, orthogonal degrees of freedom to space-time when the latter is embedded in a larger manifold. [Similar situations arise for other fields when constraints are introduced.] As a simple illustration, take a cylinder extending in the x direction and with its cross section in the y–z plane, where x is the space dimension in a $1+1$ dimensional space-time. The y and z dimensions symbolize extra degrees of freedom. Suppose a mass point is allowed to move on the surface of the cylinder $y^2+z^2=a^2$. It will in general have a spiral path, with its momentum having both x and angular components. In quantum mechanics, the angular momentum is quantized so that the energy is given by

$$E=[p^2+p_\theta^2]^{1/2},$$

$$p_\theta=n/a\,,\qquad n=0,\pm1,\pm2,\ldots$$

Now introduce a gauge field so the points on the cross section of the cylinder surface at fixed x are all gauge equivalent. The angular motion then becomes unobservable, which is equivalent to letting $a \to \infty$ in the above equation. The resulting infinite degeneracy is an unphysical one, i.e., there are no physical gauge field degrees of freedom in one space dimension.

In a lattice version, one would replace the cylinder by a set of circles repeated in the x direction with step distance b. A classical trajectory with pitch angle

$$\tan\theta = 2\pi na/b, \qquad n \neq 0$$

will be indistinguishable on the lattice from the straight one with $n = 0$. So the energy spectrum becomes different from the continuum case. In the gauged theory, on the other hand, cases with $n \neq 0$ clearly have intrinsic topological significance. One may thus add, if one wishes, an extra gauge invariant piece in the Lagrangian to give such configurations extra energy, which would not be possible in the continuum case. It therefore appears that there is a qualitative difference between continuous and discrete theories. The lattice version has richer physical potentialities. But the price one has to pay for this is the loss of Poincaré (especially Lorentz) invariance. The continuum limit $b \to 0$ would become ambiguous if one wants to retain nontrivial topological features.

(b) *Topological excitations.*

Here I will consider only those topological configurations associated with gauge fields, although similar phenomena are known to occur much more generally. One can classify the configurations by the dimensionality of the topological singularity: wall, string, pole and instanton (dim 3, 2, 1 and 0). If they are to be natural solutions of a local field theory rather than extra input, one actually needs auxiliary fields (Higgs fields) to give them a scale parameter except in the case of instantons. This is understandable from the previous discussion on lattice formulation. These solutions are thus solitons characterized by topological quantum numbers.

Much is known about topological solitons at the classical level, but their quantum theory is still at a very primitive stage. Each topological soliton is a nontrivial extremum in the function space, and one can study the quantum fluctuations around it. However, one has also to be able to handle arbitrarily many solitons, and their creation and annihilation processes. Since each soliton is well localized, one would think that it is possible to define an effective Lagrangian in which quantized fields of solitons are introduced (when one is dealing with long distance phenomena) along with the original fields. This is not an easy task. In the case of monopoles, charges and poles should be represented by respective local fields satisfying certain duality relations, but a satisfactory field theory of this kind remains to be found.

Charges and poles are related by the generic relation of Dirac

$$eg = \tfrac{1}{2}n\hbar c\,.$$

For non-Abelian gauge theories, this becomes generalized to relations between two dual gauge groups, electric and magnetic [57]. If an effective local field theory should exist, it would have to realize such relationships. Would it also enable one to carry out the renormalization of charges and poles? One expects that the two renormalizations are reciprocal so as to preserve the Dirac condition. This would imply however, that the magnetic gauge group is asymptotically non-free, when the electric gauge group is asymptotically free (and vice versa), except for the neutral case.

Important and interesting among quantum effects in topological excitations is the spectrum of quantum fluctuations, especially that of the zero modes. In general, they reflect the degeneracy, and hence the entropy, associated with a particular topological excitation. As was pointed out before, quantum effects are essentially entropy effects, so the importance of topological excitations depends critically on their entropy as well as action. In quantum chromodynamics, for example, the vacuum state is often pictured as containing large color fluctuations due to virtual poles and instantons. Quark confinement is a result of the magnetic Meissner effect of such a medium.

I will illustrate the above mentioned thermodynamic analogy with an example. In a gauge theory, the (Euclidean) quantum action integral over a 4-volume V (with appropriate boundary conditions) takes the form

$$Z = \int \exp\left[-\frac{1}{g^2}\int_V G^2 - \int_V \mathscr{L}_{\text{matter}}\right] \mathrm{D}[A/g]\,\mathrm{D}[\phi]$$
$$\equiv \exp[-F/g^2]$$

where G stands for the gauge field, A/g for the gauge potential, g for the coupling constant, and ϕ for the matter field. Then

$$F = E - g^2 S\,, \qquad E = \int_V G^2\,,$$

E being the gauge field action evaluated at a functional extremum, and S the entropy computed with a suitable gauge fixing convention. The matter field contributions are entirely in S. Clearly, F, E and g^2 play the roles of thermodynamic free energy, energy and temperature: $g^2 \to T$. Furthermore, one can somehow introduce the concept of renormalization. For example, consider the volume V to be measured in units of a scaling variable V_0, so that $V \to V/V_0$ and suppose that the free energy remains invariant under correlated changes of V and T (with fixed intrinsic degrees of freedom). The condition is thus

$$\mathrm{d}T/\mathrm{d}V = -p/S\,.$$

In particular, make an ideal gas type ansatz

$$pV = RT\,, \qquad E = \alpha RT\,, \qquad (C_p/C_V = 1 + 1/\alpha)$$

in which case the condition reads

$$-F = RT\ln(T^\alpha V/V_0) = \text{const.} = cR\,.$$

The scaling variable V_0 may be replaced with the usual running mass μ, so that $V/V_0 = k\mu^4$. In the small T and large T limits, the above relation reduces to

$$c/T \approx \ln\mu^4 \qquad (T \ll 1)$$

$$kT^\alpha \approx \mu^{-4} \qquad (T \gg 1)\,.$$

The first is the correct weak coupling behavior, whereas the second agrees with the strong coupling result (in the Hamiltonian version) if $\alpha = 2$. [Actually c and α determine the first two terms of the Callan–Symanzik β function in weak coupling expansion.]

The above exercise is to be regarded more as an observation than as a derivation. What I am trying is to seek different or alternative interpretations of renormalization. In a more formal approach, one might introduce a scale parameter μ and its conjugate as new thermodynamic variables. The statement of renormalization is then that a certain thermodynamic function is kept fixed: $d\phi = dT\, \partial\phi/\partial T + d\mu\, \partial\phi/\partial\mu = 0$. But some physical insight into the meaning of μ is needed. In the present example, one might argue that in doing the functional integration, typical configurations are thought to be made up of local excitations of certain size V_0, which contribute a volume entropy $\sim \ln(V/V_0)$ and a kinetic entropy $\sim \ln T$. It is conceivable that in a correct theory the entropy has a finite limit >0 as $T \to 0$. Incidentally, a similar calculation can be done for an instanton gas.

(c) *Chiral anomaly and relative phenomena.*

The chiral anomaly [58, 59] is a rather anomalous or pathological quantum effect which spoils renormalizability. Its physical meaning is not very clear, but it is an important effect in the sense that it touches on the general question of renormalizability and possibly serves as a clue to other problems.

In a theory in which the Lagrangian involving fermion fields have chiral invariance, the quantum action can in general produce a nonconservation of the chiral Noether current when the fermions are coupled to gauge fields. Consider a Dirac field with zero mass so that the left-handed and right-handed currents are separately conserved. In the presence of a given gauge field, one can evaluate the quantum fluctuations of the Fermi fields by finding the eigenvalues of the Dirac operator $\gamma_\mu D^\mu$ for each of the chiral components. This operator, however, converts the Lorentz spinors $D(\frac{1}{2}, 0)$ and $D(0, \frac{1}{2})$ into each other, so the eigenvalue equation must couple two fields of opposite chiralities or a field and its complex conjugate, and they must belong to the same representation of the gauge group. When such matching partners are not available, renormalization cannot be made consistent with both Lorentz and gauge invariance. This is a strong constraint on possible renormalizable theories, and can be exploited in various ways. It is remarkable that nature appears to have chosen a highly nontrivial way of satisfying the requirement, at least according to the standard SU(5) type theories.

When the renormalizability criteria are met, there can still be chiral anomalies in ungauged symmetries in the presence of CP noninvariant gauge field configurations. In other words, the gauge field behaves like having a piece of chiral (helicity) current

$$\chi_\mu = c \operatorname{tr} \varepsilon_{\mu\nu\lambda\rho}\left(A_\nu G_{\lambda\rho} + \frac{2i}{3} A_\nu A_\lambda A_\rho\right)$$

$$\partial_\mu \chi_\mu = c(\operatorname{tr} G_{\mu\nu}\, {}^*G^{\mu\nu} + 2A_\mu k^\mu),$$

$${}^*G_{\mu\nu} = \tfrac{1}{2}\varepsilon_{\mu\nu\lambda\rho} G^{\lambda\rho}, \qquad k^\mu = D_\lambda\, {}^*G^{\lambda\mu}.$$

Note that $\partial_\mu \chi^\mu$ has a contribution from monopoles. When a term $\theta c \operatorname{tr} G\, {}^*G$ is added to the Lagrangian in the so-called θ-vacuum, the monopole acquires an effective charge $2c$ due to this situation [60].

The origin of the chiral anomaly can be traced to the existence of zero eigenvalues (zero modes) of the Dirac operator. This is well known for the instanton configurations, but recently the anomaly associated with monopoles have become an important issue as they seem relevant to baryon number nonconservation [61–63]. In general, zero modes imply some kind of degeneracy or symmetry. Thus, the fermion zero modes seem related to some kind of supersymmetry. This point needs to be explored further. I will come back to it later. For the moment, my emphasis is on the fact that anomaly is a widespread phenomenon with important physical consequences.

3. Quantum mechanics of extended objects

Extended objects can arise in local field theory as collective modes of various kinds, including the topological excitations. The hadrons, for example, behave like extended systems, and this stimulated various attempts at a description of hadrons as nonlocal objects either in a phenomenological or in an intrinsic sense. Actually there is no reason to believe that the local field theory cannot describe all the phenomena, but there is a theoretical challenge in trying to find a generalization or an alternative to local field theory.

In this sense the most interesting example is the theory of strings [64], namely one-dimensional extended systems (or two-dimensional in space-time). Originally an outgrowth of the dual resonance model of hadrons, later it came to be regarded as an idealization of the long distance behavior of quantum chromodynamics. From this point of view, a string represents greatly reduced degrees of freedom of gauge fields. It also represents an idealization of magnetic flux tubes [65], which are of topological nature. In any case, the question is whether one can go backwards and arrive at a consistent quantum theory by keeping only the geometrical degrees of freedom of the ideal string.

So far this program has failed. In general, the quantization is inconsistent with covariance except when the system is embedded in a space of critical dimensionality (26 for the simple strings). Besides, there is a problem of tachyons in the mass spectrum. It is not clear that the recent elegant reformulation of string theory by Polyakov [66] will resolve these difficulties. Essentially, the underlying cause of the difficulties seems to be that a mathematical string fluctuates in an uncontrollable way. Maybe one cannot have a fully consistent quantum theory of extended systems, and only local fields exist at the fundamental level.

Nevertheless, the string theory poses three questions which can still be meaningful: (a) the string model of hadrons works up to the semiclassical level as is shown by the mass spectra (Regge trajectories) and the recovery of a Coulomb-like interquark potential as a quantum (Casimir) effect [67]. As the conventional Hamilton formalism of mass points deals with the field of geodesics, so can one define a kind of gauge (skew tensor) field describing the field of minimal surfaces. Aside from its mathematical interest, this may serve as a theory of hadrons intermediate between string model and QCD. (b) One can associate a new gauge principle and new kind of gauge field to strings, just as the conventional gauge fields are naturally associated with point sources. This Kalb–Ramond gauge principle [68] is again interesting not only for showing the existence of a hierarchy of gauge principles, but also because they may actually appear in nature in one form or another. (c) Because a string forms a generally nonflat two-dimensional space-time manifold, the string theory shares many of the features of the gravitation theory. In a recent revival [69, 83] of the ideas of Kaluza and Klein, even our space-time itself has been considered as being a submanifold in a space of higher dimensions. I will elaborate a little further on the above points (b) and (c).

The Kalb–Ramond gauge principle has a natural physical interpretation as a relativistic hydrodynamics of vortices. The antisymmetric second rank tensor potential A and the third rank field V derived from it correspond to velocity potential and velocity field of a fluid generated by (closed) vortex strings [70]. [The interpretation is not perfect because the velocity is unbounded.] They satisfy the equations

$$V_{\mu\nu\lambda} = \partial_\mu A_{\lambda\nu} + \partial_\lambda A_{\nu\mu} + \partial_\nu A_{\mu\lambda}$$

$$\partial_\lambda V_{\lambda\nu\rho} = \mathscr{S}_{\nu\rho}\,, \qquad \partial_\nu \mathscr{S}_{\nu\rho} = 0$$

where $\mathscr{S}$ represents a Dirac string field. The velocity field mediates a hydrodynamic interaction between vortex strings, and the quanta of the field are massless scalar (zero helicity) particles.

Why should nature not avail itself of this beautiful theoretical possibility? If such a field does not exist, is it because the source $\mathscr{S}$ is intrinsically nonlocal and cannot be elementary objects like charged particles? Or it may be because the field becomes massive by a spontaneous symmetry breaking, in which case the theory reduces to a description of conventional magnetic flux tubes and hadron strings. However, it is still conceivable that vortices and accompanying long range fields actually exist somewhere, at a scale which could be either microscopic or macroscopic, or even cosmological.

Turning now to the modern versions of Kaluza–Klein theory, the ordinary space-time and the space of internal symmetries (gauge groups) are considered as different subspaces of a single higher-dimensional space. In some models this differentiation of the two subspaces is a result of spontaneous symmetry breaking. The world is like the interior of a flux tube which has four large longitudinal (space-time) dimensions, and its small cross section (of Planck length scale) is the compact space of internal symmetry. The ordinary particles (< Planck mass) are confined within the tube, but there can be modes that leak out. Two comments are in order.

First of all, if this type of theory is to be meaningful, one would like to have it explain the absence of the cosmological term in gravity equations, which in the string analogy is equivalent to vanishing string tension. Therefore the dynamics must be rather different in these two cases. Secondly, various topological invariants (quantum numbers) are available to characterize a solution and ensure its stability. Thus a flux tube with two-dimensional cross section and N longitudinal dimensions has a topological invariant

$$\int F^{(2)}_{12}\, \mathrm{d}x_1\, \mathrm{d}x_2$$

where $F^{(2)}$ is an SO(2) (U(1)) gauge field, x_1 and x_2 are the coordinates of the cross section. Similarly, for cross sections of dimension $M = 4$, 6, etc., the respective invariants will be

$$\int \mathrm{tr}\, F^{(4)}_{ij} F^{(4)}_{kl} \varepsilon^{ijkl}\, \mathrm{d}^4x\,, \quad \int \mathrm{tr}\, F^{(6)}_{ij} F^{(6)}_{kl} F^{(6)}_{mn} \varepsilon^{ijklmn}\, \mathrm{d}^6x, \text{ etc.}$$

For example, a 10-dimensional space can be reduced in this way to $N = 4$, $M = 6$. For odd dimensions, one can bring in an extra factor of the potential A to form an invariant. Similar constructions using higher rank tensors also seem relevant [71].

4. Supersymmetry

Supersymmetry [72] is a beautiful but rather puzzling subject. What does it mean: Is there a simple way to understand it in terms of conventional physical concepts? The answer is not clear. But because of this novelty, there is a real possibility that supersymmetry will supply a key to the unsolved problems of particle physics, and thus become a new paradigm. So far, however, the relevance of supersymmetry has not been demonstrated clearly in spite of the intensive efforts of model building based on the idea.

The trouble is that no recognizable supersymmetry structure, even an approximate one, exists among the known particles. One might be able to make more progress if one had a better physical understanding of its origin, for example on the basis of dynamics. With this in mind, I will discuss a few model systems that seem to bear on the subject one way or another.

(a) *Supersymmetric oscillator and Wigner–Yang oscillator.*
In a nonrelativistic version of supersymmetry, the Hamiltonian is the square of a supersymmetry operator:

$$H = Q^2$$

so that the spectrum of H is nonnegative, and usually Q has zero eigenvalues. Consider, for example, a one-dimensional system with spin:

$$Q = \sigma_3 p + \sigma_1 W(x), \qquad p = -\mathrm{i}\, \mathrm{d}/\mathrm{d}x$$

$$H = Q^2 = p^2 + W^2 + \sigma_2 W'(x).$$

If $W = x$, this represents a harmonic oscillator plus a spin term, resembling the case of an electron in a magnetic field. To see if Q has zero eigenvalues, set

$$Q\psi = 0$$

or

$$\mathrm{i}\sigma_3 Q\psi = [\mathrm{d}/\mathrm{d}x - \sigma^2 W]\psi = 0,$$

$$\psi = \exp\left[\sigma_2 \int_{x_0}^{x} W\, \mathrm{d}x\right]\psi_0.$$

This function will be normalizable if the exponent $\to -\infty$ as $x \to \pm\infty$, which will be the case if W is an odd function ($W(\pm\infty) \neq 0$) and a proper sign of σ_2 is chosen. The previous harmonic oscillator satisfies this, so it has a zero eigenvalue.

The above Hamiltonian is a prototype of supersymmetric Hamiltonians widely discussed in field theory, but next consider the following. Wigner once asked the question [73]: Is the Heisenberg canonical commutation relation $[p, x] = -\mathrm{i}$ uniquely determined from the requirement that the classical and quantum equations of motion be formally identical? Wigner found that the commutation relation was not unique in the case of simple harmonic oscillator $H = \frac{1}{2}(p^2 + x^2)$. Namely one could set in general

$$[p, x] = -\mathrm{i} - \mathrm{i}F, \qquad [p, F] = [x, F] = 0$$

because then

$$\mathrm{i}\dot{x} = [x, H]_- = [[x, p]_-]_+ = 0 \qquad \mathrm{i}\dot{p} = [p, H]_- = [[p, x]_- x]_+ = -x.$$

One can realize F by setting

$$F = \alpha R , \qquad p = p_0 - A , \qquad A = \tfrac{1}{2} \mathrm{i} F/x , \qquad (p_0 = -\mathrm{i}\, \mathrm{d}/\mathrm{d}x)$$

where R is the parity operator. Since $p \pm \mathrm{i}x$ are still ladder operators, the spectrum is unchanged except that all the levels are shifted by $-\alpha/2$ (if $0 < \alpha < 1$) so the zero-point energy can be made arbitrarily small.

The Wigner–Yang Hamiltonian is not quite supersymmetric, but seems somehow related to it. The parity operator plays a role somewhat analogous to spin. The harmonic potential can be generalized to any reasonable even function. Unfortunately, however, it seems difficult to extend it further to many oscillators (except in a trivial way).

(b) *Fermionic zero modes associated with topological excitations.*

In general, classical topological solutions of field theory are accompanied by zero modes of various kinds, and these reflect the symmetry breaking aspect of the classical solutions. But what symmetries are these zero modes trying to restore? That is not always obvious. There might be hidden symmetries that are broken. Here I will restrict myself to fermionic zero modes because they may shed some light on supersymmetry, i.e., the degeneracy of mass spectra between bosonic and fermionic states. It is well known that instantons have fermionic zero modes, which are related to the index theorem and also are a cause for the anomaly. But the main focus of discussion will be on other objects like monopoles, vortices, and kinks. They again possess zero energy solutions for Dirac equations when the mass term is generated by the same Higgs mechanism that produced the topological objects [74].

This situation is very similar to that of the supersymmetric Hamiltonian discussed in (a) above. Consider in particular a Dirac Hamiltonian in one space dimension coupled to a Higgs field having two-fold degenerate vacua. It is precisely the supersymmetry operator Q if W is read as the Higgs field. For a kink, W is an odd function of x with $W \to \pm W_0$ as $x \to \pm\infty$, which satisfies the zero mode condition. In the case of vortices and monopoles, the situation is somewhat complicated. There is a non-Abelian (SU(2)) gauge field, but essentially the same mathematical structure emerges when the Dirac equation is reduced to a radial equation. Moreover, as the isospin is correlated with spatial direction through the topology, it gets incorporated into angular momentum, a remarkable phenomenon which also changes the statistics of the monopole-particle system accordingly [74].

Recently it has been argued that this coupling of spin and internal spin also leads to a baryon number anomaly [75–77]: an SU(5) monopole induced baryon number changing processes like $\mathrm{d} + \mathrm{u} \to \mathrm{e}^+ + \mathrm{u}^-$. A rough explanation of the phenomenon goes like this: Different physical states in a fermion multiplet are determined with respect to a local frame of internal spin, which varies around a monopole, and becomes ambiguous at the origin. Therefore a fermion will lose its identity if it reaches the origin, which turns out to be inevitable. One may also say that the fermionic vacuum state is ill-defined in the presence of a monopole, and the zero modes are a reflection of this situation.

The fermionic zero modes suggest that there exists some kind of supersymmetry in topological excitations, even when such symmetry is not manifestly built into the system. This point has been noted before [78], but much remains to be done. [The zero modes are somehow related to the Goldstone modes. Note that in the one-dimensional example of (a), the zero mode is in the proper Goldstone direction σ_2 relative to the symmetry breaking term $W\sigma_3$.]

(c) *Quasi-supersymmetry in nonrelativistic phenomena.*

The foregoing observations make one wonder if there are examples of zero modes or supersymmetry even in more familar nonrelativistic situations not requiring the Dirac equation. In fact, one such example involves the superconducting electrons (BCS quasi-particles) trapped in a magnetic flux tube [79].

The quasi-particle of the BCS theory obeys a two-component equation with the energy gap function $\Delta = \Delta_1 + i\Delta_2$ playing a role similar to the mass in Dirac equation. The Hamiltonian reads

$$H = \sigma_3\left\{\frac{1}{2m}(p - \sigma_3 eA)^2 - E_F\right\} + \sigma_1\Delta_1 + \sigma_2\Delta_2$$

where E_F is the Fermi energy. Around a flux tube, Δ has phase factor $e^{in\theta}$, ($n = \pm 1, \ldots$) depending on the quantized flux represented by $eA_\theta \sim n/2\rho$ (ρ = axial distance). Although the circumstances are similar to the Dirac electron case, it is not obvious that zero eigenvalues exist, because the equation is second order. Nevertheless it turns out that for odd n there are eigenvalues of order $|\Delta^2/E_F| \ll |\Delta|$ which vanish in the limit $|\Delta/E_F| \to 0$. [Incidentally, an Aharonov–Bohm effect arises for odd n as the flux is properly quantized only against Cooper pairs.]

There may be other cases where a similar situation holds. For example, one recalls that the BCS mechanism has been invoked to describe the pairing of nucleons in nuclei. Furthermore, the recent work of Iachello and Bars [80] shows that energy levels of certain even–even and even–odd nuclei are related by approximate supersymmetry relations. In Iachello's example, the nuclear excitations are built up of bosons ($j = 0$ and 2) and fermions ($j = \frac{3}{2}$, $l = 2$), forming together a representation of supersymmetry U(6/4). The equation for the fermions can be taken to be a BCS-like one, involving an eight-component wave function ($\psi_{3/2}$, $\psi^{+}_{3/2}$). The pairing functions will again go to zero toward the origin. It turns out, however, that this is similar to the even flux number case of the previous example, so one does not a priori expect low lying modes.

In the above considerations, supersymmetry is taken only in the sense that bosonic states are accompanied by certain degenerate or nearly degenerate fermionic states. Nothing is said about other states. There exists, however, another possibility for (broken) supersymmetry which does not require nontrivial topologies. This comes about when the BCS and Ginzburg–Landau theories of superconductivity are combined to describe fermionic and bosonic excitations. The corresponding Lagrangian is an analog of the σ-model with fermions in particle physics. In the case of superconductivity, the parameters of the theory are all calculable, and it has been shown [81] that the masses of the analogs of π, fermion and σ states come in the ratios $0:\Delta:2\Delta$, a fact which also has experimental support. This situation suggests a broken symmetry satisfying a mass formula

$$\sum m_{\text{Fermi}} = \sum m_{\text{Bose}}\,.$$

A similar relation appears in relativistic supersymmetry where m is replaced by m^2. On the other hand, the velocities of propagation in the case of superconductivity are not equal: $v^2_{\text{Fermi}} = 3v^2_{\text{Bose}}$.

Note added in proof: Simple boson–fermion mass relations of this kind exist in BCS-type theories in general, and suggest a possible explanation for the above-mentioned nuclear physics example.

3. Speculative outlook

In the previous section I addressed some of the problems having to do with the theoretical framework which should be relevant in tackling the more substantive questions of particle physics. I will now come to these substantive questions, or at least a few representative ones of this kind. Currently a great deal of model-building activity is going on in an effort to find a consistent unified theory of particles and forces, of which only a very small part is experimentally known. The program as a whole is highly appealing and persuasive, but a truly natural and satisfactory scheme has not been found yet. These attractive features are lost as soon as one is forced to play complicated tricks in order to fit the reality. Probably this means that even if one is on the right track, some key elements are still missing. I will avoid discussing the details of the activity, and restrict myself to asking questions of a more general nature.

1. Mass and mass spectra

The origin of mass and mass spectra is one of the oldest and most elusive problems. Actually the situation has gotten worse as the fundamental particles have proliferated, their masses showing no precise regularities. Clearly mass is a dynamical quantity governed by self-energy effects. It may arise in conjunction with spontaneous symmetry breaking of chiral symmetry or gauge invariance, but that may not account for everything. What one expects would be to derive mass formulas from dynamics in terms of a few basic parameters as has been done for composite systems like hydrogen atom and hadrons (the latter in terms of QCD and string model). So far, however, this has not been accomplished for the masses of leptons and quarks. In the grand unified theories, certain mass relations are obtained, but eventually the complexities of real life are reduced to a set of arbitrary parameters in the Higgs fields, so no essential progress is made. Moreover, a raison d'être for the several generations of fermions still remains to be found. Since divergent quantities like self-energy are a reflection of very short distance properties of dynamics, the problem of mass may be the last thing one can understand. At the same time, however, it may supply us with the keys to the short distance physics. So, leaving this subject temporarily, I will move on to the next related subject.

2. Broken symmetries

Perhaps one of the naive questions that may occur to the student of particle physics is why weak interactions seem to have so little respect for symmetries and conservation laws. In fact they do not respect any of their own symmetries except the one related to energy-momentum conservation. Why is the internal reference frame for the weak interactions so capriciously tilted relative to the strong interaction frame, giving rise to messy fermion mass matrices? Is it by God's design, or by God's mistake? Whatever the reason, the problem must be intimately linked to the previous one of mass spectra. And these properties of weak interactions and mass spectra control the complexities of the real world in a subtle but critical way. Imagine a hypothetical world where the Cabibbo angle is zero. Or a world where the neutron–proton mass difference is less than the electron mass. What if muon was heavier than pion, or if the mass of Λ hyperon was lower by 40 MeV? Such an exercise will reveal that some of our daily physical phenomena would be profoundly affected, and the course of the history of particle physics would have been different; in fact the entire history of the universe would have changed. Pursuing a similar line, one may further speculate what would happen, for example, if the fine structure constant was a little different from what it is. How would that change chemistry, biology, etc?

After these speculative exercises, perhaps one is led to conclude that enough complexities already exist at the level of fundamental particles, and particle physics is no simpler yet than the other branches of physics. It is then possible that these complexities may be due to the initial conditions of the universe, and therefore cannot be rationally explained. One might also say that the causes are hidden within the Planck length. And if such inferences are accepted, one need not despair at the difficulties of grand unification, nor should one expect an easy solution to it.

3. Phenomenology of grand unification

I have been led to think that grand unification cannot be an entirely deductive process. One has to live with a lot of phenomenology, and one has to learn from experiment. If so, it seems also natural to expect that the so-called vast desert between the currently available energies and the unification energies probably will not be a desert, but will be filled with many complex physical phenomena. To learn about them, one must explore them experimentally. During the past fifty years of particle physics, the energy of accelerators has increased by a factor of 10^6, at a steady rate, from 1 MeV to 1 TeV. If this Livingston curve [82] is blindly extrapolated to the future, it will take another century to reach the grand unification energy of 10^{15} GeV and thirty years more to reach the Planck energy of 10^{19} GeV. In a way, grand unification does not look so forbidding, but of course these estimates are totally unrealistic. So one has to look for whatever evidence is available without relying on accelerators, and probably real progress can and will be made this way even during the present century.

Viewed in this light, the three currently ongoing experiments are extremely important. These are the search for quarks, monopoles, and proton decay. If either very heavy monopoles or proton decays are experimentally detected, that will immediately confirm some basic ideas underlying the grand unification theories. On the other hand, the rather puzzling experimental results on free quarks are a potential threat to the current paradigms. Serious efforts are called for to confirm or disprove these results.

4. Beyond the Planck limit

One need not stop at the Planck limit if one is only engaged in speculations, which is actually all I have been able to do regarding higher energy physics anyway. So a few more exercises of this kind may not be out of order.

As already mentioned, I am assuming that the Planck length is where a break will have to be made from the theoretical framework known today, a break comparable to that which occurred between classical and quantum mechanics. A fundamental length scale seems inevitable for several reasons. Aside from the problem of quantum gravity, in particle physics alone one cannot keep postponing indefinitely the final resolution of the problem of mass relying on renormalizable theories. If nontrivial topologies are to be taken seriously, the dual relationship between electric and magnetic charges, for example, indicates that there may be a scale where renormalization theory breaks down. It seems plausible that in this realm the geometries of space-time as well as the matter fields become all nontrivial, and may in fact merge together, as is already envisaged in the higher-dimensional field theories. One thus imagines that the world is a topological bubble (or flux tube) of Planck size created by the big bang inside a space of higher dimensions. The fermions trapped in the bubble are the zero modes associated with the topology [83]. [Consider a Dirac equation $(\gamma \cdot D - M)\psi = 0$, where M is an operator in the extra dimensions, and has zero (or very small) eigenvalues.]

What about experimental handles? Certainly gravitational phenomena (black hole physics, quantum gravity) will be among them, but will there be others comparable to proton decay or monopoles in the case of grand unified theories? By way of simple analogy, one might speculate that the last bastions of conservation laws, i.e., those of electric charge, color, energy-momentum and spin, are eventually subject to violations, due perhaps to the existence of the other dimensions or nontrivial topology [84].

References

[1] T. Kuhn, The Structure of Scientific Revolutions (University of Chicago Press, 1962) p. 10.

[2] H. Yukawa, Proc. Phys.-Math. Soc. Japan 17 (1935) 48–57.
[3] S. Sakata, Progr. Theor. Phys. 16 (1956) 686.
[4] M. Gell-Mann, Phys. Rev. Lett. 8 (1962) 214.
[5] G. Zweig, CERN Rep. No. 8182/Th (1964) p. 401, 8419/Th (1964) p. 412.
[6] L.M. Brown, Centaurus 25 (1981) 71–132; in: Particle Physics in Japan, 1930–1950, eds. L. Brown, M. Konuma and Z. Maki, Research Inst. for Fundamental Physics (U. Kyoto) publication RIFP-407 and 408 (1980).
[7] S. Sakata, Bull. Phys.-Math. Soc. Japan (in Japanese) 16 (1943) 232–235.
[8] E. Fermi and C.N.Yang, Phys. Rev. T6 (1949) 1739.
[8a] O. W. Greenberg, Phys. Rev. Lett. 13 (1964) 598;
M.Y. Han and Y. Nambu, Phys. Rev. 139B (1965) 1006;
H. Fritsch, M. Gell-Mann and H. Leutwyler, Phys. Lett. 47B (1973) 365.
[9] N. Cabibbo, Phys. Rev. Lett. 10 (1963) 571.
[10] S.L. Glashow, J. Iliopoulos and L. Maiani, Phys. Rev. D2 (1970) 1285.
[11] M. Kobayashi and T. Maskawa, Prog. Theor. Phys. 49 (1973) 652.
[12] S. Weinberg, Phys. Rev. Lett. 19 (1967) 1264.
[13] A. Salam, in: Elementary Particles, ed. N. Svartholm (Almqvist, Forlag AB, 1968) p. 367.
[14] V. Weisskopf, Zeits. f. Physik 89 (1934) 27; 90 (1934) 817.
[15] S. Tomonaga, Bull. I.P.C.R. (Riken-Iho) 22 (1943) 545 (in Japanese); Prog. Theor. Phys. 1 (1946) 1.
[16] J. Schwinger, Phys. Rev. 74 (1948) 1439.
[17] R. Feynman, Phys. Rev. 76 (1949) 749, 769.
[18] F. Dyson, Phys. Rev. 75 (1949) 486, 1736.
[19] W. Heisenberg, Ann. Phys. 32 (1938) 20; Zeits. f. Physik 110 (1938) 251.
[20] W. Heisenberg, Zeits. f. Physik 120 (1943) 513;
R. Oehme, Theory of the Scattering Matrix, or an Introduction to Heisenberg's Papers, in: Collected Works of Werner Heisenberg (to be published 1982) MPI-PAE/PTh 48/82.
[21] A. Pais, Phys. Rev. 68 (1945) 227.
[22] S. Sakata and O. Hara, Progr. Theor. Phys. 2 (1947) 30.
[23] See, for example, K.G. Wilson and J. Kogut, Phys. Rev. 12 (1974) 75.
[24] G. 't Hooft, unpublished.
[25] D.J. Gross and F. Wilczek, Phys. Rev. D8 (1973) 3633.
[26] H.D. Politzer, Phys. Rev. Lett. 30 (1976) 1346.
[27] H. Hamber and G. Parisi, Phys. Rev. Lett. 47 (1981) 1792;
E. Marinari, G. Parisi and G. Rebbi, Phys. Rev. Lett. 47 (1981) 1795;
D. Weingarten, Phys. Lett. 109B (1982) 57.
[28] L.D. Landau, in: Niels Bohr and the Development of Physics, ed. W. Pauli (McGraw-Hill, New York, 1955) p. 52.
[29] H. Georgi and S.L. Glashow, Phys. Rev. Lett. 32 (1974) 348.
[30] J.C. Pati and A. Salam, Phys. Rev. D10 (1974) 275.
[31] H. Georgi, H.R. Quinn and S. Weinberg, Phys. Rev. Lett. 33 (1974) 451.
[32] See, for example, J.M. Blatt and V.F. Weisskopf, Theoretical Nuclear Physics (John Wiley, New York, 1952) p. 153.
[33] H.A. Jahn and E. Teller, Proc. Roy. Soc. (London) A161 (1937) 220.
[34] Y. Nambu and G. Jona-Lasinio, Phys. Rev. 122 (1961) 345;
G. Jona-Lasinio and Y. Nambu, Phys. Rev. 124 (1961) 246.
[35] J. Goldstone, Nuovo Cimento 19 (1961) 155.
[36] H. Weyl, Space-Time-Matter (Raum, Zeit und Materie, 3rd ed., 1919) (Dover edition, 1952).
[37] Th. Kaluza, Sitzungsber. d. Preuss. Akad. Wiss. (1921) 966.
[38] C.N. Yang and R. Mills, Phys. Rev. 96 (1954) 191.
[39] T.D. Lee and C.N. Yang, Phys. Rev. 98 (1955) 1501.
[40] R. Utiyama, Phys. Rev. 101 (1956) 1597.
[41] J.J. Sakurai, Ann. Phys. 11 (1969) 1.
[42] S.L. Glashow, Nucl. Phys. 22 (1961) 579.
[43] V.L. Ginzburg and L.D. Landau, Zh. Exp. Teor. Fiz. 20 (1950) 1064.
[44] P.W. Higgs, Phys. Rev. 145 (1966) 1154.
[45] J. Bardeen, L.N. Gooper and J.R. Schrieffer, Phys. Rev. 108 (1957) 1175.
[46] L.D. Faddeev and V.N. Popov, Phys. Lett. 25B (1967) 29; Kiev Report No. ITP67-36.
[47] G. 't Hooft, Nucl. Phys. B33 (1971) 173.
[48] P.A.M. Dirac, Proc. Roy. Soc. A133 (1931) 60; Phys. Rev. 74 (1948) 817.
[49] F. London, Superfluids Vol. 1 (Dover edition, 1961).
[50] Y. Aharonov and D. Bohm, Phys. Rev. 115 (1959) 485.
[51] A. Tonomura et al., Phys. Rev. Lett. 48 (1982) 1443.
[52] G. 't Hooft, Nucl. Phys. B79 (1974) 276.

[53] A.M. Polyakov, JETP Lett. 20 (1974) 194.
[54] A.A. Belavin, A.M. Polyakov, A.S. Schwartz and Yu.S. Tyupkin, Phys. Lett. 59B (1975) 85.
[55] G. 't Hooft, Phys. Rev. Lett. 37 (1976) 8; Phys. Rev. D14 (1976) 3432.
[56] G. Parisi and Y.S. Wu, Scientia Sinica 24 (1982) 81;
See also, E. Nelson, Phys. Rev. 150 (1966) 1079; Dynamical Theories of Brownian Motion (Princeton Univ. Press, 1967).
[57] P. Goddard, J. Nuyts and D. Olive, Nucl. Phys. B125 (1977) 1.
[58] J.S. Bell and R. Jackiw, Nuovo Cimento 60A (1969) 47.
[59] S.L. Adler, Phys. Rev. 177 (1969) 2426.
[60] E. Witten, Phys. Lett. 86B (1979) 283.
[61] V.A. Rubakov, Pis'ma Zh. Exp. Teor. Fiz. 33 (1981) 658 [JETP Lett. 33 (1981) 644].
[62] F. Wilczek, Phys. Rev. Lett. 48 (1982) 1146.
[63] C.G. Callan, Phys. Rev. D26 (1982) 2058.
[64] See, for example, the review article by C. Rebbi, Phys. **Reports 12 (1974)** 1.
[65] H.B. Nielsen and P. Olesen, Nucl. Phys. B61 (1973) 45.
[66] A.M. Polyakov, Phys. Lett. 103B (1981) 207.
[67] M. Lüscher, K. Symanzik and P. Weisz, Nucl. Phys. B173 (1981) 367.
[68] M. Kalb and P. Ramond, Phys. Rev. D9 (1974) 2273.
For complete literature, see P.G.O. Freund and R.I. Nepomechie, Nucl. Phys. B199 (1982) 482.
[69] Y.M. Cho and P.G.O. Freund, Phys. Rev. D12 (1975) 1711 and references therein;
L.N. Chang, I.I. Macrae and F. Mansouri, Phys. Rev. D13 (1976) 235;
E. Cremmer and J. Scherk, Nucl. Phys. B108 (1976) 409.
[70] Y. Nambu, in: Quark Confinement and Field Theory, eds. D.R. Stump and D.H. Weingarten (John Wiley, New York, 1977) p. 1.
[71] E. Cremmer, B. Julia and J. Scherk, Nucl. Phys. 76B (1978) 409;
P.G.O. Freund and M.A. Rubin, Phys. Lett. 97B (1980) 233;
F. Englert, Phys. Lett. 119B (1982) 339.
[72] See, for example, the review article by P. Fayet and S. Ferrara, Phys. Reports 32C (1977) 250.
[73] E. Wigner, Phys. Rev. 77 (1950) 711;
L.M. Yang, Phys. Rev. 84 (1951) 788;
Y. Ohnuki and S. Kamefuchi, Quantum Field Theory and Parastatistics (Univ. Tokyo Press, 1982) Chap. 23.
[74] R. Jackiw and C. Rebbi, Phys. Rev. D13 (1976) 3398;
R. Jackiw, Rev. Mod. Phys. 49 (1977) 681.
[75] V.A. Rubakov, Pis'ma Zh. Exp. Teor. Fiz. 33 (1981) 658 [JETP Lett. 33 (1981) 6442].
[76] F. Wilczek, Phys. Rev. Lett. 48 (1982) 1146.
[77] C.G. Callan, Jr., Phys. Rev. Lett. D26 (1982) 2058.
[78] For example, P. Rossi, Phys. Lett. 71B (1977) 145, and references therein.
[79] C. Caroli, P. de Gennes and J. Matricon, Phys. Lett. 9 (1964) 307;
L. Fetter and P.C. Hohenberg, in: Superconductivity Vol. 2, ed. R.D. Parks (Marcel Dekker, N.Y. 1969) p. 888.
[80] F. Iachello, Phys. Rev. Lett. 44 (1980) 772;
A.B. Balantekin, I. Bars and F. Iachello, Phys. Rev. Lett. 47 (1981) 19; Nucl. Phys. A370 (1981) 284.
[81] P.B. Littlewood and C.M. Varma, Phys. Rev. Lett. 47 (1981) 811.
[82] M.S. Livingston, High Energy Accelerators (Interscience, New York, 1958) p. 149;
W.K.H. Panofsky, Phys. Today 33 (1980) No. 6, p. 24.
[83] L. Palla, Proc. XIX Intern. Conf. on High Energy Physics, Tokyo, 1978, p. 629;
E. Witten, Nucl. Phys. B186 (1981) 412.
[84] P.G.O. Freund, private communication.

Physica 15D (1985) 147–151
North-Holland, Amsterdam

FERMION–BOSON RELATIONS IN BCS-TYPE THEORIES

Y. NAMBU
Enrico Fermi Institute and Department of Physics, University of Chicago, Chicago, IL 60637, USA

1. Introduction

Supersymmetry is a mathematical idea in search of physical relevance, but so far the search has not been successful in the area where it originated: particle physics. The present report is an outcome of my attempt to understand the physical meaning of supersymmetry and to find examples of it in more familiar low energy physics.

Iachello and collaborators [1] have observed that there is a kind of supersymmetry in nuclear physics. For certain nuclei in the mass number region of platinum, the low energy spectra of even–even nuclear species and neighboring even–odd species can be described by the same empirical formula based on group theory. I have been aware, on the other hand, that in theories of the BCS-type, there always is a simple relation between the mass (energy gap) of the basic fermion and those of the bosons (collective modes) [2]. To use the language of particle physics, the dynamically induced masses of the pion, quark and σ meson stand in the ratio 0:1:2 (subject to higher order corrections). In terms of the effective σ model (or Higgs or Ginsburg–Landau) Lagrangian, this implies that the self-coupling and the Yukawa constants are related by $\lambda = f^2$. Generic relations of this nature emerge in any BCS–Heisenberg-type four-fermion short-range interaction theory [3,11]. It is gratifying that such relations have been experimentally established in superconductors and superfluid helium 3, as I will discuss later.

My speculation, then, is that the Iachello relation in nuclear physics may also be a manifestation of the BCS mechanism which is known to account for the nuclear pairing phenomenon. An immediate question that arises is whether the BCS or Ginsburg–Landau theory has a supersymmetry of which these relations are a consequence. I do not know the answer yet. Before coming to nuclear physics, however, I will first discuss the other examples to show the origin of the relations.

2. Superconductivity

We use the two-component formalism in which the spin-up electron and spin-down hole span a τ-spin space for the quasiparticle. The free quasiparticle Hamiltonian has the form

$$H = K\tau_3 + m\tau_1, \tag{1}$$

where K is the kinetic energy measured from the Fermi surface and m is the gap parameter. (We deal only with states in the vicinity of the Fermi surface.) The charge operator is $e\tau_3$, so a τ_1 or τ_2 term will break charge conservation. The gap in the τ_1 direction will then generate a Goldstone mode in the τ_2 direction. The collective modes in the τ_1 and τ_2 directions (called amplitude and phase modes) are the analogs of σ and π mesons in particle physics.

In terms of the four-fermion interaction Lagrangian

$$G\tau_3\tau_3' \sim G\left[\left(\tau_1\tau_1' + \tau_2\tau_2' + \cdots\right)\right]$$
$$\text{(under a Fierz transformation)}, \tag{2}$$

a collective mode is determined as the pole of the

function

$$G/(1-GI) = G + G(GI) + G(GI)^2 + \ldots, \quad (3)$$

where the formal expansion indicates that the function is derived by summing to infinite order chains of "bubble" perturbation theory diagrams. I is a two-point loop integral with appropriate vertex operators, τ_1 or τ_2. Its energy (E) dependence can be displayed by a dispersion integral

$$I(z) = \int \rho(x)\,\mathrm{d}x/(x-z), \quad x > 4m^2,\ z = E^2, \quad (4)$$

where the absorptive parts ρ for τ_1 and τ_2 modes satisfy the relation

$$\rho_1(x) = \rho_2(x)(1 - 4m^2/x). \quad (5)$$

From this, it follows that

$$I_1(z = 4m^2) = I_2(z = 0). \quad (6)$$

But $GI_2(z=0) = 1$, as a result of the Ward identity corresponding to the breaking of τ_3 invariance. Thus, one sees that the τ_2 and τ_1 modes have a pole at $E = 0$ and $2m$, respectively. It looks as if these states correspond to composites of two fermions with maximum and zero binding energy. I do not have a simple physical explanation for this, but technically it depends on whether the vertex operator anticommutes or commutes with the mass operator. Another way of putting it is that, in the particle theory analog, π and σ are s- and p-wave bound states of quarks, and the extra factor in the absorptive part for the latter is just the p-wave phase space factor.

The τ_1 (amplitude) mode was experimentally detected a few years ago in certain superconductors whose gap parameters are susceptible to external forces like laser beams [4]. Actually, the mode comes out as a genuine bound state somewhat below $2m$. The theoretical explanation has been worked out along the lines sketched above [5]. The massless τ_2 (phase or Goldstone) mode, on the other hand, mixes with the τ_3 (density) mode, and turns into the plasmon by the familiar Anderson–Higgs–Englert–Brout mechanism. Both τ_1 and τ_2 modes have the same canonical velocity

$$v = v(F)/\sqrt{3}, \quad (7)$$

where $v(F)$ is the Fermi velocity. So it is a simple matter to write down an effective σ-model Lagrangian for the collective modes, which can be explicitly derived from a BCS theory. [As was mentioned in section 1, the Yukawa coupling (f) and the boson self-coupling (λ) are related as

$$\lambda = f^2 = 1/(\mathrm{d}I_2(z)/\mathrm{d}z)_{z=0}, \quad (8)$$

where $\lambda/2$ is the coefficient of the quartic term in the Lagrangian.]

3. Helium 3

Superfluid helium 3, occurring at temperatures in the millikelvin range, has two phases, A and B[6]. It is believed that the Cooper pairs of two helium atoms are formed in a triplet p state in either phase. In principle, the total angular momentum j can be 0, 1, or 2, and the pairs can condense in any state in the space spanned by these 9 states; also, recall that the degeneracy of a state of angular momentum j is $2j+1$. Actually, however, the A and B phases correspond to $j = 1$ and 0, respectively. The nine states may be expressed by a scalar S, vector V_i, or tensor T_{ik} order parameter transforming as

$$\tau_\alpha \sigma \cdot p, \quad \tau_\alpha(\sigma \times p)_i, \quad \text{or } \tau_\alpha \sigma_i p_k \quad (\alpha = 1,2), \quad (9)$$

respectively, where σ is the spin, and p is the relative momentum of the Cooper pair. As before, the τ_α's operate in the space of particle and hole components, so there are 18 collective modes altogether.

What symmetries are broken by the Cooper pairings? This question leads to another: what is the precise nature of the forces responsible for the pairing? The answer is not entirely clear, but it appears that they are not greatly different in different j channels, meaning that intrinsic spin-orbit interaction is small. The pairing in a particular j state will therefore break the conservation of total angular momentum j (if $\neq 0$) and spin (or equivalently, orbital) angular momentum, in addition to the usual violation of particle number. Each violation should generate a Goldstone mode (exact or approximate), whereas the remaining collective states should be massive. Again, these have a simple mass spectrum as shown in table I [7]. Note in particular that for a given j, the phase and amplitude modes satisfy a sum rule

$$m_1^2 + m_2^2 = 4m^2 \tag{10}$$

in terms of the fermion mass m. This is a consequence of a theorem which generalizes eq. (6):

$$aI_1(a) + (1-a)I_2(a) = \text{const} = I_2(0). \tag{11}$$

Vertices appearing in this problem generally give rise to the linear combination on the left-hand side, and the phase and amplitude modes are orthogonal and complementary in the sense $a \leftrightarrow 1 - a$.

The above theoretical predictions are roughly in agreement with observation. The phase and amplitude modes behave differently under particle–hole reflection relative to the Fermi surface. Since this symmetry is not exact, both modes can be excited by external density waves. Perhaps the above sum rule is to be more trusted than the individual mass values.

Table I
Collective modes in He 3, B phase. $H = K\tau_3 + m\tau_1\sigma\cdot p$

Mode	j	Mass	Loop integral	Comment
$\tau_1\sigma\cdot p$	0	$2m$	I_1	
$\tau_2\sigma\cdot p$	0	0	I_2	particle number Goldstone
$\tau_1(\sigma\times p)_i$	1	0	I_2	spin Goldstone
$\tau_2(\sigma\times p)_i$	1	$2m$	I_1	
$\tau_1\sigma_1 p_k$	2	$\sqrt{2/5}\,(2m)$	$\frac{2}{5}I_1 + \frac{3}{5}I_2$	
$\tau_2\sigma_i p_k$	2	$\sqrt{3/5}\,(2m)$	$\frac{3}{5}I_1 + \frac{2}{5}I_2$	

4. Nuclear physics

The BCS theory has been applied with success to account for the pairing phenomenon in nuclei [8]. Typically, the ground state of a heavy even–even nucleus is filled with singlet proton pairs and neutron pairs, whereas an even–odd nucleus has one quasi-particle added or missing. The excited states of a nucleus can be due to single-particle as well as pair excitations, the latter giving rise to collective models with $j = 0, 2, \ldots$.

In the interacting boson model of Arima and Iachello [9,1], one treats the pairs as bosons, and introduces an effective Hamiltonian with quartic self-couplings, which is determined on the basis of group theory. Thus, take $j = 0$ and $j = 2$ bosons only, considered to be degenerate in the first approximation. The relevant invariance group is then $U(6) = SU(6) \times (U(1)$. Next, break it down to a chain of subgroups. There are three distinct chains:

$$\begin{aligned} &U(6) \supset U(5) \supset SO(5)\ldots, \\ &U(6) \supset SU(4) \supset SO(6) \supset SO(5)\ldots, \\ &U(6) \supset SU(3)\ldots. \end{aligned} \tag{12}$$

The question of which chain is relevant is where physics comes in, and the answer depends on the mass number region one is considering. In the platinum region, for example, it is the second chain. At any rate, one then writes down an effective Hamiltonian as a linear combination of Casimir operators corresponding to the chain of groups in terms of bose operators. This seems to work well in general. It has been pointed out [1], moreover, that in the platinum region, the even–odd nuclei have single protons in a $D_{3/2}$ shell, for which the natural invariance group is SU(4), leading to the same chain of subgroups as for the bosons. As it turns out, the same Hamiltonian can indeed describe both even–even and even–odd cases, hence suggesting a supersymmetry [1]. Real supersymmetry, however, should involve a supergroup like U(6/4) which includes fermion–boson transitions. Such a scheme still works, although not as well as before.

Now, I come back to the scenario already developed, i.e., the BCS–Ginsburg–Landau theory, as applied to the present problem. Let us say that Cooper pairs, each formed out of nucleons in the same shell, collectively condense into a ground state, which can be excited by breaking pairs up or placing them into new states. In the case of $j = 3/2$ shell, the pair can have $j = 0$ or 2.

According to the standard methods, a nucleon should be described by combining four particle and four hole (complex conjugate) states into a 8-component wave function ψ. Then one writes an analog of the Hamiltonian, eq. (1), in this space. The ψ can form a representation of SO(8) and Spin(7) (Majorana), so there are seven anticommuting Γ matrices $\Gamma_1, \Gamma_1, \ldots \Gamma_7$. One can choose them in such a way that $\Gamma_0 = +1(-1)$ for particle (hole) states, and $\Gamma_s \equiv \Gamma_1$ behaves as a scalar (or possibly a mixture of scalar and tensor). Then the Hamiltonian will take the form

$$H = K\Gamma_0 + m\Gamma_s. \tag{13}$$

A remark is in order before proceeding further. The emergence of the parameter m is a result of interactions among many nucleons in various shells, not just in the states described by the wave function under consideration. The collective modes are generated by all of them, which contribute to the loop integral I, eq. (4). For example, $j = 1/2$ shells will contribute to $j = 0$, and $j = 5/2$ states to $j = 0, 2, 4$, collective modes. The relation between m in eq. (13) and the masses of the collective modes, therefore, is no longer so clear. On the other hand, to the extent that $j = 0$ and 2 modes are approximately degenerate while others can be ignored, it would be reasonable to suppose that the symmetry group SU(4) of the fermions under consideration and the symmetry group SO(6) of the collective modes are physically identical. So the loop integrals will still maintain the same properties as before.

There should be $6 \times 2 = 12$ collective modes, which come in six pairs $\Gamma_k, \mathrm{i}\Gamma_0\Gamma_k$ $(k = 1, \ldots 6)$ of amplitude and phase modes. Thus, Γ_s is an analog of the σ with $j = 0$ at mass $2m$, whereas the zero-mass $\mathrm{i}\Gamma_0\Gamma_1$ mode is spurious because there is actually no particle number violation in a finite system. The SO(6) symmetry, if it exists in an approximate sense, is generated by $\Gamma_k\Gamma_1 (K > 1)$. The Γ_s term in H breaks this down to SO(5), so there should be spurious near-zero modes $\Gamma_t (t \neq 1)$, and complementary modes $\mathrm{i}\Gamma_0\Gamma_t$ at mass $\approx 2m$, both with $j = 2$. Actually, the zero modes will in general couple to ordinary acoustic modes and get absorbed into them. These results are summarized in table II.

Table II
Collective modes in nuclei ($j = 3/2$ shell). $H = K\Gamma_0 + m\Gamma_s$

Mode	j	Mass	Baryon #(B)	Loop integral	Comment
Γ_s	0	$2m$	± 2	I_1	
$\Gamma_0\Gamma_s$	0	0	± 2	I_2	B Goldstone
Γ_t	2	0	± 2	I_2	SO(6) Goldstone
$\Gamma_0\Gamma_t$	2	$\sim 2m$	± 2	I_1	
Γ_0	0		0	$I_1 - I_2$	B generator
$\Gamma_s\Gamma_t$	2		0	$I_1 - I_2$	SO(6) generator
$\Gamma_t\Gamma_{t'}$	2		0	0	SO(6) generator
$\Gamma_0 - \Gamma_0\Gamma_s$	0		0	$(\mathrm{i}E/2m)I_2$	off diagonal
$\Gamma_t - \Gamma_s\Gamma_t$	2		0	$(\mathrm{i}E/2m)I_2$	elements

Γ_s: scalar, Γ_t: tensor

Finally, we will write down an effective Hamiltonian for the system. Introduce six complex boson operators $B_i(B_i^\dagger)$, transforming under the symmetry group $\mathrm{SU}(6) \times \mathrm{U}(1)$. It turns out that the proper Hamiltonian (ignoring the mixing with acoustic modes for the moment) is given by

$$H = \psi^\dagger K\Gamma_0\psi + f\left[\psi^\dagger(1+\Gamma_0)\Gamma_i\psi B_i + \text{h.c.}\right] + (f^2/2)\left[\left(B_i^\dagger B_i - c^2\right)^2 + |B_i^\dagger B_k - B_k^\dagger B_i|^2\right].$$

Note that the nonlinear terms for the B_i's are composed of Casimir operators of U(1) and SU(6). The terms involving the fermions, on the other hand, have the symmetry of $\mathrm{SU}(4) \times \mathrm{U}(1) \approx \mathrm{SO}(6) \times \mathrm{U}(1)$. The latter gets broken spontaneously to SO(5), giving rise to the above mentioned zero modes. It thus seems likely that an effective

Hamiltonian constructed along this line has properties similar to the one proposed by Iachello.

I have tried to uncover in this kind of system a genuine supersymmetry involving fermionic transformations, but so far I do not have a clear answer to that.

5. Particle physics

Finally, I will briefly come back to particle physics. Historically, the σ model in hadron physics is the oldest relevant example and is still of considerable interest in many respects, but our present concern is with mass relations. In accordance with our general results, it has been claimed [10] that the σ meson should have a mass roughly twice the constitutent quark mass, i.e., ≈ 700 MeV, although the actual σ resonance seems to occur at ≈ 900 MeV. Since the QCD interaction is rather different from the ones we are considering, it is not obvious that the present arguments apply here equally well.

A more interesting case should be the electroweak interaction [11]. Here the Higgs bosons in the Weinberg–Salam theory are analogs of the π and σ, but regarded as elementary objects. Their masses therefore are not predictable. But in composite models like Terazawa's [11], there emerge relations like ours between boson and constituent fermion masses. Similar situations may also exist in technicolor theories, subject to the uncertainties encountered in the hadronic case. Short of a specific model however, I will not indulge in further speculations at this moment.

Acknowledgements

I would like to thank Professors F. Iachello, A. Arima, I. Bars, A.J. Leggett, G. Baym, K. Levin, C.M. Varma, M. Scadron and A. Terazawa for enlightening conversations I have had at one time or another.

References

[1] F. Iachello, Phys. Rev. Lett. 44 (1980) 772; Physica 15D (1985) 85 (these proceedings).
A. Balantekin, I. Bars and F. Iachello, Phys. Rev. Lett. 47 (1981) 19; Nucl. Phys. A370 (1981) 284; A.B. Balantekin, I. Bars, R. Bijker and F. Iachello, Phys. Rev. C27 (1983) 1761.
[2] Y. Nambu and G. Jona-Lasinio, Phys. Rev. 122 (1961) 345.
[3] T. Eguchi and H. Sugawara, Phys. Rev. D10 (1974) 4257.
[4] R. Sooryakumar, M.V. Klein and R.R. Frindt, Phys. Rev. B23 (1980) 3213, 3222.
[5] P.B. Littlewood and C.M. Varma, Phys. Rev. B26 (1982) 4883.
C.A. Balseiro and L.M. Falicov, Phys. Rev. Lett. 45 (1980) 662.
[6] A.J. Leggett, Rev. Mod. Phys. 47 (1975) 331.
J. Wheatley, Rev. Mod. Phys. 47 (1975) 415.
D.M. Lee, Internat. J. Quant. Chem. 23 (1983) 1191.
K. Levin, Phys. Rep. 98 (1983) 1.
[7] P. Woelfle, Physica 90B (1977) 96; Prog. LTP 7A (1978) 191.
[8] A. Bohr, B.R. Mottelson and D. Pines, Phys. Rev. 110 (1958) 936.
S.T. Beliaev, Kgl. Dansk. Vid. Selsk. Mat-Fys. Medd. 31, no 11 (1959).
J.M. Eisenberg and W. Greiner, Microscopic Theory of the Nucleus, vol 3, (North-Holland, Amsterdam, 1972).
[9] A. Arima and F. Iachello, Phys. Rev. Lett. 35 (1975) 1069.
[10] R. Delbourgo and M. Scadron, Phys. Rev. Lett. 48 (1982) 379.
[11] H. Terazawa, Y. Chikashige and A. Akama, Phys. Rev. D15 (1977) 480.

Reprinted from Field Theory and Quantum Statistics, eds. I. A. Batalin et al.,

Field Theory of Galois' Fields†

Yoichiro Nambu

1 INTRODUCTION

The motivation for the present work comes from various sources which, however, need not be elaborated on here. I will be exploring a class of quantum field theories defined over finite sets of integers. Essentially these are the familiar Z_n lattice theories, but carried to their logical extremes.

Generally speaking, a set of integers modulo m constitutes a residue system ring Z_m of characteristic m (i.e., $ma = 0$ for any a), which is closed under addition and multiplication, but not necessarily admitting division. Actually I will introduce three kinds of Z_m's for three different physical quantities. Thus the lattice is taken to be periodic, and its coordinates x take values in Z_l, where l may be different for each space dimension. However, the length of the time dimension will be left open. Next, the field F at each lattice site takes values in another set Z_k, as does the action functional which depends on F. Finally, the partition function W (quantum or statistical) belongs to the third set Z_h, for reasons that will become clear in a moment. At any rate, one is thus dealing with mappings from Z_l to Z_k to Z_h.

Although these parameters are arbitrary integers, the case of prime numbers has a special significance. For then the sets become finite (Galois) fields which admit, like the ordinary numbers that appear in physics, all the

†Work supported in part by the National Science Foundation, Grant No PHY-83-01221, and the Department of Energy, Grant No DE FG02-84ER-45144, and the University of Tokyo where part of the work was done.

arithmetic operations. A prime characteristic will in general be denoted by p.

First of all, the partition function W must necessarily belong to a field in order to be able to define expectation values in which W appears as a denominator. It would also seem natural that the field variable F be indeed a field (the pun is unintentional!) so that, for example, Green functions may be defined as the inverses of wave operators, although this is not a mandatory requirement.

As for the lattice coordinates x, they do not have to be a field if they are regarded as forming just an affine space. However, one does need inverses if rotations and Lorentz transformations are to be considered.

The rest of the paper is organised as follows. I will first establish some basic mathematical properties of Z_m that are relevant to the physical problems on hand. Next I will consider discrete versions of heat (diffusion) equations and wave equations in $1+1$ dimension. Here my main interest concerns the recurrence time, or the Poincaré cycle, of the system. Since there are l lattice sites, on each of which the field F takes k different values, there are only $N = k^l$ distinct sets of data at any time. Thus, off hand, an initial datum is expected to reappear after an elapsed time (Poincaré cycle) P of the order of N. The question is whether this is true or not, and if it is, to determine the actual value of P for given p and l.

I will then briefly comment, without actually working out the details, on equations in higher dimensions, coordinate transformations, invariance or non-invariance properties of the system, and the partition function.

At this point it also seems appropriate to spend a few words about related literature. The problems and results concerning heat equations are related to those of the recent work of Wolfram (1983, 1985) on cellular automata. As for the general use of Galois fields in field theory, the existence of early works by Coish (1959) and Jarnefelt (1949) has come to my attention at the time of this writing. Coish mainly discusses the Lorentz and other symmetry operations defined over Galois' fields (see also Joos 1964).

2 MATHEMATICAL PRELIMINARIES

If a and b belong to Z_p, then

$$(a+b)^p = a^p + b^p \pmod{p} \tag{2.1}$$

because the coefficients of the binomial expansion for all other terms are proportional to p. Thus the exponentiation by p is distributive. Starting from $a = b = 1$, it follows that

$$n^p = n \qquad \text{or} \qquad n^{p-1} = 1 \pmod{p} \qquad 0 < n < p. \tag{2.2}$$

Any of these $p-1$ elements n can be expressed as a power of one of them, called a primitive root, which is not unique.

Let k be some power of $p: k = p^r$. For a member a of Z_k, one can write

$$a(\text{mod } k) = a(\text{mod } p) + pb$$

with some b. Applying a binomial expansion as in (2.1), one can show

$$a^{k/p}(\text{mod } k)$$
$$= [a(\text{mod } p)]^{k/p}(\text{mod } k). \tag{2.3}$$

This reduces certain problems in Z_k to those in Z_p. For example,

$$a^{k/p} = 1(\text{mod } k) \qquad \text{if } a = 1(\text{mod } p). \tag{2.4}$$

(The power k/p may be lowered depending on b.)

Consider a general Z_m, and suppose $m = m_1 m_2$, m_1 and m_2 being relatively prime to each other: $(m_1, m_2) = 1$. Then

$$Z_m = Z_{m_1} \oplus Z_{m_2}. \tag{2.5}$$

This decomposition corresponds to the mapping

$$a(\text{mod } m) \rightarrow (a(\text{mod } m_1), a(\text{mod } m_2)) \tag{2.6}$$

which may be obtained by using the fact that $xm_1 + ym_2 = 1$ for some integers x and y. The inverse mapping is also unique.

Those members a of Z_m which are relatively prime to m, $(a, m) = 1$, form a multiplicative group Z_m^* (reduced residue system). Equation (2.4) implies

$$Z_m^* = Z_{m_1}^* \otimes Z_{m_2}^*. \tag{2.7}$$

The number of elements of Z_m^* is the Euler function $f(m)$. For any member a of Z_m^*

$$a^{f(m)} = 1(\text{mod } m). \tag{2.8}$$

From (2.4) one has, in particular,

$$f(p^r) = (p-1)p^{r-1}.$$

The analogue of a plane wave $F(x) = e^{ikx}$ (eigenfunction for translation) should be a^x in our case. According to (2.1) and (2.2), then

$$F(x + f(m)) = F(x) \tag{2.9}$$

if F belongs to Z_m. Thus the existence of plane waves naturally induces a lattice size $l = f(m)$. (More generally, l may be a multiple of $f(m)$.) The same argument may be applied to the case of a partition function W which belongs to Z_h. The action functional appears as the exponent of W, so that the field F must have $k = f(h)$.

This is the reason why three different parameters l, k and h had to be introduced. However, there are still problems: if h is a prime number p as was required, then $k = p - 1$, which is not a prime number. If k is a prime number, on the other hand, then l is constrained to be $p-1$ for the existence of running waves.

These constraints and difficulties can be circumvented by means of Galois' extensions of the original fields, as will be seen later. A Galois' extension of Z_p is obtained by adjoining the roots of an algebraic equation to Z_p. The procedure is similar to the extension of real numbers to complex numbers in order to construct running waves.

3 HEAT EQUATION

I will start by defining the discrete analogues of differential operators:

$$\begin{aligned} D_x^{\pm} F(x) &= F(x \pm 1) \\ d_x^{\pm} &= D_x^{\pm} - 1 \\ D^+D^- &= D^-D^+ = 1 \\ d^+d^- &= d^-d^+ = -d^+ - d^-. \end{aligned} \tag{3.1}$$

Note the distinction between forward and backward derivatives. They are adjoints of each other under the natural scalar product

$$(F, G) = \textstyle\sum_x F(x)G(x) \tag{3.2}$$

a fact which will become relevant later. Now consider the heat equation in one space dimension

$$d_t^+ F = -c d_x^+ d_x^- F \tag{3.3a}$$

or

$$D_t^+ F = (cD_x^+ + cD_x^- - c + 1)F. \tag{3.3b}$$

Given an initial datum $F(x)$ at $t = 0$, the latter equation can immediately be iterated. For this purpose it is convenient to expand F in normal modes for $D_x^{\pm}$. However, this is not possible within Z_k unless $l = f(k)$, as was mentioned before. Therefore extend Z_k to $Z_{k,l}$ by adjoining all the solutions z (which are not already present in Z_k) of

$$z^l - 1 = 0 \ (\text{mod}\ k). \tag{3.4}$$

Taking $F(x) = z^x$, equation (3.2) becomes

$$\begin{aligned} D_t^+ F &= (cz + c/z - 2c + 1)F \\ &= u(z)F. \end{aligned} \tag{3.5}$$

Now the formulae (2.1) and (2.2), as applied to $u(z)$ and properly generalised, show that

$$u(z)^p = u(z^p)$$

which can be repeated to give

$$u(z)^{p^n} = u(z^{p^n}). \tag{3.6}$$

From now on, it is convenient to discuss three different cases separately: (i) $k = p$, a prime number, and $(p, l) = 1$; (ii) $k = p$, $(p, l) \neq 1$; and (iii) general k.

(i) Case $k = p$, $(p, l) = 1$.

Equation (2.8) implies that there is an n which is a factor of $f(l)$ and is the smallest positive integer satisfying

$$p^n = \pm 1 (\text{mod } l) \tag{3.7}$$

so that

$$u(z)^{p^n} = u(z^{p^n}) = u(z)$$

because $u(z) = u(1/z)$. This means

$$u(z)^{p^n - 1} = 1 \tag{3.8a}$$

or else

$$u(z) = 0. \tag{3.8b}$$

According to equation (3.8a), the Poincaré cycle P should be given in general by

$$mP = p^n - 1 \tag{3.9}$$

where the possible reduction factor m cannot be determined by the present reasoning alone.

The case (3.8b) merits special attention. This possibility arises if equations (3.4) and (3.8b) have common roots. An example is the case $z = -1$ (l even), and $4c + 1 = 0$ (mod p). Such a zero mode is annihilated after one time step, and will never reappear. The Poincaré cycle does not exist in the strict sense of the word.

(ii) Case $k = p$, $(p, l) \neq 1$.

One may set $l = sp^r$, $(s, p) = 1$. This time it is a matter of reducing p^n in (3.6) modulo l, or with respect to Z_l.

If $s \neq 1$, Z_l can be split into $Z_s \oplus Z_{p^r}$ according to (2.5) and (2.6). Since $(s, p) = 1$, the argument used in (i) may be applied. It is possible to find the smallest $n (\geqslant r)$ such that

$$p^n = \pm 1 (\text{mod } s) \oplus 0 \ (\text{mod } p^r). \tag{3.10}$$

This implies that p^n is an idempotent operator:

$$u(z)^{p^{2n}} = u(z)^{p^n}$$

so

$$u(z)^{p^{2n}-p^n} = 1 \tag{3.11a}$$

unless

$$u(z)^{p^n} = 0. \tag{3.11b}$$

The Poincaré cycle P is thus

$$mP = p^{2n} - p^n \tag{3.12}$$

with a possible reduction factor m. However, there can be zero modes satisfying equation (3.11b). These modes are annihilated after p^n steps (or possibly less). One may assign them a degree of nilpotency P',

$$P' \leqslant p^n. \tag{3.13}$$

If $s = 1$, on the other hand, one has

$$\begin{aligned} u(z)^{p^r} &= u(z^{p^r}) \\ &= u(1)(\text{mod } p) \end{aligned} \tag{3.14}$$

which belongs to Z_p, and is either $= 0$ or $\neq 0$. If $u(1) \neq 0$, there exists an $n < p$ such that

$$u(1)^n = 1 \qquad \text{i.e.} \qquad u(z)^{np^r} = 1 \tag{3.15}$$

so that

$$mP = np^r = nl \tag{3.16}$$

with a possible reduction factor m.

The case $u(1) = 0$ is special. It means that all the modes are annihilated after l steps. Thus the system itself is nilpotent of degree P':

$$P' \leqslant l. \tag{3.17}$$

In field theory language, the system is driven to a fixed point $F = 0$. Such a situation happens because, as was remarked before, the time translation operator is not self-adjoint.

Incidentally, the extension of Z_p to $Z_{p,l}$, $l = sp^r$, is a non-separable one since

$$z^l - 1 = (z^s - 1)^{p^r}(\text{mod } p)$$

and $Z_{p,l}$ forms only a ring. This is reflected in the fact that the eigenvalue equation

$$d_x^+ d_x^- F = eF$$

admits, for $l = p$, two solutions with $e = 0$: $F(x) = 1$ $(z = 1)$ and $F(x) = x$. The latter does not have the assumed exponential form. Similarly, for

$l = 2p$, there are two solutions with $e = 4$:

$$F(x) = (-1)^x \qquad \text{and} \qquad F(x) = x(-1)^x.$$

This indicates the existence of some subtleties when $(l, p) \neq 1$ in general.

(iii) Case of general k.

The main theorems to be used for this purpose are equations (2.3) and (2.5). Let the characteristic k of F be decomposed into its prime factors:

$$k = p_1^{n_1} \times p_2^{n_2} \times \ldots . \tag{3.18}$$

Because of the linearity of the equation at hand, and in view of equation (2.5), the problem reduces to that of separately considering each component field belonging to a different prime factor $p_i^{n_i}$. As far as the individual components are concerned, equation (2.3) then implies that the previous results on P and P' be modified by a factor $p_i^{(n_i - 1)}$, except for the cases of nilpotency as in (ii) above.

Table 3.1 The Poincaré cycle for the case $p = 17$ and $l = 3, 4, \ldots, 17$. P_{exp} is the result of computer simulation, P_{th} is the theoretical maximum expected from equations (3.9) and (3.16) and $m = P_{\text{th}}/P_{\text{exp}}$.

l	P_{exp}	P_{th}	m
3	8	$p-1$	2
4	16	$p-1$	1
5	9	p^2-1	32
6	16	$p-1$	1
7	307	p^3-1	16
8	16	$p-1$	1
9	16	$p-1$	1
10	144	p^2-1	2
11	88741	p^5-1	16
12	144	p^2-1	2
13	307	p^3-1	16
14	4912	p^3-1	1
15	10440	p^4-1	8
16	16	$p-1$	1
17	17	p	1

Table 3.1 shows the results of a computer simulation for equation (3.3) for a representative case $p = 17$ and $3 \leqslant l \leqslant 17$, $c = 1$, based on a reasonably large sample of random initial data. They confirm the formulae obtained above. The reduction factor m is generally present, but actually it is found to vary with the parameter c. I conjecture that the case $m = 1$ always exists for given p and l.

It is obvious that most of the formulae obtained so far do not depend on the precise form of $u(z)$. This means that the right-hand side of the heat equation may be replaced by more general difference expressions which depend on any number of neighbours, and yield general polynomials of z over Z_k. However, the basic results will remain unchanged.

It seems appropriate at this point to recall the work of Wolfram on cellular automata. The evolution of his system generally depends on a set of neighbours in a non-linear way. He observed in particular that there are cases where initial patterns die away after some time steps. This is similar to the nilpotent case found above.

4 THE WAVE EQUATION

Consider the discrete version of the Klein–Gordon equation in $1+1$ dimension:

$$d_t^+ d_t^- F = c d_x^+ d_x^- F + mF. \tag{4.1}$$

It is convenient to consider a plane-wave solution

$$F = z^x w^t. \tag{4.2}$$

Equation (3.18) is then reduced to

$$\begin{aligned} t(w) &\equiv w + 1/w - 2 \\ &= (cz + c/z - 2c + m) \equiv v(z). \end{aligned} \tag{4.3}$$

This is a quadratic equation for w, hence the existence of w is not guaranteed in the original $Z_{k,l}$. One needs further extensions of $Z_{k,l}$ by adjoining the solutions w of equation (4.3) for all z's, if necessary.

As before, $v(z)$ is subject to recurrence relations obtained by taking its powers. Applying this procedure to both sides of equation (4.3), one will obtain recurrence relations for $t(w)$.

First consider the case $k = p$, $(p, l) = 1$. Following equations (3.7) and (3.8) one finds

$$t(w^{p^n}) = t(w)$$

which means

$$w^{p^n} + 1/w^{p^n} = w + 1/w$$

or

$$(w^{p^n - 1} - 1)(w^{p^n + 1} - 1) = 0. \tag{4.4}$$

Thus

$$w^{p^n - 1} = 1$$

or else

$$w^{p^n+1} = 1. \tag{4.5}$$

The former case corresponds to the same Poincaré cycle found earlier, but the latter is different. Thus the overall Poincaré cycle P in the present case must be the least common multiple of the two exponents in equation (4.5)

$$mP = p^{2n} - 1 \tag{4.6}$$

where m is again an undetermined reduction factor.

The zero modes do not exist because they would imply $w = 0$, which is not possible for equation (4.3). This is due to the time reversal invariance ($w \to 1/w$) of the wave equation.

Next the case $k = p$, $(p, l) \neq 1$. Following equation (3.10), one obtains

$$t(w^l) = v(1).$$

If $v(1) = 0$, this means $w^l = 1$, so that

$$mP = l. \tag{4.7}$$

If $v(1) \neq 0$, then use the fact

$$v(1)^p = v(1)$$

which leads to

$$t(w^l)^p = t(w^{lp}) = t(w^l)$$

or

$$(w^{lp+p} - 1)(w^{lp-p} - 1) = 0 \tag{4.8}$$

hence

$$mP = (l^2 - 1)p \tag{4.9}$$

as in equations (4.4)–(4.6).

The general case of Z_m may be similarly treated as was done for heat equations.

5 OTHER PROBLEMS

Without going into details, I will comment on general problems that are forseen to arise in pursuing the current program.

5.1 Higher dimensional wave equations

It is a straightforward matter to write down wave equations in higher dimensions by replacing differential operators with difference operators (3.1). There is a question whether the basic operator should be d_x or rather D_x as is typical of lattice theories. To a degree it is a matter of convention since they differ by only a constant term. However, the real question is whether or not there is a sufficient physical correspondence between the finite field theory and the continuum field theory.

For example, there will exist in general plane-wave solutions of the type

$$F = a^x b^y c^t \text{ etc}$$

where a, b and c belong to an appropriate extension of Z_p. In terms of a primitive root g, they may be written as

$$F = g^{ix+jy+kt} \text{ etc}$$

so that i, j and k may be interpreted as momentum–energy. However, they do not satisfy the usual relations, but rather, an equation like

$$(g^i + 1/g^i) + (g^j + 1/g^j) + c = g^k + 1/g^k.$$

One is faced with two basic problems.

(*a*) Positivity and ordering. In Z_m the numbers are in circular order, so one cannot talk about large and small, or positive and negative, except that 0 and 1 play special roles. This is a serious departure from the usual notion of physical quantities. However, recall that one does not realise the roundness of the earth by local observations alone, so which of two local points are closer to you has a clearcut answer. The problem arises only if physical quantities (in appropriate units like the Planck length) become of comparable order with the characteristic which may be taken sufficiently large. Nevertheless, there is a mathematical distinction between a large characteristic and a zero characteristic, the latter meaning the usual infinite set Z.

(*b*) Non-invariance of wave equations. Clearly the lattice has no invariance under linear transformations of coordinates. In fact, it is impossible to devise a wave equation for F in terms of difference or other algebraic operations because the coordinates and momenta belong to a characteristic different from that of F.

Furthermore, there is no notion of positivity of squares in general; the sum of two squares can also be zero, for example.

It may nevertheless be meaningful to discuss linear transformations of coordinates and momenta as defined above. If they belong to a prime field Z_p, discrete rotations and Lorentz boosts can be defined over Z_p (modular representation). For example, in a two-by-two subspace one has

$$\begin{pmatrix} a & b \\ b & a \end{pmatrix} \qquad a^2 - b^2 = 1 (\text{mod } p)$$

for boosts, and

$$\begin{pmatrix} a & b \\ -b & a \end{pmatrix} \qquad a^2 + b^2 = 1 (\text{mod } p)$$

for rotations. The former has $p-1$ solutions

$$a=\tfrac{1}{2}(x+1/x) \qquad b=\tfrac{1}{2}(x-1/x)$$

if $p>2$. The latter also has the same number of solutions

$$a=\tfrac{1}{2}(x+\mathrm{i}/x) \qquad b=(x-\mathrm{i}/x)/2\mathrm{i}$$

if the i also exists in Z_p. This is the case if $p=4n+1$. It is also possible to define spinor representations. The spinors transform with one half the rotation angles, or square roots of the x's above. Thus, one has simply to extend Z_p by square roots for spinor fields.

Although the rotations and boosts can be defined in the above way, their meaning in terms of the lattice points looks rather strange, for the transformations do not preserve the concept of neighbours. Thus it would not make sense to consider only nearest neighbours in building an action functional. Herein may lie the key to the resolution of the invariance problem.

5.2 Partition function

This is perhaps the most radical part of the present program. Partition functions W must admit division, and this requires them to belong to a prime field Z_p or its extensions. The indices (exponents) of its elements form a ring because

$$(a^x)(a^y)=a^{x+y}$$
$$(a^x)^y=(a^y)^x=a^{xy} \text{ etc.}$$

Thus, an action functional A can be constructed in this way from the field variable F. However, the induced characteristic of A is $p-1$ which is not a prime number.

If a prime characteristic is wanted for A, one way to get around the problem is to generate the partition function from a primitive kth root z of unity in the algebraic extension $Z_{h,k}$. This would be the natural analogue of the quantum mechanical partition function where k is the inverse Planck constant. A thermal partition function can be obtained from it by the imaginary-time trick: replacing the time coordinate t with it provided that $\mathrm{i}=\sqrt{-1}$ exists in Z_k. More generally, one must define the ith power of a number in Z_h^* by

$$(a^{\mathrm{i}})^{\mathrm{i}}=a^{-1}.$$

The mapping $a\rightarrow a^{\mathrm{i}}$ may be viewed as a multiplicative automorphism of Z_h^* or some extension of it.

ACKNOWLEDGMENTS

I thank Professor N Nakanishi for enlightening me regarding the early literature, and T Wolf for his interest in the project and some useful comments.

Note added in proof. The following reference should be added to the work of S Wolfram: 1984 *Commun. Math. Phys.* **93** 219. This paper contains a detailed, but somewhat different, treatment of the main topics discussed in the current work. I thank Dr Wolfram for calling my attention to this and to related papers of his.

REFERENCES

Coish H R 1959 *Phys. Rev.* **114** 383
Jarnefelt J 1949 *Veroeffentlichungen des Finnischen Gaeodaetischen Institutes No* 36 (and other papers, as quoted by Coish (above))
Joos H 1964 *J. Math. Phys.* **5** 155
Wolfram S 1983 *Rev. Mod. Phys.* **55** 601
—— 1985 *Phys. Rev. Lett.* **54** 735

104 Progress of Theoretical Physics Supplement No. 85, 1985

Directions of Particle Physics*)

Yoichiro NAMBU

The Enrico Fermi Institute and Department of Physics
University of Chicago, Chicago, IL 60637, U. S. A.

§ 1. Modes of quest in particle physics

Particle physics as a subdiscipline in physics owes its origin to E. Lawrence for its experimental aspect, and to H. Yukawa for its theoretical aspect. The significance of their contributions lies not only in the particular work they did, but more importantly, in the fact that each established a basic methodology for the particle physicists to follow. These methodologies have turned out to be so powerful and fruitful that we are still following them. On the experimental side, this is most evident in the current activities going on at various accelerator laboratories all over the world. Their particle energies are now reaching a million times that of Lawrence's first cyclotron.

The purpose of my talk, however, is to concentrate on the theoretical side. In an attempt to understand the properties of nuclei with their two mysterious types of interactions, the strong and the weak, he unwittingly, so to speak, hit the tip of an iceberg which was to become the whole discipline of particle physics. I doubt that Yukawa himself, when he hypothesized his meson in 1935, would have anticipated the proliferation of elementary particles we have discovered since then.

It seems to me that Yukawa's approach had a heuristic and phenomenological tone. Because of this nature, it served as a useful guiding principle to explore uncharted and ever surprising new worlds, just as Lawrence's cyclotron and its descendants served as useful experimental tools to go with it. Yukawa's approach had both conservative and radical sides. He was conservative in the sense that he pursued the logical consequences of relativistic quantum field theory (which was in its infancy at that time) rather than vaguely anticipating a more radical solution of the problem of nuclear forces, a belief often held by his contemporaries. But he was radical in the sense that he did not hesitate to speculate on the existence of new elementary particles which had not been seen. According to him, his meson had not been seen in everyday life, not only because it had a large mass, but also because it was unstable. Strong binding and weak instability are the two peculiar features of nuclei, but one might feel uneasy to accept the notion of an unstable but elementary particle unless one has a pragmatic mind, as did Yukawa and, before him, Fermi. It is a notion we have come to accept without understanding its deep reasons even now. At any rate, Yukawa solved the problem of nuclear forces by dividing it in two inherently different parts: the theoretical framework in which to decribe nature, and the substantive question of what entities exist to be described. In the former, he embraced the basic tenets of the relativistic quantum field theory in spite of its imperfec-

*) Work supported in part by the National Science Foundation, Grant No. PHY-83-01221, and the Department of Energy, Grant No. DE FG02-84-45144.

tions. In the latter, he took the cue from nature, and asked himself what kind of particle / field should be responsible for the strong nuclear forces. He even tried to kill two birds with one stone, linking the weak interactions with the same hypothetical particle.

In the subsequent development of particle physics for the past fifty years, one may detect in the efforts of theorists two competing modes of approach which are related to the two types of questions just discussed. In this talk I would like to call them the Yukawa mode and the Dirac mode. The Yukawa mode is the pragmatical one of trying to divine what underlies physical phenomena by attentively observing them, using available theoretical concepts and tools at hand. This also includes the building and testing of theories and models. It is the standard way of doing research in all branches of science. In particle physics, the following examples come to my mind:

Quark
GUTS (grand unified theories)
Parton model
Dual resonance-string model.

The other mode, the Dirac mode, is to invent, so to speak, a new mathematical concept or framework first, and then try to find its relevance in the real world, with the expectation that (in a distorted paraphrasing of Dirac[1]) a mathematically beautiful idea must have been adopted by God. Of course the question of what constitutes a beautiful and relevant idea is where physics begins to become an art.

I think this second mode is unique to physics among the natural sciences, being most akin to the mode practiced by the mathematicians. Particle physics, in particular, has thrived on the interplay of these two modes. Among examples of this second approach, one may cite such concepts as

Magnetic monopole
Non-Abelian gauge theory
Supersymmetry.

On rare occasions, these two modes can become one and the same, as in the cases of the Einstein gravity and the Dirac equation.

§ 2. Evolution of the Lawrence-Yukawa paradigm

The Lawrence-Yukawa paradigm has been, and still is, the dominant mode of research in particle physics. As far as the experimental side is concerned, this will remain so in the forseeable future, although the big accelerators now seem to be approaching practical limits due to their cost and physical size.

Turning to the theoretical side, a kind of revolution took place in the early 70's when the above mentioned interplay of the two modes began to bear fruit. First of all, the progress in the quantum theory of non-Abelian gauge fields opened up the possibility that gauge fields are at the root of all the forces in nature. The two types of interactions that Yukawa set out to explain in terms of intermediary particles, i.e., the strong and the weak, could now be viewed, together with the classic electromagnetism, as different manifestations of gauge fields, i.e., the color $SU(3)$ and flavor $SU(2)\times U(1)$, acting on the fundamental fermions, i. e., the quarks and leptons. The ensuing grand unification of the three forces, as embodied in the GUTS, immensely appeals to our sense of the unity of natural laws. It is also interesting and satisfying that the magnetic monopole, which was originally an object of free invention by Dirac, has turned out to be a natural and inevitable consequence of GUTS.

The grand unification, however, is not a complete unification because gravity, the oldest and the most universal of the forces, is not yet included. As the next step, attempts to unify all forces including gravity have led to the revival and modern interpretations of the Kaluza-Klein theories, in which internal symmetries are reflections of the geometry and topology of higher dimensional manifolds. Thus, at least conceptually, there is a natural framework in which a unification of all forces can be envisioned. Yet this program of complete unification still suffers from some defects:

a) Gravity cannot be consistently treated as a quantum field theory. This is similar to the situation with quantum electrodynamics before renormalization theory. However, the problem goes even deeper than that, involving conceptual questions.

b) The program still remains at a phenomenological level as far as the weak interaction is concerned, because it only describes, but does not explain, the irregular and symmetry nonconserving aspects of the weak interaction.

c) Direct tests of the basic ideas can be done only at energies way beyond the reach of present accelerators. There are only a few low energy manifestations.

The last point underscores the peculiar position in which particle physics finds itself at present. There now exists a growing disparity between the capacities of theory and experiment. Theory cannot make precise predictions about phenomena just above the available energies, and accelerators cannot possibly reach the grand unification energies where predictions are clear. This is a rather unfortunate situation. Only if by luck any convincing evidence should turn up about nucleon decay, the situation would change drastically, and we could happily say that theory has leapfrogged many decades of GeV's, and fixed our bearing toward a distant but reassuring star. As of now, however, the original optimistic predictions about nucleon decay have not been confirmed, and again the lengthening time scale for experiment has become a frustrating element as is also true with the case of accelerator development.

Perhaps one should not complain too much about it. Yukawa's meson did not turn out to be just of one kind, but actually have come out in vast numbers. Even completely unexpected new leptons have showed up. It may be that we have grown too self-confident and expecting too much out of theory.

In the meantime, theory and experiment seem to be going their separate ways. Theorists have turned from accelerators to astrophysics and cosmology for guidance and testing, but this is a far less reliable or controllable means than laboratory experiments. In a way we are being forced back to the time when cosmic rays were the primary tool of particle physics.

§ 3. The rise of the Dirac mode

There is, however, yet another trend in theoretical activity which has been going on for some time. It is the rise of paradigms in the Dirac mode, and I would like to address it for a moment. The topics that falls under this category in my mind are supersymmetry and string theory.

a) Supersymmetry paradigm

If non-Abelian gauge theories and monopoles have already found their immediate physical relevance in the GUTS, such is not yet the case with supersymmetry. It seems a bit of an accident that the appearance of supersymmetry coincided with the progress in gauge theories. Gauge theories are a generalization of known physical theories, but supersymmetry is not. The situation is somewhat similar to that of the Dirac equation. In the latter case, however, it found immediate relevance.

Theoretically, supersymmetry has some very appealing features, like improved ultraviolet behavior. Perhaps more importantly, it can admit particles of all spins, from Higgs boson to graviton, to be as equals, a feat of ultimate unification. Unfortunately, the way fermions and bosons are organized in supersymmetric theories does not seem to correspond to the patterns of known fermions and bosons. So supersymmetry simply adds more unknown partners to the known ones without explaining the latter.

Unlike gauge theory, supersymmetry is not based on a conceptually simple principle. In fact it is not clear what the principle is. Without known examples, and without a clear physical principle behind it, supersymmetry by itself does not look much different from other mathematical constructs like parastatistics.

b) String paradigm

String theory has a solid phenomenological origin, and is not a result of free invention. In hadron physics where it originated, it has come to be regarded as a phenomenological substitute for quantum chromodynamics, just as the Ginzburg-Landau theory is a phenomenological substitute for the Bardeen-Cooper-Schrieffer theory of superconductivity.

But somehow string theory has acquired a life of its own. By combining the three formal ideas represented by the string, supersymmetry and Kaluza-Klein theories in a unique way, and applying them to the most speculative and inaccessible realm of physics, the recent superstring theories have suddenly found themselves serious candidates for an ultimate theory of the world.[2),3)] In the very least, it seems to offer the possibility of a finite quantum theory of gravity along with all other forces and particles.

Superstring theory presents a utopian vision of the world. The only big question is whether it is a vision of the real world or not. So far, the vision is set only at the asymptopia of Planck energies. One cannot yet say with confidence that our low energy world and the world of superstrings live in the same connected manifold. Whatever the predictive power of the superstring theory may turn out to be, there is every reason for us to explore experimentally the physics in the TeV range where something must surely happen to the weak interaction but one does not know exactly what to expect.

§ 4. Speculative comments

So far I have engaged only in philosophical characterizations of the ongoing trends of particle physics. I would like now to present a few ideas of my own. My intention is to explore alternative ways, as different as possible from the standard ones, to look at some unsolved problems (in the perverse hope that I may be proven wrong!).

My primary interest in this regard concerns the weak interaction in general, and the question of fermion generations and mass spectra in particular. I am struck by the fact that among the four types of interactions, the weak interaction alone is the one that does not respect symmetries. Under the current thought, the mass spectra of quarks and leptons are intimately related to their weak interactions. It is these seemingly capricious mass spectra that make the world look so complicated. Without the problem of mass, the world might be much simpler, but at the same time very boring.

Physicists engaged in model building take pains to conform to the real world without questioning the latter. It is instructive, however, to reflect on how the nature of the real world, down to the existence of life itself, may be critically dependent on some very subtle properties of the Kobayashi-Maskawa fermion mass matrix. If a model builder were not concerned about fitting the real world too well, but just followed his natural logic, the predicted world probably would be vastly simpler than the real one. Perhaps he would first assume the mass matrix to be diagonal, in which case each generation would be stable, and there would exist many exotic forms of matter. Even if he took into account the possibility of generation mixing and worked out other fine details, would he correctly conclude that the up quark is lighter than the down quark, in such a way that the neutron-proton mass difference comes out greater than the electron mass so that free neutrons decay ? Without this, the world we know would not exist and the human beings would not be around to pose such questions.

What I am driving at is the question whether or not our world is really predictable down to its minute and subtle details. As pointed out above, the most serious difficulties come from the weak interaction sector. A natural answer to the existence of fermion generations with complex spectra would be that the fermions are composite objects with substructure. Yet a naive notion of compositeness does not seem to apply here, because a) the mass spectra are too irregular and too sparse, and b) there are no signs of a substructure ; in fact the good agreement of the Weinberg-Salam theory with experiment and the absence of processes like $\mu e \gamma$ decay cannot easily be reconciled with compositeness.

So let me for a moment try to be as radical as possible, and challenge the notion of the laws of physics in the sense commonly understood by physicists, namely that there is a unique Lagrangian from which all the laws of physics are derived and the properties of the physical world determined. This could be done in various different contexts. For example, one could say that one really has to integrate over all possible theories of the world according to the quantum mechanical principle. Actually this may not be much different from what we usually do, if a particular theory, the "right" one, has a dominant probability among all theories.

In a more serious vein, one could ask whether the laws of physics are intimately bound up with the evolution of the universe, influenced not only by the initial conditions, but also by the subsequent evolutionary processes themselves. In a way I am suggesting the biological evolution as a possible model for physical evolution.

One would have to be more specific, however, so let me entertain the idea that the term "generation" means more than just an analogy. Is it at all possible that the generations of quarks and leptons have "evolved" one after another in some sense, that each generation is "born", so to speak, at the corresponding energy (or length) scale of an expanding universe, its properties being influenced, but not necessarily deterministically fixed, by what already exists?

Biological evolution apparently is made possible by the vast degrees of freedom residing in complex molecules. If translated to particle physics, this might again bring back the compositeness issue. Are lower mass generations more complex than the higher ones? This is hardly likely although the opposite might be true. So what I should mean would be that the constants like mass are really dynamical quantities that were selected, with some degree of chanciness, from among other possibilities in the course of the universal evolution.

We do not know how many generations really exist and why. But let us consider, for example, the following hypothesis. Fermions of one generation with a characteristic mass m will exist in abundance during the cosmological expansion if m is less than the temperature. Assume that these fermions will beta decay by creating a new generation with lower mass if the decay rate is higher than the rate of expansion at that moment.

The beta decay rate goes like

$$\sim m^5/M_W{}^4,$$

where M_W is the W boson mass. On the other hand, the expansion rate of the universe at temperature $T(\sim m)$ goes like

$$\sim T^2/M_P,$$

where M_P is the Planck mass. As the universe expands and the temperature cools down, the decay rate will decrease more rapidly than the expansion rate. These two rates become equal when T is of the order of a few MeV, which is the mass scale of the lowest and last generation. It may not be complete nonsense to read some significance into it. (The above condition is the same as that which applies to general decoupling of weak processes out of thermal equilibrium. Incidentally, if the electron decayed to an even lower generation, its lifetime should be about a week.)

I would like to end my talk with a less drastic remark than the one just given. It concerns the problem of supersymmetry again. The relativistic supersymmetry of Wess and Zumino has not yet found any direct experimental support. In the meantime, however, certain kinds of supersymmetry have turned out to be useful concepts in certain problems of quantum mechanics and statistical physics. But the real physical meaning underlying supersymmetry still remains unclear.

One of the most interesting cases of apparent supersymmetry occurs in nuclear physics. As has been analyzed by Iachello and collaborators, a group of nuclei having even or odd mass numbers form an approximate supermultiplet with regard to their low energy spectra.[4] Recently I have attempted to attribute this to a general feature of the BCS pairing mechanism, which certainly applies to nuclear physics.[5] Whether it is in fact the mechanism behind the Iachello supersymmetry or not, the BCS theory leads to rather simple relations among low energy bosonic and fermionic spectra. These relations have been experimentally confirmed in superconductors and superfluid helium 3. For example, the analogs of the pion, quark, and σ meson in a superconductor have masses in the ratio 0:1:2 (in the absence of the Coulomb interaction). In terms of the effective theory (σ model) of Ginzburg-Landau and of Gell-Mann-Levy, it means that the Yukawa and quartic self-couplings of the bosonic fields are simply related. A similar result is obtained when the formula is applied to the nuclear case.

The question is: In what sense does the BCS theory have a hidden and approximate supersymmetry? As it turns out, it is possible to express the static (non-kinetic) part of the effective Hamiltonian for a superconductor as a product of fermionic operators Q and $Q^\dagger$ as in supersymmetric quantum mechanics:

$$Q=\pi^\dagger\psi_{up}-i(\phi\phi^\dagger-c^2)\psi_{dn}^\dagger, \qquad \bar{Q}=\int Q dv,$$

$$H=\{\bar{Q},\bar{Q}^\dagger\}$$

$$=\int[\pi\pi^\dagger+(\phi\phi^\dagger-c^2)^2+\psi_{up}\psi_{dn}\phi^\dagger+\text{h.c.}]dv,$$

where the ψ's are the spin up and spin down electron fields, ϕ and $\pi^\dagger$ are a bosonic field and its canonical conjugate field representing electron pairs. (Q and $Q^\dagger$ carry a definite charge unlike quasi-particle fields.)

The difference from the case of rigorous supersymmetry lies in the fact that Q and $Q^\dagger$ do not commute with H because Q^2 and $Q^{\dagger 2}\neq 0$. But one may say that the latter are effectively proportional to $\langle\phi\rangle$, and thus the spontaneous breaking of the $U(1)$ (charge) symmetry also breaks a supersymmetry between fermions and bosons as is reflected in their mass relations.

In more general cases of the BCS mechanism, one can let the fields become multicomponent:

$$Q=\pi^\dagger\psi-iV(\phi)\psi^\dagger,$$

$$H=\{\bar{Q},\bar{Q}^\dagger\}$$

$$=\int[\mathrm{Tr}(\pi\pi^\dagger+VV^\dagger)+[\psi,V'\psi]/2+\text{h.c.}]dv,$$

$$V'=\partial V/\partial\phi\,.$$

Here the bosonic fields are matrices acting on the fermionic fields. These formulas give the rest spectra of bosons and fermions for the examples of helium and nuclear physics mentioned above.

It is not yet clear why this formulation of the BCS theory works. But one would be tempted to apply it to the problem of dynamical mass generation for the fundamental fermions, assuming that it is due to a similar mechanism. In the simplest analogy, each fermion would be accompanied by a Higgs partner of comparable mass, and their coupling would not be expected to be very weak. This certainly does not seem correct. More likely are the technicolor type theories, in which case the Higgs and some heavy fermions form partners, but this would not by itself shed light on the riddle of the mass spectra of the known fermions.

References

1) P. A. M. Dirac, Proc. Roy. Soc. **A133** (1931), 60.
2) M. Green and J. Schwartz, Phys. Lett. **149B** (1984), 117; **151B** (1985), 21.
3) D. Gross, J. Harvey, E. Martinec and R. Rohm, Phys. Rev. Lett. **54** (1985), 502.
4) F. Iachello, Phys. Rev. Lett. **44** (1980), 772; Physica **15D** (1985), 85.
5) Y. Nambu, Physica **15D** (1985), 147.

Supersymmetry and Superconductivity

*Y. Nambu**

Enrico Fermi Institute and Department of Physics
The University of Chicago, Chicago, IL 60637

ABSTRACT

The fermionic and bosonic spectra that follow from a BCS mechanism have a rather simple and universal structure. It is shown, in typical examples, that the effective Hamiltonian that reflects such a structure can be factorized in a way similar to the case of supersymmetric systems. The underlying spectrum generating superalgebras are identified.

1. Introduction

Recently I made the observation[1] that, in dynamical symmetry breaking of the BCS type, one generally gets simple relation among the gap (mass) parameters of the fermions and the composite bosons, and speculated that such relations are responsible for the apparent supersymmetry[2] in the energy levels of some nuclei. The existence of these relations have been known for a long time, but their universal character has not been fully appreciated. Moreover, it is relatively recently that experiments in superconductors and superfluid helium 3 have confirmed the theoretical results (see ref. 1).

The mass relations I am talking about are as follows. In the case of superconductors with the usual S-wave pairing, the masses of the three low energy excitations, i.e., the phase (π or Goldstone) collective mode, the quasielectron, and the amplitude (σ) mode are in the ratios 0:1:2. This is true in the weak coupling limit in which a chain of loop diagrams are summed, and the Ward identity and dispersion relations are used to derive the results. A clear understanding of the origin of the relations is still lacking, however.

In more complex systems line helium 3, there are a variety of collective modes, but again one can derive a generalized mass formula:

$$m_1^2 + m_2^2 = 4m_f^2 \,. \tag{1}$$

*Work supported in part by the NSF, PHY-83-01221, and the DOE, DE FG02-84ER 45144.

As is already mentioned above, an unanswered theoretical question is what symmetry principle, if any, underlies the mass relations, and in particular whether or not a kind of supersymmetry is inherent in the BCS mechanism. In the present article I shall show that the static (non-kinetic) part of the effective Hamiltonian for the low energy excitations of a BCS system can indeed be expressed as a bilinear form in fermionic composite fields in direct analogy to supersymmetric theories. The system is not supersymmetric, but represents only a slight and almost trivial generalization of supersymmetry. (A preliminary account appears in ref. 3.) In the following sections I will work out the mathematical details, starting from superconductivity and then going to more general cases.

2. Fermionic operators in superconductivity

The low energy excitations in a superconductor can be described by an effective Hamiltonian

$$\begin{aligned}
\bar{H} &= \int H dv \\
H &= \psi^\dagger(\epsilon\tau_3 + f\phi\tau_+ + f\phi^\dagger\tau_-)\psi \\
&\quad + (\pi\pi^\dagger + (v^2/3)\nabla\phi\cdot\nabla\phi^\dagger) + f^2(\phi\phi^\dagger - (m/f)^2)^2, \\
\psi &= (\psi_{up}, \psi^\dagger_{dn}) \,. \qquad (2)
\end{aligned}$$

Here ψ and $\psi^\dagger$ represent Bogoliubov–Valatin fermionic quasi-particle modes; ϕ and $\phi^\dagger$ stand respectively for the bosonic collective modes of electrons and holes; π and $\pi^\dagger$ are respectively the canonical conjugates to $\phi^\dagger$ and ϕ, ϵ is the electron kinetic energy measured from the Fermi energy, and v is the Fermi velocity.

The bosonic part of H is of the Ginzburg–Landau form, but with its self-coupling constant dynamically determined to be equal to the square of the Yukawa coupling constant. After the system undergoes the familiar spontaneous breaking of phase invariance (charge conservation), the phase mode $arg(\phi)$, the fermionic modes ψ and $\psi^\dagger$, and the amplitude mode $|\phi|$ acquire their masses in the ratios $0 : m : 2m$.

In the following I will completely ignore the kinetic parts of H for fermions and bosons since they are obviously very different from each other. Now define the fermionic composite fields

$$
\begin{aligned}
Q^\dagger &= \pi^\dagger \psi_{up} - iV\psi^\dagger_{dn}, \\
Q &= \pi\psi^\dagger_{up} + iV\psi_{dn}, \\
V &= f(\phi\phi^\dagger - (m/f)^2) .
\end{aligned} \tag{3}
$$

It is then straightforward to show that their anticommutator gives

$$
\begin{aligned}
\{Q(\mathbf{r}), Q^\dagger(\mathbf{r}')\} &= \delta(\mathbf{r}-\mathbf{r}')H_s(\mathbf{r}), \\
H_s &= \pi\pi^\dagger + V^2 + \psi_{up}\psi_{dn}\partial V/\partial\phi + \psi^\dagger_{dn}\psi^\dagger_{up}\partial V/\partial\phi^\dagger .
\end{aligned} \tag{4}
$$

H_s is nothing but the static part of H. Equation (4) is to be compared with the case of supersymmetric quantum mechanics:

$$
\begin{aligned}
Q = (p - iV(x))\psi, \quad Q^\dagger &= (p + iV(x))\psi^\dagger, \\
\{Q, Q^\dagger\} = p^2 + V(x)^2 + V(x)'[\psi^\dagger, \psi] &= 2H .
\end{aligned} \tag{5}
$$

Beyond the rough correspondence $p \to \pi$, $x \to \phi$, there is a difference in the degrees of freedom of both bosonic and fermionic operators. This difference leads to the consequence that, whereas Q and $Q^\dagger$ commute with H in the case of Eq. (5) because $Q^2 = Q^{\dagger 2} = 0$, these properties are not shared by the operators in Eq. (4). Since spin-up and spin-down states are not treated symmetrically in Q and $Q^\dagger$, I will also introduce their time-reversed counterparts

$$
\begin{aligned}
Q^{\dagger\prime} &= -\pi^\dagger\psi_{dn} - iV\psi^\dagger_{up} , \\
Q' &= -\pi\psi^\dagger_{dn} + iV\psi_{up} .
\end{aligned} \tag{6}
$$

Each of the four operators in Eqs. (3) and (4) carries a definite charge and spin. The pair (Q, Q') forms a spin doublet with charge -1 (electronlike), and the pair $(Q^{\dagger\prime}, -Q^\dagger)$ form s a doublet with charge $+1$ (holelike). The anticommutators among them are found to be (omitting spatial delta functions)

$$
\begin{aligned}
\{Q, Q\} &= -2f\phi S_+ , \\
\{Q', Q'\} &= 2f\phi S_- , \\
\{Q, Q'\} &= -2f\phi S_3 , \\
\{Q, Q^\dagger\} = \{Q', Q^\dagger\} &= H_s , \\
\{Q, Q^{\dagger\prime}\} &= 0, \text{ etc.}, \\
S_+ &= \psi^\dagger_{up}\psi_{dn} , \\
S_- &= \psi^\dagger_{dn}\psi_{up} , \\
S_3 &= [\psi^\dagger_{up}, \psi_{up}]/4 - [\psi^\dagger_{dn}, \psi_{dn}]/4 .
\end{aligned} \tag{7}
$$

The S's are fermion spin operators. But the algebra does not close here, eventually leading to an infinite tower. On the other hand, if V is a linearized from the beginning around the condensates $\langle\phi\rangle = mz/f$ and $\langle\phi^\dagger\rangle = mz^*/f$, $|z| = 1$, the resulting four fermionic operators and the four bosonic operators

$$\begin{aligned} &Q_0,\ Q_0^\dagger,\ Q_0',\ Q_0^{\dagger\prime}\ ; \\ &\text{and } H_0,\ S_+,\ S_-,\ S_3 \end{aligned} \tag{8}$$

form, after appropriate renormalization, a superalgebra $su(2/1)$. In general, an $su(m/n)$ algebra acting on n fermions and m bosons has an even (Lie algebra) part consisting of $su(n)$, $su(m)$ and a diagonal $u(1)$, the last being weighted in the ratio $m : n$ between fermion and boson sectors. (See, for example, ref. 4.) In the present case of $su(2/1)$, the $su(2)$ corresponds to spin, and the $u(1)$ to the Hamiltonian H_0, with a mass ratio of $1 : 2$. The zero mass boson does not appear in the algebra because, in the static approximation, it is not a true dynamical degrees of freedom. The odd part of $su(2/1)$, i.e., the Q's and $Q^\dagger$'s, act as ladder operators for H_0, connecting fermionic and bosonic sectors. Indeed one has

$$[Q_0^\dagger, \bar{H}_0] = -mzQ_0', \qquad [Q', \bar{H}_0] = -mz^*Q_0^\dagger, \quad \text{etc.} \tag{9}$$

Spontaneous charge symmetry breaking thus induces a broken supersymmetry with a spectrum generating superalgebra.

3. General forms of fermion operators

The previous example suggests that the effective Hamiltonian for a BCS-type system may in general be expressed as a bilinear form in fermionic operators. To see their structure, suppose that the fermion field ψ has $2n$ components forming the fundamental representations of $U(2n)$. (I consider only even numbers because fermions must have half-integral spins, and spin $SU(2)$ is a subgroup of $U(2n)$.) If condensates are formed in s wave pairing, they will belong to the antisymmetric representation

$$\sim \psi_i\psi_j - \psi_j\psi_i \tag{10}$$

and its complex conjugate, each having $k = n(2n-1)$ components.

Let O_a, $a = (i,j)$, $i < j$ be k antisymmetric matrices with entries ± 1 respectively at site (i,j) and (j,i), and zero elsewhere. These O's are generators of $SO(2n)$ acting on ψ or $\psi^\dagger$. Correspondingly let ϕ_a be k complex scalar matrix fields, and π_a be their canonical conjugates.

I now introduce the fermionic field Q and $Q^\dagger$:

$$Q^\dagger = \pi^\dagger \psi + iV\psi^\dagger,$$
$$Q = \psi^\dagger \pi - i\psi V^\dagger, \tag{11}$$

when the ϕ's and V are to be regarded as matrices acting on the ψ's. For simplicity's sake, the coupling constant f is taken equal to 1.

For the "superpotential" V, consider the choice

$$V = \phi\phi^\dagger - m^2 . \tag{12}$$

The Hamiltonian is generated by the anticommutators:

$$\bar{H} = \sum_n \{\bar{Q}_n, \bar{Q}_n^\dagger\}/2 = \int H dv ,$$
$$H = tr(\phi\phi^\dagger\phi\phi^\dagger)/2 - m^2 tr(\phi\phi^\dagger) + nm^4 + \psi V'^\dagger \psi/2 + \psi^\dagger V' \psi^\dagger/2,$$
$$V' = i \sum O_a [\pi_a, V] = \phi . \tag{13}$$

To see the meaning of this Hamiltonian, first note that n of the $k = n(2n-1)$ fields $\phi_{ij} = -\phi_{ji}$, for example those with

$$i, j = 1, 2; 3, 4; \ldots ; 2n-1, 2n \tag{14}$$

do not directly couple with each other in the potential function in H. Keeping only these components, H reduces to

$$H = \sum (\pi_p \pi_p^\dagger + \phi_p \phi_p^\dagger + \psi \left(\sum O_p \phi_p^\dagger\right) \psi/2 - \psi^\dagger \left(\sum O_p \phi_p\right) \psi^\dagger/2,$$
$$p = (2i-1, 2i) . \tag{15}$$

Thus the ground state has zero energy, corresponding to the condensates

$$\langle \phi_p \rangle = z_p m, \qquad \langle \phi_p^\dagger \rangle = z_p^* m, \qquad |z_p| = 1 . \tag{16}$$

With the choice of the phase factors $z_p = 1$, the fermion mass term takes the form

$$(m/2)(\psi\eta\psi - \psi^\dagger \eta \psi^\dagger), \eta = \sum O_p, \qquad \eta^2 = -1 . \tag{17}$$

Thus if the $U(2n)$ symmetry is broken down to $[SU(2)]^n$, each $SU(2)$ subspace is spanned by a complex boson field ϕ_p and a pair of fermion fields ψ_{2i-1}, ψ_{2i}, and is equivalent to the previous case of superconductivity.

The system after the symmetry breaking, however, has a larger symmetry than $[SU(2)]^n$ because of the equivalence of all the $SU(2)$ subspaces. It is in fact invariant under $Sp(2n)$. There are $2n$ massive fermions and $n(2n-1)$

massive bosons. From the last example, one would expect the relevant superalgebra to be either $su(2n/[n(2n-1)])$ or $osp(4n/[2n(2n-1)])$, the former being a subalgebra of the latter. The Hamiltonian, however, cannot be the $u(1)$ piece of $su(2n/[n(n-1)])$ for $n > 1$, because the mass ratio does not come out right. I will now proceed to check this out.

Redefine ϕ and $\phi^\dagger$ to be deviations from the condensates given in Eq. (17) with the standard choices $z = 1$;

$$\langle\phi\rangle = -\langle\phi^\dagger\rangle = m\eta\ . \tag{18}$$

(Note that $\dagger$ also implies matrix conjugation.)
Linearize Q and $Q^\dagger$ with respect to ϕ and $\phi^\dagger$; call them Q_0 and $Q_0^\dagger$, and form H_0 out of them. One finds

$$\begin{aligned} Q_0 &= \pi^\dagger\psi + (-\phi + \phi^{\dagger\prime})\eta\psi^\dagger\ , \\ Q_0^\dagger &= -\pi\psi^\dagger - \eta(-\phi^{\prime\dagger} + \phi)\psi\ , \end{aligned} \tag{19}$$

and

$$\begin{aligned} H_0 = \{Q_0, Q_0^\dagger\}/2 &= -\mathrm{tr}(\pi\eta\pi^{\dagger\prime}\eta)/2 - (m^2)\mathrm{tr}((\phi^{\dagger\prime} - \phi)\eta(\phi^{\dagger\prime} - \phi)\eta)/2 \\ &+ m(\psi\eta\psi - \psi^\dagger\eta\psi^\dagger)/2, \qquad \phi^{\dagger\prime} = \eta\phi^\dagger\eta^{-1}, \quad \pi^{\dagger\prime} = \eta\pi^\dagger\eta^{-1}. \end{aligned} \tag{20}$$

Equation (19) is written in such a way as to exhibit the invariance of H_0 under $Sp(2n)\cap U(2n)$ transformations with respect to the skew metric η: each contraction of indices is done through the intermediary of η. In other words, the choice of V given by Eq. (5) corresponds to the case of symmetry breaking $U(2n)$ to $Sp(2n)$. One might suspect that all possible anticommutators of the $4n$ Q_0's and $Q_0^\dagger$'s yield a closed superalgebra, $su(1/2n)$ which has $4n$ odd elements, and contains $sp(2n)$ as a subalgebra. This, however, is not true, as the algebra does not close.

At any rate, these anticommutators are listed in the table. The structure of the algebra for the fermion sector is simple. The $n(2n+1)$ members in a) are generators of an $sp(2n)$, and together with the $n(2n-1)$ members in b) form a $u(2n)$ algebra with $(2n)^2$ elements. The fermionic part of H_0 is a member of b) with $A = \eta$, and is the $u(1)$ piece of the algebra.

The bosonic part is rather complicated and messy. There are again the same numbers of elements as in the fermionic sector, except for the previous special case of $n = 1$, where the only element was the H_0 piece in b). For $n > 1$, however, the algebra in the bosonic sector does not close, due basically to the presence of the factor η. In fact, the commutators of Q and $Q^\dagger$ with these even elements give rise to new odd elements. Eventually the process closes on a superalgebra $osp(4n/[2n(2n-1)])$, where the $o(4n)$ is the maximal algebra for the $2n$ ψ's and $2n$ $\psi^\dagger$'s, and similarly the $sp([2n(2n-1)])$ is the maximal algebra for the $n(2n-1)$ ϕ's and $n(2n-1)$ π's of the massive modes. A difference between $osp(4n/[2n(2n-1)])$ and its subalgebra $su(n/[n(2n-1)])$ is that the total fermion number operator (as defined by $\psi^\dagger\psi$) is not in the latter.

The Hamiltonian H_0 is invariant under $Sp(2n)$, but it does not occupy a special position in the superalgebra except for the case of $n = 1$.

4. Comments

The results of the preceding sections can be summarized as follows. The static part of the effective Hamiltonian for the standard superconductor can be expressed as an anticommutator of fermion composite operators. After symmetry breaking, and in the linear approximation, one gets a spectrum generating superalgebra $su(2/1)$. The Hamiltonian is its $u(1)$ piece with a fermion-boson mass ratio of $1:2$. Prescriptions have also been given for a similar procedure for the general case of $2n$ ($n > 1$) fermions with a symmetry breaking of $U(2n)$ to $Sp(2n)$. The superalgebra is $osp(4n/[2n(2n-1)])$.

A few comments are in order regarding the results of the last section. a) There is quite a bit of arbitrariness in defining the Q operators. Out of $2n$ components of ψ and $n(2n-1)$ components of $\pi^\dagger$ one can construct $2n^2(n-1)$ tensor products, so one must also demand the terms containing the potential function V to have the same degrees of freedom. Together with their Hermitian conjugates, one thus ends up with $4n^2(2n-1)$ composite operators. Now V must be constrained by two requirements: 1) the prescribed pattern of symmetry breaking, and 2) the proper mass relations. The definition of V in Eq. (12) is a deliberately made one for the $U(2n)$ to $Sp(2n)$ breaking and a mass ratio of $1:2$. Notice that the order in which ϕ and $\phi^\dagger$ is placed in V does not make the Q operators $U(2n)$ covariant. It is covariant only under $SO(n)$, although the resulting H is miraculously $U(2n)$ invariant. A different ordering would have given different mass relations. I do not know, however, whether the solution for the Q's given here is unique or not.

b) A possible physical example of the model systems discussed here is the $SU(4/6)$ nuclear supersymmetry proposed by Iachello and collaborators.[2] In this case, $n = 2$. According to the above results, the superalgebra is $osp(8/12)$. (Its subalgebra $su(4/6)$ in the present sense is not identical with theirs.) The four fermions are the nucleons in the $j = 3/2$ shell, and the six bosons are the collective excitations with $j = 0$ and 2. The Hamiltonian H before symmetry breaking is invariant under $U(4) = U(1) \times SU(4) \sim U(1) \times SO(6)$. The six bosons in the $SO(6)$ interpretation: ϕ_s, $s = 1, \ldots, 6$, correspond to

$$\begin{aligned} &\phi_{12} \pm \phi_{34} , \\ &\phi_{23} \pm \phi_{14} , \\ &\phi_{13} \pm \phi_{24} , \end{aligned} \tag{24}$$

in the present $SU(4)$ notation. From this one immediately sees that the condensate matrix η in Eq. (17) does represent one of these $SO(6)$ states. The bosonic potential in H may be written in the $SO(6)$ notation as

$$\left(\sum_s \phi_s^\dagger \phi_s - m^2\right)^2 + \sum_{st} (\phi_s^\dagger \phi_t - \phi_t^\dagger \phi_s)^2 / 2. \tag{22}$$

Now if the ϕ's are expressed as a sum of annihilation and creation operators, and furthermore only the occupation number conserving terms are retained in each factor in Eq. (22), then Eq. (22) is indeed made up of a sum of the Casimir operators of $U(1)$ and $SO(6)$, as would be done in the interacting boson model. However, the fermion part does not have corresponding potential terms, but only Yukawa coupling terms.

After the symmetry breaking, the residual symmetry of the system is $Sp(4) \sim SO(5)$. There are six Goldstone modes and six massive modes. The fermionic part now becomes an $Sp(4)$ invariant bilinear form. Although this is a suggestive interpretation of the origin of the empirical supersymmetry relations in nuclear physics, some more work would be necessary to substantiate it. In this connection, the second order effects mediated by the Yukawa couplings in H may have to be taken into account.

c) There remain more fundamental questions: Why does the BCS mechanism have a built-in quasi-supersymmetry? Why is it possible to factorize the effective Hamiltonian in the way shown here? Is this always possible? If so, what are the rules for writing down the superpotential V for a given system? Is there a direct derivation of the factorization procedure from the Feynman diagrams that determine the effective Hamiltonian?

I would like to thank P. Freund and J. Rosner for enlightening conversations.

References

1. Y. Nambu, *Physica* **15D** (1985) 147.
2. F. Iachello, *Phys. Rev. Lett.* **44** (1980) 772; *Physica* **15D** (1985) 85, and the references cited therein.
3. Y. Nambu, to be published in the Proceedings of the International Conference MESON 50, Kyoto, 1985 (Univ. Chicago preprint EFI 85-71).
4. P. Ramond, *Physica* **15D** (1985) 25.

	type	Fermion sector	Bose sector
a)	$\{Q, SQ\} = -\{Q', S'Q'\}$	$2i\mathrm{tr}(\phi_1\pi_1' S\eta)$	$[\psi', S\eta\psi] - [\psi', \eta\psi]\mathrm{tr}(S)$
b)	$\{Q, A\eta Q'\}$	$\mathrm{tr}((\phi_1\phi_1' + \pi_1'\pi_1)S\eta)$	$\psi' A\psi' - \psi A'\psi$

$$S^t = S, \quad A^t = -A, \quad S' = -\eta S\eta, \quad A' = -\eta A\eta,$$
$$\phi_1 = \phi - \phi^{+\prime}, \quad \pi_1 = (\pi - \pi^{+\prime})/2,$$
$$\phi_1' = -\phi_1' = \eta\phi_1\eta,$$
$$\pi_1' = -\pi_1' = \eta\pi_1\eta.$$

Table. Nonzero anticommutators of Q and Q'. S and A are arbitrary symmetric and antisymmetric matrices respectively. Only massive bosonic modes ϕ_1 are retained.

THERMODYNAMIC ANALOGY IN QUANTUM FIELD THEORY*

Y. Nambu
The University of Chicago, Enrico Fermi Institute
Chicago, Illinois, 60637 USA

ABSTRACT

It is proposed that thermodynamic analogies may be useful in understanding the properties of quantum field theory. As an example, an attempt is made to translate the charge renormalization in gauge theories to a thermodynamic statement about an ideal gas.

1.

This is the 40th anniversary of that memorable year which we of the immediate postwar generation cannot forget. In the summer of 1947 when the news of the discovery of the pion in cosmic rays by the Powell group and of the discovery of the level shift in hydrogen atom by Lamb and collaborators reached Japan through scientific as well as journalistic channels, I was an assistant, and Nishijima a graduate student, at the University of Tokyo. A few years later both of us moved to the newly created Osaka City University. It was in these years that my close association with Nishijima started. Then I came over to America, and Nishijima went to Germany. But in the late Fifties we again became neighbors as he took a position at the University of Illinois. He remained there for several years until his return to Tokyo in the Sixties, which thus ended his "Wanderjahre".

* Supported in part by a U.S. National Science Foundation Grant PHY-85-21588 and a U.S. Department of Energy Grant DE FG02 84ER 45144.

Each of Nishijima's major contributions to particle physics, which he made at various locations and under various circumstances, is clearly etched in my mind: Nakano-Nishijima's relation, or the Gell-Mann-Nakano-Nishijima law[1a], was born at Osaka-City University; the two-neutrino hypothesis[1b] at Brookhaven National Laboratory; the description of composite states in quantum field theory[1c], and the textbooks[1d,e] on fundamental particles and fields at the University of Illinois. I am also familiar with Nishijima the human being, the administrator, his love of movies and travels, and so on. It is my pleasure to dedicate this piece to Kazuhiko Nishijima on the occasion of his sixtieth birthday.

2.

It is all too well known that, at least in a mathematical sense, quantum mechanics and statistical mechanics are closely related. In the functional formulation of quantum mechanics, probability amplitudes can be derived from a quantum "partition function" Z which is an integral of the action exponential over the space-time histories of classical or anticlassical variables. Similarly, thermodynamics observables can be derived from the thermodynamic partition function which is the integral of the Boltzmann factor over the phase space, or equivalently, an integral of the action exponential in a fictitious Euclidean space-time, periodic or antiperiodic in imaginary time evolution. The basic period τ of the imaginary time is related to temperature T by $\tau = \hbar/kT$.

In a more general but less precise sense, quantum mechanics and quantum field theory, after the Wick rotation to imaginary time, can be compared to statistical mechanics in a Euclidean 4-space R^4, except of course for the fact that one is dealing only with a configuration space instead of a phase space. In this correspondence, there are several ways in which the temperature may be interpreted.

First, obviously the Planck's constant is an analog of temperature as it appears universally as the denominator of the action exponent. In gauge theory however, the charge g^2 (after setting $\hbar = c = 1$) also assumes the role of temperature:

$$Z = \int \exp[-1/g^2 \bar{L}_G]\exp[-\bar{L}_M]dV \tag{1}$$

Here $\bar{L}_G$ and $\bar{L}_M$ are respectively the action for the gauge field and for the material source of the gauge field: dV stands for the usual measure of the function space, but

$$dV' \equiv \exp[-\bar{L}_M]dV \tag{2}$$

may also be interpreted as the function measure relative to the gauge field.

In a similar way, Newton's constant G in gravity theory takes the place of g^2:

$$Z = \int \exp[-1/G\ \bar{L}_E]\exp[-\bar{L}_M]dV, \tag{3}$$

where $\bar{L}_E$ is the Einstein action of the gravitational field. Finally, in string theory, it is the string tension α' that plays a similar role:

$$Z = \int \exp[-1/\alpha'\ \bar{L}_s]dV, \tag{4}$$

$\bar{L}_s$ being the world sheet action.

The thermodynamics analogy is now clear in any of such interpretations. In a system of infinite degrees of freedom, as in quantum field theory, the action exponential is usually assumed to have a sharp Gaussian maximum at the classical "on shell" value, and the tangent space around it spans the Hilbert space of physical states. The partition function then assumes the form

$$Z = \exp[-\bar{F}/T],\ \bar{F} = \bar{L}_c - T\bar{S} \tag{5}$$

$\bar{L}_c$ is the classical action which is the analog of thermodynamic energy. S is the entropy of the tangent space with respect to an essentially Gaussian measure, and marks the difference between action $\bar{L}$ and "free action" F or the difference between classical and quantum action. This interpretation is literally true in the loop (WBK) expansion of $\bar{F}$ in which $T \sim \hbar$; $T\bar{S}$ is the zero-point energy of the system. The above statement, however, is subject to some qualifications.

1) It is important to notice that $\bar{L}$ in Eq. (5), which should be defined more precisely as a statistical average $<\bar{L}>$, is not equal to $\bar{L}_c$ for finite T. The "energy" $\bar{L}$ itself should in general depend on T. In other words, the extremum of action functional would not be the same as the classical extremum at $T = 0$.. This is a thermodynamic interpretation of renormalization effects which have to be incorporated in $\bar{L}$.

2) The functional extremum is not always Gaussian. In particular, it is flat in the direction coresponding to a symmetry of the system. In a local, i.e., gauged, symmetry, this leads to an unphysical and infinite contribution to entropy which can be eliminated by gauge fixing. The system may also be infinitely degenerate with respect to a global symmetry, in which case the corresponding entropy is reduced by spontaneous symmetry breaking. Thus the entropy part of $\bar{F}$ is not so well defined as $\bar{L}$; the gauge fixing terms more properly

belong to $\bar{S}$.

3) As in statistical mechanics, there may be many disconnected extrema, and the system may undergo phase transitions from one extremum to another under changing thermodynamic parameters.

Beside temperature and entropy, there are other macroscopic order parameters that characterize the system. As in thermodynamics, one may introduce a Legendre conjugate pair of extensive and intensive parameters. The basic extensive quantity in field theory is the volume of space-time because there is no other natural way to count the number of dynamical degrees of freedom attached to each point of a continuum. One may consider any portion of space-time as a subsystem; $\bar{L}$ and $\bar{S}$ are extensive, or additive, with respect to V except for boundary effects. The system under consideration must be regarded as a grand canonical ensemble because the number of degrees of freedom is proportional to the volume. As one takes a larger chunk of space-time, more degrees of freedom are added, and the intensive parameters of the system will not change. So it is appropriate to deal with thermodynamic densities, which in general will be denoted by Roman characters without a bar.

All these arguments are formal ones, and ignores the fact that there are intrinsic divergences which have to be controlled by renormalization. What is the thermodynamic interpretation of the renormalization process?

In Ref. 2 I made an observation that the relation between the coupling constant g^2 and scale parameter m in a gauge theory is similar to that between temperature and volume of an ideal gas for fixed free energy provided one measures the volume entropy in variable units characterizing the renormalization scale(similar to the Planck's constant in phase space). Here I will give a somewhat different and more satisfactory version of the analogy.

As remarked above, quantum field theory is to be associated with a grand canonical ensemble. This means that the free energy has the expression

$$\bar{J} = -p(T, \mu)V \ , \tag{6}$$

where p is the pressure, μ is the chemical potential. For an ideal gas, $\bar{J}$ is given by

$$\bar{J} = -T \exp[(\mu - f)/T] \ ,$$

$$f = -T \ln(T^{\alpha} V) \ , \tag{7}$$

where μ is the chemical potential, f is the canonical free energy per atom with internal energy $U = \alpha T$, and $V_0 = l^4$, l being the renormalization scale length.

Consider now the Callan-Symanzik equation for a gauge field relating g^2 to the scale length parameter, and write it in the form:

$$dG = dg^2/\beta(g^2) + d \ln l = 0 ,$$

$$\beta(g^2) = b_0 g^4 + b_1 g^6 + \ldots \tag{8}$$

leading to

$$G = \int dg^2/\beta + \ln l = c_{-1}/g^2 + c_0 \ln(g^2) + \ldots + \ln l = \text{const.}$$

$$c_{-1} = -\frac{1}{b_0}, \quad c_0 = -\frac{b_1}{b_0^2} \tag{9}$$

Eqs. (7) and (9) can be brought into correspondence by the assignment

$$T = g^2, f - \mu = 4G - \ln V , \mu = -4c_{-1} , \alpha = -4c_0 , \tag{10}$$

which leads further to

$$-\overline{J}/T = pV/T = nV, n = \exp(-4G) . \tag{11}$$

Thus n, the number density, is a renormalization invariant. But, unfortunately, the chemical potential is also fixed by the theory from the beginning. In QED, $b_0 > 0, b_1 > 0$ so $\mu > 0, \alpha > 0$. In QCD, $b_0 < 0$ if n_f, the number of flavors, is < 17; $b_1 < 0$ if $n_f < 8$ and $b_1 > 0$ if $n_f > 8$. These translate into $\mu < 0$, $\alpha < 0$, and $\mu < 0$, $\alpha > 0$ respectively. For a normal Boltzmann gas α must be positive, and μ is negative at high temperature and low density. Although this analogy is not to be taken seriously, the condition involving b_1 reminds one of similar, though not identical, conditions on n_f derived by Oehme and Zimmermann[3], and further discussed by Nishijima[4]. They seem to reflect in some way on the stability of the perturbative vacuum.

The gas analogy also seems to make sense in the strong coupling (large g^2, or high T) regime. The behavior of g^2 vs. μ depends on the model, but the usual linear potential ansatz implies

$$g^2 \sim l^2, (g >> 0) . \tag{12}$$

This is consistent with

$$F \sim -g^2 \ln(g^4/l^4), \textit{ i.e.}, \alpha = 2 . \tag{13}$$

Is there any basis for all of this interesting correspondence? Perhaps these "atoms" are the gluons. The particle picture might make sense because renormalization is an ultraviolet phenomenon. One could then say that $n = \exp(-4G)$ is the 4-density of gluons.

It is not clear how these interpretations can actually be derived. The thermodynamic function F considered above does not correspond to the free energy interpretation of the action functional as defined by Eq. (5). There does exist, on the other hand, a conventional relation between the conformal (trace) anomaly of the energy-momentum tensor $T_{\mu\nu}$ and the beta function:

$$T^{\mu}_{\mu} \sim \beta . \tag{14}$$

In the present language an analogous relation might be written as

$$\partial(J/T)\partial T = -4(J/T)/\beta . \tag{15}$$

The thermodynamic considerations can also be profitably applied to the case of instanton vacuum which corresponds to a different vacuum from the normal one. For this purpose, first go back to the conventional form of L by replacing the gauge field A by gA, and write L/g^2 and F/g^2 as $L(gA)$ and $F(gA)$ respectively. There is a difference between the respective Lagrange densities due to the Jacobian:

$$L(A)/g^2 \rightarrow L' = L(gA)/g^2 - N_G \ln(g) \tag{16}$$

Here N_G represents the (infinite) degrees of freedom of the field A per unit volume, and may be regarded as belonging to the entropy part of F. This should not change the values of other thermodynamic observables.

Since the Gaussian degrees of freedom around the instanton are decreased from the normal case by the presence of zero modes, the difference in entropy between the instanton and the normal vacuum is then given by the difference of the second term of Eq. (16) integrated over space-time, i.e.,

$$\bar{S}_I - \bar{S}_N = -n \ln g . \tag{17}$$

$n (= 8)$ being the number of zero modes.[5]

3.

I end this article on an even more speculative note. The present ideas may be useful in gravity and string theories as well. For example, the problem of the cosmological constant may be reduced to that of entropy, which in turn leads one to suspect the significance of the third law of thermodynamics in our analogy. (I thank David Zoller for raising these questions.) One might speculate the possibility of a superquantum mechanics which would replace the Boltzmann statistics used here and realize the third law in quantum field theory.

References

1a. T. Nakano and K. Nishijima, Prog. Theor. Phys. 10 (1953) 581.

b. K. Nishijima, Phys. Rev. 108 (1957) 907.

c. K. Nishijima, Phys. Rev. 111 (58) 995.

d. "Fundamental Particles" (published by Benjamin, New York, 1963).

e. "Fields and particles" (published by Benjamin, New York, 1969).

2. Y. Nambu, Phys. Reports 104 (1984) 237.

3. R. Oehme and W. Zimmermann, Phys. Rev. D21 (1980) 471, 1661; R. Oehme, Phys. Lett. 132B (1983) 114.

4. K. Nishijima, Prog. Theor. Phys. 75 (1984) 1221.

5. G. 't Hooft, Phys. Rev. D14 (1976) 3432.

1

BCS MECHANISM, QUASI-SUPERSYMMETRY, AND FERMION MASSES*

Y. Nambu
The Enrico Fermi Institute and Department of Physics
University of Chicago, Chicago, IL 60637

ABSTRACT

The characteristic features of the BCS mechanism, one mathematical and the other physical, are explored in relativistic field theories. The first is called quasi-supersymmetry among fermions and composite bosons, and various examples can be found at various energy scales. The other has to do with the notion of a hierarchical chain of symmetry breakings and that of a self-sustaining (bootstrap) mechanism. The latter is applied to electroweak unification, and some predictions are made.

1. INTRODUCTION

The BCS theory is thirty-one years old. The great impact it has given to particle as well as condensed matter physics in the years that followed is now a matter of history. But as you will see, there is still something new to be learned from the original BCS theory, by which I mean a dynamical symmetry breaking by a short range (four-fermion type) interaction. I would like to report on some results of my work that started a few years ago.

First I will give an account of the general features of the BCS theory in a nutshell. These features can then be captured by a Ginzburg-Landau effective Hamiltonian of a restricted type, which in turn is shown to possess a broken or what I call quasi-supersymmetry.

These considerations lead to a set of questions, both mathematical and physical. On the mathematical side, I search for examples of relativistic models which satisfy quasisupersymmetry conditions. They include chiral flavor dynamics, the SU(3) × SU(2) × U(1) and the SU(5) unification.

On the physical side, I discuss two scenarios; one is a cascading chain of symmetry breakings (tumbling), and the other a bootstrap mechanism in which symmetry breaking sustains itself among a set of effective fields without a need to refer to the underlying substratum. I will then apply the last idea to the problem of the fermion masses with some predictions.

* Work supported in part by the National Science Foundation Grant No. PHY-85-21588.

2. BASIC FEATURES OF THE BCS THEORY

A BCS mechanism has two energy scales, high and low. The high energy scale reflects that of the constituents of Cooper pairs. The low energy scale is the pairing energy, or the energy gap parameter. There are fermionic and bosonic excitations at the low energy scale: the Bogoliubov-Valatin quasifermion, the Goldstone ("pi") boson, and the Higgs ("sigma") boson. Their masses (energy gaps) m_f, m_π and m_σ satisfy universal ratios 0:1:2 in the weak coupling limit.[1]

The same low energy properties can also be reproduced by an effective Ginzburg-Landau-Gell-Mann-Levy (GL)Hamiltonian involving these fermion and boson fields, with Yukawa couplings and a Higgs potential.[2] It contains two parameters: the sigma condensate c (pion decay constant) representing the (usually) high energy scale, and a dimensionless Yukawa coupling G which is the ratio of the low and high energy scales (since $m_f = Gc$). The Higgs self-coupling is equal to G^2 in order to satisfy the mass ratio constraint. In nonrelativistic theories, c is given by $c^2 = N/4$, where N is the density of states of the constituents at the Fermi surface.

What is the origin of the mass relations and the broken supersymmetry that they imply? This is not clear, although one can formally derive the relations on the basis of phase (gauge) and amplitude (scale) Ward identities, the former being exact and the latter approximate.[1] (An alternative way is to rely on dispersion relations.[1,2]) One can also define fermionic operators Q, similar to the supersymmetry operators, out of which the GL Hamiltonian can be constructed. Consider the static limit in which all kinetic energies are ignored.[4] [4] In the case of superconductors, let

$$Q = \Pi^\dagger \psi_{up} + iW\psi_{dn}^\dagger ,$$
$$W = G(\Phi^\dagger\Phi - c^2) . \tag{1}$$

Here ψ_{up} and ψ_{dn} represent respectively spin up electrons and spin down holes, and Φ and Π are the complex field of Cooper pairs and its canonical conjugate. Now one easily checks that

$$\overline{H}_{st} = \{\overline{Q}, \overline{Q}^\dagger\} ,$$
$$H_{st} = \Pi^\dagger\Pi + W^2 + G(\psi_{up}^\dagger\psi_{dn}^\dagger\Phi + \psi_{dn}\psi_{up}\Phi^\dagger). \tag{2}$$

where H_{st} is precisely the static GL Hamiltonian density containing a Higgs potential W^2 and Yukawa couplings. (The bar indicates spatial integral.) Q and $Q^\dagger$ look like a Bogoliubov mixture, but carry a definite charge. They are not nilpotent, so there is no strict supersymmetry. When W is linearized around its minimum, however, Q, Q^* generate a superalgebra su(2/1) consisting of two fermions (spin 1/2) and one boson (sigma). The pi mode acts as a central charge.

This decomposition of H_{st} works in general. For example, with $n(>2)$ fermions and $m = n(n-1)/2$ complex composite bosons, one defines n-component Q and $Q^\dagger$ in which the boson fields Φ and Π are $n \times n$ antisymmetric matrices acting on the fermion field. The superalgebra generated is $osp(2n/2m)$.

My student Mukerjee and I have used these results in an interpretation of the Interacting Boson Model (IBM).[5] The spin 0 and 2 bosons of the IBM are collective excitations of Cooper pairs of nucleons that have filled up the shell states. A GL Hamiltonian is written down for bosons and fermions with a symmetry dictated by that of the degeneracy of active valence shell states. In a typical case, $n = 4$ ($j = 3/2$ shell), $m = 6$. The starting symmetry of SU(4) (~ O(6)) spontaneously breaks to Sp(4) (~ O(5)). One collects all the tree graphs for boson-boson, boson-fermion, and fermion-fermion scatterings, then performs a projection to a basis diagonal in nucleon number, as is appropriate for a finite system. The results are expressible in terms of basic nuclear parameters only, and reproduce key features of the empirical IBM formulas.

3. RELATIVISTIC QUASI-SUPERSYMMETRY

Although the meaning of the static quasi-supersymmetry found in BCS-GL theories is not clear, it naturally leads to the next question: Can it be implemented in relativistic field theories? I will show that it can.

A prototype is obtained by emulating the BCS superconductivity. Let ψ and ψ^c be a two component Weyl spinor and its charge conjugate, and construct

$$Q = (\Pi^\dagger\sigma_0 \pm \nabla_i\Phi^\dagger\sigma_{i\pm})\psi + iW\psi^c, \ (\sigma_0 = 1;\ \sigma_i\pm = \pm\sigma_i\ ,\ i = 1,2,3) \tag{3}$$

where the complex (or two real) Bose fields Φ and Π are the same as before. Then

$$\{\overline{Q}, \sigma_{\mu\pm}\overline{Q}^\dagger\}/2 = P_\mu,\ \mu = 0, \ldots 3\ , \tag{4}$$

gives the total energy and momentum. In particular,

$$P_0 = \overline{H}, H = \Pi^\dagger\Pi + |\nabla\Phi|^2 + W^2 \pm i\psi^\dagger\sigma\cdot\nabla\psi + G(\Phi\psi^\dagger\psi + \Phi\psi^{\dagger c}\psi)/2\ . \tag{5}$$

is the proper Hamiltonian for a fermion coupled to a Higgs field that gives a Majorana mass with the standard BCS mass ratios.

4

It is crucial in this construction that the bosonic and fermionic degrees of freedom are equal (= 2). The boson kinetic energy receives its weight from the fermions, and the fermion kinetic energy receives its weight from the bosons. Note also that Q and $Q^\dagger$ would be nilpotent if $W = 0$, giving an ordinary free supersymmetric system of spin 0 and 1/2. So one is dealing here with a deformation of supersymmetry.

This suggests that a (1/2, 1) supersymmetry may also be emulated. In fact

$$Q = \sigma \cdot F_{\pm}\psi, \; F_{\pm} = (E \pm iB)/\sqrt{2}, \tag{6}$$

will do, where B and E are the usual U(1) gauge field components. It leads (in the $A_0 = 0$ gauge) to

$$\{\overline{Q}, \overline{Q}^\dagger\}/2 = \overline{H}, \; H = (E^2 + B^2)/2 \mp i\psi^\dagger\sigma \cdot \nabla\psi \, . \tag{7}$$

The two cases can be combined to handle a Dirac fermion. Thus let

$$Q_l = (\Pi^\dagger - \sigma \cdot D\Phi^\dagger)\psi_r + (\eta W + \sigma \cdot F_-)\psi_l \, ,$$

$$Q_r = (\Pi^\dagger + \sigma \cdot D\Phi)\psi_l + (\eta W + \sigma \cdot F_+)\psi_r , \; |\eta| = 1 \, . \tag{8}$$

Here ∇ is replaced by a covariant derivative D. One finds that either pair $(Q_l, Q_l^\dagger)$ or $(Q_r, Q_r^\dagger)$ gives a Hamiltonian for a Dirac fermion coupled to a U(1) gauge field and a U(1) × U(1) Higgs field. A few remarks are in order.

1) The suffix l or r implies left- or right-handed spinor (0, 1/2) or (1/2, 0), which goes with the $\mp$ sign. Switching ψ_l and ψ_r in Eq. (8) would make Q and $Q^\dagger$ behave like components of (1, 1/2) or (1/2, 1), and reverse the sign of fermion kinetic terms. There seems nothing intrinsically wrong about this.

2) The anticommutator of $Q(x)$ and $Q^\dagger(x')$ must be constructed by inserting a Wilson line between them to get the correct gauge-invariant results.

3) The Yukawa terms get contributions from both the Higgs terms $[\Pi, W^2]$ and the gauge field terms $[E, D\phi]$. Their relative chiral phase (η) is arbitrary, and the BCS mass relation between m_f and m_σ is replaced by a more general one that also involves the vector meson mass m_V.

Putting aside the details for now, I will list some models theories that are compatible with quasi-supersymmetry just by the matching of the number n of degrees between bosons and fermions, allowing for the possibility of some bosons giving negative weights to fermions.

a) $[U(1) \times U(1)]_G \times U(1)_g$: $n = 4$. (G and g stand respectively for global and local symmetries.) The above example Eq. (8). 1 Dirac fermion. The bosons are S(1) (scalar singlet), P(1) (pseudoscalar), and V(1) (U(1) gauge vector with 2 helicities.)

b) $[U(1) \times U(1)]_G \times SU(2)_g$: $n = 8$ with 2 Dirac fermions, S(1), P(1), and V(3) (isotriplet).

c) $[SU(2) \times SU(2]_g$: $n = 8$ with 2 Dirac fermions, -S(1), -P(3), V(3), and A(3) (axial vector. (-S means weight -1.)

d) $[U(1) \times U(1)]_G \times [SU(2) \times SU(3)]_g$: $n = 24$. 2 flavors and 3 colors of quarks, a gauge octet $V(8)$ added to the bosons in a).

e) $[SU(2) \times SU(2) \times SU(3)]_g$: $n = 24$, similarly obtained from c). This shows a compatibility between chiral dynamics at the nucleon level and at the quark level, reminiscent of the 't Hooft consistency for chiral anomaly.[6] Furthermore, there seems to be a close relation between these models and the hidden symmetry schemes of Bando et al.[7]

f) $[SU(3) \times SU(2) \times U(1)]_g$ (color and electroweak): $n = 32$. 1 generation of fermions vs. gauge bosons V(8), V(3), V(1), and 2 sets of Higgs H(4).

g) $SU(5)_g$: $n = 96$ with 3 generations of 16 = (1 + 5 + 10) chiral fermions vs. gauge bosons V(24), and 2 sets of adjoint Higgs H(24).

An important physical assumption implied here is that, at each energy scale, there should be a set of fields to satisfy the quasi-supersymmetry constraints, but some of them, especially the Higgs, are composite fields so they cease to exist at a higher energy scale. For example, g) does not contain the Higgs of f) that would have made the fermions superheavy. This is in line with the "naturalness" philosophy of 't Hooft.[6]

4. BOOTSTRAP SYMMETRY BREAKING

The quasi-supersymmetry explored above is so far of purely mathematical nature. I will now turn to more physical propositions inspired by the BCS theory. Suppose a symmetry breaking at a high energy constituent level gives rise, among other things, to a sigma boson at the low energy scale. The latter, being a scalar, will induce attraction between the quasi-fermions, which in turn may generate a second generation symmetry breaking, and so on. Such a scenario in particle model building has been called tumbling.[8]

I would like to claim, however, that an example of tumbling already exists in nuclear physics. It is generally accepted that quantum chromodynamics causes chiral symmetry breaking and quark mass generation leading to hadron dynamics. In particular, the sigma meson supplies strong attraction between nucleons, which not only makes the formation of nuclei possible, but also is responsible for nuclear pairing. In fact one can estimate the pairing energy fairly well from basic nuclear parameters and the gap equation.[9] A somewhat similar and perhaps more familiar example is the ordinary superconductivity, in the sense that the phonons responsible for Cooper pairing are the Goldstone modes for the breaking of translation (and rotation) invariance in solids.

Instead of a cascading hierarchy of symmetry breakings, one can also envisage a theoretical possibility of bootstrap: The sigma boson and the rest are responsible for their own existence. Even if there may well be a substratum at a higher energy scale, it may not be necessary to look beyond what one sees at the low energy scale; the dynamics may be self-consistent among the latter. Hadron dynamics may have this property[10] in view of the 't Hooft type consistency mentioned above. But how does one clearly formulate the bootstrap condition?

I will discuss this with the concrete example of the Weinberg-Salam electroweak unification. The mass m_i of the ith fermion is generated by the Yukawa coupling f_i of the Higgs as (Fig. 1)

$$m_i = f_i c \ . \tag{9}$$

But there are also tadpole diagrams (Fig. 2a) which represent a Higgs-induced attractive interaction between fermions, i.e., a bootstrap mechanism just mentioned. The tadpole also couples to the Higgs and gauge boson (W and Z) loops, Figs. 2b and 2c. These are all quadratically divergent, going with a cut-off parameter Λ like

$$\Lambda^2 - m^2 \ln\Lambda^2 \ . \tag{10}$$

If the cut-off is interpreted as a higher energy scale, the bootstrap must mean independence (at least insensitivity) of the result on Λ, i.e., cancellation of divergences among the three contributions of Figs. 2. Two points of view are possible here: a weak condition (quadratic divergences only) and a strong one (both quadratic and logarithmic divergences). At any rate, the quadratic and logarithmic conditions respectively lead to the following two equations

$$I. \quad \Sigma\, m_f^2 = 3/2\, m_W^2 + 3/4\, m_Z^2 + 3/8\, m_H^2$$

$$II.\ \ \Sigma m_f^4 = 3/2\ m_W^4 + 3/4\ m_Z^4 + 3/8\ m_H^4\ . \tag{11}$$

The condition I is analogous to one that would follow from supersymmetry, since a weakening of the degree of divergences is a characteristic feature of supersymmetry. Here, however, there are no a priori constraints on the number of fields or the values of individual coupling constants and masses. An important consequence of I is that the sum of Yukawa couplings are bounded from below by the known gauge couplings. If the sum is dominated by the top quark, one finds $m_t \geq 70$ Gev .

In ordinary or quasi-supersymmetry that is pertinent to the present problem, however, one requires two Higgs. They must strongly mix so that each can couple to the top quark as the dominant term. In this case Eq. (11) has to be written down for each Higgs with proper modifications.

The weak condition I is a secure one to impose. It is gauge independent; the quadratic divergences in the Higgs self-energy is also eliminated by the same condition.

Let us now turn to the condition II that would be required by a strict adherence to bootstrap. But then there are the usual self-energy terms (Fig. 3) which are also logarithmically divergent, and add to the condition II cross terms in the masses. Keeping only the dominant top quark, I and II together should then determine m_H and m_t from the known m_W and m_Z. The following results emerge.

a. Eq. (11) reduces to a quadratic equation in $(m_H/m_W)^2$ which has no real solutions. With addition of the self-energies of Fig. 3, however, it appears that one is essentially at the critical point of the existence of degenerate solutions

$$m_t \sim m_W \sim 80 \text{ Gev},\ m_H \sim 100 \text{ Gev}. \tag{12}$$

(Although m_t is rather stable, solutions for m_H are sensitive to changes in the parameters.)

b. I have also considered the case of two Higgs doublets. In a model with the Peccei-Quinn symmetry, for example, the four degenerate (neutral and charged) Higgs and an axion modify Eq. (11) to

$$I'.\ \ m_t^2 \sim 13/24\ m_H^2 + 1/2\ m_W^2 + 1/4\ m_Z^2\ ,$$

$$II'.\ \ m_t^4 \sim 1/2\ m_H^4 + 1/2 m_W^4 + 1/4 m_Z^4\ . \tag{13}$$

One finds two solutions

$$m_H \sim 180 \text{ Gev},\ \ m_t \sim 110 \text{ Gev}\ ;$$

$$m_H \sim 40 \text{ Gev}, \; m_t \sim 80 \text{ Gev} \tag{14}$$

where minor contributions from Fig. 3 have been taken into account.

I close with a brief discussion of the entire mass matrix. The weak consistency condition is common to all fermions, but the strong one is not. The non-tadpole diagrams spoil the universality of the condition II. There may be two possible ways to proceed. One is to stay with the weak condition. In other words, the masses depend logarithmically on a cut-off, as in the BCS theory. The other is to enforce II for all fermions with the ansatz

$$m_i = (f_i + \delta_i)c \; , \tag{15}$$

where f_i is the Higgs coupling, and δ_i is a correction term. In either case one is led to a set of coupled consistency relations (gap equations) involving masses and mixing angles. Important in this regard are the up-down (and quark-lepton) difference of electroweak charges, and the coupling between different fermions mediated by the W and the charged Higgs.

The strong bootstrap condition is similar to the assumption that low energy theory is at a critical point of renormalization group equations since the latter also implies absence of logarithmic dependences of physical parameters on the cut-off. Perhaps it is not surprising that Zimmermann and collaborators have obtained, on the basis of renormalization group arguments, results which are not very different from Eq. (12).[11] In a different context, the work of Fritzsch[12] and that of Kaus and Meshkov[13] on the mass matrix are related to the present one as they also seem to be inspired by the BCS theory.

This is a revised version of the talk given at the conference. I thank Professors G. E. Brown, J. L. Rosner, H. Fritzsch, S. Meshkov, Dr. C. Savoy, and Mr. R. Rosenfeld for useful discussions and comments.

REFERENCES

1. Y. Nambu, and G. Jona-Lasinio, Phys. Rev. **122** (1961) 345. P. B. Littlewood and C. Varma, Phys. Rev. **B26** (1982) 4883.
2. Y. Nambu, Physica **150** (1985) 147.
3. Y. Nambu, Phys. Rev. **117** (1960) 648. C. Varma, Ref. 1.
4. Y. Nambu, in *Rationale of Beings*, Festschrift in honor of G. Takeda (eds. K. Ishikawa et al., publ. World Scientific, Singapore, 1986), p. 3; EFI 85-86.
5. Y. Nambu and M. Mukerjee, Phys. Lett. **B209** (1988) 1.

6. G. 't Hooft, in *Recent Developments in Gauge Theories* (Plenum Press, London), p. 135.
7. M. Bando, T. Kugo and K. Yamawaki, Nucl. Phys. **B259** (1985) 493.
8. S. Raby, S. Dimopoulos and L. Susskind, Nucl. Phys. **B169** (1980) 373.
9. Y. Nambu, to be published in *Festi-Val - Festschrift for Val Telegdi*, (ed. K. Winter, publ. North-Holland, Amsterdam, 1988).
10. G. E. Brown, private communication.
11. J. Kugo, K. Sibold and W. Zimmermann, Nucl. Phys. **B259** (1985) 331.
12. H. Fritzsch, talk at this conference.
13. P. Kaus and S. Meshkov, Cal Tech preprint CALT-68-1492.

10

FIG. 1.

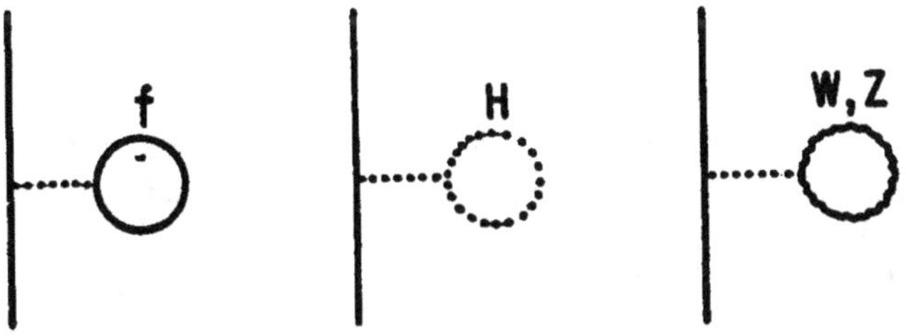

a. b. c.

FIG. 2

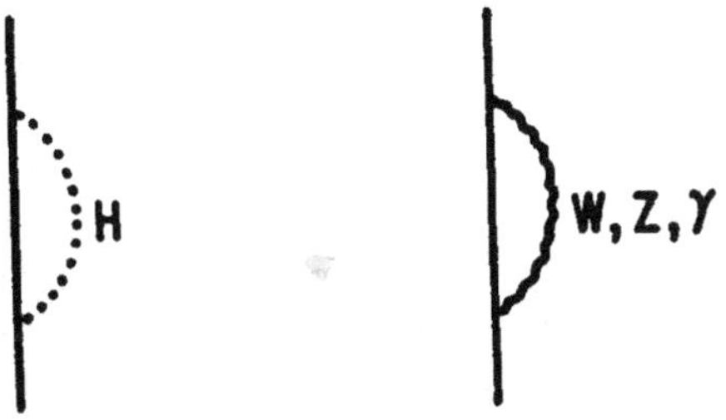

FIG. 3

Reprinted from ANNALS OF PHYSICS Vol. 191, No. 1, April 1989

BCS and IBM

MADHUSREE MUKERJEE AND YOICHIRO NAMBU

Department of Physics and The Enrico Fermi Institute, University of Chicago, 5640 S. Ellis Avenue, Chicago, Illinois 60637

Received December 5, 1988

The BCS theory of fermionic pairing and condensation is used to understand the interacting boson model. Results from BCS are incorporated into an effective Hamiltonian that after symmetry-breaking and second-order corrections yields an IBM-type Hamiltonian with coefficients determined by well-known nuclear constants. The $O(6)$ and $O(5)$ chains are shown to be largely of spontaneous origin. Supersymmetry aspects of the model are also discussed.

1. INTRODUCTION

The interacting boson model [1–7] is well established as a unified description of collective levels of heavy nuclei. The BCS theory of fermionic pairing and condensation [8] is likewise a tour-de-force with wide-ranging applications, not the least of which is to the nucleus [9]. The work we describe here unifies these two paradigms of modern nuclear physics so that the former becomes the inevitable manifestation of the latter, and the latter the cause of the former. BCS theory provides the microscopic rationale for the interacting boson model [10].

The interacting boson model (or IBM for short) postulates that two types of bosons, of spin 0 and spin 2 (known as the s and d bosons) describe collective motions of heavy nuclei. These form a Hilbert space of $U(6)$ symmetry. The bosons are identified with nucleon pairs, so that half the number of nucleons outside major closed shells gives the $U(6)$ quantum number. Some nuclei seem to have the symmetry of a diagonal subgroup chain of $U(6)$. For these, the IBM provides a beautifully simple description of the energy levels and transition matrix elements, in a prescription that unifies whole regions of the table of isotopes. In the end, however, the IBM is a phenomenological theory, and its parameters (other than the crucial one of which representation applies to which nucleus) have to be obtained from experiment.

In contrast, BCS is a microscopic theory; when described as a theory of spontaneous symmetry-breaking as in this application, it has no free parameters. It allows two kinds of bosonic excitations–the π, which are oscillations along the rim of the "Mexican hat" and are massless (see Fig. 1), and the σ, which are radial oscillations and massive. In addition, there are fermionic excitations which break up

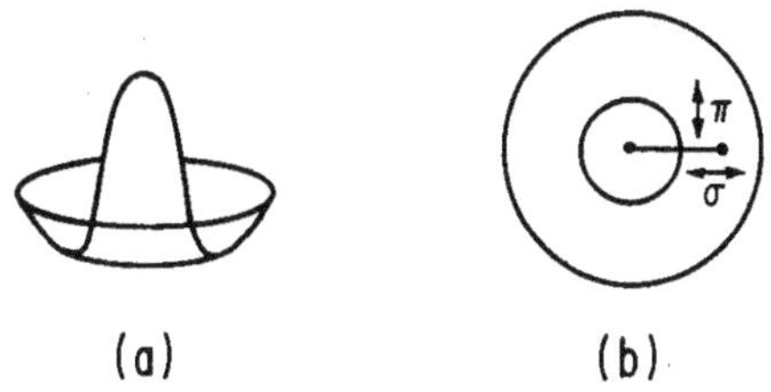

FIG. 1. The Mexican hat, from the side (a) and the top (b) showing the radial (sine) and phase (π) oscillations.

the Cooper pairs and promote the fundamental fermions above the BCS sea. BCS theory implies mass relations between these bosonic and fermionic excitations of the system. These relations can be mimicked by an appropriate choice of coupling coefficients in the equivalent Ginsburg–Landau [11] type effective theory, thus determining *at this effective level* all the parameters. After symmetry-breaking and second-order corrections the Hamiltonian of the system can be written as a sum of Casimirs (of the IBM type) with coefficients determined by well-known nuclear constants [10].

In our application of BCS to the nucleus, we have to come to terms with the finiteness in both volume and particle number of our system. We work in infinite nuclear matter with a symmetry that comes from shell structure (which is caused by finiteness!). Since we use a discrete basis, the Hamiltonian is written without a spatial integral and diagrams are calculated in the limit of zero kinetic energy. In the end we correct for finiteness by an overall fuzz-factor C—this makes our results quantitavely uncertain by about a factor of two while retaining the precision of the infinite medium in the *relative* sizes of the different terms.

The problem of the fixed number of particles is also serious, since the BCS ground state is a superposition of states with different particle number. Indeed, the BCS sea must be looked upon as existing in *a fictitious, unphysical* Hilbert space to which the Hilbert spaces of real nuclei are tangent spaces, So we perform the calculations in the BCS system and in the end project to fixed-particle-number subspaces. The problem of finiteness is central to any theory of nuclear matter that hopes to illuminate real nuclei. Our prescriptions are imperfect but (for our purposes) inescapable answers to the special problems that the nucleus presents us with.

This article is divided into seven parts. The first recapitulates the ideas of BCS theory relevant to this work; for example, the mass relations and the couplings in the equivalent Ginsburg–Landau picture. The second sets up the problem in the nucleus, using the $O(6)$ subgroup chain for specificity. The third details calculations for the bosonic interactions. The fourth gives our adaptations for finiteness for the bosonic part and comparisons to the IBM for the $O(6)$ chain. The fifth part returns to the fermions. After a short discussion of quasi-supersymmetry in the BCS system we compute the fermionic interactions and list the results. The sixth section describes the $SU(3)$ subgroup (the $U(5)$ chain does not fit naturally into the BCS scenario and we will not deal with it other than to point out that an $O(6)$ theory with explicit symmetry-breaking looks very similar to the $U(5)$ chain). Finally we end with discussions, conclusions, and last but not least, speculations.

2. MODES AND MASSES IN BCS

Let us begin with the BCS Lagrangian with the usual four-fermion interaction [8, 12–14]:

$$H(x)=\sum_{s=\downarrow,\uparrow}[\psi_s^\dagger(x)\,\varepsilon_p\psi_s(x)]-2g\psi_\uparrow^\dagger(x)\,\psi_\downarrow^\dagger(x)\,\psi_\downarrow(x)\,\psi_\uparrow(x). \tag{1}$$

Here $\psi_\uparrow$ and $\psi_\downarrow$ are spin-$\frac{1}{2}$ fermions, ε_p is the kinetic energy, and g is a fundamental coupling which goes to zero outside a shell of thickness Λ centered at the Fermi surface.

This can be rewritten using the two-component (Nambu) notation [12–14]

$$\Psi(x)=\begin{pmatrix}\psi_\uparrow(x)\\ \psi_\downarrow^\dagger(x)\end{pmatrix} \tag{2}$$

as

$$H(x)=\Psi^\dagger\,\varepsilon_p\tau_3\Psi-g\,\Psi^\dagger\tau_3\Psi\Psi^\dagger\tau_3\Psi+\text{constant terms}. \tag{3}$$

To linearise, we make the BCS ansatz that there is a Cooper pair condensate and $\langle\psi\psi\rangle$, $\langle\psi^\dagger\psi^\dagger\rangle\neq 0$:

$$\phi\sim\Psi^\dagger\tau_1\Psi. \tag{4}$$

Here we choose only one linear combination $\langle\psi^\dagger\psi^\dagger\rangle+\langle\psi\psi\rangle$ to be non-zero since the other (proportional to τ_2) can be obtained by a rotation in the Nambu space (this is called the "phase" rotation). Now we write the Hamiltonian as the sum of "free" and "interacting" parts:

$$\begin{aligned}H&=H_0+H_I,\\ H_0&=\Psi^\dagger\,\varepsilon_p\tau_3\Psi+\Psi^\dagger\tau_1\Psi\Delta,\\ H_I&=-g\,\Psi^\dagger\tau_3\Psi\Psi^\dagger\tau_3\Psi-\Psi^\dagger\tau_1\Psi\Delta\end{aligned} \tag{5}$$

and calculate diagrams using the "bare" propagator

$$iG(p)=i\frac{(p_0+\tau_3\varepsilon_p+\tau_1\Delta)}{(p_0^2-\varepsilon_p^2-\Delta^2+i\varepsilon)}. \tag{6}$$

Self-consistency requires that we choose Δ so that H_I gives no further self-energy corrections proportional to τ_1. This gives us the gap equation with Δ the mass of the fermionic excitations. (The gap energy is that required to break up a Cooper pair and is therefore 2Δ). To see how the four-fermion term can give contributions proportional to τ_1 we use the Fierz identity for Pauli matrices

$$(\tau_3)_{ij}(\tau_3)_{kl} = \frac{1}{2}\sum_{A=1}^{4} \tau^A_{kj}(\tau_3\tau^A\tau_3)_{il} \tag{7}$$

where $\tau^A = (\tau_1, \tau_2, \tau_3, 1)$ to write

$$-\tfrac{1}{2}\Psi^\dagger\tau_3\Psi\Psi^\dagger\tau_3\Psi$$
$$= \tfrac{1}{4}[\Psi^\dagger\Psi\Psi^\dagger\Psi + \Psi^\dagger\tau_3\Psi\Psi^\dagger\tau_3\Psi - \Psi^\dagger\tau_1\Psi\Psi^\dagger\tau_1\Psi - \Psi^\dagger\tau_2\Psi\Psi^\dagger\tau_2\Psi]. \tag{8}$$

The gap equation now reads (Fig. 2)

$$g\int\frac{d^4p}{(2\pi)^4}\operatorname{Tr}[\tau_1 G(p)] + i\Delta = 0, \tag{9}$$

or, doing the p_0 integral,

$$1 = g\int_{p_F - p_{\Delta/2}}^{p_F + p_{\Delta/2}} \frac{d^3p}{(2\pi)^3}\frac{1}{(\varepsilon_p^2 + \Delta^2)^{1/2}}, \tag{10}$$

where the integration is over a shell around the Fermi surface of thickness $2p_{\Delta/2}$.

Hence we can make the approximation

$$\frac{d^3p}{(2\pi)^3} = \frac{4\pi p^2\,dp}{(2\pi)^3} = \frac{p_F}{2\pi^2}p\,dp = \frac{N_F}{2}\,d\varepsilon_p, \tag{11}$$

where N_F is the Fermi density (including the spin degeneracy $\nu = 2$) to write for the gap equation

$$1 = \frac{gN_F}{2}\int_{\varepsilon_F - \varepsilon_{\Delta/2}}^{\varepsilon_F + \varepsilon_{\Delta/2}} \frac{d\varepsilon_p}{(\varepsilon_p^2 + \Delta^2)^{1/2}}. \tag{12}$$

To find the resonances in the fermion–fermion interaction we now calculate the bubble chains in Fig. 3 with τ_1 or τ_2 at the vertices and look for the poles. These describe the amplitude and phase (Goldstone and Higgs) modulations, respectively,

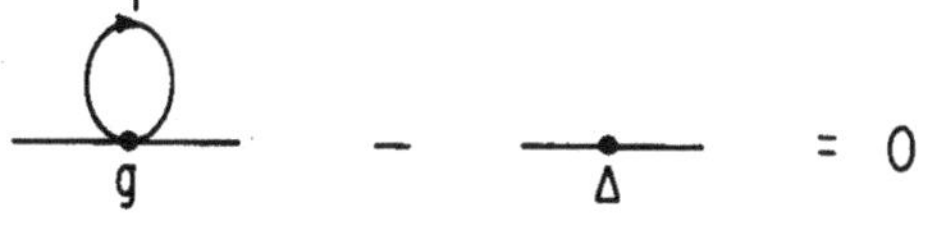

FIG. 2. Mass contributions from H_I.

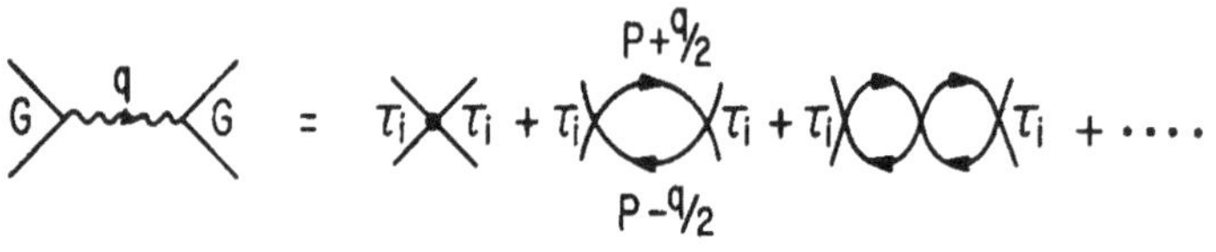

FIG. 3. Repeated fermion-fermion interactions and the equivalent meson-exchange diagram. For $\tau_i = \tau_1$ we get the σ meson and for $\tau_i = \tau_2$ the π meson.

since the condensate has been chosen to lie in the direction τ_1. An alternate approach has been described by Nambu [12] and Littlewood and Varma [15]: the poles in the gauge-invariant vertex functions are used to find the collective modes. The connection with the Ward identities is then more explicit. The π and σ modes are seen to be manifestations of symmetries of the Hamiltonian (3) under $SU(1, 1)$ rotations generated by τ_3, $i\tau_1$, and $i\tau_2$.

The matrix element for the bubbles is given by

$$M = g/(1 - J_{\pi,\sigma}(q)), \tag{13}$$

where

$$J_\pi(q) = ig \cdot \int \frac{d^4p}{(2\pi)^4} \mathrm{Tr}[\tau_2 G(p+q/2)\, \tau_2 G(p-q/2)] \tag{14}$$

and

$$J_\sigma(q) = ig \cdot \int \frac{d^4p}{(2\pi)^4} \mathrm{Tr}[\tau_1 G(p+q/2)\, \tau_1 G(p-q/2)] \tag{15}$$

are the integrals for a single bubble. Note that the poles of M are at

$$J_{\pi,\sigma}(q) = 1; \tag{16}$$

this is the condition for the existence of a mode.

To calculate the integrals we resort to dispersion theory techniques. Using the expression for the propagator we find, for ε_q set to zero (i.e., for zero momentum transfer),

$$J_{\pi,\sigma}(q) = 2ig \int \frac{d^4p}{(2\pi)^4} \frac{[(p_0^2 - q_0^2/4) - \varepsilon_p^2 \mp \Delta^2]}{[(p_0 - q_0/2)^2 - \varepsilon_p^2 - \Delta^2 + i\varepsilon)][(p_0 + q_0/2)^2 - \varepsilon_p^2 - \Delta^2 + i\varepsilon]} \tag{17}$$

where the $-$ and $+$ signs are for the π and σ modes, respectively. We do the p_0 integral and pick up the residues at the two poles

$$p_0 = \pm q_0/2 - (\varepsilon_p^2 + \Delta^2)^{1/2} + i\varepsilon \tag{18}$$

to get (writing $q_0 = \omega$)

$$J_\pi(q_0^2 = \omega^2) = g \int_{p_F - p_{A/2}}^{p_F + p_{A/2}} \frac{d^3p}{(2\pi)^3} \frac{1}{(\varepsilon_p^2 + \Delta^2)^{1/2}} \left[1 - \frac{\omega^2}{(\omega^2 - 4(\varepsilon_p^2 + \Delta^2))}\right]. \tag{19}$$

The expression for $J_\sigma(\omega^2)$ is similar, with the ω^2 in the numerator replaced by $\omega^2 - 4\Delta^2$. Note that

$$1 = J_\pi(\omega^2 = 0) = J_\sigma(\omega^2 = 4\Delta^2) \tag{20}$$

by the gap equation. Hence the poles of M are at $\omega^2 = 0$ and $4\Delta^2$ for the π and σ modes, respectively. The masses of the modes are therefore [16]

$$m_\pi = 0, \qquad m_\sigma = 2\Delta. \tag{21}$$

(For $\varepsilon_q \neq 0$, we find instead of the above the dispersion relation [15], $\omega^2 = m_{\pi,\sigma}^2 + v_F^2 q^2/3$). Using Eq. (11) and then the substitution $\kappa = 2(\varepsilon_p^2 + \Delta^2)^{1/2}$, we get

$$J_{\pi,\sigma}(\omega^2) = \frac{gN_F}{2} \int_{4\Delta^2}^{\Lambda^2} \frac{d\kappa^2}{(\kappa^2 - \omega^2)(1 - 4\Delta^2/\kappa^2)^{\pm 1/2}}, \tag{22}$$

where the $\pm$ signs refer to the π and σ modes respectively and $\Lambda^2 \gg 4\Delta^2$ is the shell-thickness cutoff.

To evaluate the effective coupling of the σ and π modes we write (from Fig. 3)

$$-\frac{G_{\pi,\sigma}^2}{\omega^2 - m_{\pi,\sigma}^2} = \frac{g}{1 - J_{\pi,\sigma}(\omega^2)} \tag{23}$$

and expand $J_{\pi,\sigma}$ about the mass-shell value of ω^2, namely $\omega^2 = 0, 4m^2$ for π, σ. Using the gap equation again, we get

$$G_{\pi,\sigma}^2 = g(dJ_{\pi,\sigma}/d\omega^2)^{-1} \tag{24}$$

with the derivative evaluated on the mass shell. We find

$$G^2 = G_{\pi,\sigma}^2 \simeq 8\Delta^2/N_F, \tag{25}$$

since the cutoff $\Lambda^2 \gg 4\Delta^2$. Here we have approximated the spectral density $(1 - 4m^2/\kappa^2)^{\pm 1/2} \simeq 1$. We need an approximation of this kind to use the sigma model in which the π and σ modes are identical prior to symmetry breaking.

If we evaluate $G_{\pi,\sigma}$ exactly at $\omega^2 = 0$ we get

$$\begin{aligned} G_\pi^2 &= 8\Delta^2/2N_F, \\ G_\sigma^2 &= 3(8\Delta^2)/2N_F \end{aligned} \tag{26}$$

so that the average of the two couplings at this momentum transfer is still G^2. However, at arbitrary ω^2, this might not be true. This difference of the π and σ couplings is the primary problem with modelling the BCS model by the σ model.

The above results carry over to the analysis of BCS systems in larger Hilbert spaces. In general the Fermi density is given by

$$N_F = \frac{\nu p_F^2}{2\pi^2 v_F}, \tag{27}$$

where ν is the spin degeneracy. With this allowance for the enlarged space, the main results Eqs. (21) and (25) remain unchanged.

3. BCS IN THE NUCLEUS

We concentrate on the case of $O(6)$ symmetry since it is the most amenable to a BCS interpretation. The symmetry is dictated by that of the valence shell; while this assumption is not essential to the description of the even–even nuclei it is important to the even–odd nuclei (as we shall see, however, the supersymmetry aspects of this model do not coincide with those in the extensions of the IBM). We have a symmetry dictated by the shell structure and a discrete basis, and so we write the Hamiltonian without an integral; however, we also work in an infinite medium in the limit of zero momentum. In any case, consider the valence fermions to be in a spin $\frac{3}{2}$ shell. These pair up to form spin 0 and spin 2 bosons. So the starting symmetry is that of the valence shell, that is, $U(4) \sim O(6)$. We also impose a $U(1)$ symmetry for particle number conservation:

$$\begin{aligned} H_0 = {} & \sum_n \pi_n^\dagger \pi_n + 2G^2 \left[\sum_n \phi_n^\dagger \phi_n - c^2 \right]^2 \\ & - G^2 \sum_{nm} [\phi_n^\dagger \phi_m - \phi_m^\dagger \phi_n]^2 + \Psi^\dagger \Gamma_7 K \Psi \\ & + G/\sqrt{2} \sum_n [\Psi^\dagger (1+\Gamma_7)\Gamma_n \Psi \phi_n + \Psi^\dagger (1-\Gamma_7)\Gamma_n \Psi \phi_n^\dagger]. \end{aligned} \tag{28}$$

In the above ϕ_n, $\phi_n^\dagger$ with $n = 0, 1, \ldots, 5$ are composite bosonic fields ($\phi \sim \psi\psi$) and π_n, $\pi_n^\dagger$ their canonical conjugates. The first term is the usual bosonic kinetic energy, the second an $U(6)$-invariant symmetry-breaking piece, and the third has $O(6) \times U(1)$ invariance ($L_{mn} = -i(\phi_m^\dagger \phi_n - \phi_n^\dagger \phi_m)$ is like an $O(6)$ angular momentum, and $\frac{1}{2} \sum L_{mn}^2$ is therefore the Casimir invariant). Ψ represents nucleon states in the valence shell (considered in this infinite medium approach as the Fermi surface) and is written as an eight-component Majorana fermion $\Psi = (\psi_{3/2}, \psi_{1/2}, \ldots, \psi_{-3/2}, \psi_{3/2}^\dagger, \ldots, \psi_{-3/2}^\dagger)/\sqrt{2}$. The Γ_n's are 8×8 Clifford algebra matrices. That is, they transform as vectors under $O(6)$, anticommute among themselves, and serve to couple the Ψ's together to form spin 0 and spin 2 composites. $\Gamma_7 = \prod_n \Gamma_n = (1, 1, 1, 1, -1, -1, -1, -1)$ is used to construct the kinetic energy piece for the fermions since it gives positive and negative energies to the particles and holes, respectively. The other Γ's are off-diagonal. The last two terms in Eq. (28) are arranged to conserve particle number; that is, they only allow terms like $\psi\psi\phi^\dagger$ (and not, for instance, $\psi\psi\phi$). The coefficients are chosen, as we shall see, to reproduce the results of the last section.

We allow the spin 0 boson to develop a vacuum expectation value and call the excitations around vacuum π and σ modes in anticipation of their respective masses:

$$\begin{aligned} \phi_0 &= c + (\sigma_0 + i\pi_0)/\sqrt{2} \\ \phi_i &= (\pi_i + i\sigma_i)/\sqrt{2}, \qquad i = 1, ..., 5. \end{aligned} \tag{29}$$

Then our Hamiltonian becomes (omitting the kinetic terms for simplicity)

$$\begin{aligned} H_0 = {} & 4G^2c^2 \sum_n \sigma_n^2 + \sqrt{2}Gc\,\Psi^\dagger \Gamma_0 \Psi \\ & + 2\sqrt{2}G^2 c\,\sigma_0 \sum_n (\sigma_n^2 + \pi_n^2) + 4\sqrt{2}G^2 c \sum_i \sigma_i(\sigma_0\sigma_i - \pi_0\pi_i) \\ & + G^2/2 \sum_n (\sigma_n^2 + \pi_n^2)^2 + G^2 \sum_{ij} (\pi_i\sigma_j - \sigma_i\pi_j)^2 \\ & + 2G^2 \sum_i (\sigma_0\sigma_i - \pi_0\pi_i)^2 \\ & + G\Psi^\dagger(\Gamma_0\sigma_0 + i\Gamma_7\Gamma_0\pi_0 + \Gamma_i\pi_i + i\Gamma_7\Gamma_i\sigma_i)\Psi. \end{aligned} \tag{30}$$

In the above $n = 0, 1, ...5$ and $i = 1, 2, .., 5$. The first two are the mass terms, the next two vertex (interaction) terms, the following three quartic and the last Yukawa terms. The σ mass is

$$m_\sigma = 2\sqrt{2}Gc \tag{31}$$

and the fermion mass

$$\Delta = \sqrt{2}Gc. \tag{32}$$

As promised, $m_\sigma = 2\Delta$. We further find the value of c using Eqs. (25) and (32) to be

$$c = \sqrt{N_F}/4. \tag{33}$$

The overall scale is fixed by the Yukawa coupling G of Eq. (25), with N_F given by Eq. (27).

The task ahead of us is to evaluate perturbation corrections to the above Hamiltonian. Since H_0 is an effective Hamiltonian that reproduces only the low-energy characteristics of BCS, it is not meaningful to carry out perturbation theory to all orders. Diagrams with internal loops depend on high-energy (and hence untrustworthy) components of the fields. We accordingly restrict ourselves to two-particle interactions and consider only tree graphs to second order; further, we include only energy-conserving (on-shell) processes. The mass relations between various fields select a subset of tree diagrams that meet the latter (on-shell) requirement. On adding these corrections to H_0 we obtain a *new* effective Hamiltonian that will be compared to the IBM. Conceptually, the corrected Hamiltonian is analogous to the S-matrix or the optical potential; its matrix elements give physical quantities directly. Unlike the S-matrix, however, scattering causes energy shifts for a finite system. Thus our results will carry a volume factor $1/V$. In the following section we detail the bosonic corrections.

4. Bosonic Interactions

We write the π and σ fields in terms of creation and annihilation operators, remembering that we will work in the limit of zero momentum:

$$\begin{aligned} \sigma_n &= (a_n + a_n^\dagger)/\sqrt{2m_\sigma V} \\ \pi_n &= (b_n + b_n^\dagger)/\sqrt{2m_\pi V}. \end{aligned} \tag{34}$$

In the above V is the volume of the nucleus and the pion has been assigned a small mass m_π which will later be taken to be zero. (We anticipate problems with convergence–more on this shortly). A typical diagram is shown in Fig. 4; a σ is exchanged between two other σ's. The contribution from this diagram is

$$\frac{(2\sqrt{2}G^2c)^2}{(-m_\sigma^2)}\, 9(4a_i^\dagger a_0^\dagger a_i a_0)\frac{1}{(2Vm_\sigma)^2} \tag{35}$$

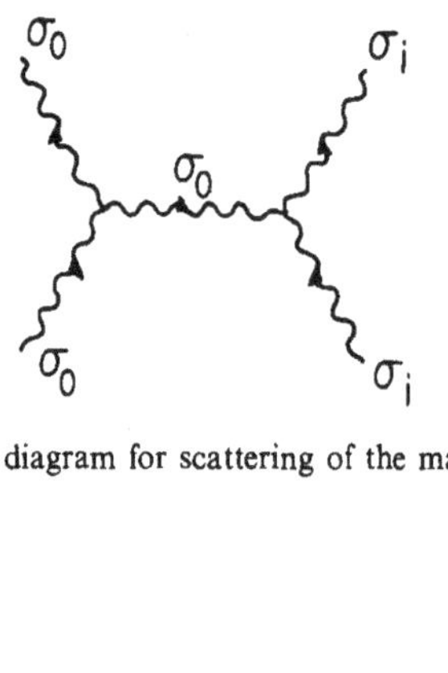

FIG. 4. A typical diagram for scattering of the massive σ mesons.

where the first term is the coupling (with the propagator in the dénominator) and the last term is from the normalisation of the a's. The coefficient of the a's has the dimensions of energy per unit volume:

$$\frac{G^4c^2}{m_\sigma^4 V^2}=\frac{1}{c^2V^2}. \tag{36}$$

It is interesting to compare this coefficient with that of the quartic terms in H_0. A typical quartic term, when expressed in terms of the a's (we leave out particle-number-nonconserving terms) is

$$\frac{2G^2}{(2Vm_\sigma)^2}\, a_i^\dagger a_0^\dagger a_i a_0 \tag{37}$$

and has the coefficient

$$\frac{1}{16V^2c^2} \tag{38}$$

which we note has not only the same dimensions but also the same parametric form as the term obtained from the diagram. In fact the overall coefficient turns out to have the same $1/c^2V^2$ dependence for all the diagrams and all the quartic terms.

Let us now turn to the diagrams involving the massless π bosons, with which we expect divergence problems. The contribution from the diagram in Fig. 5, for example is

$$(2\sqrt{2}G^2c)^2\, b_0^\dagger b_0^2 \frac{1}{(2Vm_\pi))^2}\frac{1}{4m_\pi^2-m_{\sigma^2}}. \tag{39}$$

The last term is from the propagator, where for reasons that will become clear, we have been careful to include the small effect of the π mass. Expanding it in powers of m_π^2/m_{σ^2} to first order we get for the above

$$-\frac{(2\sqrt{2}G^2c)^2}{4V^2}\, b_0^{\dagger 2}b_0^2\left(\frac{1}{m_\pi^2 m_\sigma^2}+\frac{4}{m_\sigma^4}\right). \tag{40}$$

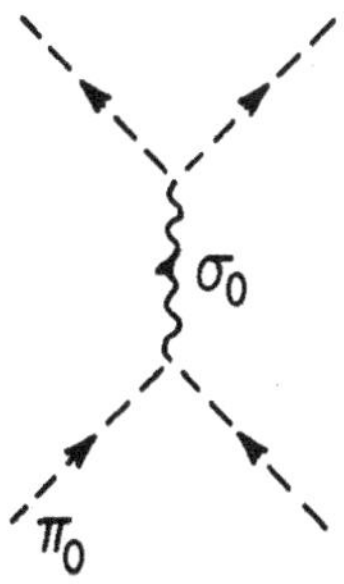

FIG. 5. A typical diagram for scattering of π mesons.

In the last bracket we see two terms: one has m_π^2 in its denominator and is divergent, while the second is finite and has the same overall coefficient as the usual σ terms. We find that when we add together all the contributions from the first- and second-order tree diagrams, *all the divergent terms cancel identically*. This propitious event is not a coincidence but attests to the self-consistency of the BCS theory as embodied in our Hamiltonian [17]. Be that as it may, it is most welcome as it takes away the specter of renormalisation with its additional cutoff parameter. Now we observe that the a's can be grouped into terms which are invariants of $U(6)$, $O(6)$, and $O(5)$:

$$
\begin{aligned}
N_6(\sigma) &= \sum a_m^\dagger a_n \\
O_6(\sigma) &= -1/2 \sum (a_m^\dagger a_n - a_n^\dagger a_m)^2 \\
O_5(\sigma, \pi) &= \sum (a_i^\dagger a_j - a_j^\dagger a_i)(b_i^\dagger b_j - b_j^\dagger b_i)
\end{aligned}
\tag{41}
$$

and so on. We write the final result in terms of these invariants and multiply by a factor V to go from units of energy per unit volume to energy:

$$
\begin{aligned}
H_b = m_\sigma N_6(\sigma) + 1/(8c^2V)[&-3N_6(\sigma)^2 - 3O_6(\sigma) + 2O_5(\sigma) \\
&- N_6(\pi)^2 - O_6(\pi) + 2O_5(\pi) + 2O_5(\sigma, \pi)].
\end{aligned}
\tag{42}
$$

The self-energies have not been included because of the problem of overcounting– the loop integrals were incorporated into the underlying BCS theory. The above is our result for the bosonic case in the infinite medium. In the following section we describe the effects of finiteness in volume (V) and particle number on the Hamiltonian H_b.

5. Finiteness

Let us first look at the problem central to the application of BCS to most systems, that of fixed and finite particle number. The bosonic modes have well-defined interpretations in the infinite medium as fluctuations in the Cooper pairs. For a nucleus in the mass region $A \sim 200$, the coherence length can be found from the dispersion relation following Eq. (21): $\lambda = v_F/\sqrt{3}m_\sigma$. Using the empirical value of the energy gap ($= 2\Delta = m_\sigma \sim 1.5$ Mev) and the Fermi momentum $p_F \sim 240$ Mev, we find λ to be ~ 20 fm. This is larger than the radius of ~ 7 fm in this region. So the interpretation of the modes leaves something to be desired.

We deal with the finite particle number problem by projection of the Hamiltonian from an energy diagonal basis to substates diagonal in particle number.

Since the σ mode describes amplitude oscillations ($\sigma \sim \Psi^\dagger \tau_1 \Psi \sim \psi\psi + \psi^\dagger\psi^\dagger$) and the π mode phase oscillations ($\pi \sim \Psi^\dagger \tau_2 \Psi \sim i(\psi^\dagger\psi^\dagger - \psi\psi)$) we can write

$$\begin{aligned} a &= (u+v)/\sqrt{2} \\ b &= (u-v)/i\sqrt{2}, \end{aligned} \tag{43}$$

where u annihilates a pair of particles and v a pair of holes, and keep only u (or v) terms. (Nucleons above and holes below a major shell can both be considered "particles" in this picture.) Thus for the finite nucleus the π and σ modes become identified and acquire a common mass of $m_\sigma/2$. The Hamiltonian becomes

$$H'_b = 1/2m_\sigma N_6 + (1/8c^2V)[-N_6^2 - O_6 + (3/2)O_5] \tag{44}$$

where the Casimirs are now functions of the one variable u and have the same form listed in Eq. (41); i.e., $O_6 = O_6(u)$ and so on.

Now we set $m_{\sigma/2}$ equal to the pairing energy and get $m_\sigma \sim 3$ Mev. This reduces the coherence length by half from that in the infinite medium. Using Eq. (27) for $N_F, p_F \sim 240$ MeV and degeneracy $v=4$, we find from Eq. (33)

$$c \sim 53 \text{ MeV} \tag{45}$$

for the value of the condensate. Then Eq. (32) gives

$$G = 0.02 \tag{46}$$

which justifies the perturbation theory we have used.

To correct for the finite volume effect we note that the momentum uncertainty Δp reduces the propagator in our tree diagrams relative to the infinite volume case. For example, the propagator

$$\frac{1}{m_\sigma^2} \to \frac{1}{m_\sigma^2 + \Delta p^2} = \frac{1}{m_\sigma^2}\left(\frac{1}{1+\Lambda^2/R^2}\right) \sim \frac{1}{3m_\sigma^2}, \tag{47}$$

where R is the nuclear radius. Hence in this example the reduction factor C is $\sim 1/3$. C should actually be different for different diagrams but we choose to preserve the integrity of our infinite volume results by applying it as an overall correction factor:

$$H = 1/2m_\sigma N_6 + (C/8c^2V)[-N_6^2 - O_6 + (3/2)O_5]. \tag{48}$$

This is the Hamiltonian that is to be compared to the IBM. The coefficient

$$\frac{C}{8c^2V} \sim \frac{40C}{A} \text{ MeV} \sim 200C \text{ keV} \tag{49}$$

for $A \sim 200$. In the IBM the phenomenological coefficients in the platinum region vary from -30 to -50 keV for the O_6 term and from 40 to 60 keV for the O_5 term. Equation (48) reproduces these average results for $C \sim 1/4$. The dependence on A is consistent with the larger coefficients for ^{134}Ba (-74 keV and 87 keV, respectively).

A caveat remains. We have ignored the coupling of the π and σ modes to density modes. This occurs via fermion bubbles which we evaluate near the Fermi surface. The π_0 mode in both protons and neutrons is expected to couple to the isospin density ($\sim \Gamma_7$) and get absorbed into a plasmon mode. With a plasmon energy of ~ 10 MeV, the π_0 contributions would be suppressed. The π_i modes, on the other hand, would not disappear even though they can couple to quadrupole deformations ($\sim \Gamma_0 \Gamma_i$).

Our analysis applies to protons and neutrons separately. BCS theory has no mechanism for their mixing; they could, however, mix by coupling to density modes. In this case the degeneracy v would effectively double and the coefficient in Eq. (49) would be reduced by half. Also, we have no mechanism for generating the $O(3)$ piece spontaneously. There must always be an intrinsic $O(3)$ piece present in this and, by extension, in the $U(3)$ Hamiltonian.

6. Fermions and Supersymmetry

In order to demonstrate the supersymmetry inherent in the BCS formalism [18] let us express the $O(6)$ to $O(5)$ symmetry-breaking in the equivalent language of $SU(4)(\sim O(6))$ breaking to $\mathrm{Sp}(4)(\sim O(5))$. As before, we have six bosonic fields, their conjugate momenta, and four fermionic fields corresponding to a spin $\frac{3}{2}$ object. We represent the bosonic fields as the six independent components of a 4×4 antisymmetric matrix $\mathbf{\Phi}$ and the conjugate momenta similarly as $\mathbf{\Pi}$. The fermions form a four-component spinor $\mathbf{\Psi}$. Then the SU(4) BCS Hamiltonian equivalent to Eq. (28) is

$$H = \mathrm{Tr}[\mathbf{\Pi}^\dagger \mathbf{\Pi}] + 2G^2 \mathrm{Tr}[(\mathbf{\Phi}^\dagger \mathbf{\Phi} - c^2)^2] + \sqrt{2}G[\mathbf{\Psi}\mathbf{\Phi}^\dagger\mathbf{\Psi} + \mathbf{\Psi}^\dagger \mathbf{\Phi} \mathbf{\Psi}^\dagger]. \tag{50}$$

In the above, $\mathbf{\Psi}$ and $\mathbf{\Psi}^\dagger$ are taken to be row (or column) matrices according as they are to the left (or right) of $\mathbf{\Phi}$. Symmetry-breaking to the ground state

$$\mathbf{\Phi} \to c\eta, \tag{51}$$

$$\eta = \begin{pmatrix} 0 & -1 & 0 & 0 \\ 1 & 0 & 0 & 0 \\ 0 & 0 & 0 & -1 \\ 0 & 0 & 1 & 0 \end{pmatrix} \in \mathrm{Sp}(4) \tag{52}$$

leads us to an equivalent description of the nucleus as in Section 2.

Let us form the hybrid spinors

$$\mathbf{Q} = \mathbf{\Pi}^\dagger \mathbf{\Psi} + i\sqrt{2}G(\mathbf{\Phi}^\dagger\mathbf{\Phi} - c^2)\mathbf{\Psi}^\dagger,$$
$$\mathbf{Q}^\dagger = \mathbf{\Psi}^\dagger\mathbf{\Pi} - i\sqrt{2}G\mathbf{\Psi}(\mathbf{\Phi}^\dagger\mathbf{\Phi} - c^2). \tag{53}$$

Then it is easily seen that

$$H = \mathrm{Tr}\{\mathbf{Q}, \mathbf{Q}^\dagger\}, \tag{54}$$

where the curly bracket denotes anticommutation. So the Hamiltonian can be written in a supersymmetric form. The algebra after linearisation corresponds to that of osp $(\frac{8}{12})$ [18], of which the more familiar $SU(\frac{4}{6})$ is a subalgebra. However, since the $\mathbf{Q}$'s are not nilpotent, we do not have a true supersymmetry. This is reminiscent of the physical observation that we have two bosons per fermion, and their masses are not the same either, in contrast to the usual supersymmetry.

Since the starting Hamiltonian exhibits a kind of supersymmetry, we might expect that some relic of this will be evident even after we have corrected for second-order effects. With this hope, we calculate the fermion tree diagrams (we already have the bosonic contribution). We revert to the notation of Section 2. The calculation is simpler in this notation since the Majorana condition on the Ψ's requires that

$$[\Psi^\dagger, \mathcal{O}\Psi] = 0 \tag{55}$$

if $\mathcal{O}$ is the unit matrix or the product of three or four different Γ's. We give the fermions a small kinetic energy K to facilitate delicate corrections at the limit $K \to 0$ which we take in the end. As before, we find that ultimately divergences cancel and we are left only with small finite corrections from potentially divergent terms.

In order to illustrate these ideas, let us examine the pair of diagrams in Fig. 6, in which a σ meson and a fermion of incident energy $E_\Delta = \sqrt{K^2 + \Delta^2}$ scatter off each other. Their contribution, with that of their conjugate processes, is

$$\frac{iG^2}{2m_\sigma V^2}\left[\Gamma_0 \frac{1}{E_\Delta + m_\sigma - \Delta\Gamma_0 - K\Gamma_7}\Gamma_7\Gamma_i \right.$$
$$\left. + \Gamma_7\Gamma_i \frac{1}{E_\Delta - m_\sigma - \Delta\Gamma_0 - K\Gamma_7}\Gamma_0\right](a_0 a_i^\dagger + a_i a_0^\dagger), \tag{56}$$

where the fermion propagator is written in a matrix form reminiscent of Eq. (6) (E_f is the energy of the propagating fermion):

$$G_\Delta(K) = \frac{1}{E_f - \Delta\Gamma_0 - K\Gamma_7}. \tag{57}$$

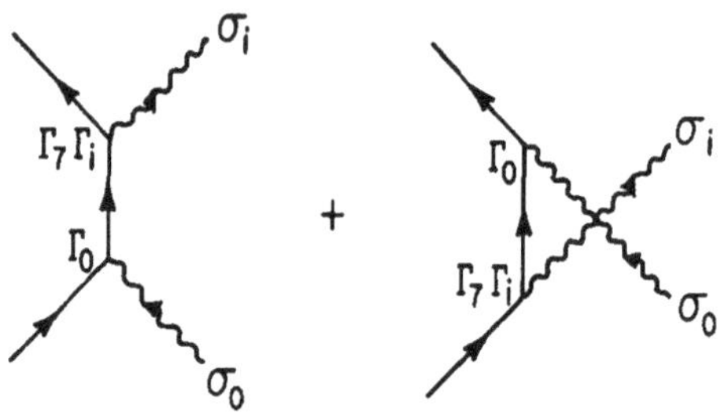

FIG. 6. Typical meson–fermion scattering diagrams.

After some manipulation Eq. (56) yields as the result before projection

$$\frac{1}{K}\times(-i)\frac{G^2}{2m_\sigma V^2}\Gamma_0\Gamma_i(a_0a_i^\dagger+a_ia_0^\dagger). \tag{58}$$

In spite of the K in the denominator, this expression is actually finite because the numerator $\Gamma_0\Gamma_1$ is also of order K for on-shell states. Incidentally, if we had not allowed for a small but non-vanishing K, the diagrams would have cancelled by the Majorana condition above. Thus we see there is a discontinuity at $K=0$ and extreme care has to be taken to retain all the terms till after the projection, when we can take the limit $K\to 0$. Note also that the coefficient $1/c^2V^2$ of the fermionic contribution is the same as for the bosons. The sum of the terms of the above mixed boson-fermion type (henceforth called $\mathbf{L}\cdot\mathbf{s}$ type for reasons that will be clear soon) are

$$\frac{G^2}{m_\sigma^2V^2}\Gamma_i\Gamma_j(a_i^\dagger a_j-b_i^\dagger b_j)+iG^2\left[\frac{3\Gamma_7\Gamma_i}{m_\sigma^2V^2}-\frac{\Gamma_0\Gamma_i}{2m_\sigma KV^2}\right](a_0^\dagger a_i+a_i^\dagger a_0). \tag{59}$$

The fermion mass terms are fairly straightforward, so we will not detail the calculation.

The projection to particle-number-diagonal space is somewhat tricky. We first project the operators $\mathcal{O}$ onto states of positive (or negative) energy and then onto positive (or negative) particle number and sum over the two projections:

$$\mathcal{O}\to\sum_{\pm}\frac{1\pm\Gamma_7}{2}\frac{|E|\pm H}{2|E|}\mathcal{O}\frac{|E|\pm H}{2|E|}\frac{1\pm\Gamma_7}{2}. \tag{60}$$

This yields, for example,

$$\begin{aligned}\Gamma_0\Gamma_i&\to\Gamma_0\Gamma_iK/2\Delta,\\ \Gamma_7\Gamma_i&\to\Gamma_0\Gamma_i/2,\end{aligned} \tag{61}$$

and so on. Adding together all the second-order contributions and putting in the volume correction C we get as the final result for the fermionic pieces

$$H_{bf'}+H'_f=\frac{1}{2}\Delta N_f+\frac{C}{8c^2V}\left[-\frac{1}{2}N_fN_6+\sum_i L_{0i}s_{0i}+\frac{1}{4}\sum_i s_{0i}s_{0i}\right]. \tag{62}$$

In the above we have made a phase rotation $a_0 \to ia_0$ in order to write the bosonic pieces in the standard form L_{0i} for angular momenta. The Γ's have been similarly disguised as spin: $s_{0i}=(i/2)\,\Gamma_0\Gamma_i$. By this notation we have betrayed our fond hope that the fermionic contributions can be coupled to the bosonic contributions Eq. (48) in such a way as to yield Casimirs of Spin (6) and Spin (5):

$$\begin{aligned} C_{\text{Spin}(5)} &= 1/2\sum_{ij}(L_{ij}+s_{ij})^2, \\ C_{\text{Spin}(6)}-C_{\text{Spin}(5)} &= O_6-O_5+2\sum_i L_{0i}s_{0i}. \end{aligned} \tag{63}$$

However, this is not to be. The fermionic contributions turn out to be too small [19, 20] by about a factor of 2 because of extensive cancellations. This is clear from a glance at Eq. (59)—the first term drops out completely because of the equivalence of the a's and b's after projection.

The result is not surprising because there was no strict supersymmetry between bosons and fermions from the beginning. At the same time, we are reminded that our σ-model Hamiltonian is in fact an approximation to the underlying BCS theory. In particular, there is a difference in BCS in the coupling of the σ and π modes to the fermions which we have neglected in our model. It is also possible, of course, that the assumed fermion space is inadequate and needs to be enlarged to describe the platinum region.

7. The $SU(3)$ Chain

Let us now turn to the other prominent IBM chain, that of $U(6)\supset U(3)\supset O(3)$ which describes rotational nuclei. The $U(3)$ to $O(3)$ symmetry-breaking is also of spontaneous origin, as the following analysis shows. We assume a starting $U(3)$ symmetry and adapt it to the BCS system with the help of a pseudospin [21, 22]: in particular, our valence fermions consist of six ($l=1$, $s=\frac{1}{2}$) states which couple into spin singlets and spin 0 and 2 bosons (the spin here simply takes care of the antisymmetry of the fermions):

$$(1\times\tfrac{1}{2})\otimes(1\times\tfrac{1}{2})=(1\times 1=2\oplus 0)\otimes(\tfrac{1}{2}\times\tfrac{1}{2}=0). \tag{64}$$

The bosons can be represented as the six independent components of 3×3 symmetric matrices $\mathbf{\Phi}$ and the fermions as the spinor $\Psi=(\psi_{1\uparrow}\psi_{0\uparrow}\psi_{-1\uparrow}\psi^\dagger_{1\downarrow}\psi^\dagger_{0\downarrow}\psi^\dagger_{-1\downarrow})$ in analogy whith Eq. (2) (the subscripts are the z-components of l and s).

The Hamiltonian is written in a somewhat mixed notation as

$$H_{U(3)} = \mathrm{Tr}[\boldsymbol{\Pi}^\dagger \boldsymbol{\Pi}] + 4G^2\,\mathrm{Tr}[\boldsymbol{\Phi}^\dagger \boldsymbol{\Phi}\boldsymbol{\Phi}^\dagger \boldsymbol{\Phi}] - \frac{16c^2G^2}{3}\mathrm{Tr}[\boldsymbol{\Phi}^\dagger \boldsymbol{\Phi}] + G\Psi^\dagger[\Gamma_m\sigma_m + \Gamma'_m\pi_m]\Psi. \tag{65}$$

In the above, the Yukawa piece is given in terms of real fields σ and π, where

$$\boldsymbol{\Phi} = (\sigma + i\pi)/\sqrt{2}, \tag{66}$$

with σ and π further expanded as

$$\sigma = \sum_m \sigma_m \lambda_m/\sqrt{2}, \qquad \pi = \sum_m \pi_m \lambda_m/\sqrt{2}. \tag{67}$$

Here λ_m, $m = 0, 1, \ldots, 5$ are the Gell–Mann matrices of SU(3) (or any unitarily equivalent set) with $\lambda_0 = \sqrt{2/3}\mathbf{1}$. The λ's are normalised so that the trace of their square is two. The spinors are coupled to the bosons using the 6×6 matrices

$$\Gamma_m = \tau_1 \otimes \lambda_m = \begin{pmatrix} 0 & \lambda_m \\ \lambda_m & 0 \end{pmatrix}$$
$$\Gamma'_m = \tau_2 \otimes \lambda_m = \begin{pmatrix} 0 & -i\lambda_m \\ i\lambda_m & 0 \end{pmatrix}, \tag{68}$$

where the τ's are Pauli matrices. The Γ's are not Clifford algebra matrices but serve to couple the fermions into pairs such as $\psi_{1\uparrow}\psi_{1\downarrow}$ etc.

We set the condensate proportional to the unit matrix

$$\langle \boldsymbol{\Phi} \rangle = \sqrt{2/3}c\mathbf{1} \tag{69}$$

and note that the mass conditions Eq. (21) are satisfied, with

$$m_\sigma = 4\sqrt{2/3}Gc. \tag{70}$$

Finally we break up the σ and π fields as

$$\sigma = (a + a^\dagger)/\sqrt{2m_\sigma V}$$
$$\pi = (b + b^\dagger)/\sqrt{2m_\pi V} \tag{71}$$

and calculate the diagrams. For the bosonic diagrams we do not need to expand the fields in terms of the λ matrices, but we do have to be careful with divergences. As before we take $m_\pi \to 0$ in the end and divergent diagrams cancel. The following definitions for the Casimirs of $U(3)$ and $O(3)$ are used to rewrite the result for the bosonic part:

$$\begin{aligned} U_3(\sigma) &= \sum_i \mathrm{Tr}[a^\dagger \lambda_i a]\,\mathrm{Tr}[a^\dagger \lambda_i a], \qquad \lambda_i \in U(3) \\ O_3(\sigma) &= 4\sum_i \mathrm{Tr}[a^\dagger \lambda_i a]\,\mathrm{Tr}[a^\dagger \lambda_i a], \qquad \lambda_i \in O(3). \end{aligned} \tag{72}$$

In the above, the antisymmetric λ_i are the elements of the $O(3)$ subgroup of $U(3)$. The $U(3)$ Casimir is three times smaller than in the IBM, and the $O(3)$ Casimir is given an extra factor of four to conform to the usual convention.

The following identities are similarly useful for the fermionic calculations:

$$\begin{aligned} [\Gamma_j, \Gamma_i]\, a_j^\dagger a_i &= -\Gamma_k\, \mathrm{Tr}[a^\dagger \lambda_k a - a\lambda_k a^\dagger] \\ \{\Gamma_j', \Gamma_i'\} b_j^\dagger b_i &= \Gamma_k'\, \mathrm{Tr}[b^\dagger \lambda_k b + b\lambda_k b^\dagger]. \end{aligned} \tag{73}$$

We identify the above as $\mathbf{L}\cdot\mathbf{s}$ type generators. $\mathbf{L}$ and $\mathbf{s}$ are defined as

$$\begin{aligned} L_i(\sigma) &= \mathrm{Tr}\left[a^\dagger \frac{\lambda_i}{2} a \pm a \frac{\lambda_i}{2} a^\dagger\right], \\ s_i &= \Gamma_i/2, \end{aligned} \tag{74}$$

where the $\pm$ signs are for the symmetric ($\in U(3)/O(3)$) and antisymmetric ($\in O(3)$) λ's respectively. With these definitions, and after projection, the result is

$$\begin{aligned} H'_{U_3} = {} & 1/2m_\sigma N_6 + 1/2\Delta N_f + 3C/32c^2V[-U_3 + 3/8O_3 \\ & -3\mathbf{L}\cdot\mathbf{s}(U_3) + 2\mathbf{L}\cdot\mathbf{s}(O_3) - 4\mathbf{s}\cdot\mathbf{s}(U_3/O_3)]. \end{aligned} \tag{75}$$

The condensate $c = 65$ MeV ($\nu = 6$). As noted earlier, the $U(3)$ Casimir here is three times as small as in the IBM. So in the IBM notation the ratio of the coefficients of U_3 and O_3 is $-\frac{8}{9}$. This agrees roughly with the available data [4] in which the ratio varies from about -1 to $-\frac{1}{3}$. The $1/A$ dependence is also consistent. It is to be expected that as in the $O(6)$ case an intrinsic $O(3)$ piece is always present.

The $U(5)$ chain is not easily amenable to a BCS interpretation. While a pseudospin picture with ($l = 2, s = \frac{1}{2}$) and a 5×5 symmetric bosonic matrix seems natural, the spin 4 part would have to be suppressed to conform to the IBM. It could alternately be regarded as an $O(6)$ chain with explicit $U(5)$ symmetry-breaking.

8. Conclusions and Speculations

We find that two important group chains of the IBM are of the spontaneous symmetry-breaking type, reflecting an underlying BCS mechanism. The magnitude of the resulting Hamiltonian is derived from ballpark values of the nuclear density and pairing energy while the relative sizes of its different terms are uniquely determined by the infinite-medium field theory. Heavier nuclei have smaller Hamiltonians—they scale as $1/A$, A being the atomic number.

This is a parameter-free theory, and its limitations directly reflect the limitations of the basic assumptions. The fuzziness of the overall scale results from the adaptation of infinite-medium results to the finite medium. The relative sizes of the different terms are taken for precision to be that of the infinite medium. However, since the fuzzfactor C is actually different for different diagrams, the *relative* sizes of the terms are also a little fuzzy. In addition, there are questions about the accuracy with which the σ-model reflects the BCS theory—as discussed earlier, the Yukawa coupling constant G is not necessarily constant or even the same for σ's and π's except in the limit of zero momentum-transfer.

On the plus side, the overall results are insensitive to details of the starting Hamiltonian and even to details of the adaptations to finiteness. They are applicable to all nuclei of the type discussed, with the single identifier A changing the scale of the effect for different nuclei. While the fermionic results depend on the assumption of a valence shell with a specific symmetry, the bosonic part is determined only by the starting symmetry (whether $O(6)$ or $U(3)$); the valence symmetry provides merely the rationale for the latter. The spontaneous effects are, in short, surprisingly robust. On the whole, BCS seems to be the secret of IBM.

One might remark on the coincidence that the coherence length of the Cooper pairs is so close to the nuclear radius—does this mean then that they are one and the same thing, and that the nucleus is a set of Cooper pairs held together by surface tension? In this case, there remains only one basic parameter, the saturation nuclear density. The pairing energy would then die as $1/A^{1/3}$ and disappear for an infinite medium. The relevance of BCS theory in the nucleus inspite of the large coherence length leads one to ask if other aspects of BCS might also be important. For example, are there vortices in nuclei? One of the effects of a vortex is that it allows rotational spectra about an axis of symmetry. This it does by effectively breaking the symmetry about the axis—all directions are no more equivalent but distinguished by an integrable phase. While the anomalous spectra would probably have a shell-model explanation, a BCS description would be simple and universal.

Finally, it is curious that the two different cases of symmetry-breaking that occur in the nucleus, the chirality pairing in pion physics and the nucleon pairing we have been discussing, have similar values for their condensate. Perhaps in some devious fashion this derives from their common root, QCD.

ACKNOWLEDGMENTS

We thank Professors A. Arima, F. Iachello, and Dr. Lee Brekke for helpful discussions. This work is supported by NSF Grant PHY-85-21588.

References

1. A. Arima and F. Iachello, *Phys. Rev. Lett.* **35** (1975), 1069.
2. A. Arima and F. Iachello, *Phys. Rev. Lett.* **40** (1978), 385.
3. A. Arima and F. Iachello, *Ann. Phys.* **99** (1976), 253.
4. A. Arima and F. Iachello, *Ann. Phys.* **111** (1978), 201.
5. A. Arima and F. Iachello, *Ann. Phys.* **123** (1979), 436.
6. F. Iachello and I. Talmi, *Rev. Mod. Phys.* **59**, No. 2 (1978), 339.
7. R. F. Casten and D. D. Warner, *Rev. Mod. Phys.* **60**, No. 2 (1988)), 389.
8. J. Bardeen, L. N. Cooper, and J. R. Schreiffer, *Phys. Rev.* **108**, No. 5 (1957), 1175.
9. A. Bohr, B. R. Mottelson, and D. Pines, *Phys. Rev.* **110** (1958), 936.
10. Y. Nambu and M. Mukerjee, *Phys. Lett.* **209**, No. 1 (1988), 1.
11. V. L. Ginsburg and L. D. Landau, *JETP* **20** (1950), 1064.
12. Y. Nambu, *Phys. Rev.* **117**, No. 3 (1960), 648.
13. J. R. Schreiffer, "Theory of Superconductivity," Frontiers of Physics, Benjamin, New York, 1964.
14. D. Lurie, *in* "Quantum Fluids, Proceedings, Batsheva Seminar, Technion-Israel Institute of Technology, Haifa" (N. Wiser and D. J. Amit, Eds.), North-Holland, New York, 1970.
15. P. B. Littlewood and C. M. Varma, *Phys. Rev. B* **26**, No. 9 (1982), 4883.
16. Y. Nambu, *Physica D* **15** (1985), 147.
17. P. Langacker and H. Pagels, *Phys. Rev. D* **8** (1973), 4595. H. Pagels, *Phys. Rep.* **16** (1975), 219.
18. Y. Nambu, *in* "The Rationale of Beings," Festschrift honouring G. Takeda, (K. Ishikawa *et al.*, Ed.), World Scientific, Singapore, 1986.
19. A. B. Balantekin, I. Bars, and F. Iachello, *Phys. Rev. Lett.* **47** (1981), 19.
20. A. B. Balantekin, I. Bars, and F. Iachello, *Nucl. Phys. A* **370** (1981), 284.
21. K. T. Hecht and A. Adler, *Nucl. phys. A* **137** (1969), 129.
22. A. Arima, M. Harvey, and M. Shimizu, *Phys. Lett. B* **30** (1969), 517.

Dynamical Symmetry Breaking*

Y. Nambu

Enrico Fermi Institute and Department of Physics,
University of Chicago, Chicago, IL 60637, USA

ABSTRACT

An overview is given concerning the concept of dynamical symmetry breaking and its examples in condensed matter, nuclear, and particle physics, including some speculations about the nature of the Higgs field in the Standard Model of electroweak unification.

1. Introduction

In my student days, Yoshio Nishina was one of those exalted names we talked about in awe. He was at that time heading cosmic ray and nuclear physics groups at Riken, the famous Institute for Physical and Chemical Research, which had played a unique role in the development of science and industry in Japan during the period between the two world wars. Having few, if any, professors in my university of Tokyo to teach us particle physics, we students used to frequent the weekly Riken seminars run by Nishina and his theoretical colleague S. Tomonaga. It was in this way that I was initiated into cosmic ray physics; I learned, for example, how Nishina's group was engaged in measuring cosmic ray intensities underground and over the Pacific Ocean. I also learned at first hand the mode of operation of the great school of theorists of the time, represented by people like Tomonaga, Yukawa, and Sakata, as they were developing their ideas about the cosmic ray mesons. I remember Tomonaga, at one of those seminars, reading Sakata's communication to him, in which Sakata was proposing that the cosmic ray "meson" (now called muon) and Yukawa's meson (now called pion) were different particles. However, this paper will concern certain theoretical ideas which have little direct connection with Nishina.

*Work supported in part by the NSF; PHY 90 0086

Springer Proceedings in Physics, Volume 57
Evolutionary Trends in the Physical Sciences Eds.: M. Suzuki and R. Kubo

2. Symmetry and Symmetry Breaking

The symmetry principle occupies an important place in our pursuit of physical laws, but it is not my intention to give an exhaustive discussion of the symmetry principle in general. Besides, my talk will inevitably reflect the fact that I am a particle theorist. When Yukawa created a theoretical paradigm with his meson theory in his search for the origins of nuclear forces, the concept of symmetry did not play any role. His paradigm, which I would like to call the Yukawa mode [1], was to hypothesize that behind new phenomena there are new particles in terms of which one can explain the former; the pursuit of particle physics is the pursuit of new particles. When a subfield of physics like particle physics was in its exploratory stage, this turned out to be a highly effective methodology. In fact it has remained so up to the present. But in the meantime the symmetry principle has also proven its power and importance as the field matured. In recent years we have seen a gradual emergence and even dominance of what I call the Einstein mode, in which theoretical principles drive the direction of particle physics. A key elemnt of this mode is the symmetry principle.

The purpose of this paper is to address one particular aspect of the symmetry principle, namely the dynamical, or spontaneous, breaking of symmetries. But the symmetry principle as it appears in modern physics has many facets, some of which had not been recognized before the recent developments in quantum field theory. So it seems appropriate for me to first give a brief summary of these various facets.

a) Symmetry gives a sense of esthetic beauty to physics and the natural world it describes. In mathematical terms, a symmetry essentially means a group of congruent operations under which the laws of physics are unchanged. The group may be continuous or discrete, and implies an associated conservation law which is respectively additive or multiplicative. Finding patterns of symmetry is highly useful in discovering regularities and conservation laws; conversely symmetry serves as a guiding principle in our search for a unified description of physical laws.

b) There are global symmetries as well as local, or gauged, symmetries. Wigner once remarked, according to my recollection, that there are two ways of establishing conservation laws: one by finding selection rules that apply between the initial and final states of a process, and the other by directly measuring the conserved charges by the fields they produce. Certainly this is a good characterization of the distinction between global and gauged symmetries. A gauged symmetry is richer and more restrictive than a global one in the sense that the former is in fact an infinite product of symmetries referring to each point of space-time. The Einstein gravity also belongs to this category. (Among the global symmetries one may include the so-called dynamical symmetries, like those found in the Keplerian and the harmonic motion, which are symmetries in the phase space, and are outside of the Noether theorem.)

c) It is often emphasized that physics consists of physical laws in local and differential form, plus the boundary and initial conditions which are subject to independent physical considerations. Symmetries usually refer to the former, but not necessarily to the latter. However, there are cases in which the topology of the physical space is coupled to that of the group manifold in question, so the boundary and initial conditions become an integral part of the symmetry. Topological considerations have led to concepts like solitons, monopoles, strings and instantons.

d) A symmetry implies degeneracy. In general there are multiplets of equivalent states related to each other by congruence operations. They can be distinguished only relative to a weakly coupled external environment which breaks the symmetry. Local gauged symmetries, however, cannot be broken this way because such an external environment is not allowed (a superselection rule), so all states are singlets, i. e., the multiplicities are not observable except possibly for their global part.

e) In reality global symmetries may be perfect or only approximate, leading to strict or approximate conservation laws. There may be a hierarchy of approximate symmetries, and often the patterns with which symmetries are broken are as meaningful and pleasing as the symmetries themselves. A symmetry may be so blatantly violated that it is the asymmetry rather than the symmetry that is interesting and significant. Parity violation in weak processes is an example.

f) There can be clashes of symmetries: different interactions may have different symmetries which are in conflict, and this conflict becomes the prominent feature of certain phenomena, again as in the case of the weak vs. the strong interactions, where their symmetry axes are tilted with respect to each other, so to speak.

g) A symmetry and the associated conservation law that are strict in classical theory may be violated by a quantum anomaly, i. e., the symmetry in question may be valid only "on shell", but not in the entire function space of fields over which the quantum action is defined. The chiral anomaly is a prominent example of it. Absence of anomalies is thought to be a necessary condition for a renormalizable theory based on gauged symmetries.

h) Symmetries inherent in the physical laws may be dynamically and spontaneously broken, i.e., they may not manifest themselves in the actual phenomena. The rest of the paper will address this topic in more detail.

3. Dynamical (Spontaneous) Symmetry Breaking

The fact that crystals, molecules and atoms exhibit symmetries as well as asymmetries seems to have caught the attention of physicists already when the group theory was being developed by mathematicians in the last century. According to Radicati [2], Pierre Curie [3] was one of the first physicists to discuss the aspects of symmetries and asymmetries in a modern language, mostly in the properties of crystals and of their responses to external forces. Be that as it may, a comprehensive historical review is not intended here.

The spontaneous breakdown of symmetries as a general concept is of more recent origin, although it predates the term coined by Baker and Glashow [4] in the 60's. The name is too long, and does not represent its content very adequately, but it has stuck for lack of a better one. It also appears that there exist subtle nuances in the way different people understand its meaning. Sometimes the term dynamical symmetry breaking is used as opposed to spontaneous symmetry breaking to denote dynamical mechanisms which are not immediately apparent. But in my opinion such a distinction is irrelevant. It is always a dynamical question whether a symmetry breaks or not. The two terms may be used interchangeably. Each term has its merits, but I will mainly use the word dynamical, and furthermore give it a rather narrow meaning. It does not include asymmetries of small finite systems like molecules (the Jahn-Teller effect). I will start with my definition of dynamical symmetry breaking, relying on concepts taken from group theory, statistical mechanics and quantum field theory.

As already mentioned, a symmetry implies degeneracy of energy eigenstates. Each multiplet of states forms a representation of a symmetry group G. Each member of a multiple is labeled by a set of quantum numbers for which one may use the generators and Casimir invariants of the chain of subgroups, or else some observables which form a representation of G. It is a dynamical question whether or not the ground state, or the most stable state, is a singlet, the most symmetrical one.

Consider now a system with a large number of degrees of freedom N, and the ground state is either degenerate or asymptotically degenerate so that its multiplicity grows and the energy splittings go to zero with increasing N. Usually one has in mind a uniform medium, where N is proportional to the number of constituents, and the spatial extent of the medium also grows with N. In the limit $N \to \infty$ (the thermodynamic limit) one may choose any particular state belonging to the degenerate multiplet, and call it the ground state of the medium. The quantum numbers of the state are infinite, but one may define their densities per unit volume, and call them order parameters.

Physical phenomena that can happen in this medium span a Hilbert space of states including the ground state under consideration. This space, however, is only a subspace of the Hilbert space of the system one had when N was finite. This is because the other ground states cannot be reached from the present one by means of local perturbations that operate only on a subset of its constituents. The two ground states are infinitely orthogonal, so to speak. The effective Hilbert space is one built on the present ground state by exciting it by local perturbations only. The system behaves as if the ground state was nondegenerate, but had reduced symmetry. Its symmetry (if one remains) is that of the subgroup H of G that leaves the order parameters invariant. The order parameters as a representation of G then belong to the coset space G/H.

According to the above characterization, the emergence of a superselection rule that reduces the Hilbert space is the essence of dynamical breaking of a symmetry as I would like to define it. It is crucial that N goes to infinity, but the symmetry may only be asymptotic, and the degeneracy need not be infinite. The familiar example of a double-well potential density [5] [7]

$$V(\phi) = G^2(\phi^2 - v^2)^2 \tag{1}$$

for a real scalar field $\phi(x)$ has two minima $\phi = \pm v$. If the number N of points in space is considered to be finite, the twofold degeneracy will be lifted by tunneling. For an asymmetric state centered around one of the two minima, the kinetic energy that causes tunneling is a symmetry-restoring agent, but it becomes ineffective as $N \to \infty$, so no mixing will take place between the two degenerate states.

In the case of a continuous symmetry, the large-N limiting behavior becomes more subtle. If the real field ϕ in the above example is replaced by a complex one, one has a $U(1)$ symmetry. The order parameter $< \phi >$ then is determined only up to an arbitrary phase angle $\theta(mod\ 2\pi)$ that labels the degenerate vacua. Two vacua corresponding to two distinct θ's are orthogonal, but the phase θ may be regarded a field, and local variations of θ from the given constant value will generate excited states. If the region of variation becomes large and the wave length of its Fourier transform also becomes proportionately large, one approaches a constant (and nonlocalized) variation, which amounts to a transition to a different vacuum, hence no change in energy. From this argument one infers that there will be a normal mode, the Goldstone mode, of excited states which have no energy gap in the long wavelength limit. In relativistic theories, the Goldstone mode behaves as a relativistic massless particle.

The above statements about the existence of symmetry breaking and associated gapless modes have some exceptions. Basically it has to do with the effectiveness of symmetry-restoring forces, i. e., how big the barrier is between broken symmetry configurations. In the case of continuous symmetry, there is no potential barrier, only a kinetic barrier. As a result, the Goldstone mode can exhibit an infrared instability, i. e., its large wavelength zero-point fluctuations wash out the order parameters and restore a single symmetric ground state. This can happen in low-dimensional media.

Another notable exception is when the symmetry is a gauged one. If the complex field ϕ in the above example is coupled to a $U(1)$ gauge field, the phase θ is a gauge parameter. Fixing it to a constant breaks gauge invariance. As θ turns into the dynamical Goldstone field, it couples to the gauge field which is also gapless. The mixing of two gapless modes then lifts their degeneracy, and gives rise to a massive mode with three polarizations, the plasmon mode. One may also say that the gauge field causes long range correlations between constituents so one cannot gently modulate the order parameter; such a modulation gets shielded.

Often quoted examples of dynamical symmetry breaking are ferromagnetism, crystal formation, and superconductivity. In an isotropic Heisenberg ferromagnet, the total spin is conserved. Dynamics favors neighboring spins to be parallel, so the ground state of the system has maximum spin pointed in some direction. The symmetry breaks from $SU(2)$ (or $O(3)$) to $U(1)$ (or $SO(2)$), the latter being the rotation group around the chosen axis. Which axis the system chooses depends on the initial and boundary conditions or on the environment. A typical procedure is to impose a weak magnetic field which then is gradually switched off. The Goldstone mode is the spin wave (polarization perpendicular to the magnetization axis) belonging to the coset Lie algebra $o(3)/o(2)$.

A crystal is said to violate the Euclidean incariance $O(3) \times T(3)$ down to a discrete subgroup, i. e., the space group of the crystal, because one imagines it to be fixed in space. The kinetic energy of the center of mass motion in the Hamiltonian of the crystal is a symmetry recovering agent, but it vanishes in the infinite mass limit, so one can localize the system by an infinitesimam force, breaking momentum and angular momentum conservation. Because the Euclidean group is a semidirect product, the Goldstone modes corresponding to o(3) and t(3) are coupled, and one ends up having only three modes, the isotropic longitudinal and transverse sound waves in the long wavelength limit.

Superconductivity, as described by the Bardeen, Cooper and Schrieffer (BCS) theory [6] as well as by its predecessor, the Ginsburg-Landau (GL) theory [7], is a nontrivial example of spontaneous symmetry breaking. The BCS theory is a microscopic, and hence more fundamental, description than the GL theory which is a phenomenological representation of the former. But both have served as the prototype of theories for various phenomena in condensed matter, nuclear and particle physics. The essence of the BCS theory is the Cooper pair formation: the pairing of an indefinite number of electrons of opposite spin near the Fermi surface due to a phonon-induced attraction. It leads to a nonzero pair correlation function $< \psi_{up}\psi_{dn} >$, and analog of the magnetization in ferromagnets. This complex pair field carries electric charge, and its phase is the gauge parameter.

The notion of the dynamical symmetry breaking with its characteristic properties as defined above first emerged in an attempt to resolve the question of gauge invariance in the BCS theory [8]. The concept of degenerate vacuua and the analogy to ferromagnetism had also been invoked in an earlier work of Heisenberg [9]. In his nonlinear theory of elementary particles, it was assumed that some internal quantum numbers like isospin and strangeness were not the intrinsic attributes of the elementary fermion field, but were spurions (a sort of nondynamical Goldstone mode with zero momentum and energy) picked up by particles from degenerate state vectors of the world acting as a reservoir.

4. The BCS Mechanism

By BCS mechanism I mean here the formation of a Cooper pair condensate as an order parameter, due to an attractive interaction between fermions, typically a short range one. Some salient features of the mechanism are [10, 11]:

a) There are fermionic and bosonic excitation modes. The order parameter causes mixing of fermions of opposite charges (particle and hole) leading to an energy gap in the dispersion relation. The bosons are collective states of fermion pairs, and come in two kinds, the "π" or "Goldstone" mode, and the "σ" or "Higgs" mode, corresponding respectively to modulations of the phase and the modulus of the pair field.

b) There are two energy scales: that of the dynamics of the constituents and that of the energy gap which is usually lower than the first. The latter is dynamically determined by a gap equation as a nonperturbative solution. In the short range and weak interaction limit, the fermion and boson modes satisfy simple mass relations:

$$m_\sigma : m_f : m_\pi = 2 : 1 : 0, \tag{2a}$$

$$m_1^2 + m_2^2 = 4m_f^2. \tag{2b}$$

The second relation, of which the first is a special case, applies to a pair of extra bosonic modes (m_1 and m_2) that exist in p-wave pairs like in ^{3}He.

c) There are induced interactions among those modes. They are controlled by a single coupling parameter which represents the ratio of the gap and the constituent energy scale. One can therefore translate the low energy contents of the BCS mechanism into an equivalent and restricted Ginzburg-Landau-Gell-Mann-Lévy system [12] which contain phenomenological fermion and complex boson ("Higgs") fields, one dimensionless coupling, and one mass scale (the vacuum expectation value $v = <\sigma>$ of the "Higgs" field σ).

d) There exists what I call quasi-supersymmetry [13], of which the mass formula (2a) is a consequence. It means that the static part of the GL Hamiltonian can be factorized in terms of fermionic composite operators as in supersymmetric quantum mechanics and quantum field theory. More specifically, these operators are spatial integrals of the densities

$$Q = \Pi\psi + W(\phi)\psi^\dagger, \text{ and its Hermitian conjugate } Q^\dagger . \tag{3}$$

Here Π and ϕ are canonical conjugates expressed as $n \times n$ matrix fields multiplying an n-component fermion field ψ; W is the square root of the Higgs potential: $V = tr\ W^2$. These fermionic operators give rise to a spectrum generating superalgebra. The physical origin of the quasi-supersymmetry underlying the BCS mechanism is not clear, but it is possible to generalize quasi-supersymmetry to a relativistic quantum field theory in which the Poincaré part of the superalgebra can be realized among a set of fermion, Higgs, and gauge fields.

Among examples of the BCS mechanism are superconductivity, superfluidity in ^{3}He, and the nucleon pairing effects in nuclei. Bosonic modes satisfying the mass relations of Eq. (2) have been found in superconductors and ^{3}He [14].

As for the nucleon pairing, I have recently claimed [15, 11] that the Interacting Boson Model [16] of nuclear excitations may be interpreted basically as a GL description of the BCS mechanism at work in nuclei. There is a caveat to be made here, however. Nuclei are finite systems so the concept of spontaneous symmetry breaking does not apply in the literal sense, but the latter may nevertheless offer a reasonably good picture of the dynamics involved. The near degeneracy of a multiplet of nucleon valence shell states dictates the corresponding degeneracy of bosonic pair states of spin 0 and 2, and forms the basis of a GL Hamiltonian in terms of nucleon and boson fields. In a typical example, a $U(1) \times SU(4)(\sim O(6))$ symmetry of the six complex bosons is broken spontaneously to $Sp(2)(\sim O(5))$ after forming a condensate in one of the pairs states, and thereby breaks the baryon number $U(1)$. To take care of the fact that finite nuclei do not actually break $U(1)$, one projects the broken symmetry states onto unbroken ones, thus eliminating spurious modes. The boson-boson interaction obtained in this way reproduces the corresponding part of the phenomenological formula of the IBM fairly well in terms of the density and volume of the nucleus only.

In particle physics, the chiral dynamics of hadrons consisting of massive quarks, pion, sigma meson and others, is generally interpreted as a realization of the BCS mechanism in QCD, although not of the short range and weak coupling variety of the previous examples. Each massless quark field has a chiral (γ_5) invariance. The masses of quarks generated spontaneously by gluonic interactions are usually referred to as "constituent masses". The pion, which is pseudoscalar, is essentially the accompanying Goldstone boson, but it is not strictly massless because the chiral invariance is broken by small "current masses" already present due to electroweak interactions. Furthermore, chiral symmetry is in general anomalous, so one does not expect massless or nearly massless bosons to exist except for a particular linear combination of chiral transformations for which the anomalies cancel.

In a similar fashion, it is often thought that the Standard Model of electroweak unification may in fact be the low energy effective form of a more fundamental dynamical theory in which the Higgs field is a composite object. Recently I have suggested that the Higgs field is not formed out of new heavy fermions as in the technicolor theory, but rather of the top and antitop quarks. This will be discussed below.

5. Tumbling and Bootstrap

I now come to bring up some new theoretical possibilities related to spontaneous symmetry breaking. One is known by the name tumbling [17]; I will call the other bootstrap.

In the BCS mechanism a massless fermion field acquires a mass, and composite bosons are created at the same time. Consider a set of fundamental fields having chiral invariance, for example in a grand unified theory, which is valid at a large energy scale E_1. Suppose the chiral symmetry is broken, and a mass scale $E_2 < E_1$ is created. The various composite bosons can be exchanged (in the "t-channel") between the fermions. If the induced interaction is attractive, it may trigger a second round of Cooper pair formation (in the s-channel) generating a new mass scale $E_3 < E_2$. This will occur most readily in a channel in which the attraction is the most attractive. In principle the process can be repeated any number of times, leading to a hierarchy of mass scales. Such a possibility of "tumbling" has been explored in model building in particle physics. The process of tumbling, however, already exists in known phenomena. One such example is the chain: crystal formation to superconductivity, for the phonons are the Goldstone bosons resulting from crystal formation, and they in turn become the agent of Cooper pair formation in superconductors. One might even ask if the process can be continued one step further.

Another example of tumbling is found in nuclear physics [18]. First the QCD of quarks produces the mass scale of the hadrons (of the order of the so-called Λ parameter, at which the QCD interaction becomes large enough to cause chiral symmetry breaking). The nucleons and various mesons are thereby generated. The exchange of the σ meson between nucleons is attractive, and makes it possible to form nuclei out of nucleons, especially because the σ field, being a neutral scalar, can be coherently enhanced in a many-nucleon system. One might say in a nutshell that the σ is largely responsible for the existence of nuclei and for their basic properties like the shell structure, the spin-orbit interaction and the pairing, the last of which corresponds to the second stage of tumbling.

In contrast to the tumbling chain of symmetry breakings of descending energy scales, the idea of bootstrap is that the chain is circular and self-sustaining. It refers to the theoretical possibility that the Higgs field is both the cause and the effect of a BCS mechanism at the same time. The concept is similar to Chew's bootstrap hypothesis in hadron dynamics [19]. In its most general form, his bootstrap implied a duality of s- and t-channels so that the hadrons were in effect composites made out of each other. The duality principle has found its mathematical realization in the Veneziano model and the subsequent string theory.

The bootstrap BCS mechanism could in principle occur in many systems, e.g., (high T_c?) superconductors, but the specific hypothesis I have proposed concerning the $SU(2) \times U(1)$ electroweak unification means the following [20]. In the Standard Model of Salam and Weinberg, a complex doublet Higgs field is introduced to trigger a spontaneous breaking of $S(2) \times U(1)$ down to a $U(1)$ subgroup corresponding to electromagnetism. Physically realized particles are a massless gauge boson, i.e. the photon, for electromagnetism, three massive gauge bosons $W^{\pm}$ and Z^0 for the weak interactions, massive quarks and leptons (except possibly for the neutrinos), and the scalar Higgs boson. Their masses are given by

$$m_i = g_i v, \tag{4}$$

where the g_i's are appropriately defined coupling constants; $v = 246$ GeV is the vacuum expectation value (order parameter) of a component of the Higgs field, which sets the overall mass scale.

The Standard Model has so far proven remarkably accurate in describing the experimental data concerning the weak and electromagnetic interactions. No discrepancies or indications of new phenomena going beyond the model have been seen. Two input parameters, i. e., the electric charge and the Fermi constant, and a mixing angle determine v and the two gauge couplings, and thereby fix the W and Z masses, now known to be 80 GeV and 91 GeV respectively. On the other hand, the fermion and Higgs boson masses depend on Yukawa- and self-couplings of the Higgs fields which are arbitrary, so the model has no predictive power in this regard.

Recall now that, in the BCS mechanism, there was a simple 2 to 1 mass ratio between the fermion and the Higgs (σ) boson. If only one degree of freedom out of the many fermions in the Standard Model participated in this mechanism, one would expect this ratio to hold, up to renormalization corrections (which tend to reduce the ratio). It also implies that the Higgs field is a phenomenological substitute for the bound states of the particular fermion pairs in question, just as it was the case with the chiral dynamics of hadrons.

Even if this interpretation were correct, one would not know the agent that caused the Cooper pairing. Perhaps it originated in a grand unified theory which involved extra gauge fields and other degrees of freedom hidden from us at the electroweak energies. But there is another possibility, namely the bootstrap. In this case, the stronger the Yukawa coupling of a fermion, the stronger the pairing interaction, and the larger the fermion mass. Therefore one may say the heaviest one among the leptons and quarks is responsible for the formation of the Higgs field and breaking of $S(2) \times U(1)$ to $U(1)$. It is also consistent with the fact that the yet undiscovered top quark appears to be very heavy compared to all the other fermions. Current lower limits are 89 GeV for the top quark, and 40 GeV for the Higgs boson. In short, the Higgs boson is a bound state of top and antitop quarks, and their Yukawa coupling is $\gtrsim 1/3$. The Higgs boson mass should be roughly double the top mass (or less). A similar idea has been proposed also by Miransky, Yamawaki and Tanabashi [21].

The precise formulation of the bootstrap mechanism has some latitudes. Bardeen, Hill, and Lindner [22] start from a local limit of Higgs-exchange interaction between the top quark fields (as in the Fermi form of the weak interaction), and apply the BCS formalism in the standard manner, which lead to a gap equation with a quadratic divergence. The results depend on the cut-off parameter L, but in general the masses are rather large ($m_t \gtrsim 200$ GeV, and m_H is somewhat larger than m_t).

The approach I have taken [20] starts from the Standard Model as is, but treats the vacuum expectation value v as a dynamical one. Namely v is the expectation value of the potential, the so-called tadpole potential, acting on the fermion and giving it a mass due to the Higgs echange with the zero-point fermion, Higgs and the gauge fields in the vacuum. (Usually the tadpoles are regarded as a correction to a given v, and to be renormalized away, but here it is the whole contribution.) This sets up a gap equation for v since the tadpoles themselves are proportional to v, but with quadratically divergent coefficients. This is interpreted to mean that the Higgs theory is only a phenomenological representation which breaks down at the cut-off energy Λ and has to be replaced by a more fundamental theory of less divergent nature. On the other hand, the bootstrap hypothesis means that the low energy effective theory is closed and self-consistent by itself, and should not be sensitively dependent on the hidden underlying dynamics symbolized by an extra cut-off parameter. Thus one demands that the quadratic divergences cancel each other among the fermion, Higgs and gauge fields tadpole contributions. In this way, one gap equation splits into two equations, the quadratic and the remaining logarithmic part. The quadratic cancellation condition had been proposed by Veltman [22].

The two equations are constraints on v/Λ and the various coupling constants. Given v/Λ and the gauge couplings, one can determine the two unknown parameters, Yukawa coupling and the self-coupling of the Higgs field, and hence determine m_H and m_t. These equations have the form

$$\Sigma\ c_i g_i^2 Z_i^{-1} = 0\ ,\ \ \Sigma\ c_{i'} g_i^4 Z'^{-1}_i\ ln(\Lambda/m_i) = g_H^2\ . \tag{5}$$

Here the c_i's and c'_i's are numerical weights; the Z_i's and Z'_i's are renormalization constants and related quantities for the various fields contributing to the tadpoles, regarded as functions involving powers of $g_i^2 \ln(\Lambda/v)$, g_i^2 and $\ln g_i$. In the lowest approximation, one may set all the Z's equal to 1, and solve for the top and Higgs couplings g_t and g_H (or m_t and m_H) in terms of the gauge coupling's (or m_W and m_Z). One finds two sets of solutions: $m_t \sim 80$ GeV, $m_H \lesssim 60$ GeV, and $m_t \gtrsim 120$ GeV, $m_H \gtrsim 200$ GeV. Their exact values depend on Λ, getting larger for smaller Λ. The low mass solution seems incompatible with experiment. For values of Λ of the order of the Planck mass and less, the high mass solution actually gives considerably larger masses than the lower limits. However, the renormalization corrections seem appreciable, although they have not been evaluated.

It remains to be seen whether or not the basic assumptions concerning a bootstrap mechanism as the origin of symmetry breaking and mass generation in the electroweak interactions will hold up experimentally. Their tests mainly lie in the prediction of the top quark and Higgs boson masses, and the absence of early deviations from the Standard Model.

A more ambitious program would address the origin of the entire mass matrix of the fermions. From the viewpoint of the bootstrap, it is interesting that the top quark plays a special role and is by far the heaviest fermion. In fact W, Z, t and H seem to belong to the same natural mass scale of the weak interactions. So the problem is why the other fermions are so light. One has already a hierarchy problem at the current energy range. It might not be unreasonable to expect that these small masses are higher order corrections to the basic BCS mechanism proposed here. At any rate, understanding the fermion hierarchy might give one a clue to the hierarchy problem at higher energy scales.

I have benefited from informative discussions with Laurie Brown on historial literature concerning symmetry and symmetry breaking. This work was supported in part by the NSF; PHY 90-00386.

REFERENCES

[1] Y. Nambu, Prog. Theor. Phys. Suppl. **85**, 104 (1985); in *Particle Accelerators*, 1990, (Gordon and Breach, New York, 1990), Vol. 26, p. 1.

[2] L. A. Radicati, in *Symmetries in Physics (1600-1990)*, Proc. 1st International Meeting on the History of Scientific Ideas, Catalonia, ed. M. G. Doncel et al. (Univ. Autonoma de Barcelona, 1987), p. 197.

[3] For more information about the historical aspects, see, for example: Y. Nambu, Fields and Quanta 1, **33** (1970) and *The Past Decade in Particle Theory*, ed. E. C.G. Sudarshan et al. (Gordon and Breach, 1933), p. 33. *Dynamical Gauge Symmetry Breaking*, ed. E. Farhi and R. Jackiw (World Scientific, Singapore, 1982).

[4] M. Baker and S. L. Glashow, Phys. Rev. **128**, 2462 (1962).

[5] J. Goldstone, Nuovo Cimento **19**, 155 (1961).
P. Higgs, Phys. Rev. **145**, 1156 (1966).

[6] J. Bardeen, L. N. Cooper, and J. R. Schrieffer, Phys. Rev. **106**, 162 (1957).

[7] V. L. Ginsburg and L. D. Landau, Zh. Exp. Teor. Fiz. **20**, 1064 (1950).

[8] P. W. Anderson, Phys. Rev. **110**, 827 (1958); **112**, 1900 (1958).
G. Rickayzen, Phys. Rev. **111**, 817 (1958); Phys. Rev. Lett. **2**, 91 (1959).
Y. Nambu, Phys. Rev. 117, 648 (1960).

[9] H. P. Duerr, W. Heisenberg, H. Mitler, S. Schlieder and K. Yamazaki, Z. f. Naturf. **14a**, 441 (1959).

[10] Y. Nambu and G. Jona-Lasinio, Phys. Rev. **122**, 345 (1961); **124**, 246 (1961).

[11] Y. Nambu, Physica **15D**, 147 (1985); "Mass Formulas and Dynamical Symmetry Breaking", U. Chicago preprint 90-37, to appear in a festscrift in honor of S. Okubo.

[12] M. Gell-Mann and M. Levy, Nuovo Cimento **16**, 705 (1960)

[13] Y. Nambu, in *Rationale of Beings*, Festschrift in honor of G. Takeda, ed. R. Ishikawa et al. (World Scientific, Singapore, 1986), p. 3; "Supersymmetry and Quasi-Supersymmetry", U. Chicago preprint EFI 89-30, to appear in a festschrift in honor of Murray-Gell-Mann.

[14] See [11] and the references quoted therein.

[15] Y. Nambu and M. Mukerjee, Phys. Lett. **209**, 1 (1988); M. Mukerjee and Y. Nambu, Ann. Phys. **191** (1989).

[16] A. Arima and F. Iachello, Phys. Rev. Lett. **35**, 1069 (1975); Ann. Phys. **99**, 253 (1976).

[17] S. Dimopoulos, S. Raby, and L. Susskind, Nuc. Phys. **B169**, 493 (1980).

[18] Y. Nambu, in *Festi-Val*, Festschrift for Val Telegdi, ed. K. Winter (Elsevier Science Publishers B. V., 1988), p. 181; and the second reference in [11].

[19] G. F. Chew, Proc. 1960 International Conference on High Energy Physics, ed. E. C. G. Sudarshan et al. (U. Rochester/Interscience Publishers, 1960), p. 273.

[20] Y. Nambu, in *New Theories in Physics*, Proc. XI Warsaw Symposium on Elementary Particle Physics, ed. Z. A. Ajduk et al. (World Scientific, Singapore, 1989), p. 1; *1988 International Workshop on New Trends in Strong Coupling Gauge Theories*, ed. M. Bando et al. (World Scientific, Singapore, 1989), p. 2.; "New Bootstrap and the Standard Model", U. Chicago preprint EFI 90-46, to appear in a festschrift in honor of D. Dalitz; "More on Bootstrap Symmetry Breaking", U. Chicago preprint EFI 90-69, to appear in the Proceedings of the 1990 International Workshop on Strong Coupling Gauge Theories and Beyond, Nagoya.

[21] V. Miransky, M. Tanabashi, and K. Yamawaki, Mod. Phys. Lett. **A4**, 1043 (1989); Phys. Lett. **B221**, 177 (1989).

[22] W. Bardeen, C. Hill, and M. Lindner, Phys. Rev. **D41**, 1647 (1990).

[23] M. Veltman, Acta Phys. Polon. **B12**, 437 (1981).

YOICHIRO NAMBU

Date of birth:	18 January 1921
Birthplace:	Tokyo, Japan
U. S. Citizenship:	31 March 1970

Education

B. S.	University of Tokyo 1942
D. Sc.	University of Tokyo 1952

Positions

1950–1956	Osaka City University — Professor
1952–1954	Institute for Avdanced Study — Member
1954–1956	University of Chicago — Research Associate
1956–1958	University of Chicago — Associate Professor
1958–1990	University of Chicago — Professor
1991–	Professor Emeritus, University of Chicago

Honors

Dannie Heinemann Prize for Mathematical Physics, 1970

Member of National Academy of Sciences, 1971

Member of American Academy of Arts and Sciences, 1971

J. Robert Oppenheimer Prize, 1976

Order of Culture, Government of Japan, 1978

United States National Medal of Science, 1982

Honorary Member of Japan Academy, 1984

Max Planck Medal, 1985

P.A.M. Dirac Medal, ICTP, Trieste, 1986

Wolf Prize, Government of Israel, 1995

List of Publications

1948

1. On the Relativistic Formulation of the Perturbation Theory,
Prog. Theor. Phys. **3**, 444.

1949

2. The Level Shift and the Anomalous Magnetic Moment of the Electron,
Prog. Theor. Phys. **4**, 82.

3. Second Configuration Space and Third Quantization,
Prog. Theor. Phys. **4**, 96.

4. Effect of the C-Meson Field on the Anomalous Magnetic Moment of the Electron (with Z. Koba and T. Tati),
Prog. Theor. Phys. **4**, 99.

5. On the Method of Third Quantization, I and II,
Prog. Theor. Phys. **4**, 331; 339.

1950

6. A Note on the Eigenvalue Problem in Crystal Statistics,
Prog. Theor. Phys. **5**, 1.

7. The Use of the Proper Time in Quantum Electrodynamics I
Prog. Theor. Phys. **5**, 82.

8. On the Electromagnetic Properties of Mesons (with T. Kinoshita),
Prog. Theor. Phys. **5**, 307.

9. Derivation of the Interaction Potential from Field Theory,
Prog. Theor. Phys. **5**, 321.

10. Force Potentials in Quantum Field Theory,
Prog. Theor. Phys. **5**, 614.

11. On the Interaction of Mesons with the Electromagnetic Field II (with T. Kinoshita),
Prog. Theor. Phys. **5**, 749.

1951

12. On the Nature of V-Particles I and II (with K. Nishijima and Y. Yamaguchi),
Prog. Theor. Phys. **6**, 615; 619.

13. Meson–Nucleon Scattering (with Y. Yamaguchi),
Prog. Theor. Phys. **6**, 1000.

1952

14. On Lagrangian and Hamiltonian Formalism,
Prog. Theor. Phys. **7**, 131.

15. An Empirical Mass Spectrum of Elementary Particles,
Prog. Theor. Phys. **7**, 595.

1954

16. The Collective Description of Many-Particle System (A Generalized Theory of Hartree Fields) (with T. Kinoshita),
Phys. Rev. **94**, 598.

1955

17. Structure of the Scattering Matrix,
Phys. Rev. **98**, 803.

18. Structure of Green's Function in Quantum Field Theory I,
Phys. Rev. **100**, 394.

1956

19. Structure of Green's Function in Quantum Field Theory II,
Phys. Rev. **101**, 459.

20. Renormalization Constants,
Phys. Rev. **101**, 1183.

21. Application of Dispersion Relations to Low Energy Meson–Nucleon Scattering (with G.F. Chew, M.L. Goldberger and F.E. Low),
Phys. Rev. **106**, 1337.

22. Relativistic Dispersion Relation Approach to Photomeson Production (with G.F. Chew, M.L. Goldberger and F.E. Low),
Phys. Rev. **106**, 1345.

1957

23. Possible Existence of a Heavy Neutral Meson,
Phys. Rev. **106**, 1366.

24. Dispersion Relations for Nucleon-Nucleon Scattering (with M.L. Goldberger and R. Oehme),
Ann. Phys. **2**, 226.

25. Parametric Representations of General Green's Functions,
Il Nuovo Cimento X **6**, 1064.

1958

26. Dispersion Relations for Form Factors,
Il Nuovo Cimento X **9**, 610.

27. Dispersion Theory Treatment of Pion Production in Electron–Nucleon Collisions (with S. Fubini and V. Wataghin),
Phys. Rev. **111**, 329.

1960

28. Quasi-Particles and Gauge Invariance in the Theory of Superconductivity,
Phys. Rev. **117**, 648.

29. Axial Vector Current Conservation in Weak Interactions,
Phys. Rev. Lett. **4**, 380.

30. Electro-production of *pi*-Mesons (with R. Blankenbecler, S. Gartenhous and R. Hugg),
Il Nuovo Cimento X **17**, 775.

31. Anomalous Thresholds in Dispersion Theory I (with R. Blankenbecler),
Il Nuovo Cimento X **18**, 595.

1961

32. Odd $\Lambda\Sigma$ Parity and the Nature of the $\pi\Lambda\Sigma$ Coupling (with J.J. Sakurai),
Phys. Rev. Lett. **6**, 377.

33. A 'Superconductor' Model of Elementary Particles and Its Consequences, in the *Proceedings of the Midwest Conference on Theoretical Physics*, eds. F.J. Belinfante, S.G. Garten haus and R.W. King, Purdue University, Lafayette, Indiana, April 1–2, 1960, p. 1.

34. Dynamical Model of Elementary Particles based on an Analogy with Superconductivity I (with G. Jona-Lasinio),
Phys. Rev. **122**, 345.

35. Dynamical Model of Elementary Particles based on an Analogy with Superconductivity II (with G. Jona-Lasinio),
Phys. Rev. **124**, 246.

36. Possible Bound $\Sigma - \Lambda$ System,
(with E. Shrauner), *Il Nuovo Cimento X* **21**, 864.

1962

37. Chirality Conservation with Soft Pion Production (with D. Lurié),
Phys. Rev. **125**, 1429.

38. Soft Pion Emission Induced by Electromagnetic and Weak Interactions (with E. Shrauner),
Phys. Rev. **128**, 862.

39. Magnetic Field Dependence of the Energy Gap in Superconductors (with San Fu Tuan),
Phys. Rev. **128**, 2622.

40. Rare Decay Modes of the $\omega(\eta)$ Meson (with J.J. Sakurai),
Phys. Rev. Lett. **8**, 79.

1963

41. Double Phase Representation of Analytic Functions (with M. Sugawara),
Phys. Rev. **131**, 2335.

42. High-Energy Behavior of Total Cross-Sections (with M. Sugawara),
Phys. Rev. **132**, 2724.

43. Magnetic Field and Phase Transition in Thin Film Superconductors (with San Fu Tuan),
Phys. Rev. Lett. **11**, 119.

44. Non-shrinking Diffraction Scattering (with M. Sugawara),
Phys. Rev. Lett. **10**, 304.

45. K* (725) and the Strangeness-changing Currents of Unitary Symmetry (with J.J. Sakurai),
Phys. Rev. Lett. **11**, 42.

46. Self-Consistent Models of Strong interactions with Chiral Symmetry (with P. Pascual), *Il Nuovo Cimento X* **30**, 354.

1964

47. Considerations on the Magnetic Field Problem in Superconducting Thin Films (with San Fu Tuan),
Phys. Rev. **133**, A1.

48. Axial Vector Mesons (with P. Freund),
Phys. Lett. **12**, 248.

49. Broken SU(x) × SU(3) × SU(3) × SU(3) Symmetry of Strong Interactions (with P. Freund),
Phys. Rev. Lett. **12**, 714.

50. Mass and Coupling Constant Formulas in Broken Symmetry Schemes (with P. Freund),
Phys. Rev. Lett. **13**, 221.

51. Magnetic Field and Phase Transition in Superconducting Thin Films (with San Fu Tuan),
Rev. Mod. Phys. **36**, 288.

52. Quasi-elementary Massless Bosons Associated with the Quantum Electrodynamics of Johnson, Baker and Willey,
Phys. Lett. **9**, 214.

53. Outlook of Elementary Particle Physics,
in *Nature of Matter*, ed. Luke C.L. Huan, BNL F88(T-360), p. 55.

1965

54. Three-Triplet Model with Double SU(3) Symmetry (with M.-Y. Han),
Phys. Rev. **B139**, 1006.

55. Dynamical Symmetries and Fundamental Fields, in the *Proceedings of Coral Gables Conference 1965*, eds. B. Kursunoglu, A. Perlmutter and I. Sakmar (W. H. Freeman and Co.), p. 274.

56. Broken SU(3) × SU(3) × SU(3) × SU(3) Symmetry (with P. Freund),
Ann. Phys. **32**, 201.

57. Triplets, Static SU(6), and Spontaneously Broken Chiral SU(3) Symmetry, in *Proceedings of the International Conference on Elementary Particles*, in Commemoration of the Thirtieth Anniversary of Meson Theory, Kyoto, 1965, ed. Y. Tanikawa (Kyoto Publication Office) p. 131; General Discussion, *ibid.*, p. 327.

58. Axial Vector Meson and the Hyperfine Structure of Hydrogen (with S. Fenster and R. Koberle),
Phys. Lett. **19**, 513.

59. Baryon Structure and Electromagnetic Properties,
(with S. Fenster), *Prog. Theor. Phys. Suppl.*, Extra No., 250.

60. Electromagnetic Properties of the Baryon (Hyperfine Structure of Hydrogen), *1965 Tokyo Summer Lecture*, Part II, ed. G. Takeda (Syokabō Publishing), p. 87.

1966

61. A Systematics of Hadrons in Subnuclear Physics, in *Preludes in Theoretical Physics*, eds. A. De-Shalit, H. Feshbach and L. van Hove (North-Holland), p. 133.

62. Coupling Constant Relations for $1^{\pm}$ and Induced $0^{\pm}$ Mesons (with J. Cronin),
Il Nuovo Cimento **A41**, 380.

63. Nonleptonic Decays of K-Mesons (with Y. Hara),
Phys. Rev. Lett. **16**, 875.

64. Nonleptonic Decays of Hyperons (with Y. Hara and J. Schechter),
Phys. Rev. Lett. **16**, 380.

65. Relativistic Wave Equations for Particles with Internal Structure and Mass Spectrum,
Prog. Theor. Phys. Supp. **37** and **38**, 368.

1967

66. Infinite Multiplets, in the *Proceedings of 1967 International Conference on Particles and Fields*, eds. C.R. Hagen *et al.* (Interscience), p. 347.

67. Infinite Component Wave Equations with Hydrogenlike Mass Spectra,
Phys. Rev. **160**, 1177.

1968

68. Magnetic Moments and Charge Radii for States Described by an Infinite Component Wave Equation (with S.P. Rosen),
Prog. Theor. Phys. **40**, 5.

69. Quantum Electrodynamics in Nonlinear Gauge,
Prog. Theor. Phys. Suppl., Extra No., 190.

70. Scalar Fields Coupled to the Trace of the Energy Momentum Tensor (with P. Freund),
Phys. Rev. **174**, 1741.

71. S-Matrix in Semiclassical Approximation,
Phys. Lett. **B26**, 626.

72. Relativistic Groups and Infinite Component Fields, talk at *Nobel Symposium on Elementary Particle Theory*, Goteborg, May, 1968, ed. N. Svaartholm (Almquist and Wiksell), p. 105.

1970

73. Asymptotic Behavior of Partial Widths in the Veneziano Model of Scattering Amplitudes (with Paul Frampton), in *QUANTA, A Collection of Scientific Essays dedicated to Gregor Wentzel*, eds. P.G.O. Freund, C. Goebel and Y. Nambu (University of Chicago Press), p. 33.

74. Quark Model and the Factorization of the Veneziano Amplitude, in the *Proceedings of the International Conference on Symmetries and Quark Models*, Wayne University, 1969, ed. R. Chand (Gordon and Breach), p. 269.

75. Statistical Approach to the Veneziano Model (with L.-N. Chang and P. Freund),
Phys. Rev. Lett. **24**, 628.

76. Axial-Vector Form Factor of Nucleon Determined from Threshold Electropion Production (with M. Yoshimura),
Phys. Rev. Lett. **24**, 25.

77. Symmetry Breakdown and Small Mass Bosons,
Fields and Quanta 1, 33 (1970), also published in *The Past Decade in Particle Theory*, eds. E.C.G. Sudarshan and Y. Ne'eman (Gordon and Breach, 1973), p. 33.

78. Duality and Hadrodynamics,
notes prepared for Copenhagen High Energy Symposium, August 1970 (unpublished).

1971

79. Electromagnetic Currents in Dual Hadrodynamics,
Phys. Rev. **D4**, 1195.

1972

80. Gauge Conditions in Dual Resonance Models (with F. Mansouri),
Phys. Lett. **B39**, 375.

81. Use of Regulator Fields in Dual Resonance Models (with J. Willemsen), submitted to *XVI International Conference on High Energy Physics, 1972*, eds. J.D. Jackson and A. Roberts, National Accelerator Laboratory, 1972, number 110, unpublished.

82. Chiral Symmetries and Current Algebras,
Lecture at International Summer School, Erice, Italy, 1972 (unpublished).

1973

83. Generalized Hamiltonian Dynamics,
Phys. Rev. **D7**, 2405.

84. Models Concerning the Chemical Structure of Hadrons, in the *Proceedings of Tokyo Symposium on High Energy Physics*, July.

85. Elementary Particle Physics in Perspective,
Butsuri **28**, 452 (in Japanese).

1974

86. Three Triplets, Paraquarks and "Colored" Quarks (with M.-Y. Han),
Phys. Rev. **D10**, 675.

87. Quarks, Strings and Gauge Fields, in the *Proceedings of Johns Hopkins Workshop on Current Problems in High Energy Particle Theory, 1973.*

88. Strings, Monopoles and Gauge Fields,
Phys. Rev. **D10**, 4262.

1975

89. Dynamics of the Zweig–Iizuka Rule and a New Vector Meson Below 2 GeV/c^2 (with P. Freund),
Phys. Rev. Lett. **34**, 1645.

1976

90. Magnetic and Electric Confinement of Quarks, talks given at Topical Conference on Extended Systems, Ecole Normale Superieure, June, 1975; *Phys. Rep.* **C23**, 250.

91. Diquark Color Excitation and the Narrow Resonances (with M.-Y. Han),
Phys. Rev. **D14**, 1459.

1977

92. Strings, Vortices and Gauge Fields, in *Quark Confinement and Field Theory*, eds. D.R. Stump and D.H. Weingarten (John Wiley and Sons), p. 1.

93. Description of Hadronic Structure, in the *Proceedings of the International School on Elecrtro- and Photonuclear Reactions, 1976*, eds. S. Costa and C. Shaerf; also published in *Quark Models and Hadronic Structure*, Physics Workshop Series, 1, ed. G. Morpurgo (Plenum Press), p. 281.

94. Monopoles, Strings and Instantons, in *Five Decades of Weak Interactions, Proceedings of CCNY Conference, 1977*, ed. N. P. Chang *Ann. of the N.Y. Acad. of Sci.* **74**, 294.

95. Elementary Particles,
Butsuri **32**, 11 (in Japanese).

96. String-Like Configurations in the Weinberg–Salam Theory,
Nucl. Phys. **B130**, 505.

1978

97. Some Topological Configurations in Gauge Theories,
Int. J. Theor. Phys. **17**, 287.

98. Remarks on the Topology of Gauge Fields, in *New Frontiers in High Energy Physics, Proceedings of the Orbis Scientiae*, University of Miami, January 1978, eds. Kursunoglu and Perlmutter (Plenum Press), p. 632.

1979

99. Topological Problems in Gauge Theories, talk at *UCLA Symposium in Honor of Julian Schwinger on the Occasion of His 60th Birthday, Physica* **A96**, 89.

100. QCD and the String Model, *Phys. Lett.* **B80**, 372.

101. Concluding talk at the *XIX International Conference on High Energy Physics, Tokyo, 1978*; in the *Proceedings of the Conference*, eds. S. Homma *et al.* (Phys. Soc. Japan), p. 971.

1980

102. Quark Confinement: the Cases for and Against, in *To Fulfill a Vision: Jerusalem Einstein Centennial Symposium on Gauge Theories and Unification of Physical Forces*, ed. Y. Ne'eman (Addison and Wesley), p. 118.

103. Hamilton–Jacobi Formalism for Strings,
Phys. Lett. **B92**, 327.

1981

104. Effective Abelian Gauge Fields,
Phys. Lett. **B102**, 149.

105. Strings and Vortices,
VPI Conference on Weak Interactions as Probes of Unification, AIP Conference Proceedings, Particles and Fields subseries No. 23, p. 633.

106. *Quark* (Kodansha Publishers) in Japanese.

107. Theory of Strings, in the *Proceedings of the 1981 INS Symposium on Quark and Lepton Physics*, Tokyo, Japan, eds. K. Fujikawa *et al.* (Institute of Nuclear Study, Univ. of Tokyo), p. 347.

1982

108. Many Quark Problem in Two Dimensional Gauge Theory (with B. Bambah), *Phys. Rev.* **D26**, 2871.

109. One-Dimensional Quark Gas (with B. Bambah and M. Gross), *Phys. Rev.* **D26**, 2875.

1983

110. Concluding Remarks, *Lecture Notes in Physics*, **176**; *Proceedings of the International Symposium on Gauge Theory and Gravitation*, Nara, Japan, 1982, eds. K. Kikkawa *et al.*, (Springer), p. 277.

111. Magnetic Monopoles and Related Topics, in the *Proceedings of the Topical Conference*, June, 1982, eds. T. Eguchi and Y. Yamaguchi (World Scientific), p. 3.

112. The Activity of the Tomonaga Group up to the Time of the 1947 Shelter Island Conference (with K. Nishijima), in *Shelter Island II, Proc. of the 1983 Shelter Island Conference on Quantum Field Theory and the Fundamental Problems of Physics*, eds. R. Jackiw *et al.* (MIT Press), p. 367.

1984

113. Concluding Remarks, in the *Proceedings of the Solvay Conference on Higher Energy Physics*, Austin, Texas, November 1982; *Phys. Rep.* **104**, 237.

114. Quantum Mechanics: Prospects and Problems, in the *Proceedings of the Symposium on Foundations of Quantum Mechanics in Light of New Technology*, Tokyo, August, 1983, eds. Kamefuchi *et al.* (Physical Society of Japan), p. 363.

115. Superconductivity and Particle Physics, talk at International Conference on Low Temperature Physics, Karlsruhe, Germany, August, 1984; *Physica* **B126**, 328.

116. The Turbulent Ether, *Symmetries in Particle Physics, Symposium in Honor of F. Gürsey's 60th Birhtday*, eds. I. Bars, A. Chodos and C.-H. Tze (Plenum Press), p. 9.

117. Fermions Living in a Space of Lie Groups, in *From SU(3) to Gravity, Festschrift in Honor of Yuval Ne'eman*, eds. E. Gotzman and G. Tauber (Cambridge Univ. Press), p. 45.

1985

118. Topological Excitations in Physics, *Lectures in Applied Mathematics*, **21**, 3; (*Proceedings of 1982 Mathematical Seminar*, American Mathematical Society).

119. Fermion–Boson Relations in BCS-Type Theories, in *Supersymmetry in Physics, Proceedings of the Los Alamos Workshop on Supersymmetry in Physics*, December, 1983, eds. A. Kosteleck'y and D.K. Campbell (North-Holland); also published as *Physica* **D15**, 147.

120. Some Theoretical Problems in Particle Physics, *Phys. Blätter* **41**, 173.

121. Directions of Particle Physics, *Proceedings of the Kyoto International Symposium, MESON50*, 1985, eds. M. Bando, R. Kawabe and N. Nakanishi, *Prog. Theor. Phys. Suppl.* **85**, 104.

122. Gauge Principles, Vector Meson Dominance and Spontaneous Symmetry Breaking, *International Symposium on Particle Physics in the 1950s*, Fermilab, May.

1986

123. Supersymmetry and Superconductivity, in *Rationale of Beings, Festschrift in Honor of Gyo Takeda*, eds. K. Ishikawa *et al.* (World Scientific), p. 3.

1987

124. Field Theory of Galois Fields, in *Field Theory and Quantum Statistics, Essays in Honor of the Sixtieth Birthday of E.S. Fradkin*, eds. I.A. Batalin *et al.* (Adam Hilgar), p. 625.

125. Thermodynamic Analogy in Quantum Field Theory, in *Wanderings in Physics, Festschrift in Honor of K. Nishijima*, eds. K. Kawarabayashi and A. Ukawa (World Scientific), p. 167.

1988

126. BSC Mechanism and the Interacting Boson Model (with M. Mukherjee), *Phys. Lett.* **B209**, 1.

127. Dynamical Symmetry Breaking in Nuclei, in *Festi-Val, Festschrift in Honor of Valentine Telegdi's 60th Birthday*, ed. K. Winter (North-Holland), p. 181.

128. BCS Theory, Quasi-Supersymmetry and Nuclear Physics, (*Encuentro Latino-Americano de Fisica de Alta Energia*, Universidad Federal Santa Maria, Valparaiso, Chile, December 1987, eds. N. Bralic *et al.*, Universidad Fedederal Santa Maria), Sciencia **165–166**, 141.

1989

129. The BCS Theory and the Sigma Model Revisited, in *Themes in Particle Physics*, dedicated to 70th Birthday of Julian Schwinger, eds. S. Deser and R.J. Finkelstein (World Scientific).

130. BCS Mechanism, Quasi-Supersymmetry and Fermion Masses, in the *Proceedings of XI Warsaw Symposium on Elementary Particle Physics, New Theories in Physics*, Kazimierz, Poland, May 1988, eds. Z. Ajduk *et al.* (World Scientific), p. 1.

131. Quasi-Supersymmetry, Bootstrap Symmetry Breaking and Fermion Masses, in the *Proceedings of the 1988 International Workshop on New Trends in Strong Coupling Gauge Theories*, Nagoya, August 1988 (World Scientific), p. 3.

132. BCS and IBM (with M. Mukherjee),
Ann. Phys. **191**, 143.

1990

133. Particle Physics since Lawrence and Yukawa, Particle Acclerators, **26** p. 89: *Proceedings of the 14th International Conference on High Energy Accelerators*, Part I, eds. Y. Kimura *et al.* (Gordon and Breach).

134. Model Building Based on Bootstrap Symmetry Breaking, in the *Proceedings of the 1989 Workshop on Dynamical Symmetry Breaking*, Nagoya, December 1989, eds. T. Muta and K. Yamawaki (Nagoya University), p. 1.

135. Mass Formulas and Dynamical Symmetry Breaking, in *From Symmetries to Strings: Forty Years of Rochester Conferences, A Symposium to Honor Susumu Okubo on His 60th Year*, May 1990, ed. A. Das (World Scientific), p. 1.

136. New Bootstrap and the Standard Model, *Plots, Quarks & Strange Particles, Proceedings of the Dalitz Conference*, Oxford, July 1990, eds. I.J.R. Aitchison, C.H. Llewellyn Smith and J.E. Paton (World Scientific), p. 56.

137. More on Bootstrap Symmetry Breaking, in the *Proceedings of the 1990 International Workshop on Strong Coupling Gauge Theories and Beyond*, Nagoya, Japan, July 1990, eds. T. Muta and K. Yamawaki (World Scientific), p. 3.

138. Summary of Personal Recollections of the Tokyo Group, *Elementary Particle Theory in Japan, 1935–1960* — Japan–US Collaboration on the History of Particle Physics (Lake Yamanaka, 1985), eds. L. Brown *et al.*, *Prog. Theor. Phys. Suppl.* **105**, 111.

1991

139. Supersymmetry and Quasi-supersymmetry, in *Elementary Particles and the Universe, Essays in Honor of Murray Gell-Mann*, ed. J.H. Schwarz (Cambridge Univ. Press), p. 89.

140. Theoretical Perspective in High Energy Physics, in *Proceedings of the 25th International Conference on High Energy Physics*, Singapore, eds. K.K. Phua and Y. Yamaguchi (South East Theoretical Association), p. 51.

141. Dynamical Symmetry Breaking, in *Evolutionary Trends in the Physical Sciences*, Proceedings of the Yoshio Nishina Centennial Symposium, Tokyo, Japan, December 1990, eds. M. Suzuki and R. Kubo (Springer-Verlag), p. 51.

142. Majorana's Infinite Component Wave Equation, *A Commentary to be published in the Collected Papers of E. Majorana*, ed. V. Telegdi.

143. Old Wine in New Bottles — Mass Spectra, Bootstrap and Hierarchy, talk given at *E. C. G. Sudarshan's Contributions to Theoretical Physics*, ed. A. Gleeson.

1992

144. Dynamical Symmetry Breaking: An Overview, in the *Proceedings of the International Workshop on Electroweak Symmetry Breaking*, Hiroshima, Japan, November, 1991, eds. W.A. Bardeen *et al.* (World Scientific).

145. Cosmic Rays and Particle Physics, talk at the *International Workshop on Super-high Energy Hadronic Interactions* at Waseda University, October 1991.

1993

146. Spontaneous Symmetry Breaking and the Origin of Mass, in *Some New Trends on Fluid Mechanics and Theoretical Physics*, in *Honor of Professor Pei-Yuan Chou's 90th Anniversary*, 1992, eds. C.C. Lin and Ning Hu (Peking University Press), p. 97.

147. Nonlocal Separable Solutions of the Inverse Scattering Problem (with T. Gherghetta), *Int. J. Mod. Phys.* **A8**, 3163.

1994

148. Feedback Effects in Superconductors (with T. Gherghetta), *Phys. Rev.* **B49**, 49.

www.ingramcontent.com/pod-product-compliance
Lightning Source LLC
Chambersburg PA
CBHW061739210326
41599CB00034B/6732
* 9 7 8 9 8 1 0 2 2 4 2 0 2 *